Medical Image Analysis and Informatics: Computer-Aided Diagnosis and Therapy

Medical Image Analysis and Informatics: Computer-Aided Diagnosis and Therapy

Edited by

Paulo Mazzoncini de Azevedo-Marques
Arianna Mencattini
Marcello Salmeri
Rangaraj M. Rangayyan

CRC Press
Taylor & Francis Group
Boca Raton London New York

CRC Press is an imprint of the
Taylor & Francis Group, an **informa** business

CRC Press
Taylor & Francis Group
6000 Broken Sound Parkway NW, Suite 300
Boca Raton, FL 33487-2742

First issued in paperback 2019

ISBN-13: 978-1-4987-5319-7 (hbk)
ISBN-13: 978-0-367-87628-9 (pbk)

Visit the Taylor & Francis Web site at
http://www.taylorandfrancis.com

and the CRC Press Web site at
http://www.crcpress.com

We dedicate this book
with gratitude and admiration
to medical specialists and clinical researchers
who collaborate with engineers and scientists
on computer-aided diagnosis and therapy
for improved health care.

Paulo, Arianna, Marcello, and Raj

Contents

Foreword on CAD: Its Past, Present, and Future

Computer-aided diagnosis (CAD) has become a routine clinical procedure for detection of breast cancer on mammograms at many clinics and medical centers in the United States. With CAD, radiologists use the computer output as a "second opinion" in making their final decisions. Of the total number of approximately 38 million mammographic examinations annually in the United States, it has been estimated that about 80% have been studied with use of CAD. It is likely that CAD is beginning to be applied widely in the detection and differential diagnosis of many different types of abnormalities in medical images obtained in various examinations by use of different imaging modalities, including projection radiography, computed tomography (CT), magnetic resonance imaging (MRI), ultrasonography, nuclear medicine imaging, and other optical imaging systems. In fact, CAD has become one of the major research subjects in medical imaging, diagnostic radiology, and medical physics. Although early attempts at computerized analysis of medical images were made in the 1960s, serious and systematic investigations on CAD began in the 1980s with a fundamental change in the concept for utilization of the computer output, from automated computer diagnosis to computer-aided diagnosis.

Large-scale and systematic research on and development of various CAD schemes was begun by us in the early 1980s at the Kurt Rossmann Laboratories for Radiologic Image Research in the Department of Radiology at the University of Chicago. Prior to that time, we had been engaged in basic research related to the effects of digital images on radiologic diagnosis, and many investigators had become involved in research and development of a picture archiving and communication system (PACS). Although it seemed that PACS would be useful in the management of radiologic images in radiology departments and might be beneficial economically to hospitals, it looked unlikely at that time that PACS would bring a significant clinical benefit to radiologists. Therefore, we thought that a major benefit of digital images must be realized in radiologists' daily work of image reading and radiologic diagnosis. Thus, we came to the concept of computer-aided diagnosis.

In the 1980s, the concept of automated diagnosis or automated computer diagnosis was already known from studies performed in the 1960s and 1970s. At that time, it was assumed that computers could replace radiologists in detecting abnormalities, because computers and machines are better at performing certain tasks than human beings. These early attempts were not successful because computers were not powerful enough, advanced image processing techniques were not available, and digital images were not easily accessible. However, a serious flaw was an excessively high expectation from computers. Thus, it appeared to be extremely difficult at that time to carry out a computer analysis of medical images. It was uncertain whether the development of CAD schemes would be successful or would fail. Therefore, we selected research subjects related to cardiovascular diseases, lung cancer, and breast cancer, including for detection and/or quantitative analysis of lesions involved in vascular imaging, as studied by H. Fujita and K.R. Hoffmann; detection of lung nodules in chest radiographs by M.L. Giger; and detection of clustered microcalcifications in mammograms by H.P. Chan.

Our efforts concerning research and development of CAD for detection of lesions in medical images have been based on the understanding of processes that are involved in image readings by radiologists. This strategy appeared logical and straightforward because radiologists carry out very complex and difficult tasks of image reading and radiologic diagnosis. Therefore, we considered that computer algorithms should be developed based on the understanding of image readings, such as how radiologists can detect certain lesions, why they may miss some abnormalities, and how they can distinguish between benign and malignant lesions.

Regarding CAD research on lung cancer, we attempted in the mid-1980s to develop a computerized scheme for detection of lung nodules on chest radiographs. The visual detection of lung nodules is well-known as a difficult task for radiologists, who may miss up to 30% of the nodules because of the overlap of normal anatomic structures with nodules, i.e., the normal background in chest images tends to camouflage nodules. Therefore, the normal background structures in chest images could become a large obstacle in the detection of nodules, even by computer. Thus, the first step in the computerized scheme for detection of lung nodules in chest images would need to be the removal or suppression of background structures in chest radiographs. A method for suppressing the background structures is the difference-image technique, in which the difference between a nodule-enhanced image and a nodule-suppressed image is obtained. This difference-image technique, which may be considered a generalization of an edge enhancement technique, has been useful in enhancing lesions and suppressing the background not only for nodules in chest images, but also for microcalcifications and masses in mammograms, and for lung nodules in CT.

At the Rossmann Laboratories in the mid-1980s, we had already developed basic schemes for the detection of lung nodules in chest images and for the detection of clustered microcalcifications in mammograms. Although the sensitivities of these schemes for detection of lesions were relatively high, the number of false positives was very large. It was quite uncertain whether the output of these computerized schemes could be used by radiologists in their clinical work. For example, the average number of false positives obtained by computer was four per mammogram in the detection of clustered microcalcifications, although the sensitivity was about 85%. However, in order to examine the possibility of practical uses of CAD in clinical situations, we carried out an observer performance study without and with computer output. To our surprise, radiologists' performance in detecting clustered microcalcifications was improved significantly when the computer output was available. A paper was published in 1990 by H.P. Chan providing the first scientific evidence that CAD could be useful in improving radiologists' performance in the detection of a lesion. Many investigators have reported similar findings on the usefulness of CAD in detecting various lesions, namely, masses in mammograms, lung nodules and interstitial opacities in chest radiographs, lung nodules in CT, intracranial aneurysms in magnetic resonance angiography (MRA), and polyps in CT colonography.

The two concepts of automated computer diagnosis and computer-aided diagnosis clearly exist even at present. Therefore, it may be useful to understand the common features and also the differences between CAD and automated computer diagnosis. The common approach to both CAD and automated computer diagnosis is that digital medical images are analyzed quantitatively by computers. Therefore, the development of computer algorithms is required for both CAD and computer diagnosis. A major difference between CAD and computer diagnosis is the way in which the computer output is utilized for the diagnosis. With CAD, radiologists use the computer output as a "second opinion," and radiologists make the final decisions. Therefore, for some clinical cases in which radiologists are confident about their judgments, radiologists may agree with the computer output, or they may disagree and then disregard the computer. However, for cases in which radiologists are less confident, it is expected that the final decision can be improved by use of the computer output. This improvement is possible, of course, only when the computer result is correct. However, the performance level of the computer does not have to be equal to or higher than that of radiologists. With CAD, the potential gain is due to the synergistic effect obtained by combining the radiologist's competence with the computer's capability, and thus the current CAD scheme has become widely used in practical clinical situations.

With automated computer diagnosis, however, the performance level of the computer output is required to be very high. For example, if the sensitivity for detection of lesions by computer were lower than the average sensitivity of physicians, it would be difficult to justify the use of automated computer diagnosis. Therefore, high sensitivity and high specificity by computer would be required for implementing automated computer diagnosis. This requirement is extremely difficult for researchers to achieve in developing computer algorithms for detection of abnormalities on medical images.

The majority of papers related to CAD research presented at major meetings such as those of the RSNA, AAPM, SPIE, and CARS from 1986 to 2015 were concerned with three organs–chest, breast, and colon–but other organs such as brain, liver, and skeletal and vascular systems were also subjected to CAD research. The detection of cancer in the breast, lung, and colon has been subjected to screening examinations. The detection of only a small number of suspicious lesions by radiologists is considered both difficult and time-consuming because a large fraction of these examinations are normal. Therefore, it appears reasonable that the initial phase of CAD in clinical situations has begun for these screening examinations. In mammography, investigators have reported results from prospective studies on large numbers of patients regarding the effect of CAD on the detection rate of breast cancer. Although there is a large variation in the results, it is important to note that all of these studies indicated an increase in the detection rates of breast cancer with use of CAD.

In order to assist radiologists in their differential diagnosis, in addition to providing the likelihood of malignancy as the output of CAD, it would be useful to provide a set of benign and malignant images that are similar to an unknown new case under study; this may be achieved using methods of content-based image retrieval (CBIR). If the new case were considered by a radiologist to be very similar to one or more benign (or malignant) images, he/she would be more confident in deciding that the new case was benign (or malignant). Therefore, similar images may be employed as a supplement to the computed likelihood of malignancy in implementing CAD for a differential diagnosis.

The usefulness of similar images has been demonstrated in an observer performance study in which the receiver operating characteristic (ROC) curve in the distinction between benign and malignant microcalcifications in mammograms was improved. Similar findings have been reported for the distinction between benign and malignant masses, and also between benign and malignant nodules in thoracic CT. There are two important issues related to the use of similar images in clinical situations. One is the need for a unique database that includes a large number of images, which can be used as being similar to those of many unknown new cases, and another is the need for a sensitive tool for finding images similar to an unknown case.

At present, the majority of clinical images in PACS have not been used for clinical purposes, except for images of the same patients for comparison of a current image with previous images. Therefore, it would not be an overstatement to say that the vast majority of images in PACS are currently "sleeping" and need to be awakened in the future for daily use in clinical situations. It would be possible to search for and retrieve very similar cases with similar images from PACS. Recent studies indicated that the similarity of a pair of lung nodules in CT and of lesions in mammograms may be quantified by a psychophysical measure which can be obtained by use of an artificial neural network trained with the corresponding image features and with subjective similarity ratings given by a group of radiologists. However, further investigations are required for examining the usefulness of this type of new tool for searching similar images in PACS.

It is likely that some CAD schemes will be included together with software for image processing in workstations associated with imaging modalities such as digital mammography, CT, and MRI. However, many other CAD schemes will be assembled as packages and will be implemented as a part of PACS. For example, the package for chest CAD may include the computerized detection of lung nodules, interstitial opacities, cardiomegaly, vertebral fractures, and interval changes in chest radiographs, as well as the computerized classification of benign and malignant nodules. All of the chest images taken for whatever purpose will be subjected to a computerized search for many different types of abnormalities included in the CAD package, and, thus, potential sites of lesions, together with relevant information such as the likelihood of malignancy and the probability of a certain disease, may be displayed on the workstation. For such a package to be used in clinical situations, it is important to reduce the number of false positives as much as possible so that radiologists will not be distracted by an excessive number of these, but will be prompted only by clinically significant abnormalities.

Radiologists may use this type of CAD package in the workstation for three different reading methods. One is first to read images without the computer output, and then to request a display of the computer output before making the final decision; this "second-read" mode has been the condition that the Food and Drug Administration (FDA) in the United States has required for approval of a CAD system as a medical device. If radiologists keep their initial findings in some manner, this second-read mode may prevent a detrimental effect of the computer output on radiologists' initial diagnosis, such as incorrectly dismissing a subtle lesion because of the absence of a computer output, although radiologists were very suspicious about this lesion initially. However, this second-read mode would increase the time required for radiologists' image reading, which is undesirable.

Another mode is to display the computer output first and then to have the final decision made by a radiologist. With this "concurrent" mode, it is likely that radiologists can reduce the reading time for image interpretations, but it is uncertain whether they may miss some lesions when no computer output was shown, due to computer false negatives. This negative effect can be reduced if the sensitivity in the detection of abnormalities is at a very high level, which may be possible with a package of a number of different, but complementary CAD schemes. For example, although two CAD schemes may miss some lung nodules and other interstitial opacities on chest radiographs, it is possible that the temporal subtraction images obtained from the current and previous chest images demonstrate interval changes clearly because the temporal subtraction technique is very sensitive to subtle changes between the two images. This would be one of the potential advantages of packaging of a number of CAD schemes in the PACS environment.

The third method is called a "first-read" mode, in which radiologists would be required to examine only the locations marked by the computer. With this first-read mode, the sensitivity of the computer software must be extremely high, and if the number of false positives is not very high, the reading time may be reduced substantially. It is possible that a certain type of radiologic examination requiring a long reading time could be implemented by the concurrent-read mode or the first-read mode due to economic and clinical reasons, such as a shortage of radiologist manpower. However, this would depend on the level of performance by the computer algorithm, and, at present, it is difficult to predict what level of computer performance would make this possible. Computer-aided diagnosis has made a remarkable progress during the last three decades by numerous investigators around the world, including those listed in the footnote* and researchers at the University of Chicago. It is likely in the future that the concept, methods, techniques, and procedures related to CAD and quantitative image analysis would be applied to and used in many other related fields, including medical optical imaging systems and devices, radiation therapy, surgery, and pathology, as well as radiomics and imaging genomics in radiology and radiation oncology. In the future, the benefits of CAD and quantitation of image data need to be realized in conjunction with progress in other fields including informatics, CBIR, PACS, hospital

* Faculty, research staff, students, and international visitors who participated in research and development of CAD schemes in the Rossmann Laboratory over the last three decades have moved to academic institutions worldwide and continue to contribute to the progress in this field. They are H. P. Chan, University of Michigan; K.R. Hoffmann, SUNY Buffalo; H. Yoshida, MGH; R. M. Nishikawa, K. T. Bae, University of Pittsburgh; N. Alperin, University of Miami; F. F. Yin, Duke University; K. Suzuki, Illinois Institute of Technology; L. Fencil, Yale University; P. M. Azevedo-Marques, University of São Paulo, Brazil; Q. Li, Shanghai Advanced Research Institute, China; U. Bick, Charite University Clinic, Germany; M. Fiebich, University of Applied Sciences, Germany; B. van Ginneken, Radbound University, The Netherlands; P. Tahoces, University of Santiago de Compostella, Spain; H. Fujita, T. Hara, C. Muramatsu, Gifu University, Japan; S. Sanada, R. Tanaka, Kanazawa University, Japan; S. Katsuragawa, Teikyo University, Japan; J. Morishita, H. Arimura, Kyushu University, Japan; J. Shiraishi, Y. Uchiyama, Kumamoto University, Japan; T. Ishida, Osaka University, Japan; K. Ashizawa, Nagasaki University, Japan; K. Chida, Tohoku University, Japan; T. Ogura, M. Shimosegawa, H. Nagashima, Gunma Prefectural College of Health Sciences, Japan.

information systems (HIS), and radiology information systems (RIS). Due to the recent development of new artificial intelligence technologies such as a deep learning neural network, the performance of the computer algorithm may be improved substantially in the future, but will be carefully examined for practical uses in complex clinical situations. Computer-aided diagnosis is still in its infancy in terms of the development of its full potential for applications to many different types of lesions obtained with various diagnostic modalities.

Kunio Doi, PhD

Preface

Medical Imaging, Medical Image Informatics, and Computer-Aided Diagnosis

Medical imaging has been well established in health care since the discovery of X rays by Röntgen in 1895. The development of computed tomography (CT) scanners by Hounsfield and others in the early 1970s brought computers and digital imaging to radiology. Now, computers and digital imaging systems are integral components of radiology and medical imaging departments in hospitals. Computers are routinely used to perform a variety of tasks from data acquisition and image generation to image visualization and analysis (Azevedo-Marques and Rangayyan 2013, Deserno 2011, Dhawan 2011, Doi 2006, Doi 2007, Fitzpatrick and Sonka 2000, Li and Nishikawa 2015, Rangayyan 2005, Shortliffe and Cimino 2014).

With the development of more and more medical imaging modalities, the need for computers and computing in image generation, manipulation, display, visualization, archival, transmission, modeling, and analysis has grown substantially. Computers are integrated into almost every medical imaging system, including digital radiography, ultrasonography, CT, nuclear medicine, and magnetic resonance (MR) imaging (MRI) systems. Radiology departments with picture archival and communication systems (PACS) are totally digital and filmless departments. Diagnosis is performed using computers not only for transmission, retrieval, and display of image data, but also to derive measures from the images and to analyze them.

Evolutionary changes and improvements in medical imaging systems, as well as their expanding use in routine clinical work, have led to a natural increase in the scope and complexity of the associated problems, calling for further advanced techniques for their solution. This has led to the establishment of relatively new fields of research and development known as medical image analysis, medical image informatics, and computer-aided diagnosis (CAD) (Azevedo-Marques and Rangayyan 2013, Deserno 2011, Dhawan 2011, Doi 2006, Doi 2007, Fitzpatrick and Sonka 2000, Li and Nishikawa 2015, Rangayyan 2005, Shortliffe and Cimino 2014). CAD is defined as diagnosis made by a radiologist or physician using the output of a computerized scheme for image analysis as a diagnostic aid (Doi 2006, 2007). Two variations in CAD have been used in the literature: CADe for computer-aided detection of abnormal regions of interest (ROIs) and CADx for computer-aided diagnosis with labeling of detected ROIs in terms of the presence or absence of a certain disease, such as cancer.

Typically, a radiologist using a CAD system makes an initial decision and then considers the result of the CAD system as a second opinion; classically, such an opinion would have been obtained from another radiologist. The radiologist may or may not change the initial decision after receiving the second opinion, be it from a CAD system or another radiologist. In such an application, the CAD system need not be better than or even comparable to the radiologist. If the CAD system is designed to be complementary to the radiologist; the symbiotic and synergistic combination of the radiologist with the CAD system can improve the accuracy of diagnosis (Doi 2006, 2007).

In a more radical manner, one may apply a CAD system for initial screening of all cases and then send to the radiologist only those cases that merit attention at an advanced level; the remaining cases may be analyzed by other medical staff. While this process may be desirable when the patient population is large and the number of available medical experts is disproportionately small, it places heavier reliance and responsibility on the CAD system. Not all societies may accept such an application where a computational procedure is used to make an initial decision.

Medical image informatics deals with the design of methods and procedures to improve the efficiency, accuracy, usability, and reliability of medical imaging for health care. CAD and content-based image retrieval (CBIR) are two important applications in medical image informatics. CBIR systems are designed to bring relevant clinically established cases from a database when presented with a current case as a query. The features and diagnoses associated with the retrieved cases are expected to assist the radiologist or medical specialist in diagnosing the current case. Even though CBIR systems may not suggest a diagnosis, they rely on several techniques that are used by CAD systems and share some similarities. In this book, we present a collection of chapters representing the latest developments in these areas.

Why Use CAD?

At the outset, it is important to recognize the need for application of computers for analysis of medical images. Radiologists and other medical professionals are highly trained specialists. Why, when, and for what would they need the assistance of computers? Medical images are voluminous and bear intricate details. More often than not, normal cases in a clinical set up or details within a given image overwhelmingly outnumber abnormal cases or details. Regardless of the level of expertise and experience of a medical specialist, visual analysis of medical images is prone to several types of errors, some of which are listed in Table 1. The application of computational techniques could address some of these limitations, as implied by Table 2.

The typical steps of a CAD system are as follows:

1. Preprocessing the given image for further analysis
2. Detection and segmentation of ROIs
3. Extraction of measures or features for quantitative analysis
4. Selection of an optimal set of features
5. Training of classifiers and development of decision rules
6. Pattern classification and diagnostic decision making

Table 3 shows a simplified plan as to how one may overcome some of the limitations of manual or visual analysis by applying computational procedures.

The paths and procedures shown in Table 3 are not simple and straightforward; neither are they free of problems and limitations. Despite the immense efforts of several researchers, the development

TABLE 1 Causes of Various Types of Errors in Visual Analysis of Medical Images

Causes	Types of error
Subjective and qualitative analysis	Inconsistency
Inconsistencies in knowledge and training, differences in opinion, personal preferences	Inter-observer error
Inconsistent application of knowledge, lack of diligence, environmental effects and distraction, fatigue and boredom due to workload and repetitive tasks	Intra-observer error

TABLE 2 Comparison of Various Aspects of Manual versus Computer Analysis of Medical Images

Manual Analysis	Computer Analysis
Inconsistencies in identifying landmarks or ROIs	Consistent application of established rules and methods
Errors in localization of landmarks due to limited manual dexterity	High numerical precision and computational accuracy
Extensive time and effort for manual measurement of intricate details	High speed of computation
Limitations in the precision and reproducibility of manual measurement and calculations	Ease of repeatability and reproducibility
Effects of distraction, fatigue, and boredom	Immunity to effects of work environment, fatigue, and boredom

TABLE 3 Techniques and Means to Move from Manual to Computer Analysis of Medical Images

Move From	Via	To
Qualitative analysis	Computation of measures, features, and attributes using digital image processing techniques	Quantitative analysis
Subjective analysis	Development of rules for diagnostic decision making using pattern classification techniques	Objective analysis
Inconsistent analysis	Implementation of established rules and robust procedures as computational algorithms	Consistent analysis
Inter-observer and intra-observer errors	Medical image analysis, medical image informatics, and CAD	Improved diagnostic accuracy

and clinical application of CAD systems encounter several difficulties, some of which are listed below:

- Difficulty in translating methods of visual analysis into computational procedures
- Difficulty in translating clinical observations into numerical features or attributes
- Difficulty in dealing with large numbers of features in a classification rule: curse of dimensionality
- Substantial requirements of computational resources and time
- Need for large sets of annotated or labeled cases to train and test a CAD system
- Large numbers of false alarms or false positives
- Difficulty in integrating CAD systems into established clinical workflows and protocols

The World Health Organization (WHO), in its 58th World Health Assembly held in Geneva in 2005, recognized the potential of application of information and communication technologies (ICT) as a way to strengthen health systems and to improve quality, safety, and access to care. Despite recent advances, there are, as yet, many difficulties in improving the utilization of ICT in the healthcare environment. Different ways to use diverse technologies, lack of widely adopted data communication standards, and the intersection of multiple domains of knowledge are some of the issues that must be overcome in order to improve health care worldwide.

These introductory paragraphs do not offer solutions: They lead us toward the latest developments in the related fields presented by leading researchers around the world who have contributed the chapters in the book.

Organization of the Book

The chapters in this book represent some of the latest developments in the fields related to medical image analysis, medical image informatics, CBIR, and CAD. They have been prepared by leading researchers in related areas around the world. Unlike other books in related areas, we have chosen not to limit the applications covered by the chapters to imaging of certain parts of the body (such as the brain, the heart, or the breast) or to certain diseases (such as stroke, coronary artery disease, or cancer). Instead, the range of applications is from the head to toe, or craniocaudal, to use an imaging term. Several different medical imaging modalities and techniques related to CAD and image informatics are included. The chapters should appeal to biomedical researchers, medical practitioners, neuroscientists, ophthalmologists, dentists, radiologists, oncologists, cardiologists, orthopedic specialists, gastroenterologists, pathologists, computer scientists, medical physicists, engineers, informatics specialists, and readers interested in advanced imaging technology and informatics, and assist them in learning about a broad range of latest developments and applications in related areas.

In Chapter 1, Reiche et al. present an approach for segmentation and characterization of white matter lesions in fluid-attenuated inversion recovery (FLAIR) MR images. They describe the rationale for use of the FLAIR modality, as well as the problem of noise in MRI and its effect on reliable segmentation.

Hatanaka and Fujita present, in Chapter 2, several methods for CAD of multiple diseases via analysis of retinal fundus images. Their methods serve the purposes of segmentation of blood vessels and measurement of vessel diameter, as well as detection of hemorrhages, microaneurysms, large cupping in the optic disc, and nerve fiber layer defects.

Oloumi et al. present, in Chapter 3, several algorithms for CAD of retinopathy in premature infants. Gabor filters and morphological image processing methods are formulated to detect and analyze the vascular architecture of the retina. It is shown that measures related to the thickness and tortuosity of blood vessels, as well as the openness of the major temporal arcade, can assist in CAD of retinopathy of prematurity.

Chapter 4 by Roychowdhury et al. presents methods for image segmentation and measurement of the thickness of sub-retinal layers in optical coherence tomography (OCT) images. The importance of denoising as a preprocessing step in the segmentation process is analyzed. The results of the algorithm presented for multiresolution iterative sub-retinal surface segmentation are shown to be useful for the assessment of macular diseases.

Muramatsu et al. present techniques for CAD with dental panoramic radiographs in Chapter 5. The techniques presented address several clinically important issues, including detection of carotid artery calcifications for screening for arteriosclerosis, detection of radiopacity in maxillary sinuses, and quantitative analysis of periodontal diseases.

In Chapter 6, Pérez-Carrasco et al. introduce the problem of diagnosis of burn wounds. They describe several methods for burn diagnosis, including segmentation, feature extraction, estimation of depth, measurement of surface area, and automatic classification of burns.

Gutierrez et al. present, in Chapter 7, the state of the art of noninvasive cardiac imaging for diagnosis and treatment of cardiovascular diseases. The authors show how cardiac image segmentation plays a crucial role and allows for a wide range of applications, including quantification of volume, localization of pathology, CAD, and image-guided interventions.

In Chapter 8, Pezeshk et al. address issues related to databases for training and testing CAD algorithms. In order to overcome practical difficulties and limitations that often severely constrain the number of cases one may be able to acquire in a CAD study, Pezeshk et al. describe methods to insert a lesion or tumor selected from an available case into other available images so as to increase the number of cases. The various techniques and transformations described in this chapter facilitate blending of an original lesion into its recipient image in several ways to accommodate natural variations in shape, size, orientation, and background.

Koenigkam Santos and Weinheimer investigate, in Chapter 9, the topic of diagnosis of diffuse lung diseases. They discuss clinical applications of CAD for emphysema, airway diseases, and interstitial lung diseases. Furthermore, they describe methods for computerized detection and description of airways and lung parenchyma in CT images.

In Chapter 10, Mencattini et al. present several methods and measures to characterize and detect bilateral asymmetry in mammograms. Their procedures include landmarking mammograms, segmenting matched pairs of regions in mammograms of the left and right breasts of an individual, and deriving features based on the semivariogram and structural similarity indices. The methods are demonstrated to be effective and efficient in CAD of bilateral asymmetry and breast disease.

In Chapter 11, Chan et al. discuss the impact of the digital breast tomosynthesis (DBT) imaging technique on breast cancer detection. The authors describe the characteristics of DBT and present state-of-the-art approaches that address this topic. In addition, the authors analyze the advantages and disadvantages of a CAD approach applied to DBT in relation to standardized and approved digital mammography.

Nogueira-Barbosa and Azevedo-Marques investigate, in Chapter 12, CAD methods for spinal abnormalities with radiographic images, CT, and MRI. They study clinical applications such as detection and classification of vertebral body fracture, as well as characterization of intervertebral disc degeneration.

Yuan and Meng present, in Chapter 13, techniques to capture images of the gastrointestinal tract using imaging and data transmitting devices packaged in a capsule that may be swallowed. Furthermore, they present image processing, feature extraction, coding, and pattern classification techniques to detect ulcers.

In Chapter 14, Virmani and Kumar present CAD applications in the diagnosis of diseases of the liver and show how noninvasive methods can enhance the results of clinical investigation. They demonstrate that ultrasonographic measurements that characterize the structure of soft tissue are potentially useful tissue signatures because important features of diffuse and focal liver diseases are indicated by disruptions of the normal tissue architecture.

Cipriani Frade et al. study, in Chapter 15, the topic of dermatological ulcers. They propose color image processing methods for analysis of dermatological images in the context of CBIR.

Chapter 16 by Boyd and Lagacé presents a detailed study on *in vivo* bone imaging with micro-computed tomography, quantitative CT (QCT), and other specialized imaging modalities. The authors present a multifaceted discussion on the physiological and structural characteristics of bone, bone loss and osteoporosis, and analysis of bone density and other parameters that could be useful in diagnosis.

Borotikar et al. present methods of statistical shape modeling for augmented orthopedic surgery and rehabilitation in Chapter 17. Their procedures include building image-based bone models, registration of images, derivation of patient-specific anatomic references, and modeling of shape.

In Chapter 18, Knight and Khademi present color image processing and pattern recognition techniques for analysis of histopathology images. They describe methods to detect and characterize nuclei and related features, and demonstrate the effectiveness of their measures in the recognition of tissue patterns related to breast cancer.

Chandra et al. describe, in Chapter 19, methods to obtain images using microwaves. They illustrate how image reconstruction and radar techniques may be used to obtain medical images that could assist in the detection of brain tumors.

In Chapter 20, Traina et al. present a CBIR system designed to locate and retrieve mammographic images from a database that are similar to a given query image. The authors introduce concepts of and criteria for similarity and diversity to facilitate searching for and resolving nearly duplicate images.

Chapter 21 by Deserno and Reichertz gives an overview of health informatics for clinical applications of CAD. Several paradigms, models, and concepts related to informatics are described and shown to be important in moving CAD from research laboratories toward application to patient care.

We are confident that you will find the chapters interesting, intriguing, and invigorating.

References

Azevedo-Marques, P. M. and Rangayyan, R. M. 2013. *Content-Based Retrieval of Medical Images: Landmarking, Indexing, and Relevance Feedback*. San Francisco, CA: Morgan & Claypool.

Deserno, T. M. (Ed.) 2011. *Biomedical Image Processing*. Berlin, Germany: Springer.

Dhawan, A. P. 2011. *Medical Image Analysis*, 2nd ed., New York: IEEE and Wiley.

Doi, K. 2006. Diagnostic imaging over the last 50 years: Research and development in medical imaging science and technology. *Physics in Medicine and Biology*, 51(13):R5–R27, June.

Doi, K. 2007. Computer-aided diagnosis in medical imaging: historical review, current status and future potential.*Computerized Medical Imaging and Graphics*, 31(4–5):198–211.

Fitzpatrick, J. M. and Sonka, M. (Eds.) 2000. *Handbook of Medical Imaging, Volume 2. Medical Image Processing and Analysis*. Bellingham, WA: SPIE.

Li, Q. and Nishikawa, R. M. (Eds.) 2015. *Computer-Aided Detection and Diagnosis in Medical Imaging*. Boca Raton, FL: CRC Press.

Rangayyan, R. M. 2005. *Biomedical Image Analysis*. Boca Raton, FL: CRC Press.

Shortliffe, E. H. and Cimino, J. J. (Eds.) 2014. *Biomedical Informatics: Computer Applications in Health Care and Biomedicine*. Berlin, Germany: Springer.

Paulo Mazzoncini de Azevedo-Marques
(University of São Paulo, Brazil; pmarques@fmrp.usp.br)

Arianna Mencattini
(University of Rome Tor Vergata, Rome, RM, Italy; mencattini@ing.uniroma2.it)

Marcello Salmeri
(University of Rome Tor Vergata, Rome, RM, Italy; salmeri@ing.uniroma2.it)

Rangaraj Mandayam Rangayyan
(University of Calgary, Calgary, Alberta, Canada; ranga@ucalgary.ca)

Acknowledgment

We thank the authors of the chapters for contributing their research work for publication in this book. It was an enjoyable learning experience to review the articles submitted and a pleasure to work with experts in the related topics around the world. We offer special thanks to Dr. Kunio Doi, popularly referred to as the "Father of CAD," for writing the foreword for the book. We thank the staff of Taylor & Francis Group, CRC Press, for their assistance in publication of this work.

Our research work in related topics over the past several years has been supported by many grants from the following agencies, and we are grateful to them: the São Paulo Research Foundation (FAPESP), the National Council for Scientific and Technological Development (CNPq); Financing of Studies and Projects (FINEP); the Foundation to Aid Teaching, Research, and Patient Care of the Clinical Hospital of Ribeirão Preto (FAEPA/HCRP) of Brazil; and the Natural Sciences and Engineering Research Council of Canada (NSERC).

Editors

Paulo Mazzoncini de Azevedo-Marques is an associate professor of medical physics and biomedical informatics with the Internal Medicine Department, University of São Paulo (USP), School of Medicine, in Ribeirão Preto, SP, Brazil. In the 1990s he worked on medical imaging quality control. He held a research associate position at the University of Chicago in 2001, where he worked on medical image processing for computer-aided diagnosis (CAD) and content-based image retrieval (CBIR), under the supervision of Professor Kunio Doi. He is the coordinator of the Medical Physics facility at the University Medical Center at Ribeirão Preto Medical School – USP. His main research areas are CAD, CBIR, Picture Archiving, and Communication System (PACS) and Radiomics.

Arianna Mencattini is an assistant professor at the Department of Electronic Engineering, University of Rome Tor Vergata, Italy. She is a member of the Italian Electrical and Electronic Measurement Group. At present, she holds a course on Image Processing, Master Degree in Electronic Engineering. Her main research interests are related to image processing techniques for the development of computed assisted diagnosis systems, analysis of speech and facial expressions for automatic emotion recognition, and design of novel cell tracking algorithms for immune-cancer interaction analysis. She is the principal investigator of project PainTCare, Personal pAIn assessemeNT by an enhanced multimodAl architecture, funded by University of Rome Tor Vergata, for the automatic assessment of pain in post-surgical patients, and team member of the Project Horizon 2020 PhasmaFOOD: Portable photonic miniaturised smart system for on-the-spot food quality sensing. Currently, she is author of 80 scientific papers.

Marcello Salmeri is an associate professor at the Department of Electronic Engineering, University of Rome Tor Vergata, Italy. He is a member of the Italian Electrical and Electronic Measurement Group and IEEE. At present, he is coordinator of the Electronic Engineering courses and delegate of engineering for orientation and tutoring of students. His research interests include signal and image processing, theory, applications, and implementations of fuzzy systems, pattern recognition. Currently, he is author of about 110 papers in the fields of electronics, measurement, and data analisys. He has collaborated with many companies in the fields of Electronics and ICT.

Rangaraj M. Rangayyan is a professor emeritus of electrical and computer engineering at the University of Calgary, Canada. His research areas are biomedical signal and image analysis for computer-aided diagnosis. He has been elected Fellow of the IEEE, SPIE, American Institute for Medical and Biological Engineering, Society for Imaging Informatics in Medicine, Engineering Institute of Canada, Canadian Medical and Biological Engineering Society, Canadian Academy of Engineering, and the Royal Society of Canada. He was recognized with the 2013 IEEE Canada Outstanding Engineer Medal.

Contributors

Begoña Acha-Piñero
Signal Theory and Communications Department
University of Seville
Seville, Spain

Paulo Mazzoncini de Azevedo-Marques
University of Sao Paulo
Sao Paulo, Brazil

Ilangko Balasingham
Department of Electronics and Telecommunication
Norwegian University of Science and Technology
Trondheim, Norway
and
Intervention Center
Oslo University Hospital
University of Oslo
Oslo, Norway

Bhushan Borotikar
Le département Image et traitement de l'information
IMT Atlantique
and
Laboratoire de Traitement de l'Information Médical (LaTIM)
Brest, France

Steven K. Boyd
McCaig Institute for Bone and Joint Health
University of Calgary
Calgary, Alberta, Canada

Sylvain Brochard
Laboratoire de Traitement de l'Information Médical (LaTIM)
and
CHRU de Brest
Brest, France

Valérie Burdin
Le département Image et Traitement de l'Information
IMT Atlantique
and
Laboratoire de Traitement de l'Information Médical (LaTIM)
Brest, France

Guilherme Ferreira Caetano
University of Sao Paulo
School of Medicine of Ribeirao Preto
Department of Internal Medicine
Divison of Dermatology
and
Herminio Ometto University
Biomedical Science Postgraduate Program
Sao Paulo, Brazil

Luiz Olmes Carvalho
University of São Paulo
São Paulo, Brazil

Paola Casti
Department of Electronic Engineering
University of Rome Tor Vergata
Rome, Italy

Heang-Ping Chan
Department of Radiology
University of Michigan
Ann Arbor, Michigan

Rohit Chandra
Ericsson AB
Kista, Sweden

Marco Andrey Cipriani Frade
University of Sao Paulo
School of Medicine of Ribeirao Preto
Department of Internal Medicine
Division of Dermatology
Sao Paulo, Brazil

Alceu Ferraz Costa
University of São Paulo
São Paulo, Brazil

Thomas M. Deserno
Peter L. Reichertz Institute for Medical Informatics (PLRI) of TU Braunschweig and Hannover Medical School,
38106 Braunschweig, Germany

Éderson Dorileo
Center of Imaging Sciences and Medical Physics
Internal Medicine Department
Ribeirão Preto Medical School
University of São Paulo
São Paulo, Brazil

Lucio Fernandes Dutra Santos
University of São Paulo
São Paulo, Brazil

Anna L. Ells
Division of Opthalmology
Department of Surgery
Cumming School of Medicine
University of Calgary
Calgary, Alberta, Canada

Hiroshi Fujita
Department of Electrical, Electronic & Computer Engineering and
Department of Intelligent Image Information, Graduate School of Medicine
Gifu University
Gifu, Japan

Enjie Ghorbel
Institut de recherche en systèmes électroniques embarqués (IRSEEM)
Rouen, France

and

Ecole des Mines de Douai
Douai, France

Marco A. Guiterrez
Heart Institute
University of Sao Paulo
Sao Paulo, Brazil

Lubomir M. Hadjiiski
Department of Radiology
University of Michigan
Ann Arbor, Michigan

Takeshi Hara
Department of Electrical, Electronic and Computer Engineering and
Department of Intelligent Image Information
Graduate School of Medicine
Gifu University
Gifu, Japan

Yuji Hatanaka
Department of Electronic Systems Engineering
School of Engineering
University of Shiga Prefecture
Hikone, Japan

Tatsuro Hayashi
Media Co., Ltd, Bunkyo-ku
Tokyo, Japan

Akitoshi Katsumata
Department of Oral Radiology
Asahi University School of Dentistry
Mizuho, Japan

April Khademi
Image Analysis in Medicine Laboratory
Ryerson University
Toronto, Ontario, Canada

Jesse Knight
Image Analysis in Medicine Laboratory
University of Guelph
Guelph, Ontario, Canada

Dara D. Koozekanani
Associate Professor, Department of Ophthalmology and Visual Neurosciences Minneapolis, Minnesota

Vinod Kumar
Jaypee University of Information Technology
Shimla, India

Pierre-Yves Lagacé
McCaig Institute for Bone and Joint Health
University of Calgary
Calgary, Alberta, Canada

Mathieu Lempereur
Laboratoire de Traitement de l'Information Médical (LaTIM)
and
CHRU de Brest
Brest, France

Maysa M. G. Macedo
Heart Institute
University of Sao Paulo
Sao Paulo, Brazil

Agma Juci Machado Traina
University of São Paulo
São Paulo, Brazil

Arianna Mencattini
Department of Electronic Engineering
University of Rome Tor Vergata
Rome, Italy

Max Q.-H. Meng
Department of Electronic Engineering
The Chinese University of Hong Kong, Hong Kong

Alan R. Moody
Department of Medical Imaging
University of Toronto
Toronto, Ontario, Canada

Ramon A. Moreno
Heart Institute
University of Sao Paulo
Sao Paulo, Brazil

Chisako Muramatsu
Department of Electrical, Electronic and Computer Engineering
Gifu University
Gifu, Japan

Tinashe Mutsvangwa
Department of Biomedical Engineering
University of Cape Town
Cape Town, South Africa

Ram M. Narayanan
Department of Electrical Engineering
The Pennsylvania State University
University Park, Pennsylvania

Marcos Vinícius Naves Bedo
University of São Paulo
São Paulo, Brazil

Marcello H. Nogueira-Barbosa
Ribeirão Preto, Medical School
University of São Paulo, Brazil

Faraz Oloumi
Director, Aurteen Inc.
Calgary, Alberta Canada

Keshab K. Parhi
Professor, Dept of Electrical and Computer Engineering, University of Minnesota, Minneapolis, Minnesota

Glauco Vítor Pedrosa
University of São Paulo
São Paulo, Brazil

José-Antonio Pérez-Carrasco
Signal Theory and Communications Department
University of Seville
Seville, Spain

Nicholas Petrick
Division of Imaging, Diagnostics, and Software Reliability
US Food and Drug Administration
Silver Spring, Maryland

Aria Pezeshk
Division of Imaging, Diagnostics, and Software Reliability
US Food and Drug Administration
Silver Spring, Maryland

Rangaraj M. Rangayyan
Department of Electrical and Computer Engineering
Schulich School of Engineering
University of Calgary
Calgary, Alberta, Canada

Marina S. Rebelo
Heart Institute
University of Sao Paulo
Sao Paulo, Brazil

Brittany Reiche
Image Analysis in Medicine Laboratory
University of Guelph
Guelph, Ontario, Canada

Michael Reinsbach
NuHealth - Ophthalmology Resident, Nassau University Medical Center East Meadow, New York

Christian Roux
Institut Mines-Telecom
Paris, France

Sohini Roychowdhury
Sr. Deep Learning Engineer
Volvocars R&D
Mountain View, California
and
Affiliate Assistant Professor
University of Washington
Seattle, Washington

Berkman Sahiner
Division of Imaging, Diagnostics, and Software Reliability
US Food and Drug Administration
Silver Spring, Maryland

Marcello Salmeri
Department of Electronic Engineering
University of Rome Tor Vergata
Rome, Italy

Ravi K. Samala
Department of Radiology
University of Michigan
Ann Arbor, Michigan

Anderson G. Santiago
Heart Institute
University of Sao Paulo
Sao Paulo, Brazil

Marcel Koenigkam Santos
School of Medicine of Ribeirao Preto
University of Sao Paulo
Sao Paulo, Brazil

Carmen Serrano-Gotarredona
Signal Theory and Communications Department
University of Seville
Seville, Spain

Eric Stindel
Laboratoire de Traitement de l'Information Médical (LaTIM)
and
CHRU de Brest
Brest, France

Caetano Traina
University of São Paulo
São Paulo, Brazil

Jitendra Virmani
CSIR-CSIO
Chandigarh, India

Jun Wei
Department of Radiology
University of Michigan
Ann Arbor, Michigan

Oliver Weinheimer
Department of Diagnostic and Interventional Radiology
Heidelberg University
Heidelberg, Germany

Yixuan Yuan
Department of Electronic Engineering
The Chinese University of Hong Kong,
Hong Kong

Huiyuan Zhou
Department of Electrical Engineering
The Pennsylvania State University
University Park, Pennsylvania

1

Segmentation and Characterization of White Matter Lesions in FLAIR Magnetic Resonance Imaging

Brittany Reiche

Jesse Knight

Alan R. Moody

April Khademi

1.1 Introduction

Acute ischemic stroke is described as the sudden interruption of blood flow to the brain that results in the deprivation of oxygen and nutrients to the cells; and stroke duration directly increases the risk of permanent brain damage. According to Statistics Canada, a government agency commissioned with the production of statistics to analyze all aspects of life in Canada, strokes were the third leading cause of death in Canada in 2011 (Statistics Canada, 2011), and they represent a 3.6 billion dollar a year burden on the economy in associated health costs and lost wages (Public Health Agency of Canada, 2011).

Physicians are now looking at Magnetic Resonance Images (MRI) to identify precursors to strokes. There is a strong relationship between white matter lesions (WML) and risk of stroke, as well as correlations with Alzheimer's disease (Oppedal et al., 2015), multiple sclerosis (Grossman and McGowan, 1998), and vascular dementia (Hajnal et al., 1992). It has been noted that the prevalence of WML increases with age, and that the lesions are more common and extensive in those who already have cardiovascular risk factors or symptomatic cerebrovascular disease. WML are best seen in Fluid-Attenuated Inversion Recovery (FLAIR) MR images, manifesting as hyperintense objects distributed throughout the white matter, and this imaging modality has enhanced discrimination of ischemic pathology (Malloy et al., 2007). The total volume of these lesions are an important prognostic indicator for stroke risk (Altaf et al., 2006).

Traditionally, WML volume measurements are obtained by manual delineation; however, this is known to be laborious and subject to inter- and intra-observer variability. Automated image analysis techniques are a better alternative as they can segment WML accurately, efficiently, and consistently (Khademi et al., 2012). These methods are also ideal for large databases, as images can be processed quickly and without user intervention, in a way not feasible with manual processing. This is particularly important because technological advances have given way to the consolidation of large image repositories for multi-center studies. By analyzing this quantity of data, results will have more statistical significance and power (Suckling et al., 2014). However, due to the multi-center nature of this data, there is greater variability in image quality, contrast, and resolution. Methods developed for automatic segmentation must be able to account for these variations in order to be robust.

Many automatic segmentation methods have already been developed and generally fall into two categories: model-based or nonparametric. Model-based approaches tend to use intensity-based pixel classification with the Expectation–Maximization (EM) algorithm (Santago and Gage, 2003; Cuadra et al., 2002), where the model is constructed using a Gaussian Mixture Model (GMM). The results from these techniques are promising; however, they are based on the assumption that the underlying intensity distributions are Gaussian and also require estimates of distribution parameters. These assumptions lead to inaccurate segmentations in images from multi-coil MR scanners, as intensity distributions may be non-Gaussian and/or nonstationary (Khademi et al., 2009a). Also, the signal values of pathologies, like WML, do not follow a known distribution and cannot easily be handled by model-based approaches. Nonparametric techniques attempt to overcome the use of these assumptions by processing co-registered, multi-modality datasets (i.e., T1, T2, PD, FLAIR) (Anbeek et al., 2004; Lao et al., 2006; de Boer et al., 2007) to perform segmentation. These modalities are subsets of the MRI modality, where the resultant images have varying contrast qualities based on different parameter settings at image acquisition. This eliminates the requirement of assumed distributions, but increases the cost of image acquisition (multiple modalities per patient), computational complexity, and introduces registration error, reducing the appeal of this approach.

Current manual analysis of WML focuses on using total lesion volume in order to draw conclusions about the relation of the pathology to patient outcomes; however, other features not readily defined and measured by a human observer could hold more information about the disease. For example, shape characteristics of the lesions could be measured using quantitative image analysis techniques and then could be used to describe disease in a novel manner. Specifically, robust shape features could differentiate between types of WML, as well as characteristics associated with other neurodegenerative diseases (Khademi et al., 2014).

Currently, WML are divided into two classes: periventricular WML (PVWML) and deep WML (DWML). The pathophysiology of many diseases associated with WML can be better understood if the mechanisms causing PVWML and DWML can be characterized (Woong Kim et al., 2008). Because the visual appearance of these lesions differs, this chapter investigates whether shape analysis techniques can be used to differentiate between the two WML classes.

In order to differentiate between the two types of WML, the calculation of shape features must be reliable and robust. As such, this chapter is focused on the robust and highly accurate segmentation of WML. This is a difficult task, as MR images are degraded by a range of artifacts, such as acquisition noise, intensity inhomogeneity, intensity non-standardness, and partial volume averaging (PVA).

To combat these sources of variability, a model-free, efficient approach to WML segmentation is presented. The algorithm focuses on the correction of PVA, yielding lesions which are segmented at subvoxel precision to produce boundaries that are ideal for reliable shape analysis. Initially, the fundamentals of MRI and the FLAIR modality, specifically, are examined; this is followed by the sources of variability present in MRI. A framework for exploratory noise analysis is presented, which works to identify the characteristics of image noise. A novel pipeline for image standardization is also explained. Next, methods for PVA-based WML segmentation and shape metric calculations are shown. Finally, results demonstrate non-Gaussian noise in MR images and the necessity of image standardization. The results of WML segmentation in both ischemic stroke and Alzheimer's subjects are detailed, and classification using shape characterization is explored.

1.2 Background

In order to understand the challenges of segmenting WML in MR images, a foundation in imaging physics is required. Therefore, this section begins by introducing the principles of MRI. Attention is given to the FLAIR modality specifically, which is the leading-edge standard for imaging, segmenting and characterizing WML (Hajnal et al., 1992).

1.2.1 MRI Fundamentals

MRI provides high-contrast images of anatomical structures without the use of ionizing radiation. This modality has become the standard for visualizing neuropathologies, particularly those characterized by the prevalence of WML, such as in cases of ischemic stroke (Brant-Zawadzki et al., 1996), multiple sclerosis (Grossman and McGowan, 1998), and Alzheimer's disease (Oppedal et al., 2015).

MRI is based on the principle of nuclear magnetic resonance (NMR), which is concerned with the behavior of nuclei under the influence of a strong magnetic field. Clinical MRI typically examines hydrogen atoms (protons) in tissues with large water and fat composition. Hydrogen atoms within the body have NMR characteristics that are influenced by tissue type and pathology, allowing the imaging of anatomical structures and lesions (Hornak, 2014).

In the absence of a magnetic field, protons in the body are randomly oriented, and the net magnetic effect of the orientations is zero. When placed in a large magnetic field, a proportion of the atoms align in the direction of this field, yielding a nonzero net magnetization. The signal is then produced by perturbing the spins from their alignment using a radio frequency (RF) pulse.

Following a RF pulse with enough energy to tip the protons away from the strong magnetic field, the magnetization will recover in the original direction. This rate of recovery is characterized by the T1 time constant; however, this differs from the rate of magnetization decay in the transverse direction, which has a time constant T2. With the knowledge of these time constants, the acquisition parameters of the scan can emphasize the image gray levels, according to the T1 and T2 properties of the tissues.

Given that the brain is composed primarily of water and fat, the soft tissue structures are imaged with good clarity.

Gradients are then applied in the X, Y, and Z directions to encode spatial information in their respective directions. The signal is acquired using a receiver coil, which is positioned at a right angle to the main magnetic field, and the positional information is then encoded into a Fourier-domain representation of the image. The inverse Fourier transform of this data is then taken to produce the image.

Image resolution is determined by the area that is excited by the gradient field. If a small area is excited, the pixel resolution in the resultant images will be finer. Slice thickness is affected in a similar way. If a large slice is selected, the resolution in this direction will be low. Images with very high resolution can be acquired by exciting very small volumes; however, this requires more excitations to obtain the images, which significantly increases scan time.

1.2.2 Fluid Attenuation Inversion Recovery (FLAIR) MRI

For the identification of pathologies in tissues, T2-weighted images are often used, as fatty tissues appear dark, and water-filled tissues appear bright. When a scan is taken of a subject with a pathology, the gray and white matter appear as dark gray, and the pathology-affected area lights up due to the edema caused by inflammation from the pathology (MacKay et al., 2014). However, the cerebral spinal fluid (CSF) also appears bright and reduces the ability to discriminate between CSF and lesions along their boundaries in ventricular and cortical regions.

This is remedied by using an inversion recovery (IR) sequence to null the signal from the CSF; the sequence is also known as FLAIR. This, combined with an extended echo time (TE), produces heavily T2-weighted sequences without bright CSF. It has been demonstrated that FLAIR provides greater contrast between the brain and lesions, particularly where CSF and PVA artifacts confound segmentation in traditional T2-weighted data, that is at the edge of the hemispheres and at the interface of white matter and gray matter areas (Essig et al., 1998).

Figure 1.1 shows an example of a T2-weighted scan, as well as its corresponding FLAIR image. In (a), it can be seen that the brain tissue shows up as darker than the surrounding CSF. In (b), the CSF signal was nulled during acquisition, giving the brain boundaries crisper edges and emphasizing WML. This makes the identification of WML easier, as the edges are more defined and there is more contrast between the lesions and the surrounding brain tissue.

It was first concluded that the T2-weighted hyperintensities in the white matter arose from unmyelinated or sparsely myelinated nerve fibers (De Coene et al., 1992), which are characteristic of disease.

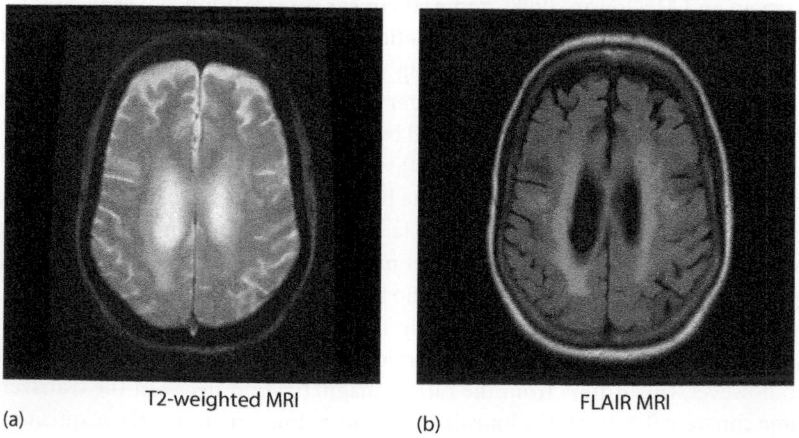

FIGURE 1.1 A T2 image (a) and the corresponding FLAIR image (b).

This sequence provides high lesion contrast within the white matter itself, indicating pathologies in the tissue (Essig et al., 1998). By analyzing FLAIR images, quantitative measurements of WML can be taken to help monitor the progress of neurological disease and aid in prevention and management of these diseases.

1.3 Challenges of Segmenting FLAIR MRI

When acquiring an MR image, there are several factors that reduce the quality of the image by introducing artifacts; these include acquisition noise, intensity inhomogeneity, intensity non-standardness, and partial volume averaging (PVA). Without these factors, images would have unique pixel values assigned to each class, and image segmentation could be done simply with peak detection and thresholding techniques. These artifacts may not appear significant to a human viewer, but can pose challenges for machine-learning algorithms.

The commonly used mathematical model of noise in MR images considers the ideal image, $f(x_1, x_2)$, where x_1, x_2 represent the pixel intensities at different locations in the image, distorted by bias field, $\beta(x_1, x_2)$, and acquisition noise, $n(x_1, x_2)$:

$$y(x_1,x_2) = f(x_1,x_2) \times \beta(x_1,x_2) + n(x_1,x_2), \tag{1.1}$$

Yielding the resultant image, $y(x_1, x_2)$. Noise and bias field are discussed in Sections 1.3.1 and 1.3.2, respectively. The other sources of variability, intensity non-standardness, and PVA cannot be modeled with this equation. These artifacts are discussed in Sections 1.3.4 and 1.3.5, respectively. Variability between images in a database is further increased by the variations of scanning parameters during image acquisition, as discussed in Section 1.3.3. Although these artifacts may not be significant to a human observer, they may present obstacles to algorithms for automatic analyses. By suppressing and/or removing these artifacts, automatic algorithms can be simplified and made more robust (Zhuge and Udupa, 2009; Palumbo et al., 2011; De Nunzio et al., 2015).

1.3.1 Acquisition Noise

Additive acquisition noise poses a significant challenge for automated WML segmentation, as it changes the intensity profile of each tissue class. With noise, image classes are no longer discernible in the image histogram. This reduces the contrast between tissue classes.

To reduce the effects of acquisition noise, many works have investigated MRI noise characteristics and have tuned processing frameworks to reflect these properties. This requires a knowledge of the acquisition process, as noise in the output image is inherently related to the method by which the image was acquired. Single-and multi-coil scanners are most commonly used, and these approaches create substantially different noise properties in their resultant images, as discussed below.

1.3.1.1 Single-Coil Noise Models

There has been extensive work done to model and remove noise in conventional MRI. Models of the imaging physics have revealed that noise is Gaussian at the receiver coil of the scanner, and that transforming the received data to the image domain via the Fourier transform yields a Rician distribution (Dietrich et al., 2008), which simplifies to a Rayleigh in background regions. For high signal-to-noise ratio (SNR) images, the noise distribution can be approximated by a Gaussian. There are three noise assumptions that are generally employed for conventional (single-coil) MR images: (1) the noise field is assumed to have a Gaussian, Rician, or Rayleigh probability density function (PDF); (2) this distribution is assumed to be stationary (not varying with spatial location); and (3), that the individual noise pixels are assumed to be spatially uncorrelated (Thunberg and Zetterberg, 2007; Dietrich et al., 2008). These assumptions underpin many popular neuroimage classification algorithms, including the GMM-based approaches used in the Statistical Parametric Mapping (SPM) toolbox (Ashburner and Friston, 2005; Cuadra et al., 2005). Also, the hidden Markov Random Field model employed in the widely used FMRIB Software Library (FSL, by the FMRIB Analysis Group) toolbox was developed and tested using simulated images containing uncorrelated and stationary Gaussian noise (Zhang et al., 2001).

1.3.1.2 Noise in Parallel MRI

A novel and more modern image acquisition technique, known as parallel MRI (PMRI), produces images using a very different acquisition and reconstruction workflow than single-coil systems. In PMRI, multiple sensor coils are employed to facilitate Fourier-domain subsampling, yielding two- to four-fold reductions in scan time. Advanced reconstruction methods are used to interpolate the missing data either in the image domain, as in SENSE (SENSitivity Encoding), or in the Fourier domain, as in GRAPPA (GeneRalized Autocalibrating Partially Parallel Acquisition) (Thunberg and Zetterberg, 2007). For this reason, many scan sequences are expected to be replaced by multi-coil parallel MRI acquisition in the coming years (Deshmane et al., 2012).

Unfortunately, PMRI systems do not have the same noise characteristics as single-coil systems. For instance, noise fields in PMRI have been shown to vary significantly over the image plane and may not follow typical distributions, such as a Gaussian (Aja-Fernandez et al., 2014; Dietrich et al., 2008; Khademi, 2012). For this reason, the use of methods designed for single-coil images on PMRI is invalid and liable to give erroneous results. While there have been considerable efforts to model PMRI noise, they are almost always theoretical or require per-scan coil sensitivity profiles. Moreover, this approach is limited by the prevalence of proprietary reconstruction algorithms, which cannot be modeled. Rather, a more robust solution in the face of modern MR noise variability would be the use of models that make minimal assumptions about image noise.

1.3.2 Intensity Inhomogeneity

Intensity inhomogeneity, also known as bias field, refers to the artifact that introduces a variation of signal intensities within an acquired image, caused by poor RF coil uniformity. This artifact is known to affect image analysis algorithms, including those used for segmentation and tissue classification. In a standard 1.5T MR scanner, the magnitude of the intensity variation can potentially exceed 30% of the signal's actual value (Hah et al., 2014). During image acquisition, the RF coil is placed as close as possible to a standardized location relative to the desired region for imaging. This is done to maximize SNR, but inevitable imperfections in the RF coil and any deviation from the nominal position of the coil cause a low-pass filtering effect of the Fourier domain. When the image is calculated by taking the inverse Fourier transform of the raw data, the result corresponds with a pixel-wise multiplication of the ideal image with a bias. This artifact may be imperceptible to a human observer, but it is a challenge for an automated algorithm that relies on mathematical relationships in order to perform the same analyses.

1.3.3 Scanning Parameters

In addition to variability in reconstruction methodologies, images from multi-center databases have a wide range of acquisition parameters. Even for the same scan protocol, the magnetic field strength, repetition time (TR), echo time (TE), and inversion time (TI) can differ between institutions and vendors; therefore, image quality and contrast levels may still vary. The TR, TE, and TI are known to affect contrast levels, and TE and TR are directly related to the SNR (Bydder and Young, 1985). TR affects the image contrast, where a long TR yields T2-weighted images, and a short TR gives T1-weighted images. A long TR allows for all of the protons to fully relax back into alignment with the main magnetic field. Reducing this time could prevent the protons of some tissues from fully relaxing before the next measurement, lowering the SNR in the next scan. This can affect the contrast between neighboring tissues.

A longer TE can reduce the signal strength, as the protons are more likely to come out of phase, whereas a shorter TE reduces the amount of dephasing that can occur, yielding a higher SNR. Magnetic field strength is also related to image quality, as increased magnetic strength results in a higher signal relative to constant acquisition noise. These parameters all have an effect on the characteristics of resultant images, and can create variability between images within the same database. Algorithms for automated analyses must be able to account for these differences.

1.3.4 Intensity Non-Standardness

Intensity non-standardness refers to the lack of standardization of the MR image intensity scale. For instance, if a subject has two scans done on the same region, with the same scanner, with the same protocol, the resultant images will yield different pixel intensities (Nyul and Udupa, 1999). This is a challenge for fully automated methods, as there is no consistency to the intensities of the tissues that the algorithm is trying to identify (Palumbo et al., 2011).

1.3.5 Partial Volume Averaging (PVA)

The ideal pixel contains the signal of one object class. PVA is the effect where a pixel represents more than one object class in an image, which is common for pixels that lie on the boundary of an object. PVA blurs the intensity distinction between the tissue classes in image histograms and leaves the object boundaries looking unclear. These mixture voxels can lead to a 30%–60% error in the volume measurement of complex brain structures (Samsonov and Johnson, 2004). PVA-affected pixels have intensities that are linearly dependent on the proportion of each tissue in the pixel (Khademi et al., 2009b). In an example where two tissue classes are present in a pixel at a spatial coordinate of $x = (x_1, x_2) \in Z^2$, the voxel intensity is determined by the proportion of Tissue 1 present at x, as in:

$$Y_{12}(\mathbf{x}) = \alpha(\mathbf{x}) \cdot Y_1(\mathbf{x}) + (1 - \alpha(\mathbf{x})) \cdot Y_2(\mathbf{x}), \tag{1.2}$$

where:

$Y_{12}(\mathbf{x})$ is the resultant intensity of the PVA voxel

$Y_1(\mathbf{x})$ is the intensity of the first tissue where $Y_1 \sim p_1(y)$

$Y_2(\mathbf{x})$ is the intensity of the second tissue where $Y_2 \sim p_2(y)$ and α (commonly referred to as the tissue fraction) is the proportion of the first tissue present in the PVA voxel where $\alpha \in [0, 1]$.

The amount of PVA in an image is dependent on the slice thickness and resolution parameters during image acquisition. Thick slices will have a more noticeable PVA effect, and because of this, the histograms of each tissue cannot be described by a single intensity value, but will instead occupy a range of values that overlap with neighboring classes. This PVA effect, along with the image noise that is inherent to MRI, makes it difficult to accurately segment imaged tissues, since it is difficult to determine the boundaries of image objects.

1.4 Framework for Exploratory Noise Analysis on Modern MR Images

Given the variability in noise characteristics associated with parallel MRI acquisition, as well as other factors, it is prudent to explore image noise in a database before employing automated image analysis. This can determine whether images contradict the assumptions of potential downstream models and can identify the correct de-noising procedures. Note that it is typically impossible to characterize the noise within the imaged object, as $f(x_1, x_2)$ is almost always unknown. Instead, the image background (where $f(x_1, x_2) = 0$) is isolated and used to quantify noise properties.

The three important noise field attributes for investigation are: (1) whether the probability distribution varies significantly over the image space (non-stationarity), (2) whether these distribution(s) agree with common models (e.g., Gaussian PDF) and, (3) whether there is spatial correlation in the signal. The statistical tests described below test each of these features, while making minimal assumptions about the distributions, and they represent a framework for diligent consideration of image noise before choosing and designing appropriate image analysis methods.

1.4.1 Testing for Stationarity

Stationarity should be tested first because if there is evidence of non-stationarity, the global PDF will not be representative of the noise in the image. If the noise is stationary, the distribution, or parameters of the distribution, will vary as a function of space.

To test for stationarity without assuming any underlying distribution, we compare the distribution of several patches in the image background. We consider background noise as it is not contaminated by tissue signal. First, image slices are subdivided into patches, and only patches completely within the background mask are considered. Ideally, each background patch is compared with all other background patches, and all pairs are tested for distribution similarity.

A good comparative test for patch pairs is the two-sample Kolmogorov-Smirnov (KS) Test. This tests the null hypothesis that the data in both samples (image patches) come from the same underlying distribution, but makes no assumptions about the parameterization of that distribution. The test statistic is the mathematical supreme of the difference between the cumulative distribution functions (Kvan and Vidakovic, 2007). Non-stationarity is indicated if the majority (>50%) of comparisons reject the null hypothesis of equal distributions.

1.4.2 Testing for Common Distributions

Even if the noise is found to be non-stationary, it may still be distributed according to a known distribution, just with parameters that vary over the image volume (Aja-Fernandez et al., 2014). For this reason, each hypothesized distribution should also be tested using patches. To test a patch for a given distribution, the distribution parameters are first estimated using the maximum likelihood. Then, a KS Test is used to test whether patch data come from the optimally fit distribution, found by using the estimated parameters (Kvan and Vidakovic, 2007). Distributions that were investigated included Gaussian, Rician, Rayleigh, and Weibull.

1.4.3 Testing for Spatial Correlation

Lastly, it is important to investigate the presence of local signal correlation. An extension of Mantels test for spatial clustering, the 2D Spatial Correlation Test (2DSCT), has been shown to indicate this feature without making assumptions about the distribution of the data (Khademi, 2012). The test computes a correlation statistic, M_2, for the gray-level differences between pixels and the spatial distances between those pixels. For a patch of image Z, with a dimensional spatial indexing variable $s = (x_1, x_2)$, M_2 is calculated using:

$$W_{ij} = \left\| s_i - s_j \right\|, \tag{1.3}$$

$$U_{ij} = | Z(s_i) - Z(s_j) |, \tag{1.4}$$

$$M_2 = \sum_{i=1}^{N} \sum_{j=1}^{N} W_{ij} U_{ij}. \tag{1.5}$$

The statistic for the original arrangement, M_2^{obs}, is compared with the distribution of M_2 values for random rearrangements of the same data, and the hypothesis of no correlation is then rejected if M_2^{obs} is sufficiently extreme; a Z-test was used to make the comparison, and a significance level of $\alpha = 0.05$ is used.

1.5 Standardization and Brain Extraction

In order to account for the variability present in large-scale studies, a standardization framework was developed to suppress the effects of MRI artifacts, as presented in Reiche et al. (2015).

To reduce the effects of high-frequency acquisition noise, a low-pass filter was applied. Bias field was reduced using a method similar to Zhong et al. (2014), where the image was divided by a low-pass filtered version of itself. This low-pass filtered image is representative of the slowly varying artifact, and this approach was successful at removing the multiplicative noise. To account for intensity non-standardness, a histogram-matching algorithm (Reinhard et al., 2001) was used to transform the histograms into the same space by matching the mean and standard deviation of the database images to those of a template image (Winkler et al., 2012).

To demonstrate the benefits of standardization, brain extraction was performed on standardized and non-standardized images. Methods are further described in Reiche et al. (2015).

1.6 PVA Quantification and WML Segmentation

This section details the methods used to quantify PVA, as well as perform segmentation of WML using a PVA-based approach. In many current works, algorithms designed to segment structures in neurological MRI use the EM approach to perform intensity based tissue classification (Cuadra et al., 2002; Santago and Gage, 2003; Dugas-Phocion et al., 2003) using assumed distributions, which are generally Gaussian (Cuadra et al., 2002; Lin et al., 2004; Ballester et al., 2002). An extension on this parametric approach is Markov random fields, which impose spatial constraints (Lin et al., 2004; Van Leemput et al., 2003).

Although results from these approaches are promising, they make assumptions regarding the underlying distributions of the images, which we have noted is not a valid assumption when dealing with multi-coil scanners. Pathology also presents an obstacle, as the distribution of diseased tissue does not follow a known distribution.

Nonparametric classification approaches attempt to avoid the use of models by processing co-registered, multi-modality datasets (i.e., T1, T2, PD) (Anbeek et al., 2004; Lao et al., 2006; de Boer et al., 2007). By using additional modalities, the dependence on model parameters is reduced, but the costs of image acquisition (multiple modalities per patient), computational complexity, and potential for registration errors are increased.

Model-based approaches are not easily applied to pathology segmentation, as the distributions of disease in an image are not easily modeled. In addition, neurological MRI often have non-Gaussian or unknown noise properties, meaning that approaches that assume normality will be inaccurate. To overcome these obstacles, the current work uses a model-free, adaptive PVA modeling approach for the robust segmentation of WML in FLAIR MRI. It is computationally efficient and only requires the FLAIR modality, as it exploits a novel mathematical relationship between edge information and PVA for robust PVA quantification and tissue segmentation. The following subsection details this approach, which has been introduced in Khademi et al. (2012) and Khademi et al. (2014).

1.6.1 The PVA Model

The PVA in neurological MRI usually contains the mixture of two tissue types (Ballester et al., 2002), as seen in Equation 1.2. When analyzing WML in FLAIR, there are three pure tissue classes in the ideal model: CSF, GM/WM, and WML. Assuming standardized images (each class has its own discrete intensity), an ideal multi-class PVA model is constructed from Equation 1.2:

$$Y_{12}(\mathbf{x}) = \alpha_{12}(\mathbf{x}) \cdot I_1 + (1 - \alpha_{12}(\mathbf{x})) \cdot I_2, \tag{1.6}$$

$$Y_{23}(\mathbf{x}) = \alpha_{23}(\mathbf{x}) \cdot I_2 + (1 - \alpha_{23}(\mathbf{x})) \cdot I_3, \tag{1.7}$$

where $Y_{12}(\mathbf{x})$ and $Y_{23}(\mathbf{x})$ are the intensities of PVA voxels in the WML-brain and brain-CSF boundaries, respectively. The brightest tissue (WML) is denoted by I_1 where $I_1 > I_2 > I_3 \geq 0$

The parameter α_{12} describes the percentage of the voxel that is made up of WML and its quantification is required for robust WML volume computation. Finding an image-based estimate of α is ideal, as this does not rely on models or multiple modalities. In the ideal image, pure classes appear as constant, while mixture (PVA) classes occur over specific intensity intervals in the edges between pure classes. From this, it was concluded that intensity and edge strength features can discriminate between pure and mixture classes (Khademi et al., 2012). The remainder of this section examines edge content to model PVA (by estimating α).

1.6.2 Edge-Based PVA Modeling

To examine the edge content of PVA regions, the gradient of the ideal signal model can be calculated, resulting in:

$$Y'_{12} = \alpha'_{12} \cdot (I_1 - I_2), \qquad Y'_{23} = \alpha'_{23} \cdot (I_2 - I_3), \tag{1.8}$$

where α' is the change in tissue fraction, and dictates how the proportion of a tissue changes as a function of space. Solving for the change in tissue fraction, α', results in two PVA quantifiers for each PVA region:

$$\alpha'_{12} = \frac{Y'_{12}}{I_1 - I_2}, \qquad \alpha'_{23} = \frac{Y'_{23}}{I_2 - I_3}. \tag{1.9}$$

Because edge information is quantified by a gradient, each α' is a class-specific, normalized representation of the edge information in the PVA voxels (edgy regions). The current work focuses on an edge-based estimate for α', and uses it to decode α, the partial volume fraction, which describes the proportion of each tissue in every voxel. This is a novel innovation that holds the potential to change the way that PVA is handled in MR image analysis algorithms. It does not depend on an intensity distribution model, or multiple modalities, but is based on the mathematical relationship that describes the PVA voxels in term of their respective edge content.

1.6.3 Fuzzy Edge Model

To estimate $\alpha'(\mathbf{x})$, a fuzzy technique based on the cumulative distribution function (CDF) of the gradient (Khademi et al., 2009b; Meer and Georgescu, 2001), is employed. First, the traditional magnitude of the gradient, $g = \|\nabla_y\|$ is calculated by using the Sobel operator. Based on the PDF of the gradient $p_G(g)$, the CDF is found and used as an estimate for the edge information in the image:

$$\alpha'(g) = \text{Prob}(G \leq g) = \sum_{i=0}^{g} p_G(i), \tag{1.10}$$

where $\alpha'(g) \in [0, 1]$. This nonlinear, probabilistic modeling of the edge information quantifies the "certainty of edge presence." This parameter is expressed as a function of the gradient (to be used to approximate $\alpha'(\mathbf{x})$). $\alpha'(g)$ is then mapped back to the spatial domain: $\alpha'(g) \rightarrow \alpha'(\mathbf{x})$.

As shown in Khademi et al. (2009b) and Khademi et al. (2010), this fuzzy edge measure assigns large and similar values to significant edges, while suppressing irrelevant ones, and is used to represent PVA.

Although such a nonlinear mapping function localizes PVA in ideal images and demonstrates the motivation for an edge-based approach, the local nature of the gradient operator allows noise to severely degrade its performance. Since not all noise can be removed by denoising MRI, a new estimate for α is developed, based on the global edge content instead to combat noise.

1.6.4 Global Edge Description

PVA occurs over specific intensity ranges, and at high edge values. These two features are coupled to arrive at a denoised version of the fuzzy edge metric. They are combined through the conditional PDF of $\alpha'(\mathbf{x})$, for a particular intensity y by:

$$p_{\alpha'(\mathbf{x})|Y}(\alpha'(\mathbf{x}) = a \mid Y = y) = \frac{\#\text{voxels with } \alpha'(\mathbf{x}) = a \mid Y = y}{\#\text{voxels with } Y = y}, \tag{1.11}$$

where $0 \leq \alpha \leq 1$, $0 \leq y \leq y_{max}$, a is the realization of $\alpha'(\mathbf{x})$ and y_{max} is the maximum gray level in the image. This PDF quantifies the distribution of the edge information $\alpha'(\mathbf{x})$ for a specific gray level y. Generally, in flat regions (pure tissues), there is clustering in the PDF for low edge values at corresponding intensities. Across anatomical boundaries (PVA), high edge values dominate the PDF for the respective intensities.

To approximate $p_{\alpha'(\mathbf{x})|Y}(a|y)$, a kernel density estimator is used to create a smooth construction of the frequency distribution (Khademi et al., 2009a). Given a sample of fuzzy edge values $\alpha'(\mathbf{x}_1)$, $\alpha'(\mathbf{x}_2)$, $\cdots$ $\alpha'(\mathbf{x}_n)$, for a specific intensity y, the result is the summation of a series of kernels for all sample data:

$$p_{\alpha'(\mathbf{x})|Y}(a \mid y) = \frac{1}{nh_n} \sum_{i=1}^{n} K\left(\frac{a - \alpha'(\mathbf{x}_i) \mid y}{h_n} \right), \tag{1.12}$$

where:

$K(\cdot)$ is a Gaussian kernel function
h_n is the standard deviation of the Gaussian
n is the number of pixels with intensity y

To discriminate between pure and PVA voxels, two bin locations at $a = 0$ and $a = 1$ are used to solve Equation 1.12. The result is a conditional PDF that describes edge presence for each gray level y for the non-edge class (pure) and edge class (PVA), respectively. The probability distribution for the non-edgy regions is described by:

$$p_{\alpha'(\mathbf{x})|Y}(a = 0 \mid y) = \frac{1}{nh_n} \sum_{i=1}^{n} K'(\alpha'(\mathbf{x}_i) \mid y), \tag{1.13}$$

The probability a voxel lies in a PVA region (significant edge) is found by:

$$p_{\alpha'(\mathbf{x})|Y}(a = 1 \mid y) = \frac{1}{nh_n} \sum_{i=1}^{n} K'(1 - \alpha'(\mathbf{x}_i) \mid y). \tag{1.14}$$

Estimation of the PDF this way automatically classifies the voxel y as belonging to either the pure tissue or PVA class. Pure regions correspond to maxima in $p\ (a = 0|y)$ and minima in $p\ (a = 1|y)$, whereas minima in $p\ (a = 0|y)$ and maxima in $p\ (a = 1|y)$ indicate with high likelihood that these voxels belong to a PVA region.

To determine the global estimate of $\alpha'(y)$, the conditional expectation operator is used. It offers the best prediction of α' given that the intensity is y in the mean square error (MSE) sense. The result is an enhanced edge map $\alpha'(\mathbf{y})$:

$$\begin{aligned} \alpha'(y) &= \sum_{\forall a} a \cdot p_{\alpha'(\mathbf{x})|Y}(a \mid y), \\ &= 0 \cdot p_{\alpha'(\mathbf{x})|Y}(a = 0 \mid y) + 1 \cdot p_{\alpha'(\mathbf{x})|Y}(a = 1 \mid y), \\ &= p_{\alpha'(\mathbf{x})|Y}(a = 1 \mid y). \end{aligned} \tag{1.15}$$

The last line of the equation above indicates that quantification of PVA content is directly proportional to the probability that a voxel is located on an edge.

1.6.5 Estimating α

To decode $\alpha(y)$, some regions of $\alpha'(y)$ are retained, while others are discarded. Recall that the maxima of $p(a = 1|y)$ dictate which voxels y are most likely PVA (maximally edgy), while the minima are correlated to voxels y from pure tissue classes. Ideally, in pure tissues regions, there should be no edge information, but noise creates "artificial" edginess, causing the minima of $\alpha'(y)$ to be nonzero in these regions. To account for the relative nature of $\alpha'(y)$, an adaptive threshold is applied.

An adaptive threshold that retains voxels most likely to contain mixture components are computed for the left and right side of each PVA pulse, where t_L and t_R are the left and right thresholds, respectively. The minima and maxima values are easily found with peak finding. For two pixel classes, an array of thresholds is defined:

$$T = \left(t_{23}^{L}, t_{23}^{R}, t_{12}^{L}, t_{12}^{R}\right), \tag{1.16}$$

where t_{23}^{L}, t_{23}^{R} are applied to the first PVA pulse (CSF-brain PVA), t_{12}^{L}, t_{12}^{R} to the second pulse (brain-WML PVA):

$$\alpha_T'(y) \rightarrow \alpha'(y) > T, \tag{1.17}$$

$$\alpha_T'(y) \rightarrow p\left(a = 1 \mid y\right) > T. \tag{1.18}$$

The nonzero regions correspond to the adaptively found, class-specific estimates for $\alpha'_{\mathrm{jk}}(y)$, where $\alpha_T'(y) = \alpha_{23}'(y) \cup \alpha_{12}'(y)$.

To decode the tissue fraction $\alpha_{\mathrm{jk}}(y)$, each refined PVA pulse $\alpha'_{\mathrm{jk}}(y)$ is integrated over the corresponding intensity values:

$$\alpha_{jk}(y) = \frac{\int_{y_k}^{y} \alpha_{jk}'(t)dt}{\int_{y_k}^{y_j} \alpha_{jk}'(t)dt}, \quad y_k < y \leq y_j, \tag{1.19}$$

where the denominator is a normalizing constant. The final estimate for the proportion of the tissue parameter is the union of these two PVA regions: $\alpha(y) = \alpha_{23}(y) < \alpha_{12}(y)$. This technique automatically detects which gray levels correspond to PVA, and also how much of each tissue is present within these voxels.

1.6.6 WML Segmentation

As the PVA map $\alpha(y)$ dictates how tissue classes are mixing, it can be easily modified to get a class membership function for the WML.

If y_2 and y_1 denote the starting and ending gray level values for $\alpha_{12}(y) > 0$, respectively, the class membership $\xi_{WML}(y)$ for the WML class may be found by:

$$\xi_{WML}(y) = \begin{cases} \alpha_{jk}(y), & y_2 \leq y < y_1, \\ 1 & y \geq y_1. \end{cases} \tag{1.20}$$

A value of "1" is assigned to voxels that are pure WML, a "0" for voxels that are not part of the lesion class, and $0 < \xi < 1$ for PVA voxels (mixture of brain and WML tissues).

To generate binary WML segmentations, $B(x_1, x_2)$, an optimal threshold was determined and applied to class memberships by thresholding the maps at various values $\tau \in (0, 1)$, and comparing the masks to ground truth data with the Dice Similarity Coefficient, *DSC*. Methods for creating ground truth masks can be found in Section 1.8.2. The threshold that yielded the highest average segmentation accuracy was applied to all other membership maps to yield the final binary segmentation. Boundaries of these objects, $s(k) = x(k), y(k)$, where $K = 1{:}\ N$ can be created using boundary tracing methods.

Figure 1.2 shows plots of the *DSC* over all thresholds and the average *DSC* for all images, respectively. 195 images were used in the validation set. The *DSC* was high for lower thresholds. The optimal threshold τ^* was found to be 0.15, suggesting that expert WML segmentation includes the majority of PVA voxels in WML. Similar results were found by analyzing the *ROC* curves as well. Class membership functions for all images were then thresholded using τ^* to yield lesion segmentations.

1.7 Shape Analysis

This section will describe shape analysis methods that can be applied to objectively and quantitatively analyze the shape signatures of different types of WML using the binary masks, $B(x_1, x_2)$, and their boundaries, $s(k)$.

Shape analysis techniques refer to a set of image processing tools that focus on characterizing segmented objects based on their shape. Shape analysis has typically been used for object recognition and matching, boundary filtering, and general shape characterization (Loncarica, 1988). These methods will be used to differentiate between different types of WML.

There are currently two classifications for WML: periventricular white matter lesions (PVWML) and deep white matter lesions (DWML). DWML are characterized as being close to the periphery of the brain (i.e., "deep" in the WM), and the PVWML are described as being alongside the ventricles. A possible dissimilarity between pathogenic mechanisms of the PVWML and DWML may provide clues for understanding the pathophysiology of the many diseases associated with WML (Woong Kim et al., 2008).

Shape descriptors are calculated from segmented objects; therefore, the boundaries of the WML must accurately represent the pathology. Because the proposed segmentation method was PVA-based, the lesions are segmented with subvoxel precision, and can yield WML objects with very accurate boundaries that are ideal for shape analysis.

Two approaches to shape analysis are used in this work: external (boundary-based) and global methods. The external techniques focus on analyzing the boundary of an object, while global methods analyze the geometric shape of the object. This section discusses the shape analysis methods that will be used to discriminate between DWML and PVWML.

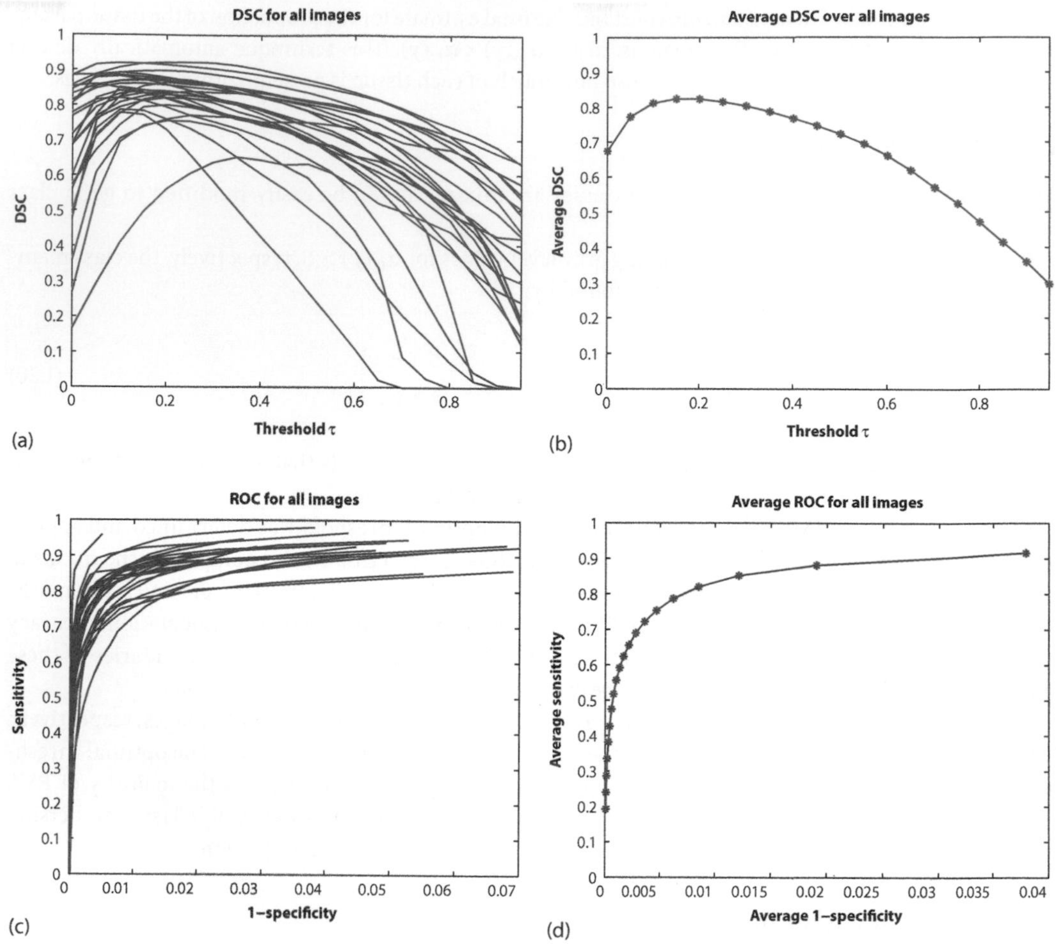

FIGURE 1.2 (a) DSC as a function of threshold τ for all patients, (b) Average DSC curve, which shows optimal performance is achieved with $\tau = 0.15$. (c) ROC plots for all patients, and (d) Average ROC graph.

1.7.1 Boundary-Based Techniques

In boundary-based methods, the contour of the object is represented by its spatial coordinates x_k, y_k, where $k = 0, 1, \cdots, N-1$ for a boundary with N pixels. The kth pixel along the contour is described with two parametric equations: $x(k) = x_k$ and $y(k) = y_k$, which can be described jointly using complex notation: $s(k) = x(k) + iy(k)$.

A series of features are extracted from the boundary representation, known as Fourier Descriptors (FD). They are computed by taking the Fourier transform of $s(k)$, yielding $S(f)$, where the length of the sequence is determined by the number of points used to compute the Fast Fourier Transform (FFT). Since the Fourier transform is linear:

$$S(f) = \mathcal{F}(x(k) + iy(k)), \tag{1.21}$$

$$S(f) = S_x(f) + iS_y(f), \tag{1.22}$$

where $S_x(f)$ and $S_y(f)$ are the spectral coefficients of the x and y coordinates Usually, the magnitude spectrum is used to analyze the overall shape and defines the final set of FD as

$$z(f) = \| S(f) \| = \sqrt{S_x(f).^2 + S_y(f)^2}. \tag{1.23}$$

Fourier descriptors $z(f)$ are a series of spectral components that describe the way a contour changes as the boundary is traversed. For example, a bumpy contour generates a spectrum that is dominated by high-frequency content. Conversely, a smooth boundary would have little high-frequency content. The descriptors $z(f)$ for values of f close to zero will describe an approximate shape and the higher frequencies will describe detail.

The invariance properties of the Fourier descriptors are related to the properties of the Fourier transform, in terms of translation, rotation, scale, and start point. A description of these properties can be found in Khademi et al. (2014). To ensure that the shape descriptors of the WML are robust, the following operations were performed to make the FD invariant: for translation, $z(0) = 0$; for rotation and start point, $z(f) = \|S(f)\|$, and $z(f) = z(f)/z(1)$ for scale invariance. Boundaries were also resampled to make fair comparisons between boundaries of different sized lesions.

1.7.2 Global Shape Metrics

Global shape metrics quantify the geometric shape of the WML. These metrics are attractive because they are not sensitive to boundary definitions and they also have a more intuitive meaning.

Each metric is computed based on the binary segmentation mask for each WML, called $B(x_1, x_2)$. Using the binary masks, several global shape metrics are considered for WML shape analysis and are briefly described below:

1.7.2.1 Circularity

The simplest formula to quantify the circularity of an object is found by:

$$Circ = \frac{P^2}{A}, \tag{1.24}$$

where:

P is the perimeter

A is the area of the WML in $B(x_1, x_2)$

1.7.2.2 Haralick's Circularity

Haralick's measure of circularity is computed as:

$$Circ_H = \frac{\mu_R}{\sigma_R}, \tag{1.25}$$

where:

R is the Euclidean distance computed between every boundary point and the centroid

μ is the mean of these distances

σ is the standard deviation

This metric is like circularity, but is less sensitive to noise and digitization artifacts.

1.7.2.3 Elongatedness

This describes how elongated the WML are. This measure is dependent on the central moments of an object and is calculated as the ratio of the lengths of the axes that describe the best fit ellipse. Further explanation can be found in Khademi et al. (2014).

1.7.2.4 Solidity

This describes the proportion of WML pixels that are contained in the corresponding convex hull (CH):

$$sol = \frac{n(WML \cap CH)}{n(CH)}, \tag{1.26}$$

where:

$n(WML \cap CH)$ is the number of WML pixels inside the convex hull
$n(CH)$ is the total number of pixels that make up the convex hull

1.8 Results

This section describes the experimental results of exploratory noise analysis, image standardization, WML segmentation, and shape characterization. Exploratory noise analysis summarizes the results of statistical testing of PMRI image noise and demonstrates the variations in noise distributions in PMRI images. The effects of artifacts and image standardization on segmentation tasks are demonstrated, and WML segmentation algorithm performance is judged. Lastly, shape characterization was used to classify lesions as PVWML and DWML, and classification accuracy was quantified using the Misclassification Rate (MCR).

1.8.1 Evaluation Metrics

To quantify segmentation accuracy, binary masks generated by the WML segmentation algorithm were compared to ground truth data using the *DSC* and the Receiver Operator Characteristic (*ROC*) curve. The amount of overlap between a segmented object and the ground truth was measured using the *DSC*:

$$\text{DSC}(A,B) = \frac{2\,|A(\mathbf{x}) \cap B(\mathbf{x})|}{|A(\mathbf{x})| + |B(\mathbf{x})|}, \tag{1.27}$$

where $A(\mathbf{x})$ and $B(\mathbf{x})$ are binary masks for the segmentation and ground truth. To generate a *ROC* curve, sensitivity and specificity were also calculated. Sensitivity, *sens*, is defined as the true positive rate:

$$sens = \frac{TP}{TP + FN}, \tag{1.28}$$

where TP and FN are the number of true positives and false negatives, respectively. Specificity, *spec*, describes the true negative rate:

$$spec = \frac{TN}{TN + FP}, \tag{1.29}$$

where: TN and FP are the number of true negatives and false positives, respectively. The TP, FP, TN, and FN are all defined on a voxel-by-voxel basis, using binary masks of the ground truth and segmentation result.

The accuracy of lesion classification was quantified using the Misclassification Rate, which is the percentage of lesions incorrectly classified.

1.8.2 Experimental Data

There were twelve images investigated through exploratory noise analysis. They were acquired in the axial plane on a 3T Philips Achieva, which is a multi-coil scanner. As a result, it is expected that the images are inflicted with non-Gaussian noise. The imaging parameters are as follows: pixel bandwidth of 250°, 90° flip angle, 12 bit, $0.42857 \times 0.42857 \times 3$ mm^3 voxel dimensions, TR/TE/TI = 9000/125/2800 ms, FOV 240×240 mm. FLAIR data from twenty-four patients was used to examine and quantify the WML segmentation performance and lesion characteristics. These images were acquired in the axial plane on a 1.5T GE Signa Excite. This is also a multi-coil scanner, causing non-Gaussian noise characteristics in the images. The imaging parameters are as follows: pixel bandwidth of 97.65°, 90° flip angle, 16 bit, $0.5 \times 0.5 \times 5$ mm^3 voxel dimensions, TR/TE/TI = 8000/128/2000 ms, FOV 180×240 mm.

In addition, two subjects from the Alzheimer's Disease Neuroimaging Initiative (ADNI) Database were used for analysis (Alzheimer's Disease NeuroImaging Initiative, 2015). These images were acquired in the axial plane on a 3T Siemens Verio, which is also multi-coil. The imaging parameters are as follows: pixel bandwidth of 222°, 150° flip angle, 12 bit, $0.8594 \times 0.8594 \times 5$ mm^3 voxel dimensions, TR/TE/TI = 9000/91/2500 ms, FOV 220×220 mm.

These images will highlight the benefits of the current approach over traditional model-based approaches as these images have non-Gaussian noise characteristics and pathology which cannot be accurately accounted for with Gaussian intensity distribution models.

Validation of algorithms on real images requires the manual segmentation of the objects being detected. To generate the ground truth data, a radiologist used the Sedeen Image Viewer (Pathcore, 2015), a specialized software platform, to outline WML objects in each image.* The estimated and ground truth WML masks will be compared using various metrics. Ground truth masks were not available for the ADNI images, and the resultant segmentations from this data are used to demonstrate the robustness of the algorithm across scanner manufacturer, scanner parameters, and differing pathologies.

1.8.3 Exploratory Noise Analysis

The analysis pipeline that has been described in Section 1.4 is demonstrated on 12 SENSE-reconstructed images, having size $560 \times 560 \times 50$, as represented in Figure 1.3. The background was isolated using 2-class

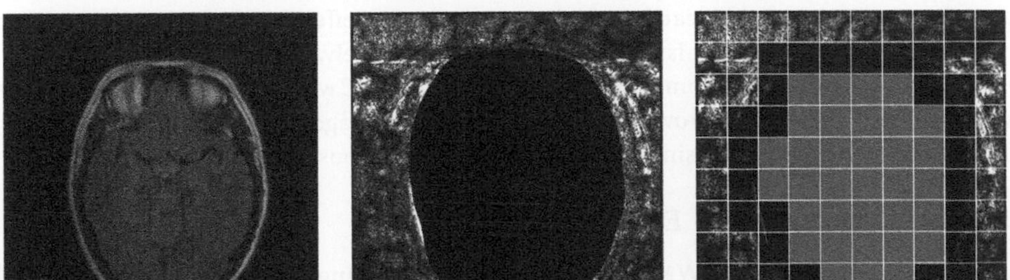

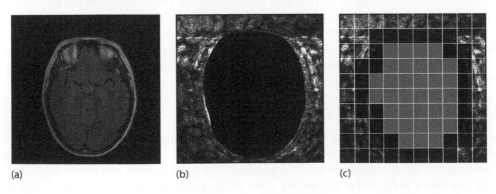

FIGURE 1.3 (a) Original PMRI FLAIR image, (b) corresponding background signal at 100 contrast scale, and (c) patched background signal.

* http://www.pathcore.ca/sedeen/

k-means segmentation on a low-pass filtered version of each image, followed by morphological dilation of the background mask by 2% of the image width. This ensures highly specific selection of background voxels, at the expense of sensitivity, as some voxels adjacent to the head are excluded. Patch sizes of $56 \times 56 \times 1$ (2D) were used throughout, and an example is shown in Figure 1.3c. Uncolored patches are completely background and therefore used for testing; red patches contain noise signal which was not tested due to incomplete overlap with the background mask; yellow patches were completely head signal.

The summary for each noise test is shown in Table 1.1. Shown are the percentages of patch tests for which the hypotheses of single-coil or classical assumptions could not be rejected at $\alpha = 0.05$. The hypotheses are: (1) Stationarity: data distributions are equal for all patches; (2) PDF Match: data are distributed according to the tested distribution; and (3) Uncorrelation: there is no spatial correlation in the patch data. For the uncorrelated data, 1,000 patches were used for computational efficiency.

The results of the noise characterization tests indicate that the three classical noise assumptions are invalid for all 12 images in the tested database (Table 1.1). The images demonstrate significant non-stationarity: 96% of patch-to-patch comparisons rejected the hypothesis of equal distributions. This is easily understood by visual analysis of the PDFs of the data, as shown in Figure 1.4a. A large variability in the probability functions of the data can be seen. Similarly, patch data appeared to rarely conform to any common distribution, with the most consistently plausible distribution being the Weibull (Figure 1.4b); however, this only agrees with about 10% of tested patches. Finally, 88% of the patches showed evidence of spatial correlation using the 2DSCT.

In single-coil MRI, noise is assumed to be stationary, uncorrelated, and follow a consistent distribution over the entire image volume. These assumptions have been incorporated into many modern MRI analysis algorithms. However, parallel MR images have fundamentally different noise characteristics from single-coil images, and many of these violate the single-coil assumptions. The implications of these findings suggest that a number of Gaussian-based model approaches may no longer be applicable to segmentation tasks on PRMI images.

Due to the variability of PMRI reconstruction algorithms and their proprietary secrecy, it is almost impossible to adequately model noise in a large clinical database. Rather, exploratory analysis should be used to understand the noise characteristics in the database, so that the appropriate algorithms can be used. For instance, in a sample of 12 SENSE-reconstructed FLAIR images, the overwhelming evidence suggests that image noise is non-stationary, spatially correlated, and not distributed according to common distributions.

1.8.4 Standardization

Given the number of artifacts in MRI (Section 1.3) that pose challenges in WML segmentation, the need for a robust standardization protocol is obvious. A sample of brain extraction before and after standardization is show in Figure 1.5. As it can be seen, by suppressing the effects of various artifacts, the rate of false positives in the segmentation is greatly reduced. Quantitatively, it was found that standardization increased segmentation overlap from 65.6% to 82.8% using the *DSC* when compared to manual segmentations (Reiche et al., 2015). This proves that standardization of images in large databases allows for the simplification of automatic processing methods, and increased robustness.

1.8.5 WML Segmentation Evaluation

This section details the results of WML segmentation. Preprocessing of the images included standardization and brain extraction (Section 1.5). Manual segmentation was performed on 25 randomly chosen images. Sample manual segmentations, corresponding PVA quantification and binary segmentations of WML are shown in Figure 1.6. The interior of the WML (pure class) were classified with high probability, and the boundaries received a decreasing value, due to PVA.

The average *DSC*, sensitivity, and specificity for all images at the optimal threshold value, τ^*, are shown in Table 1.2. These results demonstrate good segmentation performance over a variety of patients and lesion loads.

TABLE 1.1 Summary of Exploratory Noise Analysis Framework for 12 Images

Image no.		1	2	3	4	5	6	7	8	9	10	11	12	Mean
# Tested patches %		2050	2498	2409	2408	2500	1917	1871	2092	2093	2113	2459	2468	2240
Stationary		3.8	5.2	6.7	2.8	1.7	4.2	5.6	1.4	8.5	5.5	4	2.7	4.34
M% PD match	Gaussian	0	0	0.2	0	0	0	1.2	0	0.2	0.1	0	0	0.14
	Rician	0.6	0.2	0.3	0.6	0.7	0.1	1.3c	0.5	0.8	0.5	0.3	0.2	0.51
	Rayleigh	0.3	0.1	0	0.4	0.3	0.1	1	0.4	0.5	0.2	0.2	0.2	0.31
	Weibull	11.3	7.5	5.6	12.3	14.8	9.4	10	18.1	6.6	7.7	10.7	11.2	10.4
% Uncorrelated*		11.3	12.9	14.2	10.3	9.4	13.4	11.6	9.7	17.7	13.4	12	10.9	12.23

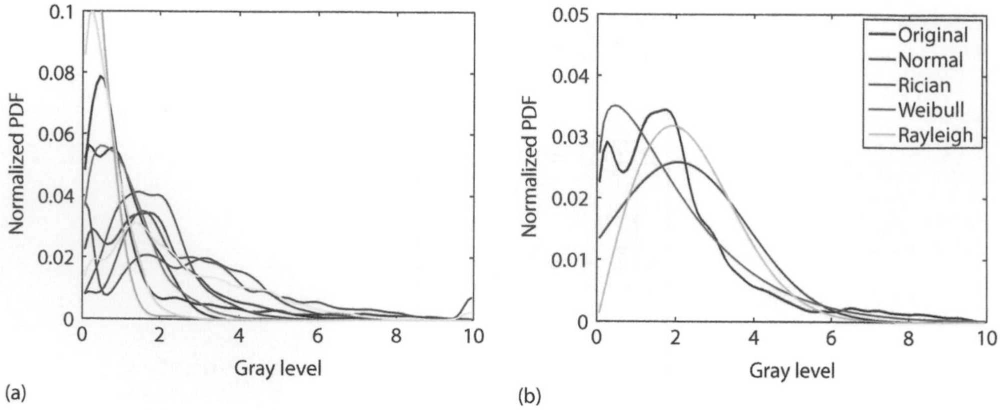

FIGURE 1.4 Sample patch cumulative distribution functions (a) from 10 randomly selected patched and (b) from a single patch, and the corresponding best fit distributions.

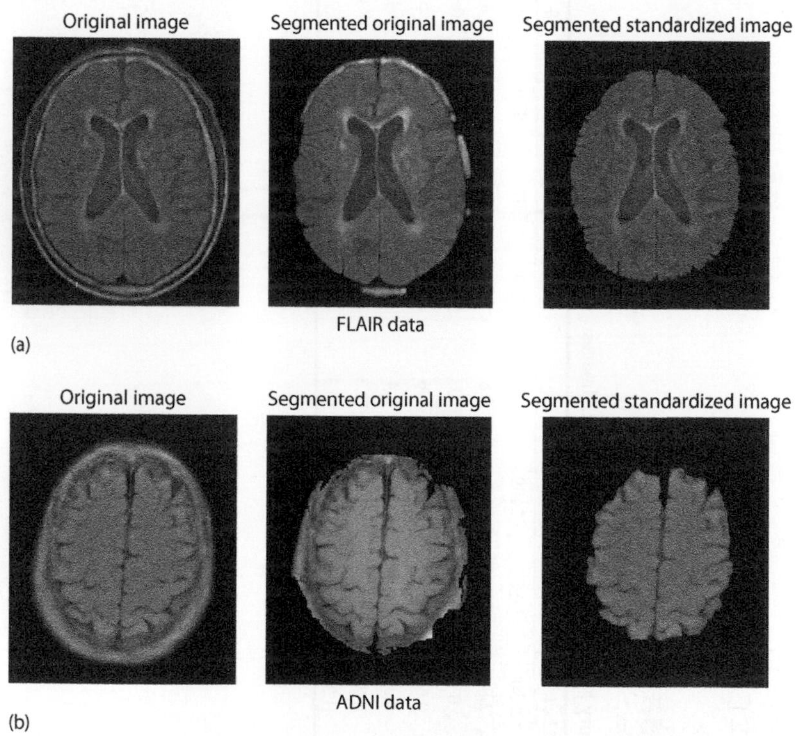

FIGURE 1.5 Effects of image standardization on brain extraction algorithms.

The algorithm was also applied to data from the ADNI database, and the results are shown in Figure 1.7. As it can be seen, the algorithm was robust to changes in magnetic field strength, scanner parameters, and pathology. As previously discussed, these variations cannot be modeled, and this further proves the necessity of model-free segmentation algorithms.

Regardless of the previously mentioned artifacts, the algorithm yielded high segmentation accuracy. In addition, the algorithm used information calculated directly from a single modality, proving

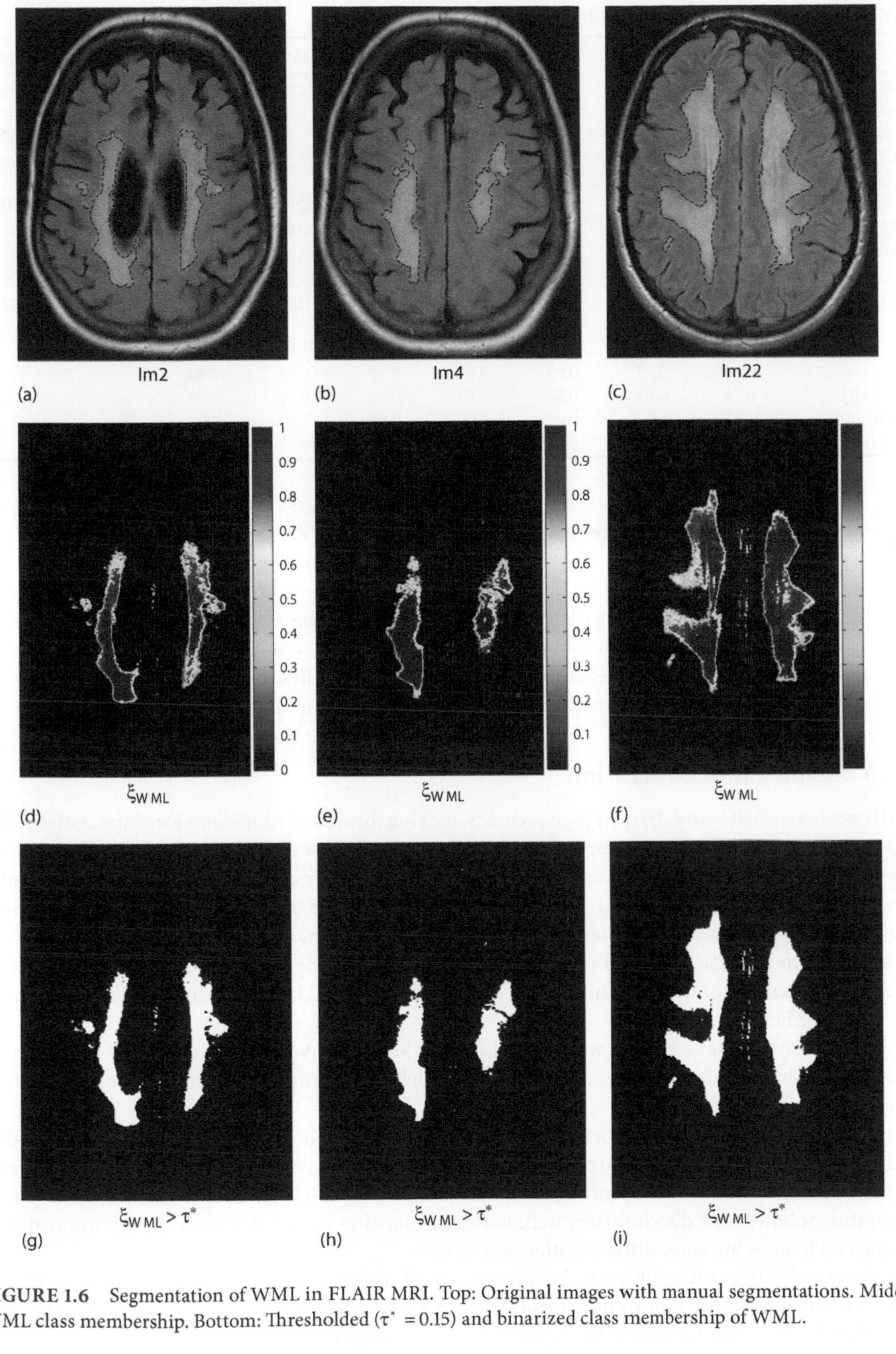

FIGURE 1.6 Segmentation of WML in FLAIR MRI. Top: Original images with manual segmentations. Middle: WML class membership. Bottom: Thresholded ($\tau^* = 0.15$) and binarized class membership of WML.

TABLE 1.2 Results for Real FLAIR MRI with WML using $\tau^* = 0.15$

	IM1	IM2	IM3	IM4	IM5	IM6	IM7	IM8	IM9
DSC	0.84	0.90	0.82	0.92	0.83	0.81	0.74	0.74	0.79
Sensitivity	0.82	0.89	0.84	0.95	0.89	0.88	0.84	0.74	0.77
Specitivity	0.98	0.99	0.99	0.99	0.98	0.99	1.00	1.00	0.99
Vol. (mL)	8.89	8.76	2.51	6.14	4.87	2.24	0.38	0.28	2.34
	IM10	**IM11**	**IM12**	**IM13**	**IM14**	**IM15**	**IM16**	**IM17**	**IM18**
DSC	0.78	0.84	0.84	0.81	0.85	0.86	0.87	0.87	0.88
Sensitivity	0.73	0.80	0.81	0.76	0.89	0.86	0.87	0.85	0.83
Specitivity	1.00	1.00	1.00	0.99	1.00	0.98	0.99	0.99	0.99
Vol. (mL)	1.55	0.21	1.81	**3.41**	1.35	11.29	4.62	5.20	7.42
	IM19	**IM20**	**IM21**	**IM22**	**IM23**	**IM24**	**IM25**		
DSC	0.84	0.77	0.89	0.92	0.87	0.83	0.50		
Sensitivity	0.75	0.68	0.83	0.89	0.83	0.77	0.78		
Specificity	1.00	1.00	0.99	0.99	0.98	1.00	0.99		
Vol. (mL)	3.89	2.71	6.84	18.78	20.33	2.02	0.37		

efficiency. This is advantageous over traditional approaches, as it does not depend on predetermined intensity distribution models. It is also robust to the effects of different pathologies and does not use multiple modalities to segment lesions.

Results from this FLAIR data with WML demonstrate that the algorithm can robustly segment WML. Due to the fractional membership maps, WML segmentation is completed with subvoxel accuracy, giving highly accurate representations of lesion edges. With these highly accurate segmentations, automatic shape analysis can be conducted with ease, as discussed in the following section.

1.8.6 Shape Characterization

WML were extracted with extremely high accuracy, making them ideal candidates for shape analysis. The binary masks $B(x_1, x_2)$ and the corresponding boundaries (x_k, y_k) for several WML are shown in Figure 1.8.

Five images containing WML were selected for shape analysis, yielding 48 lesions for experimentation. Postprocessing was performed on the segmentations to tidy up the binary masks as to only analyze relevant lesions. Operations included false positive reduction from the periphery of the brain (pixels missed in brain extraction), as well as the removal of lesions with the largest diameter being lesser than three millimeters or comprised of nine pixels or less. This removed irrelevant and spurious noise from the WML segmentation results.

As previously mentioned, WML were labeled as PVWML and DWML. These class labels, in conjunction with the shape analysis features extracted from each lesion, are used for the automatic classification of lesion type.

In Figure 1.9, global shape metrics are shown per lesion and coded differently to highlight the discriminatory behavior of each feature in classifying DWML and PVWML. The circularity metric demonstrated a high amount of discriminatory power, whereas Haralick's circularity measure was not as successful. Solidity also discriminates well, but the elongation metric did not robustly discriminate between all lesions, but some differentiation can be seen.

A supervised, linear discriminate classifier was used, where a leave-one-out training and testing approach was applied. This allows us to identify which features robustly discriminate between PVWML and DWML. The performance of the classifier is evaluated using the MCR. Several experiments were conducted to test the individual discrimination power of each feature. First, each feature was analyzed individually. Next, the global features and boundary-based features were grouped into their own respective sets. The last experiment used all proposed features together for classification.

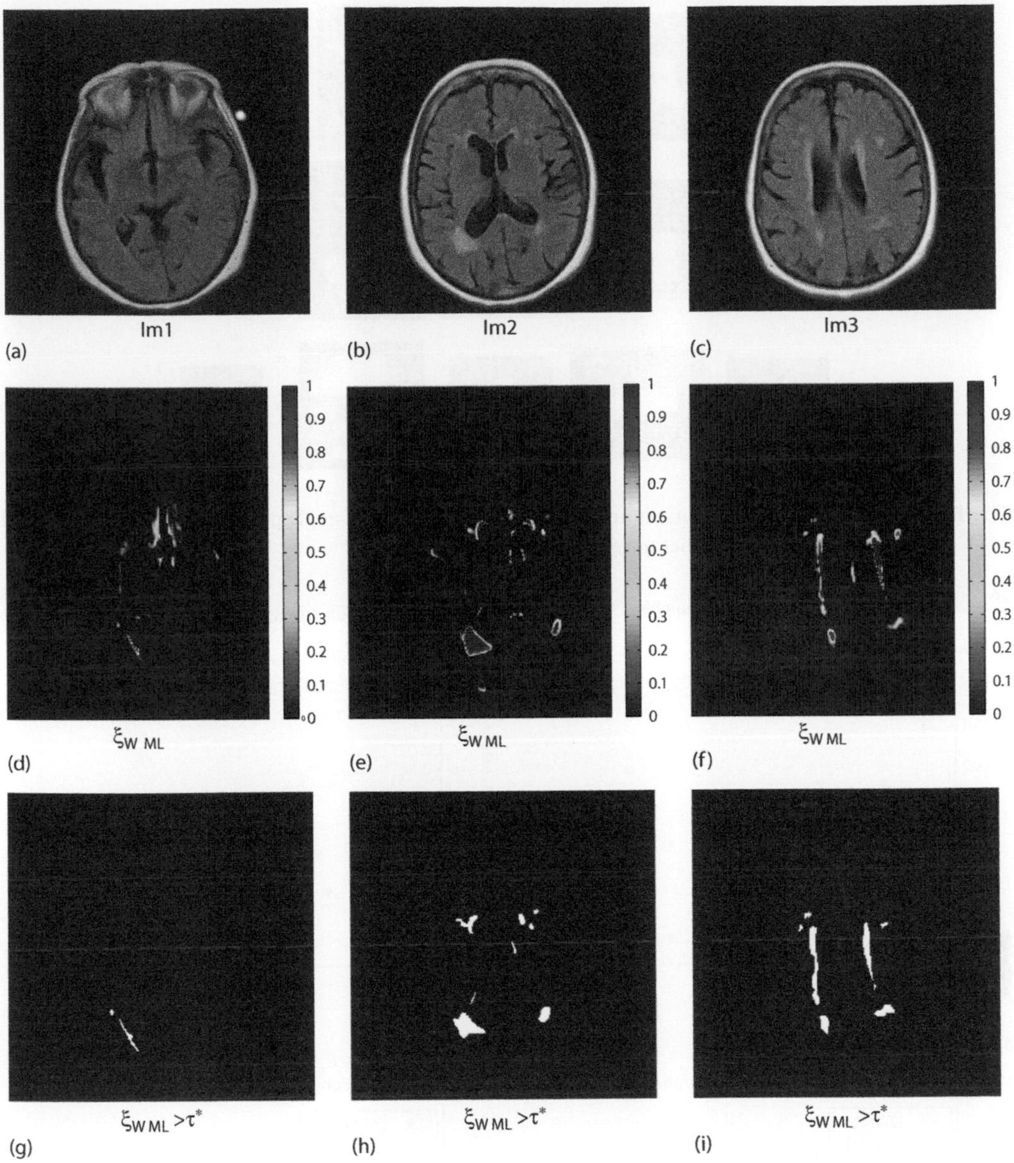

FIGURE 1.7 Segmentation of WML in FLAIR MRI from the ADNI database. Top: Original images with manual segmentations. Middle: WML class membership. Bottom: Thresholded ($\tau^* = 0$) and binarized class membership of WML.

As it can be seen in Figure 1.10, the circularity and elongation metrics outperform all other features with MCR of 10% and 27%, respectively. Using all global features, and the use of all features also yielded good results, with MCR of 21% and 25%, respectively. It should be noted that the use of a single feature (circularity) gave the best results. This demonstrates the great utility and promise that shape analysis techniques hold for neurological disease quantification and classification. Future work will involve the application of these methods on larger databases in order to further analyze performance and for other applications.

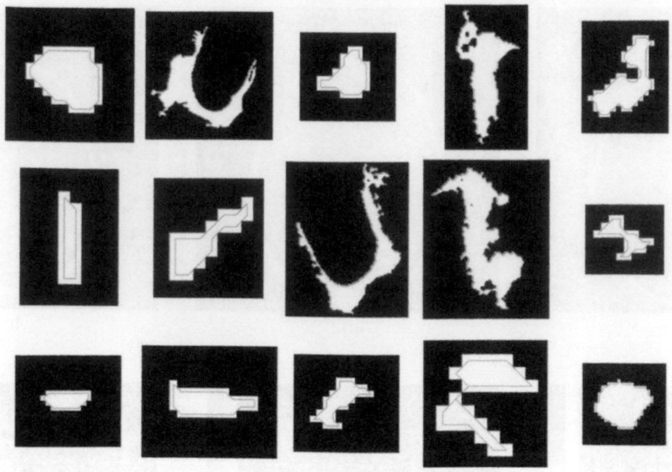

FIGURE 1.8 WML segmentation masks and detected boundaries for a single patient. Lesions are numbered 1 through 15, starting at the top left corner traversing each row.

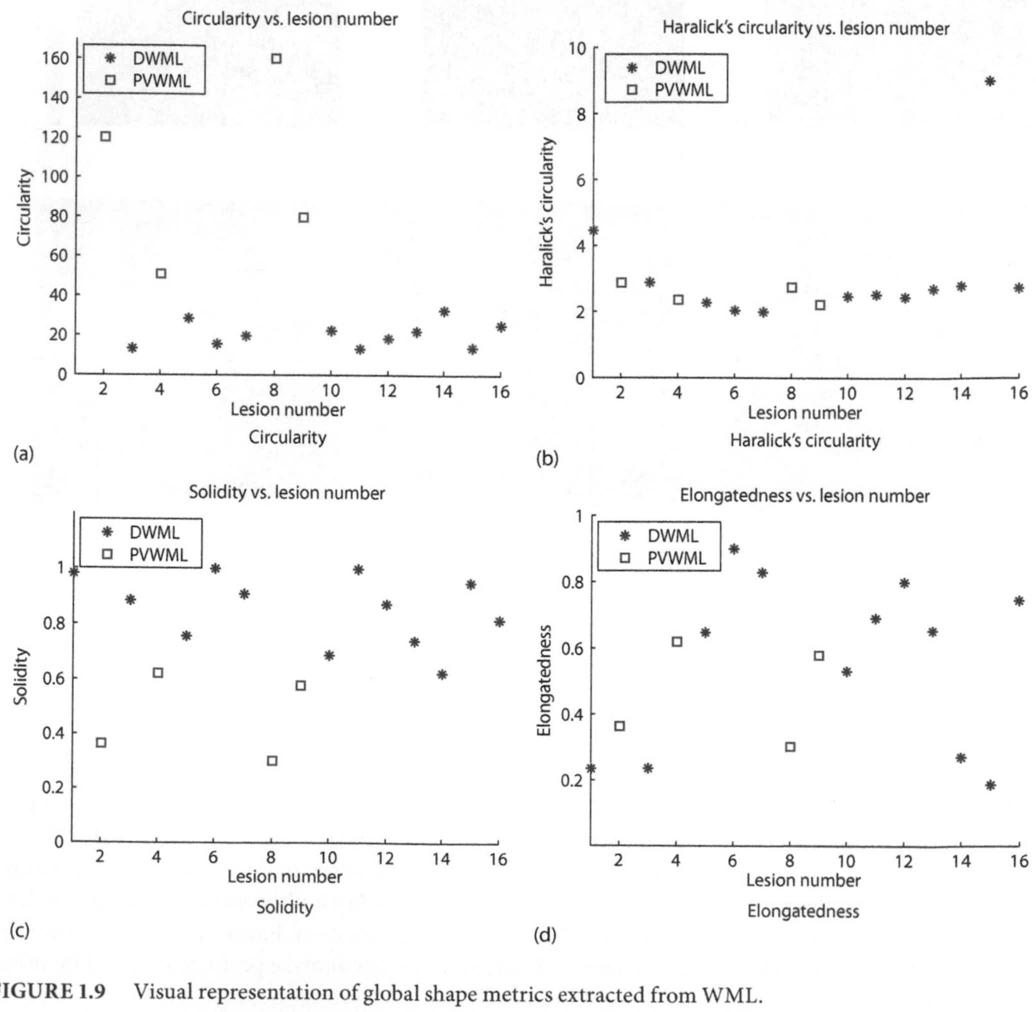

FIGURE 1.9 Visual representation of global shape metrics extracted from WML.

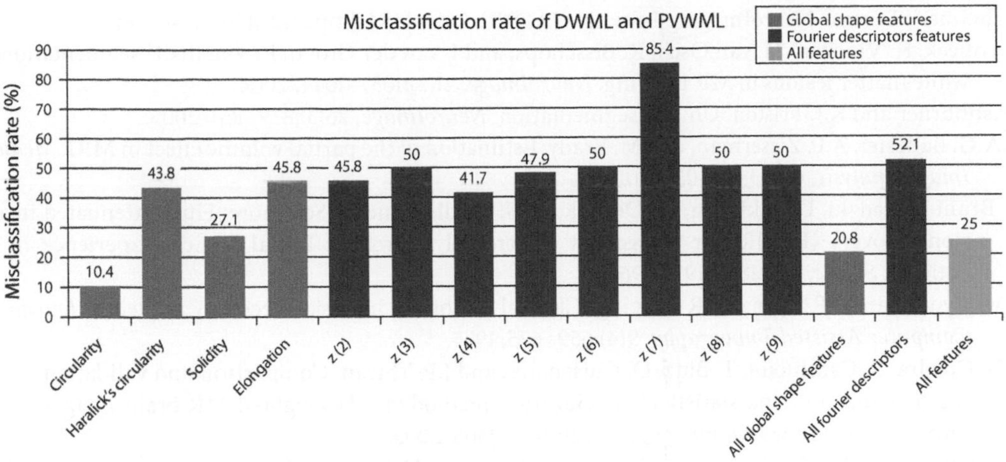

FIGURE 1.10 Misclassification results using different features.

Conclusion

The goals of this work were to demonstrate a robust approach to the segmentation of WML and to also investigate quantitative shape features for discriminating between two classes of lesions. Using a novel exploratory noise analysis approach, it was shown that noise distributions in modern MRI scanners are not stationary, nor do they reliably follow Gaussian, or other known distributions. For these reasons, segmentation algorithms that rely on Gaussian Mixture Models may not be applicable to images acquired on multi-coil scanners. Nonparametric approaches to lesion segmentation are also not ideal, as they require multiple modalities, which increases acquisition costs, computational power, and potential for registration error. The proposed segmentation algorithm does not rely on assumptions of noise distributions, and only requires the FLAIR modality, making it superior to the aforementioned approaches.

In addition to the complex distributions of noise in MR images, other sources of variability include different scanner manufacturers and scan parameters, intensity inhomogeneity, and intensity non-standardness. As repositories of FLAIR data grow, automatic methods of analyses must be robust to the effects of these parameters. A framework for image standardization was also presented, and demonstrates that a simple preprocessing step can increase segmentation accuracy. By reducing the variability between images in a database, automated algorithms can be simplified and made more robust.

Lastly, shape characterization was used to classify lesions as PVWML or DWML. With this automatic classification, combined with other quantitative properties of the lesions, large-scale studies of WML can identify interesting features that are correlated with the progression of disease. With this analysis, neuropathologies that manifest with WML can finally be better understood, and interventions and treatments can be optimized for these patients.

References

S. Aja-Fernandez, G. Vegas-Sanchez-Ferrero, and A. Tristan-Vega. Noise estimation in parallel MRI: GRAPPA and SENSE. *Magnetic Resonance Imaging*, 32(3): 281–290, April 2014.

N. Altaf, L. Daniels, P. Morgan, J. Lower, J. Gladman, S. MacSweeney, A. R. Moody, and D. Auer. Cerebral white matter hyperintense lesions are associated with unstable carotid plaques. *European Journal of Vascular and Endovascular Surgery*, 31: 8–3, 2006.

Alzheimer's Disease NeuroImaging Initiative (ADNI). 2015. URL http://adni.loni.usc.edu/.

P. Anbeek, K. Vincken, M. van Osch, R. Bisschops, and J. van der Grond. Probabilistic segmentation of white matter lesions in MR imaging. *NeuroImage*, 21(3):1037–1044, 2004.

J. Ashburner and K.J. Friston. Unified segmentation. *NeuroImage*, 26(3):839–851, 2005.

M.A.G. Ballester, A.P. Zisserman, and M. Brady. Estimation of the partial volume effect in MRI. *Medical Image Analysis*, 6(4):389–405, 2002.

M. Brant-Zawadzki, D. Atkinson, M. Detrick, W.G. Bradley, and G. Scidmore. Fluid-attenuated inversion recovery (FLAIR) for assessment of cerebral infarction: Initial clinical experience in 50 patients. *Stroke*, 27: 1187–1191, 1996.

G.M. Bydder and I.R Young. MR imaging: Clinical use of the inversion recovery sequence. *Journal of Computer Assisted Tomography*, 9(4):659–675, 1985.

M.B. Cuadra, .L Cammoun, T. Butz, O. Cuisenaire, and J.P. Thiran. Comparison and validation of tissue modelization and statistical classification methods in Tl-weighted MR brain images. *IEEE Transactions on Medical Imaging*, 24(12):1548–1565, 2005.

M.B. Cuadra, B. Platel, E. Solanas, T. Butz, and J.P. Thiran. Validation of tissue modelization and classification techniques in Tl-weighted MR brain images. *Lecture Notes in Computer Science (LCNS)*, 2488: 290–297, 2002.

R. de Boer, F. van der Lijn, H.A. Vrooman, M.W. Vernooij, M.A. Ikram, M.M.B. Breteler, and W.J. Niessen. Automatic segmentation of brain tissue and white matter lesions in MRI. *IEEE International Symposium on Biomedical Imaging (ISBI)*: 652–655, 2007.

B. De Coene, J.V. Hajnal, P. Gatehouse, D.B. Longmopre, S.J. White, A. Oatridge, J.M. Pennock, I.R. Young, and G.M. Bydder. MR of the brain using fluid-attenuated inversion recovery (FLAIR) pulse sequences. *American Journal of Neuroradiology*, 13(6):1555–1564, 1992.

G. De Nunzio, R. Cataldo, and A. Carl. Robust intensity standardization in brain magnetic resonance images. *Journal of Digital Imaging*, 28(6): 727–37, December 2015.

D. Deshmane, V. Gulani, M.A. Griswold, and N. Seiberlich. Parallel MR imaging. *Journal of Magnetic Resonance Imaging*, 36:55–72, 2012.

O. Dietrich, J.G. Raya, S.B. Reeder, M. Ingrisch, M.F. Reiser, and S.O. Schoenberg. Influence of multichannel combination, parallel imaging and other reconstruction techniques on MRI noise characteristics. *Magnetic Resonance Imaging*, 26: 754–762, 2008.

G. Dugas-Phocion, M.A.G. Ballester, G. Malandain, C. Lebrun, and N. Ayache. Improved EM-based tissue segmentation and partial volume effect quantification in multi-sequence brain MRI, *Lecture Notes in Computer Science (LCNS)* 3216, 26–33, 2003.

M. Essig, H. Hawighorst, S.O. Schoenberg, R. Engenhart-Cabillic, M. Fuss, J. Debus, I. Zuna, M.V. Knopp, and G. van Kaick. Fast fluid-attenuated inversion-recovery (FLAIR) MRI in the assessment of intraaxial brain tumors. *Journal of Magnetic Resonance Imaging*, 8(4):789–798, 1998.

R.I. Grossman and J.C. McGowan. Perspectives on multiple sclerosis. *American Journal of Neuroradiology*, 19: 1251–1265, 1998.

T.T.T. Hah, J.Y. Kim, and S.H. Choi. White matter hyperintensities extraction based T2-FLAIR MRI using non-local means filter and nearest neighbor algorithm. In *2014 International Conference on IT Convergence and Security (ICITCS)*, 1–4, 2014. Beijing, China.

J.V. Hajnal, D.J. Bryant, L. Kasuboski, P.M. Pattany, B. De Coene, P.D. Lewis, J.M. Pennock, A. Oatridge, I.R. Young, and G.M. Bydder. Use of fluid attenuated inversion recovery (FLAIR) pulse sequences in MRI of the brain. *Journal of Computer Assisted Tomography*, 16(6):841–844, 1992.

J.P. Hornak. The Basics of MRI, 2014.

A. Khademi. Medical image processing techniques for the objective quantification of pathology in magnetic resonance images of the brain. Presented at University of Toronto, 2012.

A. Khademi, D. Hosseinzadeh, A. Venetsanopoulos, and A.R. Moody. Nonparametric statistical tests for exploration of correlation and nonstationarity in images. *International Conference on Digital Signal Processing (DSP)*: 1–6, 2009a. Santorini-Hellas, Greece.

A. Khademi, A.R. Moody, and A. Venetsanopoulos. Accurate pathology segmentation in FLAIR MRI for robust shape characterization. *Advances of Shape Analysis in Medical Image Analysis*: 187–228, 2014.

A. Khademi, A. Venetsanopoulos, and A.R. Moody. Automatic contrast enhancement of white matter lesions in FLAIR MRI. *IEEE International Symposium on Biomedical Imaging:* 322–325, 2009b.

A. Khademi, A. Venetsanopoulos, and A.R. Moody. Edge-based partial volume averaging estimation in FLAIR MRI with white matter lesions. *IEEE Engineering in Medicine and Biology Conference*, 6114–6117, 2010.

A. Khademi, A. Venetsanopoulos, and A.R. Moody. Robust white matter lesion segmentation in FLAIR MRI. *IEEE Transactions on Biomedical Engineering*, 59(3):860–871, 2012.

P.H. Kvan and B. Vidakovic. Goodness of Fit. In *Nonparametric Statistics with Applications to Science and Engineering*, 81–114. Wiley, New Jersey, 2007.

Z. Lao, D. Shen, A. Jawad, B. Karacali, D. Liu, E.R. Melhem, and R.N. Bryan. Automated segmentation of white matter lesions in 3d brain MRI, using multivariate pattern classification. *IEEE International Symposium on Biomedical Imaging (ISBI)*, 307–310, 2006.

P. Lin, Y. Yang, C.X. Zheng, and J.W. Gu. An efficient automatic framework for segmentation of MRI brain image. *International Conference on Computer and IT*, 896–900, 2004. Wuhan, China.

S. Loncarica. A survey of shape analysis techniques. *Pattern Recognition*, 31(8): 983—1001, 1988.

A. MacKay, C. Laule, D.K.B. Li, S.M. Meyers, B. Russell-Schulz, and I.M. Vavasour. Magnetic resonance techniques for investigation of multiple sclerosis. *XIII Mexican Symposium on Medical Physics*, 1626: 22–35. AIP Conference Proceedings, 2014.

P. Malloy, S. Correia, G. Stebbins, and D.H. Laidlaw. Neuroimaging of white matter in aging and dementia. *The Clinical Neuropsychologist*, 21: 73–109, 2007.

P. Meer and B. Georgescu. Edge detection with embedded confidence. *IEEE Transactions on Pattern Analysis and Machince Intelligence PAMI*, 23(12): 1351–1365, 2001.

G.L. Nyul and J.K. Udupa. On standardizing the MR image intensity scale. *Magnetic Resonance in Medicine*, 42: 1072–1081, 1999.

K. Oppedal, T. Eftestl, K. Engan, M.K. Beyer, and D. Aarsland. Classifying dementia using local binary patterns from fifferent regions in magnetic esonance images. *International Journal of Biomedical Imaging*, 2015:1–14, 2015.

D. Palumbo, B. Yee, P. O'Dea, S. Leedy, S. Viswanath, and A. Madabhushi. Interplay between bias field correction, intensity standardization, and noise filtering for t2-weighted mri. In *Engineering in Medicine and Biology Society, EMBC, 2011 Annual International Conference of the IEEE*, 5080–5083, 2011. Boston, MA.

Public Health Agency of Canada. Tracking Heart Disease and Stroke in Canada: Stroke Highlights 2011, 2011.

B. Reiche, A.R. Moody, and A. Khademi. Effect of image standardization on FLAIR MRI for brain extraction. *Signal Image and Video Processing*, Volume 9, supplement 1, pp. 11–16, 2015.

E. Reinhard, M. Ashikhmin, B. Gooch, and P. Shirley. Color transfer between images. *IEEE Computer Graphics and Applications*, 21(5): 34–41, 2001.

A. Samsonov and C. Johnson. Noise-adaptive nonlinear diusion ltering of MR images with spatially varying noise levels. *Magnetic Resonance in Medicine*, 52:798–806, 2004.

P. Santago and H. Gage. Statistical models of partial volume effect. *IEEE Transactions on Image Processing*, 4(11):1531–1540, 2003.

Statistics Canada. Leading Causes of Death, by Sex, 2011.

J. Suckling, J. Henty, C. Ecker et al., Are power calculations useful? A multicentre neuroimaging study: Validation of the voxel-based power calculations. *Human Brain Mapping*, 35(8): 3569–3577, 2014.

P. Thunberg and P. Zetterberg. Noise distribution in SENSE- and GRAPPA-reconstructed images: A computer simulation study. *Magnetic Resonance Imaging*, 25: 1089–1094, 2007.

K. Van Leemput, F. Maes, D. Vandermeulen, and P. Suetens. A unifying framework for partial volume segmentation of brain MR images. *IEEE Transactions on Medical Imaging*, 22(1):105–119, 2003.

A.M. Winkler, P. Kochunov, and D.C. Glahn. FLAIR Templates. 2012. URL http://glahngroup.org.

K. Woong Kim, J.R. MacFall, and M.E. Payne. Classification of white matter lesions on magnetic resonance imaging in the elderly. *Biological Psychiatry*, 64(4):273–280, 2008.

Y. Zhang, M. Brady, and S. Smith. Segmentation of brain MR images through a hidden markov random field model and the expectation-maximization algorithm. *IEEE Transactions on Medical Imaging*, 20(1):45–57, 2001.

Y. Zhong, D. Utriainen, Y. Wang, Y. Kang, and E. Mark Haacke. Automated white matter hyperintensity detection in multiple sclerosis using 3d T2 FLAIR. *International Journal of Biomedical Imaging*, 2014: 1–7, 2014.

Y. Zhuge and J.K. Udupa. Intensity standardization simplifies brain MR image segmentation. *Computer Vision and Image Understanding*, 113(10): 1095–1103, 2009.

2

Computer-Aided Diagnosis with Retinal Fundus Images

Yuji Hatanaka

Hiroshi Fujita

2.1 Introduction

In this chapter, image analysis methods for computer-aided diagnosis (CAD) based on retinal fundus images are described. The retina is the only area that allows direct and noninvasive observation of its blood vessels. Funduscopy is effective in the diagnosis of several diseases, such as glaucoma, diabetic retinopathy, and age-related macular degeneration (AMD). Moreover, by funduscopy, physicians may diagnose not only ophthalmological diseases but also systemic hypertension. The retina is located near the brain; thus, some conditions of the cerebrovascular system may be diagnosed by funduscopy [1]. There is a possibility that retinal arteriosclerosis may be a sign of cerebral hemorrhage. Thus, a patient may have a cerebral hemorrhage, a subarachnoid hemorrhage, or a hypertensive encephalopathy if they have a retinal hemorrhage. Since funduscopy is very useful in ophthalmology, neurosurgery, and cardiology, the retina is assessed in several medical departments. Therefore, a retinal CAD system can be useful for physicians in several medical departments. Arteries, veins, and other retinal anatomy are visualized in a retinal image as shown in Figure 2.1. Retinal condition is described by Keith-Wagner-Barker classification or Scheie

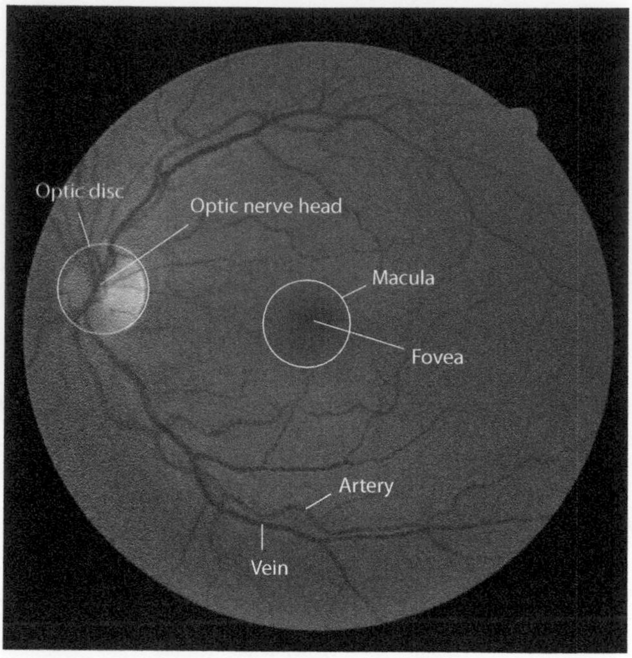

FIGURE 2.1 Structures of the retina seen in a fundus image.

TABLE 2.1 Scheie Classification

Grade	Hypertension (H)	Arteriolosclerosis (S)
0	No changes	Normal
1	Barely detectable arteriolar narrowing	Barely detectable light reflex changes in arterial wall, or arteriolar-venous nicking
2	Obvious arteriolar narrowing with focal irregularities	Obvious increased light reflex changes, or arteriolar-venous nicking
3	Grade 2 plus hemorrhages or exudates	Copper wire arteries, or arteriolar-venous nicking
4	Grade 3 plus papilledema	Silver wire arteries

classification. Detailed Scheie classification is shown in Table 2.1. It is divided into two diseases, hypertension and arteriosclerosis. The signs of low-grade diseases are related to blood vessels; thus, blood vessel segmentation techniques are very important and are described in Section 2.2. Section 2.3 introduces artery diameter measurement as a parameter to assist diagnosis of hypertensive retinopathy. Diabetic retinopathy and glaucoma are leading causes of blindness. CAD techniques for diabetic retinopathy and detection of hemorrhages and microaneurysms are described in Section 2 .4. In Section 2.5, techniques for detection of large cupping in the optic disc and nerve fiber layer defects are described for CAD of glaucoma.

2.2 Blood Vessel Segmentation

CAD systems for glaucoma, diabetic retinopathy, and cardiovascular diseases require blood vessel segmentation. The retinal image databases STARE (Structured Analysis of the Retina) [2] and DRIVE (Digital Retinal Images for Vessel Extraction) [3,4] have been publicly available for several years, and the performances of many algorithms have been compared using these databases. Retinal blood vessel extraction algorithms are divided into three categories: matched filtering, morphological processing, and supervised methods using multiscale features.

2.2.1 Matched Filtering

Blood vessels are enhanced by using a matched filter that convolves a 2-D kernel with a retinal image. Chaudhuri et al. proposed a method using 12 different templates which were used to search for vessel segments along all possible directions [5]. The profile of the cross section of a blood vessel resembles a Gaussian curve. Thus, they modeled the matched filters with Gaussians. Zolfagharnasab et al. proposed a method that used the Cauchy distribution instead of the Gaussian function for improvement of the performance of matched filters in the detection of retinal vessels [6]. Hatanaka et al. proposed a double-ring filter [7]. An original double-ring filter is structured by an inner ring region and an outer ring region. When the image resolution is low, the double-ring filter is structured by an inner square region (region A) and an outer square region (region B), as shown in Figure 2.2. It outputs the contrast by calculating the difference of the means in regions A and B. To process images from the DRIVE database, the widths of regions A and B were set as 1 and 9 pixels, respectively, and the green component of the color retinal image was used for calculation of the output of the double-ring filter. Figure 2.3c shows the filtered image obtained by a double-ring filter for the retinal image in Figure 2.3a.

2.2.2 Morphological Image Processing

Morphological processing is based on geometry. The main morphological operations are dilation and erosion. In addition, the combined operations are closing and opening. "Closing" is a dilation followed by an erosion, and "opening" is an erosion followed by a dilation. The gray-level profile of the cross section of a blood vessel is concave in shape. Zana et al. proposed a method based on morphological filters combined with cross-curvature evaluation and linear filtering [8]. The method used vessel-like structures with examples of blood vessels.

2.2.2.1 Double-Ring Filter and Black Top-Hat Transformation

Muramatsu et al. combined two methods, a double-ring filter and a black top-hat transformation [9]. A black top-hat transform is defined as the difference between the result of morphological "closing" and the original image. For the DRIVE database, the structuring element for the black top-hat transformation was defined as a circle with a diameter of 15 pixels, which is approximately equal to the diameter of the thickest vessel. The green component of the color retinal image was processed by "closing." As shown in Figure 2.4c,e, the valley regions formed by blood vessels are filled by "closing." By calculation of the difference between the "closing" image and the original one, the blood vessel regions are enhanced as shown in Figure 2.4c,f [10]. The outputs from the two filters are summed as shown in Figure 2.3e. The

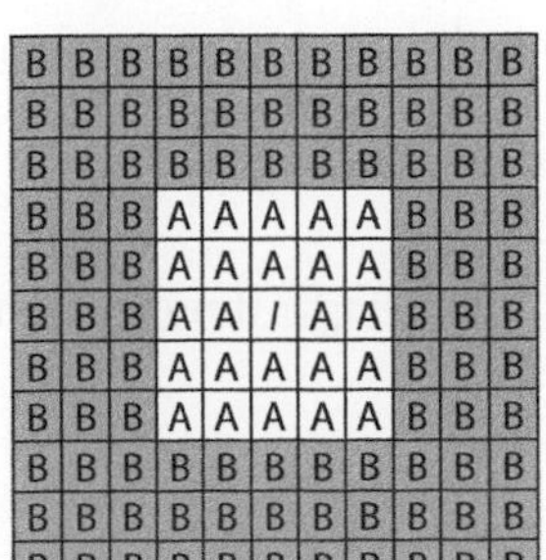

FIGURE 2.2 Structure of a double-ring filter.

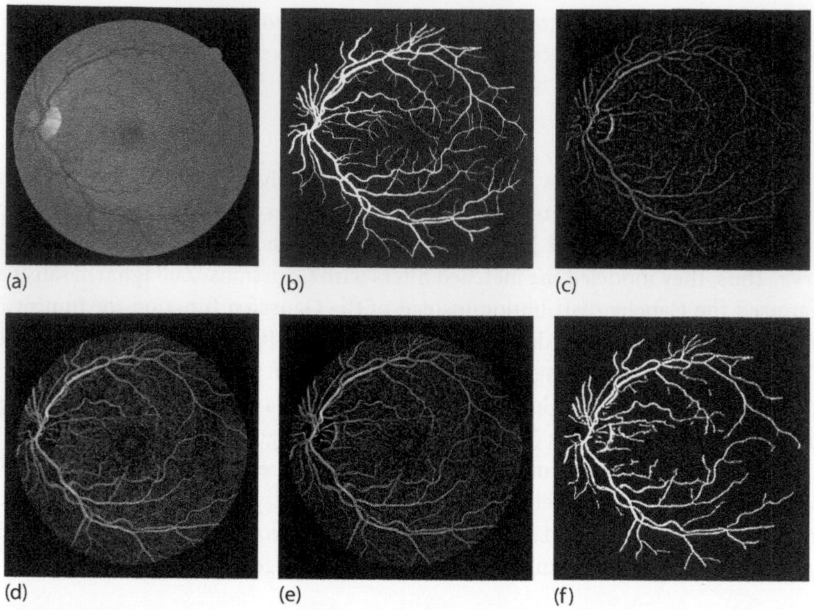

FIGURE 2.3 Example of blood vessel segmentation using a double-ring filter combined with a black top-hat transformation (a) Green component image. (b) Ground truth for blood vessel segmentation. (c) A filtered image using a double-ring filter. (d) A filtered image using a black top-hat transformation. (e) A filtered image using a combination of a double-ring filter and black top-hat transformation. (f) Binarized image from (e).

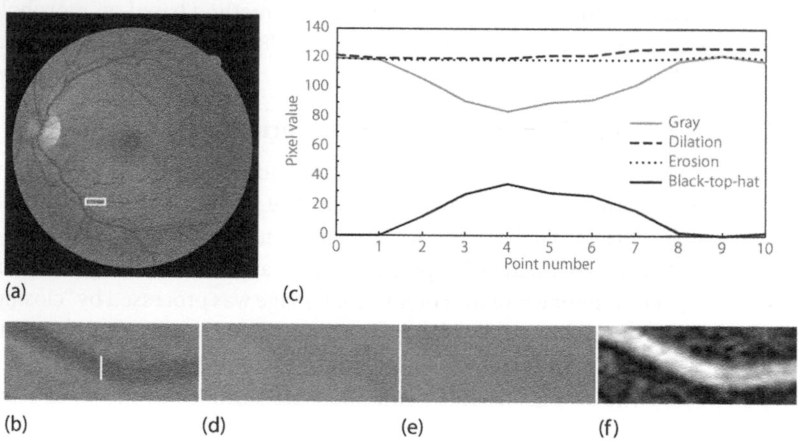

FIGURE 2.4 Process of black top-hat transformation. (a) Green component image. (b) Enlarged image of the blood vessels marked by a rectangle in (a). (c) Pixel profiles of (b)–(f) marked by a line in (b). (d) Dilation from (b). (e) Erosion from (d). (f) Black top-hat transformed image.

blood vessel regions are determined by applying a threshold, and small regions are removed as false positives (Figure 2.3f).

2.2.3 Supervised Methods Using Multiscale Features

There are several types of blood vessels with different widths and curvatures. Thus, blood vessel segmentation methods that use multiscale frequency analysis, multiscale pattern analysis, and related methods are effective. Blood vessels and other regions are classified by using supervised machine learning with methods such as neural networks, the support vector machine (SVM), and k-nearest neighbors. Niemeijer et al. classified blood vessels using the k-nearest neighbor (kNN) classifier with a Gaussian matched filter and its first-order and second-order derivatives at multiple scales [3]. Miri et al. proposed a method based on image enhancement using a curvelet transform [11]. Curvelets can represent the edges along curves much more efficiently than traditional wavelets. Wavelet transform is not useful for wave decomposition on oblique, long, and narrow regions. On the other hand, ridgelet transform differentiates in the arbitrary direction. However, ridgelet transform is not useful for wave decomposition on isotropic regions. Curvelets is a method based on wavelet transform, followed by ridgelet transform to serve both types of regions. They first applied the wavelet transform to the given retinal image, and then the high-frequency components were processed by using the ridgelet transform. Oloumi et al. used multiscale vesselness measures, Gabor filters, line operators, and matched filters, and classified pixels using a multilayer neural network [12]. Soares et al. segmented retinal vessels by using a method based on each pixel's feature vector that was composed of the pixel's intensity and the results of the two-dimensional Gabor wavelet transform at multiple scales [13]. Staal et al. proposed a method based on segmentation of image ridges that used the kNN classifier with properties of patches and line elements [4]. Ricci et al. classified blood vessels by using an SVM with two orthogonal line detectors [14].

2.2.3.1 Artificial Neural Network Using High-Order Local Autocorrelation

Hatanaka et al. proposed an automated blood vessel segmentation method using an artificial neural network (ANN) with high-order local autocorrelation (HLAC) derived from the vessel-enhanced images by black top-hat transformation [15]. HLAC is a method based on the relation of neighboring pixels and is effective for many image recognition applications, including facial recognition, chest nodule detection, and human sensing. HLAC is appropriate for center-shifted, hot-spot pattern feature extraction. A general template-matching technique depends strongly on a preset model, but HLAC does not require a preset model. Thus, HLAC is expected to be effective for blood vessel extraction. However, HLAC has directional dependency. By transforming the region of interest (ROI) to a polar coordinate image, HLAC features become rotation-invariant. The *N*th order HLAC is calculated by the following autocorrelation:

$$R_N^T(a_1, a_2, \ldots, a_N) = \sum_r I(r)I(r+a_1)\ldots I(r+a_N) \tag{2.1}$$

where:

- $I(r)$ is a pixel value of the image
- T is the local pattern number (as shown in Figure 2.5)
- r $= (x, y)'$ (the apostrophe denotes the transpose) is position vector
- a_i are the displacement vectors
- x and y are coordinates in the image

By varying the parameter value N, Equation 2.1 can take many forms. N was limited to ($N = 0, 1, 2$) for blood vessel detection. In general, the window size of HLAC is limited to 3 × 3 pixels, and T is limited to 35 ($T = 1, 2, \ldots, 35$) by reducing the equivalent patterns, which are equivalent by the shift. The 35 types of general patterns are shown in Figure 2.5a. Hatanaka et al. extended the window size to 5 × 5 and 7 × 7 pixels (as shown in Figure 2.5b), and they defined 105 HLAC patterns to detect retinal blood vessels. All HLAC patterns are calculated in the image transformed by black top-hat (as shown in Figure 2.3d), and their features are input into an ANN with four features. The four features are the maximum value, minimum value, mean, and standard deviation of the pixel values in the ROI. The ANN is a feed-forward, three-layered network with twelve middle-layer units.

Hatanaka et al. compared five kinds of images, including the output of the ANN (as shown in Figure 2.6b), the Gabor filter output (as shown in Figure 2.6c), the double-ring filter output [7] (as shown

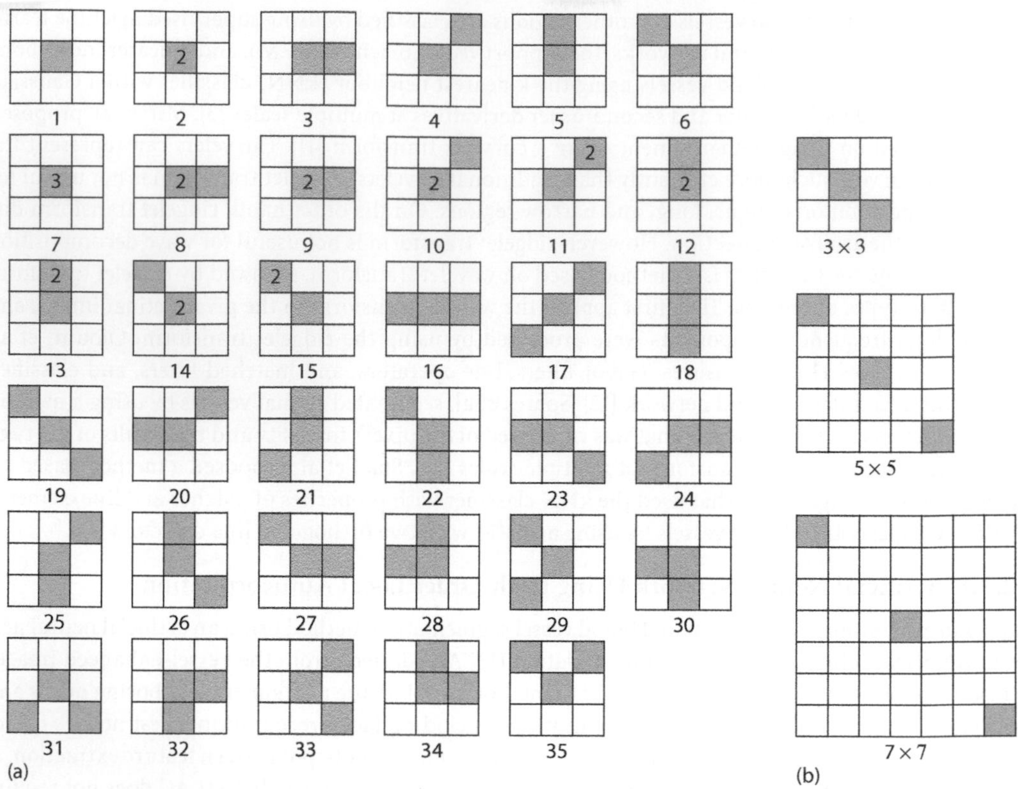

FIGURE 2.5 High-order local autocorrelation (HLAC) patterns. (a) Thirty-five basic patterns. (b) Extended pattern from No. 19 in (a).

in Figure 2.3c), the result of black top-hat transformation [10] (as shown in Figure 2.3d), and the green component of the color retinal image. Cross correlations between the five images were low. To improve the performance of blood vessel segmentation, the above five images were input to the next ANN (ANN2), which is a feed-forward, three-layered network with five middle-layer units. The Gabor filter is represented as follows:

$$g(x,y)=\exp\left(-\frac{x'^2+\gamma^2 y'^2}{2\sigma^2}\right)\cos\left(2\pi\frac{x'}{\lambda}+\varnothing\right) \tag{2.2}$$

where:

$$x' = x\cos\theta + y\sin\theta$$

$$y' = -x\sin\theta + y\cos\theta$$

$$\sigma = \frac{K}{\pi}\lambda\frac{2^b+1}{2^b-1}, K=\sqrt{\frac{\ln 2}{2}} \tag{2.3}$$

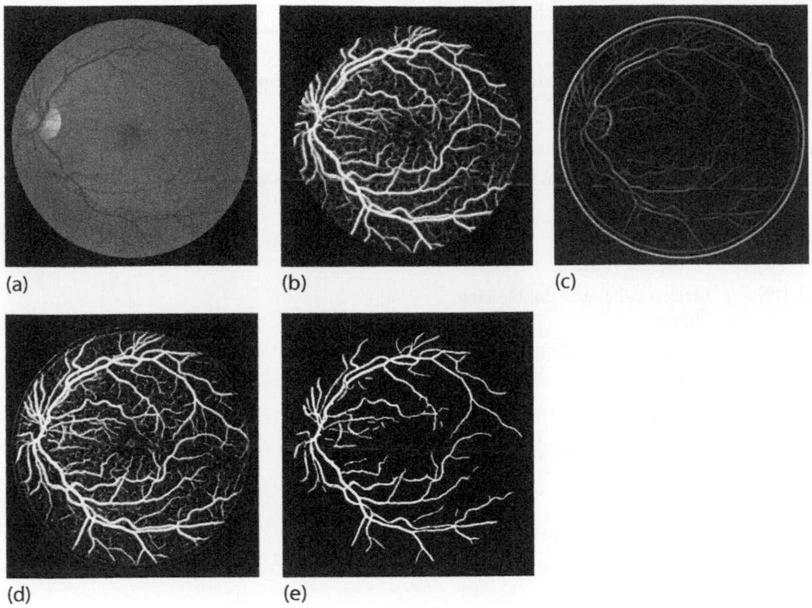

FIGURE 2.6 Example of blood vessel segmentation using HLAC features. (a) Green component image. (b) Output of first artificial neural network (ANN) based on HLAC features in black top-hat transformed image. (c) Output of Gabor filter. (d) Output of next ANN (ANN2) using four outputs of first ANN (Figure 2.6 (b)), black top-hat transformation (Figure 2.3d), double-ring filter (Figure 2.3c) and Gabor filter (Figure 2.6 (c), and pixel value in green component image (Figure 2.6 (a). (e) Binarized image from (d).

For the DRIVE database, $\lambda = 12$, $\gamma = 1$, $\varnothing = 0$, and $b = 1$ were used, and the filtered image takes the maximum value of $g(x,y)$ by setting $\theta = 0, \pi/12, 2\pi/12, \ldots, 11\pi/12$. The parameters of the double-ring filter and black top-hat transformation were the same as described in Section 2.2.2.1. Figure 2.6d shows an example of the output of ANN2. By using a thresholding technique and removing small regions, a binarized image was obtained as shown in Figure 2.6e.

2.2.4 Comparison of Performances of Blood Vessel Segmentation Methods

A total of 11 blood vessel segmentation methods and segmentation by a human observer were compared using the DRIVE database [3,4]. The DRIVE database consists of 20 training images and 20 test images. Table 2.2 shows the results of each method, comparing the sensitivity, specificity, accuracy, and area under the curve (AUC) based on receiver operating characteristic (ROC) analysis. Methods using supervised machine learning showed high performance. The performances of several methods were close to that of the human expert.

2.3 Hypertensive Retinopathy

Symptoms of systemic cardiovascular diseases also appear in retinal fundus images. Hypertension causes changes in the retinal artery-to-vein diameter ratio (AVR). Decreasing AVR is associated with risks of stroke and myocardial infarction. Arterio-venous nicking (AVN) is associated with hypertension and cardiovascular diseases such as stroke. AVN is a phenomenon where a vein is compressed by an artery and becomes narrow on both sides of an arterio-venous crossing (AVC). In addition, retinal blood vessel occlusion is related to systemic vascular disease.

TABLE 2.2 Comparison of Performance for Blood Vessel Segmentation Using DRIVE

Proposer	Main technique	Sensitivity	Specificity	Accuracy	Area under ROC Curve
Human observed [3]	—	0.776	0.973	0.947	—
Chaudhuri et al. [5]	2D Gaussian matched filter	—	—	0.877	0.788
Zolfagharnasab et al. [6]	Cauchy distribution matched filter	0.624	0.971	—	—
Zana et al. [7]	Morphological filter and cross-curvature	0.697	—	0.938	0.898
Muramatsu et al. [10]	Black top-hat and double-ring filter	—	—	—	0.918
Niemeijer et al. [3]	Gaussian matched filter and kNN	0.715	—	0.942	0.929
Miri et al. [11]	Curvelet transform	0.735	0.980	0.946	—
Oloumi et al. [12]	Gabor filter, etc., and multilayer perceptron	—	—	—	0.961
Soares et al. [13]	Gabor wavelet	0.728	0.979	0.947	0.961
Staal et al. [4]]	Image ridge features and kNN	0.719	0.977	0.944	0.952
Ricci et al. [14]	Two orthogonal line detectors and SVM	—	—	0.956	0.956
Hatanaka et al. [15]	HLAC, black top-hat, Gabor filter, double-ring filter and neural network	—	—	—	0.960

2.3.1 Artery-Vein Ratio Measurement

2.3.1.1 Artery-Vein Ratio Measurement Method of Niemeijer et al.

Niemeijer et al. proposed methods for AVR measurement [16]. They determined the AVR measurement zone between half-disc and one-disc diameters from the optic disc margin by applying the protocol proposed by Knudtson et al. [17]. They segmented the blood vessels by using a k-nearest neighbor (kNN) classifier with the Gaussian matched filter and its first-order and second-order derivatives at multiple scales [3]. They then obtained the centerlines of the blood vessels by applying a skeletonization algorithm [18]. After detecting the location of the optic disc, they set the region of interest (ROI) in the retinal image. The vessel widths were then measured in the measurement zone. Finally, the vessels were classified as arteries or veins, and the AVR was determined. The mean unsigned error was 0.06 (SD 0.04) with a mean AVR of 0.67 for 40 images from the INSPIRE-AVR database [19].

2.3.1.2 Artery-Vein Ratio Measurement Method of Muramatsu et al.

Muramatsu et al. [10] determined the AVR measurement zone between quarter-disc and one-disc diameters from the optic disc margin according to a Japanese study (as shown in Figure 2.7a). They segmented blood vessel regions by using a double-ring filter [7] combined with a black top-hat transform [8]. Optic disc locations and approximate disc diameters were then determined using an active contour method to determine the AVR measurement zone [20]. Thin vessels with diameters less than two pixels were removed, since AVR is usually measured on large vessels. The remaining blood vessels were partitioned to blood vessel segments by detecting bifurcations and vessel crossings as described in Section 2.3.2. Each centerline pixel of the blood vessel segments in the measurement zone was then classified as belonging to an artery or vein by using a linear discriminant classifier. For features of artery-vein classification, they used six features, including the three original color components (red, green, and blue), and three measures of contrast in the color channels. Their method selected the temporal artery-vein (AV) pair, since they used images in which the macula existed in the center. Small vessel segments were

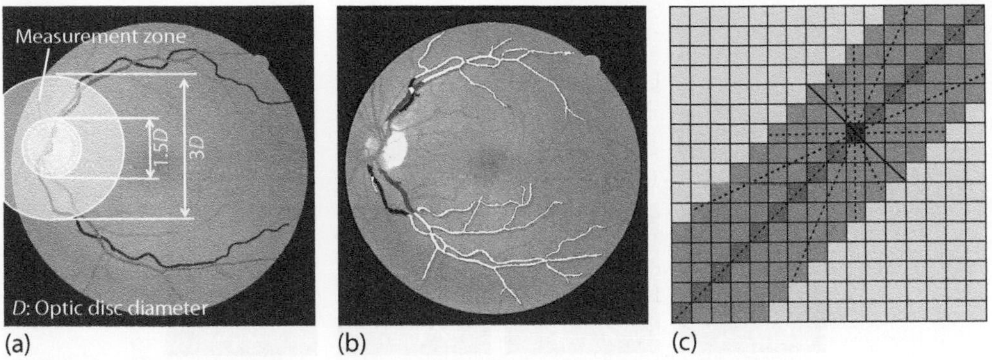

FIGURE 2.7 Artery-to-vein diameter ratio (AVR) measurement. (a) Determination of AVR measurement zone. The circle with dotted line shows an outline of the optic disc. The translucent rim region shows the AVR measurement zone. The gray lines and black lines show ground truth of arteries and veins, respectively. (b) Result of selected AV pairs. (c) Vessel diameter measurement. The gray boxes and dark gray boxes show vessel region and the centerline, respectively. The black box shows pixel of interest and the black line shows the shortest path through the pixel of interest inside the vessel region. The length of the black line was defined as the vessel diameter at the pixel of interest.

removed and then the vessel segments within the limits of direction determined experimentally were selected as candidates of AV pairs. Figure 2.7b shows an example of selected AV pairs. The measurement of vessel diameter is shown in Figure 2.7c. The vessel diameters were determined by the length of the shortest path through the centerline pixels inside the vessel region. The AVRs were determined at these pixels and then averaged for the upper and lower AV pairs. Thirty-six out of 40 vessel pairs were correctly identified in the test set of DRIVE. Although the mean error in the AVRs with respect to those based on the ground truth was 0.11, the mean error in vessel diameter was less than 1 pixel.

2.3.1.3 Improvement of Artery-Vein Selection

Hatanaka et al. improved the AV selection method of Muramatsu et al. [10]. AVR was measured on an AV pair running side by side. However, the previous method could not segment an AV pair coming in contact with each other, since the method did not consider such an AV pair. Figure 2.8a,b show an example of such a case, and the previous method [10] incorrectly recognized the AV pair as a vein (as shown in Figure 2.8c). Thus, Hatanaka et al. proposed a method for classification of such an AV pair [21]. They determined the AVR measurement zone, which is the same as that determined by Muramatsu et al. The absorbance of oxygenated hemoglobin is significantly lower than that of reduced hemoglobin in the red-light band (600–750 nm). The contrast of an artery is low in the red component of a color fundus image since arterial blood contains more oxygenated hemoglobin than venular blood. Figure 2.9a,b

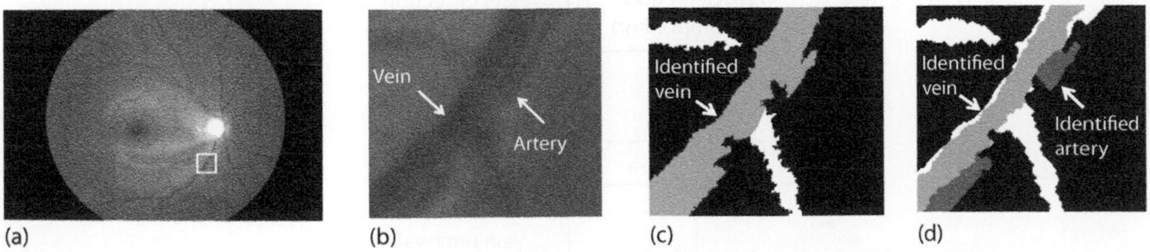

FIGURE 2.8 Result of main artery and vein selection. (a) Example of artery and vein coming in contact with each other. (b) Enlarged image of the blood vessel marked by a rectangle in (a). (c) Artery and vein selection using a method in a study by Muramatsu et al. [10]. The gray region shows an identified vein. An artery region was not identified. (d) Result of artery (dark gray) and vein (gray) selection using a method in a study by Hatanaka et al. [21].

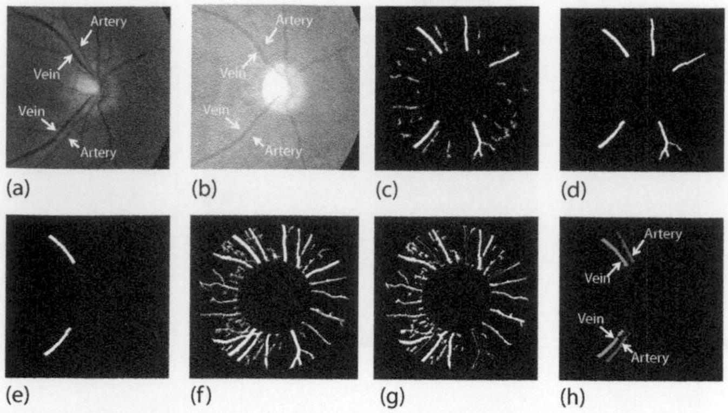

FIGURE 2.9 Main artery and vein selection using a method in a study by Hatanaka et al. [21]. (a) Green component image. (b) Red component image. The contrast of arteries is lower than that in (a). (c) Vein candidates extracted using the image in (b). (d) Vein candidates selected. (e) Final main vein selected. (f) Blood vessels extracted using the image in (a). (g) Artery candidates selected by removing the veins. (h) Final selection of main arteries and main veins.

shows examples of green and red component images. Therefore, Hatanaka et al. proposed a method based on identification of veins in the red channel of a color image and identification of arteries in the green channel of a color image [21]. Figure 2.10 shows the flowchart of the method.

They selected only veins by using a black top-hat transformation and a double-ring filter with the red component of a color retinal image. However, the red component contains high levels of noise and parts of an artery as shown in Figure 2.9c. Thus, small candidates are removed as noise, and veins are selected by using a linear discriminant classifier with eight features, which are pixel values and contrast in the red, green, and blue components, the values of black top-hat transformation, and the double-ring filter output for the green component. Figure 2.9d shows the result of the selected veins. To select the main veins for AVR measurement, crossing-points and bifurcations of blood vessels were detected by using the method described in the next section. Vein candidates were divided into branch segments at crossing-points and bifurcations. If several vein candidates ran side by side, the thickest candidate was determined as the main vein. Figure 2.9e shows the result of the selection of the main veins.

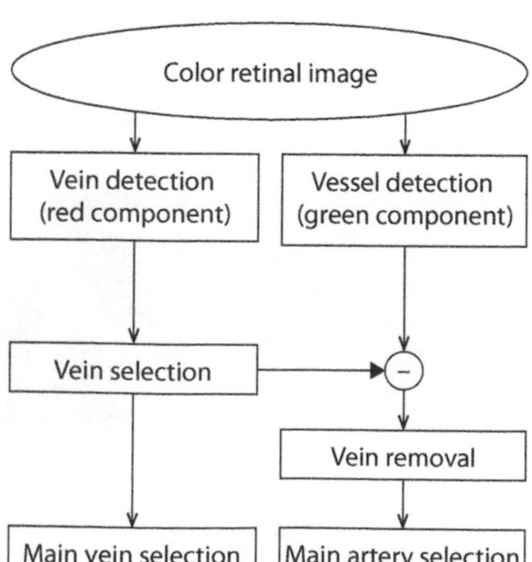

FIGURE 2.10 Flowchart of the method for selection of the main arteries and veins.

For selection of the main arteries, vessel candidates were first extracted by using a black top-hat transformation and double-ring filter with the green component image (as shown in Figure 2.9f). By removing the vein candidates (Figure 2.9d) from the vessel candidates, the artery candidates were obtained, as shown in Figure 2.9g. The main arteries were selected by using a decision tree model with three features. By comparison of an artery candidate and the nearest candidate, the difference in mean diameter and difference in area were defined as two features. Another feature was the difference between an artery candidate direction and the nearest main vein direction, where a direction was defined using the center of the optic disc and the center of an artery candidate or main vein. Figure 2.9h shows the final result with the main arteries and main veins selected for AVR measurement.

This method and the previous method [10] were compared by using 22 images, including 44 main artery segments and 44 main vein segments. Whereas the previous method identified 70.5% of main arteries and 80.0% of main veins, the improved method identified 77.3% and 97.8%, respectively. Figure 2.8d shows the result using this method. The previous method could not identify a main artery and a main vein, whereas the present method succeeded.

2.3.2 Arterio-Venous Nicking

Nguyen et al. proposed a semiautomated AVN grading method [22]. Vessels were first extracted using the multiscale line detection method. To detect AVCs, the vessel skeleton was extracted from the segmented image using binary morphological operators. When the number of neighbor pixels in the skeleton image was four, the point was defined as an AVC [7]. The AVC detection method was then performed to localize all AVC locations. The connected vessel was divided into four branches at each detected AVC. Each of two crossing branch pairs was classified as an artery or vein by using the pixel values of the green component image; the branch pair with lower pixels was defined as an artery. The vein diameters were measured for analysis of AVN severity. Using 47 trimmed images with AVC, the sensitivity achieved was 77%, and the specificity was 92%.

Hatanaka et al. proposed an automated method for AVN detection [23]. They first segmented the blood vessels by using a double-ring filter [7] (as shown in Figure 2.2). AVCs were then detected by using the ring filter as shown in Figure 2.11d. If over four vessels existed on the ring filter, the point of interest was determined as the AVC candidate. A pair of points on the ring filter in which their connected line crosses the interest point was determined as belonging to the same vessel. For example, A1-A2 and B1-B2 in Figure 2.11d are identified as same vessels, respectively. A point without a corresponding point was determined to be a false point. The ring filter incorrectly detected many false AVCs, which were removed by the feature analysis. The box regions just outside the ring filter were set, and the means of the pixel values of the boxes in the green and blue components were calculated. Referring to Figure 2.11e, if the difference between the means of boxes A1 and A2 was large, A1 and A2 were determined to be branches of different vessels, and the AVC was removed. The means of the pairs of boxes were also calculated. If the difference of means A1-A2 and B1-B2 was large, one pair was determined as a false vessel and the related AVC was removed. For a true AVC, the pair with the larger mean in A1-A2 and B1-B2 was determined to be an artery. V_1 and V_2 were then calculated for detection of the AVN. The vein on the ring was determined to be position P1, and the position with a half radius from the center of the ring on the vein was determined to be position P2 (as shown in Figure 2.11f). V_1 and V_2 were calculated as diameters at P1 and P2, respectively. Finally, the point of interest was determined to be AVN when the V_2/V_1 ratio was under 0.8, which was set experimentally. The method was tested using 24 retinal images, which included 27 AVNs. Among these AVNs, 85% (23/27) were detected as ANC with 7.2 false positives per image (including 36 normal AVC). Using an automated V_2-to-V_1 ratio measurement, 70% (16/23) of AVNs were classified correctly. However, only 59% (16/27) of the total AVNs were detected automatically with 1.7 false positives. Therefore, fully automated AVN detection remains a difficult problem.

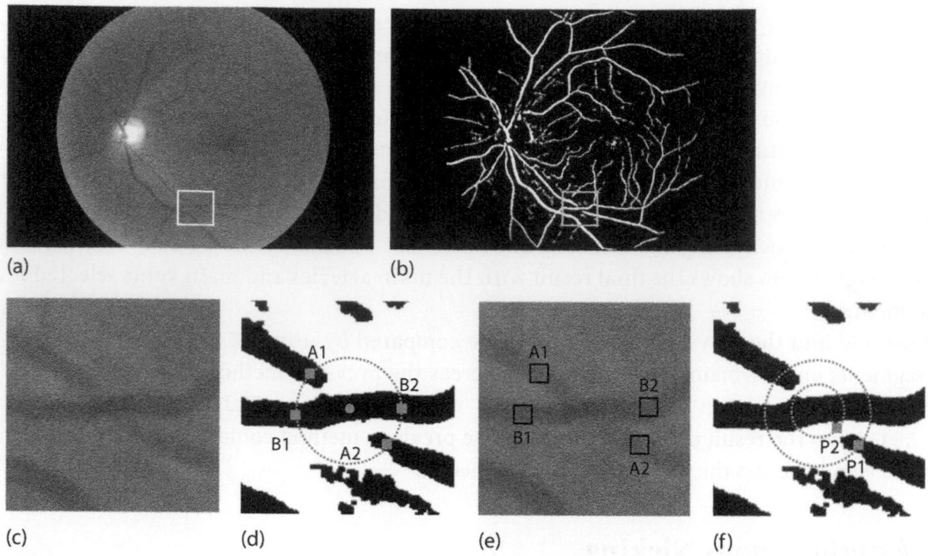

FIGURE 2.11 Identification of arterio-venous nicking. (a) Green component image. (b) Blood vessel segmentation. (c) Enlarged image of region marked by a rectangle in (a). (d) Enlarged image of the blood vessel region marked by a rectangle in (b). Dot ring shows a ring filter. The circular point shows the center of the filter. Boxes show the points crossing the blood vessel and ring filter. (e) The pixel means of four boxes in the green and blue components were calculated for removing false AVC and identification of veins. (f) Determination of venous width V_1 and V_2 on position P1 and P2. The vein on the ring was determined to be position P1, and the position with the half radius from the center of the ring on the vein was determined to be position P2.

2.4 Diabetic Retinopathy

Diabetic retinopathy (DR) is the leading cause of vision loss, but it can be prevented by early detection and treatment. The findings related to DR in a retinal fundus image are microaneurysms (MA), hemorrhages, exudates, and neovascularization.

2.4.1 Hemorrhage Detection Methods

Li et al. developed a hemorrhage detection method using a kNN classifier with many features [24]. They applied two feature selection methods, which were the filter approach and the wrapper approach [25]. They evaluated the method using the MESSIDOR (Methods to evaluate segmentation and indexing techniques in the field of retinal ophthalmology) database [26]. The AUC via ROC analysis was 0.96. Rocha et al. also proposed methods for detection of red and bright lesions based on an SVM with Speeded-Up Robust Features (SURF) [27]. Using MESSIDOR, AUCs for detection of red and bright lesions were 0.72 and 0.89, respectively.

2.4.2 Microaneurysm Detection Methods

An MA appears as a dark dot on a non-contrast retinal image. Its diameter may be just a few pixels, and hence, it is difficult for a human observer to detect MAs. To serve as retinal image databases for MA detection, ROC (Retinopathy Online Challenge) [28], DIARETDB1 (Standard Diabetic Retinopathy Database Calibration level 1) [29], and eOphtha [30] have been made publicly available, and the performances of many algorithms have been compared using these databases.

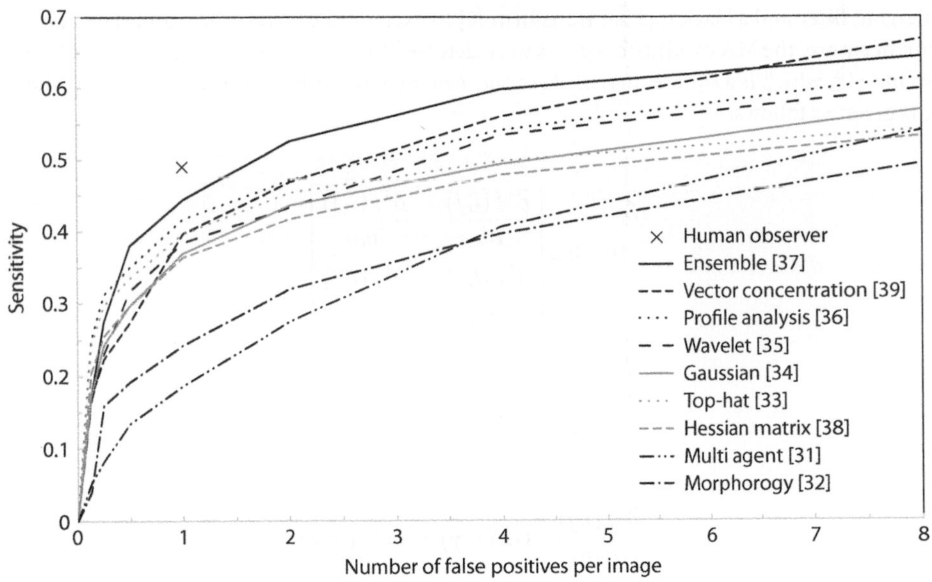

FIGURE 2.12 FROC curves of nine methods using the retinopathy online challenge database [28].

The ROC database contains 50 training images and 50 test images, in which each set includes images with three different resolutions. The gold standard for 336 MAs was determined by four experts using the training images. However, the gold standard for the test images is not provided. The 50 training images present three different types of images with different resolutions, which contain 22 images of 768 × 576 pixels, three images of 1058 × 1061 pixels, and 25 images of 1389 × 1383 pixels. The test images include 343 MAs, determined by three experts, but the gold standard locations in the test images are not available. The 50 test images present three different types of images with different resolutions, which contain 22 images of 768 × 576 pixels, six images of 1058 × 1061 pixels, and 22 images of 1389 × 1383 pixels. The field of view of the training and test images is 45°. Many groups have been developing automated methods for MA detection based on multi-agents [31], morphological operators [32,33], multiscale Gaussian correlation filtering [34], template matching on wavelet translated images [35], analysis of profiles on MA [36], ensemble-based MA detector [37], eigenvalue analysis of the Hessian matrix [38], and gradient vector concentration [39] using the ROC database. Figure 2.12 shows the FROC (Free-response Receiver Operating Characteristic) curves of eight methods using the ROC database. The ensemble-based method [37] shows the best performance in Figure 2.12, but this method combined several MA detectors. The method using the MA profile [36] was the best among the simple MA detectors for false positives per image under 2.0. Gradient vector concentration [39] detected MAs the most over four false positives per image, although this tended to cause an incorrect detection of a part of a capillary as an MA. Eigenvalue analysis of the Hessian matrix [38] and gradient vector concentration [39] are described in the following sections.

2.4.2.1 MA Detection Based on Eigenvalue Analysis of the Hessian Matrix [38]

There are differences in brightness and contrast in retinal fundus images. In order to reduce the adverse effects on image processing due to these differences, the images were preprocessed with gamma correction and contrast enhancement. The contrast between an MA and the surrounding retinal area is the highest in the green channel of the color image, and hence, the green channel of the color image was used for blood vessel extraction and MA detection. A low-pass filter based on the fast Fourier transform was applied to the green channel for decreasing noise. The blood vessels were extracted by combining a double-ring filter and a black top-hat transform [8] to exclude them from the candidate regions of the MA. Furthermore, the MA candidate regions were detected by using an eigenvalue analysis based on the Hessian matrix, which is a square matrix of second-order partial differentiation functions. The Hessian matrix is given as follows:

$$H(i,j)=\begin{bmatrix} \frac{\partial^2 f(i,j)}{\partial^2 i} & \frac{\partial^2 f(i,j)}{\partial i \partial j} \\ \frac{\partial^2 f(i,j)}{\partial j \partial i} & \frac{\partial^2 f(i,j)}{\partial^2 j} \end{bmatrix}$$

$$\frac{\partial^2 f(i,j)}{\partial^2 i} = G_{xx}(x,y) * I(x,y)$$

$$\frac{\partial^2 f(i,j)}{\partial i \partial j} = G_{xy}(x,y)(x,y) * I(x,y)$$

$$\frac{\partial^2 f(i,j)}{\partial j \partial i} = G_{yx}(x,y) * I(x,y)$$

$$\frac{\partial^2 f(i,j)}{\partial^2 j} = G_{yy}(x,y) * I(x,y)$$

where:

*	is the convolution operator
$I(x,y)$	is the preprocessed image
$G_{xx}(x,y)$, $G_{xy}(x,y)$, $G_{yx}(x,y)$, and $G_{yy}(x,y)$	are second-order partial derivative functions of the Gaussian function $G(x,y)$ in each direction.

The Gaussian function $G(x,y)$ is

$$G(x,y)=\frac{1}{2\pi\sigma^2}\exp\left(-\frac{x^2+y^2}{2\sigma^2}\right)$$

where,

$$G_{xx}(x,y)=\frac{x^2-\sigma^2}{2\pi\sigma^6}\exp\left(-\frac{cx^2+y^2}{2\sigma^2}\right)$$

$$G_{xy}(x,y)=\frac{xy}{2\pi\sigma^6}\exp\left(-\frac{x^2+y^2}{2\sigma^2}\right)$$

$$G_{yx}(x,y)=G_{xy}(x,y)$$

$$G_{yy}(x, y) = \frac{y^2 - \sigma^2}{2\pi\sigma^6} \exp\left(-\frac{x^2 + y^2}{2\sigma^2}\right)$$

where σ is a parameter to determine the scale of the Gaussian function.

In this study, σ is 3.0 (pixels) and the convolution area is a square region within 3σ of the pixel of interest. With the first two eigenvalues of the Hessian matrix, it is possible to classify the shape of the intensity curve surface. The shape index is the classification index. By using the two eigenvalues ($\lambda_1 \geq \lambda_2$), the shape index *S* is given by

$$S = \begin{cases} -\frac{2}{\pi} \tan^{-1} \frac{(\lambda_1 + \lambda_2)}{\lambda_1 - \lambda_2} & (\lambda_1 \neq \lambda_2) \\ -1 & (\lambda_1 = \lambda_2 > 0) \\ 1 & (\lambda_1 = \lambda_2 < 0) \end{cases}$$

where $-1 \leq S \leq 1$. When *S* has a small value, the possibility of MA is high.

The MA candidate regions included many false positives. Therefore, in order to remove the false positives, 48 image features were calculated. These image features are listed in Table 2.3. These features included (1) area; (2) degree of circularity; (3) length-to-width ratio; (4–6) mean value of the candidate lesion in each of the red, green, and blue components; (7–9) difference between the maximum and minimum pixel values of the candidate lesion in each of the red, green, and blue components; (10–12) contrast; (13) similarity of blood vessel; (14) number of labels in binarized image; (15) nearest distance from blood vessels; (16, 17) ordinate and abscissa with origin located at the center

TABLE 2.3 The List of 48 Image Features for Reduction of False Positives

Number	Features
1	Area
2	Circularity
3	Length-to-width ratio
4–6	Mean pixel value of the candidate region in red, green, and blue channel components
7–9	Difference between the maximum and minimum pixel value of the candidate region in red, green, and blue channel components
10–12	Contrast between peripheral region and candidate region in red, green, and blue channel components
13	Similarity to blood vessel
14	Number of candidate regions in region of interest
15	Nearest distance from blood vessels
16, 17	Ordinate and abscissa from optic disc
18–23	Mean output value of two double-ring filters in red, green, and blue channel components
24–36	Haralick features obtained from the co-occurrence matrix in green-channel component: angular second moment, contrast, correlation, variance, inverse difference moment, sum average, sum variance, sum entropy, entropy, difference variance, difference entropy, and two kinds of information measure of correlation
37–38	Weszka features obtained from the difference statistics in green-channel component: angular second moment and mean
39–48	Galloway features obtained from the run length matrices (0 and 90°) in green-channel component: short runs emphasis, long runs emphasis, gray-level non uniformity, run length non uniformity, and run percentage

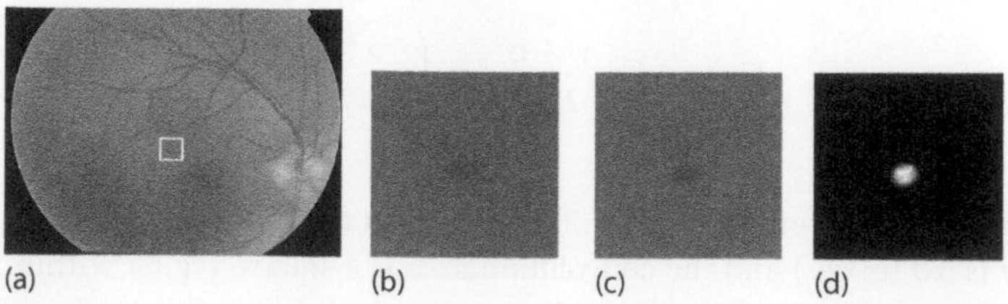

FIGURE 2.13 Example of retinal images with a microaneurysm. (a) Green component image, and (b) enlarged view of the white box in image (a). There is a microaneurysm at the center of the image (b). (c) Preprocessed image. (d) Results of eigenvalue analysis. Here, if $S = -1$, pixel value is 255, and if $S \geq 0$, pixel value is 0. Otherwise, pixel value is $-S \times 255$.

of the optic disc; (18–23) mean values of double-ring filter in red, green, blue values; (24–36) twelve of Haralick features from co-occurrence matrix in green values [40]; (37, 38) two of Weszka features from gray-level difference statistics in green values [41]; (39–48) each five of vertical and horizontal Galloway features from run length matrix in green values [42]. On the basis of these features, false positives were preliminarily removed using the rule-based method. Moreover, an SVM classifier based on features selected by a principal component analysis (PCA) was trained to differentiate between MAs or false positives. The PCA was applied to reduce dimensionality of the vectors that serve as inputs to the SVM. The eigenvectors corresponding to the cumulative contribution ratio that exceeds 95% were used for the SVM.

The MA image was emphasized by the shape index in Figure 2.13. The shape index of the proposed method can classify the shape of the intensity curve surface, i.e., the classification of a cup and rut is possible. Hence, the shape index is effective in detecting dark lesions in the retinal images.

2.4.2.2 MA Detection Based on Gradient Vector Concentration [39]

The retinal image background was corrected, the blood vessels were extracted, and noise was reduced as described in Section 2.4.2.1. In a typical MA, almost all density gradient vectors point to the center of the MA. Thus, MAs were detected by using a measure of concentration by density gradient vector.

Pixel values of retinal images were inverted for calculation of vector concentration at this stage. A typical MA is round, thus the direction component of the density gradient vector points to the center of the MA, and the density gradient vectors concentrate near the center of the MA. Hatanaka et al. [43] proposed the concentration measure as follows:

$$C_1(i,j) = \sum_{x,y \in R} d(x,y) f(x,y)$$

where $C_1(i, j)$ was defined by the cumulative values in a circular mask region R as shown in Figure 2.14, and was calculated by the density gradient from a Sobel filter. In the equation, $d(x, y)$ represents the intensity component of the density gradient as shown in Figure 2.14, and $f(x, y)$ is given by the following:

$$f(x,y) = \cos\left[\theta(x,y) - m \cdot \sin\left\{2\theta(x,y) - \pi\right\}\right]$$

where:

$\theta(x, y)$ is shown in Fig. 2.14, and m is an adjustment factor

$C_1(i, j)$ depends on $d(x, y)$, and the point concentration was not considered

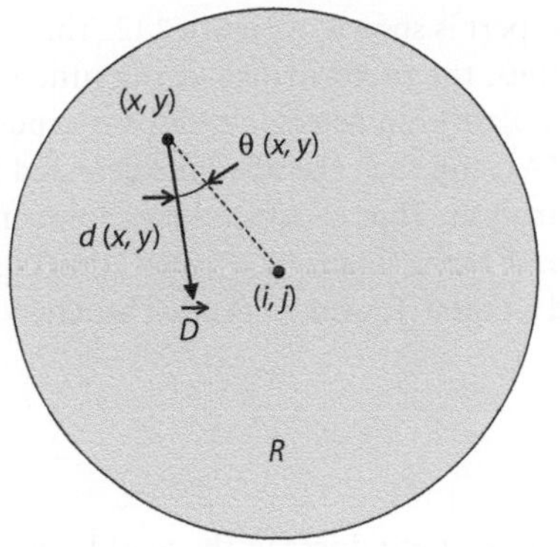

FIGURE 2.14 The calculation area for density gradient vector concentration.

Thus, $C_1(i, j)$ incorrectly enhanced the exudates and the edges of the optic nerve head. Hatanaka et al. [43] proposed a new concentration $C_2(i, j)$ as an improvement of $C_1(i, j)$.

$$C_2(i,j)=\sum_{x,y\in R} d(x,y)(f(x,y)\times\left\{\frac{1}{n}\sum_{x,y\in R} f(x,y)\right\}^2$$

The first term of $C_2(i, j)$ represents the gradient vector concentration, the second term represents the point concentration, and n is the number of pixels in R.

Figure 2.15a shows an example of a retinal image, and Figure 2.15b shows the MA, which looks like a dark spot. Figure 2.15b–e, f–i show examples of the filtered MA region and false positive region by using $C_1(i, j)$, $C_2(i, j)$ and the second term of $C_2(i, j)$. Every concentration enhanced the MA region, as shown in Figure 2.15, but the point concentration incorrectly enhanced many normal tissues, as shown in Figure 2.15e,i. Although $C_2(i, j)$ tended to have low sensitivity for normal regions, as shown in Figure 2.15h, $C_1(i, j)$, incorrectly enhanced the normal region with color shading. Irregular edges exist in a normal region, thus $C_1(i, j)$ tended to affect such edges. The point concentration does not include the intensity component, therefore, it controls the undesired influence of the intensity component in $C_2(i, j)$. In Figure 2.12, when the number of false positives was 8.0, the sensitivity of the proposed method was considered the best among nine methods.

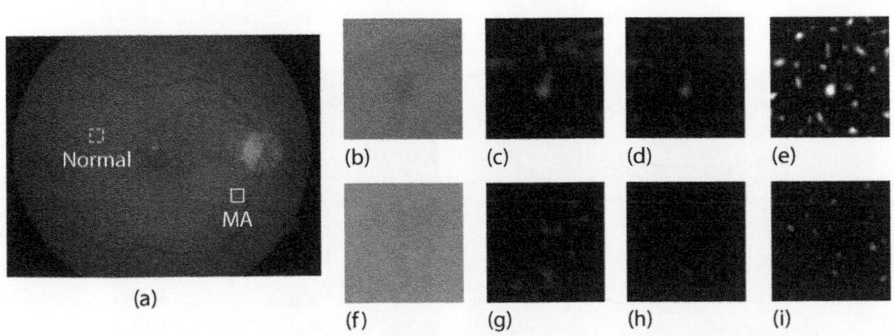

FIGURE 2.15 Example of microaneurysm and normal region. The solid line box shows an example of an MA, and an enlarged image is shown in (b). The broken line box shows an example of a normal region, and an enlarged image is shown in (f). (c)–(e) images are filtered versions of the image in (b) using the density gradient vector concentration $C_2(i, j)$, the previous vector concentration $C_1(i, j)$, and the point concentration (second term of $C_2(i, j)$), respectively. (g)–(i) images are filtered versions of the image in (f).

The performance of a human expert is shown in Figure 2.12. This result shows that detection of MAs in the test image was difficult. Thus, the performance of the nine methods was less than that of the expert. In general, a computerized algorithm detects a small dark spot as an MA. The proposed method also pre-detected such spots as MAs, and then classified the detected MA candidates into the MAs and false positives by using a texture analysis. Thus classification of low contrast MAs and false positives was difficult in the low false positives rate range (Figure 2.12). Detection of an MA that the expert overlooked was also difficult for the proposed method. Researchers must find new features that are based on clinical knowledge.

2.5 Glaucoma

Glaucoma is the second leading cause of blindness in the world. Once the optic nerve is damaged and the visual field deteriorates, it cannot recover. Therefore, early detection of glaucoma is important for halting disease progression and preventing total blindness. However, due to the lack of symptoms and the slow progression in its early stages, most patients do not consult a doctor until glaucoma reaches advanced stages. Retinal imaging is an easy and effective diagnostic exam. Retinal image findings that indicate glaucoma include large cupping, rim loss, rim notching, peripapillary atrophy (PPA), and nerve fiber layer defects (NFLDs). Methods for detection of large cupping and NFLDs are described in the following sections.

2.5.1 Large Cupping Detection

Large cupping is found by measuring the vertical optic cup-to-disc ratio (CDR), and rim loss is found by measuring the rim-to-disc ratio (RDR). The vertical CDR was determined using a method described by Gloster et al. [44]:

$$\text{CDR} = C_v / D_v$$

where:

C_v is the vertical cup diameter
D_v is the vertical disc diameter

C_v and D_v were determined by the distances between the top and bottom levels of these diameters, as shown in Figure 2.16a. The higher the CDR is, the higher the glaucoma risk is. In the upper region

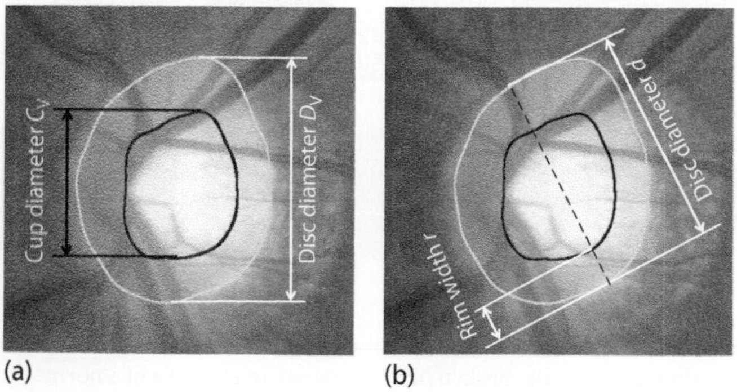

FIGURE 2.16 Determination of (a) vertical optic cup-to-disc ratio (CDR) and (b) rim-to-disc ratio (RDR). The white and black lines show OD and cup outlines, respectively.

(11–1 o'clock) or the lower region (5–7 o'clock) of the optic disc (OD), the narrowest dimension r of the rim divided by the diameter d of the OD at that point was determined as the RDR [44].

$$\text{RDR} = r/d$$

The lower the RDR is, the higher the glaucoma risk is.

Most of the proposed algorithms for this purpose consist of four steps: detection of OD location, OD segmentation, cup segmentation, and CDR or RDR measurement. Figure 2.16 shows an example of cup and disc contours drawn by an expert. Several groups, including Joshi et al., Hatanaka et al., and Cheng et al., proposed methods for OD and cup segmentation [45–47]. OD segmentation methods by Joshi et al. [45] and Hatanaka et al. [46] were similar; therefore, the method proposed by Hatanaka et al. is presented below.

2.5.1.1 OD Segmentation [20]

The OD can be identified as a bright region on a retinal image. In order to reduce computational time, the approximate OD locations were identified, and regions of interest (ROIs) that included the OD were extracted from the retinal images, as shown in Figure 2.17b. The potential location of an OD was identified through the p-tile thresholding method, using the red channel of RGB color images. If a region with the maximum area in the binarized image had a degree of circularity larger than an arbitrary threshold,

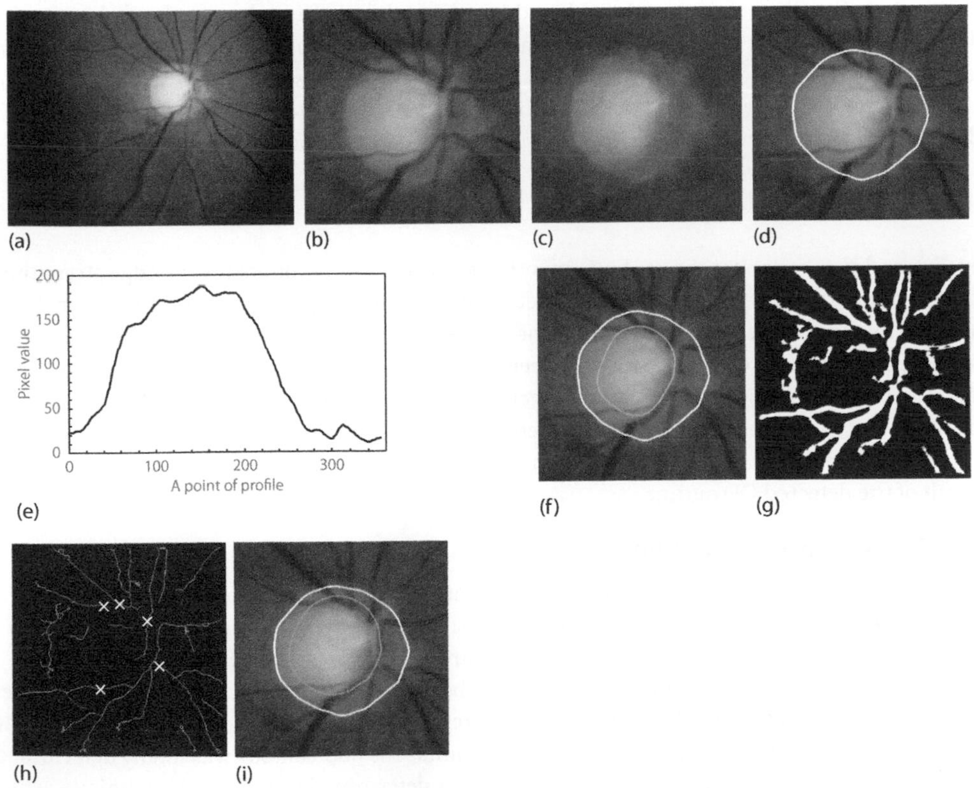

FIGURE 2.17 An example of OD and cup outline detection. (a) Green component image. (b) ROI of OD. (c) "Blood-vessel-erased" image. (d) Result of OD outline detection. (e) The vertical profile in the OD region. (f) The initial cup outline (gray contour). The white contour shows the OD outline. (g) The binarized image of blood vessels obtained by using the gradient vector concentration. (h) Result of detection of bends in blood vessels (x marks). (i) Result of the final cup outline (gray contour).

the region was considered to be the approximate OD region, and an ROI was extracted approximately at the center of the region. Otherwise, the next brightest region was identified by repeating the process.

Blood vessels on the OD may interfere with the precise determination of the OD region. In order to reduce this issue, a "blood-vessel-erased" image, as shown in Figure 2.17c, was created [10]. The pixels on the blood vessel regions were first extracted by using a black top-hat transformation on green-channel images, and their pixel values were then inpainted by the weighted average of the surrounding non blood vessel pixels, whose weights were inversely proportional to the distance between the two pixels.

The expected OD outline was identified in radial directions from the center of the ROI by active contour modeling. The initial candidate points of the OD outline were selected on the basis of the edges identified by applying the Canny edge detector on the red channel of the blood-vessel-erased images. Assuming that the ODs are approximately round in shape, the edges with a gradient directed perpendicularly to the radial direction were suppressed by

$$A' = A\cos(\theta - \varphi)$$

where:

A is the amplitude of the gradient

θ and φ are the radial and edge directions, respectively

The edges in the radial direction are sustained, while the edges perpendicular to the radial direction are eliminated. The edges of the remaining blood vessels can be suppressed using this technique. Based on the detected edges, the median distance from the center to the closest edges was determined in 48 directions. In each direction, a point on the detected edges, which was close to the median value to the edge candidates in the previous and next directions, was selected as the initial candidate point. These candidate points were adjusted by energy minimization based on the following equation:

$$E = W_{\text{dist}}E_{\text{dist}} + W_{\text{curv}}E_{\text{curv}} + W_{\text{edge}}E_{\text{edge}}$$

where E_{dist} and E_{curv} represent the contour smoothness, while E_{edge} accounts for the edge strength with the corresponding weight factors, which were determined empirically using the training cases. E_{dist} was determined by the average distance between candidate points in the current and next directions at both sides, and E_{curv} was based on the distance between the candidate point in the current direction and the middle point of the candidates in the next directions. At each iteration, eight neighborhood pixels of the candidate pixel in the previous iteration were searched. If the cumulative energy in all directions became lower than that of the previous iteration, the candidate points were updated. Figure 2.17d shows the result of the detected OD outline.

2.5.1.2 Optic Cup Segmentation

Chen et al. proposed methods for OD and cup segmentation using superpixel classification [45]. They used histograms and center-surround statistics to classify each superpixel as OD or not OD. Moreover, the location information was also included in the feature space for cup segmentation.

The cup was segmented by considering blood vessel bends as a part of cup contour [46,47]. Joshi et al. [46] detected vessel bend candidates within the bright region by curvature, and the potential bends were selected by finding the best-fitting circle to the set of points. The cup contour was finally determined by local spline fitting. In their method, the bright region was determined using a p-tile technique, which is rather a simple method. Thus, Hatanaka et al. determined the initial cup region (pallor region) by using a line profile technique [47]. The related procedures are given below.

Figure 2.17e shows the vertical profile in the OD region. A profile was then obtained around the center of gravity of the disc region that was extracted automatically. The blood vessel regions that were not erased affected the profile. Thus, 10 profiles were obtained around the center of gravity of the disc region, and profiles were then averaged so that the result would not depend on the specific line selected. This profile was smoothed, and impulses, if any, were removed in order to reduce the effect of noise and the remaining blood vessels. Subsequently, the profile was divided at the top point into two parts. The area under the mean pixel values on the profile was determined as the search area for detection of the cup edge. The cup edge was then determined by using the zero-crossing method in the search area. Profiles were obtained in 10-degree intervals around the center of gravity of the disc region. In other words, 36 points of cup edges were obtained by this process experimentally. The initial cup outline was finally determined by the spline interpolation method based on the 36 points. Figure 2.17f shows an example of an initial cup outline.

The black top-hat transform could not detect blood vessel regions with low contrast. Thus, such blood vessels were detected by using the concentration feature determined from the density gradient [43]. The concentration feature $C_1(i, j, r)$ is described in Section 2.4.2.2. In cup segmentation, the radius r of circle mask R (as shown in Figure 2.14) was set to 9, 11, and 13 pixels, and the maximum value of $C_1(i, j)$ is determined as the final concentration $C(i, j)$.

$$C(i,j) = \max_{r=9,11,13} C_1(i,j,r)$$

The retinal image was filtered by $C(i, j)$, and then the blood vessels were detected by a p-tile method, as shown in Figure 2.17g. The blood vessel bends were detected by tracking the blood vessels from the OD outline to the initial cup outline. The bends were divided into two types, visible bend and invisible bend going backward. The visible bends were detected by the k-curvature [48]. K-curvature is the angle on the blood vessel using three points: the interest point P_0; the previous point $P_{-1,}$ located k pixels away from P_0; and the next point P_{+1}, located k pixels away from P_0. The invisible bends look like endpoints on the OD. Thus, such endpoints were detected as bends. The cup edge candidates were updated by using the bends. The outline of the cup was finally determined using the spline interpolation method. Figure 2.17h shows an example of bend detection, and (i) shows the final cup outline by updating based on the bends.

This method for measurement of CDR and RDR was evaluated on the basis of the manual outlines drawn by two ophthalmologists who are experts in glaucoma diagnosis. Two ophthalmologists classified 50 retinal images as glaucoma and normal, and 44 retinal images were agreed upon by both. Therefore, 44 retinal images were used in this test, including 32 glaucoma images. The results of CDR and RDR were analyzed by using ROC analysis. The AUC values obtained with CDR and the RDR were 0.966 and 0.936, respectively.

2.5.2 Nerve Fiber Layer Defect Detection

NFLD is one of the earliest signs of glaucoma on retinal images. However, there are not enough ophthalmologists specialized in glaucoma diagnosis, and diagnostic findings by different ophthalmologists are varied. Therefore, automated detection of NFLDs is required for early and efficient diagnosis of glaucoma. Muramatsu et al. proposed a method for NFLD detection based on multi-step candidate detection and false positive removal techniques [49,50], which is described as follows.

In their study, Muramatsu et al. used retinal images with 768×576 pixels. NFLDs are depicted with the highest contrast in the green-channel image. Therefore, an original color fundus image was converted to a gray-scale image by using the green-channel image as shown in Figure 2.18b. In order to exclude blood vessel regions from potential candidates, a "blood-vessel-erased" image was created [10]. The major blood vessels were extracted using black top-hat transformation, and the pixels corresponding to the detected vessel regions were inpainted by the surrounding non-vessel pixels. Subsequently, the images were transformed in the same way as the polar transformation but using elliptical functions so that the directions of the retinal nerves are roughly vertically aligned as shown in Figure 2.18c. The brightness in retinal images is often uneven due to the imaging condition.

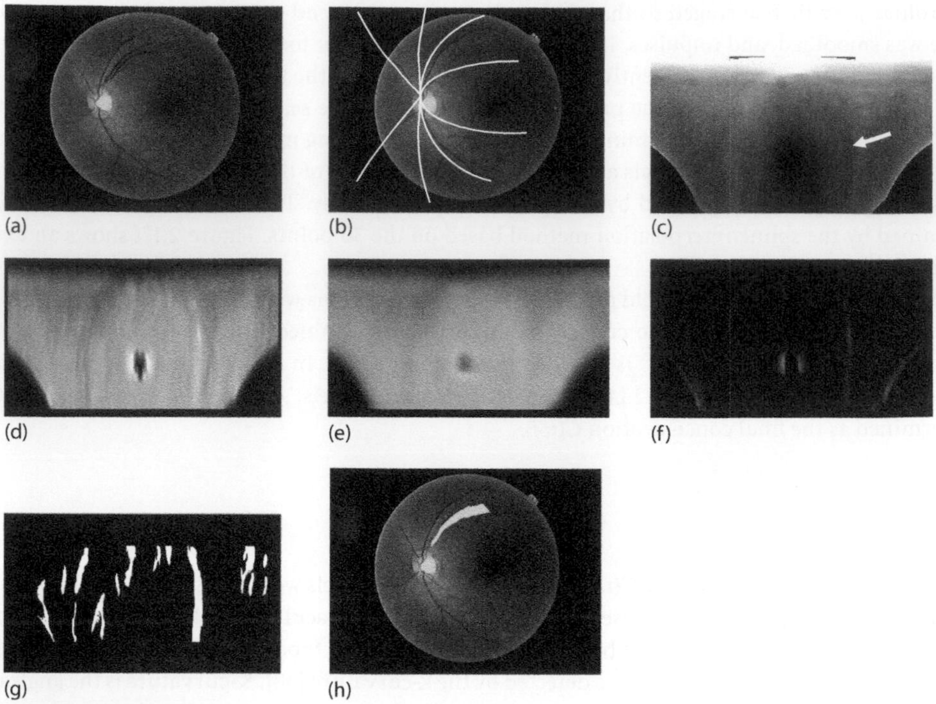

FIGURE 2.18 An example of nerve fiber layer defects (NFLD) detection. (a) Retinal image with a NFLD identified by an ophthalmologist. The two lines specify the extent of the defect. (b) The "blood-vessel-erased" image created by vessel detection and image interpolation. Retinal image with a set of elliptic lines used in the image transformation. The lines represent the estimated paths of retinal nerves. (c) Modified polar transformed green-channel image. White arrow shows the NFLD of (a). (d) NFLD enhanced image using Gabor filters, (e) NFLD attenuated image using a smoothing filter, (f) result of subtraction of (d) from (e), and (g) NFLD candidate regions obtained by thresholding. (h) Result of NFLD detection by this method.

Background brightness was corrected in order to facilitate the detection step. The background trend was estimated by pixel averaging using the red channel image, and the fraction of the inverted background value was added to the green-channel image. Moreover, an un-sharp mask filter was applied for detecting narrow lesions, and small noise enhanced as a by-product was reduced using the median filter.

Gabor filters with four sizes were applied to enhance NFLD regions. The filter sizes were 11×31, 13×31, 21×31, and 41×61 pixels, and the standard deviations of the Gaussian functions were 3 and 10, 3 and 10, 6 and 10, and 12 and 20 pixels, respectively. These four filters were applied to a preprocessed image, and by selecting the maximum output of the four filters in each pixel, the NFLD enhanced image was created, as shown in Figure 2.18d. A smoothing operation was then performed to create the NFLD attenuated image, as shown in Figure 2.18e. The initial NFLD candidate regions were binarized by thresholding the subtraction image of the two. Figure 2.18f shows the subtraction image and (g) shows the first step detection procedure. The thickness of the normal retinal nerve layer varies around the OD, and, accordingly, the intensity of the retina gradually changes around the OD. Therefore, the second detection technique was based on adaptive thresholding in order to detect local intensity changes that were possibly NFLDs. The third detection method was based on k-means clustering after excluding the OD. The mean pixel value and area were determined for each cluster. The candidate pixels were detected using a selective thresholding technique in each cluster. The results of the three methods were combined by retaining the

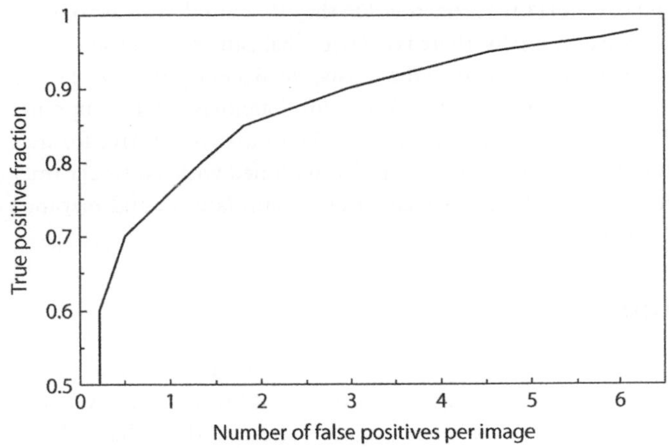

FIGURE 2.19 Free-response receiver operating characteristic (FROC) curve for NFLD detection.

candidate regions detected by the first method if they intersected with the pixels detected by the second and third detection methods.

By focusing on the difference in shape between the NFLDs and blood vessels, false positives were removed. The pixel near the centerline of the blood vessel has a tendency of having black intensity peaks, whereas pixel values inside the NFLD have a tendency to be more uniform. The point of the maximum depth (minimum pixel value) in each cross section of the candidate regions was detected. If the points were stretched like a line, the candidate region was considered a blood vessel. On the other hand, the points on an NFLD region are distributed randomly. Therefore, the candidates with high likelihood measures were removed as false positives.

Moreover, false positives were removed by using an LDA classifier with 11 features. Six features included the area, vertical length, mean pixel values in the original and filtered images, and contrast in the original and filtered images. The contrast is defined as the difference in the average pixel values in a candidate region and its surrounding background region.

Five more features were determined from the original images. The seventh feature was the difference in widths of a candidate region on the OD side and the temporal side (distal to OD). NFLDs can be observed in a fan-like shape, whereas diameters of vessels remain almost constant or decrease with increasing distance from the OD. The eighth feature was the positional feature based on the probabilistic atlas created with the training cases that represents the frequent sites of NFLDs. The other three features were based on the pixel values and contrast of the candidate regions. As the blood vessel likelihood measure, these features were intended to distinguish between NFLD regions with relatively constant pixel values and false positive regions with larger variations in pixel values.

This method was tested on 261 cases that included 130 cases with 203 NFLDs. Figure 2.19 shows the FROC curve obtained by this method. If a candidate region overlapped with the gold standard NFLD region, it was considered to be a true positive region. The initial detection sensitivity was 82% with 6.0 false positives per image. By applying a false positive removal procedure, the sensitivity was 80% with 1.36 false positives.

2.6 Three-Dimensional Eye Devices

This chapter described techniques limited to a two-dimensional, non-contrast retinal image. For diagnosis of glaucoma, other eye examination devices have been developed. The Heidelberg Retina Tomography (HRT) system is a confocal scanning laser ophthalmoscope that can create a three-dimensional (3D) image and is effective for diagnosis of large cupping. However, it is not suited for screening of retinal diseases, since the field of view (FOV) is narrow. On the other hand, the stereo retinal camera can obtain 2D and 3D color images. Additionally, there is a device that can create mosaic images automatically and is effective for screening, detailed glaucoma diagnosis, etc. Recently, the use of optical coherence tomography (OCT) has been increasing, which is effective for diagnosis of large cupping, since it can provide 3D images. It can also measure nerve fiber layer thickness and is effective for diagnosis of NFLD. The FOV of OCT is narrow, but the latest OCT device is combined with a retinal camera feature. Therefore, OCT will become popular as a clinical device for eye examinations, and methods for CAD of the eye using multiple devices must be developed.

2.7 Conclusion

This chapter introduced several component techniques for retinal CAD. In particular, the blood vessel segmentation step is the most important component in retinal CAD, since it is necessary for almost all applications. Several blood vessel segmentation methods have shown high performance with the DRIVE database. To use them in other CAD applications, they need to be evaluated with a large clinical or screening database. Moreover, several methods for diagnosis of diabetic retinopathy and glaucoma, which are the first and second leading causes of blindness, were introduced in this chapter. MA is an early sign of diabetic retinopathy and has low contrast in non-contrast retinal images. Therefore, ophthalmologists generally identify MAs by using fluorescent retinal images. On the other hand, the ROC database, including non-contrast retinal images, has become available for comparison of the performance of several algorithms. Therefore, many MA detection algorithms have been published. However, the performance of all algorithms has not reached that of experts yet. The ROC database includes many invisible MAs, and novel algorithms for detection of such MAs must be developed. Algorithms for detection of large cupping are necessary for diagnosis of glaucoma in its early stages. Many researchers are continuing to develop various techniques for retinal CAD.

References

1. Vuong, L.N., Thulasi, P., Biousse, V., Garza, P., Wright, D.W., Newman, N.J., and Bruce, B.B. 2015. Ocular fundus photography of patients with focal neurologic deficits in an emergency department. *Neurology* 85:256–62.
2. Hoover, A., Kouznetsova, V., and Goldbaum, M. 2000. Locating blood vessels in retinal images by piece-wise threshold probing of a matched filter response. *IEEE Transactions on Medical Imaging* 19:203–10.
3. Niemeijer M., Staal J.J., van Ginneken B., Loog M., and Abramoff, M.D. 2004. Comparative study of retinal vessel segmentation methods on a new publicly available database. *In Conference Proceedings of SPIE* 5370:648–56.
4. Staal, J.J., Abramoff, M.D., Niemeijer, M., Viergever, M.A., and van Ginneken, B. 2004. Ridge based vessel segmentation in color images of the retina. *IEEE Transactions on Medical Imaging* 23:501–9.
5. Chaudhuri, S., Chatterjee, S., Katz, N., Nelson, M., and Goldbaum, M.S. 1989. Detection of blood vessels in retinal images using two dimensional matched filters. *IEEE Transactions on Medical Imaging* 8:263–9.
6. Zolfagharnasab, H., and Naghsh-Nilchi, A.R. 2014. Cauchy based matched filter for retinal vessels detection. *Journal of Medical Signals and Sensors* 4:1–9.
7. Hatanaka, Y., Nakagawa, T., Aoyama, A., Zhou, X., Hara, T., Fujita, H., Kakogawa, M., Hayashi, Y., Mizukusa, Y., and Fujita, A. 2005. Automated detection algorithm for arteriolar narrowing on fundus images. In *Conference Proceedings of the 27th IEEE Engineering in Medicine and Biology Society* 286–9. Shanghai, China.

8. Zana, F., and Klein, J. 2001. Segmentation of vessel-like patterns using mathematical morphology and curvature evaluation. *IEEE Transactions on Image Processing* 10:1010–19.
9. Muramatsu, C., Hatanaka, Y., Iwase, T., Hara, T., and Fujita, H. 2011. Automated selection of major arteries and veins for measurement of arteriolar-to-venular diameter ratio on retinal fundus images. *Computerized Medical Imaging and Graphics* 35:472–80.
10. Nakagawa, T., Suzuki, T., Hayashi, Y., Mizukusa, Y., Hatanaka, Y., Ishida, K., Hara, T., Fujita, H., and Yamamoto, T. 2008. Quantitative depth analysis of optic nerve head using stereo retinal fundus image pair. *Journal of Biomedical Optics* 13:064026.
11. Miri, M.S., and Mahloojifar, A. 2011. Retinal image analysis using curvelet transform and multistructure elements morphology by reconstruction. *IEEE Transactions on Biomedical Engineering* 58:1183–92.
12. Oloumi, F., Rangayyan, R., and Ells, A. 2014. *Digital Image Processing for Ophthalmology: Detection and Modeling of Retinal Vascular Architecture*. Morgan & Claypool Publishers. San Rafael, CA.
13. Soares, J.V., Leandro, J.J., Cesar Június, R.M., Jelinek, H.F., and Cree, M.J. 2006. Retinal vessel segmentation using the 2-D Gabor wavelet and supervised classification. *IEEE Transactions on Medical Imaging* 25:1214–22.
14. Ricci, E., and Perfetti, R. 2007. Retinal blood vessel segmentation using line operators and support vector classification. *IEEE Transactions on Medical Imaging* 26:1357–65.
15. Hatanaka, Y., Samo, K., Tajima, M., Ogohara, K., Muramatsu, C., Okumura, S., and Fujita, H. 2016. Automated blood vessel extraction using local features on retinal images. In *Conference Proceedings of SPIE* 9785: 97852F. San Diego, CA.
16. Niemeijer, M., Xu, X., Dumitrescu, A.V., Gupta, P., van Ginneken, B., Folk, J.C., and Abramoff, M.D. 2011. Automated measurement of the arteriolar-to-venular width ratio in digital color fundus photographs. *IEEE Transactions on Medical Imaging* 30:1941–50.
17. Knudtson, M.D., Lee, K.E., Hubbard, L.D., Wong, T.Y., Klein, R., and Klein, B.E.K. 2003. Revised formulas for summarizing retinal vessel diameters. *Current Eye Research* 27:143–9.
18. Rockett, P.I. 2005. An improved rotation-invariant thinning algorithm. *IEEE Transactions on Pattern Analysis and Machine Intelligence* 27:1671–4.
19. INSPIRE-AVR. Iowa normative set for processing images of the retina artery vein ratio. Available: http://www.medicine.uiowa.edu/eye/Datasets/
20. Muramatsu, C., Nakagawa, T., Sawada, A., Hatanaka, Y., Hara, T., Yamamoto, T., Fujita, H. 2011. Automated segmentation of optic disc region on retinal fundus photographs: comparison of contour modeling and pixel classification methods. *Computer Methods and Programs in Biomedicine* 101:23–32.
21. Hatanaka,Y. Tachiki, H., Okumura, S., Ogohara, K., Muramatsu, C., Fujita, H. 2016. Artery and vein diameter ratio measurement based on improvement of arteries and veins segmentation on retinal images. In *Conference Proceedings of the 38th IEEE Engineering in Medicine and Biology Society* 1336–9. Orlando, FL.
22. Nguyen, U.T., Bhuiyan, A., Park, L.A., Kawasaki, R., Wong, T.Y., Wang, J.J., Mitchell, P., and Ramamohanarao, K. 2013. An automated method for retinal arteriovenous nicking quantification from color fundus images. *IEEE Transactions on Biomedical Engineering* 60:3194–203.
23. Hatanaka, Y., Muramatsu, C., Hara, T., and Fujita, H. 2011. Automatic arteriovenous crossing phenomenon detection on retinal fundus images. In *Conference Proceedings of SPIE* 7963:79633V. Orlando, FL.
24. Tang, L., Niemeijer, M., Reinhardt, J.M., Garvin, M.K., and Abràmoff, M.D. 2013. Splat feature classification with application to retinal hemorrhage detection in fundus images. *IEEE Transactions on Medical Imaging* 32:364–75.
25. Kohavi, R., and John, G. 1997. Wrappers for feature subset selection. *Artificial Intelligence*, 97:272–324.
26. Decencière, E., Zhang, X., Cazuguel, G., Lay, B., Cochener, B., Trone, C., Gain, P., Ordonez, R., Massin, P., Erginay, A., Charton B., and Klein, J.C. 2014. Feedback on a publicly distributed image database: The MESSIDOR database. Short Research Communication. *Image Analysis and Stereology* 33: 231–34.

27. Rocha, A., Carvalho, T., Jelinek, H.F., Goldenstein, S., and Wainer, J. 2012. Points of interest and visual dictionaries for automatic retinal lesion detection. *IEEE Transactions on Biomedical Engineering* 59: 2244–53.
28. Niemeijer, M., van Ginneken, B., Cree, M.J. et al., 2010. Retinopathy online challenge: automatic detection of microaneurysms in digital color fundus photographs. *IEEE Transactions on Medical Imaging* 29:185–95.
29. Kauppi, T., Kalesnykiene, V., Kamarainen, J. K., Lensu, L., Sorri, I., Raninen, A., Voutilainen, R., Pietila, J., Kalviainen, H., and Uusitalo, H. 2007. The DIARETDB1 diabetic retinopathy database and evaluation protocol. In *Conference Proceedings of the British Machine Vision Conference* 15. 1–10. Warwick, UK.
30. Decencière, E., Cazuguel, G., Zhang, X., et al., 2013. TeleOphta: Machine learning and image processing methods for teleophthalmology. *IRBM* 34:196–203.
31. Pereira, C., Veiga, D., Mahdjoub, J., Guessoum, Z., Gonçalves, L., Ferreira, M., Monteiro, J. 2014. Using a multi-agent system approach for microaneurysm detection in fundus images. *Artificial Intelligence in Medicine* 60:179–88.
32. Ram, K., Joshi, G.D., and Sivaswamy, J. 2010. A successive clutter rejection based approach for early detection of diabetic retinopathy. *IEEE Transactions on Biomedical Engineering* 58:664–73.
33. Niemeijer, M., van Ginneken, B., Staal, J.J., Suttorp-Schulten, M.S.A., and Abràmoff, M.D. 2005. Automatic detection of red lesions in digital color fundus photographs. *IEEE Transactions on Medical Imaging* 24:584–92.
34. Zhang, B., Karray, F., Li, Q., and Zhang, L. 2012. Sparse representation classifier for microaneurysm detection and retinal blood vessel extraction. *Information Sciences* 200: 78–90.
35. Quellec, G., Lamard, M., Josselin, P.M., Cazuguel, G., Cochener, B., Roux, C. 2008. Optimal wavelet transform for the detection of microaneurysms in retina photographs. *IEEE Transactions on Medical Imaging* 27:1230–41.
36. Lazar, I., and Hajdu, A. 2013. Retinal microaneurysm detection through local rotating cross-section profile analysis. *IEEE Transactions on Medical Imaging* 32:400–7.
37. Antal, B., and Hajdu, A. 2012. An ensemble-based system for microaneurysm detection and diabetic retinopathy grading. *IEEE Transactions on Biomedical Engineering* 59:1720–26.
38. Adal, K.M., Sidibe, D., Ali, S., Chaum, E., Karnowski, T.P., and Meriaudeau, F. 2014. Automated detection of microaneurysms using scale-adapted blob analysis and semi-supervised learning. *Computer Methods and Programs in Biomedicine* 114:1–10.
39. Hatanaka, Y., Okumura, S., Ogohara, K., Inoue, T., Muramatsu, C., and Fujita, H. 2015. Automated detection of microaneurysms using improved density gradient vector concentration on retinal fundus images. In *Conference Proceedings of the 2015 Joint Conference of the International Workshop on Advanced Image Technology and the International Forum on Medical Imaging in Asia*, Paper 217. Tainan, Taiwan.
40. Haralick, R., Shanmugam, K., and Dinstein, I. 1973. Textural features for image classification. *IEEE Transactions on Systems, Man, and Cybernetics* 3:610–20.
41. Weszka, J.S., Dyer, C.R., and Rosenfeld, A. 1976. A comparative study of texture measures for terrain classification. *IEEE Transactions on Systems, Man, and Cybernetics* 6:269–85.
42. Galloway, M.M. 1975. Texture analysis using gray level run lengths. *Computer Graphics and Image Processing* 4:172–9.
43. Hatanaka, Y., Hara, T., Fujita, H., Kasai, S., Endo, T., and Iwase, T. 2001. Development of an automated method for detecting mammographic masses with a partial loss of region. *IEEE Transactions on Medical Imaging* 20:1209–14.
44. Gloster, J., and Parry, D.G. 1974. Use of photographs for measuring cupping in the optic disc. *British Journal of Ophthalmology* 58:850–62.
45. Cheng, J., Liu, J., Xu, Y., Yin, F., Wong, D.W., Tan, N.M., Tao, D., Cheng, C.Y., Aung, T., and Wong, T.Y. 2013. Superpixel classification based optic disc and optic cup segmentation for glaucoma screening. *IEEE Transactions on Medical Imaging* 32:1019–32.

46. Joshi, G.D., Sivaswamy, J., and Krishnadas, S.R. 2011. Optic disk and cup segmentation from monocular color retinal images for glaucoma assessment. *IEEE Transactions on Medical Imaging* 30: 1192–1205.
47. Hatanaka, Y., Nagahata, Y., Muramatsu, C., Okumura, S., Ogohara, K., Sawada, A., Ishida, K., Yamamoto, T., and Fujita, H. 2014. Improved automated optic cup segmentation based on detection of blood vessel bends in retinal fundus images. In *Conference Proceedings of the 36th IEEE Engineering in Medicine and Biology Society* 126–9. Chicago, IL.
48. Rosenberg, B. 1972. The analysis of convex blobs. *Computer Graphics and Image Processing* 1:183–92.
49. Muramatsu, C., Hayashi, Y., Sawada, A., Hatanaka, Y., Hara, T., Yamamoto, T., and Fujita, H. 2010. Detection of retinal nerve fiber layer defects on retinal fundus images for early diagnosis of glaucoma. *Journal of Biomedical Optics* 15:016021.
50. Muramatsu, C., Ishida, K., Sawada, A., Hatanaka, Y., Yamamoto, T., and Fujita, H. 2016. Automated detection of retinal nerve fiber layer defects on fundus images: False positive reduction based on vessel likelihood. In *Conference Proceedings of SPIE* 9785:97852L. San Diego, CA.

3

Computer-Aided Diagnosis of Retinopathy of Prematurity in Retinal Fundus Images

Faraz Oloumi

Rangaraj M. Rangayyan

Anna L. Ells

3.1 Introduction

3.1.1 Retinopathy of Prematurity

Retinopathy of prematurity (ROP) is a complex disease that affects the process of development of the retina and retinal vasculature in preterm infants. ROP has been identified as the leading cause of preventable childhood blindness [1]; it is estimated that at least 50,000 children worldwide are suffering from blindness caused by ROP [2]. Based on a study conducted in 1993 in the U.S., it was reported that there are about 30,000 preterm babies born each year with a birth weight of 500–1249 g, out of whom 1000 are estimated to progress to a level of ROP that requires treatment (threshold ROP) [3]. An analysis [4] of a New York state patient database over a four-year span indicated that the incidence of ROP among newborn infants was 1 in 511. An incidence of 81.5% of any stage of ROP was reported by Gunn et al. [5] in premature infants 23–25.6 weeks of gestational age (GA), based on a screening study conducted over an 18-year period in Australia.

3.1.1.1 Plus Disease

Considering that ROP advances rapidly in the first few weeks of life, timely identification of signs of ROP is crucial for diagnosis and clinical management of the affected infants. In recent years, early detection of threshold ROP has been observed to be highly correlated with the presence of plus disease, which is a designation and an indicator of a severely progressing phase of ROP [6–11]. Plus disease is manifested in terms of vascular changes, including increased vessel thickness and tortuosity in the posterior part of the retina [12]. The posterior area is loosely defined as a circular area centered on the optic nerve head (ONH) and reaching up to the macular region [13]. A change in the openness of the major temporal arcade (MTA), the thickest branch of venules, has also been observed as a sequela of ROP, as well as an indicator of compromised structural integrity of the macular region [1,2,14–16]. The presence of plus disease is now considered to be the main indicator for the need for treatment. Even though the stage and zone of ROP are important markers of the severity of the disease, plus diagnosis is almost always associated with cases of acute ROP requiring treatment [7–10].

Plus disease is clinically defined and diagnosed by visual qualitative comparison of the patient's retina (using a funduscope) in relation to a standard retinal fundus photograph [1] that exhibits abnormal levels of thickness and tortuosity. However, the standard photograph is believed to be atypical because it shows more vascular dilation and less tortuosity as compared to most cases with plus disease [7]. Also, the standard photograph possesses a narrow field of view (FOV) and does not reveal possible tortuous vessels in the periphery of the retina.

The current clinical definition and diagnosis of plus disease are subjective. Furthermore, it is likely that no optimal visual reference standard exists for the diagnosis of plus disease, as shown by disagreement even among recognized experts [8,10,17]. Chiang et al. [8] demonstrated that, among 22 recognized ROP experts who performed diagnosis of plus disease using 34 fundus images of preterm infants based on two-level classification (without plus, with plus), the experts agreed on the diagnosis of only 22% of the images (seven out of 34). Considering such facts, it can be concluded that there is a great need of methods for computer-aided diagnosis (CAD) of plus disease, which can lead to timely diagnosis and effective treatment of ROP.

3.1.1.2 CAD of Plus Disease

Various semiautomated, computer-aided procedures have been designed to perform diagnosis of plus disease by quantification of changes in the tortuosity and thickness of retinal vessels [9,10,18–25]. Such studies have demonstrated that methods for CAD are capable of discriminating between cases with and without plus disease as accurately as experts. To the best of the authors' knowledge, all studies that have performed diagnosis of plus disease via quantification of vascular changes have employed methods that involve manual marking of vessel segments to be analyzed, or manual selection and correction of parts of automatically detected vessels, to include only the desired or selected vessels for further analysis. This may not be feasible in a clinical or teleophthalmological setting due to time and resource constraints. Considering the measurement of tortuosity, all studies seen in the literature have limited the area of analysis to the posterior of the retina; however, peripheral-vessel tortuosity has been shown to be more correlated to the presence of plus disease as compared to posterior-vessel tortuosity [11,26]. Despite the clinical importance of abnormal changes in the openness of the MTA, it has been quantified, in various manners, in only three studies reported in the literature dealing with ROP [14–16].

The main aim of the present work is CAD of plus disease by quantitative feature extraction related to changes in the thickness of the MTA, tortuosity of vessels, and the openness of the MTA, in retinal fundus images of preterm infants via digital image processing techniques. Feature selection and pattern classification methods are employed to determine if the computed features or their combinations can lead to accurate diagnosis of plus disease.

The present work provides an overall view of several methods developed for quantification of the thickness and openness of the MTA, as well as vascular tortuosity. The specific details of these methods have been documented in detail in previous related publications [16,27–33].

3.1.2 Computer-Aided Analysis of Retinal Fundus Images

3.1.2.1 Detection of Retinal Vasculature

There has been an extensive amount of research performed on the development of methods for enhancement, detection, and segmentation of retinal vessels by considering various physiological properties of vessels; these properties include being piecewise-linear oriented features, possessing negative contrast with respect to the background, possessing an inverted Gaussian cross-sectional intensity profile, and presenting a branching tree pattern. In addition to enabling quantitative analysis of the vasculature, vessel detection facilitates localization of certain additional retinal features, such as the macula and the ONH, in relation to the vascular structure [34–36].

Even though vessel detection is a major step in the present work as a prerequisite to quantitative measurement of vascular features, vessel detection is not the focus of the present work; for details regarding this step, please refer to related publications [29,30]. The vessel detection methods available in the literature fall within several general categories that include, but are not limited to, the use of (1) template matching via matched filters [37,38], (2) vessel ridge, edge, or contour detection and tracking [39,40], (3) directional filters [27,41], (4) gray-scale variation and local thresholding [42,43], (5) morphological operators [44,45], (6) Hessian-based gray-scale gradient information [46,47], (7) multi-feature analysis [48,49], and (8) probabilistic models [50,51]. See Oloumi et al. [29] for comparative analysis of a few retinal vessel detection methods.

3.1.2.2 Measurement of the Openness of the MTA

Wilson et al. [14] defined the angle of insertion of the MTA as follows: the center of the ONH and the fovea are manually marked by two independent observers. A line is drawn through the manually marked centers of the ONH and the fovea; this is the retinal raphe. A line perpendicular to the retinal raphe is drawn from the fovea until it intersects the inferior and superior temporal arcades (ITA and STA). From the points of intersection, two lines are drawn to the center of the ONH. The temporal arcade angle (TAA) is defined as the sum of the inferior and the superior arcade angles. In a related follow-up study by Wong et al. [15], semiautomated measurements were made of four different angles of the temporal and the nasal venules and arterioles. The procedures required manual editing of automatically detected vessels; this step required 10–15 min per image. As compared to the previous related study of Wilson et al. [14], the angles were measured using reference points selected closer to the center of the ONH.

The published methods to measure the angle of insertion of the MTA may not properly reflect the changes that occur in the structure of the MTA, as they define the openness of the MTA based on only three points. Furthermore, only the location of the vertex of the arcade angle has been consistently defined as the center of the ONH; the locations of the other two points have been defined in different manners. Even though the structure of the MTA has been used to estimate the ONH and the macula in previously reported works, only Tobin et al. [35] modeled the arcade for parameterization of its openness; however, they used the openness parameter only to draw the parabolic model on the image. It should be noted that identifying the fovea in fundus images of preterm infants is challenging even for experts, as shown by Chiang et al. [52].

The results of quantification of the openness of the MTA using the method of Wong et al. [15] and a modeling approach proposed by Oloumi et al. [28] are presented and compared in Section 3.4.1.

3.1.2.3 Measurement of the Thickness of Vessels

Methods used for measurement of vessel thickness generally fall into two main categories. The first category of methods initially attempts to detect and segment vessels, then obtains pixels associated with vessel centerlines and edges, and finally computes a thickness measure based on the distance between the associated edges and centerline pixels [18,53]. The second category of methods employs vessel detection techniques via a modeling approach; the models are mainly Gaussian-based since the intensity profile of a vessel at a given vessel centerline pixel resembles an inverted Gaussian curve [19,54]. The spread or standard deviation (STD) of the Gaussian model used to detect the vessel could then be used to estimate the width of the vessel.

Using semiautomated methods, Heneghan et al. [18] obtained a binary image of retinal vessels via multiscale analysis of the second-order derivative of the original intensity images, combined with a two-level thresholding strategy. Heneghan et al. measured the vessel width by extending a line from the two opposite sides of a given vessel pixel until it reached the boundary of the vessel and noting the length of the line. This procedure was repeated for lines of various orientations originating from the same pixel. The vessel width at the given pixel was taken as the minimum distance obtained for all orientations.

Fiorin and Ruggeri [53] used a web-based software package to draft manually the centerline of a retinal vessel segment. Canny's edge detection method was used to obtain a set of vessel-edge pixels around the selected centerline. Two edge curves were then estimated on either side of the selected centerline by fitting a cubic spline to the previously detected edge pixels. The width of the selected vessel segment at each pixel was defined as the distance along the normal at a centerline pixel between the two estimated edge curves.

Wilson et al. [54] estimated vessel width using Gaussian models and a contrast measure based on the magnitude response of Laplacian-of-Gaussian (LoG) models obtained at vessel centerline pixels. Wallace et al. [19] determined a vessel width estimate at each detected ridge point using the spread of a LoG model based on tracking and extraction of skeletons of manually selected vessels via multiscale ridge detectors.

Martínez-Pérez et al. [42,55] developed a semiautomated image analysis software called Retinal Image multiScale Analysis (RISA) for detection, segmentation, and measurement of vessel statistics in retinal images. RISA uses a region-growing algorithm to segment (binarize) blood vessels based on two features of edge strength and ridge strength. RISA estimates the width of each vessel segment as the total area (in pixels) of the segment, divided by its length. RISA requires manual user correction at the stage of detection of branching points, as well as manual input regarding the vessel segment to analyze and to distinguish between venular and arteriolar branches.

Locating vessel edges based on binarization of the intensity image obtained using a vessel detection algorithm is affected by the spatial resolution of the image and is prone to error due to sampling, the spatial scale of the model used, as well as the threshold value used to binarize and segment vessels [18]. Furthermore, a Gaussian model is a continuous curve of infinite extent that does not contain a clear-cut point to indicate the exact location of the vessel edges; hence, such estimation of vessel width using the STD of Gaussian-based models is also prone to error. Such a method for estimation of width requires an assumption regarding suitable, yet arbitrary, weighting of the STD of the Gaussian model to estimate vessel width [56]. Methods that use a vessel's centerline for measurement of width are prone to error in case the centerline is not precisely in the middle of the vessel; it is especially challenging to guarantee such a requirement. Furthermore, such methods require accurate detection and subpixel representation of vessel edges. In addition, the methods available in the literature for measurement of vessel width in case of plus disease either do not distinguish between venular and arteriolar branches at all [18] or require a manual setup to distinguish between them [9,10,20,25]; the mentioned methods obtain the width measurement for only manually selected vessel segments.

3.1.2.4 Measurement of Tortuosity of Vessels

Although there have been several measures of tortuosity proposed in the literature, tortuosity does not have a specific mathematical definition. The proposed methods can be typically divided into three categories: length-to-chord (LTC) measures [9,19,54], angle-based measures [20,57], and curvature-based measures [58,59].

It should be noted that measures of tortuosity are typically obtained using a skeletal representation of the vasculature, which can either be drafted manually using a specially designed graphical user interface (GUI) or obtained using image analysis algorithms in an automated or semiautomated environment. All of the methods presented in the literature for measurement of tortuosity in the presence of plus disease have used manually segmented or selected vessels.

The simplest definition of tortuosity is the LTC measure, which is the ratio of the true (geodesic) length, or arc-length, of a vessel segment to the length of the line connecting the segment's end points (chord) [9,19]. The drawback of the LTC measure is that it does not account for the possible changes in the curvature of a vessel segment; hence, an arched segment and a sinusoidal segment of the same true length and chord length may lead to the same tortuosity values which is undesirable. Some studies have proposed modified definitions of the LTC measure to overcome this inherent limitation [54].

The angle-based tortuosity measures are derived by defining a set of vectors of fixed length connecting various points on a given vessel segment and using the angles formed at the tip and tail of two consecutively connected vectors to compute a tortuosity measure. Gelman et al. [20] defined tortuosity as the sum of all angles (SOA) between a set of vectors, as previously explained, normalized by the total length of the segment. However, Gelman et al. did not specify the length of the vectors used to obtain the angles and noted that the proposed methods would not work on a segment without any branching points.

By defining the angle of a given pixel based on the coordinates of the current and previous pixels, Poletti et al. [57] proposed an angle-based measure of tortuosity. The tortuosity measure was computed as the sum of the squared angle changes, divided by the total length of the vessel segment, where the local change in angle was defined as the difference in the inverse tangent of the angle of the current and previous pixels. It should be noted that such a definition of the angle at a pixel based on its 8-connected neighboring pixels will only provide angle measures with increments that are integral multiples of 45°.

Hart et al. [58] originally proposed the use of curvature to define a measure of tortuosity using skeletonized retinal vessel segments, including the total curvature and total squared curvature measures. Grisan et al. [59] separated vessel segments into curved and linear parts using curvature and then assigned each half of a given linear subsegment to its previous and next curved subsegments. Grisan et al. defined the tortuosity of the entire segment as the sum of the LTC of each subsegment normalized by the true length of the entire segment and weighted based on the number of subsegments (number of sign changes in the curvature). Grisan et al. noted that their proposed measure would provide misleading values if a subsegment does not contain any nonlinear parts.

The tortuosity measures that use the SOA approach are dependent on the length of the vectors used for a vessel segment to define the angles; short vector lengths may result in large angle values, whereas long vector lengths may miss small local variations. Ideal representation of the skeletons (or centerlines) of vessels is a difficult task that is prone to discretization errors as well as inaccuracies in detection and segmentation of vessels; defining the orientation of a given vessel pixel using the coordinates of such a skeletal representation of vessels is also prone to errors.

3.2 Database of Retinal Fundus Images and the Evaluative Setup

3.2.1 The Telemedicine for ROP in Calgary Database

The Telemedicine for ROP In Calgary (TROPIC) database [60] is a private collection of retinal fundus images of preterm infants that is used in the present work for CAD of plus disease. Written consent was obtained from the parents of the patients to capture and use the images. The tenets of the Declaration of Helsinki were followed while compiling the database. The images of the TROPIC database were captured using the RetCam II camera equipped with a wide-angle ROP lens (130°). The RetCam II–130 images have a size of 480 × 640 pixels and are considered to possess a poor spatial resolution of about 30 μm per pixel [61].

In total, 110 images from 41 preterm patients (16 females, 25 males) were selected from the database for the present study. In most cases, there are five different images available for each eye of each patient from each visit, representing different retinal fields to provide, collectively, an almost-complete photographic documentation of the retina. In each case, the image with the highest visibility of the entire vasculature was chosen. Images were not selected based on overall quality and/or vessel-to-background contrast. Nineteen of the 110 selected images are from patients diagnosed with plus disease (stages 2 and 3 of ROP), and 91 show no signs of plus disease (stages 0, 1, 2, and 3 ROP). At most, two images from the same patient were included for the same stage of ROP (one image from each eye). Multiple images of the same eye from the same patient were included only if the ROP stages were different at the time of imaging. All diagnoses were performed by a retinal specialist (A. L. Ells) at the time of clinical examination using a funduscope and the patient's clinical records; the diagnosis was not based solely on the RetCam images of the patient. Patients corresponding to 90 of the 110 images were diagnosed with no ROP or with stages 1 or 2 ROP (30 images per category), and patients corresponding to 20 images were diagnosed with Stage 3 ROP.

A training set of 10 images, including five without and five with signs of plus disease was also formed. All abnormally tortuous vessel segments in the five training images corresponding to plus disease were manually identified by the retinal specialist. The images of the training set are mutually independent of the test set described in the previous paragraph, even though they are drawn from the same population of patients.

Figure 3.1 displays two images from the TROPIC database. Part (a) of the figure shows an example of an image without signs of plus disease or ROP. Part (b) of the same figure illustrates an image with signs of Stage 3 ROP, as well as plus disease at the time of imaging. The case with ROP clearly demonstrates dilation of the MTA and tortuosity of several vessels.

Table 3.1 provides the mean and STD for the birth weight (BW), GA, and chronological age (CA) of the patients. As observed in previously published clinical studies [5,62], GA is a statistically significant indicator of the risk of ROP.

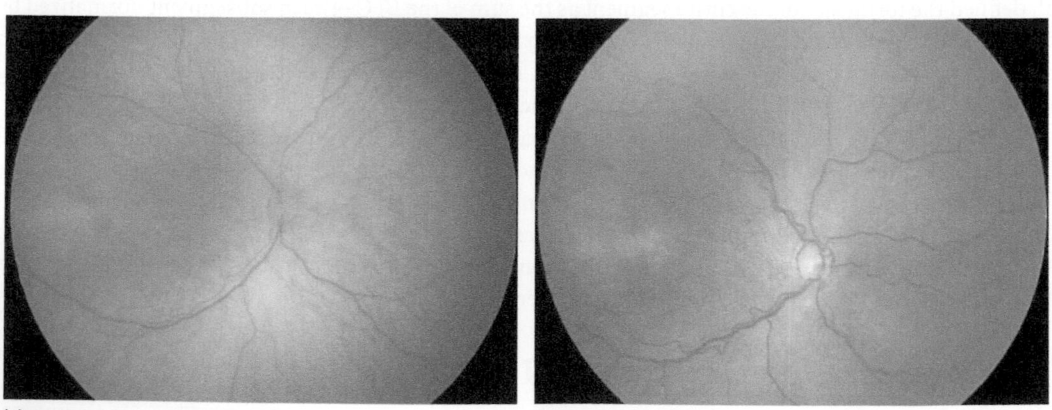

(a) (b)

FIGURE 3.1 Two images from the TROPIC database, where (a) illustrates no signs of plus disease or ROP, and (b) diagnosed with signs of ROP (not visible in the present FOV) as well as plus disease (tortuous and dilated vessels).

TABLE 3.1 The Mean and STD of BW, GA, and CA, in Grams (g), Weeks, and Days, respectively, for Normal Preterm Infants as well as Patients Diagnosed with Plus Disease

Parameter	Without Plus Mean ± STD ($n=91$)	With Plus Mean ± STD ($n=19$)	p-value
BW(g)	818.00 ± 210.78	815.89 ± 203.71	>0.1[NS]
GA (weeks)	26.73 ± 1.88	24.95 ± 1.77	<0.001***
CA (days)	71.05 ± 23.67	69.84 ± 13.00	>0.1[NS]

Note: The p-values indicating the statistical significance between the means of the two classes are also provided.

***$p < 0.001$.

NS stands for not significant.

3.2.2 Evaluation of the Diagnostic Accuracy of the Results

The most suitable method for assessment of the results of feature extraction and classification for the purpose of diagnostic decision making, as applied to MTA thickness (Section 3.4.2) and openness measurement (Section 3.4.1), as well as the total length of tortuous vessels in a given image (Section 3.4.3), is analysis of the receiver operating characteristics (ROC) [63].

The area under the ROC curve, A_z, provides a quantitative measure of the overall level of classification, the performance of the underlying feature, and the trade-off between sensitivity and specificity. In the present work, the ROCKIT software [64] is used to perform ROC analysis for evaluation of the results of diagnostic classification. The A_z value provided by ROCKIT is based on the area under the binormal fit to the ROC curve.

3.2.3 Feature Selection and Pattern Classification

When dealing with a set of features to perform CAD of a certain disease, it would be of interest to determine whether any single feature or a subset of the computed features will yield a more precise diagnosis than any other feature or combination of features. It is possible for various features to provide complementary information when combined together through pattern classification methods and lead to higher diagnostic accuracy than any individual feature on its own. Pattern classification methods of logistic regression (LR) [65], naïve Bayes [66], and multilayer perceptrons (MLPs) [66] were employed in the present work.

Most pattern classification methods require training. It would be ideal to perform training using a dedicated set of data; however, considering the limited number of available cases in this work (110 cases from the TROPIC database), it is required to use all the available dataset to perform the training as well as the testing steps. One practical approach to this situation is the k-fold cross-validation method, in which the given dataset is divided into k parts (folds). One fold is kept aside while the rest of the data are used to train the chosen classifier. The trained classifier is then applied to the fold kept aside [66,67] in the testing or validation step; the procedure is repeated over all folds.

The wrapper method is a feature selection technique that employs supervised learning along with classification algorithms, such as those explained above, to select the most discriminatory combination of features based on a given set of data [67]. Two common search methods of best-first [67] and exhaustive search [66] were employed with the wrapper method to select the set of features that provides the highest level of discrimination between cases with and without plus disease. The exhaustive search method traverses the entire search space and examines all possible combinations of features to determine the feature sets with the highest accuracy of discrimination, also based on the cross-fold-validation method. In all instances, 11 folds (10 images each) were used for cross-validation.

The results of feature selection by applying the wrapper method, a given classifier, a specific search method, and k-fold cross-validation are presented in terms of the number of times any given feature was selected out of the k steps [67]. In the present work, the features that were selected more than half the time ($\geq k/2$), using the proposed setup, were used for pattern classification using the same classifier as with the wrapper method.

3.3 Methods for Quantitative Analysis of Retinal Vasculature

3.3.1 Detection of Retinal Vessels

In this work Gabor filters, which are sinusoidally modulated Gaussian functions, are used for detection of vessels. Gabor filters provide optimal localization in both the frequency and space domains. A bank of K Gabor filters may be obtained by rotating the main Gabor filter response over the range $(-\pi/2, \pi/2)$ with the step size of π/K. For a given pixel, the maximum output value over all of the K filters is saved as the Gabor-magnitude response at that particular pixel; the corresponding angle is recorded as the Gabor-angle response. For simplicity of design, variables τ and l are used to represent the average thickness and elongation of the vessels to be detected in the image. For more information on Gabor filters and other methods for detection of blood vessels in retinal fundus images, as well as preprocessing steps required in application of Gabor filters, please refer to the related studies [27,29,30].

3.3.2 Segmentation of the Detected Vessels

Gabor filters provide the results in terms of a gray-scale image of the vessels, which can be thought of as vessel-strength images. Any further analysis of the retinal vasculature would require the vessel-strength image to be transformed to an image that indicates what is (unity-valued) or is not an object of interest (zero-valued) through binary representation of the image. The binarized vasculature image may require to be processed further in order to obtain the required measurements or information, which is usually achieved through morphological image processing methods.

Considering the ground truth data provided for the 20 training images of the DRIVE database, on the average, only 13% of the FOV of a retinal image is covered by vessel pixels. In fact, it has been observed that the gray-scale intensity values of the result of vessel detection are unimodal and skewed toward zero [29,33]; there is no clear separation between the pixels belonging to blood vessels from the background pixels. Thresholding such gray-scale output images of vessel detection methods with high accuracy is a difficult task, especially since most automated thresholding methods assume a bimodal histogram or a certain level of separation between the foreground and background pixel values.

Several automated thresholding methods, including Otsu's method [68], a moment-preserving thresholding method [69], the Ridler-Calvard thresholding method [70], and an entropy-based thresholding method [71] were explored in a related study [29]. Based on the results of the study [29], it can be concluded that accurate and precise segmentation of retinal vessels for both adult and pediatric cases requires more sophisticated and complex thresholding techniques that adaptively adjust to local variations in each given image.

In the present work, the moment-preserving method was used as part of the methods for measurement of the thickness of the MTA. For the purpose of measurement of the openness of the MTA and tortuosity of vessels, a manual threshold value was set by the user using a threshold slider for each image using a GUI.

3.3.3 Measurement of the Openness of the MTA

3.3.3.1 Measurement of the Arcade Angle

Measurement of the TAA using the method of Wong et al. [15] requires several inputs from the user and is executed in this work as follows [16]:

1. Mark, manually, the center of the ONH and draw a circle with radius $r = 60$ or 120 pixels on the image.
2. Mark, manually, the points of intersection of the circle with the STA and ITA.
3. Measure the TAA as the angle between the three manually marked points, where the center of the ONH is the vertex of the angle.

For the present work, this procedure was implemented with a GUI [72]. The procedure results in one parameter (feature) to quantify the openness of the MTA.

3.3.3.2 Parabolic Modeling of the MTA

It is possible to model the MTA as single parabola or the STA and ITA as two separate semiparabolas via the generalized Hough transform (GHT) [73], as shown in our previous works [16,28,74]. The following list provides an overall step-by-step representation of the various stages of the method designed for measurement of the openness of the MTA via parabolic modeling using the GHT. For specific details of each stage, please refer to the associated works [16,28,74].

3.3.3.2.1 Detection and Segmentation of the MTA Skeleton [27,30]

1. Apply Gabor filters to the inverted green-channel image using a large τ value.
2. Divide the Gabor-magnitude response image into its superior and inferior parts with reference to the center of the ONH.
3. Binarize the Gabor-magnitude response using a threshold slider.
4. Skeletonize the binary image of the vasculature.
5. Remove small disconnected segments using the area-open morphological operation.

3.3.3.2.2 Application of the GHT [28,30]

1. Indicate whether the image is of the left or right eye.
2. Rotate the skeleton images by 180° if the image is of the right eye.
3. Restrict the skeleton image horizontally from $0.25 \times$ ONHW in the nasal direction to $2 \times$ ONHW in the temporal direction with respect to the center of the ONH, where ONHW is the average width of the ONH for preterm infants.
4. Apply the GHT to the superior and inferior skeleton images separately for dual-parabolic modeling of the STA and ITA.
5. Rotate the Hough space by 180°, if the image is of the right eye.
6. Obtain, separately, the openness parameters of the best-fitting semi-parabolas for the STA and ITA.

This procedure results in a single parameter, a_{MTA}, or two parameters, a_{STA} and a_{ITA} to represent the openness of the MTA, STA, and ITA, respectively.

3.3.4 Measurement of the Thickness of the MTA

The process of measurement of the thickness of the MTA consists of several stages including detection, segmentation, and extraction of the MTA skeleton, as well as detection and interpolation of the edges of the MTA, and ultimately, computation of the width of the MTA at all available pixels along its skeleton. The various stages of the entire procedure designed for measurement of the width of the MTA are provided in list form below. For specific details of the procedures and methods please see our related publication [31].

3.3.4.1 Detection, Segmentation, and Extraction of the MTA Skeleton [31,75]

3.3.4.1.1 Preprocessing

1. Mark the approximate location of the center of the ONH.
2. Apply Gabor filters to the inverted green-channel image with $\tau = 10$ pixels, $l = 1.2$, and $K = 45$ [27].
3. Restrict the Gabor-response images horizontally from 0.25 × ONHW in the nasal direction to 2 × ONHW in the temporal direction with respect to the center of the ONH.
4. Remove a circular region with diameter = 1 × average ONHW, centered at the center of the ONH, from the Gabor-response images.
5. Divide the Gabor-response images into their superior and inferior parts.
6. Normalize the superior and inferior Gabor-magnitude-response images and apply gamma correction.
7. Binarize the normalized Gabor-magnitude-response images using the moment-preserving thresholding method [69].
8. Skeletonize the resulting binary images.
9. Remove the effects of the boundary of the ONH from the skeleton images.
10. Apply morphological operations of pruning and area-open to remove short branches and unconnected segments.

3.3.4.1.2 Selecting the Seed Labels

1. Find the branching points of the skeleton.
2. Remove areas of 3 × 3 pixels centered on the branching points to break the skeleton into disconnected segments.
3. Obtain and store the labels of segments that are present in a neighborhood of 13 × 13 pixels around each branching point in a matrix.
4. Search an annular region around the center of the ONH for seed-label candidates, and select the labels with the highest and second-highest average Gabor-magnitude response as the first and second seed labels.

3.3.4.1.3 Extracting the MTA Skeleton Segments

1. Start with the first-selected seed label, and select the next segment as the neighboring label with the highest average Gabor-magnitude response.
2. Repeat step (1) until no more labels are left.
3. Check the total length of the selected segments; if it is less than a pre-specified limit, use the second seed label and perform step (2).

3.3.4.1.4 Postprocessing

1. Construct a binary image of the extracted arcade consisting of the selected segments in the previous stage.
2. Combine the extracted STA and ITA to obtain a binary image of the MTA.

3.3.4.2 Segmentation of the MTA Skeleton into Linear Parts

3.3.4.2.1 Sequencing the Skeleton Pixels

1. Start at one of the end points of the current segment.
2. Move along sequentially one pixel at a time in a 3 × 3-pixel neighborhood.

3. Record the position of each traversed pixel and its associated Gabor-angle response in a vectorial format.
4. Repeat steps (2–3) until the other end point is reached.

*3.3.4.2.2 **Analyzing the Gabor-Angle Sequence***

1. Compute the statistical measure of median absolute deviation (MAD) [76] in a window of length 7 pixels, applied to the previously sequenced Gabor-angle response.
2. Analyze the MAD measure as follows:
   a. Obtain the backward and forward derivatives (d_b and d_f) of the MAD measure.
   b. Set values of $d_b < 0$ and $d_f > 0$ to 0.
   c. Set $d_{combined} = d_b + d_f$.
   d. Analyze $d_{combined}$ for possible patterns in the derivative values.
   e. Determine the breaking points if any one of the expected patterns is detected.
   f. Break apart the segment if any breaking point is determined in the previous step.
   g. Assume the entire segment to be linear if no breaking point is determined.

3.3.4.3 Interpolation of Vessel Edges

1. Apply Canny's edge detection method to the inverted green-channel image.
2. Select two separate sets of nearest edge pixels in Canny's output to every linear segment or subsegment of the extracted MTA on either side.
3. Fit two separate first-order functions to the two sets of edge-pixel candidates for every linear segment or subsegment to obtain a continuous-form interpolation of the two edges.
4. Represent the two edge lines as $Y_{e_{1,2}} = m_{e_{1,2}} X_{e_{1,2}} + C_{e_{1,2}}$, where the subscripts $e_{1,2}$ indicate edge lines 1 and 2, respectively, and the variables $m_{e1,2}$ and $C_{e1,2}$ represent their corresponding slopes and y-intercepts.

3.3.4.4 Measurement of the Width of the MTA

1. Obtain the slope of the normal, n, at a given skeleton pixel, i, as $m_{n_i} = -1/\tan[\phi(i)]$, where $\phi(i)$ is the Gabor-angle response at the skeleton pixel i.
2. Define the normal line using the coordinates of the given skeleton pixel, (x_i, y_i), and its corresponding slope, as $Y_{n_i} = m_{n_i} X_{n_i} + C_{n_i}$, where $C_{n_i} = y_i - (m_{n_i} x_i)$ is the y-intercept.
3. Obtain the coordinates of the two points of intersection as $x_{i_{1,2}} = \left(C_{n_i} - C_{e_{1,2}}\right) / \left(m_{e_{1,2}} - m_{n_i}\right)$ and $y_{i_{1,2}} = m_{e1,2} x_{i_{1,2}} + C_{e_{1,2}}$.
4. Compute vessel width at the given pixel i as $w_i = \sqrt{(x_{i_1} - x_{i2})^2 + (y_{i_1} - y_{i2})^2}$.

The MTA extraction algorithm was originally developed based on adult retinal images from the DRIVE database and was validated by obtaining the mean-distance-to-the-closest-point measure from the tracked MTA to the hand-drawn traces of the MTA, as delineated by a retinal specialist (A. L. Ells) [75]. The results showed a low average MDCP error of about 2.3 pixels for the 40 images of the DRIVE database [75], indicating that the methods are reliable for detection and extraction of the MTA.

The marking of the center of the ONH is the only manual input in the entire procedure used for vessel width measurement and may be replaced by an automated procedure [77].

Given the fact that vascular dilation has been observed to occur close to the ONH (posteriorly) in the presence of ROP [1], the subsequent analysis of each image was limited to the horizontal range from 0.25× ONHW nasal to 2× ONHW in the temporal direction with respect to the center of the ONH.

The pixels in the skeleton that represent the boundary of the ONH were removed by first obtaining the difference between the radial angle (originating from the center of the ONH) and the Gabor angle for skeleton pixels within an annular region (the area between two circles with diameters = 1× and 2× ONHW). If the angle difference was more than 60°, the associated pixel was removed [29].

The annular search area used to find seed-label candidates was adaptively increased to ensure that there were at least two labels present for comparison for each of the STA and ITA parts; the procedure makes an exception to this criterion if only a single label is available in the image.

When selecting the label with the highest average Gabor-magnitude response as the next potential segment of the arcade being analyzed, the algorithm ensures that the distance between the ending point of the current label and the starting point of the next potential label is no more than 10 pixels. The algorithm checks the total length of the selected segments and how far they extend collectively, with respect to the horizontal size of the image, as the exit criteria; specific details are provided in the related publication [31].

The selected MTA segments are analyzed in terms of their associated Gabor-angle information to determine whether each selected segment is approximately linear or not. If a segment is not linear, breaking points are determined to separate the given segment into a set of linear subsegments.

The algorithm obtains the MAD measure of the previously obtained Gabor-angle sequence and sets all nonzero values in the MAD measure equal to unity to avoid detection of local changes and emphasize the global variations in the orientation of the vessel segment under analysis.

The obtained MAD measure needs to be analyzed further to detect any possible significant variations in the orientation of the segment being evaluated and, subsequently any possible cutting points. The analysis is performed as follows:

Given a variation in the orientation of a vessel segment, there are three possibilities and patterns to be detected: (1) A change that starts and stops within a sequence, indicated by a leading positive and a trailing negative derivative. (2) A change that starts from the first pixel in a sequence and ends within it, indicated by only a negative derivative (no leading positive value). (3) A change that starts within a sequence, but does not end, indicated by only a positive derivative (no trailing negative value). Any one of these conditions indicates a possible cutting point on the segment under analysis. Any single index obtained from cases (2) and (3) was taken as a cutting point. In case (1), the median point between the two indices obtained was taken as the cutting point.

3.3.5 Measurement of Tortuosity of Vessels

Similar to the methods for quantification of the openness and thickness of the MTA, measurement of tortuosity requires a skeletal representation of vessels. In this work, the Gabor-angle information associated with the segmented vessel segments is used to define an angle-variation-based tortuosity (AVT) measure, as described below.

3.3.5.1 Detection and Segmentation of Vessel Skeletons [32]

1. Obtain the Gabor-magnitude and Gabor-angle-response images using the inverted green-channel image.
2. Normalize the Gabor-magnitude-response image to the range [0,1].
3. Binarize the normalized Gabor-magnitude-response image using a threshold slider via the GUI.
4. Remove 8-connected segments shorter than the maximum length (in pixels) as specified by the user using the area-open procedure.
5. Skeletonize the resulting binary image.
6. Remove an elliptical area from the skeleton image centered at the center of the ONH as previously marked by the user.
7. Remove spurs of 5 pixels or shorter in length.
8. Detect branching points on the skeleton and remove an area of 3×3 pixels centered at each branching point to separate all vessel segments.
9. Remove remaining disconnected segments of length less than 7 pixels.

3.3.5.2 Analysis of Variation in the Orientation of Vessel Segments

1. Detect and number all disconnected vessel segments obtained in the previous step.
2. Label the skeleton pixels in sequence:
   a. Ensure that the segment under analysis has exactly two end points.
   b. Start at one of the end points of the current segment.
   c. Move along sequentially one pixel at a time in a 3×3-pixel neighborhood.
   d. Record the position of each traversed pixel and its associated Gabor-angle response in a vectorial format.
   e. Repeat steps (b-c) until the other end point is reached.
3. Compute the statistical measure of MAD [76] in a window of length 7 pixels applied to the sequenced Gabor-angle response, ϕ.
4. Mark a pixel on the segment under analysis as being a linear only if all normalized MAD measures in a 7-pixel-long window centered at the current pixel are zero.
5. Obtain skeleton image of only linear segments as indicated by the normalized MAD measure.
6. Obtain skeleton image of nonlinear vessel segments by subtracting the image of linear segments from the original skeleton image.

3.3.5.3 Computation of Vascular Tortuosity per Subsegment

1. Compute a local tortuosity index (LTI) as $\mathrm{LTI}(p)=\{\sin[\phi(p)-\phi(p-1)]-\sin[\phi(p)+\phi(p+1)]\}/2$, where p, $p-1$, and $p+1$ are the current, previous, and next pixels, respectively, along a nonlinear vessel segment.
2. Compute the AVT measure over a nonlinear vessel subsegment as $\mathrm{AVT}=\frac{1}{P}\sum_{p=1}^{P}\mathrm{LTI}(p)$ [32], where P is the total number of pixels in the given subsegment.

The training set of images was used to compare the AVT measures for the abnormally tortuous and normal vessels, as marked by the retinal specialist, to obtain a suitable threshold. Based on this analysis, a threshold of $t=0.07$ was determined. Each nonlinear vessel segment with AVT > t was marked as being abnormally tortuous.

Providing one average or overall measure of tortuosity for an image, or even for each quadrant of an image, may be misleading if the tortuous segments are not sufficiently long; two highly tortuous but short segments in two quadrants should not lead to a positive diagnosis. To avoid such a situation, the total length of each abnormally tortuous vessel segment was obtained by its associated chain-code [78] representation. The length of each segment was computed as the number of even codes plus $\sqrt{2}$ multiplied by the number of odd codes multiplied by the pixel size of 30 μm. The total length of the abnormally tortuous vessel segments in the entire FOV was used as a feature to discriminate between cases with and without plus disease.

3.4 Results

The following sections provide illustrative results of the application of the methods for quantification of the openness and thickness of the MTA, as well as vascular tortuosity, as applied to cases from the TROPIC database. Overall results of statistical and diagnostic analysis of each feature, as well as feature selection and pattern classification, are also provided.

3.4.1 Openness of the MTA

Figures 3.2 and 3.3 show the results of single- and dual-parabolic modeling as well as the measurement of the TAA using a circle of radius $r=60$ pixels for two images from the TROPIC database; one image

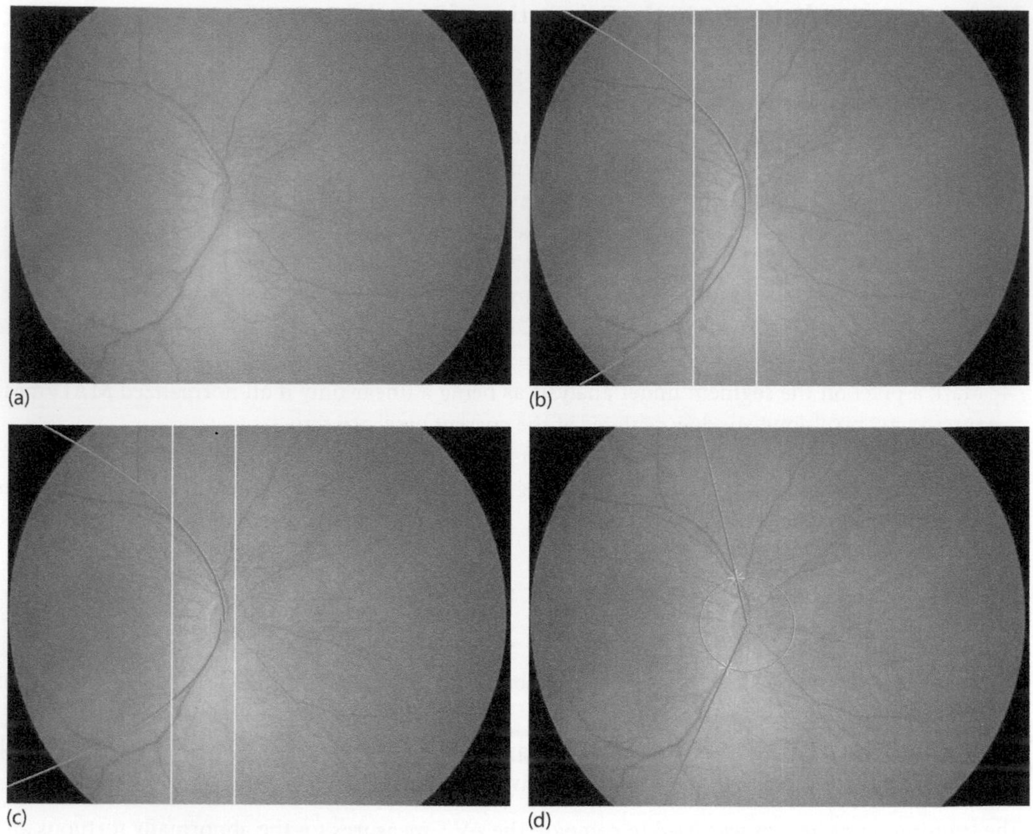

FIGURE 3.2 (a) Image 1701 of the TROPIC database which does not show any sign of plus disease. (b) Single-parabolic model with $a_{MTA}=60$. (c) Dual-parabolic model with $a_{STA}=66$ and $a_{ITA}=40$. (d) TAA measured using a circle of radius $r=60$ pixels with TAA=141.99°. The vertical lines in yellow show the extent of the arcade used in the modeling procedure.

contains no signs of plus disease (Figure 3.2) and the other shows signs of plus disease (Figure 3.3). The ITA portion of the dual-parabolic model in Figure 3.2(c) is providing an accurate fit close to the ONH, whereas the MTA model in part (b) is providing an average fit to the ITA. The TAA obtained using the circle of radius $r=60$ pixels is providing a measure close to the posterior pole. The parameters of Gabor filters used for detection of vessels were $\tau=10$ pixels, $l=2$, and $K=30$. The parameters of the parabolic models and the TAA have smaller values for the case with plus disease than for the case without the disease.

3.4.2 Thickness of the MTA

Figure 3.4 demonstrates the application of the methods for measurement of the width of the MTA to a case without plus disease from the TROPIC database. It can be seen from the results in part (b) that the originally extracted STA and ITA are, indeed, made of up of several linear segments and are not linear by themselves. Parts (c) and (d) of the figure present separate zoomed-in views of the actual width measurements of the STA and ITA, respectively.

Figure 3.5 illustrates the application of the methods to a case diagnosed with plus disease.

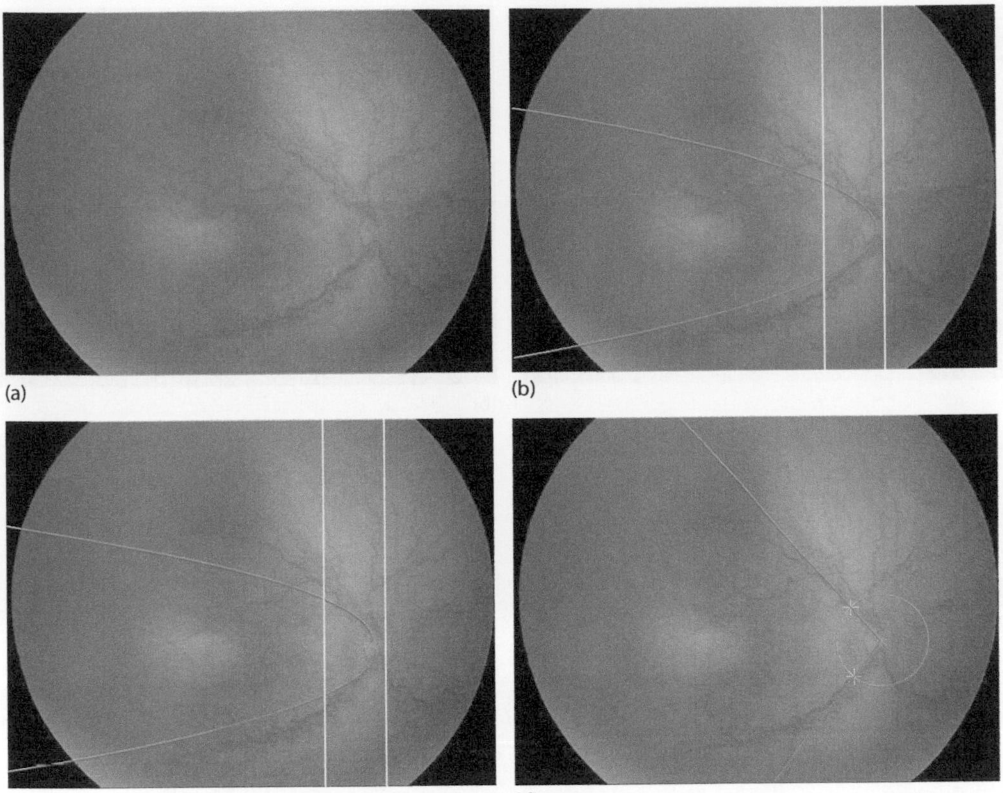

(a) (b) (c) (d)

FIGURE 3.3 (a) Image 3602 of the TROPIC database of a patient diagnosed with plus disease. (b) Single-parabolic model with $a_{MTA}=14$. (c) Dual-parabolic model with $a_{STA}=12$ and $a_{ITA}=13$; both models are providing fits close to the posterior pole. (d) TAA measured using a circle of radius $r=60$ pixels with TAA = 100.28°.

Visual comparison of the width lines illustrates that the MTA is thicker, as compared to the case without plus disease in Figure 3.4; this observation is confirmed by the average MTA width measured for the two images (91.0 μm for the case without plus disease, compared to 161.2 μm for the plus case).

The MTA extraction algorithm failed to detect all or most of a half arcade (STA or ITA, but not both) in a total of seven images out of the 110 analyzed. Such failure is mainly due to poor quality of images and/or poor contrast between vessels and the background.

3.4.3 Tortuosity of Vessels

Figures 3.6 and 3.7 illustrate the results of applying the methods for analysis of tortuosity to an image without any signs of plus disease from the TROPIC database. A total length of 0.33 mm of the vessel segments was found to be abnormally tortuous in the entire FOV based on the predetermined AVT threshold of $t=0.07$.

Figure 3.8 illustrates the results of applying the methods to an image with signs of plus disease from the TROPIC database. A total length of 26.35 mm was found to be abnormally tortuous in the entire FOV based on the predetermined AVT threshold. Note that not all of the steps of the procedure are presented in this figure.

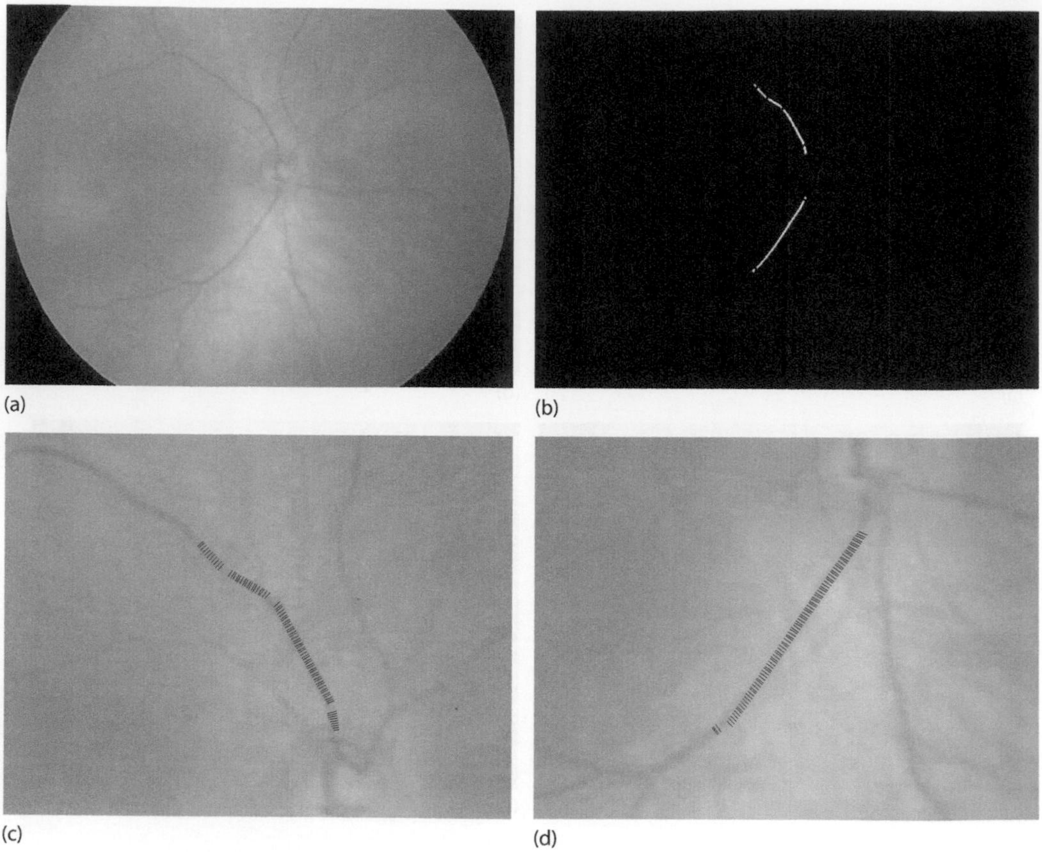

FIGURE 3.4 (a) The original color image 2306 of the TROPIC database that shows no signs of plus disease. (b) The linear subsegments of the tracked STA and ITA; the skeleton pixels have been dilated using a disk of radius one pixel for better illustration. Illustration of the calculated vessel widths for (c) STA (enlarged) and (d) ITA (enlarged), respectively. The length of each normal line (in blue) represents the vessel width at that exact location. The average calculated MTA width for this image is 109.4 μm (3.65 pixels).

3.4.4 Statistical and Diagnostic Results

Table 3.2 shows the results of statistical analysis as well as the A_z and the associated asymmetric confidence interval, CI_a, values indicating the overall diagnostic accuracy of all of the seven features studied in the present work for the purpose of CAD of plus disease. The features are the TAA measured using circles of radii $r=60$ and 120 pixels; openness parameters of the single- and dual-parabolic models of the MTA, STA, and ITA, respectively; the average width of the MTA; and the total length of tortuous vessels in the entire FOV of the image. Based on the A_z values, it can be concluded that the measure of tortuous-vessel length in the entire FOV leads to the highest diagnostic accuracy as compared to all of the other features derived.

3.4.5 Feature Selection and Pattern Classification

As mentioned in Section 3.2.3, the wrapper method was used with three classifiers and two search methods to determine whether combinations of the different features computed in the present work based on the openness, thickness, and tortuous-vessel length could lead to better diagnostic classification and higher discrimination between cases with and without plus disease, as compared to the use of each single feature separately.

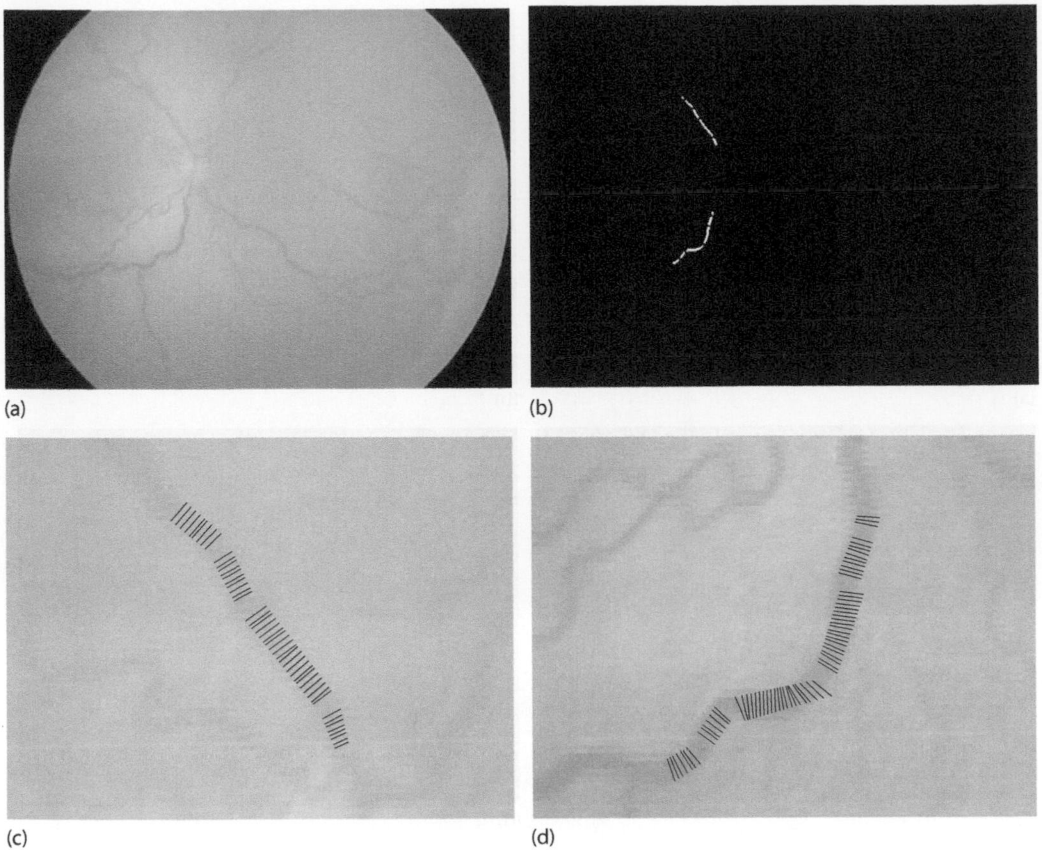

FIGURE 3.5 (a) The original color image 2904 of the TROPIC database that shows signs of plus disease. (b) The linear subsegments of the tracked STA and ITA. Illustration of the calculated vessel widths for (c) STA (enlarged) and (d) ITA (enlarged), respectively. The average calculated MTA width for this image is 161.2 μm (5.37 pixels). The image in part (b) has been dilated using a disk of radius one pixel for better illustration.

In all instances, the tortuous-vessel length over the entire FOV was selected in 100% of the folds of cross-validation; only the naïve Bayes classifier selected other features as well, including the openness parameter of the single-parabolic model and the two TAA measures.

Pattern classification was performed using the results of feature selection via the naïve Bayes model using the two search methods. However, in both instances, the results of pattern classification did not yield higher A_z values as compared to using the total tortuous-vessel length alone (A_z=0.98).

Various other methods were tested for feature selection and classification, but the results are not provided here as they were the same. All methods always included the total tortuous-vessel length with at least 90% selection rate during cross-validation. None of the methods that selected any additional feature along with total tortuous-vessel length led to improved diagnostic results.

3.5 Discussion

3.5.1 Openness of the MTA

The diagnostic performance (in terms of A_z) of the parameters of the single- and dual-parabolic models obtained in the present work is comparable to that provided by the TAA measures obtained based on the method of Wong et al. [15] via the GUI. However, the TAA measures provide better statistical

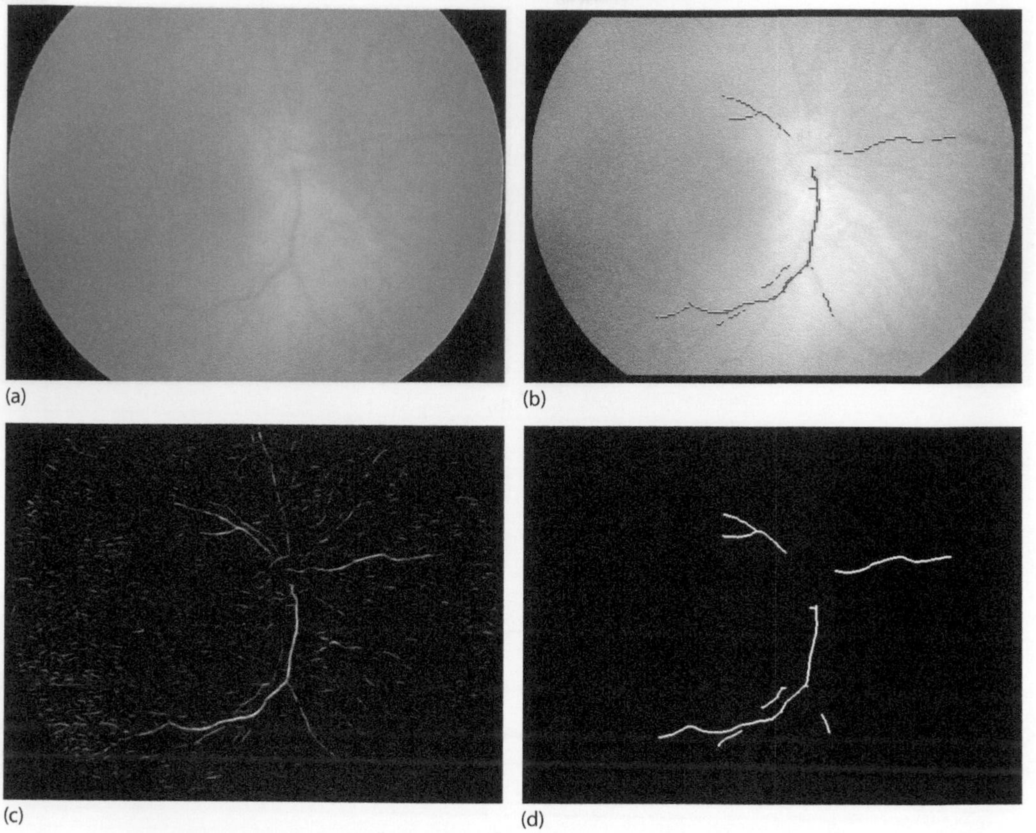

FIGURE 3.6 (a) Image 403 of the TROPIC database showing no signs of plus disease. (b) The vessel orientation for every third pixel on selected vessels is shown by the blue needles. (c) The vessel-strength image obtained using the inverted green-channel of the image in part (a). (d) The skeleton image after thresholding the image in part (c) at 0.025 of the normalized intensity value and removal of 8-connected segments having fewer than 60 pixels each. The image in part (d) has been morphologically dilated using a disk of radius one pixel for better visual representation.

performance as compared to the parameters of the parabolic models. All of the studied measures show a similar trend: there is a decrease in the openness of the MTA in the presence of plus disease.

If the angle of the retinal raphe (the line passing through the center of the ONH and the fovea) is large with respect to the horizontal axis of the image, it could lead to a much larger openness parameter for one of the dual-parabolic models than the other. Large variations in the openness parameters of the parabolic models could be the reason for the high SE and relatively large p-values for the results of parabolic modeling as shown in Table 3.2. It may be possible to estimate the retinal raphe as the principal axis [79] of the skeleton of the MTA and correct for any rotation that might exist in the image.

Upon close inspection, it becomes clear that, first, the STA and the ITA are asymmetric, and second, more accurate modeling of each arcade may be possible by applying models based on higher-order curves than parabolic (second-order) curves. A high-order curve fitting method may provide more accurate results in terms of modeling and parameterization of the MTA, at the cost of an increased number of parameters.

Using the results of MTA tracking, the same single- and dual-parabolic modeling procedures were applied to the 110 tracked MTAs and ROC analysis was performed. The results, as assessed visually, showed close fits to the MTA. However, the related parameters indicated poorer discrimination between cases with and without plus disease. This result may be further indication that restructuring of the

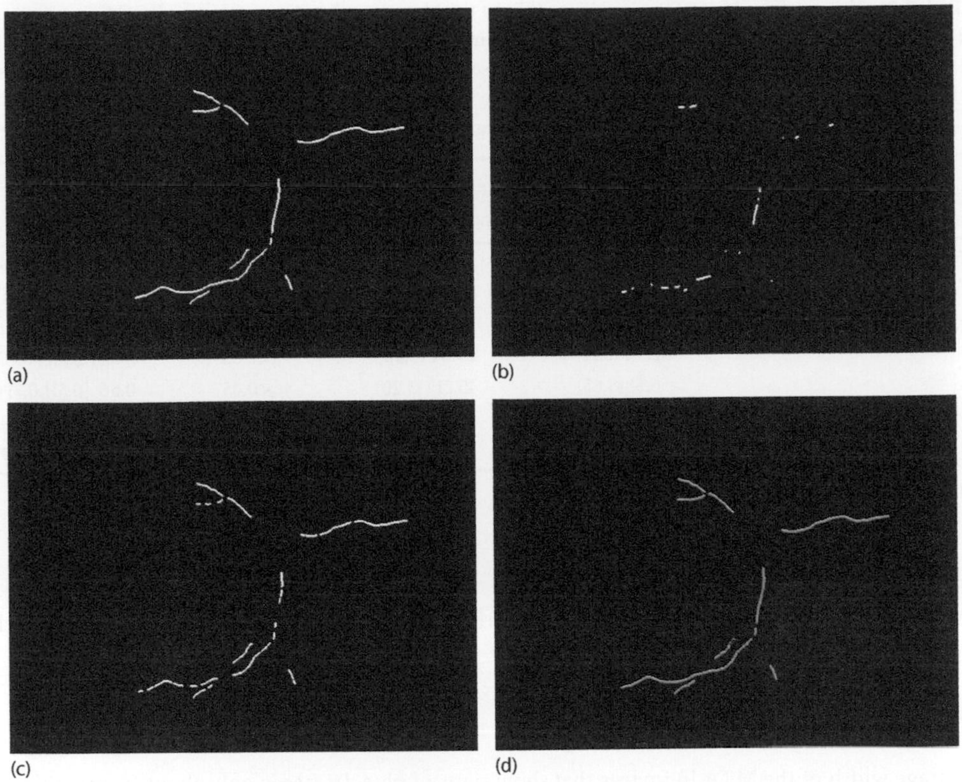

FIGURE 3.7 (a) The skeleton image in Figure 3.6d after removal of the ONH area and the branching points. (b) The skeleton image of the linear vessel subsegments obtained using the MAD measure. (c) Skeleton image of the nonlinear vessel subsegments obtained by subtracting the image in part (b) from the image in part (a). (d) Color-coded skeleton image distinguishing the abnormally tortuous vessel segments, if any, in red from the other vessel segments in green, using the AVT threshold of 0.07; in this case, a total length of 0.33 mm was found to represent abnormal tortuosity. Images in parts (a)–(d) have been morphologically dilated using a disk of radius one pixel for better visual representation in the figure.

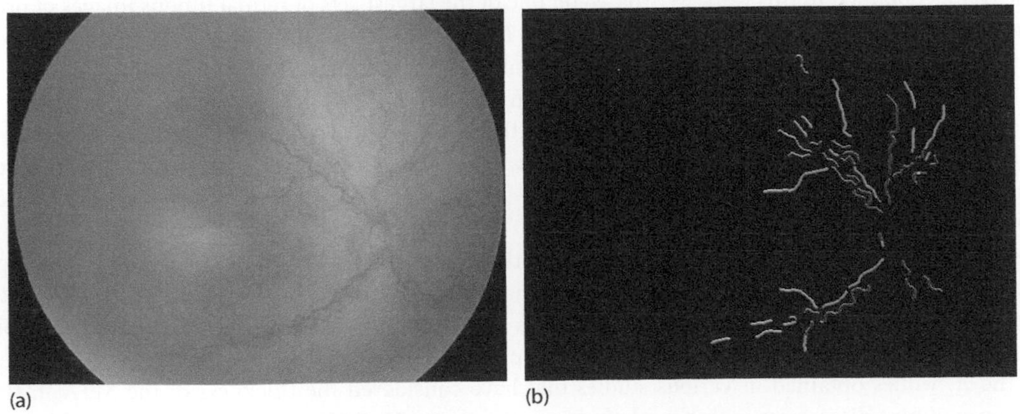

FIGURE 3.8 (a) Image 3602 of the TROPIC database showing signs of plus disease. (b) Color-coded skeleton image distinguishing the abnormally tortuous vessel segments in red from the other vessel segments in green. A total length of 26.35 mm of the nonlinear vessel segments was found to be abnormally tortuous in the entire FOV.

TABLE 3.2 Values of the Mean, Standard Error (SE) of the Mean, p-value, Area under the ROC Curve (A_z), and the Associated 95% $(\alpha = 0.025)$ Asymmetric Confidence Interval (CI_a), Obtained in the Discrimination of 19 Cases with Plus Disease versus 91 Cases without Plus Disease using the TAA with Radii of $r = 60$ and 120 pixels, the Parameter of the Single-Parabolic Model ($|a_{MTA}|$), the Parameters of the Dual-Parabolic Models ($|a_{STA}|$ and $|a_{ITA}|$), the Average Width of the MTA, and the Total Length of Tortuous Vessels in the Entire FOV of the Image

	Without Plus	With Plus		
Parameter	Mean (SE) $(n = 91)$	Mean (SE) $(n = 19)$	p-value	A_z, CI_a
$TAA_{r=60}$	132.14 (1.55)	119.19 (3.86)	<0.01**	0.73, [0.589, 0.844]
$TAA_{r=120}$	115.25 (1.53)	104.88 (3.12)	<0.01**	0.69, [0.560, 0.805]
$\|a_{MTA}\|$	42.31 (2.55)	34.89 (8.67)	<0.01**	0.67, [0.513,0.801]
$\|a_{STA}\|$	50.44 (5.51)	29.84 (3.56)	<0.05*	0.70, [0.560, 0.808]
$\|a_{ITA}\|$	56.51 (5.65)	45.21 (13.20)	<0.05*	0.66, [0.511,0.790]
MTA width (μm)	110.6 (1.94)	125.0 (3.97)	<0.001***	0.75, [0.632, 0.844]
Tortuous-vessel length (mm)	1.48 (0.17)	16.53 (2.42)	<0.001***	0.98, [0.910,0.997]

$^*p < 0.05$, $^{**}p < 0.01$, and $^{***}p < 0.001$.

MTA, referred to as the change in the openness of the MTA, may be more complex than defined clinically at this time.

3.5.2 Thickness of the MTA

The average width of the MTA in images that show signs of plus disease is only about 15 μm wider than the average width of the MTA in the cases without plus disease that were analyzed. The median of the width measurements shows a slightly larger difference between the two classes (about 22 μm). The difference between the two measures is likely due to the fact that the median is less sensitive to the presence of outliers, as compared to the mean. The differences between the mean and median MTA width measurements for normal images, as compared to images with plus disease, were found to be statistically extremely significant with $p < 0.001$ in both cases. A total of 13,878 width measurements were obtained over the 110 images, leading to 126 measurements per image, on the average.

Retinal images of preterm infants vary substantially in terms of pigmentation, contrast, and quality. The results obtained in this work were based on one of the largest sets of retinal fundus images of preterm infants in the literature used for such a study. The TROPIC database is not limited to only images with high quality and contrast. The methods presented for measurement of the MTA width are robust and work well in the presence of large variations in the characteristics of the images used; the results show a statistically highly significant difference in the width of the MTA between the two classes and good accuracy in the detection of plus disease. It would be of interest to test the methods with more images, including more cases with plus disease, as well as against combined diagnosis of the same set of images by multiple experts.

The thresholding step has a significant effect on the skeleton-extraction step. The wide range of pigmentation and contrast that exists in preterm retinal images could cause the thresholding step not to provide the best possible results. Analysis of various automatic thresholding methods, and their combinations may lead to more suitable methods for this purpose [29].

The A_z values obtained in various studies that have considered the thickness of the MTA in the presence of plus disease are within a relatively small range (0.75–0.82). The methods presented in this work perform as well as the state-of-the-art methods available in the literature without the need for user-correction of the results and discrimination in selection of images to be analyzed based on quality, which all of the other mentioned methods required.

Except for the step requiring the location of the center of the ONH, the methods described are automated. In the future, the method of Rangayyan et al. [77], which performs phase portrait analysis of the Gabor-angle information, could be adapted for pediatric retinal fundus images to detect automatically the center of the ONH, which could make the entire procedure of vessel width measurement fully automated.

The range of changes in the width of retinal vessels that is observed in later stages of ROP prior to the presence of plus disease is designated as pre-plus [1]. Based on this observation, it can be expected that an increase in vessel width should be observed as the ROP stage progresses prior to the diagnosis of plus disease. Indeed, the differences in vessel width between stages 0 and 3 ROP, stages 1 and 3 ROP, and stages 2 and 3 ROP were found to be statistically highly significant, when excluding cases with plus disease. It would be of interest to test this observation further and strengthen the statistical analysis using more cases that show no signs of plus disease.

The combined application of tracking the MTA, use of the MAD measure for analysis of variations in vessel orientation, and geometrical methods enables measurement of the width of the MTA to subpixel accuracy. Blurring and image artifacts due to movement may introduce errors in measurement of the width even when using the methods presented in this work. A washout effect may be introduced when averaging the width measure over all available segments in an image or over the entire length of a segment. To avoid such effects, it may be of interest to use a training set where abnormally dilated vessel segments are marked by an expert, to determine a threshold to detect abnormally dilated vessel segments, and to use the total length of such segments for diagnosis.

Analysis of variation of the width of a specific segment of the MTA over time (longitudinally) using the methods describe in this work may provide useful diagnostic information regarding the presence and development of plus disease, as well as the effects of treatment.

3.5.3 Tortuosity in Vessels

The total-tortuous-vessel-length criterion described in this work combines the clinical definition of plus disease with practical understanding of a sufficiently tortuous vessel. The results indicate high performance in diagnosis of plus disease using a single feature with $A_z = 0.98$.

The methods described are fully automated, except for the binarization step, the initial removal of small segments, and marking of the center of the ONH. No single automated thresholding method provides consistent results for binarization of all images due to the variable nature of the retinal images of preterm infants, including varying pigmentation, blurring, and low vessel-to-background contrast. Combination of the results of multiple thresholding methods may lead to better binarization results [80]. The method of Rangayyan et al. [77] will be adapted in the future to detect automatically the center of the ONH.

All studies that have performed diagnosis of plus disease using tortuosity have provided a single measure of tortuosity for either each selected vessel segment [9,10,20], the entire image [18], or for each quadrant of the image [19]. However, since tortuosity is not formally defined, numerical representations of its quantitative measurement may not be meaningful to an ophthalmologist. Use of the AVT measure to detect abnormally tortuous vessels and then to measure and provide their total length is more practical from a CAD point of view, and more meaningful to an ophthalmologist.

Almost all of the studies that have dealt with quantification of tortuosity have limited the area of analysis to the posterior of the retina. However, peripheral-vessel tortuosity has been shown to be more correlated to the presence of plus disease as compared to posterior-vessel tortuosity [11]. Given such limitations, it is beneficial to obtain any necessary thresholding parameters using an independent training set of images. In fact, based on the results obtained in the present study, it is questionable whether posterior-arteriolar tortuosity is more important in diagnosis of plus disease than peripheral-venular tortuosity. The methods described in this work are capable of detecting all tortuous vessels regardless of location.

All studies that have performed diagnosis of plus disease via quantification of vessel tortuosity have employed methods that involve manual marking of vessel segments, or manual selection and/or correction of parts of automatically detected vessels to include only the desired vessels for further analysis. The methods presented in this work are capable of distinguishing tortuous vessels in a given image without any manual selection and/or correction.

The use of a training set is advantageous in finding a suitable threshold to indicate abnormal tortuosity. Obtaining the total length of abnormally tortuous vessels eliminates the use of arbitrary and unitless values of tortuosity; tortuous-vessel length in millimeters is easily comprehensible. Furthermore, by breaking apart all available vessels segments and by using the length of the abnormally tortuous vessels, it is possible to avoid any washout effect that may result from averaging the measure of tortuosity over the entire length of a segment, including its branches or parts that may not be tortuous, or averaging over all vessel segments in the entire image. Using the derivatives of the sequenced MAD measure, it should also be possible to determine 180° changes in the orientation (a twist) and to use the total number of such changes in vessel orientation as a weight to modify the AVT measure.

3.5.4 Feature Selection and Pattern Classification

Considering the results of feature selection (Section 3.4.5), it may be concluded that, based on the methods and materials used, the measure of total-tortuous-vessel length alone, which provides the highest diagnostic accuracy in CAD of plus disease based on the A_z value, may be the single most important feature in the diagnosis of plus disease; no combination of the features studied led to better diagnostic accuracy.

The results indicate that the total tortuous-vessel length over the entire FOV alone can lead to the best CAD performance in the detection of plus disease with the database used (TROPIC). The MTA-thickness measure derived from images of higher quality and resolution may lead to better results. Larger databases with a large range of image characteristics, and the need for robust performance may call for the use of multiple diagnostic features.

3.6 Conclusion

The methods described in this work can aid in quantitative assessment, objective evaluation, and accurate diagnosis of plus disease in retinal fundus images of preterm infants in a clinical setting. Among the three signs associated with plus disease (increased tortuosity, thickness, and openness) the total length of tortuous vessels indicated the highest level of diagnostic accuracy and provided more reliable and accurate results, as compared to the remaining six computed features. No linear/nonlinear combination of the obtained features led to a higher diagnostic result as compared to the use of total-tortuous-vessel length alone. The methods may be used in a clinical or teleophthalmological setting for quantitative and accurate diagnosis of plus disease and ultimately may lead to efficient clinical management and treatment of ROP.

Acknowledgment

This work was supported by the Natural Sciences and Engineering Research Council of Canada. We thank April Ingram (Ells Retina Center) for help with the TROPIC images and Dr. Eliana Silva de Almeida (Federal University of Alagoas) for help with statistical analysis.

References

1. International committee for the classification of retinopathy of prematurity. The international classification of retinopathy of prematurity revisited. *Archives of Ophthalmology*, 123:991–999, 2005.
2. J. S. Grewal and A. R. Fielder. Sequelae of retinopathy of prematurity. *Paediatrics and Child Health*, 22(1):19–24, 2011.

3. J. Javitt, R. D. Cas, and Y. Chiang. Cost-effectiveness of screening and cryotherapy for threshold retinopathy of prematurity. *Pediatrics*, 91(5):859–866, 1993.
4. M. F. Chiang, R. R. Arons, and J. T. Flynn. Incidence of retinopathy of prematurity from 1996 to 2000: Analysis of a comprehensive New York state patient database. *Ophthalmology*, 111(7):1317–1325, 2004.
5. D. J. Gunn, D. W. Cartwright, and G. A. Gole. Incidence of retinopathy of prematurity in extremely premature infants over an 18-year period. *Clinical and Experimental Ophthalmology*, 40:93–99, 2012.
6. Early treatment for retinopathy of prematurity cooperative group. Revised indications for the treatment of retinopathy of prematurity: Results of the early treatment for retinopathy of prematurity randomized trial. *Archives of Ophthalmology*, 121:1684–1696, 2003.
7. B. V. Davitt and D. K. Wallace. Plus disease. *Survey of Ophthalmology*, 54(6): 663–670, 2009.
8. M. F. Chiang, L. Jiang, R. Gelman, Y. E. Du, and J. T. Flynn. Interexpert agreement of plus disease diagnosis in retinopathy of prematurity. *Archives of Ophthalmology*, 125(7):875–880, 2007.
9. R. Gelman, L. Jiang, Y. E. Du, M. E. Martínez-Pérez, J. T. Flynn, and M. F. Chiang. Plus disease in retinopathy of prematurity: Pilot study of computer-based and expert diagnosis. *Journal of American Association for Pediatric Ophthalmology and Strabismus*, 11(6):532–540, 2007.
10. S. Koreen, R. Gelman, M. E. Martínez-Pérez, L. Jiang, A. M. Berrocal, D. J. Hess, J. T. Flynn, and M. F. Chiang. Evaluation of a computer-based system for plus disease diagnosis in retinopathy of prematurity. *Ophthalmology*, 114(12):59–67, 2007.
11. K. M. Keck, J. Kalpathy-Cramer, E. Ataer-Cansizoglu, S. You, D. Erdogmus, and M. F. Chiang. Plus disease diagnosis in retinopathy of prematurity: Vascular tortuosity as a function of distance from optic disk. *Retina*, 33(8): 1700–1707, 2013.
12. R. A. Saunders, E. C. Bluestein, R. B. Sinatra, M. E. Wilson, J. W. O'Neil, and P. F. Rust. The predictive value of posterior pole vessels in retinopathy of prematurity. *Journal of Pediatric Ophthalmology and Strabismus*, 32(2):82–85, 1995.
13. B. Cassin and M. L. Rubin. *Dictionary of Eye Terminology*. Triad Publishing Company, Gainsville, FL, 6th edn, 2012.
14. C. M. Wilson, M. Theodorou, K. D. Cocker, and A. R. Fielder. The temporal retinal vessel angle and infants born preterm. *British Journal of Ophthalmology*, 90:702–704, 2006.
15. K. Wong, J. Ng, A. L. Ells, A. R. Fielder, and C. M. Wilson. The temporal and nasal retinal arteriolar and venular angles in preterm infants. *British Journal of Ophthalmology*, 95(12):1723–1727, 2011.
16 F. Oloumi, R. M. Rangayyan, and A. L. Ells. Quantification of the changes in the openness of the major temporal arcade in retinal fundus images of preterm infants with plus disease. *Investigative Ophthalmology and Visual Science*, 55(10):6728–6735, 2014.
17. D. K. Wallace, G. E. Quinn, S. F. Freedman, and M. F. Chiang. Agreement among pediatric ophthalmologists in diagnosing plus and pre-plus disease in retinopathy of prematurity. *Journal of American Association for Pediatric Ophthalmology and Strabismus*, 12(4):352–356, 2008.
18. C. Heneghan, J. T. Flynn, M. O'Keefe, and M. Cahill. Characterization of changes in blood vessels width and tortuosity in retinopathy of prematurity using image analysis. *Medical Image Analysis*, 6(1): 407–429, 2002.
19. D. K. Wallace, J. Jomier, S. R. Aylward, and M. B. Landers, III. Computer-automated quantification of plus disease in retinopathy of prematurity. *Journal of American Association for Pediatric Ophthalmology and Strabismus*, 7:126–130, 2003.
20. R. Gelman, M. E. Martínez-Pérez, D. K. Vanderveen, A. Moskowitz, and A. B. Fulton. Diagnosis of plus disease in retinopathy of prematurity using retinal image multiscale analysis. *Investigative Ophthalmology and Visual Science*, 46(12):4734–4738, 2005.

21. D. K. Wallace, Z. Zhao, and S. F. Freedman. A pilot study using ROPTool to quantify plus disease in retinopathy of prematurity. *Journal of American Association for Pediatric Ophthalmology and Strabismus,* 11(4):381–387, 2007.
22. Z. Zhao, D. K. Wallace, S. F. Freedman, and S. R. Aylward. A tool for computer-aided diagnosis of retinopathy of prematurity. In *Proceeding of SPIE Medical Imaging,* 6915: pages 69152W-1-69152W-7, 2008.
23. S. C. Johnston, D. K. Wallace, S. F. Freedman, T. L. Yanovitch, and Z. Zhao. Tortuosity of arterioles and venules in quantifying plus disease. *Journal of American Association for Pediatric Ophthalmology and Strabismus,* 13(2):181–185, 2009.
24. A. E. Kiely, D. K. Wallace, S. F. Freedman, and Z. Zhao. Computer-assisted measurement of retinal vascular width and tortuosity in retinopathy of prematurity. *Archives of Ophthalmology,* 128(7):847–852, 2010.
25. J. P. Thyparampil, Y. Park, M. E. Martinez-Perez, T. C. Lee, D. J. Weissgold, A. M. Berrocal, R. V. P. Chan, J. T. Flynn, and M. F. Chiang. Plus disease in retinopathy of prematurity: Quantitative analysis of vascular change. *American Journal of Ophthalmology,* 150(4):468–475, 2010.
26. E. Ataer-Cansizoglu, V. Bolon-Canedo, J. P. Campbell, A. Bozkurt, D. Erdogmus, J. Kalpathy-Cramer, S. Patel, K. Jonas, R. V. P. Chan, S. Ostmo and M. F. Chiango. Computer-based image analysis for plus disease diagnosis in retinopathy of prematurity: Performance of the i-ROP system and image features associated with expert diagnosis. *Translational Vision Science and Technology,* 4(6):1–12, 2015.
27. R. M. Rangayyan, F. J. Ayres, F. Oloumi, F. Oloumi, and P. Eshghzadeh-Zanjani. Detection of blood vessels in the retina with multiscale Gabor filters. *Journal of Electronic Imaging,* 17(2): 023018–023018-7, 2008.
28. F. Oloumi, R. M. Rangayyan, and A. L. Ells. Parabolic modeling of the major temporal arcade in retinal fundus images. *IEEE Transactions on Instrumentation and Measurement,* 61(7):1825–1838, 2012.
29. F. Oloumi, A. K. Dhara, R. M. Rangayyan, and S. Mukhopadhyay. Detection of blood vessels in retinal fundus images (invited paper). *Computer Science Journal of Moldova,* 22(2):155–185, 2014.
30. F. Oloumi, R. M. Rangayyan, and A. L. Ells. *Digital Image Processing for Ophthalmology: Detection and Modeling of the Retinal Vascular Architecture.* Morgan & Claypool, 2014. San Rafael, CA.
31. F. Oloumi, R. M. Rangayyan, P. Casti, and A. L. Ells. Computer-aided diagnosis of plus disease via measurement of vessel thickness in retinal fundus images of preterm infants. *Computers in Medicine and Biology,* 66:316–329, 2015.
32. F. Oloumi, R. M. Rangayyan, and A. L. Ells. Computer-aided diagnosis of plus disease in retinal fundus images of preterm infants via measurement of vessel tortuosity. In *Engineering in Medicine and Biology Society (EMBS) 37th Annual International Conference of the IEEE,* Milan, Italy, August 2015.
33. F. Oloumi. Computer-aided Diagnosis of Plus Disease via Quantitative Analysis of the Vascular Architecture in Retinal Fundus Images of Preterm Infants. Ph.D. thesis, Schulich School of Engineering, University of Calgary, 2015.
34 A. D. Fleming, K. A. Goatman, S. Philip, J. A. Olson, and P. F. Sharp. Automatic detection of retinal anatomy to assist diabetic retinopathy screening. *Physics in Medicine and Biology,* 52:331–345, 2007.
35. K. W. Tobin, E. Chaum, V. P. Govindasamy, and T. P. Karnowski. Detection of anatomic structures in human retinal imagery. In *IEEE Transactions on Medical Imaging,* 26(12):1729–1739, 2007.
36. H. Ying and J. C. Liu. Automated localization of macula-fovea area on retina images using blood vessel network topology. In *IEEE International Conference on Acoustics Speech and Signal Processing,* pages 650–653, March 2010. Dallas, TX.

37 S. Chaudhuri, S. Chatterjee, N. Katz, M. Nelson, and M. Goldbaum. Detection of blood vessels in retinal images using two-dimensional matched filters. *IEEE Transactions on Medical Imaging*, 8(3):263–269, 1989.
38. T. Chakraborti, D. K. Jha, A. S. Chowdhury, and X. Jiang. A self-adaptive matched filter for retinal blood vessel detection. *Machine Vision and Applications*, 26(1): 55–68, 2015.
39. J. Staal, M. D. Abràmoff, M. Niemeijer, M. A. Viergever, and B. van Ginneken. Ridge-based vessel segmentation in color images of the retina. *IEEE Transactions on Medical Imaging*, 23(4):501–509, 2004.
40. X. Li and W. G. Wee. Retinal vessel detection and measurement for computer-aided medical diagnosis. *Journal of Digital Imaging*, 27(1):120–132, 2014.
41. J. V. B. Soares, J. J. G. Leandro, R. M. Cesar Jr., H. F. Jelinek, and M. J. Cree. Retinal vessel segmentation using the 2-D Gabor wavelet and supervised classification. *IEEE Transactions on Medical Imaging*, 25(9): 1214–1222, 2006.
42. M. E. Martinez-Perez, A. D. Hughes, A. V. Stanton, S. A. Thorn, A. A. Bharath, and K. H. Parker. Retinal blood vessel segmentation by means of scale-space analysis and region growing. *In Conference Proceedings of Medical Image Computing and Computer-Assisted Intervention-MICCAI99*, Vol. 1 90–97, 1999. Cambridge, UK.
43. S. Belhadi and N. Benblidia. Automated retinal vessel segmentation using entropic thresholding based spatial correlation histogram of gray level images. *International Arab Journal of Information Technology (IAJIT)*, 12(5):441–447, 2015.
44. F. Zana and J. C. Klein. Segmentation of vessel-like patterns using mathematical morphology and curvature estimation. *IEEE Transactions on Image Processing*, 10(7): 1010–1019, 2001.
45. A. Fathi and A. R. Naghsh-Nilchi. General rotation-invariant local binary patterns operator with application to blood vessel detection in retinal images. *Pattern Analysis and Applications*, 17(1):69–81, 2014.
46. A. F. Frangi, W. J. Niessen, K. L. Vincken, and M. A. Viergever. Multiscale vessel enhancement filtering. In *Medical Image Computing and Computer-Assisted Intervention - MICCAI98*, volume 1496 of *Lecture Notes in Computer Science*, 130–137. Springer, Berlin, Germany, 1998.
47. M. N. Salem, A. S. Salem, and A. K. Nandi. Segmentation of retinal blood vessels based on analysis of the Hessian matrix and clustering algorithm. In *15th European Signal Processing Conference (EUSIPCO 2007)*, 428–432, Poznan, Poland, September 2007.
48. C. A. Lupascu, D. Tegolo, and E. Trucco. FABC: Retinal vessel segmentation using AdaBoost. *IEEE Transactions on Information Technology in Biomedicine*, 14(5):1267–1274, September 2010.
49. D. Marin, A. Aquino, M. E. Gegundez-Arias, and J. M. Bravo. A new supervised method for blood vessel segmentation in retinal images by using gray-level and moment invariants-based features. *IEEE Transactions on Medical Imaging*, 30(1):146–158, 2011.
50. D. Kaba, C. Wang, Y. Li, A. Salazar-Gonzalez, X. Liu, and A. Serag. Retinal blood vessels extraction using probabilistic modelling. *Health Information Science and Systems*, 2:1–10, 2014.
51. Z. Xiao, M. Adel, and S. Bourennane. Bayesian method with spatial constraint for retinal vessel segmentation. *Computational and Mathematical Methods in Medicine*, ID 401413:9, 2013.
52. M. F. Chiang, P. J. Thyparampil, and D. Rabinowitz. Interexpert agreement in the identification of macular location in infants at risk for retinopathy of prematurity. *Archives of Ophthalmology*, 128(9): 1153–1159, 2010.
53. D. Fiorin and A. Ruggeri. Computerized analysis of narrow-field ROP images for the assessment of vessel caliber and tortuosity. In *33rd Annual International Conference of the IEEE Engineering in Medicine and Biology Society*, 2622–2625, Boston, MA, 2011.
54. C. M. Wilson, K. D. Cocker, M. J. Moseley, et al. Computerized analysis of retinal vessel width and tortuosity in premature infants. *Investigative Ophthalmology and Visual Science*, 49(1): 3577–3585, 2008.

55. M. E. Martínez-Pérez, A. D. Hughes, A. V. Stanton, S. A. Thorn, N. Chapman, A. A. Bharath, and K. H. Parker. Retinal vascular tree morphology: A semi-automatic quantification. *IEEE Transactions on Biomedical Engineering,* 49(8): 912–917, 2002.
56. L. Gang, O. Chutatape, and S. M. Krishnan. Detection and measurement of retinal vessels in fundus images using amplitude modified second-order Gaussian filter. *IEEE Transactions on Biomedical Engineering,* 49(2): 168–172, 2002.
57. E. Poletti, E. Grisan, and A. Ruggeri. Image-level tortuosity estimation in wide-field retinal images from infants with retinopathy of prematurity. In *34th Annual International Conference of the IEEE Engineering in Medicine and Biology Society,* 4958–4961, San Diego, CA, 2012.
58. W. E. Hart, M. Goldbaum, B. Cote, P. Kube, and M. R. Nelson. Measurement and classification of retinal vascular tortuosity. *International Journal of Medical Informatics,* 53(2–3): 239–252, 1999.
59. E. Grisan, M. Foracchia, and A. Ruggeri. A novel method for the automatic grading of retinal vessel tortuosity. *IEEE Transactions on Medical Imaging,* 27(3):310–319, 2008.
60. P. L. Hildebrand, A. L. Ells, and A. D. Ingram. The impact of telemedicine integration on resource use in the evaluation ROP analysis of the telemedicine for ROP in Calgary (TROPIC) database. *Investigative Ophthalmology and Visual Science,* 50: E-Abstract 3151, 2009.
61. D. J. De Silva, K. D. Cocker, G. Lau, S. T. Clay, A. R. Fielder, and M. J. Moseley. Optic disk size and optic disk-to-fovea distance in preterm and full-term infants. *Investigative Ophthalmology and Visual Science,* 47(11): 4683–4686, 2006.
62. A. L. Ells, M. Hicks, M. Fielden, and A. D. Ingram. Severe retinopathy of prematurity: Longitudinal observation of disease and screening implications. *Eye,* 19(12): 138–144, 2005.
63. Metz C. E. Basic principles of ROC analysis. *Seminars in Nuclear Medicine,* VIII(4): 283–298, 1978.
64. ROCKIT. Metz ROC Software. radiology.uchicago.edu/page/metz-roc-software.
65. J. Neter, M. H. Kutner, C. J. Nachtsheim, and W. Wasserman. *Applied Linear Statistical Models.* Irwin, Chicago, IL, 4th edn, 1990.
66. R. O. Duda, P. E. Hart, and D. G. Stork. *Pattern Classification.* Wiley, New York, 2nd edn, 2001.
67. R. Kohavi and G. H. John. Wrappers for feature subset selection. *Artificial Intelligence,* 97(1): 273–324, 1997.
68. N. Otsu. A threshold selection method from gray-level histograms. *IEEE Transactions on Systems, Man, and Cybernetics,* SMC-9: 62–66, 1979.
69. W.-H. Tsai. Moment-preserving thresolding: A new approach. *Computer Vision, Graphics, and Image Processing,* 29(3): 377–393, 1985.
70. T. W. Ridler and S. Calvard. Picture thresholding using an iterative selection method. *IEEE Transactions on Systems, Man, and Cybernetics,* 8: 630–632, 1978.
71. J. N. Kapur, P. K. Sahoo, and A. K. C. Wong. A new method for gray-level picture thresholding using the entropy of the histogram. *Computer Vision, Graphics, and Image Processing,* 29: 273–285, 1985.
72. F. Oloumi, R. M. Rangayyan, and A. L. Ells. A graphical user interface for measurement of temporal arcade angles in fundus images of the retina, In *Proceeding of IEEE Canada 25th Annual Canadian Conference on Electrical and Computer Engineering (CCECE)* 4 pages on CD-ROM, Montreal, QC, Canada, April 2012.
73. J. Illingworth and J. Kittler. A survey of the Hough transform. *Computer Vision, Graphics, and Image Processing,* 44: 87–116, 1988.
74. F. Oloumi, R. M. Rangayyan, and A. L. Ells. Computer-aided diagnosis of proliferative diabetic retinopathy via modeling of the major temporal arcade in retinal fundus images. *Journal of Digital Imaging,* 26(6): 1124–1130, 2013.
75. F. Oloumi, R. M. Rangayyan, and A. L. Ells. Tracking the major temporal arcade in retinal fundus images. In *Proceeding of IEEE Canada 27th Annual Canadian Conference on Electrical and Computer Engineering (CCECE),* pages 1–5, Toronto, ON, Canada, May 2014.
76. T. Pham-Gia and T.L. Hung. The mean and median absolute deviations. *Mathematical and Computer Modelling,* 34(7–8): 921–936, 2001.

77. R. M. Rangayyan, X. Zhu, F. J. Ayres, and A. L. Ells. Detection of the optic nerve head in fundus images of the retina with Gabor filters and phase portrait analysis. *Journal of Digital Imaging*, 23(4):438–453, 2010.
78. H. Freeman. On the encoding of arbitrary geometric configurations, *IRE Transactions on Electronic Computers*, , EC-10(2): 260–268, 1961.
79. R. M. Rangayyan. *Biomedical Image Analysis*. 641–642, CRC Press, Boca Raton, Florida, 2005.
80. R. Medina-Carnicer, A. Carmona-Poyato, R. Muñoz Salinas, and F. J. Madrid-Cuevas. Determining hysteresis thresholds for edge detection by combining the advantages and disadvantages of thresholding methods. *IEEE Transactions on Image Processing*, 19(1): 165–173, 2010.

4

Automated OCT Segmentation for Images with DME

Sohini Roychowdhury

Dara D. Koozekanani

Michael Reinsbach

Keshab K. Parhi

4.1 Introduction

Diabetic macular edema (DME) is a leading cause of vision loss in patients with diabetes. The World Health Organization estimates that by the year 2020, there will be 75 million blind people and 314 million partially blind people in the world [1]. While treatments are available, including intravitreal injections and macular laser therapy, not all patients respond to these. Currently, there are no reliable methods for predicting patient response to therapy. Therefore, analysis of the patient images prior to treatment may allow the development of measures to predict patient response. Computer-aided diagnostic (CAD) systems enable automated detection of ophthalmic pathological sites, monitoring the progression of pathology, and can guide follow-up treatment processes. Optical Coherence Tomography (OCT) images have been widely used to assess macular diseases, and they have enabled analysis of the extent of *disorganization* in the sub-retinal layers due to DME [2,3]. Sub-retinal layer disorganization refers to the variation in the underlying retinal microstructure due to the presence of cystoid regions or to disruptions in the cellular architecture of the sub-retinal layers due to pathology [4]. For each patient's eye under analysis, a stack of images centered at the macula are acquired, such that reconstruction of the sub-retinal surfaces from the OCT image stacks aid localization of disease-related abnormalities in the retinal microstructure. In this work, a CAD system is presented that automatically segments sub-retinal surfaces and layers in OCT image stacks from normal patients and abnormal ones with DME, such that thickness maps corresponding to the sub-retinal layers can be further analyzed for their clinical relevance to the severity of DME.

Existing studies on OCT images from patients with DME by Maalej et al. [5] have shown that clinical edema causes the tomographic thickness to increase by 1.33 times the normal thickness values. Also, significant relationship has been found between retinal thickness and visual acuity [5]. Other studies by Otani et al. [6] have shown strong correlation of visual acuity with the photoreceptor inner and outer segment junction integrity, and lower correlation between cystoid macular edema and visual acuity. Another study by Lattanzio et al. [7] shows that the macular thickness in eyes tends to increase with the severity of diabetic retinopathy and macular edema. In this work, we propose an automated CAD system that estimates the sub-retinal layer thicknesses in healthy and pathological OCT images with DME with correlation coefficient $r \geq 0.7$. Such an automated system can lead to faster and more efficient detection and treatment of patients with DME if the macular thickness for the particular patient is found to exceed the clinically acceptable levels. Additionally, such an automated system can be used for clinical trials to analyze more metrics other than the macular thickness for their clinical relevance to visual acuity.

The proposed automated OCT segmentation algorithm involves two key steps. In the first step, additive noise introduced by the imaging systems is removed from the images. In the second step, the denoised images are subjected to model-based segmentation. In prior efforts for denoising OCT images in the first step, OCT image noise suppression has been accomplished by weighted averaging in 3-D space by Mayer et al. [8], discrete wavelet transforms by Adler et al. [9], and dual-tree wavelet transforms by Chitchian et al. [10]. In another work by Fang et al. [11], multiscale, sparsity-based, tomography denoising is performed by subsampling the noisy OCT image, followed by dictionary parameter training, k-means clustering, and final image restoration. It is observed that the sparsity-based denoising approach by Fang et al. [11] is most useful in signal retrieval from noise in un-averaged OCT frames from Bioptigen SDOCT systems. However, images obtained from Spectralis OCT (Heidelberg Engineering, Germany) have built-in stabilization and averaging systems that produce significantly high SNR in the OCT images. In this work, we analyze images from Spectralis OCT and compare the denoising performances of the proposed Wiener de-convolution algorithm with Fourier-domain-based noise variance estimator, with the complex dual-tree wavelet transform method proposed by Chitchian et al. [10]. This comparative analysis demonstrates the importance of image noise removal on automated OCT segmentation algorithms.

For the second step of automated sub-retinal surface and layer segmentation, some prior works rely on edge detection methods searching for peaks of gradient changes in column wise pixel intensities followed by interpolation by Koozekanani et al. [12], or edge classification by Ishikawa et al. [13] and Bacgi et al. [14]. Such edge detection methods, however, suffer from large segmentation errors when the image pixel intensities are inconsistent and in the presence of underlying pathology. Active contours-based methods have been explored by Mishra et al. [15] with a two-step kernel-based optimization scheme for sub-retinal layer segmentation. Another well-known graph-cut segmentation method separates a 3D composite image into sub-retinal layers by finding a minimum-cost closed set in a 3D graph by Garvin et al. [16]. Modifications to this method have been used to segment the optic nerve head by Antony et al. [17], and an alternative segmentation approach using dynamic programming have been explored by Chiu et al. [18] and Lee et al. [19]. Further, multi-resolution graph search has been applied for segmentation of up to 3 significant sub-retinal surfaces for DME images by Abhishek et al. [20], and for segmenting up to 12 surfaces in OCT images with pigment epithelial detachment by Shi et al. [21]. The major drawback of such graph-based 3D segmentation methods is that they suffer from high computational complexity and are restrictive in their assumptions. Another well-known, freely available OCT image segmentation system (OCTSEG) by Mayer et al. [8] utilizes second-order edge detection followed by fifth-order polynomial fitting for automated segmentation of the sub-retinal surfaces. Although this method results in fast OCT segmentation in normal images, the inner sub-retinal surfaces are incorrectly identified in abnormal OCT images with pathology.

This work makes two key contributions. The first contribution is the comparative analysis of the existing wavelet-transform-based denoising method with a novel Fourier-domain, structural, error-based denoising method presented in our previous work [22]. In this previous conference paper, the concept of noise parameter estimation for Wiener deconvolution was described. In this work, we analyze the importance of image denoising on automated sub-retinal surfaces and layer segmentation process. Our analysis demonstrates that the wavelet-based denoising approach loses most sub-retinal surfaces by over-smoothing the noisy OCT images, thereby incurring approximately two times more error in segmenting sub-retinal surfaces and layers when compared to the proposed Fourier-domain-based denoising method. The second contribution is a novel multi-resolution iterative sub-retinal surface segmentation algorithm that is adaptive and robust to normal OCT images and images with DME. Also, the proposed method is comparatively analyzed with the OCTSEG system on normal and abnormal OCT images for automated sub-retinal surface segmentation performance.

The organization of this chapter is as follows. In Section 4.2, the method and materials used to analyze the proposed automated segmentation system are presented. In Section 4.3 the experimental results of automated sub-retinal surface and layer segmentation are presented. Conclusions and discussion are presented in Section 4.4.

4.2 Materials and Method

The two key steps for automated denoising and segmentation of the sub-retinal layers from OCT images are described in the following subsections. From all the OCT data image stacks each OCT image is treated as a separate stand-alone image while denoising and segmenting the sub-retinal surfaces. This operation of separately segmenting each image without any information regarding the adjacent images from the OCT image stack is different from the existing graph-based segmentation approaches in [16]. Once each image is denoised, seven sub-retinal surfaces (i.e., Surfaces 1 to 7) and six sub-retinal layers that are then segmented, as shown in Figure 4.1. The segmented sub-retinal layers extend from the inner limiting membrane (ILM, Surface 1) to the Bruch's Membrane (BM, Surface 7) surface. The automatically segmented sub-retinal layers as shown in Figure 4.1 are: the Nerve Fiber Layer (NFL, between Surfaces 1 and 2), Inner Plexiform Layer (IPL) and Ganglion Cell Layer (GCL) combined (IPL/GCL, between Surfaces 2 and 3), Inner Nuclear Layer and outer plexiform layer combined (INL, between Surfaces 3 and 4), Outer Nuclear Layer (ONL, between Surfaces 4 and 5), Photoreceptor Inner/Outer Segment (IS/OS, between Surfaces 5 and 6), and Retinal Pigment Epithelium (RPE, between Surfaces 6 and 7). Additionally, to analyze the average retinal thickness, two more combined layers are analyzed as such: the Inner layer, which combines the NFL, IPL/GCL, INL and ONL (between Surfaces 1 and 5), and the Outer layer, which combines the IS/OS and RPE layers (between Surfaces 5 and 7).

The data for assessing the performance of automated OCT segmentation is obtained using the Heidelberg Spectralis Imaging system at the Department of Ophthalmology of the University of Minnesota. OCT image stacks from normal and abnormal eyes with DME are obtained such that each image is averaged over 12–19 frames. Each OCT image has a resolution of 5.88 μm/pixel along the length and 3.87 μm/pixel along the width and [1024 × 496] pixels per image. To evaluate the performance of the proposed segmentation system, each image is manually segmented by a human expert for all the

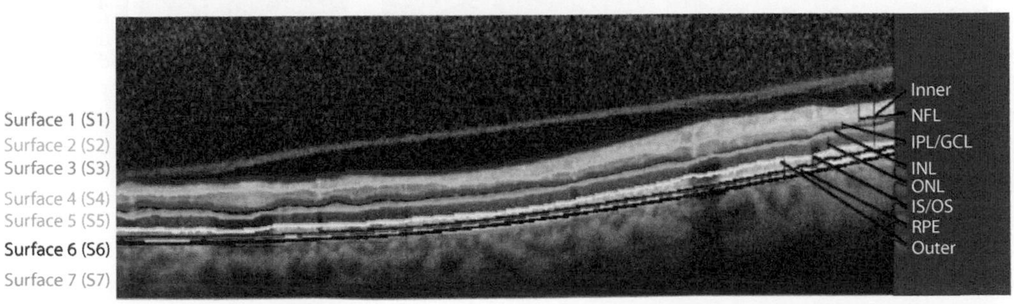

FIGURE 4.1 The manually segmented sub-retinal surfaces and layers in OCT images. The 7 sub-retinal surfaces are color coded as Surface 1 (red), Surface 2 (cyan), Surface 3 (pink), Surface 4 (green), Surface 5 (yellow), Surface 6 (black), Surface 7 (blue). The sub-retinal layers are: NFL, IPL/GCL, INL, ONL, IS/OS, inner and outer layers.

sub-retinal surfaces. The following two sets of OCT image stacks are obtained corresponding to normal eyes and eyes with pathology:

1. Normal set: This image set comprises of OCT image stacks of the macular region from 10 healthy eyes with 19–25 images per OCT stack. A total of 203 normal OCT images are collected in this set.
2. Abnormal set: This image set comprises of macular OCT image stacks from 15 patients with DME and 19–33 images per OCT stack. A total of 357 images with sub-retinal cysts and layer disorganization are collected in this set. The Surface 6, which is not affected by DME, is not manually marked for images from this dataset.

4.2.1 Automated Denoising

Additive noise removal from OCT images is achieved based on the inherent assumption that the additive noise is Gaussian in nature. The statistical model for a noisy image (I) follows $\psi(I) = \psi(H) . \psi(I_d) + \psi(N)$, where I is the noisy OCT image, H is a 2-D point spread function (PSF), I_d is the noiseless image, N is the additive Gaussian noise, and ψ denotes a particular frequency-based or texture-based transform. Some existing OCT image denoising methods in [9,10] achieve image denoising using the above statistical model with the wavelet transform. In this work, we estimate additive noise using the Fourier transform as function ψ [22].

The proposed system utilizes the Wiener deconvolution algorithm that reduces the least-square error in Fourier domain [23] as shown in Equation 4.1.

$$\psi(I_d) = \left[\frac{1}{\psi(H)} \frac{|\psi(H)|^2}{|\psi(H)|^2 + \frac{\Sigma_N}{\Sigma_{I_d}}} \right] \psi(I), \tag{4.1}$$

where Σ_N and Σ_{I_d} represent the power spectral density of noise and the original image, respectively. Assuming a Gaussian PSF (H) with zero mean and variance 10^{-3} [23], the best estimate of noise to signal ratio (Σ_N / Σ_{I_d}) will provide the best estimate of a denoised image. It is known that for OCT images the strength of additive noise varies with the number of averaged frames and across imaging systems. Thus, it is imperative to estimate the additive noise variance of each OCT image separately.

In Figure 4.2, we observe that the absolute Fourier-domain representation of the noisy image ($|\psi(I)|$) has two prime components: the central frequency spectrum structure that contains information regarding the sub-retinal microstructure (C), and the extra frequency components occupying high-frequency regions that appear due to the additive noise (E). Since the position of the frequency components in

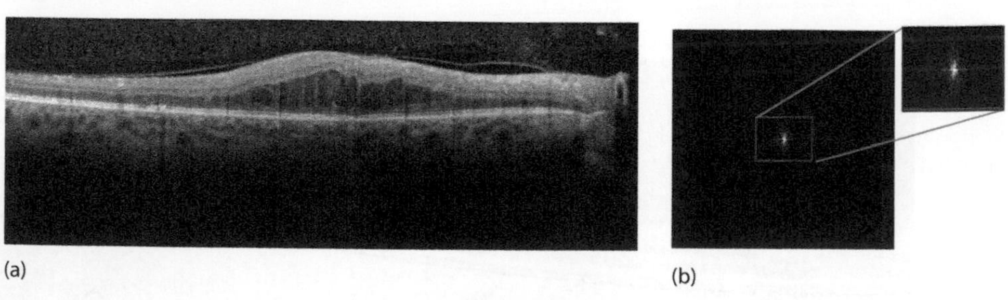

FIGURE 4.2 Fourier-domain spectrum composition for OCT images. (a) Original noisy image (I), (b) binary image of the Fourier-domain representation ($\psi'(I)$). The region within the red box is the central frequency spectrum structure that contains information regarding the sub-retinal microstructure. C is an image with components within in the red box only and all other pixels outside the box are zero. All remaining high-frequency components outside the red box are high-frequency terms introduced by noise and are represented by image E.

the Fourier domain signify the sub-retinal components and the noise components, a binary image ($\psi'(I)$) can be generated from the Fourier domain representation of the noisy image $\psi'(I)$, such that $\psi'(I) = |\psi(I)| > 0$, and $\psi'(I) = C + E$. For an ideal denoised image, the central frequency spectrum must be conserved while the extra frequency terms introduced by noise must be suppressed. The challenge of separating the noise components from the central frequency components lies in the fact that the exact position of the sub-retinal and noisy components in the absolute Fourier-domain representation are unknown and variable across images.

We observe that absolute Fourier-domain representations of the denoised images ($|\psi(I_{d_k})|$) can also be separated into the central frequency and the extra frequency components as $(|\psi(I_{d_k})|) = C_k + E_k$, where I_{d_k} is estimated by Wiener deconvolution using additive Gaussian noise estimate $N(0, 10^{-k})$. For an image with spatial configuration of $[n_1 \times n_2]$ pixels, analysis of the Fourier-domain noise component image (E_k) demonstrates, $\sum_{i=1}^{n_1} \sum_{j=1}^{n_2} E_k(i,j) \approx 0$ for Wiener deconvolution with high-noise variances (around 10^{-1} to 10^{-8}), and $\sum_{i=1}^{n_1} \sum_{j=1}^{n_2} E_k(i,j) > 0$ for deconvolution with low-noise variances (around 10^{-10} to 10^{-15}). Here, (i,j) refers to the frequency domain image pixels. The Fourier-domain central frequency structure of the denoised images (C_k) with different noise variances is shown in Figure 4.3.

For the best estimate of additive noise variance, a Fourier-domain structural error metric (e) is defined in Equation 4.2 and 4.3. A feasibility range for the convex error metric "e" can be identified by the range of values for a constant a.

$$e_k = \sum_{i=1}^{n_1} \sum_{j=1}^{n_2} \left[\psi'(I) - a(|\psi(I_{d_k})| . \psi'(I)) + \psi(I_{d_k})^2 \right](i,j)$$

$$\text{Since, } |\psi(I_{d_k})| = C_k + E_k, \psi'(I) = C + E, \tag{4.2}$$

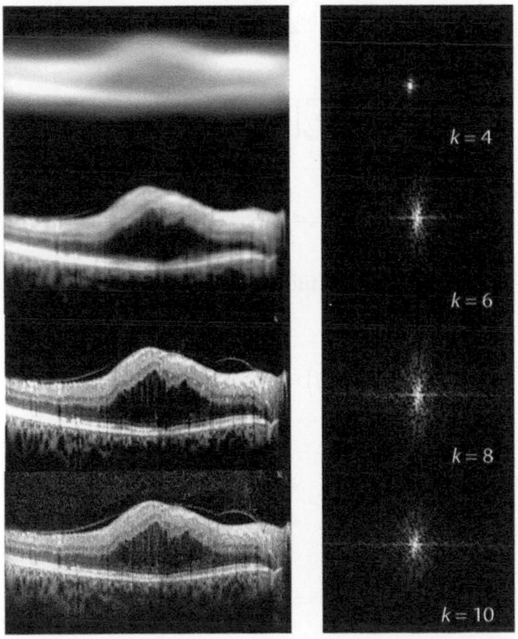

FIGURE 4.3 Denoised OCT images by wiener deconvolution with varying additive noise variance as 10-k. (a) Denoised image. (b) The respective absolute central frequency spectrum structure C_k in the Fourier-domain representation of the denoised image when k varies as k = (4,6,8,9) and the additive Gaussian noise variance varies as 10^{-4}, 10^{-6}, 10^{-8}, 10^{-9}, respectively. As estimated additive noise variance decreases from 10^{-4} to 10^{-9}, the structure of C_k becomes more similar to that of C, and additive noise components E_k increases.

$$e_k = \sum_{i=1}^{n_1}\sum_{j=1}^{n_2}\left[C + E - a(C.C_k + E.E_k) + C_k^2 + E_k^2\right](i,j) \tag{4.3}$$

Next, the trends of the structural error are analyzed in Equations 4.4 through 4.10 by considering the situations when the estimated noise variance is high, in Equations 4.5 through 4.7), and low, in Equations 4.8 and 4.9, respectively. From Figure 4.3, it is evident that as the estimated Gaussian noise variance used for Wiener deconvolution iteratively decreases (as 10^{-k}, $k = 1,2,\ldots15$), the central components C_k of the denoised image become more similar to C, and the extra additive noise components increase. The iteration at which this Fourier-domain error is minimum corresponds to the best estimate of the additive Gaussian noise and Wiener deconvolution of noisy OCT image I using this best noise variance estimate yields the best denoised image I_d.

$$e_k - e_{k+1} = \sum_{i=1}^{n_1}\sum_{j=1}^{n_2}\left[a\left(C.C_{k+1} - C.C_k + E.E_{k+1} - E.E_k\right) + C_k^2 - C_{k+1}^2 + E_k^2 - E_{k+1}^2\right](i,j). \tag{4.4}$$

$$\text{High} - \text{noise variance}: E_{k+1} \approx E_k, \sum_{i=1}^{n_1}\sum_{j=1}^{n_2} E_k(i,j) \approx 0, \tag{4.5}$$

$$\text{Since } C_k, E_k \in [0,1] \Rightarrow \forall(i,j), \text{ and if } a \geq 1,$$

$$a.C.C_k(i,j) \geq C_k^2(i,j), \sum_{i=1}^{n_1}\sum_{j=1}^{n_2}\left[a.C.C_{k+1} - C_{k+1}^2\right](i,j)$$
$$\geq \sum_{i=1}^{n_1}\sum_{j=1}^{n_2}\left[a.C.C_k - C_k^2\right](i,j). \tag{4.6}$$

$$\Rightarrow e_k - e_{k+1} \geq 0 \Rightarrow e \text{ has decreasing trend.} \tag{4.7}$$

$$\text{Low} - \text{noise variance}: \; C_{k+1} \approx C_k, \text{ and } \forall(i,j)$$

$$\sum_{i=1}^{n_1}\sum_{j=1}^{n_2} E.E_{k+1}(i,j) \approx \sum_{i=1}^{n_1}\sum_{j=1}^{n_2} E.E_k(i,j), \tag{4.8}$$

$$\sum_{i=1}^{n_1}\sum_{j=1}^{n_2} E_{k+1}^2(i,j) \geq \sum_{i=1}^{n_1}\sum_{j=1}^{n_2} E_k^2(i,j) \geq 0,$$

$$\Rightarrow e_k - e_{k+1} \leq 0 \Rightarrow e \text{ has increasing trend.} \tag{4.9}$$

$$\text{Substituting Equation (4.4) and (4.8) in Equation (4.8) we have}: \forall(i,j),$$

$$a\left(\left[E.E_{k+1} - E.E_k\right](i,j)\right) + \left[E_k^2 - E_{k+1}^2\right](i,j) \leq 0$$
$$\Rightarrow a \leq \left[E_{k+1} + E_k\right](i,j) \Rightarrow a \leq 2. \tag{4.10}$$

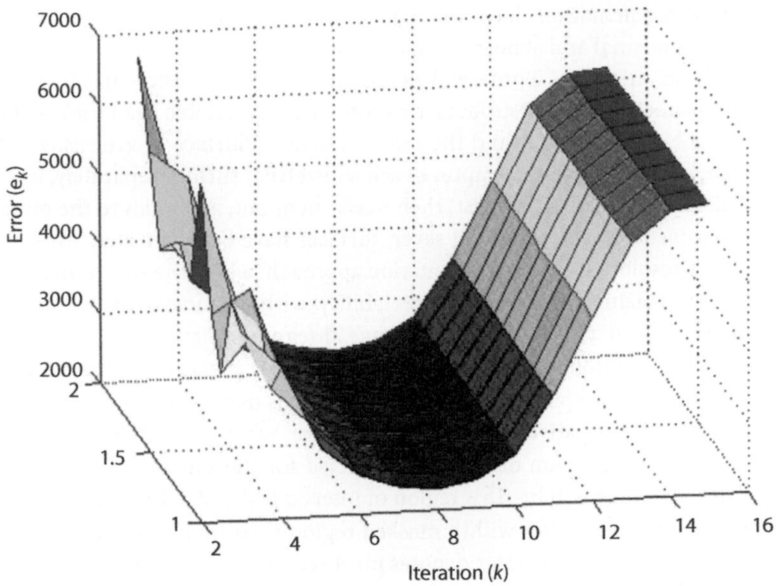

FIGURE 4.4 Fourier-domain error e_k as $k = (1, 2...15)$ and $a = [1,1.1,1.2....2]$. Minimum error occurs at $a = 1$ and $k = 8$. Thus, best estimate of additive noise for this image is $N(0, 10^{-8})$.

To ensure iterative optimality of e_k, it is necessary that $[1 < a < 2]$. Figure 4.4 shows the Fourier-domain error e as iteration k and constant a varies. This particular image has low variability with respect to a. For all OCT images denoised using the proposed approach, the error e_k is minimized with respect to a and k to obtain the best denoised image I_d.

To comparatively assess the importance of noise removal on automated OCT image segmentation, we selected a baseline method of image denoising by the wavelet tree-based approach in [10,24]. Wavelet-based denoising approaches are known to estimate clean signal coefficients from the noisy ones using Bayesian estimators [24]. While wavelet coefficients with large magnitudes represent edges or textures, coefficients with small magnitude represent the smooth background regions. Thus, OCT image denoising using the wavelet shrinkage method by Chitchian et al. [10] is achieved in four steps that include: wavelet transform on the noisy image I, estimation of adaptive threshold using the wavelet transform, denoising image I using the estimated threshold, and finally, inverse wavelet transform on the denoised image. The most critical step in this wavelet-based denoising approach is the adaptive threshold estimation. While a low threshold will not produce significant denoising, a high threshold will estimate a signal with a large number of zero coefficients that will correspond to an over-smoothened image with no edges or textures. To facilitate optimal threshold estimation, soft-thresholding is performed as described by Wagner et al. [24]. This method of OCT image denoising using complex dual-tree wavelet transform is referred as "CDWT" in the following sections. The performances of OCT image segmentation and sub-retinal layer thickness map estimation using the proposed denoising approach and the wavelet-based denoising approach are compared. The pixel intensities of the denoised image (I_d) are scaled in the range [0,1] before automated segmentation.

4.2.2 Automated Segmentation

Every denoised image (I_d) is subjected to multi-resolution iterative high-pass filtering to extract 7 sub-retinal surfaces [22]. Several OCT image segmentation methods that involve edge detection followed by segment linking or gradient search methods have been developed [12]. However, most such methods fail in the presence of sub-retinal disorganization owing to pathology. Since the proposed method of sub-retinal surface segmentation does not require interpolation between partially detected line segments, it is robust to normal and abnormal images with pathology.

Each sub-retinal layer in the OCT images has a distinct pixel intensity that is different from the adjacent layers. Also, some sub-retinal surfaces are more easily detectable than the others; for instance, the top surface of the NFL (Surface 1) and the IS/OS junction (Surface 5) are easily detectable when compared to the other surfaces. Thus, to capture each sub-retinal surface separately, it is imperative to look for the most distinct edges/surfaces first, then mask them out, and analyze the remaining regions for the next most-significant surface until all seven surfaces have been detected. This idea is the basis for the proposed multi-resolution image segmentation approach, where the sub-retinal surfaces are segmented iteratively in decreasing order of their detectability, with the most detectable surface (Surface 1) segmented first and the least detectable surface (Surface 2) segmented at last.

The proposed multi-resolution surface extraction algorithm proceeds in six iterative steps, that is $l = [1,2, ...6]$. In each iterative step l, a different source image is used. The source image I_{S_l} refers either to the denoised image or the negative of the denoised image, whichever enhances the detectability of the surface of interest. Next, a region of interest is selected for the source image based on the region where the most distinctive edge will lie. This region of interest is denoted by a masked region G_l. Next, the region in source image I_{S_l} that lies within masked region G_l is high-pass filtered in different ways as shown in Equation 4.11. The "o" operator denotes pixel-wise multiplication. The filtered image (I_k) is then subjected to pixel-level thresholding (Δ_k) and the regions in the remaining image are subjected to a particular decision criterion (ϕ_l) to extract a binary image (I_η) with one or more regions in it that satisfy the segmentation decision criterion in Equation 4.12. The top or bottom surface ($Surf_i$) of the region/regions in image I_η is determined as the segmented surface (S_l) in Equation 4.13.

$$\forall l = \left[1,2...6\right], I_l = HPF_l\left(I_{s_l} \circ G_l\right) \tag{4.11}$$

$$\Rightarrow I_l = I_{s_l} \circ G_l - LPF_l\left(I_{s_l} \circ G_l\right).$$

$$I_\eta = \phi_l\left(\Delta_l\left(I_l\right)\right). \tag{4.12}$$

$$S_l = Surf_l\left(I_\eta\right). \tag{4.13}$$

Corresponding to each iterative step (l), the choice of source image (I_{Sl}), mask for the region of interest (G_l), filter (HPF_l/LPF_l), threshold (Δ_t), decision criterion (ϕ_t) and surface of interest ($Surfi$) are given in Table 4.1. The decision criterion (ϕ_t) for selecting a particular region of interest from the thresholded binary image $\Delta_t(I_l)$ can vary from the region with maximum area to the region with maximum major axis length as shown in Table 4.1.

For an image from the abnormal dataset OCT stack, the proposed iterative surface segmentation algorithm is shown in Figure 4.5. The denoised image obtained by the proposed approach (I_d) is shown in Figure 4.5a. The image I_1 obtained after high-pass filtering in iteration $l = 1$ is in Figure 4.5b. Thresholding I_1 results in the detection of Surface 1 and the choroidal segment. In Figure 4.5c, negative source image $1 - I_d$ in iteration $l = 2$ is shown within the region of interest marked by G_2 that extends between the Surface 1 and the choroid segment. In Figure 4.5d, the image obtained after high-pass filtering and thresholding the image in Figure 4.5c is shown. The region with maximum major axis length is extracted into image I_{r2}. The top surface of this region is Surface 5, and the bottom surface is Surface 7. In Figure 4.5e, the image obtained in iteration $l = 4$ after high-pass filtering and thresholding is shown. The region with maximum major axis length is selected in image I_{r4}, and the bottom surface of this region is Surface 4. Two more iterations are performed to extract all seven surfaces. In Figure 4.5f, automated segmentations achieved at the end of six iteration steps by the proposed method are shown.

TABLE 4.1 Iterative Parameters for Sub-Retinal Surface Segmentation using the Proposed Multi-Resolution High-Pass Filtering Method

Iteration (l)	Source (I_{Sl})	Mask (G_l)	Filter (HPF_l/LPF_l)	Threshold (Δ_t)	Criterion (ϕ_t)	$Surf_i$
1	I_d	Whole image	$HPF_1 = $ 2-D Gaussians, horizontal and vertical	Regiongrow (I_1in [0, 255], seed = 255, threshold = {230–235})	Arg max (area)	$S_1 =$ top (I_{r1}), choroid = bottom (I_{r1})
2	$1 - I_d$	Between S_1 and choroid	$LPF_2 =$ average [25 × 25]	Pixel values < 0	Argmax (major axis length)	$S_5 =$ top (I_{r2}), $S_7 =$ bottom (I_{r2})
3	$1 - I_d$	Between S_5 and S_7	$HPF_3 =$ contrast enhancement	Pixel values > Otsu's threshold estimated on I_3	—	$S_6 =$ bottom (I_{r3})
4	$1 - I_d$	Between S_1 and S_5	$LPF_4 =$ average [25 × 25]	Pixel values > Otsu's threshold estimated on I_4	Argmax (major axis length)	$S_4 =$ bottom (I_{r4})
5	$1 - I_d$	Between S_1 and S_4	$LPF_5 = $ 2-D Gaussian, horizontal [10 × 10]	Pixel values > Otsu's threshold estimated on I_5	Arg max (area)	$S_3 =$ bottom (I_{r5})
6	I_d	Between S_1 and S_3	$LPF_6 =$ average [25 × 25]	Pixel values > Otsu's threshold estimated on I_6	Arg max (area)	$S_2 =$ bottom (I_{r6})

Note: These parameters are used in Equations 4.11 through 4.13 to obtain seven sub-retinal surfaces in six iterations

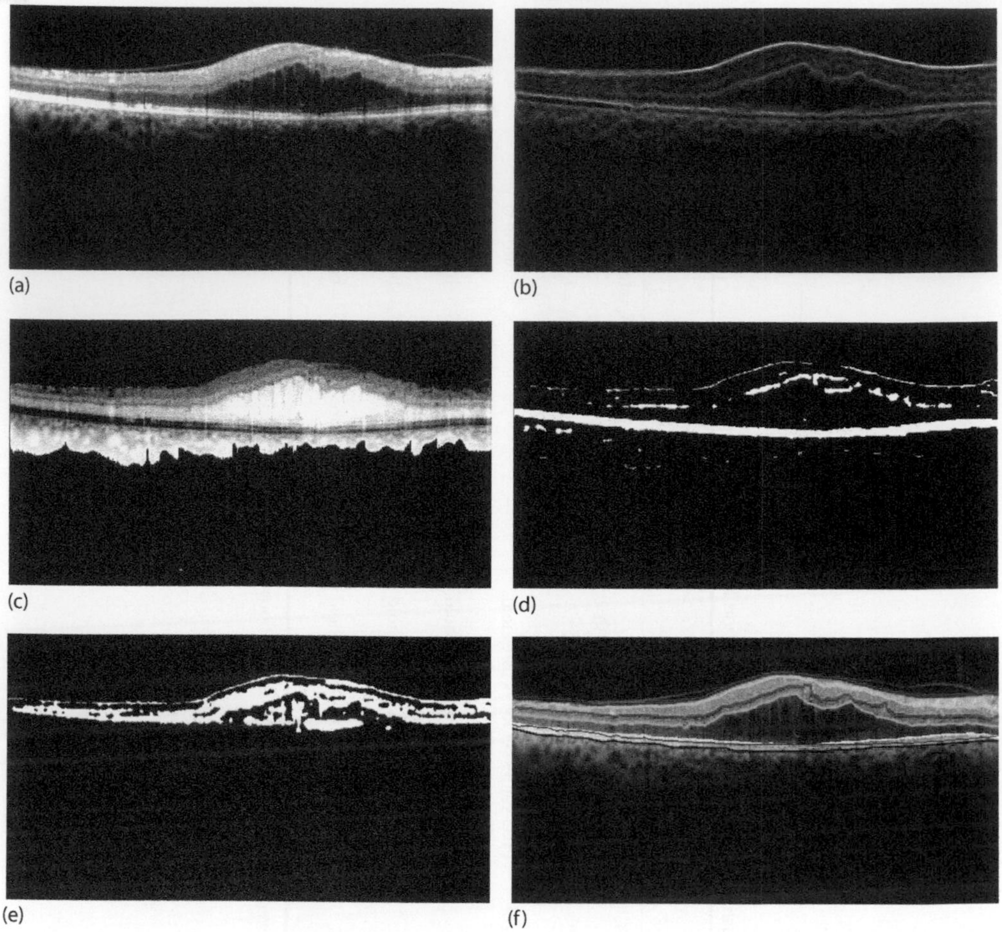

FIGURE 4.5 Examples of results at each step of the proposed iterative multi-resolution seg-mentation algorithm on an abnormal OCT image.

We observe that due to the multi-resolution nature of the proposed segmentation procedure, the inner sub-retinal surfaces (S_2 to S_4) are correctly detected in spite of the sub-retinal disorganization that occurs due to the presence of large cystoid regions. It is noteworthy that in certain OCT images, like the ones with the fovea as shown in Figure 4.1, while the Surface 1 is always continuous, other surfaces, such as Surfaces 3 or 2, may appear discontinuous. To enable the detection of complete surfaces in such cases, an exception rule is applied to detect all surfaces, apart from Surface 1, that have chances of appearing discontinuous. For this exception rule, the Surface 1 from iteration $l = 1$ is first detected and all its x-coordinates are noted. Next, from iteration two through six, for each extracted surface, if the length of the surface is less than 75% of the length of Surface 1, then another region satisfying the criterion (ϕ_l) apart from the region already selected is added to image I_{r1}. The final detected surface becomes the combination of surfaces from all the regions thus detected in image I_{r1}.

4.3 Experiments and Results

The performance characteristics of automated segmentation of sub-retinal surfaces and sub-retinal layers are evaluated in three sets of experiments. In the first experiment, the performance of image denoising is analyzed for normal and abnormal OCT image stacks. In the second experiment, the error in segmenting the sub-retinal surfaces is analyzed for normal and abnormal images. In the third experiment, the mean sub-retinal layer thickness estimates obtained using automated algorithms are compared against the manual segmentation estimates. The impact of denoising on automated segmentation is evaluated by comparing the segmentation performance of the proposed Fourier-domain-based denoising approach with the CDWT approach followed by the proposed multi-resolution iterative segmentation algorithm for sub-retinal surface and layer segmentation.

4.3.1 Performance of Automated Denoising

For this experiment, the denoised OCT images are scaled in [0, 255] and then analyzed for pixel intensities. The performance metrics for evaluating the improvement in image quality by automated denoising are defined in terms of the image foreground region and background region. For an OCT image, the region of interest is the retinal microstructure that lies between the ILM (Surface 1) and BM (Surface 7). The foreground region extends from 10 pixels above the manually segmented Surface 1 through 50 pixels below manually segmented Surface 7 to include the retinal microstructure and the choroidal region. The remaining regions in each image are selected as the background. Next, μ_f, σ_f and μ_b, σ_b are estimated as mean and standard deviation of pixel values in the foreground and background regions, respectively. The spatial denoising metrics are defined as global signal-to-noise ratio (SNR) in Equation 4.14, global contrast to noise ratio (CNR) in Equation 4.15 and peak SNR (PSNR) in Equation 4.16.

$$SNR = 20\log_{10}\frac{\mu_f}{\sigma_b}. \tag{4.14}$$

$$CNR = \frac{|\mu_f - \mu_b|}{\sqrt{0.5(\sigma_f^2 + \sigma_b^2)}}. \tag{4.15}$$

$$\forall[i',j'] \in I, I_d,$$

$$PSNR = 20\log_{10}\left(\frac{\max_{i',j'} I(i',j')}{\sqrt{\frac{1}{n_1.n_2}\sum_{i'=1}^{n_1}\sum_{j'=1}^{n_2}\left(I(i',j') - I_d(i',j')\right)^2}}\right) \tag{4.16}$$

The foreground and background regions for an abnormal OCT image and the denoising performance metrics using the proposed Fourier-domain error-based method and the wavelet-based CDWT approach are shown in Figure 4.6. Here, we observe that the proposed denoising method retains the surface edges and contrast variation in the foreground region, while it significantly reduces image noise from the background region, when compared to the denoised image using CDWT.

The comparison between denoising metrics using the proposed method and the CDWT method are shown in Table 4.2. Here, we observe that for OCT images from the normal and abnormal datasets, the proposed denoising method improves the image SNR by 12–13 dB, while the CDWT method improves the SNR by 3–4 dB only. The improvements in CNR fraction by the proposed and CDWT method are about 0.5 and 0.1, respectively. However, the CDWT method achieves PSNR about 4.5 dB greater than the proposed method. The PSNR of the proposed method is smaller than the CDWT method because of the denominator in Equation 4.16 that denotes the similarity in structure between the denoised and noisy image, and a more similar structure will ensure a small value for this quantity. The proposed method alters the structure of the noisy image in the background region due to the underlying Fourier-domain error-based Wiener deconvolution strategy, while the CDWT method conserves the image

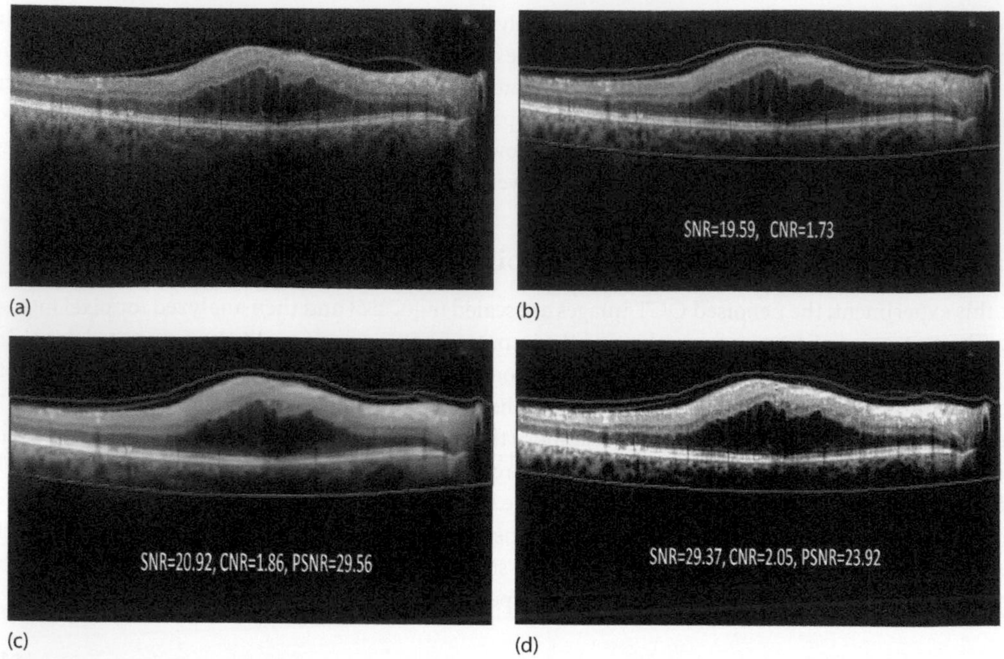

FIGURE 4.6 Example of automated OCT image denoising. (a) Noisy image (I), (b) the foreground region lies within the region bordered by the red boundary. All other regions are the background. (c) Denoised image by CDWT method. (d) Denoised image by the proposed method.

structure, and hence the difference in PSNR. The impact of the gain in SNR by the proposed denoising method versus the gain in PSNR by the CDWT method on automated sub-retinal surface segmentations are analyzed in the next experiments.

4.3.2 Sub-Retinal Surface Segmentation Error

The absolute errors incurred in segmenting each sub-retinal surface using the proposed de-noising method or CDWT method followed by the proposed segmentation method on the normal and abnormal OCT images are shown in Tables 4.3 and 4.4, respectively. The mean and standard deviation in the absolute errors between the automatically segmented surface and the manually segmented surfaces for all OCT images under analysis are evaluated in μm. In Table 4.3, the segmentation errors incurred by

TABLE 4.2 Performance of OCT Image Denoising Using the Proposed Method versus the CDWT Denoising Method Evaluated on Normal and Abnormal OCT Image Stacks

Image	SNR (dB)	CNR	PSNR (dB)
	Normal set		
Noisy image	17.52 ± 1.77	2.11 ± 0.51	—
CDWT denoising	21.03 ± 1.87	2.20 ± 0.54	27.91 ± 1.39
Proposed denoising	30.17 ± 5.19	2.58 ± 0.53	23.56 ± 0.51
	Abnormal set		
Noisy image	16.92 ± 1.74	1.79 ± 0.105	—
CDWT denoising	19.32 ± 1.16	1.91 ± 0.08	28.08 ± 1.84
Proposed denoising	29.23 ± 2.23	2.32 ± 0.11	23.67 ± 0.38

TABLE 4.3 Mean and Standard Deviation of Sub-Retinal Surface Segmentation Error Using the Proposed Method, and the CDWT Method Compared to the Performance of Existing Methods on Normal OCT Images

Surface	Proposed	CDWT	Bagci et al. 14	Chiu et al. [18]	Yang et al. [25]
S1	1.22 ± 0.96	8.45 ± 3.92	4.3 ± 0.8	2.8 ± 2.3	2.2 ± 0.7
S2	5.80 ± 4.47	14.77 ± 4.34	5.7 ± 0.7	2.5 ± 2.0	4.3 ± 1.0
S3	2.78 ± 1.53	13.38 ± 6.14	5.3 ± 0.5	3.2 ± 2.4	3.2 ± 0.9
S4	4.55 ± 2.15	9.55 ± 3.89	6.1 ± 1.0	4.8 ± 3.4	—
S5	1.71 ± 1.38	4.27 ± 2.17	8.8 ± 1.2	3.8 ± 2.9	—
S6	2.51 ± 2.35	12.90 ± 3.13	4.3 ± 1.1	2.8 ± 2.4	—
S7	2.05 ± 2.17	12.58 ± 5.01	5.5 ± 1.0	3.2 ± 2.8	2.4 ± 1.2

Note: These Errors are computed in μm

TABLE 4.4 Mean and Standard Deviation of Sub-Retinal Surface Segmentation Error Using the Proposed Method, and the CDWT Denoising Method on Abnormal OCT Images

Surface	Proposed	CDWT
S1	4.02 ± 4.82	10.97 ± 5.64
S2	26.99 ± 10.04	22.57 ± 5.69
S3	19.97 ± 7.49	23.64 ± 12.18
S4	16.83 ± 8.77	20.95 ± 11.59
S5	6.81 ± 6.42	8.02 ± 2.29
S7	3.51 ± 2.86	7.05 ± 1.78

Note: These errors are computed in μm

three existing segmentation algorithms on the normal images are also shown for comparison. The absolute surface segmentation error was estimated by Bagci et al. [14] on 14 SD-OCT images, by Chiu et al. [18] on 100 B-scans, and by Yang et al. [25] on 38 image scans, respectively.

In Table 4.3, we observe that on normal OCT images, the proposed denoising and segmentation method incurs 1–6 μm of error across all the sub-retinal surfaces. This is significantly better than the CDWT denoising method followed by proposed segmentation that results in 4–14 μm of error. Also, the proposed method has lower segmentation error when compared to other segmentation algorithms on all surfaces other than Surface 2, which is the least distinctive edge. Thus, on normal OCT images, the proposed method has relatively better sub-retinal surface segmentation performance when compared to CDWT or existing algorithms.

For images from the abnormal set, the surface segmentation errors by the proposed method and CDWT method are shown in Table 4.4. Since pathological changes due to DME do not affect the IS/OS layer (Surface 6), this sub-retinal surface was not manually segmented. For all the other sub-retinal surfaces, we observe a significant increase in the segmentation error for Surfaces 2, 3, 4 and 5 for the abnormal OCT images when compared to the normal images. This increase in segmentation error is unavoidable since the disorganization caused by pathology significantly alters the inner retinal layers, making the inner surfaces significantly less distinct, even for manual segmentation. In Table 4.4, we also observe that for all surfaces other than Surface 2, the CDWT method incurs more error on abnormal images than the proposed approach. This happens because for abnormal images with large cysts, the Surface 2 may appear very close to Surface 1, in which case false edges may get detected by the proposed denoising method, causing the segmentation error to increase. In such cases the CDWT approach smoothens the false edges close to the Surface 1, thereby resulting in less error for Surface 2, but in turn increasing the error for segmenting Surfaces 3 through 7 due to this smoothing action.

Since the error in Surface 2 is significantly high for abnormal OCT images using the proposed segmentation algorithm, the sub-retinal layer thicknesses on such abnormal images will be analyzed for the NFL and IPL layers combined (NFL + IPL, between Surfaces 1 and 3), the INL, ONL and Outer layer (between Surfaces 5 and 7). It is noteworthy that, although the performance of automated surface segmentation deteriorates on images with pathology, the proposed denoising and segmentation method has more reliable and repeatable performance for inner sub-retinal surface segmentation when compared to the OCTSEG system as shown in Figure 4.7. The proposed method is capable of reliable estimation of the INL and ONL in image with cysts and sub-retinal disorganizations.

4.3.3 Sub-Retinal Layer Thickness Estimation

Once the sub-retinal surfaces are segmented, the next step is to estimate average sub-retinal layer thicknesses. The mean and standard deviation in the sub-retinal layer thicknesses for normal and abnormal OCT images are shown in Tables 4.5 and 4.6, respectively. In these tables, the mean thicknesses of every sub-retinal layer segmented manually are compared with the automated sub-retinal layer thickness estimation results using the proposed method and the CDWT approach. The sub-retinal layer segmentation distribution is analyzed in terms of the average layer thickness, mean correlation coefficient "r" and R^2 statistic between the manual and automated thickness measurements made for each image at every horizontal pixel location. A higher value of "r" and R^2 statistic for each OCT image denotes better

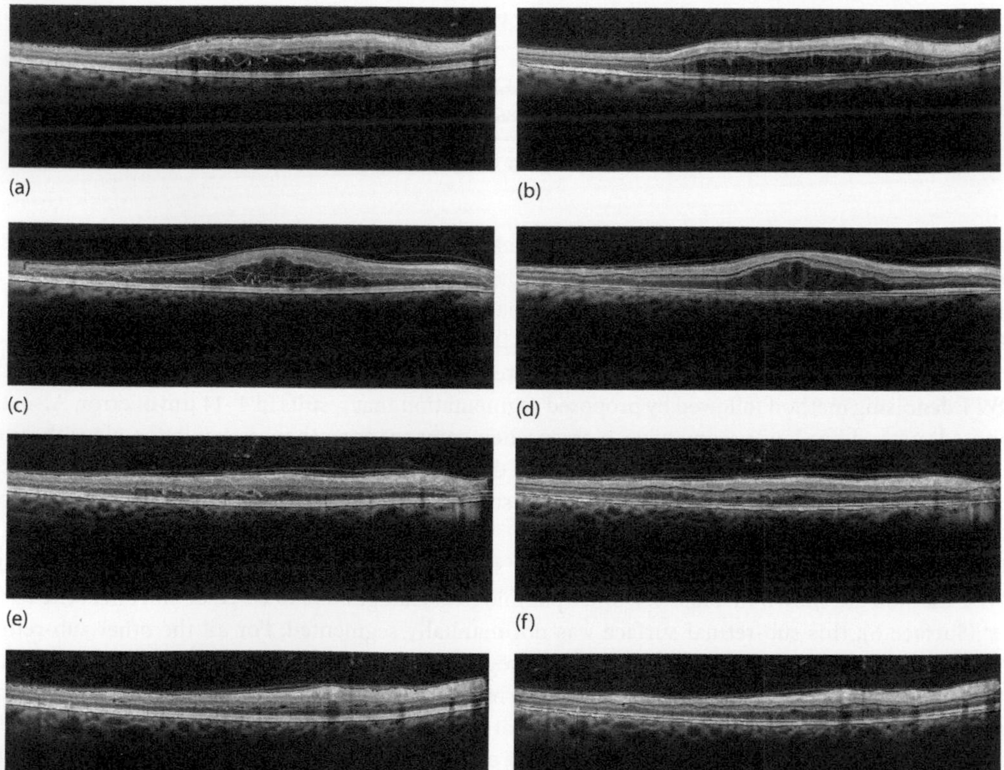

FIGURE 4.7 Comparative assessment of sub-retinal surface segmentation using the proposed method and the OCTSEG system on images with DME. Left column represents the automated segmentation produced by OCTSEG. Right column represents automated segmentation produced by the proposed method.

TABLE 4.5 Mean and Standard Deviation of Sub-Retinal Layer Thickness in Normal Images Measured in μm

Layer	Manual	Proposed (r, R^2)	CDWT (r, R^2)
NFL	52.4 ± 8.7	52.6 ± 11.5(0.86,0.7)	49.4 ± 12.2(0.62,0.4)
IPL/GL	72.7 ± 5.6	73.3 ± 9.3(0.79,0.6)	61.9 ± 15.6 (0.49,0.2)
INL	57.42 ± 7.4	59.9 ± 10.3(0.83,0.7)	72.2 ± 4.9(0.39,0.1)
ONL	74.97 ± 6.0	73.1 ± 6.8(0.83,0.7)	70.5 ± 11.0(0.39,0.1)
IS/OS	38.02 ± 10.5	38.2 ± 11.0(0.77,0.6)	20.3 ± 5.9(0.36,0.1)
RPE	35.01 ± 6.6	35.5 ± 10.0(0.75,0.6)	42.0 ± 8.3(0.33,0.1)
NFL+IPL	225.9 ± 9.1	226.9 ± 9.9(0.98,0.9)	226.4 ± 6.39(0.43,0.2)
Inner	244.6 ± 10.8	245.8 ± 11.1(0.98,0.9)	250.3 ± 8.9(0.32,0.1)
Outer	64.5 ± 7.1	65.1 ± 8.64(0.82,0.7)	53.70 ± 6.3(0.39,0.1)

TABLE 4.6 Mean and Standard Deviation of Sub-Retinal Layer Thickness in Abnormal Images Measured in μm

Layer	Manual	Proposed(r,R^2)	CDWT(r,R^2)
NFL+IPL	111.5 ± 15.9	129.4 ± 19.7(0.78,0.6)	135.6 ± 29.7(0.75,0.6)
INL	62.7 ± 10.6	60.9 ± 15.0(0.73,0.5)	76.9 ± 14.5(0.36,0.1)
ONL	89.9 ± 15.3	81.0 ± 13.9(0.71,0.5)	70.9 ± 11.4(0.23,0.1)
Inner	265.6 ± 33.1	274.9 ± 34.6(0.92,0.8)	274.0 ± 61.3(0.90,0.8)
Outer	60.9±8.44	57.4 ± 14.4(0.74,0.6)	49.7 ± 10.6(0.32,0.1)

sub-retinal layer segmentation performance. Table 4.5, we observe that for OCT images from the normal set, the proposed method incurs up to 2 μ*m* error in estimating the mean sub-retinal layer thickness with r, R^2 consistently greater than 0.75 and 0.6, respectively. However, the CDWT method incurs 3–18 μm error in estimating the sub-retinal layer thickness with significantly low r, R^2 metrics. The p-values for the proposed denoising segmentation system and the CDWT denoising followed by the proposed segmentation system are $p < 0.001$ and $p < 0.01$, respectively. This analysis shows that the errors between the proposed denoising approach and CDWT approach for mean sub-retinal layer thicknesses and distributions of sub-retinal layer thickness estimations are significantly higher than the errors in segmenting sub-retinal surfaces. In Table 4.6, we observe that the proposed method incurs 2–18 μm error in estimating the mean sub-retinal layer thickness with r, R^2 statistics consistently greater than 0.7 and 0.5, respectively. On the other hand, the CDWT method incurs 12–24 μm error for mean thickness estimation with (r, R^2) statistics ranging from (0.3, 0.1) to (0.9, 0.8).

Conclusion and Discussion

In this work we have presented a CAD system that denoises and segments seven sub-retinal surfaces and six sub-retinal layers in the retinal microstructure for normal OCT images from healthy patients and in abnormal images from patients with DME in less than 35 s per image. The proposed system is implemented on a 2.53 GHz Intel Core i3 and 3 GB RAM Laptop system using MATLAB®. Also, the sub-retinal layer thickness distributions from all images belonging to an OCT image stack are combined to generate thickness maps that aid automated assessment of the progression in DME with time.

We analyze the performance of OCT image noise removal based on the proposed Wiener deconvolution method that estimates noise parameters using a Fourier-domain structural error metric. The proposed denoising method is compared to wavelet-transform-based CDWT method by Chitchian et al. [10]. Our analysis shows that the proposed Fourier-domain-based denoising method improves the image SNR by more than 12 dB, and it retains the sub-retinal surface edges within the retinal microstructure while suppressing the noise significantly in the image background, thereby incurring a PSNR of about 23.5 dB. The CDWT method on the other hand conserves the sub-retinal structure of the image foreground and background by smoothing the sub-retinal edges, thereby achieving a high PSNR of about 28 dB but an SNR enhancement of less than 5 dB. Further analysis shows that the CDWT denoising method incurs more error while automated segmentation of sub-retinal surfaces, when compared to the proposed denoising method. Thus, the proposed Fourier-domain-based denoising method is an important first step for the proposed sub-retinal surface and layer segmentation algorithm.

Another key observation is that the segmentation performance of the proposed denoising and segmentation algorithm deteriorates significantly from normal OCT images to abnormal ones, that is errors in mean sub-retinal thickness estimations are 0–2 μm for normal images and 2–18 μm for abnormal images. However, the CDWT denoising method followed by the proposed segmentation significantly deteriorate the estimations of the distributions of sub-retinal thicknesses (low "r" value), but the estimations of mean sub-retinal layer thicknesses do not deteriorate as much, that is errors in mean sub-retinal thickness estimations are 3–18 μm for normal images and 12–24 μm for abnormal images. These observations demonstrate that the proposed segmentation algorithm is robust and can be combined with other denoising approaches to estimate mean sub-retinal layer thickness in normal and abnormal images.

The proposed sub-retinal surface segmentation algorithm exhibits significant reliability (in terms of correlation coefficient (r) and R^2 metric) in extracting sub-retinal layer thicknesses for normal and abnormal OCT image stacks. Such a CAD system can be useful for detecting and monitoring the progression of pathology in DME patients, and in guiding clinical research directed toward treatment protocols. Future efforts will be directed toward developing CAD systems for automated segmentation of OCT image stacks from patients with other vision threatening pathologies that cause variations in sub-retinal layer thicknesses such as Glaucoma and Macular Telangiectasia (MacTel).

References

1. W. H. Organization, Vision 2020 action plan for 2006 to 2011 planning meeting. Geneva. 11 July 2006.
2. S. Roychowdhury, D. Koozekanani, S. Radwan, and K. Parhi, Automated localization of cysts in diabetic macular edema using optical coherence tomography images, In *35th Annual International Conference of the IEEE Engineering in Medicine and Biology Society (EMBC)*, Osaka, Japan. July 2013, 1426–1429.
3. G. Wilkins, O. Houghton, and A. Oldenburg, Automated segmentation of intrareti-nal cystoid fluid in optical coherence tomography, *IEEE Transactions on Biomedical Engineering*, 59, 4, 1109–1114, 2012.
4. C. Gandorfer, A. Haritoglou, A. Gandorfer, and A. Kampik, Retinal damage from indocyanine green in experimental macular surgery, *Investigative Ophthalmology and Visual Science*, 44, 1, 316–323, 2003.
5. A. Maalej, W. Cheima, K. Asma, R. Riadh, and G. Salem, Optical coherence to-mography for diabetic macular edema: Early diagnosis, classification and quantitative assessment, *Journal of Clinical and Experimental Ophthalmology*, 2, S2-004, 1:6, 2012.
6. T. Otani, S. Kishi, and Y. Maruyama, Patterns of diabetic macular edema with optical coherence tomography, *American Journal of Ophthalmology*, 127, 6, 688–693, 1999.
7. R. Lattanzio, Macular thickness measured by optical coherence tomography (oct) in diabetic patients, *European Journal of Ophthalmology*, 12, 6, 482–487, 2002.
8. M. A. Mayer, J. Hornegger, C. Y. Mardin, and R. P. Tornow, Retinal nerve fiber layer segmentation on fd-oct scans of normal subjects and glaucoma patients, *Biomedical Optics Express*, 1, 5, 1358–1383, 2010.

9. D. C. Adler, T. H. Ko, and J. G. Fujimoto, Speckle reduction in optical coherence tomography images by use of a spatially adaptive wavelet filter, *Optics Letters*, 29, 24, 2878–2880, 2004.
10. S. Chitchian, M. A. Mayer, A. R. Boretsky, F. J. van Kuijk, and M. Motamedi, Retinal optical coherence tomography image enhancement via shrinkage denoising using double-density dual-tree complex wavelet transform, *Journal of Biomedical Optics*, 17, 11, 116 004–116 009, 2012.
11. L. Fang, S. Li, Q. Nie, J. A. Izatt, C. A. Toth, and S. Farsiu, Sparsity based denoising of spectral domain optical coherence tomography images, *Biomedical Optics Express*, 3, 5, 927–942, 2012.
12. D. Koozekanani, K. L. Boyer, and C. Roberts, Retinal thickness measurements from optical coherence tomography using a markov boundary model, *Transactions on Medical Imaging*, 20, 9, 900–916, 2001.
13. H. Ishikawa, D. M. Stein, G. Wollstein, S. Beaton, J. G. Fujimoto, and J. S. Schuman, Macular segmentation with optical coherence tomography, *Investigative Ophthalmology and Visual Science*, 46, 6, 2012–2017, 2005.
14. M. Bagci, M. P. N. Shahidi, R. Ansari, M. Blair, N. P. Blair, and R. Zelkha, Thickness profiles of retinal layers by optical coherence tomography image segmentation, *American Journal of Ophthalmology*, 146, 5, 679–687, 2008.
15. A. Mishra, A. Wong, K. Bizheva, and D. A. Clausi, Intra-retinal layer segmentation in optical coherence tomography images, *Optics Express*, 17, 26, 23 719–23 728, 2009.
16. M. Garvin, M. Abramoff, R. Kardon, S. Russell, X. Wu, and M. Sonka, Intraretinal layer segmentation of macular optical coherence tomography images using optimal 3-d graph search, *IEEE Transactions on Medical Imaging*, 27, 10, 1495–1505, 2008.
17. J. Antony, M. D. Abramoff, M. M. Harper, W. Jeong, E. H. Sohn, Y. H. Kwon, R. Kardon, and M. K. Garvin, A combined machine-learning and graph-based framework for the segmentation of retinal surfaces in sd-oct volumes, *Biomedical Optics Express*, 4, 12, 2712–2728, 2013.
18. S. J. Chiu, X. T. Li, P. Nicholas, C. A. Toth, J. A. Izatt, and S. Farsiu, Automatic segmentation of seven retinal layers in sdoct images congruent with expert manual segmentation, *Optics Express*, 18, 18, pp. 19 413–19 428, 2010.
19. J. Y. Lee, S. J. Chiu, P. P. Srinivasan, J. A. Izatt, C. A. Toth, S. Farsiu, and G. J. Jaffe, Fully automatic software for retinal thickness in eyes with diabetic macular edema from images acquired by cirrus and spectralis systems, *Investigative Ophthalmology and Visual Science*, 54, 12, 7595–7602, 2013.
20. A. Abhishek, T. Berendschot, S. Rao, and S. Dabir, Segmentation and analysis of retinal layers (ilm amp; rpe) in optical coherence tomography images with edema,In *IEEE Conference on Biomedical Engineering and Sciences (IECBES)*, Kuala Lumpur, Malaysia. 2014, 204–209.
21. F. Shi, X. Chen, H. Zhao, W. Zhu, D. Xiang, E. Gao, M. Sonka, and H. Chen, Auto-mated 3-d retinal layer segmentation of macular optical coherence tomography images with serous pigment epithelial detachments, *IEEE Transactions on Medical Imaging*, 34, 2, 441–452, 2015.
22. S. Roychowdhury, D. D. Koozekanani, and K. K. Parhi, Automated denoising and segmentation of optical coherence tomography images, In *Asilomar Conference on Signals, Systems and Computers*, Pacific Grove, CA, USA. November 2013, 258–262.
23. Y. Liu, Y. Liang, G. Mu, and X. Zhu, Deconvolution methods for image deblurring in optical coherence tomography, *Journal of the Optica* Society *of America*, 26, 1, 72–77, 2009.
24. C. Wagner and I. Selesnick, Double density wavelet software. [Online]. Available: http://taco.poly.edu/selesi/DoubleSoftware/
25. Q. Yang, C. A. Reisman, Z. Wang, et al., Automated layer segmentation of macular OCT images using dual-scale gradient information, *Opics Express*, 18, 20, 21293–21307, 2010.

5

Computer-Aided Diagnosis with Dental Images

Chisako Muramatsu

Takeshi Hara

Tatsuro Hayashi

Akitoshi Katsumata

Hiroshi Fujita

5.1 Introduction

Early studies on computerized analysis of dental images utilized intraoral radiographs. The subjects of such analysis include temporal subtraction, caries detection, periapical disease detection and classification, and treatment follow-up evaluation. Some basic techniques such as apex localization and tooth segmentation are also investigated for the purposes of preprocessing for lesion detection and forensic identification. With the widespread use of digital dental panoramic radiographs (DPRs), a number of studies on computerized analysis of DPRs have been reported. While image quality of DPRs is generally inferior to intraoral radiographs, their advantage is a broad field of view enabling the examination of the whole dental region as well as the upper and lower jaw. The coverage of extra-dental regions broadened the subjects of computerized analysis from dental diseases to systemic diseases.

In this chapter, computerized analysis of intraoral radiographs is briefly introduced. However, the main topics are the methodologies of quantitative analysis and computerized disease diagnosis with DPRs. Computer-aided diagnosis (CAD) in which medical image interpreters, in this case oral radiologists and dentists, make diagnosis using the results of computerized analysis can improve the accurate and efficient diagnosis of oral diseases. Moreover, they can explain to patients the possible risk and suggest more detailed examinations for systemic diseases based on the findings on DPRs. Computerized analysis of dental computed tomography (CT) is not discussed in this chapter. Readers should refer to other publications for this topic [1–4].

5.2 CAD with Intraoral Radiographs

There are three types of intraoral radiographic techniques commonly used in dental practice [5]. Figure 5.1 shows three types of images. A periapical image is obtained to examine an entire tooth and its surrounding structures. A bitewing image shows crowns of both upper and lower teeth, but not apices. An occlusal image is used to examine the maxilla and mandible by providing an extensive view.

Bitewing images are frequently examined for detection of caries. A commercial CAD system for caries detection, called Logicon system, is available from Carestream Dental LLC. The software provides the likelihood of a selected tooth being carious using a neural network trained with histologically known image datasets [6, 7]. Like a mammography CAD system, the level of the acceptable number of false positives (FPs) can be adjusted to clinicians' preference [8, 9].

Image subtraction techniques can be used for detection of temporal differences in a pair of images, such as the appearance of a new lesion or effect of treatment [10–12]. For intraoral radiographs, standardization and reproducibility of projection geometries of two radiographs may be more difficult than for other common applications of subtraction radiography, such as chest radiography. Image reconstruction techniques for manipulating an image with arbitrary projection geometry to the projection geometry of a reference image have been proposed [13, 14]. In addition, methods for correction of gray-level distribution have also been investigated [15].

Van der Stelt et al. proposed a rather simple image processing technique to segment bone and tooth regions for computerized analysis of bone lesions [16]. The method includes three steps: image encoding, which creates the density map; preprocessing for noise reduction and contrast enhancement; and segmentation by finding the largest contrast transition. In their later study, Mol and van der Stelt have investigated a computer-aided image interpretation method for diagnosis of periapical bone lesions [17]. The method starts with a manual identification of an arbitrary point on the root. The system automatically analyzes the root, and its apical end is located using a cubic polynomial. Subsequently, texture analysis is performed to identify a disturbance of trabecular pattern due to a periapical bone lesion. When a lesion is found, the final component is the edge detection for estimation of its size.

Carmody et al. proposed an image classification method using machine learning and visual search information [18]. Periapical images of four periapical disease states, including no disease (normal tooth), mild disease (widened periodontal ligament space), moderate disease (destruction of lamina dura), and severe disease (resorption of bone in the periapical area), were included. The images were reviewed by observers with different experience and their eye motions were recorded. A classifier was trained

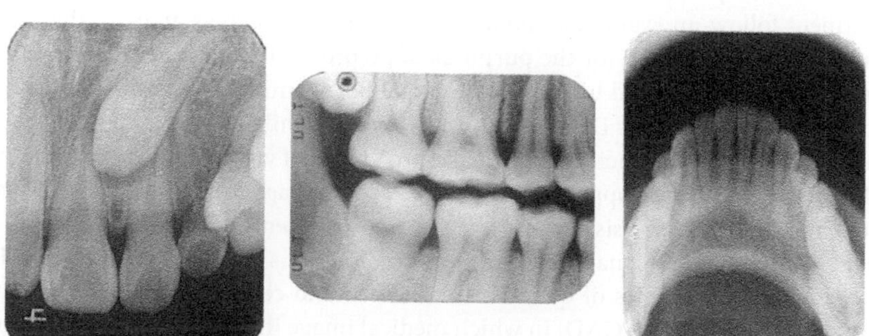

(a) (b) (c)

FIGURE 5.1 Three types of intraoral radiographs. (a) Periapical image, (b) bitewing image, and (c) occlusal image.

with regions of interest (ROIs) sampled at the observers' fixations. The classification performance of the machine was superior to those of observers and of the machine trained with random samples.

Sutthiprapaporn et al. studied a CAD system for evaluation of alveolar bone grafting [19]. Occlusal images of pre-operation and different time periods after bone grafting were analyzed. The morphological skeletonization process was applied to compare the texture of non-cleft side and cleft side of the same images. They found that the difference in trabecular patterns on the non-cleft and cleft sides was decreased at 2–7 months post-operation. Dental radiographs with such a system may be a potential alternative to CT images for evaluating grafting failure and for determination of the appropriate period for re-grafting.

Other research groups have investigated some basic techniques for localization and segmentation of teeth as potential preprocessing steps for CAD systems. Wu et al. proposed an automated method for detection of apical roots on periapical images using Haar-like features with the AdaBoost classification algorithm [20]. Huang et al. studied a tooth isolation method on bitewing images as a preprocessing step for dental diagnosis and identification systems [21]. Using horizontal integral projection, the upper and lower jaws were first separated, and vertical separations were found by vertical integral projection with adaptive windowing. Li et al. proposed a tooth segmentation method using the variational level set [22]. Using the patches classified by a support vector machine as an initial contour, coupled level set segmentation was performed, and the result was used for an uncertainty map to indicate possible areas of bone loss and decay.

Jain et al. investigated a dental image matching method for human identification [23]. Each tooth was isolated by integral projection, and the contours of the crown and root parts were extracted separately. First, a radial scan from the center of the crown was performed to separate the inner (tooth) and outer (background) pixels by probability maximization. The root boundary was searched from both ends of the crown on the basis of the location and pixel intensity. Forensic identification was performed by matching the contour shape with a rigid transformation.

5.3 Dental Panoramic Radiographs (DPRs)

DPRs are the most widespread examination in dental clinics in developed countries. More than 17 and 10 million DPRs are taken annually in the United States and Japan, respectively [24, 25]. As shown in Figure 5.2, DPR is obtained by a pair of x-ray tube and detector rotating around an examinee. A DPR has a wide coverage depicting not only the upper and lower teeth, but also other structures around the

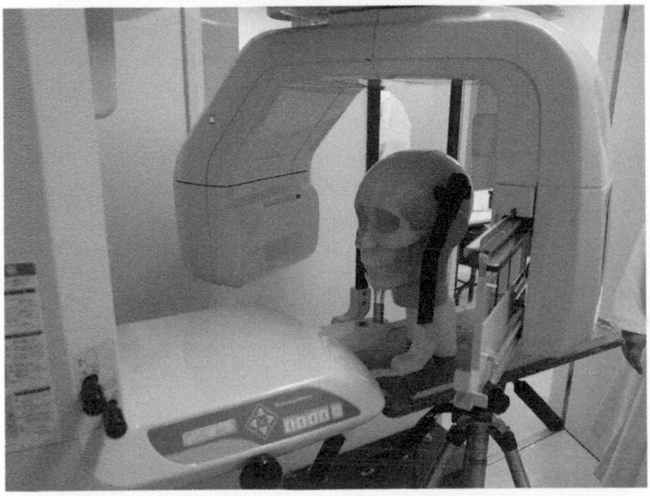

FIGURE 5.2 Dental panoramic radiography examination.

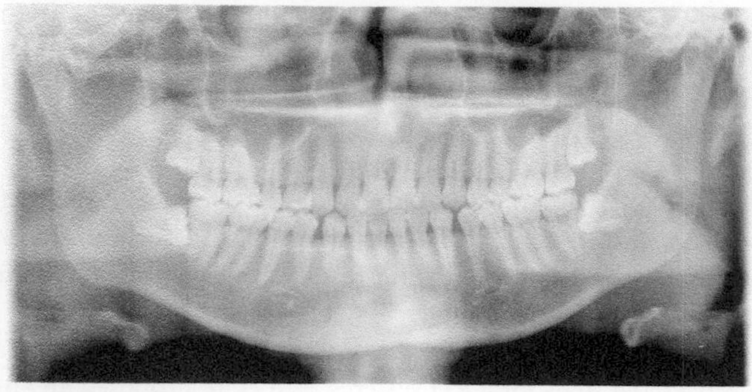

FIGURE 5.3 Dental panoramic radiograph.

mouth and neck, such as the maxilla, mandible, temporomandibular joint, maxillary sinus, and cervical vertebrae, as shown in Figure 5.3. A DPR is a kind of tomography; thus, the image is blurry compared to intraoral radiographs and has a small magnification effect. However, it allows the examination of the whole dental region, as well as the extra-dental regions, which have been attracting the attention of researchers in recent years.

5.4 Measurement of Mandibular Cortical Width for Osteoporosis Screening

Osteoporosis, which can be represented by low bone mass and structural deterioration of the bone tissue, is considered one of the major public health problems. In the aging world, an increase in its incidence rate is expected to continue, as its disease rate increases with age [26]. A report estimated that 200 million or more women are affected by osteoporosis worldwide [27]. Osteoporosis is generally asymptomatic, and bone strength is gradually weakened unnoticeably. It can increase the susceptibility to fractures in the hip and vertebra by a small force. There were an estimated nine million osteoporotic fractures worldwide in 2000, a part of which could be related to a cause of death and disability of patients [28, 29]. Early diagnosis is expected to reduce the number of severe fracture and, consequently, the medical cost; however, the screening rate is not very high due to the asymptomatic nature. The establishment of an effective framework for screening is an urgent issue.

As mentioned earlier, a DPR depicts an entire mandibular bone. Studies have reported that mandibular cortical width (MCW) on the DPR is significantly correlated with bone mineral density (BMD) in the hip, lumbar spine, and forearm [30–32]. Figure 5.4 presents DPRs with enlarged views of the cortical bone of osteoporotic and non-osteoporotic patients. The measurement of MCW for osteoporotic risk assessment during general dental examinations may be a benefit to patients. However, DPRs are generally used only for diagnosis of dental conditions, and dentists are not paying attention to the mandible. In addition, manual MCW measurement would be laborious because of a large number of exams. Therefore, computerized measurement can assist efficient and effective secondary screening.

5.4.1 Related Works

Several research groups have investigated computerized methods for measurement of MCW. Arifin et al. proposed a CAD system for MCW measurement by identifying the cortical margin on the basis of the gradient [33]. Using two manually identified points, the inner and outer margins were traced according to the perpendicularity of the gradient obtained from linear regression at every point along the line and the gradient measured between the mental foramens to the appropriate point. In their subsequent study, the cortical margin was automatically determined using various image processing techniques

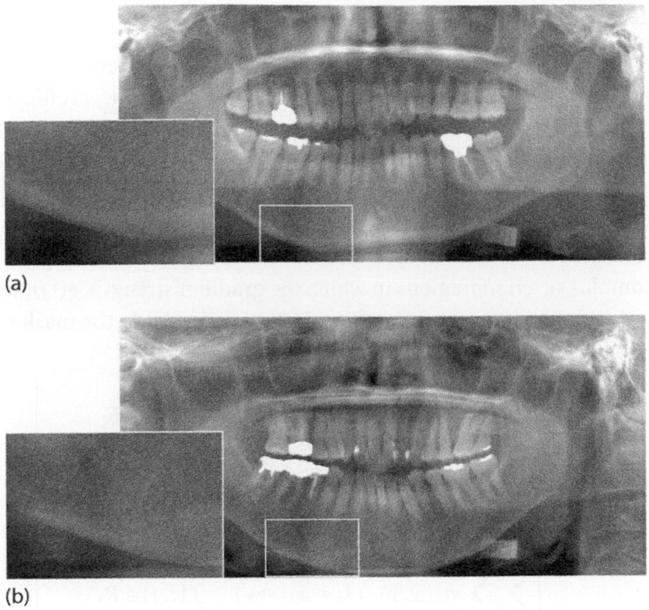

FIGURE 5.4 Thinning of mandibular cortical bone observed on dental panoramic radiographs. (a) Osteoporotic subject and (b) non-osteoporotic subject.

after manual extraction of ROIs [34]. They first applied histogram equalization to enhance cortical bone followed by a thresholding method [35] to remove background. A high-pass filter was then applied to enhance the inner cortical margin. For delineation of the inner margin, an eight-neighborhood distance function was employed, whose output pixel values represent the distance from the object's margin. The pixels with the maximum values represent the axis of the cortical bone and were traced by dynamic programming. Finally, the margins were obtained with a rolling disc. The MCW was measured by fitting the boundary with a second-order polynomial function. In a recent study, they proposed a new scheme to improve the measurement in images with noise [36]. A clustering algorithm was applied to the MCW histogram with automatic selection of the optimal number of clusters. By maximizing the ratio of elements of the largest and second largest clusters, erroneous MCW values were removed. Classification between osteoporotic and normal subjects was performed using a support vector machine (SVM) with moment-based features of the MCW histogram.

Allen et al. proposed a computerized method for MCW measurement without and with manual intervention in detecting the upper and lower cortical margins [37]. The method uses an active shape model (ASM) [38], which was trained with landmark points manually placed on cortical borders. In a semiautomated mode, four reference points on the lower mandible edge at the right and left mental foramina and the antegonion were provided by an expert. They reported that detection of the lower edge was relatively successful in both modes; however, lateral misalignment occurred in the unconstrained-fit (automatic) mode along the edge of the mandible because of the absence of strong features at the mental foramen and antegonion landmarks. This result may cause the system to obtain the measurement at different locations on the mandibular bone. In their subsequent studies, Roberts et al. [39, 40] proposed a hybrid method using the ASM and an active appearance model (AAM) [41]. By this method, an inferior border of the mandible was first detected by using the ASM. Subsequently, the AAM was applied on the left and right sides separately for searching the lateral position and the superior border. They also proposed measures for assessing failure of the automatic search, for which the conventional semiautomatic mode with manual four-point initialization would be applied.

5.4.2 Methodology

We have proposed an automated scheme for MCW measurement by detecting the mandibular contour using the similar contour model and determining the inner border using profile analysis [42–44]. The overall scheme is outlined in Figure 5.5. The method starts with the automated segmentation of lower mandibular contour [42]. Potential edges corresponding to the mandibular contour were detected using the Canny edge detector [45]. Because mandibles are generally depicted in particular shapes on DPRs, anticipated edges with specific directions and locations were detected using a mandibular mask, as in Kirsch's method [46]. Figure 5.6(a) shows the mask created by overlapping the 100-sample DPRs with manually defined contours. It contains seven subregions in which the gradient strength, $g(i,j)$, was determined using the following equations and the filter kernels for a pixel of interest, $f(i,j)$ in the mask region R_n.

$$g(i,j)=\begin{cases} \sum_{x=-1}^{1}\sum_{y=-1}^{1} h_1(x,y)\cdot f(i+x,j+y), & f(i,j)\in R_1 \\ \sum_{x=-1}^{1}\sum_{y=-1}^{1} h_2(x,y)\cdot f(i+x,j+y), & f(i,j)\in R_3 \\ \sum_{x=-1}^{1}\sum_{y=-1}^{1} h_3(x,y)\cdot f(i+x,j+y), & f(i,j)\in R_4 \\ \sum_{x=-1}^{1}\sum_{y=-1}^{1} h_4(x,y)\cdot f(i+x,j+y), & f(i,j)\in R_5 \\ \sum_{x=-1}^{1}\sum_{y=-1}^{1} h_5(x,y)\cdot f(i+x,j+y), & f(i,j)\in R_7 \\ \max_{m=1,2}\sum_{x=-1}^{1}\sum_{y=-1}^{1} h_m(x,y)\cdot f(i+x,j+y), & f(i,j)\in R_2 \\ \max_{m=4,5}\sum_{x=-1}^{1}\sum_{y=-1}^{1} h_m(x,y)\cdot f(i+x,j+y), & f(i,j)\in R_6 \end{cases} \tag{5.1}$$

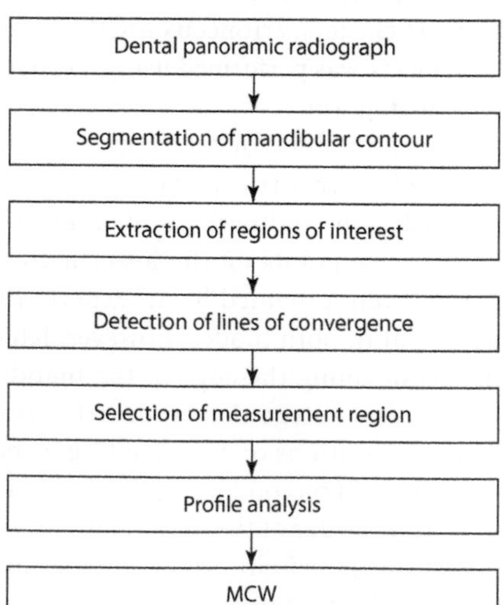

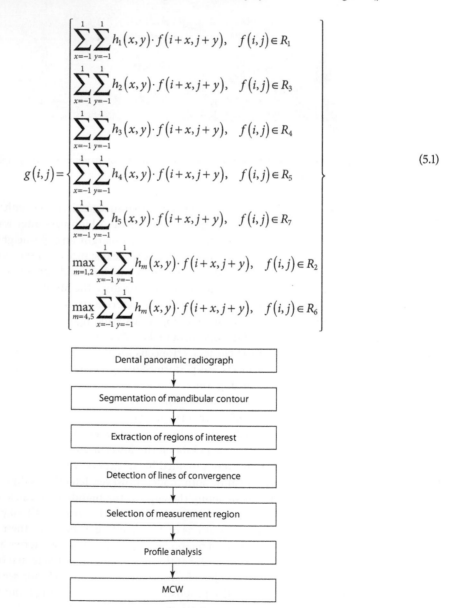

FIGURE 5.5 Flowchart of MCW measurement scheme.

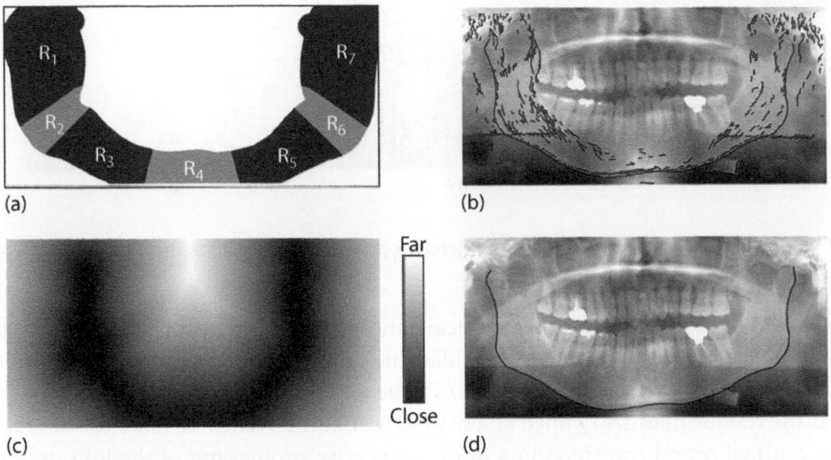

FIGURE 5.6 Segmentation of the mandibular contour. (a) An edge detection mask, (b) potential edges detected by Canny filter, (c) distance image, and (d) contour segmentation result.

$$h_1 = \begin{bmatrix} -3 & -3 & +5 \\ -3 & 0 & +5 \\ -3 & -3 & +5 \end{bmatrix} \quad h_2 = \begin{bmatrix} -3 & +5 & +5 \\ -3 & 0 & +5 \\ -3 & -3 & -3 \end{bmatrix} \quad h_3 = \begin{bmatrix} +5 & +5 & +5 \\ -3 & 0 & -3 \\ -3 & -3 & -3 \end{bmatrix}$$
$$h_4 = \begin{bmatrix} +5 & +5 & +3 \\ +5 & 0 & -3 \\ -3 & -3 & -3 \end{bmatrix} \quad h_5 = \begin{bmatrix} +5 & -3 & -3 \\ +5 & 0 & -3 \\ +5 & -3 & -3 \end{bmatrix} \tag{5.2}$$

Figure 5.6(b) shows the detected edges by the Canny edge detector with the above kernels.

At this point, many unrelated edges may be present. For removing irrelevant edges, the most probable edge was selected in each subregion. The longest horizontal edges in R_3 and R_5, and the longest vertical edges in R_1 and R_7, were selected because the mandibular edges are relatively reliably detected in these regions. In the remaining regions, the edges closest to the selected ones in the neighboring regions were retained. The final determination of the contour was performed using an active contour model (ACM) [47]. The distance transform was applied to the edges detected in the previous step. The output image has pixel values representing the distance from the nearest edge, as shown in Figure 5.6(c). Using the distance image, the most similar contour model from the manual reference contours was selected as the initial model in the ACM process. Figure 5.6(d) shows the final contour fitted to the study case by iteratively adjusting the control points. Each contour model contains two reference points below the right and left mental foramina corresponding to the MCW measurement positions. After the model fitting by ACM, the adjusted reference point locations were employed as the points of extracting ROIs.

In a few early studies [42, 43], MCWs were measured on 21 profiles obtained perpendicular to the contour around each reference point. However, accurate measurement is sometimes difficult when cortical bone is eroded and/or overlapped with other tissue, such as a hyoid bone. Thus, optimal profiles for MCW measurement were selected by residue line analysis [44]. First, 101 profiles of 100 pixel length were obtained along the contour line around a reference point, and they were arranged horizontally to create an ROI of 100 pixels by 101 lines (pixels). Figures 5.7(a)–(c) show a reference point, profile lines perpendicular to the fitted tangential lines to the contour line, and the ROI, respectively.

When osteoporosis has progressed, the inner margin of the cortical bone can become eroded, which may be indicated by the appearance of horizontal lines. For detection of high risk cases, we defined the

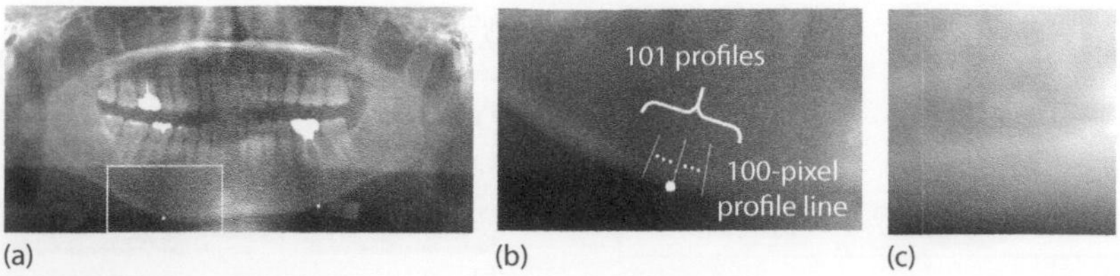

FIGURE 5.7 Extraction of a region of interest. (a) Reference points of measurement, (b) an enlarged view with an illustration of profiles, and (c) extracted ROI.

MCW as the width of the dense part of the cortical bone excluding the eroded part. A line convergence filter [48] was applied to detect the potential residue lines and an axis line corresponding to the peak of the cortical bone, as illustrated in Figures 5.8(c), (d). The axis line is defined as the lowermost line in the ROI, whereas the residue lines are defined as any lines within the vertical distance of 20 pixels from the axis line. An optimal region corresponding to 15 consecutive profiles out of the 101 profiles for MCW measurement was selected, using the result of line detection according to the following rules.

1. Region where the line corresponding to the peak of cortex (cortical axis) is present
2. Region where the residue line(s), if any, is present
3. Region where the average contrast in 15 consecutive profiles is the largest

The contrast is defined as the difference in pixel value at the axis and the minimum pixel value between the axis and residue lines, if they are present, or the pixel at the distance of 20 pixels from the axis, if no residue line is present. Figure 5.8(e), (f) present the selected regions for MCW measurement.

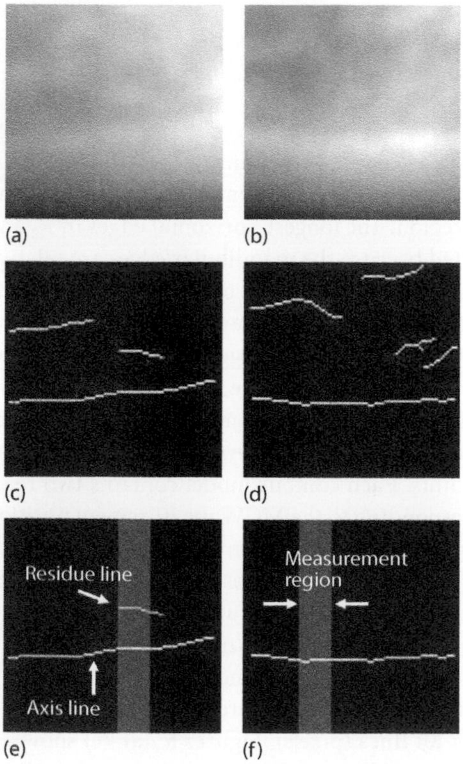

FIGURE 5.8 Determination of MCW measurement regions. (a), (b) right and left ROIs, (c), (d) horizontal lines of convergence, and (e), (f) selection regions for MCW measurement.

After the region was selected, MCW was determined on each of the 15 profiles (columns). From the peak of the profile, an average downslope of every pixel in the search range was computed as

$$G_{ave,i} = \left\{ \frac{\sum_{j=s_i}^{e_i} a_{i,j} - a_{i,j-1}}{n_i} \middle| a_{i,j} > a_{i,j-1} \right\}, \tag{5.3}$$

where:

$a_{j,j}$ is the pixel value at the (i,j) in the selected region of the ROI
s_i and e_i are the starting and ending points, respectively, of the border search range
n_i is the number of pixels satisfying the condition of $a_{i,j} > a_{i,j-1}$

The starting point is at the peak of the profile (axis of cortex), and the ending point is either at the residue line or at s–20. The inner border of the cortical bone is determined as the pixel closest to s_i that satisfies $a_{i,j} - a_{i,j-1} > G_{ave,i}$. The distance between the inner border pixel and the inferior border (contour) is considered as the MCW, and the average of 15 MCWs is determined as the MCW of the right/left side.

5.4.3 Evaluation

The proposed method was applied to 100 DPRs, including 26 osteoporotic cases. The MCW measurement result was compared with the manual measurement by a dental radiologist. The correlation coefficient between the MCWs by the dental radiologist and the proposed method was 0.87. For detection of osteoporotic cases, the area (AUC) under the receiver operating characteristic (ROC) curve was 0.95, and the sensitivity and specificity with the threshold MCW of 3.0 mm were 84.6% and 90.5%, respectively. The method was also applied to an independent dataset of 437 DPRs, including 21 suspected osteoporotic cases obtained at seven community dental clinics with five different panoramic imaging units. The image quality and imaging parameters vary considerably for these cases. Based on ROC analysis, an AUC of 0.93 was obtained for these cases. The sensitivity and specificity were 90.5% and 75.7%, respectively, with the same threshold value of 3.0 mm. These results indicate the potential usefulness of the proposed method for secondary osteoporotic screening as an added value to general dental examinations with no additional cost.

5.5 Texture Analysis of Cortical Bone for Osteoporotic Risk Assessment

The decrease in MCW is considered an index of osteoporosis, as described in the previous section. However, it is not a perfect measure. Another possible sign is the cortical erosion observed as porous cortical bone. Klemetti et al. have suggested the classification of change in the cortex into three groups [49]:

1. The endosteal margin of the cortex was even and sharp on both sides.
2. The endosteal margin showed semilunar defects (lacunar resorption) or seemed to form one to three layers of endosteal cortical residues on one or both sides.
3. The cortical layer formed heavy endosteal cortical residues and was clearly porous.

They found that the mean bone mineral density (BMD) values in the femoral neck and lumber spine are significantly different between groups 1 and 3. However, they concluded that this index was not sufficient in estimating the osteoporotic risk, especially with the limited samples they used. Taguchi et al. investigated the mandibular cortical index (MCI) proposed by Klemetti et al. and reported that the classification was useful in identifying women with osteoporotic risk [50]. The utility of the erosion index, however, could be dependent on the reliability of the subjective classification suggested by intra-observer agreement. Computerized classification may assist in consistent evaluation of the condition of the cortex.

5.5.1 Related Works

Several groups have applied intensity, morphologic, and fractal analyses to intraoral and panoramic radiographs for assessment of structural changes in alveolar cancellous bone [51–55]. Fractal analysis was found potentially useful for identifying subjects with osteoporotic changes in some studies [47, 51, 52], whereas it was considered to have no strong association with BMD and/or osteoporotic risk in other studies [53–55]. More recently, several groups investigated advanced computerized methods for evaluating cortex erosion on DPRs. Nakamoto et al. proposed an automated method to determine cortical erosion using a morphological skeletonization method [56, 57]. With detection of the lower mandibular edge by the Canny filter and manual identification of reference points on the inferior border of the mandible under the second premolar, ROIs of 100 × 400 pixels were extracted. The skeletonization was performed by a morphological opening operation to detect locally bright regions. The major skeleton segments were detected by Otsu's thresholding method [58]. If the image contained one clear segment, the image was considered as having thick cortical bone. If two or more segments with various lengths were detected, the image was considered as having endosteal cortical residues.

Roberts et al. proposed a computerized classification method for women with low BMD based on texture features [59]. For this study, a semiautomated method was employed to obtain ROIs at the cortex between the mental foramen and the antegonion. Gray-level co-occurrence matrices (GLCM) [60] were computed, with the direction normal to the cortical border and distance parameter of 2–8 pixels. Haralick texture features were determined, and a random forest classifier [61] was employed for detection of cases with MCI of 2 and 3 and those with low BMDs at the femoral neck and lumbar spine. They found that the texture features were useful for classification of osteoporotic individuals, and the performance was even better if the features were combined with the MCW. Kavitha et al. performed a similar study, in which GLCMs were computed with 4 directions (0, 45, 90 and 135°) and distance of 1 pixel, and the Haralick features were inputted to a SVM for classification [62]. They also concluded that the combination of texture features and MCW was superior in detecting cases with low BMD compared to the texture features or MCW on their own.

5.5.2 Methodology

MCI could be an effective index for detection of osteoporotic patients. However, a clear distinction between classes 2 and 3 can be difficult. Roberts et al. [59] found higher classification ability of texture features than that of the MCI categories subjectively assessed by experts in identifying osteoporotic subjects. This fact may partially be due to subjective and coarse categorization of the MCI. We have also investigated an automated method to classify cases into three MCI groups and faced difficulty in the classification of group 2, which may be partly owing to the ambiguity of class definitions [63]. For improving the classification and providing the degree of osteoporotic risk, we investigated a computerized method to estimate the continuous rate of mandibular cortical erosion degree (MCD) [64].

In order to estimate MCD, we first obtained the gold standard of MCD determined subjectively by a dental radiologist. The condition of the cortex was evaluated by the expert on a continuous scale from clear and sharp to severely eroded. The overall scheme is shown in Figure 5.9. The steps for segmentation of the mandibular contour, extraction of ROIs, and detection of lines of convergence are the same as the method for MCW measurement, except for the ROI size and the parameters of the line convergence filter. The details can be found in Muramatsu et al. [64]. Figure 5.10 shows the ROIs and the result of line detection.

Based on the detected lines, an ROI was divided into three regions of an outer cortex region (OCR), a cortex margin region (MR), and an inner trabecular bone region (ITR). The OCR was determined as

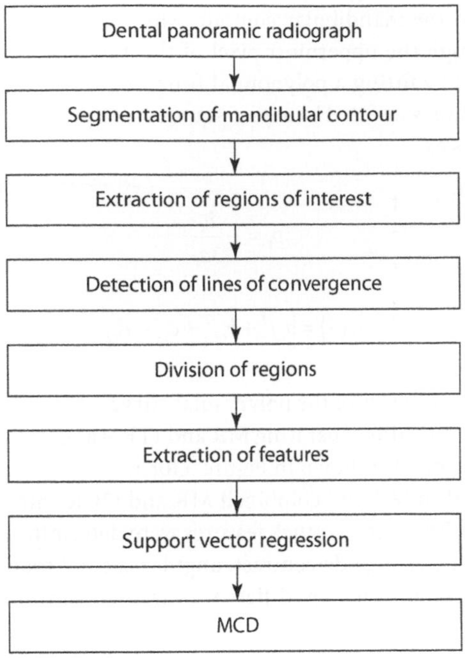

FIGURE 5.9 Flowchart of MCD determination scheme.

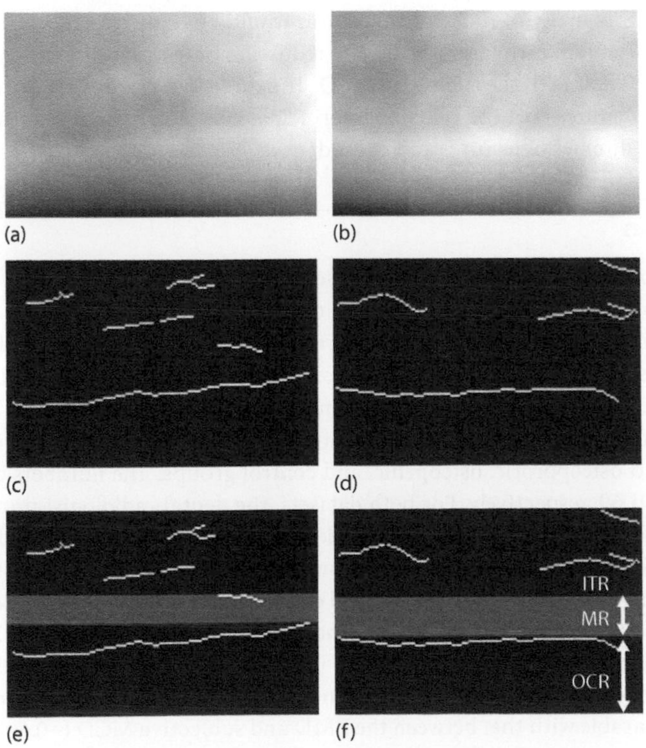

FIGURE 5.10 Division of ROIs for texture analysis. (a), (b) right and left ROIs, (c), (d) lines of convergence, and (e), (f) three regions of OCR, MR and ITR.

a rectangular region between the mandibular contour, which corresponds to the bottom of the ROI, and the horizontal line through the uppermost pixel of the axis line. The line of separation between MR and ITR was determined by fitting a polynomial function to each profile (column) of the ROI. A third-order polynomial function was fitted by a least-squares method, and the point of inflection, p_i, was determined by the coefficients as,

$$p_i = -\frac{b_i}{3a_i} \tag{5.4}$$

$$\hat{f}_i(j) = a_i j^3 + b_i j^2 + c_i j + d_i, \tag{5.5}$$

where a_i, b_i, c_i and d_i are the coefficients of the polynomial curve

The position, j, of the horizontal line separating MR and ITR was determined as the mean of p_i on 101 profiles. Examples of three regions are shown in Figure 5.10e,f.

The GLCMs were computed in MR and combined MR and OCR with 4 directions of 0, 45, 90, and 135° and a distance of 5 pixels. Thirteen Haralick features were determined using the 8 GLCMs yielding 104 features. The Haralick features include contrast, angular second moment, correlation, inverse difference moment, variance, difference entropy, difference variance, sum entropy, sum variance, entropy, sum average, and information measures of correlation 1 and 2, as defined in Haralick et al. [60]. We also determined the original features: the area of MR, the number of line component pixels in MR, and the ratio of the average pixel values at the line components in OCR and MR. With inclusion of the MCW as a feature, the total number of features was 108.

The MCD was estimated using support vector regression (SVR) with the radial basis function kernel. It was expected that many of the texture features would be highly correlated. Furthermore, some features may not be useful in estimating MCD. For reduction of features, the utility of each feature was assessed by the correlation with the subjective MCD. The threshold of 0.3 was employed empirically for removing unrelated features. For the remainder, principal component analysis (PCA) was applied for further reducing the number of features and redundancy.

5.5.3 Evaluation

The proposed method was evaluated with two datasets. The first dataset is the same dataset used for the MCW study described in the previous section and consisted of 99 DPRs: one was excluded due to the atypical cortical morphology. This dataset was used as the training set. The second dataset consists of 99 DPRs whose patients had undergone x-ray absorptiometry measurements at the distal forearm. The second dataset was used as the test dataset. The DPRs in the two datasets were obtained with different panoramic imaging units, allowing independent evaluation. Based on the BMD data, the cases in the test set were categorized into osteoporotic, osteopenic, and control groups. The numbers of cases in the three groups are 12, 18, and 69, respectively. For both datasets, the dental radiologist provided the subjective MCD based on the images without any other clinical information. The training dataset was used for reduction of image features, determination of eigenvectors by PCA, and selection of SVR parameters.

The correlation between the subjective MCD and estimated MCD by the proposed method was moderate (0.617) for the test dataset, indicating the potential usefulness of texture analysis for characterizing cortical erosion; however, there remains room for improvement compared with that for the training dataset of 0.841. The correlation coefficient between the BMD and MCD by the proposed method was 0.619, which is comparable with that between the BMD and subjective MCD (−0.600).

For detecting osteoporotic cases, the AUC using the estimated MCD was 0.901, which was comparable with that using the subjective MCD (0.874) and slightly higher than that using the MCW alone. The AUC for detecting high risk cases, that is osteoporotic and osteopenic cases, by the proposed MCD was 0.880, which is slightly higher than those by the subjective MCD (0.833) and MCW alone (0.835). None of these was statistically significantly different from another. These results indicate the potential usefulness of the proposed MCD for osteoporotic risk assessment. The result could be improved by using a more balanced dataset. The test cases used in this study have the BMD measurement done at the distal forearm, which may not be strongly related to the risk of severe osteoporotic fractures. The utility of the method should be evaluated with a larger dataset with the BMD measurement in the lumbar spine and/or femoral neck.

5.6 Detection of Carotid Artery Calcifications for Arteriosclerosis Screening

According to the health statistics provided by the Centers for Disease Control and Prevention, heart diseases and cerebrovascular diseases are the first and fifth leading causes, respectively, of deaths in the United States in 2013 [65]. Similarly, they are the second and fourth leading causes, respectively, in Japan in 2014 based on the statistics reported by the Ministry of Health, Labour, and Welfare [66]. The risk of these asymptomatic deadly diseases may be predicted by the presence of carotid artery calcifications (CACs) [67]. Studies have reported that DPRs can depict CACs and could become a possible alternative screening method for atherosclerosis [68–71]. However, CACs may not be clearly visible on DPRs, especially with the common window setting for dental examinations, as shown in Figure 5.11. In addition, regions outside the oral region are generally not paid attention to by dentists. Computerized detection can play an important role in such a situation.

5.6.1 Related Works

To our knowledge, a limited number of groups have studied automated detection of CACs on DPRs. Shinjo et al. proposed a computerized method for detection of CACs based on local contrast [72, 73]. First, ROIs of 300 × 300 pixels on the lower right and left regions between the throat and spine are selected manually as the candidate regions. Fuzzy image enhancement was applied to detect edges of the calcification regions, and FP regions were removed based on size [72]. In their later studies [73], after the background trend was corrected by subtracting the approximated plane, calcification candidates were detected using k-means clustering. FPs were removed using local features.

5.6.2 Methodology

We have proposed an automatic CAC detection scheme using a top-hat filter [74, 75]. A flowchart of the proposed scheme is shown in Figure 5.12. For computational efficiency and excluding sources of FPs,

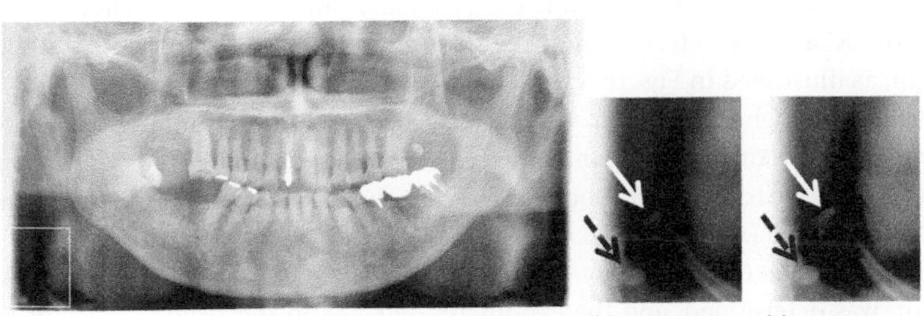

FIGURE 5.11 Visibility of CACs on DPR. (a) Original DPR with the ROI position, (b) visible and hardly visible CACs on normal window setting, (c) improved visibility by windowing adjustment. Solid arrows showing the visible CAC and dotted arrows showing the hardly visible CAC.

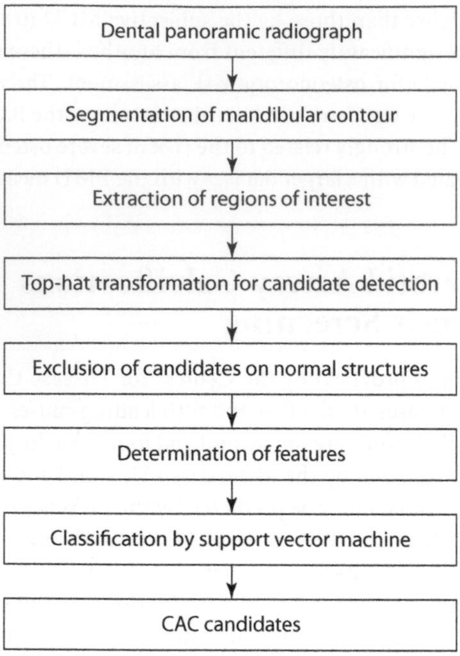

FIGURE 5.12 Flowchart of CAC detection scheme.

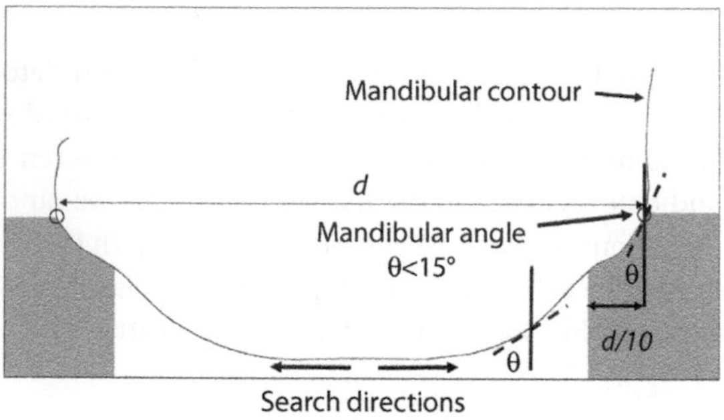

FIGURE 5.13 Searching of the mandibular angle by slope of contour for determination of ROIs.

ROIs including the areas of carotid arteries were extracted. Using the result of automated segmentation of the mandibular contour [42], the locations of mandibular angles were searched on the basis of the contour slope. The ROIs were determined as regions inferior and posterior to the angles with a prespecified margin, as illustrated in Figure 5.13.

Regions with locally high pixel values were detected using a top-hat filter with a circular element approximately 3 mm in diameter. The candidate regions include many FPs, especially on the cervical vertebrae. To exclude vertebral regions, locations of anterior edge were searched. Because the vertical edges of vertebrae are not clearly depicted on DPRs, horizontal edges of vertebra bodies were detected by the Sobel filter. Based on the number of detected edge components in each column of the image, the anterior position was determined, and the candidates detected in the region posterior to the position were excluded.

For each candidate region that remained, 12 image features were determined, including the area, average pixel value, pixel value variance, contrast, width, height, ratio of width and height, circularity, irregularity, relative locations in the horizontal and vertical directions, and normalized distance to the mandibular contour. Using these features, a rule-based FP reduction step was performed. The remaining candidates were classified into CACs and FPs by SVM. The features input to the SVM were selected by testing all possible combinations, and five features, including the average pixel value, contrast, circularity, relative horizontal location, and normalized distance from the contour, were employed.

5.6.3 Evaluation

The proposed method was evaluated using 100 DPRs, including 34 cases with CACs. The number of CACs in 34 images varied from one to nine, and the total number of CACs was 94. The locations of CACs were marked by a dental radiologist using CT images as reference. Using a leave-one-out cross-validation method, the sensitivity of the proposed method was 90% with 2.5 FPs per image. With 80% sensitivity, the number of FPs was 1.2 per image. Based on the fact that most FPs can be easily dismissed by a dental radiologist, the sensitivity of 90% is considered relatively successful. However, further reduction of FPs is preferable.

We have explored for possible approaches to reduce FPs and looked into the locations of carotid arteries [76]. For not missing any CACs, the ROIs were set relatively large and, in many cases, included unrelated regions. It was found that the number of FPs could be reduced by 20% if manually selected ROIs were used. The locations of the arteries can change with the imaging conditions. Accordingly, the appearances of the vertebrae and the mandibular contour vary with the patients' positioning, such as the head inclination. By detecting these structures, ROIs could be personally adjusted and FPs may be reduced.

Since most CACs are ignored in general dental practice, a system to prompt possible CACs would be beneficial. Many CACs are not easily identifiable even if dentists are paying attention to the arteries, as shown in Figure 5.11b. Our proposed scheme with the sensitivity of 90% has potential usefulness in assisting dentists for detection of CACs and in screening patients with risk of arteriosclerosis. Further improvement and evaluation with a large, independent dataset are needed.

5.7 Detection of Radiopacity for Diagnosis of Maxillary Sinusitis

Maxillary sinuses are one of the four paired air spaces around the nose, called paranasal sinuses. These sinuses occasionally cause inflammation and, if left untreated, may become chronic. Therefore, early detection and treatment are desired. Maxillary sinuses are located just below the eyes and can be depicted in DPRs. When inflamed, it can be observed with elevated gray levels, as shown in Figure 5.14. However, these findings can be very subtle, and dentists who are not paying attention to these regions may not notice them. Automated detection of sinusitis can be useful to draw attention of dentists to the suspected sinus, and they may suggest further examinations.

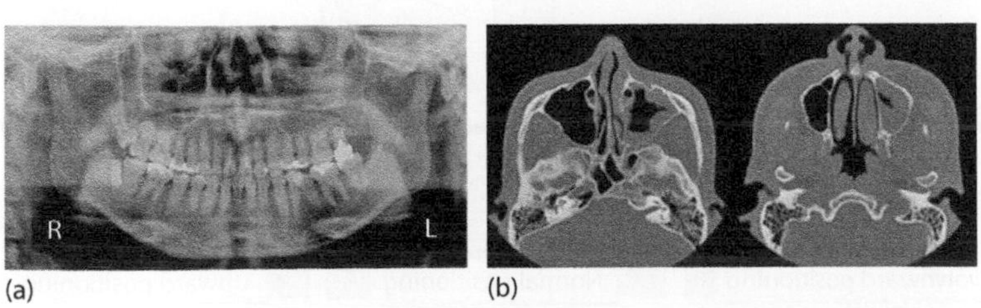

FIGURE 5.14 A case with maxillary sinusitis. (a) DPR and (b) the corresponding CT image.

5.7.1 Related Work

To our knowledge, only one group has attempted quantitative analysis of maxillary sinuses, not on DPRs, but on x-ray images of paranasal sinuses [77]. They compared different quantities, such as the ratio and difference in copper-equivalent values of the orbit and nasal regions. ROIs were manually set at approximately the centers of the maxillary sinus and orbit, and an entire region of the sinus was also sampled. They found that the ratio of copper-equivalent values of the sinus ROI and orbit ROI provided the best quantitative values corresponding to the five clinical grades for diagnosis of sinusitis.

5.7.2 Methodology

The target of this study is detection of unilateral maxillary sinusitis. Thus, our proposed method for detection of opacities is based on contralateral comparison [78]. It consists of three steps: (1) contralateral subtraction, (2) ROI determination, and (3) pixel value comparison. First, the mandibular contour was obtained by the method described earlier in Section 5.4.2 [42]. Registration of an original image and the horizontally flipped image was carried out on the basis of the mutual information of the contours. Arithmetic subtraction of the pixel values in the flipped image from the corresponding ones in the original image was performed to create the contralateral subtraction image.

The locations of sinus ROIs are determined on the basis of other anatomic structures, and their size varies according to the 4 sides of the ROIs. Figure 5.15 shows the ROI localization process. The superior wall approximately corresponds to the Frankfurt plane, which connects the upper end of an external acoustic pore and the lower end of an orbit. Detection of these structures, however, is difficult. Instead, we detect the head of the mandible based on the steep gradient from the head to the mandibular fossa. The horizontal line through the upper end of the mandibular head is considered the upper border of the ROIs. The inferior border is set using the hard palate, which is depicted with relatively high pixel values

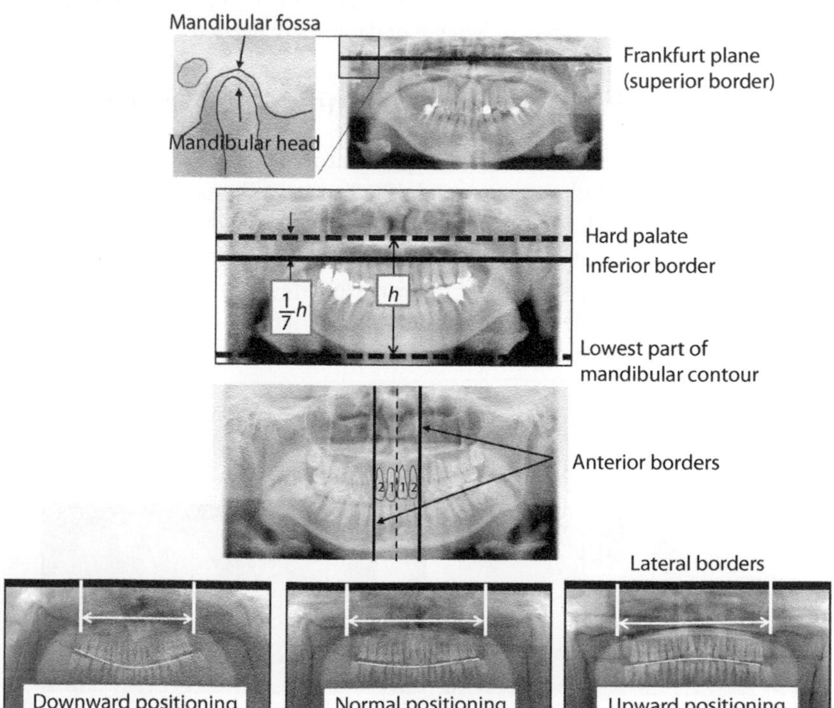

FIGURE 5.15 Localization of maxillary sinus ROIs.

in the center of DPRs. The hard palate is detected using the vertical profile of the image, and a seventh of the distance between the palate and the mandibular contour below the palate is determined as the position of lower border.

The anterior wall is approximately located next to the lateral incisors. Because some or all teeth are missing in some cases, average tooth sizes are used for determination of the central ends. Finally, the posterior wall location was determined according to the patients' positioning. In general, patients are positioned so that the Frankfurt plane is parallel to the ground. However, if a patient's head is declined, the posterior wall is observed slightly inward to the normal position. On the other hand, if a patient's head is inclined upward, it is observed slightly lateral to the normal position. The head position is estimated on the basis of the curvature of the occlusal line, and the location of the posterior wall is computed using their relationship.

The average pixel values in the right and left ROIs on a subtraction image are determined. If the difference in the average pixel values is greater than the pre-specified threshold value, the side with the higher pixel value is considered abnormal.

5.7.3 Evaluation

The proposed method was evaluated using two databases: One includes 59 cases with 19 cases of abnormal sinuses, and another includes 39 cases with 19 cases of abnormal sinuses. Two datasets were obtained at different institutions with different panoramic imaging units. All abnormal images have unilateral disease, and the diagnosis was confirmed by a dental radiologist based on other modalities such as CT. By visual evaluation, the proposed method was able to locate ROIs successfully in the images with variable positioning.

Using ROC analysis, AUCs for distinguishing between normal and abnormal cases were 0.86 and 0.82 for the two datasets. Two observer performance studies were performed for testing the utility of the CAD system. In the first study [79], 82 cases, including 34 abnormal cases, were evaluated by 13 observers without and with CAD. The 13 observers included nine oral radiology residents and four dentistry students. The AUC was improved from 0.68 without CAD to 0.73 with CAD. The difference was statistically significant ($p = 0.042$). In the second study [80], 49 cases of maxillary sinusitis and 49 control cases were evaluated by 12 dentists with less than two years of experience and four expert oral and maxillofacial radiologists. For the inexperienced group, the sensitivity was improved from 63.4% to 71.6%, whereas the specificity was also improved from 68.6% to 75.3% using CAD. The AUC without CAD was 0.728, which was increased to 0.78 with CAD. For the expert group, the sensitivities without and with CAD were 74.5% and 76.0%, and the specificities were 85.2% and 86.2%, respectively. The AUCs improved from 0.871 to 0.897 with the computer assistance. The differences in AUCs were statistically significant for both groups ($p < 0.001$ for both).

These results indicate the potential utility of the proposed method in detecting radiopacity in maxillary sinuses and also the usefulness of the CAD system in diagnosis of maxillary sinusitis on DPRs by dentists and dental radiologists. The limitation of the method is that only unilateral abnormality can be detected. For detection of bilateral abnormalities, differences in pixel values with respect to reference values may be used. Although ROI arrangement was improved by use of anatomical structures and estimation of patients' positioning, there is still some disagreement between automatic and manual placement. A method for further refinement of ROI localization is needed. In some cases, it is difficult to detect abnormalities even with manual ROIs. Additional techniques to improve sensitivity are expected in the future.

5.8 Quantitative Analysis of Periodontal Diseases

Periodontal disease is a typical dental disease that affects many adults. In clinical practice, the disease is generally assessed by inserting a probe into a gap between a tooth and the adjacent gum and measuring the depth. This examination may be painful and can cause bleeding. Radiographic examination can be an alternative choice to evaluate the level of resorption and may be effective in reducing subjective variation induced by probing. Conventionally, intraoral radiographs were used for evaluation of periodontal disease.

Although image quality is better with intraoral radiographs, image acquisition and evaluation for all teeth take time and effort; DPR on the other hand, enables examination of the total condition of all teeth.

The alveolar levels on DPRs are generally assessed by visual evaluation. Manual measurement on each tooth is labor intensive, and a quantitative method is not yet established. For evaluation of alveolar bone resorption, Schie et al. suggested the alveolar bone loss (ABL) index, which is the ratio of decrease of alveolar bone level to the distance between the cement-enamel junction and the tooth apex [81]. Computerized quantification of ABL can be helpful for consistent and efficient evaluation of alveolar bone resorption.

5.8.1 Methodology

In DPRs, it is difficult to detect the location of the cement-enamel junction. Thus as a surrogate to ABL, alveolar bone ratio (ABR), which is the ratio of the height of the alveolar bone level to the height of the tooth, was proposed. We have developed fully automated [82] and semiautomated [83] methods for quantitative measurement of ABR.

In the automatic mode, three series of processes are applied to detect three lines corresponding to tooth occlusion, alveolar bone, and root apices. The flowchart is shown in Figure 5.16. First, preprocessing is performed for noise reduction and image size reduction. Using the result of mandibular contour segmentation, ROIs are determined by sliding the contour upward. For detection of the occlusion line, a horizontal Gabor filter is applied, followed by binarization to detect tooth crown edges. Concurrently, the phase congruency (PC) map [84] is created. By taking the intersection of the two, occlusion feature points are detected, and FP points are removed using an ROI mask.

For detection of the alveolar line, the PC map is binarized. Possible FP feature points are removed on the basis of the area and location. In detection of root apices, multidirectional Gabor filters are applied according to the positions of roots. After the filtered image is binarized, FP points are removed on the basis of the area and location. For each set of detected feature points, a second-order polynomial curve is fitted using a least-squares method. Finally, ABR is computed using three lines.

In the semiautomated mode, the points corresponding to the top of the tooth crown and root apex or apices for a two-rooted tooth are manually identified on each tooth. Using these points, the image is transformed to a standardized dentition format, in which teeth are vertically aligned in a rectangular

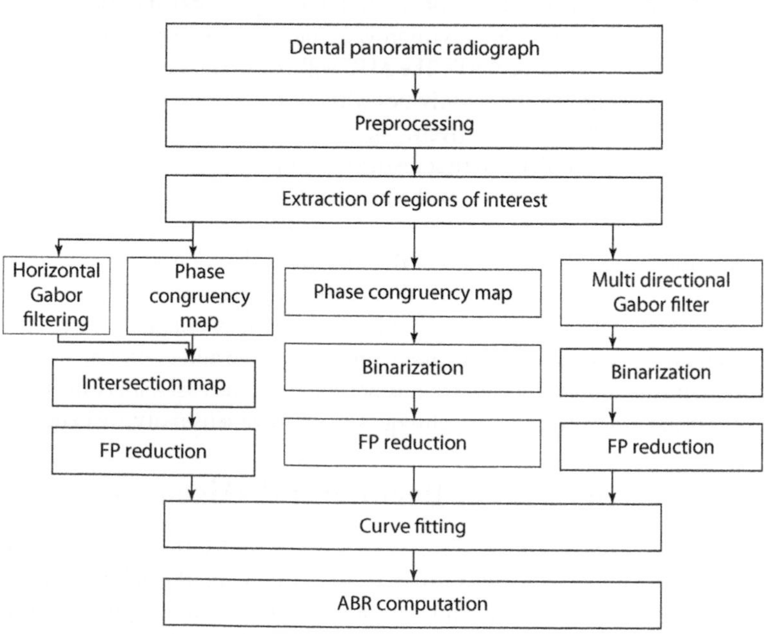

FIGURE 5.16 Flowchart of automated alveolar bone resorption measurement method.

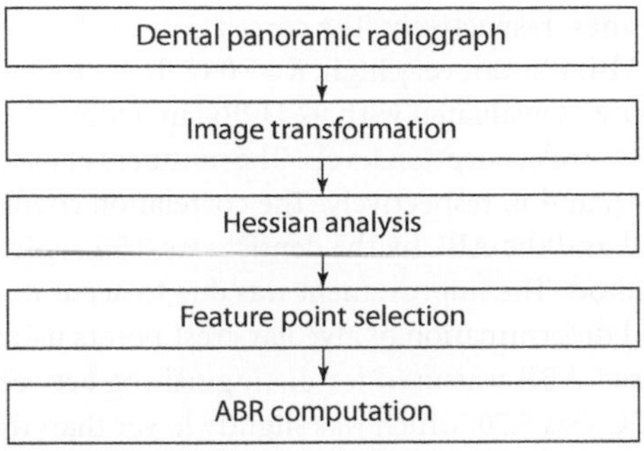

FIGURE 5.17 Flowchart of semiautomated alveolar bone resorption measurement method.

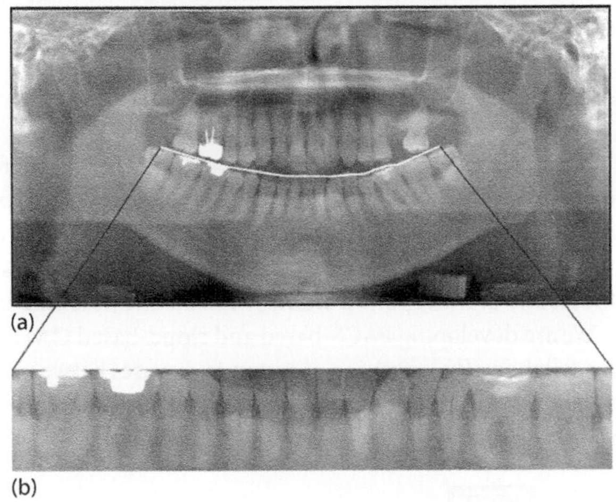

FIGURE 5.18 Standard dentition image transformation. (a) Original DPR, and (b) standard dentition image.

image. Figure 5.17 shows the flowchart of the proposed method. Between each pair of crown and apex points, 10 additional control points are placed equidistantly. Using these points, the curved tooth region is transformed to a rectangular region of aligned teeth with equal height by the thin plate spline method. An example of the transformed image is shown in Figure 5.18.

In the transformed image, feature points corresponding to alveolar crests are automatically detected. The second derivatives are determined after Gaussian smoothing to compute the Hessian matrix. Based on the eigenvalues, dark spots are detected as the candidates. Using a mask with 15 subregions between teeth, one point in each subregion is selected as the crest point from the candidates on the basis of the maximum a-posteriori (MAP) estimation [85]. The probability function consists of the pixel value, gradient strength, gradient direction with respect to the reference direction, and the first and second eigenvalues of the Hessian matrix. The prior probability is based on the mean and variance of the coordinates of the gold standard crest points provided by the dentist. The candidate region that maximizes the posterior probability in each subregion is selected. The average ABR is computed using the selected points. For visual assistance, the points are connected with spline interpolation and shown on the standardized dentition image.

5.8.2 Evaluation

The proposed automated method was evaluated using 88 DPRs. The gold standard points of tooth crowns, alveolar crests, and root apices were manually identified for comparison. They also have the ABL data measured by a dentist. The mean absolute errors of three lines were 3, 6, and 9 pixels for occlusion, alveolar, and apex lines, respectively. The correlation coefficient between the ABR measured by the proposed method and ABL was not very high, $R = -0.17$. The method needs further improvement.

The semiautomated method was evaluated with 92 DPRs, including the above 88 DPRs. These cases were categorized into four grades of absorption levels. The numbers of cases in normal, mild, moderate, and severe groups are 39, 38, 11, and 4, respectively. The correlation coefficient between the ABR measured by the proposed method and the ABL by the dentist was 0.50, which was greatly improved compared with the automated method. The improvement was due to accurate determination of crown and apex points and also improved determination of alveolar crest points using Hessian analysis and MAP estimation. When the measured ABR was used for distinguishing between normal cases and cases of the other three grades, the AUC was 0.70, which was slightly lower than that by ABL (0.81) but could be useful in screening periodontal disease. The proposed method needs further improvement and should be evaluated with a larger dataset with balanced disease grades. This preliminary study indicated the potential usefulness of computerized quantitative assessment of alveolar bone resorption on DPRs and the standard dentition images for perceptive assistance.

Conclusion

In this chapter, we have introduced several studies on computerized analysis of intraoral radiographs. More detailed descriptions of the methods and discussions of the results are provided for computerized analysis of DPRs for screening and efficient diagnosis of systemic and oral diseases. Some of these methods could also be applied to dental CT images. Utilities of the CAD systems should be evaluated in clinical trials; however, the results of the reported studies indicate the potential benefit of computerized analysis of dental images. We are developing PACS-based and cloud-based CAD systems combined with teleradiology reporting for collaborative screening of systemic diseases between computer, dentists, and medical doctors (Figure 5.19). The results in the preliminary clinical trial indicate potential utility of such a screening framework.

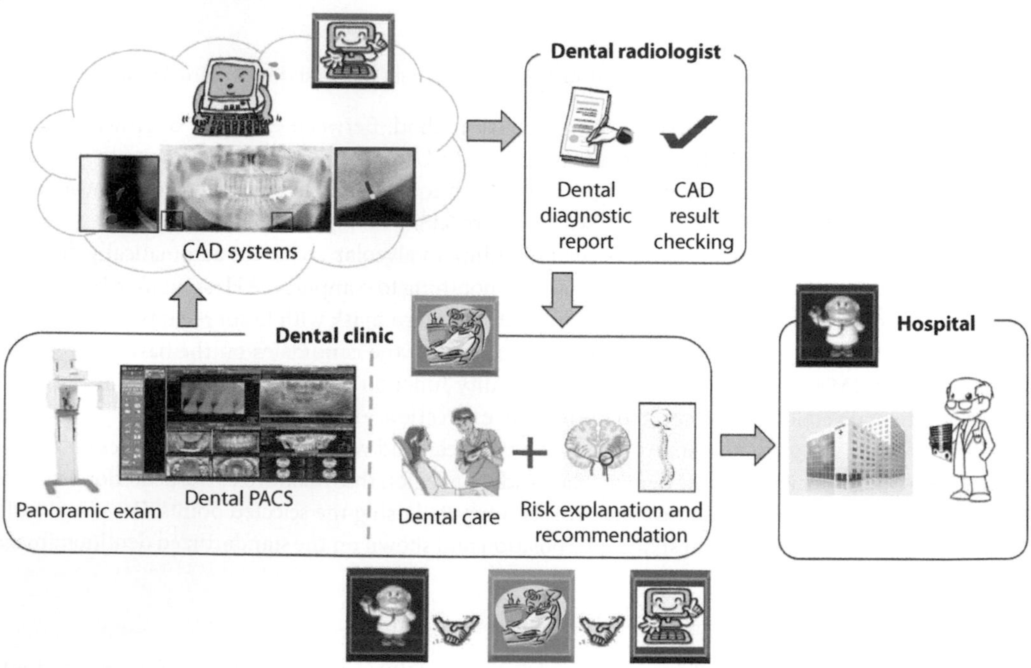

FIGURE 5.19 Conceptual diagram of collaborative screening framework between dentists, computer and medical doctors.

Acknowledgment

We are grateful to the collaborators at Asahi University, Aichi Gakuin University, and Media Co., Ltd. We also thank Asahi University Hospital staff for their contribution to this study. This study was supported in part by Grant-in-Aid for Scientific Research (b) JSPS KAKENHI Grant Number 26293402 and Grant-in-Aid for Scientific Research on Innovative Areas (Multidisciplinary Computational Anatomy), MEXT, Japan, Grant Number 26108005.

References

1. Lou L, Lagravere MO, Compton S, Major PW, Flores-Mir C, Accuracy of measurements and reliability of landmark identification with computed tomography (CT) techniques in the maxillofacial area; a systemic review, *Oral Surg Oral Med Oral Pathol Oral Radiol Endod* 104: 402–411, 2007.
2. Ji DX, Ong SH, Foong KWC, A level-set based approach for anterior teeth segmentation in cone beam computed tomography images, *Comput Biol Med* 50: 116–128, 2014.
3. Gan Y, Xia Z, Xiong J, Zhao Q, Hu Y, Zhang J, Toward accurate tooth segmentation from computed tomography images using hybrid level set model, *Med Phys* 42: 14–27, 2015.
4. Bui NL, Ong SH, Foong KWC, Automatic segmentation of the nasal cavity and paranasal sinuses from cone-beam CT images, *Int J CARS* 10: 1269–1277, 2015.
5. Farman AG, Kolsom SA, Intraoral Radiographic Techniques, dentalcare.com Continuing Education (available at http://www.dentalcare.com/media/en-US/education/ce119/ce119.pdf)
6. Yoon DC, Wilensky GD, Neuhau A, Manukian N, Gakenheeimer DC, Quantitative dental caries detection system and method. US Patent, 5742700, 1998.
7. Price JB, *A Review of Dental Caries Detection Technologies*, PennWell Corporation: Tulsa, 2013.
8. Tracy KD, Dykstra BA, Gakenheimer DC, Scheetz JP, Lacina S, Scarfe WC, Farman AG, Utility and effectiveness of computer-aided diagnosis of dental caries, *Gen Dent* 59: 136–144, 2011.
9. Gakenheimer DC, Farman TT, Farman AG, Benjamin SD, Huysing PDA, Chang HJC, Liao MYT, Tu HPY, Advancements in automated dental caries detection using DICOM image files, *International Congress Series* 1281: 1250–1255, 2005.
10. Grondahl HG, Grondahl K, Subtraction radiography for the diagnosis of periodontal bone lesions, *Oral Surg Oral Med Oral Pathol* 55: 208–213, 1983.
11. Grondahl HG, Grondahl K, Webber RL, A digital subtraction technique for dental radiography, *Oral Surg Oral Med Oral Pathol* 55: 96–102, 1983.
12. Van der Stelt PF, Modern radiographic methods in the diagnosis of periodontal disease, *Adv Dent Res* 7: 158–162, 1993.
13. Ruttimann UE, Computer-based reconstruction and temporal subtraction of radiographs, *Adv Dent Res* 1: 72–79, 1987.
14. Van der Stelt PF, Ruttimann UE, Webber RL, Determination of projection for subtraction radiographyb based on image similarity measurements, *Dentomaxillofac Radiol* 18: 113–117, 1989.
15. Ruttimann UE, Webber RL, A robust digital method for film contrast correction in subtraction radiography, *J Periodont Res* 21: 486–495, 1986.
16. Van der Stelt PF, van der Linden WJ, Geraets WGM, Alons CL, Digitized image processing and pattern recognition in dental radiographs with emphasis on the interdental bone, *J Clin Periodontol* 12: 815–821, 1985.
17. Mol A, van der Stelt PF, Application of computer-aided image interpretation to the diagnosis of periapical bone lesions, *Dentomaxillofac Radiol* 21: 190–194, 1992.
18. Carmody DP, McGrath SP, Dunn SM, van der Stelt PF, Schouten E, Machine classification of dental images with visual search, *Acad Radiol* 8: 1239–1246, 2001.

19. Sutthiprapaporn P, Tanimoto K, Nakamoto T, Kongsomboon S, Limmonthol S, Pisek P, Keinprasit C, Evaluation of alveolar bone grafting in unilateral cleft lip and palate patients using a computer-aided diagnosis system, *Imaging Science in Dentistry* 42: 225–229, 2012.
20. Wu Y, Xie F, Yang J, Cheng E, Megalooikonomou V, Ling H, Automatic detection of apical roots in oral radiographs, *Proc SPIE Med Imaging* 8315: 83152M1–8, 2012.
21. Huang PW, Lin PL, Kuo CH, Cho YS, An effective tooth isolation method for bitewing dental x-ray images, *Int Conf Mach Learn Cybern* 5: 15–17, 2012.
22. Li S, Fevens T, Krzyzak A, Li S, An automatic variational level set segmentation framework for computer aided dental x-ray analysis in clinical environments, *Comput Med Image Graph* 30: 65–74, 2006.
23. Jain AK, Chen H, Matching of dental x-ray images for human identification, *Pattern Recognition* 37: 1519–1532, 2004.
24. American Dental Association Survey Center, *2000 Survey of Dental Practice: Characteristics of Dentists in Private Practice and their Patients*, American Dental Association: Chicago, IL, 2007.
25. Shimano T, Suzuki Y, Sasaki T, Long-term trend of dental radiography examination in Japan – analysis on health insurance data, *Dent Radiol* 42: 9–21, 2002 (in Japanese).
26. Yamamoto I, Estimation of population of osteoporosis patients, Guideline on treatment of osteoporosis. Estimation based on results conforming to diagnostic criteria of Japan Soc for Bone and Mineral Res, *Osteoporos Jpn* 7: 10–11, 1999.
27. Kanis JA, on behalf of the World Health Organization Scientific Group, Assessment of osteoporosis at the primary health-care level. WHO Technical Report, University of Sheffield, UK, 2007.
28. Cummings SR, Melton III LJ, Epidemiology and outcomes of osteoporotic fractures, *Lancet* 359: 1761–1767, 2002.
29. Johnell O, Kanis JA, An estimate of the worldside prevalence and disability associated with osteoporotic fractures, *Osteoporos Int* 17: 1726–1733, 2006.
30. Taguchi A, Triage screening for osteoporosis in dental clinics using panorama radiographs, *Oral Dis* 16: 316–327, 2010.
31. Karayianni K, Homer K, Mitsea A, Berkas L, Mastoris M, Jacobs R, Lindh C, van der Stelt PF, Harrison E, Adams JE, Pavitt S, Devlin H, Accuracy in osteoporosis diagnosis of a combination of mandibular cortical width measurement on dental panoramic radiographs and a clinical risk index (OSIRIS): The OSTEODENT project, *Bone* 40: 223–229, 2007.
32. Alman AC, Johnson LR, Calverley DC, Grunwald GK, Lezotte DC, Hokanson JE, Diagnostic capabilities of fractal dimension and mandibular cortical width to identify men and women with decreased bone mineral density, *Osteoporos Int* 23: 1631–1636, 2012.
33. Arifin AZ, Asano A, Taguchi A, Nakamoto T, Ohtsuka M, Tsuda M, Kudo Y, Tanimoto K, Computer-aided system for measuring the mandibular cortical width on dental panoramic radiographs in identifying postmenopausal women with low bone mineral density, *Osteoporos Int*, 17: 753–759, 2006.
34. Kavitha MS, Asano A, Taguchi A, Kurita T, Sanada M, Diagnosis of osteoporosis from dental panoramic radiographs using the support vector machine method in a computer-aided diagnosis, *BMC Med Imaging* 12: 1–11, 2012.
35. Arifin AZ, Asano A, Image thresholding by histogram segmentation using cluster organization avoiding local minima, IEICE Technical Report, 104: 1–7, 2004.
36. Kavitha MS, Asano A, Taguchi A, Heo MS, The combination of a histogram-based clustering algorithm and support vector machine for the diagnosis of osteoporosis, *Imaging Sci Dent* 43: 153–161, 2013.
37. Allen PD, Graham J, Farnell DJJ, Harrison EJ, Jacobs R, Nicopolou-Karayianni K, Lindh C, van der Stelt PF, Horner K, Devlin H, Detecting reduced bone mineral density from dental radiographs using statistical shape models, *IEEE Trans Inform Technol Biomed*, 11: 601–610, 2007.

38. Cootes TF, Taylor CJ, Cooper DH, Graham J, Active shape models – their training and application, *Compute Vis Image Understand*, 61: 38–59, 1995.
39. Roberts MG, Graham J, Devlin H, Improving the detection of osteoporosis from dental radiographs using active appearance models, *Proc IEEE Int Sympo biomed Imaging* 440–443, 2010.
40. Roberts M, Yuan J, Graham J, Jacobs R, Devlin H, Changes in mandibular cortical width measurements with age in men and women, *Osteoporos Int* 22: 1915–1925, 2011.
41. Cootes TF, Edwards GJ, Taylor CJ, Active appearance models, *IEEE Trans Pattern Anal Mach Intell* 61: 38–59, 2001.
42. Matsumoto T, Hayashi T, Hara T, Katsumata A, Muramatsu C, Zhou X, Iida Y, Matsuoka M, Katagi K, Fujita H, Automated scheme for measuring mandibular cortical thickness on dental panoramic radiographs for osteoporosis screening, *Proc SPIE Med Imaging* 8315: 83152L-1–83152L-6, 2012.
43. Muramatsu C, Matsumoto T, Hayashi T, Hara T, Katsumata A, Zhou X, Iida Y, Matsuoka M, Wakisaka T, Fujita H, Automated measurement of mandibular cortical width on dental panoramic radiographs, *Int J CARS* 8: 877–885, 2013.
44. Horiba K, Muramatsu C, Hayashi T, Fukui T, Hara T, Katsumata A, Fujita H, Automated measurement of mandibular cortical width on dental panoramic radiographs for early detection of osteoporosis: Extraction of linear structures, *Med Imag Tech* 32: 342–346, 2014 (in Japanese).
45. Canny J, A computational approach to edge detection, *IEEE Trans Pattern Anal Mach Intell* 8: 679–698, 1986.
46. Kirsch RA, Computer determination of the constituent structure of biological images, *Comput Biomed Res* 4: 1315–1328, 1971.
47. Kass M, Witkin A, Terzopoulos D, Snakes: Active contour models, *Int J Comput Vis* 1: 321–331, 1988.
48. Kobatake H, Hashimoto S, Convergence index filter for vector fields, *IEEE Trans Image Process* 8: 1029–1038, 1999.
49. Klemetti E, Kolmakov S, Kroger H, Pantomography in assessment of the osteoporosis risk group, *Scand J Dent Res* 102: 68–72, 1994.
50. Taguchi A, Asano A, Ohtsuka M, Nakamoto T, Suei Y, Tsuda M, Kudo Y, Inagaki K, Noguchi T, Tanimoto K, Jacobs R, Klemetti E, White SC, Horner K, Observer performance in diagnosing osteoporosis by dental panoramic radiographs: Results from the osteoporosis screening project in dentistry (OSPD), *Bone* 43: 209–213, 2008.
51. Law AN, Bollen AM, Chen SK, Detecting osteoporosis using dental radiographs: A comparison of four methods, *JADA* 127: 1734–1742, 1996.
52. Bollen AM, Taguchi A, Hujoel PP, Hollender LG, Fractal dimension on dental radiographs, *Dentomaxillofac Radiol* 30: 270–275, 2001.
53. Shrout MK, Hildebolt CF, Potter BJ, Brunsden TKB, Pilgram TK, Dotson M, Yokoyama-Crothers N, Hauser J, Cohen S, Kardaris E, Civitelli R, Hanes P, Comparison of morphological measurements extracted from digital dental radiographs with lumbar and femoral bone mineral density measurements in postmenopausal women, *J Periodontol* 71: 335–340, 2000.
54. Southard TE, Southard KA, Lee A, Alveolar process fractal dimension and postcranial bone density, *Oral Surg Oral Med Oral Pathol Oral Radiol Endod* 91: 486–491, 2001.
55. Tosoni GM, Lurie AG, Cowan AE, Burleson JA, Pixel intensity and fractal analyses: Detecting osteoporosis in perimenopausal and postmenopausal women by using digital panoramic images, *Oral Surg Oral Med Oral Pathol Oral Radiol Endod* 102: 235–241, 2006.
56. Nakamoto T, Taguchi A, Ohtsuka M, Suei Y, Fujita M, Tsuda M, Sanada M, Kudo Y, Asano A, Tanimoto K, A computer-aided diagnosis system to screen for osteoporosis using dental panoramic radiographs, *Dentomaxillofac Radiol* 37: 274–281, 2008.
57. Nakamoto T, Mahmud Uz Z, Taguchi A, Tanimoto K, New advanced computer-aided diagnosis system to screen osteoporosis using dental panoramic radiographs, *Proc Int J CARS* 9(suppl 1) S357–365, 2014.

58. Otsu N, A threshold selection method from gray-level histograms, *IEEE Trans Syst Man Cybern* SMC-9, 1: 62–66, 1979.
59. Roberts MG, Graham J, Devlin H, Image texture in dental panoramic radiographs as a potential biomarker of osteoporosis, *IEEE Trans Biomed Eng* 60: 2384–2392, 2013.
.60. Haralick RM, Shanmugam K, Distein I, Textural features for image classification, *IEEE Trans Syst Man Cybern* 3: 610–621, 1973.
61. Breiman L, Random forests, *Mach Learn* 45: 5–32, 2001.
62. Kavitha MS, An SY, An CH, Huh KH, Yi WJ, Heo MS, Lee SS, Choi SC, Texture analysis of mandibular cortical bone on digital dental panoramic radiographs for the diagnosis of osteoporosis in Korean women, *Oral Surg Oral Med Oral Pathol Oral Radiol* 119: 346–356, 2015.
63. Horiba K, Muramatsu C, Hayashi T, Fukui T, Hara T, Katsumata A, Fujita H, Automated classification of mandibular cortical bone on dental panoramic radiographs for early detection of osteoporosis, *Proc SPIE Med Imaging* 9414: 94132J-1–94132J-6, 2015.
64. Muramatsu C, Horiba K, Hayashi T, Fukui T, Hara T, Katsumata A, Fujita H, Quantitative assessment of mandibular cortical erosion on dental panoramic radiographs for screening osteoporosis, *Int J CARS* 2016 (online).
65. National Center for Health Statistics, Health, United States, *2014: With Special feature on Adults Aged 55–64*. Hyattsville, MD, 2015.
66. Ministry of Health, Labour and Welfare, Demographic statistics in 2014 (available at: http://www.mhlw.go.jp/toukei/saikin/hw/jinkou/kakutei14/index.html)
67. Nandalur KR, Baskurt E, Hagspiel KD, Finch M, Phillips CD, Bollampally SR, Kramer CM, Carotid artery calcification on CT may independently predict stroke risk, *AJR* 186; 547–552, 2006.
68. Carter LC, Tsimidis K, Fabiano J, Carotid calcifications on panoramic radiography identify asymptomatic male patients at risk for stroke. A case report, *Oral Surg Oral Med Oral Pathol* 85: 119–122, 1998.
69. Pornprasertsuk-Damrongsri S, Virayavanich W, Thanakun S, Siriwongpairat P, Amaekchok P, Khovidhunkit W, The prevalence of carotid artery calcifications detected on panoramic radiographs in patients with metabolic syndrome, *Oral Sur Oral Med Oral Pathol Oral Radiol Endod* 108: e57–e62, 2009.
70. Ertas ET, Sisman Y, Detection of incidental carotid artery calcifications during dental examinations: Panoramic radiography as a important aid in dentistry, *Oral Surg Oral Med Oral Pathol Oral Radiol Endod* 112: e11–e17, 2011.
71. Bayer S, Helfgen EH, Bos C, Kraus D, Endling N, Mues S, Prevanlence of findings compatible with carotid artery calcifications on dental panoramic radiographs, *Clin Oral Invest* 15: 563–569, 2011.
72. Shinjo K, Muneyasu M, Fujita K, Asano A, Taguchi A, Improvement of the detection method for carotid artery calcification in dental panoramic radiographs, *Int Symposium on Intelligent Signal Processing and Communication Systems (ISPACS)*, 119–122, 2009.
73. Izumi Y, Harada H, Muneyasu M, Asano A, Uchida K, Taguchi A, Improvement of automatic detection method for carotid artery calcification in dental panoramic radiographs considering gradients of local regions, IEICE Technical Report, SIS2012-5: 23–28, 2012. (in Japanese)
74. Sawagashira T, Hayashi T, Hara T, Katsumata A, Muramatsu C, Zhou X, Iida Y, Katagi K, Fujita H, An automatic detection method for carotid artery calcifications using top-hat filter on dental panoramic radiographs, IEICE Trans Inf Syst E96-D; 1878–1881, 2013.
75. Muramatsu C, Takahshi R, Hara T, Hayashi T, Katsumata A, Zhou X, Fujita H, Toward early diagnosis of arteriosclerotic diseases: Collaborative detection of carotid artery calcifications by computer and dentists on dental panoramic radiographs, *Proc SPIE Med Imaging* 9035; 903521-1–903521-6, 2014.

76. Hattori Y, Muramatsu C, Takahashi R, Hara T, Hayashi T, Zhou X, Katsumata A, Fujita H, Automated detection for carotid artery calcifications in dental panoramic radiographs: Verification of detection performance using manual ROIs, *Medical Imaging and Information Sciences*, 32: 68–70, 2015. (in Japanese)
77. Tahashira H, Nakata S, Yoshimoto M, Kawakami T, Ohtsuka A, Quantitative analysis of the X-ray opacity of the maxillary sinus, *J Japanese Soc Radiol Technol* 52: 735–740, 1996.
78. Mori S, Hara T, Tagami M, Muramatsu C, Kaneda T, Katsumata A, Fujita H, Automated detection of abnormalities in paranasal sinus on dental panoramic radiographs by using contralateral subtraction technique based on mandible contour, *Proc SPIE Med Imaging* 8670: 86702R-1–86702R-6, 2013.
79. Hara T, Tagami M, Mori S, Kaneda T, Katsumata A, Zhou X, Muramatsu C, Fujita H, Automated detection of paranasal sinus diseases on dental panoramic X-ray image by using contralateral subtraction technique, *IEICE J* 96-D: 885–891, 2013 (in Japanese).
80. Ohashi Y, Ariji Y, Katsumata A, Fujita H, Nakayama M, Fukuda M, Nozawa M, Ariji E, Utilization of computer-aided detection system in diagnosing unilateral maxillary sinusitis on panoramic radiographs, *Entomaxillofac Radiol* 45: 2016.
81. Schie O, Waerhaug J, Loydal A, Arno A, Alveolar bone loss as related to oral hygiene and age, *J Periodontol* 30: 7–16, 1959.
82. Zhang M, Katsumata A, Muramatsu C, Hara T, Suzuki H, Fujita H, An automatic early stage alveolar-bone-resorption evaluation method on digital dental panoramic radiographs, *Proc SPIE Med Imaging* 9035: 90353G-1–90353G-9, 2014.
83. Takahashi R, Muramatsu C, Hara T, Hayashi T, Fukui T, Katsumata A, Fujita H, Semiautomatic method for measuring alveolar bone resorption level with alveolar crest line detection by thin plate spline deformation and gradient analysis on dental panoramic radiographs, IEICE Technical Report, MI2015-139, 2015.
84. Grigorescu SE, Petkov N, Kruizinga P, Comparison of texture features based on Gabor filter, *IEEE Trans Image Processing* 11: 1160–1167, 2002.
85. Bishop C, *Pattern Recognition and Machine Learning*, New York, NY: Springer Science+Business Media, LLC, p. 30, 2006.

6

CAD Tool and Telemedicine for Burns

Begoña Acha-Piñero

José-Antonio Pérez-Carrasco

Carmen Serrano-Gotarredona

6.1 Introduction

A burn injury is the result of an energy transfer that destroys skin and adjacent tissues.

Burns are often classified according to their depth into four groups (Clarke 1992).

1. First-degree burns or epidermal burns are usually red, dry, and painful. Usually, they correspond to sun burns that do not normally reach the Burn Unit and heal spontaneously.
2. Superficial dermal burns are moist, red, and weeping, and blanch with pressure. They are painful to air and temperature.
3. Deep-dermal burns are wet or waxy dry, variable in color (patchy to cheesy white to red), and do not blanch with pressure. The main characteristic of deep-dermal burns is the early development of extensive blisters.
4. Full-thickness burns are generally leathery in consistency, dry, insensate, and waxy. Full-thickness burns destroy both layers of skin (epidermis and dermis) and may penetrate more deeply into underlying structures. These wounds will not heal, except by contraction and limited epithelial migration, with resulting hypertrophic and unstable cover.

Deep-dermal burns and full-thickness burns are referred to a surgeon.

In Figure 6.1, examples of the different depths of burns are shown.

A different two-group classification can be stated:

1. Burn wounds requiring excision. These burns do not need grafts because they heal spontaneously. Superficial dermal burns belong to this group.
2. Burn wounds that need grafts. These wounds do not heal spontaneously. Deep-dermal burns and full-thickness burns belong to this type.

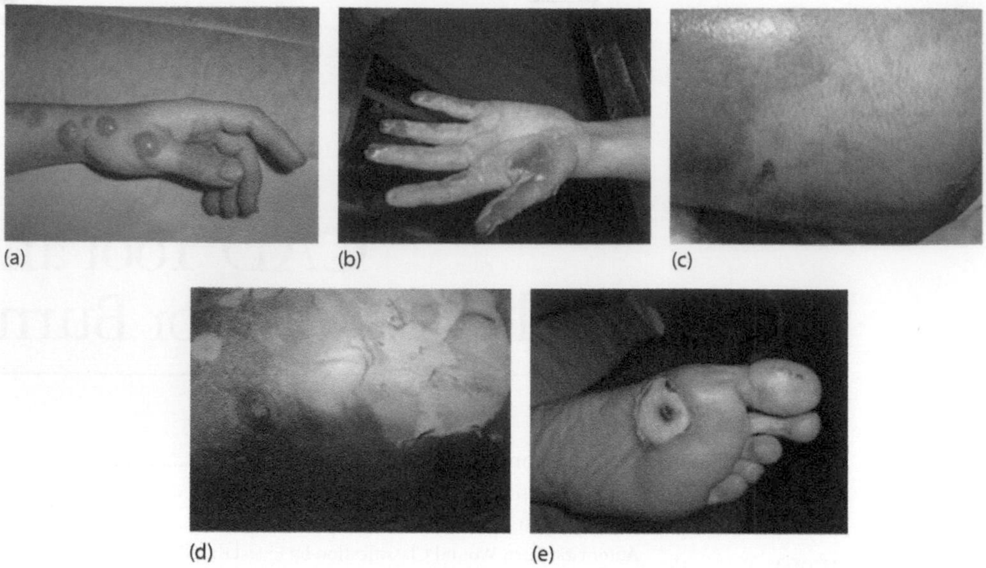

FIGURE 6.1 (a) Superficial dermal (blisters), (b) superficial dermal (red), (c) deep dermal, (d) full thickness (beige), (e) full thickness (brown).

It is very important to provide a correct burn-severity assessment of the burn wounds and the corresponding first treatments as soon as possible for a successful evolution of the wound. However, there is usually a lack of specialists in local medical centers or at the emergency room (Roa et al. 1999). According to the World Health Organization, there should be at least one bed in a Burn Unit for each 500,000 inhabitants. Therefore, this implies that only one Burn Unit covers an extensive geographic area. Thus, if a patient suffers a burn and goes to the medical center, a non-specialist will be the first to make a judgment. This non-expert doctor will contact some expert at another center via telephone communication, and this expert, according to the non-expert doctor's subjective opinion about the color, texture, and other aspects important for burn characterization, will provide a first treatment to be supplied to the patient. An incorrect first treatment may imply unnecessary surgeries, displacements of the patient, or even formation of hypertrophic scars (Jaskille et al. 2009). All these consequences involve high sanitary costs and psychological trauma for the patient and family. Thus, all these facts justify the need of telemedicine or computer-aided diagnosis (CAD) tools, which will help non-expert doctors in diagnosis. This chapter covers both issues.

Three main issues are addressed in this chapter: the feasibility of telemedicine applied to burn diagnosis, burn depth estimation from digital photographs, and burn area estimation.

The first issue is covered in Section 1.2, where it is shown that it is possible to diagnose through digital photographs sent between medical centers. Thus, the feasibility of telemedicine for burn diagnosis is demonstrated.

Digital image processing applied to burn depth estimation is described in Section 1.3. Specifically, two different methodologies are described. In Section 1.3.1, a method that is based on color and texture features extracted from digital photographs is exposed. In Section 1.3.2, a psychophysical experiment to determine subjective features that physicians employ to determine burn depth is described, and subsequently, mathematical features correlated with these subjective features are determined. Burn depth is then estimated employing these mathematical features.

Finally, in Section 1.4, a software tool for burn area estimation is described, and an extensive validation of this tool is presented.

6.2 Telemedicine

As stated in the introduction, a burn wound is classified according to parameters such as the percentage of body surface injured and the burn depth. Other parameters such as the age of the patient, sex, and the existence or absence of lung lesions (Tobiasen et al. 1982) are also important. Depth of the burn and the surface injured are usually assessed by visual inspection. If this diagnosis is to be done using an image of the burn wound instead of direct visual inspection, it is essential that the image preserves all the details and color information in order to avoid a wrong diagnosis. In this section, it is demonstrated that it is possible to diagnose burn depth from compressed digital photographs, which opens the possibility of telemedicine in burn diagnosis.

In the work by Acha et al. (2005), in order to facilitate and standardize the acquisition of the images, a protocol was developed. This protocol was validated by 12 experts of the Plastic Surgery and Burns Unit in the Virgen del Rocio Hospital, Seville (Spain), using 38 images of 22 patients with different known-burn wounds. The experts were asked to give a diagnosis, taking only into account each photograph. The results were satisfactory and proved that the image acquisition protocol was adequate for a later automatic burn assessment.

The main points of the acquisition protocol are the following:

- 40–50 cm distance between camera and patient.
- Healthy skin should appear in the image when possible (at least 20% of the whole image).
- The background should be a green/blue sheet.
- The flash must be on.
- The camera should be placed parallel to the burn.
- The different parameters were configured and set to: *ISO* (International Organization for Standardization) speed 100, exposure time 1/200 s, and aperture (*f-stop*) 20.

This last condition makes the xenon flash illumination sufficiently strong to dominate illumination in the patient room. That is very important because, in this way, color values in the picture are only influenced by the xenon flash illumination (Acha et al. 2005).

Subsequently, as described in the work by Acha et al. (2005), a colorimetric calibration algorithm has to be carried out, which allows converting RGB (Red-Green-Blue) values under the unknown main illuminant (depending on the flash used) to RGB values under D50 (Daylight D Series Illuminants, 50) illuminant. This will create an image with colors accurately representing the visual scene in RGB under D50 illuminant.

Once the image has been converted to an adequate color space, it may have to be compressed.

The methods for compression of images can be divided into two groups: methods that implement compression without losses and those with losses. The first group keeps all the information of the image, whereas the second method produces losses, making it impossible to recover the original information. However, in many cases, these losses are not perceptible by the human eye.

Those methods that perform compression without losses take benefit of the statistical redundancy of the image. The compression obtained using only these methods is not sufficient to obtain an acceptable compression rate in the number of bits, which will vary only from 2:1 to 16:1. Methods with losses perform the compression by eliminating subjective redundancies. This means that humans will not be aware of the compression implemented and the loss of information. The most common compression methods within this group are those based on the discrete cosine transformation (DCT), those based on the wavelet transformation, those using quantification of vectors, and those using fractal compression methods.

In work conducted by Roa et al. (1999), the compression JPEG (Joint Photographic Expert Group) algorithm with losses, which uses DCT, is employed. In this work, experiments to assess this compression method as valid in diagnosis were carried out. For this, 38 images of 22 different patients with burn wounds were taken. The protocol described above for taking the pictures was followed. In the study, these images were presented to a group of experts in burns. Subsequently, the experts had to answer questions concerning the diagnosis and their certainty about the diagnosis based on the quality of the image. Each image was shown by an ascendant sweeping of the quality factor (Q) of the JPEG algorithm (Van der Heijden 1994). The original images were also shown to the experts. Values fewer than 10 meant that the image presented low quality. Values of Q over 50 meant that the compressed image could not be distinguished from the original by a human observer. Two conclusions were obtained after performing the experiments. First, when the compressions were implemented using Q values over 50, the diagnosis using the compressed images coincided with those diagnoses using the original images in 90% of the cases. Second, the experts did not distinguish quality differences between the original and the compressed images. A value higher than 50 implied that the compressed images required 50 times less memory than the original ones.

This study demonstrated that an affordable camera, acquiring pictures following the protocol described, and a compression algorithm (JPEG) can be used to acquire and transmit images maintaining the original information in the picture, making possible the telediagnosis of burn wounds by experts in different locations.

6.3 Burn Depth Estimation

The depth of a burn is one of the key factors in burn classification. Depth is not easy to estimate. For inexperienced surgeons, as reported by Hlava et al. (1983), the success rate in the diagnosis is around 50%. This value increases to 64%–76% when the surgeons have a high degree of experience (Brown et al. 1998, Heimbach et al. 1984). One of the main characteristics that physicians take into account when assessing the depth of a burn wound is color. Thus, any system used to assess these wounds should preserve this property very accurately.

In the literature, there are plenty of works to analyze burn depth. These works include the use of thermographic images to evaluate burn depth (Cole et al. 1990, Romero-Mendez et al. 2009, Ruminski et al. 2007, Tylman et al. 2011, Tylman et al. 2012), terahertz pulsed imaging (Huang et al. 2007), polarization-sensitive optical coherence tomography (Pierce et al. 2003), reflectance spectrometer (Yeong et al. 2005), laser Doppler flux measurements (Monstrey et al. 2011), spectrophotometric intracutaneous analysis scope (SIAscope) (Tehrani et al. 2008), analysis of the transfer of heat through skin (Xu et al. 2009, Ng et al. 2009, Xu et al. 2010), near-infrared images and reflectance spectrophotometry (Shai et al. 2007, Bochko et al. 2010, Kaartinen et al. 2011). The problem with all these techniques is that they are very complex and expensive. For instance, in thermal images, the infrared camera should be placed in a temperature-controlled room with a maintained humidity (Acharya et al. 2012), which is an unaffordable item for most local centers.

In this section, two different techniques using simple equipment based on a camera and a computer to estimate burn wound depth are described (Serrano et al. 2005, 2015, Acha et al. 2013). In both techniques, in order to make the systems as simple as possible, images are acquired with a digital photographic camera, following the protocol described in the previous section (Roa et al. 1999), and a calibration algorithm was performed to accurately maintain the color values in the scene (Acha et al. 2005).

6.3.1 Automatic Burn Wound Classification by Extracting Features from Images

The first technique for automatic burn wound classification (Acha et al. 2005, Serrano et al. 2005) segments the burn wound and extracts some features that will be input to a neural network to classify the wound.

6.3.1.1 Segmentation Stage

A segmentation stage to separate the burn from the rest of the picture has to be performed. First, a user selects a small region inside the burn. This region is used to select the color to be segmented.

A transformation from RGB color space to CIE (Commission Internationale de l'Éclairage) L*u*v* color space (Plataniotis and Venetsanopoulos 2000) is performed. CIE L*u*v* color space is a uniform representation system where Euclidean distances are correlated with color differences in human perception (Serrano et al. 2005). Then, an anisotropic diffusion filter (Perona and Malik 1990) is applied. Subsequently, a gray-scale distance image is created by computing the Euclidean distance from each pixel in the filtered image to the color obtained in the first step. The equation that performs this operation is as follows:

$$f(n,m)=\sqrt{(L^*(n,m)-L_c^*)^2+(u^*(n,m)-u_c^*)^2+(v^*(n,m)-v_c^*)^2} \tag{6.1}$$

where L_c^*, u_c^* and v_c^* are the mean of L^*, $u^*$and v^* for the small region selected by the user.

Finally, once the distance image has been computed, a thresholding operation is performed (Acha et al. 2003). In this stage, pixels under the threshold will be selected as belonging to the burn wound.

6.3.1.2 Classification Stage

The next step is to classify the burn into its depth. Experts in burn wounds determine the depth of a wound, taking into account color and texture. In the CIE L*u*v* color space, some features, such as hue (h) and chroma (c) can be used to measure the color of the wound. Regarding texture, first-order texture features were computed, such as mean of lightness (L^*), mean of hue (h), mean of chroma (c), standard deviation of lightness (σ_L), standard deviation of hue (σ_h), standard deviation of chroma (σ_c), mean of u^*, mean of v^*, standard deviation of $u^*(\sigma_u)$, standard deviation of $v^*(\sigma_v)$, skewness of lightness (s_L), kurtosis of lightness (k_L), skewness of $u^*(s_u)$, kurtosis of u^* (k_u), skewness of $v^*(s_v)$, and kurtosis of $v^*(k_v)$.

In order to reduce the set of descriptors, Sequential Forward Selection (SFS) and Sequential Backward Selection (SBS) methods were implemented (Acha et al. 2002). The final descriptors obtained were lightness, hue, standard deviation of the hue component, u^*chrominance component, standard deviation of the v^*component, and skewness of lightness.

Finally, a Fuzzy-ARTMAP (Adaptive Resonance Theory MAP) (Carpenter et al. 1992) neural network is fed with these features and classifies the burn wound into a burn depth.

Figures 6.2 and 6.3 show the different stages of the segmentation and classification algorithm.

6.3.1.3 Results

The burn classification system was tested on 35 digital photographs acquired by physicians following the specified acquisition protocol. All the images were diagnosed by a group of plastic surgeons, belonging to the Burn Unit of the Virgen del Rocio Hospital in Seville (Spain). Both the segmentation stage and the classification step were assessed. The segmentation algorithm proposed in this chapter was tested with

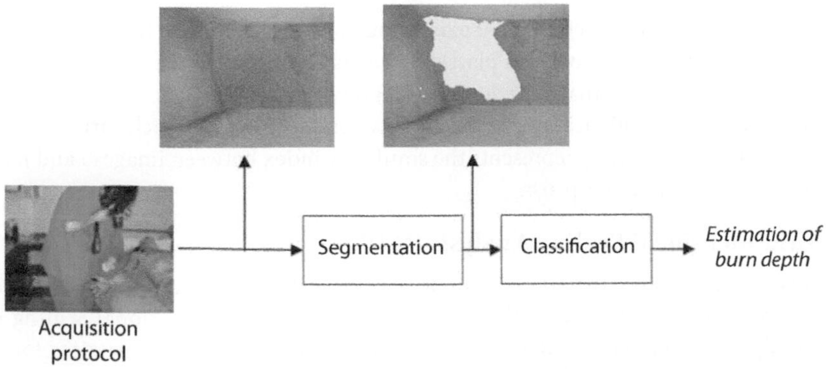

FIGURE 6.2 General scheme of the process to assess the burn depth.

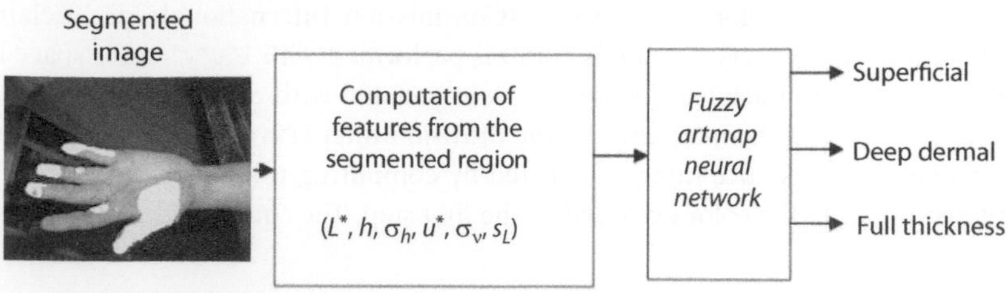

FIGURE 6.3 Scheme of the classification approach.

TABLE 6.1 Classification Results

Burn Depth	Success (%)
Superficial dermal	100
Deep dermal	84.61
Full thickness	77.77
Average	**88.57**

35 images manually segmented by five physicians from the Burn Unit of the Virgen del Rocio Hospital, from Seville (Spain). Subsequently, positive predictive value (PPV) and Sensitivity (S) value were computed. On average, both Sensitivity and PPV coefficients where over 0.8.

To assess the classification stage, 13 images with superficial dermal burns, 13 with deep-dermal burns, and 9 with full-thickness burns were used. Classification results are summarized in Table 6.1. The average success percentage was 88.57%.

6.3.2 Automatic Burn Wound Classification by Extracting Subjective Features from Burn Wounds

In work by Acha et al. (2013) and Serrano et al. (2015), a psychophysical experiment and multidimensional scaling (MDS) were used to identify meaningful features that plastic surgeons employ in their classification of burns as to their depths.

The classifiers used in this CAD system were a k-nearest neighbor classifier (KNN) and a support vector machine (SVM).

6.3.2.1 Psychophysical Experiment

In this experiment, the similarity between 190 pairs of burn images (obtained from 20 images) was measured according to the experience of eight plastic surgeons belonging to the Unit of Virgen del Rocio Hospital in Seville (Spain). A similarity index was obtained for each pair of images, in which value 0 meant no similarity and value 10 meant equivalency between the images by each surgeon. A 20 x 20 similarity matrix, D, whose element *dij* represents the similarity index between images *i* and *j*. An example of a pair of images is shown in Figure 6.4.

6.3.2.2 Multidimensional Scaling Analysis (MDS)

The aim of this step is to place the 20 images into a graphical representation space, where distances between images will be proportional to the similarity indices. Multidimensional Scaling (MDS) is a multivariate analysis method (Kruskal and Wish 1978) to analyze similarity matrices for a graphical representation of the data that would lead to a better interpretation. The input to the multidimensional scaling is the similarity matrix D 20×20 computed in the previous stage.

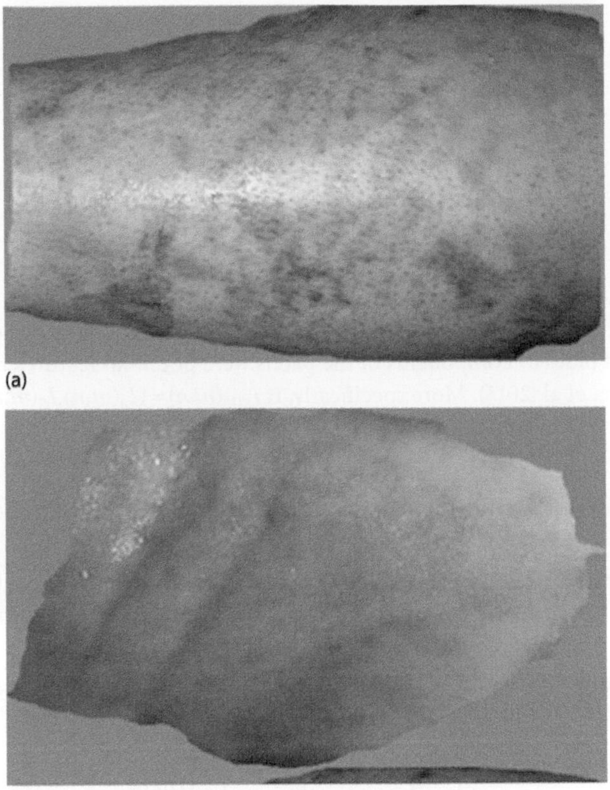

(a)

(b)

FIGURE 6.4 Example of a pair of images to be compared in terms of similarity.

The output of the MDS is a spatial representation of each image in a P-dimensional space. In other words, each image *i*, $i = 1,\ldots,20$, is assigned a position in the P-dimensional space, $X_i = (x_1,\ldots,x_P)$, where distances between images are correlated to differences between images according to their similarity indexes. In this experiment $P = 3$ was identified as the optimum space dimension (Acha 2013).

6.3.2.3 Interpretation of the Coordinates: Brainstorming

The aim of this step is to give a physical interpretation of the axes or coordinates of the MDS space. For this purpose, six experts in digital image processing analyzed the three axes ($P = 3$) of the MDS space obtained. As a result of the analysis, axis 1 was labeled as "amount of pink" and "humidity," or moisture. Axis 2 was labeled as "texture of the color." Axis 3 was considered as related to "colorfulness." Eight mathematical features were considered as possibly related to these labels.

Amount of pink. Hue is a color attribute that describes what we commonly call the color of an object. If a color pixel in an image is expressed in the L*a*b* color space (Rangayyan et al. 2011) as

$$I_{L^*a^*b^*}(m,n) = (I_{L^*}(m,n), I_{a^*}(m,n), I_{b^*}(m,n)) \tag{6.2}$$

where $I_{L^*a^*b^*}(m,n)$ represents the image in the $L^*a^*b^*$ color space, the hue average value within a burn can be estimated as (Acha et al. 2013, Rangayyan et al. 2011):

$$h = \arctan\left(\frac{s}{c}\right) \tag{6.3}$$

where

$$s = \frac{1}{|Q|} \sum_{(m,n)\in Q} \sin\left(\arctan\left(\frac{I_{b^*}(m,n)}{I_{a^*}(m,n)}\right)\right) \tag{6.4}$$

$$c = \frac{1}{|Q|} \sum_{(m,n)\in Q} \cos\left(\arctan\left(\frac{I_{b^*}(m,n)}{I_{a^*}(m,n)}\right)\right) \tag{6.5}$$

and Q represents the set of pixels belonging to the burn.

Humidity can be related to the presence of glitters in the photograph. Thus, outliers jointly in the red (R), green (G), and blue (B) components of the pixels were potential features to measure this physical characteristic (Acha et al. 2013). More specifically, if $I_{RGB}(m,n) = (I_R(m,n), I_G(m,n), I_B(m,n))$ is the RGB image,

$$o_{RGB} = \frac{1}{|Q|} \sum_{(m,n)\in Q} (I_R(m,n) > 240) \,\&\, (I_G(m,n) > 240) \,\&\, (I_B(m,n) > 240) \tag{6.6}$$

represents the number of outliers in the RGB space normalized by the area of the burn.

Texture of the color. Homogeneity of colors within a burn can be captured with first-order statistical parameters in the colorimetric coordinates of the image. To measure the color homogeneity, the third standardized moments (skewness of a* and b* components: sk_{a^*}, sk_{b^*}) and the angular variance of hue v_{h^*} (Mardia 1972, Acha et al. 2013), were chosen, with v_{h^*} computed as

$$v_{h^*} = 1 - \frac{1}{|Q^*|} \sum_{(m,n)\in Q^*} 1 - \cos\left(\arctan\left(\frac{I_{b^*}(m,n)}{I_{a^*}(m,n)}\right) - h^*\right) \tag{6.7}$$

where Q^* is the set of pixels that, not being considered outliers, belong to the burn and h^* is the hue average within the burn except for pixels that are outliers.

Colorfulness is traditionally linked to chroma, which is defined as the colorfulness of an area judged as a proportion of the brightness of a similarly illuminated white area. $C^*_{a^*b^*}$ is a numerical parameter that tries to measure this colorfulness and its calculation is (Acha et al. 2013, Rangayyan et al. 2011):

$$C^*_{a^*b^*} = \frac{1}{|Q|} \sum_{(m,n)\in Q} \sqrt{I_{a^*}(m,n)^2 + I_{b^*}(m,n)^2} \tag{6.8}$$

In addition, it was also observed that colorful sensation was also related with the histogram of the lightness component L^*: a long right tail in L^* distribution produces a vividness sensation. This has been measured according to quartile skewness (Acha et al. 2013, Brys et al. 2006), where left and right tail measures are defined relative to the median of F, denoted as $F^{-1}(0.5)$, F being a continuous univariate distribution.

$$SK_{0.25} = \frac{(F^{-1}(1-0.25) - F^{-1}(0.5)) - (F^{-1}(0.5) - F^{-1}(0.25))}{\left|F^{-1}(1-0.25) - F^{-1}(0.25)\right|} \tag{6.9}$$

Furthermore, peakedness in the distribution of the chromatic components could also be a sign of vividness. Thus, kurtosis (related to the fourth order moment) of a^* coordinate was tested as a potential feature to classify burns into their depth (Acha et al. 2013). In summary, eight features were employed as the preliminary feature set to classify burns: Chroma ($C^*_{a^*b^*}$), Outliers (O_{RGB}), hue (h), variance of hue (v_{h^*}), skewness of a*(sk_{a^*}), skewness of b*(sk_{b^*}), quartile skewness ($SK_{0.25}$) and kurtosis of a^* (k_{a^*}).

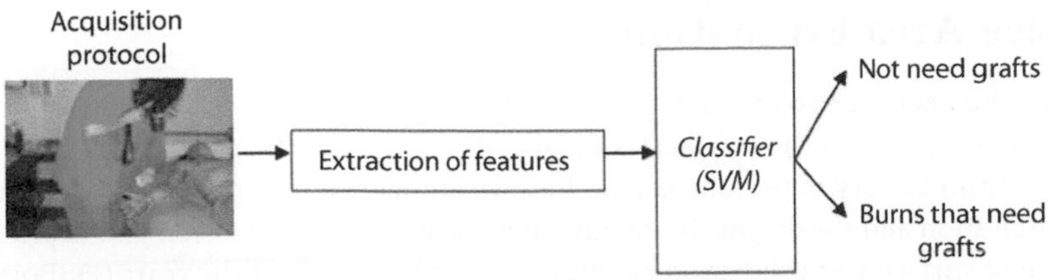

FIGURE 6.5 Scheme of the CAD tool.

Subsequently, SFS and SBS methods were employed to identify which features out of the eight numerical parameters had the most discriminant power to classify burns. For both methods, an SVM was chosen as classifier due to its advantages of accuracy, robustness to noise, or less overfitting (Acharya et al. 2012, Tomuro 2014, Cortes and Vapnik 1995).

With the application of SFS to the set of features, a set composed by four features was obtained. In the same way, applying SBS to the eight features led to a set composed by four features.

Three features were common to both SFS and SBS subsets: $C^*_{a^*b^*}$, h and k_{a^*}. The addition of a new feature to this subset slightly diminished the classification error. This feature can be either the variance of the hue component or the skewness of b^*. Both features are first-order statistics for color texture measures.

In Figure 6.5, the scheme of the CAD system can be seen.

6.3.2.4 Material

The dataset employed in this work is divided into two subsets: 20 images constitute the training dataset (used for the psychophysical experiment and during the training of the SVM), and 74 different images correspond to the test dataset (Burns BIP_US 2015). All the images were acquired with a Canon EOS 300D at the Burn Unit from Virgen del Rocio Hospital in Seville (Spain) following the protocol for acquiring the images and the calibration procedure described in previous sections. The photographs were acquired within the first 48 h of evolution of the burn. Physicians (plastic surgeons affiliated to the Burn Unit) determined their burn diagnosis by directly observing the burn at the admission in all cases and confirmed their diagnosis after 1 week or 10 days of evolution.

6.3.2.5 Results

For each image in the test database (74 images in all), the four mathematical features described in Section 1.3.2.3 were extracted after the segmentation of the wounds for each image. These features were fed to the classifier, which decides if a burn wound needs grafts or does not need grafts. A 79.73% success classification rate was attained. The achieved sensitivity is 0.97 and specificity 0.60 (see Table 6.2).

TABLE 6.2 Classification Results with the Selected Feature Subset

Success rate (%)	79.73
Sensitivity (S)	0.97
Specificity (E)	0.60
PPV	0.73
NPV	0.95

6.4 Surface Area Estimation

In this section, the analysis of burn surface area determination is analyzed. As stated in the introduction, this parameter is one of the key factors for specifying the first treatment. The burn surface area estimation constitutes a very important step in the evaluation of burned patients, because the nutrients and fluid resuscitation and the diagnosis are intimately related to the extension and depth of a burn.

The problem of surface area estimation has been analyzed throughout the years (DuBois and DuBois 1916, Berkow 1924, Boyd 1935, Lund and Browder 1944). Nowadays, the Rule of Nines (Wallace 1951) and the Rule of Palms (Lund and Browder 1944) are the methods more commonly used to estimate the surface area of a burn. The Rule of Palms consists of considering that the surface of the palms corresponds to the 1% of the total surface area. Physicians estimate the number of palms that cover the burn, and therefore, they calculate the burned area. Some modifications to this rule have decreased this percentage to 0.76% (Gehan and George 1970). The Rule of Palms is usually preferred, being more reliable than the Rule of Nines, which tends to overestimate the size of the burn and provides a larger variability (Wachtel et al. 2000, Miller-Finley et al. 1991). One of the drawbacks of these estimations is that they do not make distinctions among gender or age, or even shape or weight, except for Lund and Browder, who established two standard figures: one for children and one for adults.

Some attempts to represent Two-Dimensional (2D) and Three-Dimensional (3D) models (Lee et al. 1994a, 1994b, Yu et al. 2003) of the surface area have been developed (Sage II, SageDiagram, LLC). One of the main problems with 2D representations is that some regions cannot be represented. In (Yu et al. 2003), a system called Chang Gung Whole Body Scanner (*CGWBS*) used the axial computerized tomographic technique to compute the total body surface area in Asian groups of people. However, this system is expensive and provides estimation errors at childhood and elderly ages, as well as due to patient movements.

As far as 3D representation and estimation of burned surface area is concerned, some computer-based models have been described so far: EPRI's 3D Burn Vision (Neuwalder et al. 2002) and BurnCase 3D (Haller et al. 2009, Dirnberger et al. 2004, Dirnberber et al. 2003). These models provide good computer-user interface and allow for including clinical records. However, as it happens with the aforementioned methods, they are not adapted to the patient's constitution and have not been validated with real patients.

In the work conducted by Prieto et al. (2011), a 3D digital model (BAI) to compute the burned surface area of a patient, taking into account differences due to age, gender, weight and constitution, was built. This system was designed in C++ using OpenGL libraries and was intended to be easy to use. The Body Mass Index (BMI) was provided to classify the patient into normal, overweight, or obese. Up to 80 models accurately adapted to gender, age, weight, height, and constitution were implemented. The program uses three-dimensional triangle meshes to represent the models and allows scaling, rotation, and translation of the models.

The user represents the burns on the model with the mouse. The user can choose among green, yellow, red, or dark red color to indicate the depth of the burn (epidermal, superficial dermal burns, deep-dermal burns, and full-thickness burns, respectively). The user can choose among different well-known formulas to compute the absolute total body surface area (Prieto et al. 2011).

An example of the interface is shown in Figure 6.6.

6.4.1 Results

The system was validated using a clinical study and by implementing several assessment comparisons.

6.4.1.1 Comparison between BAI and Gold Standard

In the first comparison, estimated burned areas obtained using BAI were compared with real burn surface areas using adhesive graph paper on the patients. 37 patients of different height, age, and weight

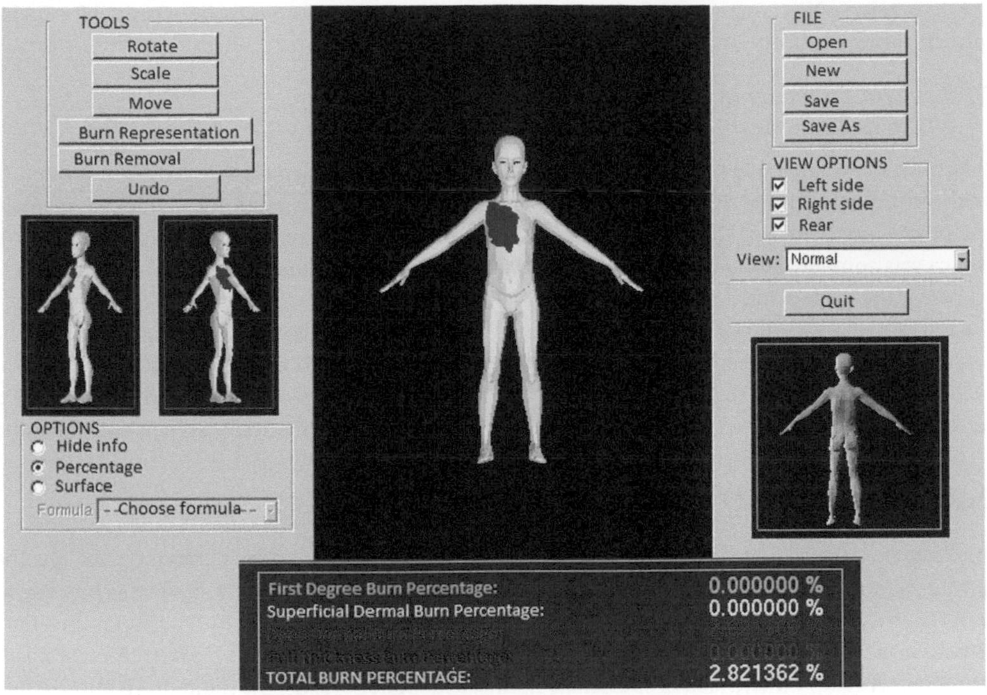

FIGURE 6.6 Example of the interface of BAI software.

participated in the study. Adhesive graph paper was used to construct patches of measured areas. The patches were attached to different parts of the body on the patients, and their burned areas were also estimated with BAI. Computation of the intraclass correlation coefficient (ICC) (Fleiss 1986) and a t-Student test were performed to measure the correlation between the BAI estimations and the real measures. The ICC obtained was 0.9918.

In the t-Student test by Prieto et al. (2011), the hypothesis that there are no significant differences between real surface values and their measurements with BAI was adopted. The parameter t obtained in the t-Student test was $t = 0.0902$. For a probability of 0.05 (95% probability of making a correct statement) and $n = 37$, $t_{0.95} = 1.648$. As $t < t_{0.95}$, the initial hypothesis can be considered as true. The p-value for the t-Student test and data resulted in $p = 0.929$.

6.4.1.2 Second Experiment: Comparison between Traditional Methods, BAI, and Gold Standard

In this evaluation, the classical Rule of Nines was assessed. For this, 12 different-sized patches were used. Four plastic surgeons and one maxillofacial surgeon from Virgen del Rocio Hospital estimated the size of the patches required by using the Rule of Nines. BAI was again used to also compute the area of the patches. Again, the ICC coefficient was computed, a t-Student test was performed for each method, and p-value was computed. In Table 6.3, ICC coefficient and p-value obtained with BAI and with the Rule of Nines applied by the 5 medical specialists are summarized. Analyzing this table, it can be observed that

TABLE 6.3 ICC and p-value Obtained When Estimating Burn Area with BAI and with the Rule of Nines Applied by 5 Different Medical Specialists

	BAI	Specialist 1	Specialist 2	Specialist 3	Specialist 4	Specialist 5
ICC	0.9990	0.8928	0.9192	0.9733	0.9656	0.8310
p-value	0.7874	0.0129	0.0061	0.0215	0.3679	0.0009

the highest value of ICC is attained with BAI and the null hypothesis can only be accepted for BAI and for plastic surgeon 4 if a typical 0.05 is accepted as significance level.

6.4.1.3 Third Experiment: Comparison between BAI and Traditional Methods in Real Cases

In this third experiment, 80 patients with burn wounds randomly chosen and aged from 9 months to 82 years participated in the study. Each time a patient arrived at the hospital, the patient was evaluated by the physicians, who used the Lund and Browder graph for burn representation. The patient's burned skin area was computed by using the Rule of Nines. BAI was also used during the first 48 h of evolution to represent the same burned skin area.

Correlation analysis and t-Student tests indicated that the hypothesis of null difference between both measures is only true for deep-dermal burn surface area estimation. There were significant differences between BAI and the Rule of Nines. Therefore, and according to the results obtained in the other two experiments, the advantage of using BAI method for burn surface estimation is shown and confirmed.

Conclusions and Discussion

In this chapter, a revision of some methods to provide burn diagnosis has been performed. As it has been stated throughout the chapter, it is essential to apply the correct first treatments as soon as possible for a successful evolution of the wound (Clarke 1992, Roa et al. 1999, Jaskille et al. 2009). Disposal of simple, noninvasive, and automatic systems for early diagnoses is very important. Moreover, in most hospital centers or emergency units, a lack of specialists in this field is latent. Thus, the main motivation of building a CAD tool should be to help non-expert physicians in local medical centers diagnose burn wounds in order to avoid a non-correct first assessment, which is crucial to the evolution of the burn and could have an important economic impact on the health system.

Two fundamental aspects in burn diagnosis are depth and surface burn area. These two factors define the kind of burn, its severity, and the time to heal, when possible.

Burn depth estimation has been deeply analyzed in Section 1.3. There are many techniques to analyze the depth of a burn. They can be integrated in two main categories. The first category encloses invasive techniques, where biopsies and posterior studies are carried out to provide a diagnosis (Watts et al. 2001). This technique, although considered the "gold standard," has some drawbacks, such as the loss of dermis in the burn, existence of variability depending on where the biopsy was acquired, and even the residual scars provoked during the acquisition. On the other hand, free of these inconveniences are noninvasive techniques (Jerath et al. 1996), including the analysis of the vascularization after the lesion by using vital colorants (Jerath et al. 1996), the use of laser Doppler imaging (Kamolz et al. 2003), or the analysis of the different spectral backscattering effects of burned skin at different burn degrees (Eisenbeiss et al. 1999). All these techniques have either logistic problems (they are expensive and may require emergency systems) and/or they require a great level of knowledge and expertise. To solve these limitations, several CAD algorithms have been described in this chapter which extracts numeric features from images of the burn wounds corresponding to physical features that physicians pay attention to for diagnosis.

Section 1.4 has focused on burn surface area estimation. The problem with most classical tools for burn area estimation is that none of them or any of other existing formulae have taken into account variables such as age, gender, or physical constitution. Even the recent developments of 2D or 3D systems to estimate burn surface area (Lee et al. 1994a, 1994b, Yu et al. 2003) have limitations in terms of providing models adapted to patient constitution. Besides, as stated previously, any tool to help in burn diagnosis should be easy and should not require a certain degree of expertise in the field, as required with the Rules of Nines and Palms, in spite of their simplicity. Thus, the availability of CAD tools such as the described BAI system constitutes a giant step in the development of inexpensive and easy-to-use tools that can be installed on computers in hospitals and emergency units, thus making it easy to compute burn area surface without higher knowledge.

References

Acha, B., C. Serrano, and J.I. Acha. 2002. Segmentation of burn images using the L*u*v* space and classification of their depths by color and texture information. *Proc SPIE* 4684:1508–15.

Acha, B., C. Serrano, J.I. Acha, and L.M. Roa. 2003. CAD tool for burn diagnosis. *Lect Notes Comput Sci* 2732:294–305.

Acha, B., C. Serrano, J.I. Acha, and L.M. Roa. 2005. Segmentation and classification of burn images by color and texture information. *J. Biomed. Opt.*10(3):1–11.

Acha, B., C. Serrano, I. Fondón, and T. Gómez-Cía. 2013. Burn depth analysis using multidimensional scaling applied to psychophysical experiment data. *IEEE Transactions on Medical Imaging* 32(6):1111–1120.

Acharya, UR., E.Y.K. Ng, J.H. Tan, and V. Sree. 2012. Thermography based breast cancer detection using texture features and support vector machine. *J Med Syst* 36(3):1503–10.

Berkow, S.G. 1924. A method of estimating the extensiveness of lesions (burns and scalds) based on surface area proportions. *Arch Surg* 8:138.

Burns BIP_US database. http://personal.us.es/rboloix/ Burns_BIP_US_database.zip (accessed March 31 2015).

Bochko, V., P. Valisuo, T. Harju, and J. Alander. 2010. Lower extremity ulcer image segmentation of visual and near-infrared imagery. *Skin Res Technol* 16(2):190–7.

Boyd, E. 1935. The *Growth of the Surface Area of the Human Body*. Minneapolis: University of Minnesota Press.

Brown, R.F., P. Rice, and N.J. Bennett. 1998. The use of laser Doppler imaging as an aid in clinical management decision making in the treatment of vesicant burns. *Burns* 24:692–8.

Brys, G., M. Hubert, and A. Struyf. 2006. Robust measures of tail weight. *Comput Stat Data Anal* 50:733–59.

Carpenter, G.A., S. Grossberg, S. Markuzon, and J.H. Reynolds. 1992. Fuzzy-ARTMAP: A neural network architecture for incremental supervised learning of analog multidimensional maps. *IEEE Trans Neural Netw* 3:698–713.

Clarke, J.A. 1992. *A Colour Atlas of Burn Injuries*. London: Chapman & Hall Medical.

Cole, R.P., S.G. Jones, and P.G. Shakespeare. 1990. Thermographic assessment of hand burns. *Burns* 16(1):60–3.

Cortes, C., and V. Vapnik. 1995. Support-vector networks. *Mach Learn* 20(3):273–97.

Dirnberger, J., M. Giretzlehner, M. Ruhmer, H. Haller, and C. Rodemund. 2003. Modelling human burn injuries in a three-dimensional virtual environment. *Stud Health Technol Inform* 94:52–8.

Dirnberger, J., M. Giretzlehner, T. Luckeneder, D. Siegl, H.L. Haller, and C. Rodemund. 2004. BurnCase 3D—realistic adaptation of 3-dimensional human body models. *LNCS* 3217:363–70.

Doi, K. 2007. Computer-aided diagnosis in medical imaging. Historical review, current status and future potential. *Comput Med Imaging Graphics* 31(4–5):198–211.

DuBois, D., and E.F. DuBois. 1916. A formula to estimate the approximate surface area if height and weight be known. *Arch Intern Med* 17:863–71.

Eisenbeiss, W., J. Marotz, and J.P. Schrade. 1999. Reflection-optical multispectral imaging method for objective determination of burn depth. *Burns* 25:697–704.

Fleiss, J.L. 1986. *Design and Analysis of Clinical Experiments*. New York: John Wiley & Sons.

Gehan, E.A., and S.L. George. 1970. Estimation of human body surface area from height and weight. *Cancer Chemother Rep* 54:225.

Haller, H.L., J. Dirnberger, M. Giretzlehner, C. Rodemund, and L. Kamolz. 2009. Understanding burns: Research project BurnCase 3D–overcome the limits of existing methods in burns documentation. *Burns* 35(3):311–7.

Heimbach, D.M., M.A. Afromowitz, L.H. Engrav, J.A. Marvin, and B. Perry. 1984. Burn depth estimation—man or machine. *J Trauma Acute Care Surg* 24:373–8.

Hlava, P., J. Moserova, and R. Konigova. 1983. Validity of clinical assessment of the depth of a thermal injury. Acta *Chir Plast* 25:202–8.

Huang, S.Y., E. Macpherson, and Y.T. Zhang. 2007. A feasibility study of burn wound depth assessment using terahertz pulsed imaging. In: *Proceedings of the 4th IEEE-EMBS International Summer School and Symposium on Medical Devices and Biosensors* 132–5.

Jaskille, A.D., J.W. Shupp, M.H. Jordan, and J.C. Jeng. 2009. Critical review of burn depth assessment techniques: Part I. Historical review. *J Burn Care Res* 30(6):937–47.

Jerath, M.R., K.T. Schomacker, R.L. Sheridan, and N.S. Nishioka. 1996. Burn wound assessment in porcine skin using indocyanine green fluorescence. *J Trauma* 46:1085–8.

Kaartinen, I.S., P.O. Valisuo, J.T. Alander, and H.O. Kuokkanen. 2011. Objective scar assessment–a new method using standardized digital imaging and spectral modelling. *Burns* 37(1):74–81.

Kamolz, L.P., H. Andel, W. Haslik, A. Donner, W. Winter, G. Meissl, and M. Frey. 2003. Indocyanine green video angiographies help to identify burns requiring operation. *Burns* 29:785–91.

Kruskal, J.B., and M. Wish. 1978. *Multidimensional Scaling. Madison*, WI: Bell Telephone Laboratories, Inc.

Lee, R.C., G. Kicska, and M.H. Mankani. 1994a. A three-dimensional computerized burn chart: Stage I: Development of three-dimensional renderings. *J Burn Care Rehabil* 15:80–3.

Lee, R.C., G. Kicska, and M.H. Mankani. 1994b. A three-dimensional computerized burn chart: Stage II: Development of three-dimensional renderings. *J Burn Care Rehabil* 15:191–2.

Lund, C.C., and N.C. Browder. 1944. The estimation of areas of burns. *Surg Gynecol Obstet* 79:352–8.

Mardia, K.V. 1972. *Statistics of Directional Data*. New York: Academic Press.

Miller-Finley, S.F., R.K. Waltman, and M. Lincks. 1991. Burn size estimate reliability: A study. *J Burn Care Rehabil* 12(6):546–59.

Monstrey, S.M., H. Hoeksema, R.D. Baker, J. Jeng, R.S. Spence, D. Wilson, and S.A. Pape. 2011. Burn wound healing time assessed by laser Doppler imaging. Part 2: Validation of a dedicated colour code for image interpretation. *Burns* 37(2):249–56.

Neuwalder, J.M., C. Sampson, K.H. Breuing, and D.P. Orgill. 2002. A review of computer-aided body surface area determination: SAGE II and EPRI's 3D Burn Vision. *J Burn Care Rehabil* 23(1):55–9.

Ng, E.Y.K., H.M. Tan., and E.H. Ooi. 2009. Boundary element method with bioheat equation for skin burn injury. *Burns* 35(7):987–97.

Perona, P., and J. Malik. 1990. Scale-space and edge detection using anisotropic diffusion. *IEEE Trans Pattern Anal Machine Intell* 12:629–39.

Pierce, M.C., R.L. Sheridan, B.H. Park, B. Cense, and J.F. De-Boer. 2003. Burn depth determination in human skin using polarization-sensitive optical coherence tomography. *Proc SPIE—Int Soc Opt Eng* 4956:263–70.

Plataniotis, K.N., and A.N. Venetsanopoulos. 2000. *Colour Image Processing and Applications*. Berlin: Springer.

Prieto, M.F., B. Acha, T. Gómez-Cía, I. Fondon, and C. Serrano. 2011. A system for 3D representation of burns and calculation of burnt skin area. *Burns* 37:1233–40.

Rangayyan, R.M., B. Acha, and C. Serrano. 2011. *Color Image Processing with Biomedical Applications*. Bellingham: SPIE Press.

Roa, L.M., T. Gomez-Cıa, B. Acha, and C. Serrano. 1999. Digital imaging in remote diagnosis of burns. *Burns* 25(7):617–24

Romero-Mendez, R., J.N. Jiménez-Lozano, M. Sen, and F.J. González. 2009. Analytical solution of the Pennes equation for burn-depth determination from infrared thermographs. *Math Med Biol* 27(1):21–38.

Ruminski, J., M. Kaczmarek, A. Renkielska, and A. Nowakowski. 2007. Thermal parametric imaging in the evaluation of skin burn depth. *IEEE Trans Biomed Eng* 54(2):303–12.

Shai, A., V. Avrin, A. Zvulunov, and I. Dinstein. 2007. Scaled healing assessment index: A novel method for measuring the degree of wound bed preparation. *Skin Res Technol* 13(3):227–35.

Serrano, C., B. Acha, T. Gómez-Cía, J.I. Acha, and R.M. Roa. 2005. A computer assisted diagnosis tool for the classification of burns by depth of injury. *Burns* 31(3):275–281

Serrano, C., R. Boloix, T. Gomez-Cia, and B. Acha. 2015. Features identification for automatic burn classification. *Burns* 41(8):1883–1890.

Tehrani, H., M. Moncrieff, and B. Philp. 2008. Spectrophotometric intracutaneous analysis: A novel imaging technique in the assessment of acute burn depth. *Ann Plast Surg* 61(4):437–40.

Tobiasen, J., J. Hiebert, and R. Edlich. 1982. A practical burn severity index. *J Burn Care* 3:229.

Tomuro, N. 2014. Neural networks and machine learning course. *DePaul College of Computing and Digital Media.* http:// condor.depaul.edu/ntomuro/courses/578/notes/SVM-overview.pdf (accessed September 30, 2014).

Tylman, W., M. Janicki, and A. Napieralski. 2011. Computer-aided approach to evaluation of burn wounds. In: *Int. Conf. Mixed Design of Integrated Circuits and Systems.* 653–6.

Tylman, W., M. Janicki, and A. Napieralski. 2012. Automatisation of computer-aided burn wounds evaluation. In: *Int. Conf. Mixed Design of Integrated Circuits and Systems.* 535–8.

Van der Heijden, H.F. 1994. *Image Based Measurement Systems: Object Recognition and Parameter Estimation.* John Wiley and Sons.

Wachtel, T.L., C.C. Berry, E.E. Wachtel, and H.A. Frank. 2000. The interrater reliability of estimating the size of burns from various burn area chart drawings. *Burns* 26(2):156–70.

Wallace, A.B. 1951. The exposure treatment of burns. *McGill MSc, Edin MB, University of Edinburgh. The Lancet* 257(6653):501–4.

Watts, A.M., M.P. Tyler, M.E. Perry, A.H. Roberts, and D.A. McGrouther. 2001. Burn depth and its histological measurement. *Burns* 27:154–60.

Xu, F., T.L. Lu, K.A. Seffen, et al. 2009. Mathematical modeling of skin bioheat transfer. *Appl Mech Rev* 62(5):50801–35.

Xu, F., P.F. Wang, M. Lin, T.J. Lu, and E.Y.K. Ng. 2010. Quantifying the underlying mechanism of skin thermal damage: A review. *J Mech Med Biol* 10(3):373–400.

Yeong, E.K., T.C. Hsiao, H.K. Chiang, and C.W. Lin. 2005. Prediction of burn healing time using artificial neural networks and reflectance spectrometer. *Burns* 31:415–20.

Yu, C.Y., Y.H. Lo, and W.K. Chiou. 2003. The 3D scanner for measuring body surface area: A simplified calculation in the Chinese adult. *Appl Ergonom* 34:273–8.

7

CAD of Cardiovascular Diseases

Marco A. Gutierrez

Marina S. Rebelo

Ramon A. Moreno

Anderson G. Santiago

Maysa M. G. Macedo

7.1 Introduction

Noninvasive cardiac imaging is an invaluable tool for the diagnosis and treatment of cardiovascular disease (CVD). Computed Tomography (CT), Single Photon Emission (SPECT), Positron Emission Tomography (PET), Ultrasound (US), Magnetic Resonance Imaging (MRI), Intravascular Ultrasound (IVUS), and Intravascular Optical Coherence Tomography (IVOCT) have been used extensively for physiologic understanding and diagnostic purposes in cardiology. These imaging technologies have greatly increased our understanding of normal and diseased anatomy. Cardiac image segmentation plays a crucial role and allows for a wide range of applications, including quantification of volume, computer-aided diagnosis, localization of pathology, and image-guided interventions. However, manual delineation is tedious, time consuming, and is limited by inter- and intra-observer variability. In addition, many segmentation algorithms are sensitive to the initialization, and, therefore, the results are not always reproducible as they are also limited by inter algorithm variability. Furthermore, the amount and quality of imaging data that needs to be routinely acquired in one or more subjects have increased significantly. Therefore, it is crucial to develop automated, precise, and reproducible segmentation methods. A variety of segmentation techniques has been proposed over the last few decades. While earlier approaches were often based on heuristics, recent studies employ techniques that are more sophisticated. However, cardiac image segmentation remained a challenge due to the highly variable nature of cardiac anatomy, function, and pathology. Furthermore, diseases, imaging protocols, artifacts, and noise heavily influence intensity distributions on images.

Therefore, many researchers are seeking techniques to deal with such constraints. The research in cardiac image segmentation ranges from the fundamental problems of image analysis, including shape modeling, motion detection, and tracking, to more applied topics such as clinical quantification and computer-aided diagnosis. In this chapter, we aim to provide an overview of automatic methods for quantification in cardiovascular images from major noninvasive modalities, such as CT, SPECT/PET, US, MRI, IVUS, and IVOCT.

7.2 X-Ray Cardiovascular Imaging

The discovery of x-rays allowed the visualization of the internal organs without invasive procedures, giving physicians an excellent tool for diagnostic and treatment. One of the main characteristics of x-rays is that it is an ionizing electromagnetic radiation and, therefore, harmful to living tissue. This leads to the concern with radiation exposure, particularly for the young and those who require repeated imaging. On the other hand, this characteristic can be used in cancer treatment to eliminate malignant cells using radiotherapy. Another defining characteristic of x-ray imaging is its high resolution.

X-rays are used in cardiology for different purposes. The interactive visualization of the interior of the patient is useful to find artery obstructions and perform interventions (e.g., x-ray angiography). It is also possible to plan surgeries, perform risk classification (e.g., calcium score, epicardial fat), detect and measure cardiac abnormalities (aortic stenosis, cardiac masses, size of aorta, ventricle wall thickness, etc.), detect and classify obstructions (stenosis and plaque composition), and calculate cardiac function (myocardial perfusion, fractional flow reserve).

In the last decades, several automatic solutions have been proposed to both segmentation and measurement of cardiac structure in CT images. In the following items, we present a five-year review of quantitative methods of CT that have been clinically validated, including some commercial software.

7.2.1 Calcium Score

Calcium score is a reliable, noninvasive screening method for estimating the overall calcified plaque burden and for identifying risk for future cardiac events (Alluri et al. 2015).

Ebersberger et al. (2013) presented the evaluation of a commercial fully automated software, COR Analyzer™ CS (Rcadia, Haifa, Israel), for calcium score measurement using contrast medium-enhanced coronary CT angiography (cCTA) for 127 patients. The coronary artery tree was segmented as described in Goldenberg et al. (2012). The calcifications are detected by their difference with a predictive model of the lumen attenuation profile for every cross section along the vessel. The authors concluded that the results obtained by the proposed method showed a high correlation with standard calcium score. The same software was analyzed by Rubinshtein et al. (2014) in a 136-patient study, which showed good correlation with the traditional score.

Isgum et al. (2010) developed an automated method for quantification of calcium score in the aorta using CT non-ECG triggered images. As atherosclerosis is a generalized process, the measurement of the aorta calcification can be used as a marker for cardiovascular disease. The authors compared the performance of the software against measurements performed by three physicians in images of 433 patients (340 for training and 93 for testing). The aorta was segmented using a multi-atlas segmentation method (Isgum et al. 2009). The calcifications were extracted using a two-stage statistic classifier and 63 different features. The authors concluded that the system results had a good correlation with those obtained from human experts.

Shahzad et al. (2013c) presented an automatic system for detection and quantification of cardiac calcium lesions. The method was validated by comparison with manual quantification, performed using the syngo calcium scoring tool (Siemens Medical Solution), using CT datasets from 366 patients, The intensity values of the CT datasets are given in Hounsfield Units (HU) and each material has a different HU value. The authors used the HU value to select calcified artery candidates by threshold (values higher than 130 HU) together with volume information of the candidates. The calcifications were detected using a classifier and 21 features from the candidates. The authors concluded that the system results achieved good sensitivity and high specificity and had a good correlation with results from human experts.

7.2.2 Pericardial/Epicardial Fat

Pericardial and epicardial fat were reported as potential parameters for cardiovascular risk stratification (Iacobellis and Willens 2009; Mahabadi et al. 2008; Ding et al. 2009). Although there was some discordance about the validity of such correlation (Larsen et al. 2015; Tanami et al. 2015), the high statistical significance of the studies that reported the correlation led to the development of semiautomatic and automatic cardiac fat measurement systems.

Coppini et al. (2010) proposed a semiautomated method that depends on the correct reorientation of the CT scan, followed by a manual determination of reference points and control points on the pericardium. The segmentation was performed by a threshold method refined by geodesic active contour. The method was tested on 10 patients but there is no ground truth comparison.

Barbosa et al. (2011) proposed a semiautomatic algorithm that uses region growing for cardiac muscle removal, followed by an angular sweep of the image to segment the pericardium. After the determination of the epicardial border, the fat was calculated as the pixels inside the epicardium with intensity between −190 and −30 HU. The method was tested in a study with 30 patients, and results showed improved reproducibility when compared to manual procedure.

Shahzad et al. (2013a) developed an automated method for epicardial fat measurement. The pericardium is defined by a multi-atlas based segmentation technique, in which manually annotated images are registered to the subject scan and fused. In the selected region, a threshold window of −200 to −30 HU is applied. The method was compared with manual segmentation in a set of 98 patients, achieving good results.

Ding et al. (2015) proposed an automated algorithm for epicardial fat volume quantification for non-contrast CT. An initial segmentation is obtained with the registration of a manually segmented multi-atlas followed by automated pericardium delineation using geodesic active contours. The range of −190 to −30 HU determines the fat. The algorithm was applied to a CT dataset of 50 patients, and the results were in good agreement to manual delineation by experts.

7.2.3 Stenosis Detection and Measurement

The high quality spatial resolution of coronary CT angiography allows the detection of the coronary tree and quantification of coronary artery stenoses. Since the adequate interpretation of coronary CT angiography requires considerable experience, the development of automatic computer-aided algorithms that can accelerate the learning curve and also provide an initial interpretation of a study is highly desirable (Dankerl et al. 2014; Goldenberg et al. 2012).

The commercial software COR Analyzer (Rcadia, Haifa, Israel) is an automatic system that reports location, type, and severity of coronary lesions. The software was evaluated in different papers that performed validation against cardiologist experts (Halpern and Halpern 2011; Kang et al. 2012; Abramowicz et al. 2013; Meyer et al. 2013; Hashoul et al. 2015), against quantitative coronary angiography (QCA) (Arnoldi et al. 2010) and invasive coronary angiography (Anders et al. 2013; Min et al. 2014). Meyer et al. and Thilo et al. (Thilo et al. 2015; Meyer et al. 2013) showed evidence of an improvement in the response of unexperienced readers using the software. The COR Analyzer is FDA approved and uses a proprietary processing algorithm derived from the work of Goldenberg et al. (2012). The main use suggested for the tool is the screening coronary artery disease in the emergency department.

Goldenberg et al. (2012) presented a CAD system for patient triage, based on coronary CT angiography. The automatic software segments the coronary artery tree, labels the major arteries and detects stenotic lesions with obstruction higher than 50%. The system also rejects low-quality images. The coronary tree segmentation is done by finding the main structures of the chest, segmenting the lung based on the air-filled areas, and, after that, finding the mediastinum. The ascending aorta is identified by using circular Hough transform. The remaining coronary tree is found by tracking tubular components using depth-first-search. The tree is then pruned to obtain only the coronary arteries. The labeling is based on a probabilistic anatomic model from a manually segmented training dataset. Finally, the detection of the stenotic lesions is performed by straightening the artery segments (curved planar reformation), and the boundary and lumen are delineated using an iterative, model-based, active contour approach. Calcified and non-calcified lesions are detected by the intensity of the voxels, and the final decision is made by a fuzzy logic system trained from manually classified images by physicians.

Boogers et al. (2010) evaluated the automatic commercial software QAngio CT 1.1 (Medis Medical Imaging Systems, Leiden, the Netherlands) with a dataset of 100 patients who also underwent invasive coronary angiography. The software uses a fast vessel-tracking algorithm to obtain 3D centerline of the coronary artery. From the centerline, it creates a stretched multiplanar-reformatted (MPR) volume and the lumen borders are detected by a model-guided minimum cost approach. From the border are calculated the cross-sectional area, diameter, minimal lumen diameter, and degree of stenosis.

Biermann et al. (2012) evaluated the semiautomatic commercial software, Syngo 3D™ (Siemens, Germany), in 40 patients using a Dual CT (Somatom, Siemens) and concluded that it improved the quality of unexperienced readers.

Kang et al. (2013) presented an algorithm that does centerline extraction, vessel classification, vessel linearization, lumen segmentation, and lesion detection. It requires three user inputs: a point at the ostium of the right coronary artery, a point at the left main coronary artery, and placing a circular region over the aortic root. 3D thinning and graph theory perform the centerline extraction, followed by artery classification using Dijkstra's shortest path algorithm and anatomical knowledge. The vessel is then linearized using a plane perpendicular to the centerline and the lumen segmentation is done with specific threshold levels for lumen and plaque. The lumen is determined by using recursive region growing in the linearized volume, and the stenosis is calculated using a knowledge-based algorithm. The system was tested in a dataset of 42 patients and the results compared to those of three expert readers, leading to a high sensitive result.

Shahzad et al. (2013b) presented a semiautomatic method for coronary artery lumen segmentation, stenosis detection, and quantification. It requires seed points at the ostium and at the distal end of each coronary artery. The centerlines are extracted using a minimal cost path, followed by a refinement step, and are used as initialization for the lumen segmentation that uses graph cuts. The diameter of the healthy lumen is then estimated by applying robust kernel regression, and the stenoses are detected and quantified by the difference between estimated and expected diameters. The algorithm was compared to results from three experienced readers and performed well for lumen segmentation and stenosis quantification, but needs improvement for stenosis automatic classification.

Dankerl et al. (2014) presented the evaluation of a commercial software, syngo.via (Siemens, Forchheim, Germany) for improving the response of inexperienced readers (medical students in the last semester of medical school) against experienced readers. The syngo.via is a semiautomatic proprietary software that reformats the images, segments coronary arteries, and provides measurement tools for stenosis quantification. The authors conclude that the software successfully assisted novice readers in interpreting coronary CTA datasets.

7.2.4 Plaque Quantification

Although there are other imaging techniques that can identify the characteristics of plaques (intravascular ultrasound, optical coherence tomography, infrared spectroscopy, angiography) they have the disadvantage of being invasive and not always being applicable to patients (Maurovich-Horvat et al. 2009). The noninvasive quantification of plaque also can provide early treatment and plaque monitoring (Maurovich-Horvat et al. 2009). One current limitation of using coronary CT angiography is the overlap of mean HU values of lipid-rich and fibrous plaques (Szilveszter et al. 2016) that sometimes hinders the proper plaque characterization.

Most vendors implemented commercial software for plaque quantification that were validated in different articles (Dey et al. 2010; Klass et al. 2010; Korosoglou et al. 2010; Kang et al. 2011; Boogers et al. 2012; Kim et al. 2013; Diaz-Zamudio et al. 2015). Nevertheless, new techniques for quantification have been recently described, and some of them are described in the following paragraphs.

Vukadinovic et al. (2012) validated a semiautomatic method for plaque measurement with 40 patients using the method described in (Vukadinovic et al. 2010). The method uses three-point initialization to segment the lumen using level set approach and determines the vessel walls using calcified regions and a GentleBoost classifier. The plaque is then classified according to its HU value. They conclude the method is comparable to manual analysis.

De Graaf et al. (2013) proposed an automated method for quantitative analysis of atherosclerosis on CT angiography. The centerline is automatically extracted using the algorithm of Yang et al. (2012), followed by the segmentation of the lumen using the method described by Boogers et al. (2012). The plaque constitution is measured by either the HU values or by an adaptive threshold based on the lumen value. Application of the proposed method was compared with IVUS Virtual Histology for a set of 57 patients. The authors concluded that there was an excellent correlation with volume and stenosis parameters, but there was still a large variability in quantitative plaque constitution.

Wei et al. (2014) used a multiscale coronary artery response-rolling balloon region growing to extract the coronary tree. Afterward, the coronaries were rectified, and a topological soft-gradient detection method was used for detecting non-calcified plaques (NCP) candidates. A classifier using 14 features performs the automatic detection of the NCP regions. The method was tested on 83 datasets and compared to manually labeled ground truth, performed by two experts.

7.2.5 Cardiac Function

Fractional flow reserve (FFR) is the ratio of maximal coronary blood flow through a stenotic artery divided by the blood flow in the hypothetical case of a normal artery. It is performed during coronary angiography, an invasive procedure, and is an important tool for revealing the ischemic potential of individual lesions (Tu et al. 2014).

Advances in computational fluid dynamics and imaging-based modeling have allowed for noninvasive calculation of FFR based on static CT images (Koo et al. 2011), denominated FFR_{CT}. The physical principles of FFR_{CT} are explained at (Taylor et al. 2013; Min et al. 2015). The most complete software for FFR_{CT} analysis is from Heart Flow (Redwood City, California) (de Araújo Gonçalves et al. 2015), which was validated in different publications (Min et al. 2012; Norgaard et al. 2014; Koo et al. 2011).

Similar measurements to FFR are: (1) the virtual fractional flow reserve (vFFR), proposed by Morris et al. (2013) and obtained from invasive coronary angiography; (2) the FFR_{QCA}, proposed by Tu et al. (2014), which is based on 3D quantitative coronary angiography (QCA) and; the virtual Functional Assessment Index (vFAI), proposed by Papaflakis et al. (2014), that is based on 3D QCA.

7.2.6 X-Ray Reviewed Methods

Table 7.1 shows a summary of the reviewed methods. It must be emphasized that neither the methods nor the applications presented in this chapter are comprehensive.

TABLE 7.1 Summary of the X-ray Methods Presented

Authors	Years	Parameter/Method	Number of Individuals
Ebersberger et al.	2013	Calcium score	127
Rubinshtein et al.	2014	Calcium score	136
Ivana Isgum et al.	2010	Calcium score	433
Shahzad, van Walsum et al.	2013	Calcium score	366
Coppini et al.	2010	Epicardial fat	10
Barbosa et al.	2011	Epicardial fat	30
Shahzad, Bos, et al.	2013	Epicardial fat	98
X. Ding et al.	2015	Epicardial fat	50
Halpern and Halpern	2011	Stenosis analysis	207
K.-W. Kang et al.	2012	Stenosis analysis	398
Abramowicz et al.	2013	Stenosis analysis	341
Meyer et al.	2013	Stenosis analysis	93
Hashoul et al.	2015	Stenosis analysis	352
Arnoldi et al.	2010	Stenosis analysis	59
Anders et al.	2013	Stenosis analysis	100
J. H. Min et al.	2014	Stenosis analysis	128
Thilo et al.	2015	Stenosis analysis	50
Goldenberg et al.	2012	Stenosis analysis	-[a]
Boogers et al.	2010	Stenosis analysis	100
Biermann et al.	2012	Stenosis analysis	40
D. Kang et al.	2013	Stenosis analysis	42
Shahzad, Kirişli, et al.	2013	Stenosis analysis	48
Dankerl et al.	2014	Stenosis analysis	61
Dey et al.	2010	Plaque quantification	70
Klass et al.	2010	Plaque quantification	35
Korosoglou et al.	2010	Plaque quantification	27
D. K. Kang et al.	2011	Plaque quantification	127
Boogers et al.	2012	Plaque quantification	51
Kim et al.	2013	Plaque quantification	57
Diaz-Zamudio et al.	2015	Plaque quantification	56
Vukadinovic et al.	2012	Plaque quantification	40
De Graaf et al.	2013	Plaque quantification	57
Wei et al.	2014	Plaque quantification	83
Koo et al.	2011	Cardiac function	103
J. K. Min et al.	2012	Cardiac function	252
Norgaard et al.	2014	Cardiac function	254
Morris et al.	2013	Cardiac function	19
Tu et al.	2014	Cardiac function	68
Papafaklis et al.	2014	Cardiac function	120

[a] Several studies—see original article.

7.3 SPECT and PET Cardiovascular Imaging

Single Photon Emission Computed Tomography (SPECT) and Positron Emission Tomography (PET) are noninvasive techniques used in the visualization and analysis of the distribution of radioactive counts, for instance within the myocardium and surrounding structures. In the last two decades, a number of techniques have been developed to aid in the classification of the SPECT and PET images. The strength and availability of these quantitative tools have, in many ways, provided a competitive advantage to nuclear cardiology, compared with other higher-resolution, noninvasive imaging modalities for the detection of coronary artery disease.

The process to obtain images in these modalities involves the detection of the radiation emitted from a patient's organ or region after the injection of a radiopharmaceutical, such as thallium-201 (^{201}TI) and technetium-99m sestamibi (^{99}Tc^{m}-MIBI) for SPECT, and F-18 (^{18}F) fluorodeoxyglucose (FDG), rubidium-82 (^{82}Rb) and nitrogen-13 (^{13}N) ammonia for PET. In SPECT, using a gamma camera, part of the emitted radiation can be used to produce an image indicating the distribution of the radionuclide in the myocardium (Garcia et al. 2000). On the other hand, PET imaging utilizes a class of radionuclide tracers that decay with the emission of a positron. The positron interacts with an electron, and their combined mass is converted into the energy of two 511 KeV photons traveling in opposite directions. A PET camera contains an array of detectors, which are able to detect the two gamma rays in coincidence. The last decade has seen the development of high performance dedicated PET cameras featuring high sensitivity (efficiency of detection) about 10–20 times higher than SPECT, high resolution, of 4–5 mm in air and about 7 mm in water, versus 20–25 mm for SPECT, high speed, and larger fields of view, 15 cm versus 10–11 cm previously (Phelps and Cherry 1998).

Temporal changes in the spatial distribution of radiopharmaceuticals can be obtained by taking multiple images over periods of time that may vary from milliseconds to hundreds of seconds. Since the resulting image, known as planar image, contains information of the projection of a radioactive volume over the detector's face, it is often difficult to determine clearly the function of tissue deep in the body. Tomographic studies obtained by taking multi-view acquisitions of the object overcome most problems caused by superposition of information in a single planar view and provides the clinician with a set of images that permits the visualization of the distribution of radioactive counts within the myocardium and surrounding structures. Defects on the distribution of some radionuclides, such as ^{201}TI , ^{99}Tc^{m}-MIBI ^{18}F, ^{82}Rb, and ^{13}N in the myocardium, indicate a muscle hypoperfusion due to obstruction of the coronary arteries

Due to the capability to acquire myocardial perfusion, the SPECT and PET studies gated to the electrocardiogram signal (ECG) provide the potential to extract additional quantitative information about cardiac function from the time-dependent motion of the myocardium. While gated studies are usually interpreted as series of two-dimensional (2D) images, there is an increasing interest in the presentation of the data in a more realistic three-dimensional (3D) form. These new displays provide an overview of the many slices in a single image and enhance the perception of spatial relationships. On the other hand, the visualization of the gated studies is complicated by the fact that 3D density images cannot be directly presented using common display devices that produce only 2D pictures. In general, some kind of transformation must be performed to the image sequence which introduces reduction of information.

A number of techniques, most of them concerned with visualization, have been developed to aid in the classification of the images. Perfusion abnormalities or defects were visually interpreted with the knowledge that normal myocardium displayed different radioactive distributions than those of abnormal myocardium (Ritchie et al. 1977; Hamilton et al. 1977; Trobaugh et al. 1978). However, it has been shown that interpretation of images by strictly visual techniques is subject to errors and inconsistencies, due to variation in the size and orientation of the heart, the low contrast of planar imaging, photon absorption, and scatter. For these reasons, assistance in diagnosis can be improved only through the development of automatic or semiautomatic methods to analyze and to quantify medical image parameters with the expectation of a significant reduction in intra-observer variability (Ficaro and Corbett 2004).

7.3.1 Planar Image Analysis

To solve the problem of inter-observer variability, several methods of quantitative analysis of the radiotracer distribution have been described (Watson et al. 1981; Berger et al. 1981; Garcia et al. 1981; Maddahi et al. 1981; Van Train et al. 1986). Watson et al. (1981) co-registered the serial images and reduced the 2D images to a series of linear radiotracer intensity profiles. Relative perfusion was then extracted from an analysis of these profiles as a function of time. Garcia et al. (1981; Maddahi et al. 1981) replaced the co-registration step by using a circumferential profile that reduced the 2D image into a line profile. The construction of the circumferential profile began with the manual definition of the center of the heart in the image. From that definition, the algorithm 13–14 sampled the myocardial tracer uptake along rays emanating from the user defined center. At 6° steps circling clockwise about the left ventricle (LV), myocardial intensity profiles were recorded. From the myocardial intensity profiles along each ray, the maximum myocardial intensity was recorded as a function of angle, reducing the 2D images to a line profile.

The advances in quantification methods by imaging patients with low likelihood for coronary artery disease allowed the construction of normal databases and the definition of mean and variance limits for normal patients. With the definition of mean and variance limits for normal perfusion, objective abnormality thresholds were defined. The use of these quantification tools and normal databases improved the diagnostic sensitivity for detecting disease in individual coronary arteries and significantly reduced the intra-observer variability for the patient's classification (Van Train et al. 1986).

7.3.2 SPECT Image Analysis

The advances in camera technology in the late 1980s allowed single photon emission computed tomography (SPECT) to be used routinely. With its inherently better image contrast, SPECT imaging gradually replaced several exams usually carried out by conventional planar imaging. Typically, a gamma-camera SPECT acquisition may consist of 64 planar views, or projections, each containing 64 × 64 image pixels and acquired at 64 discrete angles covering 360° around the patient. By the use of reconstruction algorithms, it is possible to estimate transaxial, sagittal, and coronal sectional images, creating 3D volumes or stacks of short-axis images that served as input to the quantitative software.

Although the objective of defining regional myocardial perfusion abnormalities with SPECT imaging did not change from that of planar imaging, the complexity of the problem increased significantly with the myocardium represented as a 3D volume rather than planar images.

The first effort to address this problem was carried out by Tamaki (Tamaki et al. 1984). This quantitative tomographic analysis was assessed by the circumferential profile curves of the three short-axis sections (apical, middle, and basal) of the LV. To assess the apical region, the most central, long-axis section cutting through the apex was also selected. From these images, circumferential maximal count profiles of myocardial distribution were obtained in a manner similar to that of Meade (Meade et al. 1978) and Burow et al. (1979). The center of the LV cavity was manually determined, and each tomographic image was divided into 36 radial segments at 10° intervals. The distribution of the radionuclide was determined by calculating the highest activity per pixel along each radial segment, normalizing the data to the segment with the highest counts to display a circumferential profile curve in each image. Perfusion defects were classified by use of uptake and washout distributions similar to the methodology used to quantify planar images. The results from these studies showed better sensitivity for detection of diseased coronary vessels than qualitative analysis.

Garcia et al.(1990) extended the method proposed by Tamaki et al. (1984) to the entire SPECT volume by introducing the polar map, or bull's-eye displays, for viewing circumferential profiles. The polar maps allow a quick and comprehensive overview of the circumferential samples from all slices by combining them into a single color-coded image. The points of each circumferential profile are assigned a color based on normalized count values, and the colored profiles are shaped into concentric rings. The most apical slice processed with circumferential profiles defines the center of the polar map, and each successive profile from each successive short-axis slice is displayed as a new ring surrounding the previous. The most basal slice of the LV makes up the outermost ring of the polar map. Additional maps, such as a standard deviation map that show the number of standard deviations below normal of each point in each circumferential profile, can aid in the evaluation of the study by indicating the severity of any abnormality.

In addition to tabulated defect extent values, various bull's-eye displays were developed to aid in the interpretation of the tomographic slices. The blackout polar map set those sectors in the polar map that were abnormal to the color black, while maintaining the intensity of normal sectors (Garcia et al. 1990). The blackout map provided a visual representation of the defect location and size. A second map, the defect severity map, mapped the defect abnormality in units of standard deviation below the normal mean. Both maps in combination with the vascular overlay template provided extent and severity values in each vascular region.

To account for differences among patients in the axial dimension of the heart, the slice thickness or polar map ring width was normalized so that each patient's polar map consisted of the same number of sampling points. With this technique, the polar map became the standard template for representing the 3D volume of the LV. By eliminating the anatomic variations of the LV, the polar map provided the construction of normal databases as a pattern to compare 3D perfusion distribution from patients under study. This feature has promoted the use of SPECT for the total automatic assessment of myocardial perfusion and for global and regional function, including the assessment of LV ejection fraction, LV end-diastolic (ED) and end-systolic (ES) volumes, and LV myocardial wall motion and thickening.

However, the quantification of global and regional LV parameters requires knowledge of the endocardial and, depending on the parameters being determined, the epicardial surface throughout the cardiac cycle. These surfaces may be manually assigned or automatically detected. Manual methods are subjective and time consuming. The automated methods use either boundary detection or geometric modeling. A large effort has been devoted to the analysis and segmentation of cardiac images by methods guided by prior geometric knowledge. In general, they can provide, with a limited number of global parameters, a rough shape approximation.

Faber et al. (1991) developed a general model-based surface detector for finding the four dimensional (three spatial dimensions plus time) endocardial and epicardial left ventricular boundaries for both perfusion and blood-pool SPECT studies. The model encoded LV shape, smoothness, and connectivity into the compatibility coefficients of a relaxation-labeling algorithm. From the surface estimates, volumetric data were extracted to estimate the LV ejection fraction and the regional wall motion was determined using a modified version of the centerline method. The final surfaces were visualized in a 3D rendering in which the displayed perfusion information could be presented with regional wall motion. Cauvin et al. (1993) modeled the LV as a truncated bullet, a combination of an ellipsoid and a cylinder, that is fitted to the morphological skeleton of Faber et al. (1999), which used a combination of cylindrical and spherical coordinate systems to build a discrete model of the LV in SPECT perfusion images. A radius function defined in a discrete space of longitudinal and circumferential coordinates describes the LV. For each orientation, the radius is determined by finding the position of maximal perfusion. After low-pass filtering to remove outlier radii, the radius function is mapped back to Cartesian space, where the surface is represented using triangular meshes.

Germano et al. (1995; Germano et al. 1997) introduced a fully automated algorithm that determined the LV surfaces from gated perfusion SPECT. In this approach, automatic border definition is done using thresholds of a Gaussian function pre-calibrated to a phantom. Although the spatial resolution of SPECT is too low to measure precisely the edges of the endocardial surface, the pre-calibration with a phantom and the high tolerance of the ejection fraction calculation provided highly reproducible volumetric estimates that correlated well with LV ejection fraction from other imaging modalities. This method has become popular and commercially available because it was the first to offer a fully automatic processing for gated perfusion. Figure 7.1 shows the patient's myocardial endocardial surface at end-diastole and end-systole depicted in an arbitrary orientation calculated by this method.

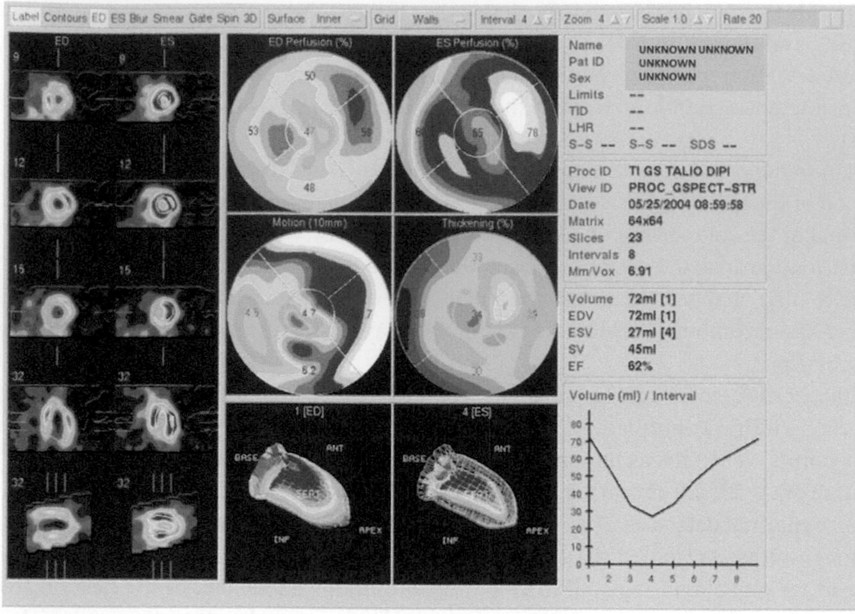

FIGURE 7.1 Screenshot from the result page of a cardiac SPECT study. Some slices are shown on the left part of the screen. The polar maps (bull's eye) are shown on the upper middle and below, the three dimensional surfaces in diastole and systole. (Screenshot of an exam report taken from the commercial software QGS, developed by Germano et al. 2007).

This display may be viewed in a dynamic fashion to visually assess wall motion showing excursion and thickening of the wall as changes in color. The color is a function of both myocardial perfusion and myocardial thickness. A myocardial segment that increases in color brightness indicates thickening and, thus, viable myocardium.

Garcia et al. (2001) proposed an expert system called PERFEX as a tool for computer-assisted diagnosis of stress/rest myocardial perfusion SPECT studies. This approach has the potential for standardizing the image interpretation process. After reviewing hundreds of studies from patients with coronary artery disease, heuristic rules were derived that best correlated the location of perfusion defects in SPECT studies. Two sets of gold standards were used to define the rules: the interpretation by human experts and the results of coronary angiography studies. These rules operate on data that is input to the PERFEX from the SPECT quantification process, which identifies defects as portions of the myocardium where normalized perfusion falls below a predetermined number of standard deviation, when compared to a normal file.

Bardinet et al. (1996) extended a geometric model based on superquadrics (Barr 1981) with the use of a free-form deformation (FFD) to estimate LV wall motion. This is accomplished by deforming the full model (superquadric and FFD) in the first frame, and modifying only the FFD in the subsequent frames. By tracking points with the same parametric coordinates along the cardiac cycle, a number of dynamic parameters like wall thickening and twisting motion can be computed.

Declerck (1998) have introduced a spatio-temporal model to segment the LV and to analyze motion from gated SPECT sequences. The model relies on a planispheric transformation that maps endocardial points in one time frame to the corresponding material points in any other frame. First, endocardial edge points are detected in all frames using a Canny-Deriche edge detector in spherical coordinates. Selected points in subsequent frames are matched to the current frame using a modification of the iterative closest point algorithm (1998). Based on corresponding point pairs, the parameters of a planispheric transformation are retrieved by least square approximation. This transformation allows describing motion parameters such as radial motion, twisting motion, and long-axis shortening.

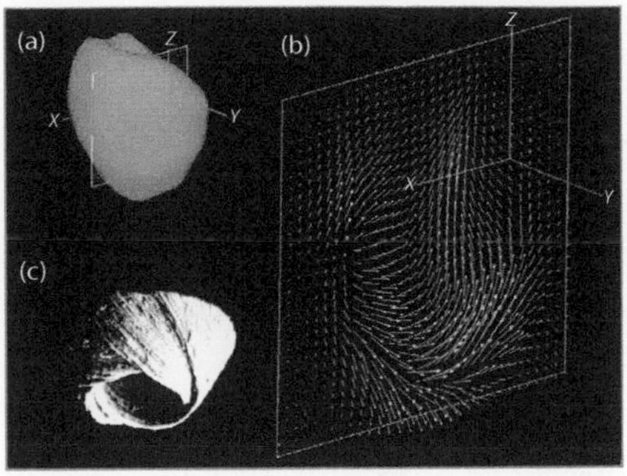

FIGURE 7.2 (a) A view of the 3D LV image of a normal heart; (b) the corresponding velocity vector field in a coronal slice at systole estimated by the gradient-based approach; and (c) the principal fiber pathways of a cow's LV.

Gutierrez et al. (2003) proposed the combination of optical method with a multi-resolution technique to estimate the myocardial Kinetic Energy (KE) obtained directly from gated SPECT images. Specifically, the method quantifies the 3D LV motion by a series of 3D velocity vector fields computed automatically for each voxel on the sequence of cardiac volumes (Figure 7.2).

The velocity estimation for each voxel is used to compute the corresponding kinetic energy, and the results are presented in a compact 2D form. The method was validated with a synthetic phantom and applied to groups of 30 volunteers and 29 patients. With the method, the cardiac condition of each subject can be studied taking the relation between the maximum and minimum values of kinetic energy observed during the cardiac cycle.

7.3.3 SPECT and PET Reviewed Methods

The strength and availability of these quantitative tools have provided in many ways a competitive advantage to nuclear cardiology, compared with other higher-resolution noninvasive imaging modalities for the detection of coronary artery disease. During the 1980s, the planar imaging of patients with low probability of coronary artery disease allowed the construction of normality databases and the definition of mean and variance limits for normal patients. With the definition of mean and variance limits for normal perfusion, objective abnormality thresholds were defined. These tools were expanded for gated SPECT, providing LV perfusion and function parameters in 2D, 3D, and 4D. Although a number of techniques have been developed to aid the classification of the SPECT images, the search for robust methods to analyze automatically the complete series of images is yet an open study area. Moreover, the comparison of the different techniques is a difficult task due to the diversity of approaches, image acquisition protocol and the lack of a "gold standard" to compare the results.

The validation of the presented techniques can be classified in three areas: (1) with no validation or only qualitative analysis; (2) with quantitative validation on patients; (3) with quantitative validation on phantoms, normal subjects, and patients. Table 7.2 shows the comparison of some techniques reported in the last 20 years. Group (1) includes papers presenting technical or methodological aspects of advanced modeling techniques (Cauvin et al. 1993; Bardinet et al. 1996; Declerck et al., 1998). Group (2) includes the evaluation performed on human volunteers and patients (Tamaki et al. 1984; Garcia et al. 1985; Faber et al. 1991; Faber et al. 1999; Garcia et al. 2001), including qualitative results in terms of cardiac functional parameters. Finally, the last group (3) includes studies performed on numerical or physical phantoms, normal subjects, and patients. They have the advantage to assess the accuracy and reproducibility of the techniques (Germano et al. 1995; Gutierrez et al. 2003). The size of test population

TABLE 7.2 Overview of Automatic Methods to Analyze and Quantify Cardiac LV Parameters by Means of SPECT and PET

Authors	Years	Parameter/Method	Number of Individuals
Tamaki et al.	1984	2D segmental analysis	104
Garcia et al.	1985	3D segmental analysis	53
Faber et al.	1991	Volume	38
Cauvin et al.	1993	Volume	—
Germano et al.	1995	Volume, ejection fraction	65
Bardinet et al.	1996	Volume, wall thickness, motion	1
Declerck et al.	1998	Motion	4
Faber et al.	1999	Volume, mass, ejection fraction	10
Garcia et al.	2001	3D segmental analysis	655
Gutierrez et al.	2003	Motion, kinetic energy	58

in most cases was small. In only five of them (Tamaki et al. 1984; Garcia et al. 1985; Germano et al. 1995; Garcia et al. 2001; Gutierrez et al. 2003), the studies were conducted on more than 50 volunteers or patients.

7.4 Ultrasound Imaging

The use of ultrasound to provide noninvasive evaluation of cardiac structure and function was a revolutionary advance in cardiac care in the late twentieth century. Development of the field of echocardiography has allowed detailed serial examinations of the development, structure, and function of the human heart in normal physiological states and in pathological conditions by means image acquisition rate in real-time and beat-to-beat. This characteristic is particularly suitable for assessing the dynamic aspects of the heart's pumping motion and perfusion. Echocardiography also provides a noninvasive method to visualize and accurately assess the function of the cardiac valves and provides a tool for the monitoring of diagnostic and therapeutic procedures (Rychik 2005).

However, ultrasound technology is based on the transmission and reflectance of a low-energy pressure wave through a medium with penetration capabilities far below those offered by other imaging modalities. Thus, the quality of the information content is often reduced by the presence of dropout, speckle, artifact, and attenuation in the far field. In addition, since the impedance mismatch between endocardium and blood is low, the endocardial border is often poorly defined or even missing.

7.4.1 M-Mode Echocardiography

M-mode, or motion-mode, echocardiography was the first type of ultrasound used for clinical cardiovascular imaging. Its use today is primarily limited to assessment of valve motion and reliable, reproducible measurements of chamber sizes and function (Rychik 2005). In M-mode echocardiography, a narrow ultrasound beam is pulsed rapidly in a single plane through the heart, and the movements of the structures in that single plane are plotted against time with very high temporal and axial resolutions.

M-mode echocardiography can be used to assess cardiac wall thickness, aortic root size, chamber sizes, or ventricular function. In general, left ventricular function is quantitated using M-mode by determining the percentage of fractional shortening of the LV, which is calculated using the following equation:

$$SF(\%) = \frac{LVEDD - LVESD}{LVEDD} \times 100 \tag{7.1}$$

where:

SF	is the shortening fraction
LVEDD	is the left ventricular end-diastolic dimension
LVESD	is the left ventricular end-systolic dimension

Normal values vary with age and range from 35% to 45% in infants to 28% to 44% in adolescents and adults (Colan et al. 1992; Gutgesell et al. 1977).

7.4.2 2D Imaging

Two-dimensional imaging provides an arc of imaging planes by employing multiple ultrasound beams to provide a cross-sectional view of the heart. Currently, 2D imaging provides the majority of information about cardiac structure and function in routine clinical studies. Two-dimensional imaging requires the presence of multiple beams of ultrasound interrogation in a single transducer, and several types of transducers are available to achieve this.

Transducers available for 2D imaging include mechanical (a sweeping or rotating ultrasound beam), phased array (multiple independently controlled sources), and linear array (a line of crystals simultaneously generating a beam of ultrasound). Today, phased-array transducers are most commonly used because of their: (1) small size, (2) ability to provide simultaneous 2D and M-mode or Doppler imaging, and (3) improved control of focal length for a more uniform image throughout the field of view (Rychik 2005).

In addition to using 2D imaging for viewing anatomical detailing, left-ventricular function can be quantitated by this mode using an estimated left-ventricular ejection fraction. This method, which has been shown to correlate well with angiographic estimates of ventricular function, takes advantage of the conical shape of the LV to estimate end-diastolic and end-systolic ventricular volumes from tracings of 2D images using Simpson's biplane rule (3). The ejection fraction is calculated as follows:

$$EF(\%) = \frac{LVEDV - LVESV}{LVEDV} \times 100 \tag{7.2}$$

where:

EF	is ejection fraction
LVEDV	is left ventricular end-diastolic volume
LV ESV	is left ventricular end-systolic volume

Normal values for ejection fraction are approximately 55%–65%, and cardiac output can be estimated by multiplying the volume ejected with each beat (stroke volume) by the heart rate using the equation:

$$CO = HR \times SV \tag{7.3}$$

where:

CO	is cardiac output
HR	is heart rate
SV	is stroke volume

7.4.3 Doppler Ultrasound

The Doppler principle, described by Christian Johann Doppler in 1843, states that the frequency of transmitted sound is altered when the source of the sound is moving (Rychik 2005). This change in frequency, or Doppler shift, also occurs when the source of sound is stationary, and the waves are reflected off a moving target, including red blood cells in the vasculature. The shift in frequency is related to the velocity of the moving target, as well as the angle of incidence, and is described by the equation:

$$f_d \equiv \frac{2(f_0)(\mathrm{v})\cos\varphi}{c} \tag{7.4}$$

where:

f_d is the observed Doppler frequency shift
f_0 is the transmitted frequency
c is the velocity of sound in human tissue at 37°C (~1560 m/s)
v is blood flow velocity
φ is the intercept angle between the ultrasound beam and the blood flow

Using this principle, Doppler ultrasound can be used to estimate the velocity of blood flow in the human heart and vasculature noninvasively. Using a modified Bernoulli equation, in which pressure drop is equal to four times the velocity squared ($4v^2$), Doppler ultrasound can also be used to estimate chamber pressures and gradients and to provide significant noninvasive hemodynamic data.

7.4.4 Ultrasound Image Analysis

Image analysis methods in ultrasound minimize the inter-observer and intra-observer variabilities and also decrease the processing time during segmentation and registration of cardiac structures. In a preliminary study, Gupta et al. (1993) demonstrated that there is a reduction of 87.5% of the segmentation time in ultrasound images using image analysis methods when compared to manual segmentation.

In general, image analysis methods involve three different steps: (1) selection of the structure of interest (segmentation); (2) extraction of image intensity based features; (3) selection of the most important characteristics and classification.

7.4.5 Selection of the Structure of Interest (Segmentation)

A variety of methods have been described in the literature for the task of selecting structures in ultrasound.

Gupta et al. (1993) proposed a semiautomatic segmentation method using deformable models, a noniterative procedure that uses user input data and dynamic programming. Mignotte and Meunier (2001) used an energy based approach, considering the statistical energy associated with an active contour method for short axis segmentation. One of the advantages presented is the fact that this approach can deal with low signal-to-noise ratio (high noise values compared to the signal values) and blurred contours. A shifted Rayleigh distribution was used to model the gray scale statistics, and the multiscale optimization proposed by Heitz et al. (1994) was used to perform the energy minimization and the maximum-likelihood (ML) algorithm. The resulting method is capable of extracting an initial contour for the endocardium. In a later work, Mignotte et al. (2001) presented an algorithm that aims to determine the boundaries of the endocardium. The proposed method was based on a Bayesian framework of deformable templates. The main characteristic of this model is that it does not require a training set; it is formulated as an optimization problem known as *maximum a posteriori problem* (MAPp) and solved using genetic algorithms. The methodology was tested over 50 short axis images.

7.4.6 Image Intensity Based Features

Image intensity-based features can be used to analyze the segmented structure and can be divided between statistical and structural features. The former provides more significant information than the latter, but structural features still present some important and interesting information. Evaluation of statistical features can be divided in two steps: first- and second-order measurements.

First-order metrics evaluate the basic statistics of a region of interest, such as standard deviation, variance, and skewness, while second-order use the joint probability of gray-level occurrences to evaluate more refined information such as homogeneity, contrast, entropy, and energy.

Sudarshan et al. (2015) present a CAD system that uses the Discrete Wavelet Transform (DWT) combined with Support Vector Machine (SVM) and compare DWT with second statistics features, aiming to use a minimum number of features. DWT achieved up to 99.5% of accuracy, 99.75% of sensitivity, and 99.25% of specificity.

7.4.7 Feature Selection, Ranking and Classification

To reduce the computational effort and time consumption for the feature classification process, some metrics must be used to select an appropriate and compact feature subset. Duda et al. (2001) demonstrated that the most reliable parameter is the Bhattacharyya distance method, which evaluates the similarity between two discrete sets of probability distributions, $p_1(x)$, the measured probability and $p_2(x)$, the ideal probability model.

Other options are the Kullback-Leibler divergence method and the Kolmogorov-Smirnov test (Angelo et al. 2007), which measures the similarity between a certain distribution and a Gaussian distribution.

Once the selection or ranking is performed, the classification of the extracted features is evaluated. This results in an estimate of lesions or an abnormalities present in the images analyzed.

Several techniques can be used in order to perform the features classification, like AdaBoost (Adaptive Booster) classifier (Freund and Schapire 1997), naive Bayes (Han et al., 2016), Gaussian mixture model (Bilmes 1998), k-nearest neighbors, and linear discriminant (Ganesan et al., 2014). This is the most important step in the development of a CAD, since it will provide a perspective of what can be expected in the image set to the radiologist. The tenfold cross-validation technique may be used in order to determine the best classification method (Duda et al., 2001). The feature set is split into 10 equal parts. Nine of them will be used to train the classifier, while the remaining one will be used to evaluate the classifier, and this process will be repeated 10 times using different parts of the training set and testing set. For each turn, the specificity, sensitivity, accuracy, and the positive predictive values (PPV) will be evaluated for each classifier, and the average of the 10 realizations provides the current performance of the classifier.

One of the most efficient classifiers is the AdaBoost method. It is a machine-learning meta-algorithm, which uses a set of many other types of learning algorithms (weak learners) in order to improve their performances by combining them with a set of weights. On the other hand, naive Bayes classifier is a very simple classifier that is based on the probability theory developed by Bayes. It considers that, given a certain class, the features considered to describe it are independent from each other. In most cases, it calculates the maximum likelihood parameters.

The Gaussian mixture model is a parametric method that seeks to maximize the likelihood parameter for a Gaussian distribution of the experimental data. The k-nearest neighbors is a nonparametric model that stores all the training set cases and classifies the new ones based on several distance metrics, such as Euclidean, Manhattan, and Minkowski distance.

Finally, the linear discriminant method is one of the fastest methods available, used in large datasets when speed may be an issue. It is a linear combination of a weighing vector ($\boldsymbol{w}$) and a feature vector ($\boldsymbol{x}$). If the linear combination ($\boldsymbol{w}^T\boldsymbol{x}$) is greater than a certain threshold, the data is considered to belong to the class c_1; if not, the data is set to the class c_2.

Some authors have successfully used some of the methods discussed here and some other new techniques in order to build CAD systems for echocardiography characterizations: Amichi and Laugier (2009) use first- and second-order statistics to identify acute myocardial contusions. In their study, a preliminary result is presented, showing that is possible to differentiate the normal heart tissue from those with acute myocardial contusion.

Some authors have used discrete wavelet transform (DWT) in order to identify MI. For example, Mojsilovic et al. (1997) presented a transthoracic echocardiography study, with unsupervised classification and DWT based on feature. Their method showed a classification rate of approximately 96%, with an average of 88.5%. In later work, Neskovic et al. (1998) described a method using DWT to detect early myocardial infarction. Their results showed up to 100% of specificity, sensibility, and accuracy depending on the time after the infarction, reaching the 100% of reliability after three weeks. In their study, a distance-based classifier was applied to 2D ultrasound transthoracic echocardiograms.

Lieback et al. (1996) used the texture of echocardiographic images in order to diagnose myocarditis, reaching 100% sensitivity.

7.4.8 Vascular Ultrasound

The carotid artery is composed of three different layers: the intima layer, the closest one to the lumen; the media layer; and the adventitia layer, as shown in Figure 7.1, along with the three main interfaces: near wall media-adventitia, far wall lumen-intima, and far wall media-adventitia. One of the first signs of cardiovascular disease is the thickening of the intima layer, which may indicate a vulnerable carotid plaque that can cause atherothrombotic events, myocardial infarction and stroke.

The main measure used in vascular ultrasound is the Intima-Media thickness (IMT), defined as the distance between the far wall lumen-intima and the far-wall media adventitia. The IMT is best visualized in longitudinal sections of the carotid (Figure 7.1).

The inter-observer and intra-observer variabilities problems are also presented in vascular ultrasound, along with a time consuming manual segmentation and analysis of the images. In order to reduce their influence in evaluating the IMT, a border detection algorithm is needed. Some authors presented methodologies that can efficiently detect the intima, media and adventitia. In general, a Gaussian filter is applied in order to blur the image and homogenize the different regions, thus, enhancing the gradient between different regions. Cheng and Jiang (2008) used dynamic programming (Scharf and Elliott 1981) in order to detect the arterial wall. The authors showed that the method, in which the user provides only the ROI, decreases the effect of the inherent ultrasound speckle noise. Liang et al. (2000) presented a multiscale dynamic programming approach to IMT segmentation. In their proposal, the image was segmented using dynamic programming from a coarse image scale to a more refined one.

Destrempes et al. (2009) proposed a probabilistic model that describes the echogenicity by three Nakagami distributions (Kolar et al., 2004) and estimates the distribution parameters using an expectation–maximization algorithm.

Gutierrez et al. (2002) propose a semiautomatic segmentation using a multiscale approach and active contour. They compared the results obtained by their Method and manual segmentation. Carvalho et al. (2015) propose a combined technique using B-Mode ultrasound, contrast enhanced ultrasound, and a non-rigid motion estimation in order to perform a lumen segmentation and wall motion estimation of the carotid (Figure 7.3).

7.4.9 Overview of the Presented Methods

Table 7.3 provides an overview of the methods discussed in this section.

7.5 Magnetic Resonance Cardiovascular Imaging

Cardiovascular magnetic resonance imaging (CMRI) is a noninvasive imaging technique based on the use of intense stationary magnetic fields to create a net magnetization in the human body, followed by the application of radiofrequency pulses, and magnetic gradient field switches. Specific combinations

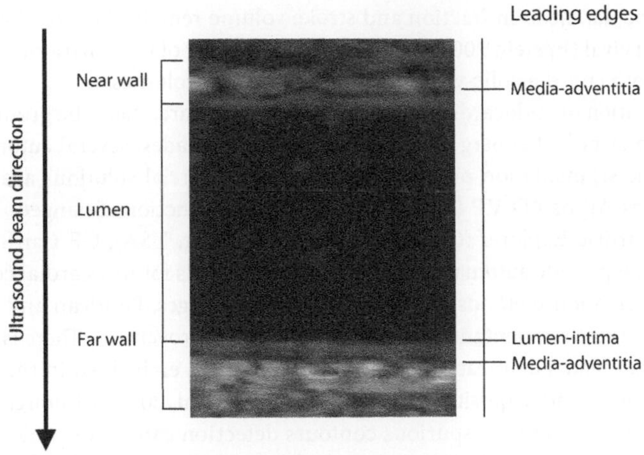

FIGURE 7.3 Carotid interfaces section. (From Gutierrez, M A et al., *Computers in Cardiology*, 359–62, 2002.)

TABLE 7.3 Overview of Automatic Methods in Ultrasound

Authors	Years	Parameter/Method	Number of Individuals
Amichi and Laugir	1996	Statistical classification	19
Mojsilovic et al.	1997	Wavelet transform/unsupervised classification	15
Neskovic et al.	1998	Wavelet transform/wall motion	19
Sudarshan et al.	2015	Discrete wavelet transform	160
Lieback et al.	1996	Myocardities	106
Cheng and Jiang	2008	Dynamic programming/arterial wall detection	23
Liang et al.	2000	Dynamic programming/IMT segmentation	50
Gutirrez et al.	2002	Active contour/IMT segmentation	30
Carvalho et al.	2015	Lumen segmentation and wall motion estimation	21

of the pulses, field switches, and acquisition times allow the generation of images describing different physical phenomena, which gives MRI its unique feature of visualizing independent anatomical and physiological features in a single modality (Brown et al. 1995). CMRI is a safe imaging modality, since there is no exposition of the patient to ionizing radiation. Besides, it has high spatial and temporal resolution, good contrast between soft tissues and blood, allows arbitrary orientations in imaging of the heart, and provides information on both anatomy and function of the cardiovascular system (Wang and Amini 2012). In the following items, we present the status of computerized aided diagnosis and quantitative methods in some applications of CMRI.

7.5.1 Left Ventricular Function Assessment

As the gold standard for the noninvasive assessment of cardiac myocardial mass, volumes, and function, CMRI exams are used in the daily routine of cardiac centers for diagnosis of cardiovascular disease (Salerno and Kramer 2013; Wieben et al., 2008). Measuring global and regional parameters of the Left LV is decisive to estimate the presence and severity of cardiac diseases and to choose the most suitable therapies. Left ventricular ejection fraction and stroke volume remain the most important parameters to predict patient survival (Epstein 2007). The correct estimation of these parameters requires an accurate calculation of both end-diastolic-volumes and end-systolic-volumes.

The correct delineation of endocardial and epicardial borders are, then, the ground tasks to compute the functional parameters (Lebenberg et al. 2015). In the last decades, several automatic solutions have been proposed to the segmentation problem. There are commercial solutions available at MRI scanners, such as Siemens Argus 4D VF (Argus 4D Ventricular Function, Erlangen, Federal Republic of Germany) Philips Cardiac Explorer (Cardiac Explorer, Andover, USA), GE Cardiac VX (CardiacVX, Milwaukee, USA), that provide automated or semiautomated tools for myocardial contour detection. In their review of segmentation methods in short axis cardiac images, Petitjean and Dacher (2011) point out that the results of these segmentation methods still need improvement. There is also indication that the techniques for automated delineation of myocardial borders embedded in these software are very sensitive to image quality and acquisition protocol (Marinol et al. 2014). Although the failure of automatic segmentation is not common, spurious contours detection can occur, which ultimately leads to extreme global-quantification outliers (Hautvast et al. 2012). Hautvast et al. (2012) performed a quantitative analysis of the time impact of myocardial contour corrections using images from a study population of 1.555 participants. They concluded that the analysis time for a single exam doubles in case of manual correction, which is relevant in clinical daily routine.

Due to the problems with the available methods, cardiac segmentation in MRI remains an active research area, with many recent contributions describing automatic or semiautomatic approaches for cardiac segmentation in MRI. Solutions include deformable models (Hu et al. 2014; Grosgeorge et al. 2011), region growing (Wang et al. 2015a), graph-based random walks (Eslami et al. 2013), dynamic programming (Liu et al. 2012), statistical models (Yun Zhu et al. 2010), and combinations of different approaches (Huang et al. 2011; Hadhoud et al. 2012; Lebenberg et al. 2015). For a comprehensive review of segmentation methods in short axis cardiac MR images, refer to the publication by Petitjean et al. (Petitjean and Dacher 2011).

7.5.2 Parametric T1 and ECV Maps

T1 relaxation time and extracellular volume (ECV) mappings are important tools for assessing a wide range of cardiomyopathies (Kellman and Hansen 2014). The composition of the individual tissues determines the T1 value, and the intensity of each voxel of the T1 map corresponds to the tissue relaxation time for each voxel of the image. (O h-Ici et al. 2014; Perea et al. 2015). The ECV is a derived measurement from T1 maps obtained before and after the injection of Gadolinium-based contrast agents (Salerno and Kramer 2013). Parametric maps of perfusion parameters or relaxation times to characterize myocardial tissues have become a viable clinical application due to improvement in CMRI technology (Salerno and Kramer 2013). The standard acquisition sequence for this application is the Modified Look-Locker (MOLLI) (Messroghli et al. 2004). The Shortened Modified Look-Locker Inversion Recovery (ShMOLLI) was created to decrease the heart-rate dependency and shorten the breath-hold (Piechnik et al. 2010) of the original MOLLI sequence. The T1 map is created by fitting a curve to the measured data inside the myocardium (Jellis and Kwon 2014). Since the images are acquired at different inversion times, registration is a mandatory step before the curve fitting to reduce the effect of motion between the images.

CVI[42] (Circle Cardiovascular Imaging, Calgary, Canada) is a widely used commercial package for creating T1 maps. Yet, as in other CMR applications, research efforts to automate the creation of the maps to improve the quantitative results are still intense. Important topics for achieving such goal are robust automatic segmentation of the myocardium and registration for correction of respiratory motion.

Roujol et al. proposed a method for non-rigid registration of the myocardium by presenting a variational framework (Roujol et al. 2014), while the approach of Xue et al. (2012) for registration was based on solving a variational energy minimization problem (Xue et al. 2012). The latter technique was applied in a clinical study by Florian et al. (2014) to correct the motion for creating T1 maps pre and post contrast, that were later used for ECV calculation (Florian et al. 2014). Some authors propose novel acquisition schemes for solving the motion correction (Weingärtner et al. 2015), while other works use new pulse sequences to provide additional information to help the registration (Xue et al. 2013).

7.5.3 Tagged CMRI

Assessment of regional wall motion allows the measurement of functional alterations of the heart and the improvement of their diagnosis and treatment (Jiang and Yu 2014; Wu et al. 2014; Xu et al. 2010). Tagged cine MRI is one technique used for quantitative measurement of local tissue deformation indicators of regional myocardial contractile function, for example, myocardial strain, strain rate, and torsion (Attili et al. 2010; Jiang and Yu 2014; Wang and Amini 2012). The first tagging technique created by Zerhouni et al. (1988) allowed the assessment of wall motion by visualizing and quantifying the movement of visible tags in MR images. Since then, several tagging imaging techniques have been established and can be broadly divided into two groups (Ibrahim 2011). The first group comprises techniques like SPAMM (Axel and Dougherty 1989), which create visual patterns of magnetization saturation—myocardial tags that can be parallel lines or grid patterns—allowing straightforward visual inspection. However, these techniques demand complicated and time-consuming algorithms for quantitative analysis, since automatic calculation of parameters requires tracking of adjacent tag intersection points (Ibrahim 2011).The second group of imaging techniques is more recent, started in 1999 with the creation of HARP (Osman et al. 1999) and DENSE (Aletras et al. 1999). This group of techniques creates images that do not show any visual markers of the tagging. Nonetheless, the motion quantification can be performed by simple and fast postprocessing algorithms, since the tracking of the tags is performed in the phase images (Jiang and Yu 2014).

Tracking of MRI tags has been performed by a number of automatic or semiautomatic techniques (Ibrahim 2011; Wang and Amini 2012). Some commercial packages, like the software HARP (Harmonic Phase Imaging, Diagnosoft, Palo Alto, CA), are available for strain analysis. Nevertheless, new techniques for quantification have been recently described using methods based on image registration (Oubel et al. 2012), time-frequency (Kause et al. 2013), deformable models (Jahanzad et al. 2015), variational techniques (Carranza-Herrezuelo et al. 2010), and nontracking-based strain estimation (Qian et al. 2011), among others.

7.5.4 Flow Assessment

2D Phase-Contrast magnetic resonance imaging (PC-MRI) is a well-established technique to visualize and quantify blood flow in the heart and great vessels. (Salerno & Kramer 2013; Santarelli et al. 2016, Stankovic et al. 2014). 4D flow MRI is the volumetric extension of this technique, whose acquisition methods encode velocity along the three spatial axes throughout the cardiac cycle, resulting in a time-resolved 3D velocity field (Stankovic et al. 2014; Santarelli et al. 2016).

The techniques used to visualize the 4D flow studies are meant to allow the qualitative evaluation of cardiovascular diseases by displaying possible changes in the normal vascular hemodynamics. Magnitude and direction of blood velocity are depicted by using color-coded vector graphs, the tangents to particle velocity vectors by using streamlines, and the paths of flow particles by using path lines (Stankovic et al., 2014; Markl et al., 2014, Santarelli et al., 2016). In addition to qualitative analysis, several authors have used quantitative parameters to describe hemodynamic behavior. The methods usually require the segmentation of the 3D vessel under study, a task that frequently is performed manually or semiautomatically (Dyverfeldt et al. 2015).

Among other parameters, wall shear stress (WSS), described as "viscous shear forces of flowing blood acting tangentially to the vessel wall" (Dyverfeldt et al. 2015), has been proposed by different authors as a quantitative parameter to describe the cardiac condition. Cibis et al. (2014) used computation fluid dynamics to calculate 3D blood-flow velocities and WSS to study carotid flow. Bieging et al. (2011) proposed a semi-quantitative WSS assessment on the entire ascending aortic wall by using a spline-based manual surface definition method followed by calculations of the velocity vectors for each surface point. Van Ooij et al. (2015) described a method that creates subject-specific "heat maps" of abnormal WSS for the study of the aorta in patients with bicuspid aortic valve disease.

Another common parameter, pulse wave velocity (PWV), is described as the "propagation speed of systolic pressure pulse in the arterial system" (Dyverfeldt et al. 2015). It may provide early atherosclerotic risk by providing information about vessel stiffness (Wentland et al. 2014). According to Dyverfelt et al. (2014), PWV estimation is usually performed by two main approaches: (1) "travel-distance," based on the distance between two vessel locations; and (2) "travel-time," the temporal shift between velocity, flow, or pressure waveforms recorded at these locations. In their work, the authors compared several methods to calculate PWV using the "travel-time" approach and applied the results to study age differences in the values of PWV. PWV was studied as a measure of stiffness of the thoracic aorta in a work by Markl et al. (2010), in which PWV was determined by transit-time methods and the linear fit from data of the entire aorta.

Turbulent kinetic energy (TKE), the mean kinetic energy measured in a region of varying velocity field (Binter et al. 2013), can be used as a measure of flow inefficiency (Zajac et al. 2014). Binter et al. (2013) calculated information about velocity distributions using Bayesian approach for multipoint velocity encoding.

Ventricular flow compartments were studied and quantified semiautomatically by Erikson et al. (2010) using a method that automatically separates the path lines into four flow compartments: Direct Flow, Retained Inflow, Delayed Ejection Flow, and Residual Volume. The method was first validated in a study with six healthy subjects and three patients and later applied in a study of 10 patients suffering from mild heart failure (Eriksson et al. 2013).

7.5.5 Carotid Plaque Quantification

MR angiographic (MRA) is a suitable method for carotid plaque quantification, as the artery is approximately stationary and superficially located, thus contributing to a precise positioning of the surface coils responsible for acquisition of the MR signal (Singh et al. 2015). Furthermore, variation of the MR pulse sequences and the use of contrast agents allow for the quantification of the stenosis grade and plaque volume, the identification of different plaque components, and the description of the carotid geometry (Singh et al. 2015). Computer-aided techniques perform automatic or semiautomatic quantification of carotid morphology and composition. Important steps for the automatic quantification of carotid plaques are segmentation of lumen and outer wall and plaque tissue characterization and classification (Singh et al. 2015).

Even though the number of publications addressing vascular segmentation is significant, the challenge remains (Lesage et al. 2009). Methods commonly use approaches like snakes, active shape/appearance model, geometric models, and level sets. In recent years, van't Klooster et al. (2012) proposed a segmentation method for the detection of the lumen and outer wall boundaries based on a deformable 3D cylindrical NURBS surface requiring minimal user interaction. After segmentation, the stenosis was graded. Hameeteman et al. (2013) combined the segmentation method proposed by van't Klooster et al. with a further learning-based correction of segmentation systematic errors to quantify the wall volume. A deformable tubular 3D Non-Uniform Rational B-Splines model was used by Suinesiaputra et al. (2012) to automatically segment the carotid lumen from MRA. Ukwatta et al. (2013) proposed a segmentation technique based on a global optimization algorithm that partitions the carotid image into the wall, lumen, and background. In a recent multi-center study, van Engelen et al. (2015) developed and tested segmentation methods—a nonlinear feature normalization approach and two transfer-learning algorithms—based on data acquired in MRI scanners from different vendors and different pulse sequences with good results.

Tang et al. (2012) proposed a method for quantifying several parameters (carotid bifurcation angle, carotid area ratio, carotid bulb size, and the vessel tortuosity) from multispectral MR images. Internal and external centerlines were found by using a minimum cost path algorithm and then used to initialize the lumen segmentation method—a topology preserving geodesic active contour. Sakellarios et al. (2012) presented a methodology for semiautomatic characterization of the plaque, in which an active contour method was used to segment the luminal and outer carotid wall surface followed by a characterization step. The characterization was based on the intensity of each plaque component in different contrast images (T1, T2) and performed by a knowledge-based algorithm that uses a fuzzy c-means clustering method. Two different modalities, MRA, and computed tomography, were used by van Engelen et al. (2014) to estimate the volume of different components of atherosclerotic plaques by using a classifier trained on histology.

7.5.6 CMRI Reviewed Methods

Table 7.4 shows a summary of the reviewed methods. It must be emphasized that neither the methods nor the applications presented in this chapter are comprehensive.

7.6 Intravascular Imaging

The *in vivo* percutaneous coronary intervention procedures were first performed by Dr. Andreas Grüntzig in 1977 for guidance based on x-ray angiography (Katouzian et al. 2012). After that, ultrasound probes or near-infrared light source were coupled to a catheter, allowing clinical intravascular visualization.

Catheter-based intravascular ultrasound and optical coherence tomography imaging of coronary arteries provide important information about the coronary lumen, wall, and atherosclerotic plaque characteristics. Both modalities are able to show 2D cross-sectional images of the coronary artery, providing visualization of lesions, which assist clinicians in diagnosis and treatment of coronary artery disease. Puri et al. (2013) includes a good review of coronary atherosclerosis using intravascular imaging (Puri et al. 2013).

7.6.1 Intravascular Ultrasound

Intravascular ultrasound (IVUS) has been clinically available for more than 25 years (Nakatani et al. 2013). Because of the large size and limited flexibility of the early ultrasound catheter, initial applications were in peripheral arteries (Pandian et al. 1990). Later, smaller and more flexible catheters enabled clinical intracoronary applications. Nissen et al. tested this new modality for coronaries in 51 *in vivo* humans, validating and demonstrating its uses as a safe and stable procedure (Nissen et al. 1991).

7.6.1.1 IVUS Features

IVUS has provided insights into *in vivo* blood vessels and assisted clinicians in characterizing the vessel walls and plaques, allowing an analysis of vessel morphology and remodeling .

IVUS places a catheter with a probe on its tip inside a vessel. While this probe rotates, it emits ultrasound pulses and receives echoes from surrounding tissues to generate 2D cross-sectional images along a vessel segment in a process called pullback. The intracoronary ultrasound imaging frequency ranges from 20 to 45 MHz, which provides an axial resolution of 70–200 μm with penetration greater than 5 mm (Elliott and Thrush 1996). The IVUS modality remains the most validated and widely used intravascular imaging technique in clinical practice. Figure 7.4 depicts the basic structures of a 2D IVUS.

TABLE 7.4 Summary of the CMRI Methods Presented

Author	Years	Parameter/Methods	Number of Individuals
Hu et al.	2014	Deformable model for LV segmentation	45
Grosgeorge et al.	2011	Deformable model for LV segmentation	59
Wang et al.	2015	Region growing for LV segmentation	45
Eslami et al.	2013	Graph-based random walks for LV segmentation	104
Liu et al.	2012	Dynamic programming for LV segmentation	45
Zhu et al.	2010	Statistical models for LV segmentation	22
Lebenberg et al.	2015	Combination of approaches for LV segmentation	45
Huang et al.	2011	Combination of approaches for LV segmentation	45
Hadhoud et al.	2012	Combination of approaches for LV segmentation	13
Houtvast et al.	2012	Combination of approaches for LV segmentation	1.555
Roujol et al.	2014	T1 mapping	29
Xue et al.	2012	T1 mapping	50
Florian et al.	2014	T1 mapping and myocardial extracellular volume fraction	49
Xue et al.	2013	T1 mapping	45
Oubel et al.	2012	Myocardial deformation analysis—movement	6
Kauze et al.	2013	Myocardial deformation analysis—movement	7
Jahanzad et al.	2015	Myocardial deformation analysis—movement	20
C-Herrezuelo et al.	2010	Myocardial deformation analysis—movement	5
Quian et al.	2011	Myocardial deformation analysis—movement	41
Cibis et al.	2014	Wall shear stress of the carotid	6
Bieging et al.	2011	Wall shear stress of the aorta	11
van Ooij et al.	2015	Wall shear stress of the aorta	23
Dyverfelt et al.	2014	Pulse wave velocity of the aorta	16
Markl et al.	2010	Pulse wave velocity of the aorta	27
Binter et al.	2013	Turbulent kinetic energy of blood	7
Erikson et al.	2010	Quantification of flow compartments	10
Erikson et al.	2013	Quantification of flow compartments	9
van't Klooster	2012	Stenosis of carotid	10
Suinesiaputra et al.	2012	Stenosis of carotid	38
Ukwatta et al.	2013	Carotid wall volume	45
van Engelen	2015	Several methods—segmentation of carotid plaque	42
Tang et al.	2012	Carotid bifurcation angle, carotid area ratio, carotid bulb size, vessel tortuosity	48
Sakellarios et al.	2012	Characterization of carotid plaque	21
Hameeteman et al.	2013	Wall volume of carotid	19
van Engelen	2014	Volume of atherosclerotic plaque	15

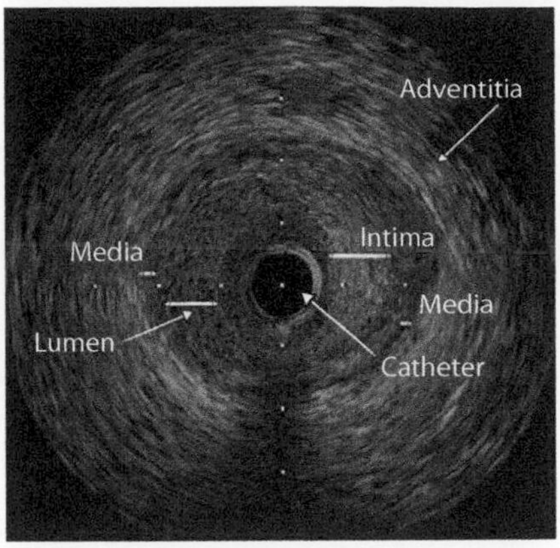

FIGURE 7.4 A single IVUS frame showing the cross section inside a coronary artery.

7.6.1.2 IVUS Computer-Aided Systems

Accurate segmentation of the borders of arterial wall layers in IVUS images provides critical information for clinical investigations such as stenosis grade and plaque detection.

It is possible to observe and measure the following structures in an IVUS image: (1) lumen; (2) intima layer; (3) media layer, which is represented as a line; (4) adventitia layer; and (5) plaques. According to Katouzian et al. (2012), automated segmentation methods for IVUS images have been reported since 1995 (Katouzian et al. 2012), while computer-aided systems have been reported for delineation of lumen vessel since 1991. Table 7.5 shows some clinical studies using computer-aided systems using IVUS images. Wenguang et al. (1991) presented a segmentation method for *in vivo* IVUS images, consisting of a tracing tool, which allows an expert to delineate the vessel lumen, and an automatic method, which computes the lumen and lesion area. (Wenguang et al. 1991). Many studies have followed this, which use semiautomated methods, employing an automated method plus a user post-correction with validation for *in vitro* and *in vivo* images (Dijkstra et al., 1999; Klingensmith et al., 2000; Sun et al., 2013). Von Birgelen et al. (1996) described an automated segmentation of a coronary lumen, using a minimum-cost algorithm and dynamic programming to identify the vessel borders and compute lumen and plaque area, requiring minimal user interaction to correct the segmentation results. Fully automated methods for segment IVUS images have also been developed (Brusseau et al. 2004; Wahle et al. 2006; Haas et al. 2000). Although post-correction is possible following the automatic processing, it was not used in the validation tests.

Another important investigation, which can be conducted using IVUS images, is determination of plaque composition. Usually, plaques consist of three different tissues: lipidic, fibrous, and calcified. Zhang et al. (1998) proposed a method to classify plaque regions into three classes: soft plaque, hard plaque, or hard plaque shadow. The authors used graphs to segment regions followed by a supervised classifier to characterize the tissues. Nair et al. (2002) developed a robust method based on the windowed Fourier transform to classify fibrous, fibrolipidic, calcified, and calcified necrotic areas using Radiofrequency (RF) data in *ex vivo* humans images. This method was later referred to as virtual histology (IVUS-VH) and became the main method to characterize tissues using IVUS pullback (König and Klauss 2007). Recently, Banach et al. (2015) reviewed nine studies using IVUS-VH with 830 patients in total and analyzed the progression of plaques and the use of statins by the patients.

TABLE 7.5 Overview of Some Computer Aided Systems for IVUS

Authors	Years	Parameters/Methods	Number of Individuals
Pandian et al.	1990	PA—luminal area	22
Nissen et al.	1991	CA—eccentricity, luminal area	51
Wenguang et al.	1991	PeA—luminal and lesion area	8
Dijkstra et al.	1999	CA—luminal area, medial-adventitial interface, longitudinal contour	—
Klingensmith et al.	2000	CA—luminal contour and medial-adventitial contour	3
Sun et al.	2013	CA—lumen and external elastic lamina contour	41
Brusseau et al.	2004	CA—luminal area	15
Wahle et al.	2006	CA—luminal and medial-adventitial contours	21
Hass et al.	2000	CA—luminal and medial-adventitial contours	29
Zhang et al.	1998	CA—luminal area, plaque composition classification	6
Nair et al.	2002	CA—plaque composition classification	51 *ex vivo*
König et al.	2007	CA—plaque composition classification	—
Banach et al.	2015	CA—plaque, luminal and external elastic membrane volume	830

Note: CA—Coronary artery; PA—Pulmonary artery; PeA—Peripheral artery.

7.6.2 Intravascular Optical Coherence Tomography IVOCT

Intravascular optical coherence tomography (IVOCT) is an *in vivo* imaging modality, in which the inner wall of blood vessels can be viewed through the introduction of an intravascular catheter with a near-infrared light source. It uses low-coherence interferometry and provides cross-sectional images of sample tissues with microscopic resolution. The acquisition of OCT cardiac images enables several new applications and the development of automatic techniques of computer vision and related areas.

Huang et al. (1991) presented the first OCT image of the human retina and coronary artery *in vitro* in 1991. Initially, only noninvasive OCT was developed for applications in ophthalmology. Later, Jang et al. (2001) were the first to present a human IVOCT, in which an intravascular catheter was inserted with a light source attached at one of its ends (Jang et al., Tearney, and Bouma 2001). IVOCT principle of acquisition is similar to IVUS, but instead of using sound, it uses the dispersion of light with near-infrared frequency signal source. This modality emerges as a new way to identify microstructures with more details (Prati et al. 2010) and superior in resolution to any other form of acquisition of *in vivo* images.

7.6.2.1 IVOCT Features

The IVOCT system uses a swept laser light source with a central wavelength of ~1300 nm and full-width half-maximum bandwidth of 50 nm resulting in a lateral resolution of ~30 μm and an axial resolution of ~12 to 15 μm in air and smaller than 20 μm in biological tissue (Ughi et al. 2012). An IVOCT image sequence consists in the capture of many frames along an arterial length of 54 mm, with a penetration depth in arteries ranges from approximately 0.1–2.5 mm. When the arterial wall is healthy, it is possible to visualize the intima and media border and part of the adventitia border, but when an artery has atherosclerotic plaque, the intima layer is not visually well established. Figure 7.5 depicts the basic structures of a 2D IVOCT. Besides having a limited penetration depth, IVOCT cannot produce images through a blood filled region, which necessitates clearing or flushing blood from the lumen (Prati et al. 2010).

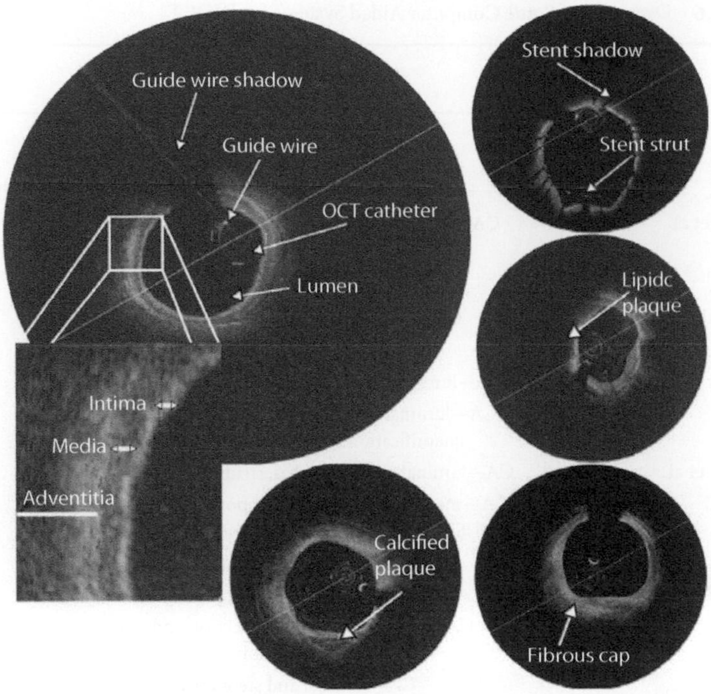

FIGURE 7.5 A single IVOCT frame showing the cross-section coronary artery. The three layers are labeled: intima, media, and adventitia.

There are at least three different OCT systems: (1) time domain (TD); (2) Fourier domain (FD); and (3) swept source (SS). The first system to be developed was the TD-OCT, which employed a variable group-delay reference arm to coherently gate backscattered light from various depths in a sample (Choma and Sarunic 2003). The FD-OCT uses a broadband light source and achieves spectral discrimination with a dispersive spectrometer in the detector arm. The third system, SS-OCT, time-encodes wave numbers by rapidly tuning a narrowband source through a broad optical bandwidth (Hou et al. 2011).

7.6.2.2 IVOCT Computer-Aided Systems

Recent studies show that IVOCT images are able to clearly show differences between the tissues surrounding the vascular wall (Kume et al. 2005; Morikawa et al. 2011). Thus, these images are useful in the evaluation of coronary atherosclerotic plaques, particularly in the evaluation of plaque rupture, erosion, and coronary thrombosis in patients with acute coronary syndrome (Kubo et al. 2011a). Furthermore, IVOCT images are capable of displaying atheroma—macrophage structures surrounding a vulnerable plaque—and to assess stent strut apposition and coverage.

Computational methods applied to IVOCT images emerged recently with the first steps toward automated characterization of vulnerable plaques (Kubo et al. 2011b; Ughi et al. 2013). Table 7.6 shows some applications using IVOCT frames mainly in a clinical phase.

The quantification of lumen cross-sectional area or the lumen volume of a frameset are important quantities to estimate. As this quantification is computationally simple, the literature has mainly focused on lumen segmentation. The work developed by Sihan et al. (2009) separated the contour of the lumen using the Canny method, frequently used in the detection of contours, and makes a series of postprocessing steps for segmenting contours, even in the presence of the catheter and in bifurcation regions. In Tsantis et al. (2012), the Hough transform was used to identify the catheter and

TABLE 7.6 Overview of Some Computer Aided Systems for IVOCT

Authors	Years	Parameters/Methods	Number of Individuals
Jang et al.	2002	CA—cap thickness	10
Ughi et al.	2012	CA—3D reconstruction	25
Kume et al.	2005	CA—intima-media thickness	18 ex-vivo
Morikawa et al.	2011	CA—intimal thickness, luminal and intimal area	37
Ughi et al.	2013	CA—plaque composition	49
Sihan et al.	2009	CA—luminal area	20
Tsantis et al.	2012	CA—luminal area, stent strut detection	10
Tung et al.	2011	CA—luminal area	4
Wang et al.	2010	CA—luminal area, calcified plaque quantification	8
Celi et al.	2014	CA—luminal area, fibrotic, fibro-calcified tissue quantification and fibrous cap thickness	10
Dubuisson et al.	2015	CA—luminal area, stent area, strut malposition rate	4
Wang et al.	2014a	CA—maximum circular unsupported surface area (MCUSA) of selected stent cells and side branch area	12
Unal et al.	2009	CA—luminal and stent area	35
Nakao et al.	2013	CA—3D reconstruction, bifurcation analysis	2
Wang et al.	2015	CA—luminal area, stent detection and 3D reconstruction	72
Xu et al.	2011	CA—luminal and stent area	9
Gerbaud et al.	2015	CA—luminal and stent area, atheroma area	42
Maejima et al.	2014	CA—fibrous cap thickness,	150
Tu et al.	2012	CA—luminal area, 3D reconstruction	37
Diletti et al.	2011	CA—luminal and plaque area, plaque classification and fibrous cap thickness	24
Gonzalo et al.	2009	CA—plaque classification and fibrous cap thickness	30
Macedo et al.	2015	CA—luminal area, bifurcation identification	9
Wang et al.	2014b	CA—luminal contour, bifurcation identification	25

Markov chains were applied to find the lumen region. Tung et al. identified the edges of vessels using a combination of expectation–maximization algorithm to graph-cuts (Tung et al. 2011).

Another IVOCT application that can be addressed computationally is the segmentation and characterization of tissues types in atherosclerotic plaques, such as calcified, lipidic, and fibrotic cap. Wang et al. (2010) applied the level set method (Sethian 1999) to identify calcified plaques in each frame. And Celi and Berti (2014) used intensity-based methods to characterize and quantify fibrous cap in order to classify the vulnerability of plaques.(Celi and Berti 2014).

The behavior of stents after their introduction into the vessel is of extreme importance for monitoring patients' stents post-introduction. The high image resolution allows IVOCT to identify the stent through a sequence of cross section acquired during the pullback, allowing a 3D reconstruction of the whole metal structure (Dubuisson et al. 2015; Unal et al. 2009; Nakao et al. 2013; Wang et al. 2014). Wang et al. (2015b) applied a method based on Bayesian network and graph searches to detect stents in 72 patients. Since the 3D reconstruction may comprise both the stent and the various tissues that constitute the blood vessel wall, some studies have developed automatic methods to segment and reconstruct the structures (stent and tissues) in three dimensions (Farooq et al. 2011; Tsantis et al. 2012; Wang et al. 2010).

IVOCT computer-aided systems can also be used for forensic tasks. Adnan et al. (Adnan et al. 2016) measured the lumen area of six *ex vivo* porcine coronaries and concluded that IVOCT is a feasible tool to perform *post-Morten* diagnosis of coronary artery disease with the potential to improve accuracy of this type of assessment.

Some studies have sought to evaluate the progress of atherosclerosis and the distribution of high-risk plaques at bifurcations using IVUS-VH and IVOCT (Gonzalo et al. 2009; Diletti et al. 2011), but the identification of side branches is performed visually. To improve this process in time, automated methods for bifurcation classification in IVOCT have been developed (Macedo et al. 2015; Wang et al. 2014).

7.6.3 Comparison between IVUS and IVOCT

Although both IVUS and IVOCT allow the visualization of the arterial wall, there are some differences between these modalities. IVUS provides a larger signal penetration and visualization of the adventitia layer, but the image is degraded considerably by speckle noise. While IVOCT provides better resolution, an important property for stent analysis, but a reduced signals penetration. These differences have been highlighted in several publications (Finn et al., 2011; Gerbaud et al. 2016; Maejima et al. 2014). Both modalities can be used in a complementary manner for a single pullback. This benefits were discussed in Li et al. (2014) and Tu et al. (2012), while Garcia-Garcia et al. analyzed the visualization of OCT compared with IVUS-VH, concluding that they are complementary to characterize tissues (García-García et al. 2009).

Conclusion

Computer-aided diagnosis in cardiology is a continuously developing discipline. One of the most challenging tasks for all modalities is cardiac image segmentation, where, despite the advances, the field remains open for new methods and strategies since there is no general solution to the problem. Other fields with intense research activity include: the search for new features to characterize normality or abnormality of cardiac function, characterization of vessel walls and plaques, and quantification and complex visualization of blood flow.

In this chapter, we have shown some of the main computer-aided diagnosis methods and applications for the widely used imaging modalities in cardiology, their main challenges and recent advances in each field.

References

Abramowicz, Alexander J, Melissa A Daubert, Vinay Malhotra, Summer Ferraro, Joshua Ring, Roman Goldenberg, Michael Kam, et al. 2013. Computer-aided analysis of 64-slice coronary computed tomography angiography: A comparison with manual interpretation. *Heart International* 8 (1): e2.

Adnan, Asif, Claire Robinson, Mike Biggs, Shiju Joseph, Bruno Morgan, Guy Rutty, and David Adlam. 2016. Optical coherence tomography of re-pressurised porcine coronary arteries: A systematic study. *Journal of Forensic Radiology and Imaging* 4:52–57.

Aletras, Anthony H, Shujun Ding, Robert S Balaban, and Han Wen. 1999. DENSE: displacement encoding with stimulated echoes in cardiac functional MRI. *Journal of Magnetic Resonance* 137 (1): 247–52.

Alluri, Krishna, Parag H Joshi, Travis S Henry, Roger S Blumenthal, Khurram Nasir, and Michael J Blaha. 2015. Scoring of coronary artery calcium scans: History, assumptions, current limitations, and future directions. Atherosclerosis 239 (1): 109–17.

Amichi, Abdelaziz, and Pascal Laugier. 2009. Quantitative texture analysis and transesophageal echocardiography to characterize the acute myocardial contusion. *The Open Medical Informatics Journal* 3: 13–18.

Anders, Katharina, Stephan Achenbach, Isabel Petit, Werner G Daniel, Michael Uder, and Tobias Pflederer. 2013. Accuracy of automated software-guided detection of significant coronary artery stenosis by CT angiography: Comparison with invasive catheterisation. *European Radiology* 23 (5): 1218–25.

Angelo, Lílian C S, Vieira, Marcelo L C, Rodrigues Sérgio L, Morelato, Renato L, Pereira, Alexandre C, Mill, José G, and Krieger, José E. 2007. Echocardiographic reference values in a sample of asymptomatic adult brazilian population. *Arquivos Brasileiros de Cardiologia* 89 (3): 184–190.

Arnoldi, Elisabeth, Mulugeta Gebregziabher, U Joseph Schoepf, Roman Goldenberg, Luis Ramos-Duran, Peter L Zwerner, Konstantin Nikolaou, Maximilian F Reiser, Philip Costello, and Christian Thilo. 2010. Automated computer-aided stenosis detection at coronary CT angiography: Initial experience. *European Radiology* 20 (5): 1160–67.

Attili, Anil K, Andreas Schuster, Eike Nagel, Johan H C Reiber, and Rob J van der Geest. 2010. Quantification in cardiac MRI: Advances in image acquisition and processing. *The International Journal of Cardiovascular Imaging* 26 (S1): 27–40.

Axel, Leon, and Lawrence Dougherty. 1989. MR imaging of motion with spatial modulation of magnetization. *Radiology* 171 (3): 841–45.

Banach, Maciej, Corina Serban, Amirhossein Sahebkar, Dimitri P Mikhailidis, Sorin Ursoniu, Kausik K Ray, Jacek Rysz, et al. 2015. Impact of statin therapy on coronary plaque composition: A systematic review and meta-analysis of virtual histology intravascular ultrasound studies. *BMC Medicine* 13 (1). 229.

Barbosa, Jorge G, Bruno Figueiredo, Nuno Bettencourt, and João M R S Tavares. 2011. Towards automatic quantification of the epicardial fat in non-contrasted CT images. *Computer Methods in Biomechanics and Biomedical Engineering* 14 (10): 905–14.

Bardinet, Eric, Laurent D Cohen, and Nicholas Ayache. 1996. Tracking and motion analysis of the left ventricle with deformable superquadrics. *Medical Image Analysis* 1 (2): 129–49.

Barr, Alan H 1981. Superquadrics and angle-preserving transformations. *IEEE Computer Graphics and Applications* 1 (1): 11–23.

Berger, Bruce C, Denny D Watson, George J Taylor, George B Craddock, Randolph P Martin, Charles D Teates, and George A Beller. 1981. Quantitative thallium-201 exercise scintigraphy for detection of coronary artery disease. *Journal of Nuclear Medicine* 22 (7): 585–93.

Bieging, Erik T, Alex Frydrychowicz, Andrew Wentland, Benjamin R Landgraf, Kevin M Johnson, Oliver Wieben, and Christopher J François. 2011. In vivo three-dimensional MR wall shear stress estimation in ascending aortic dilatation. *Journal of Magnetic Resonance Imaging* 33 (3): 589–97.

Biermann, Christina, Ilias Tsiflikas, Christoph Thomas, Bernadette Kasperek, Martin Heuschmid, and Claus D Claussen. 2012. Evaluation of computer-assisted quantification of carotid artery stenosis. *Journal of Digital Imaging* 25 (2): 250–57.

Bilmes, Jeff A. 1998. A gentle tutorial of the EM algorithm. *International Computer Science Institute.* Berkley, CA.

Binter, Christian, Verena Knobloch, Robert Manka, Andreas Sigfridsson, and Sebastian Kozerke. 2013. Bayesian multipoint velocity encoding for concurrent flow and turbulence mapping. *Magnetic Resonance in Medicine* 69: 1337–45.

Boogers, Mark J, Joanne D Schuijf, Pieter H Kitslaar, Jacob M van Werkhoven, Fleur R de Graaf, Eric Boersma, Joëlla E van Velzen, et al. 2010. Automated quantification of stenosis severity on 64-slice CT: A comparison with quantitative coronary angiography. *JACC Cardiovascular Imaging* 3 (7). 699–709.

Boogers, Mark J, Alexander Broersen, Joëlla E Van Velzen, Fleur R De Graaf, Heba M El-Naggar, Pieter H Kitslaar, Jouke Dijkstra, et al. 2012. Automated quantification of coronary plaque with computed tomography: Comparison with intravascular ultrasound using a dedicated registration algorithm for fusion-based quantification. *European Heart Journal* 33 (8): 1007–16.

Brown, Mark A, Semelka, Richard C 1995. *MRI basic principles and applications*. 1st edn. New York: Wiley.

Brusseau, Elisabeth, Chris L De Korte, Frits Mastik, Johannes Schaar, and Anton F W Van Der Steen. 2004. Fully automatic luminal contour segmentation in intracoronary ultrasound imaging -a statistical approach. *IEEE Transactions on Medical Imaging* 23 (5): 554–66.

Burow, Robert D, Malcolm Pond, A. William Schafer, and Lewis Becker. 1979. Circumferential Profiles: A New Method for Computer Analysis of Thallium-201 Myocardial Perfusion Images. *Journal of Nuclear Medicine* 20 (7): 771–77.

Carranza-Herrezuelo, Noemi, Ana Bajo, Filip Sroubek, Cristina Santamarta, Gabriel Cristobal, Andres Santos, and María J Ledesma-Carbayo. 2010. Motion estimation of tagged cardiac magnetic resonance images using variational techniques. *Computerized Medical Imaging and Graphics : The Official Journal of the Computerized Medical Imaging Society* 34 (6): 514–22.

Carvalho, Diego D B, Zeynettin Akkus, Stijn C H Van Den Oord, Arend F L Schinkel, Antonius F W Van Der Steen, Wiro J Niessen, Johan G Bosch, and Stefan Klein. 2015. Lumen segmentation and motion estimation in B-mode and contrast-enhanced ultrasound images of the carotid artery in patients with atherosclerotic plaque 34 (4): 983–93.

Cauvin, J C, J Y Boire, M Zanca, J M Bonny, J Maublant, and A Veyre. 1993. 3D modeling in myocardial 201-TL spect. *Computerized Medical Imaging and Graphics* 17 (4): 345–50.

Celi, Simona, and Sergio Berti. 2014. In-vivo segmentation and quantification of coronary lesions by optical coherence tomography images for a lesion type definition and stenosis grading. *Medical Image Analysis* 18 (7): 1157–68.

Cheng, Da-Chuan, and Xiaoyi Jiang. 2008. Detections of arterial wall in sonographic artery images using dual dynamic programming. *IEEE Transactions on Information Technology in Biomedicine : A Publication of the IEEE Engineering in Medicine and Biology Society* 12 (6): 792–99.

Choma, Michael A, and Marinko V Sarunic, Changhuei Yang, and Joseph A. Izatt. 2003. Sensitivity advantage of swept source and fourier domain optical coherence tomography. *Optics Express* 11 (18): 2183–89.

Cibis, Merih, Wouter V Potters, Frank J H Gijsen, Henk Marquering, E van Bavel, Antonius F W van der Steen, Aart J Nederveen, and Jolanda J Wentzel. 2014. Wall shear stress calculations based on 3D cine phase contrast MRI and computational fluid dynamics: A comparison study in healthy carotid arteries. *NMR in Biomedicine* 27 (7): 826–34.

Colan, Steven D, Ira A Parness, Philip J Spevak, and Stephen P Sanders. 1992. Developmental modulation of myocardial mechanics: Age- and growth-related alterations in afterload and contractility. *Journal of the American College of Cardiology* 19 (3): 619–29.

Coppini, Giuseppe, Riccardo Favilla, Paolo Marraccini, Davide Moroni, and Gabriele Pieri. 2010. Quantification of epicardial fat by cardiac CT imaging. *The Open Medical Informatics Journal* 4: 126–35.

Dankerl, Peter, Matthias Hammon, Alexey Tsymbal, Alexander Cavallaro, Stephan Achenbach, Michael Uder, and Rolf Janka. 2014. Evaluation of novice reader diagnostic performance in coronary CT angiography using an advanced cardiac software package. *International Journal of Computer Assisted Radiology and Surgery* 9 (4): 609–15.

de Araújo Gonçalves, Pedro, Gastón A Rodríguez-Granillo, Ernest Spitzer, Pannipa Suwannasom, Christian Loewe, Koen Nieman, and Hector M Garcia-Garcia. 2015. Functional evaluation of coronary disease by CT angiography. *JACC: Cardiovascular Imaging* 8 (11): 1322–35.

De Graaf, Michiel A, Alexander Broersen, Pieter H Kitslaar, Cornelis J Roos, Jouke Dijkstra, Boudewijn P F Lelieveldt, J Wouter Jukema, et al. 2013. Automatic quantification and characterization of coronary atherosclerosis with computed tomography coronary angiography: Cross-correlation with intravascular ultrasound virtual histology. *International Journal of Cardiovascular Imaging* 29 (5): 1177–90.

Declerck, Jérôme, Jacques Feldmar, and Nicholas Ayache. 1998. Definition of a four-dimensional continuous planispheric transformation for the tracking and the analysis of left-ventricle motion. *Medical Image Analysis* 2 (2): 197–213.

Destrempes, François, Jean Meunier, Marie F Giroux, Gilles Soulez, and Guy Cloutier. 2009. Segmentation in ultrasonic B-mode images of healthy carotid arteries using mixtures of nakagami distributions and stochastic optimization. *IEEE Transactions on Medical Imaging* 28 (2): 215–29.

Dey, Damini, Tiziano Schepis, Mohamed Marwan, Piotr J Slomka, Daniel S Berman, and Stephan Achenbach. 2010. Automated three-dimensional quantification of noncalcified coronary plaque from coronary CT angiography: Comparison with intravascular US 1. *Radiology* 257 (2): 516–22.

Diaz-Zamudio, Mariana, Damini Dey, Annika Schuhbaeck, Ryo Nakazato, Heidi Gransar, Piotr J Slomka, Jagat Narula, et al. 2015. Automated quantitative plaque burden from coronary CT angiography noninvasively predicts hemodynamic significance by using fractional flow reserve in intermediate coronary lesions. *Radiology* 276 (2): 408–15.

Dijkstra, J, G Koning, and J H Reiber. 1999. Quantitative measurements in IVUS images. *International Journal of Cardiac Imaging* 15 (6): 513–22.

Diletti, Roberto, Hector M Garcia-Garcia, Josep Gomez-Lara, Salvatore Brugaletta, Joanna J Wykrzykowska, Nienke van Ditzhuijzen, Robert Jan van Geuns, Evelyn Regar, Giuseppe Ambrosio, and Patrick W Serruys. 2011. Assessment of coronary atherosclerosis progression and regression at bifurcations using combined IVUS and OCT. *JACC. Cardiovascular Imaging* 4 (7): 774–80.

Ding, Jingzhong, Fang-Chi Hsu, Tamara B Harris, Yongmei Liu, Stephen B Kritchevsky, Moyses Szklo, Pamela Ouyang, et al. 2009. The Association of Pericardial Fat with Incident Coronary Heart Disease: The Multi-Ethnic Study of Atherosclerosis (MESA). *American Journal of Clinical Nutrition* 90 (3): 499–504. doi:10.3945/ajcn.2008.27358.

Ding, Xiaowei, Demetri Terzopoulos, Mariana Diaz-Zamudio, Daniel S Berman, Piotr J Slomka, and Damini Dey. 2015. Automated pericardium delineation and epicardial fat volume quantification from noncontrast CT. *Medical Physics* 42 (9): 5015–26.

Dubuisson, Florian, Emilie Péry, Lemlih Ouchchane, Nicolas Combaret, Claude Kauffmann, Géraud Souteyrand, Pascal Motreff, and Laurent Sarry. 2015. Automated peroperative assessment of stents apposition from OCT pullbacks. *Computers in Biology and Medicine* 59: 98–105.

Duda, Richard O, Peter E Hart, and David G Stork. 2001. *Pattern Classification*. New York: Wiley.

Dyverfeldt, Petter, Malenka Bissell, Alex J Barker, Ann F Bolger, Carl-Johan Carlhäll, Tino Ebbers, Christopher J Francios, et al. 2015. 4D flow cardiovascular magnetic resonance consensus statement. *Journal of Cardiovascular Magnetic Resonance: Official Journal of the Society for Cardiovascular Magnetic Resonance* 17 (1): 72.

Dyverfeldt, Petter, Tino Ebbers, and Toste Länne. 2014. Pulse wave velocity with 4D flow MRI: Systematic differences and age-related regional vascular stiffness. *Magnetic Resonance Imaging* 32 (10): 1266–71.

Ebersberger, Ullrich, Dov Eilot, Roman Goldenberg, Alon Lev, J Reid Spears, Garrett W Rowe, Nicholas Y Gallagher, et al. 2013. Fully automated derivation of coronary artery calcium scores and cardiovascular risk assessment from contrast medium-enhanced coronary CT angiography studies. *European Radiology* 23: 650–57.

Elliott, M R, and A J Thrush. 1996. Measurement of resolution in intravascular ultrasound images. *Physiological Measurement* 17 (4): 259–65.

Epstein, Frederick H 2007. MRI of left ventricular function. *Journal of Nuclear Cardiology: Official Publication of the American Society of Nuclear Cardiology* 14 (5): 729–44.

Eriksson, Jonatan, Ann F Bolger, Tino Ebbers, and Carl J Carlhall. 2013. Four-dimensional blood flow-specific markers of LV dysfunction in dilated cardiomyopathy. *European Heart Journal - Cardiovascular Imaging* 14 (5): 417–24.

Eriksson, Jonatan, Carl J Carlhäll, Petter Dyverfeldt, Jan Engvall, Ann F Bolger, and Tino Ebbers. 2010. Semi-automatic quantification of 4D left ventricular blood flow. *Journal of Cardiovascular Magnetic Resonance: Official Journal of the Society for Cardiovascular Magnetic Resonance* doi:10.1186/1532-429X-12-9..

Eslami, Abouzar, Athanasios Karamalis, Amin Katouzian, and Nassir Navab. 2013. Segmentation by retrieval with guided random walks: Application to left ventricle segmentation in MRI. *Medical Image Analysis* 17 (2): 236–53.

Faber, Tracy L, C David Cooke, Russel D Folks, Johnathan P Vansant, Kenneth J Nichols, E Gordon DePuey, Roderic I Pettigrew, and Ernest V Garcia. 1999. Left ventricular function and perfusion from gated SPECT perfusion images: An integrated method. *Journal of Nuclear Medicine* 40 (4): 650–59.

Faber, Tracy L, Ernest M Stokely, Ronald M Peshock, and James R Corbett. 1991. A model-based four-dimensional left ventricular surface detector. *IEEE Transactions on Medical Imaging* 10 (3): 321–29.

Farooq, Vasim, Bill D Gogas, Takayuki Okamura, Jung H Heo, Michael Magro, Josep Gomez-Lara, Yoshinobu Onuma, et al. 2011. Three-dimensional optical frequency domain imaging in conventional percutaneous coronary intervention: The potential for clinical application. *European Heart Journal* 34 (12): 875–85.

Ficaro, Edward P, and James R Corbett. 2004. Advances in quantitative perfusion SPECT imaging. *Journal of Nuclear Cardiology* 11 (1): 62–70.

Finn, Aloke V, Y Chandrashekhar, and Jagat Narula. 2011. IVUS and OCT: Either or survivor. *JACC. Cardiovascular Imaging* 4 (9): 1047–49.

Florian, A, A Ludwig, S Rosch, H Yildiz, U Sechtem, and A Yilmaz. 2014. Myocardial fibrosis imaging based on T1-mapping and extracellular volume fraction (ECV) measurement in muscular dystrophy patients: Diagnostic value compared with conventional late gadolinium enhancement (LGE) imaging. *European Heart Journal - Cardiovascular Imaging* 15 (9): 1004–12.

Freund, Yoav and Robert E Schapire. 1997. A decision-theoretic generalization of on-line learning and an application to boosting. *Journal of Computer and System Sciences* 55 (1): 119–39.

Ganesan, K, Acharya, UR, Chua, CK, Min, LC, Abraham, TK. 2014. Automated diagnosis of mammogram images of breast cancer using discrete wavelet transform and spherical wavelet transform features: A comparative study. *Technology in Cancer Research and Treatment*, 5: 605–15.

Garcia, Ernest, Jamshid Maddahi, Daniel Berman, and Alan Waxman. 1981. Space/time quantitation of thallium-201 myocardial scintigraphy. *Journal of Nuclear Medicine* 22 (4): 309–17.

Garcia, Ernest V, C David Cooke, Russel D Folks, Cesar A Santana, Elzbieta G Krawczynska, Levien De Braal, and Norberto F Ezquerra. 2001. Diagnostic performance of an expert system for the interpretation of myocardial perfusion SPECT studies. *Journal of Nuclear Medicine* 42 (8): 1185–91.

Garcia, Ernest V, C David Cooke, Kenneth F Van Train, Russel Folks, John Peifer, E Gordon DePuey, Jamshid Maddahi, Naomi Alazraki, James Galt, Norberto F Ezquerra, Jack Ziffer, Joseph Areeda, and Daniel S Berman. 1990. Technical aspects of myocardial SPECT imaging with technetium-99m sestamibi. *The American Journal of Cardiology* 66 (13): 23E–31E.

Garcia, E V, K F Van Train, J Maddahi, F Prigent, J Areeda, A Waxman, and D S Berman. 1985. Quantificationof rotational thaffium-201 myocardial tomography. *Journal of Nuclear Medicine* 26 (1): 17–26.

Garcia, E V, T L Faber, J R Galt, C D Cooke, and R D Folks. 2000. Advances in nuclear emission PET and SPECT imaging. *IEEE Engineering in Medicine and Biology Magazine* 19 (5): 21–33.

García-García, Héctor M, Gary S Mintz, Amir Lerman, Geoffrey Vince, Paulina Margolis, Gerrit A van Es, Marie A M Morel, et al. 2009. Tissue characterisation using intravascular radiofrequency data analysis: recommendations for acquisition, analysis, interpretation and reporting. *EuroIntervention* 5 (2): 177–89.

Gerbaud, Edouard, Giora Weisz, Atsushi Tanaka, Manabu Kashiwagi, Takehisa Shimizu, Lin Wang, Christiano Souza, et al. 2015. Multi-laboratory inter-institute reproducibility study of IVOCT and IVUS assessments using published consensus document definitions. *European Heart Journal Cardiovascular Imaging*, 7: 756–764.

Germano, G, J Erel, H Kiat, P B Kavanagh, and D S Berman. 1997. Quantitative LVEF and qualitative regional function from gated thallium-201 perfusion SPECT. *Journal of Nuclear Medicine* 38: 749–54.

Germano, Guido, Paul B Kavanagh, Piotr J Slomka, Serge D van Kriekinge, Geoff Pollard, Daniel S Berman. 2007. Quantitation in gated perfusion SPECT imaging: the Cedars-Sinai approach. *Journal of Nuclear Cardiology* 14: 433–454.

Germano, G, H Kiat, P B Kavanagh, M Moriel, M Mazzanti, H T Su, K F Van Train, and Daniel S Berman. 1995. Automatic quantification of ejection fraction from gated myocardial perfusion SPECT. *Journal of Nuclear Medicine* 36 (11): 2138–47.

Goldenberg, Roman, Dov Eilot, Grigory Begelman, Eugene Walach, Eyal Ben-Ishai, and Nathan Peled. 2012. Computer-aided simple triage (CAST) for coronary CT angiography (CCTA). *International Journal of Computer Assisted Radiology and Surgery* 7 (6): 819–27.

Gonzalo, Nieves, Hector M Garcia-Garcia, Evelyn Regar, Peter Barlis, Jolanda Wentzel, Yoshinobu Onuma, Jurgen Ligthart, and Patrick W Serruys. 2009. In vivo assessment of high-risk coronary plaques at bifurcations with combined intravascular ultrasound and optical coherence tomography. *JACC. Cardiovascular Imaging* 2 (4): 473–82.

Grosgeorge, Damien, Caroline Petitjean, Jérôme Caudron, Jeannette Fares, and Jean Nicolas Dacher. 2011. Automatic cardiac ventricle segmentation in MR images: A validation study. *International Journal of Computer Assisted Radiology and Surgery* 6 (5): 573–81.

Gupta, A, L Von Kurowski, A Singh, D Geiger, C-C Liang, M-Y Chiu, L P Adler, M Haacke, and D L Wilsont. 1993. Cardiac MR Image Segmentation Using Deformable Models. *Biomedical Image Processing and Biomedical Visualization*, 747–50.

Gutgesell, HP, M Paquet, DH Duff, and DG McNamara. 1977. Evaluation of left ventricular size and function by echocardiography. Results in normal children. *Circulation* 56 (3): 457–62.

Gutierrez, Marco A, Paulo E Pilon, SG Lage, L Kopel, RT Carvalho, and SS Furuie. 2002. Automatic measurement of carotid diameter and wall thickness in ultrasound images. In *Computers in Cardiology*, 29: 359–362. doi:10.1109/CIC.2002.1166783.

Gutierrez, Marco A, Marina S Rebelo, Sergio S Furuie, and Jose C Meneghetti. 2003. Automatic quantification of three-dimensional kinetic energy in gated myocardial perfusion single-photon-emission computerized tomography improved by a multiresolution technique. *Journal of Electronic Imaging* 12 (1): 118.

Haas, Christine, Helmut Ermert, Stephan Holt, Peter Grewe, Abderrahman Machraoui, and Jürgen Barmeyer. 2000. Segmentation of 3D intravascular ultrasonic images based on a random field model. *Ultrasound in Medicine and Biology* 26 (2): 297–306.

Halpern, Ethan J, and David J Halpern. 2011. Diagnosis of coronary stenosis with CT angiography comparison of automated computer diagnosis with expert readings. *Academic Radiology* 18 (3): 324–33.

Hameeteman, K, R van't Klooster, M Selwaness, A van der Lugt, J C M Witteman, W J Niessen, and S Klein. 2013. Carotid wall volume quantification from magnetic resonance images using deformable model fitting and learning-based correction of systematic errors. *Physics in Medicine and Biology* 58 (5): 1605–23.

Hamilton, Glen W, Gene B Trobaugh, James L Ritchie, David L Williams, W Douglas Weaver, and K Lance Gould. 1977. Myocardial imaging with intravenously injected thallium-201 in patients with suspected coronary artery disease: Analysis of technique and correlation with electrocardiographic, coronary anatomic and ventriculographic findings. *The American Journal of Cardiology* 39 (3): 347–54.

Han, Jiawei, Micheline Kamber, and Jian Pei. 2016. *Data Mining: Concepts and Techniques*, 3rd edn. MA: Elsevier.

Hashoul, Sharbell, Tamar Gaspar, David A Halon, Basil S Lewis, Yuval Shenkar, Ronen Jaffe, Nathan Peled, and Ronen Rubinshtein. 2015. Automated Computer-Assisted Diagnosis of Obstructive Coronary Artery Disease in Emergency Department Patients Undergoing 256-Slice Coronary Computed Tomography Angiography for Acute Chest Pain. *American Journal of Cardiology* 116 (7): 1017–21. doi:10.1016/j.amjcard.2015.07.014.

Hautvast, Gilion L T F, Carol J Salton, Michael L Chuang, Marcel Breeuwer, Christopher J O'Donnell, and Warren J Manning. 2012. Accurate computer-aided quantification of left ventricular parameters: Experience in 1555 cardiac magnetic resonance studies from the framingham heart study. *Magnetic Resonance in Medicine* 67 (5): 1478–86.

Heitz, F, P Perez, and P Bouthemy. 1994. Multiscale minimization of global energy functions in some visual recovery problems. *CVGIP Image Understanding* 59 (1): 125–34.

Hou, Randy, Tho Le, Septimiu D Murgu, Zhongping Chen, and Matt Brenner. 2011. Recent advances in optical coherence tomography for the diagnoses of lung disorders. *Expert Review of Respiratory Medicine* 5 (5): 711–24.

Hu, Huaifei, Zhiyong Gao, Liman Liu, Haihua Liu, Junfeng Gao, Shengzhou Xu, Wei Li, and Lu Huang. 2014. Automatic segmentation of the left ventricle in cardiac MRI using local binary fitting model and dynamic programming techniques. *PloS One* 9 (12): e114760. doi:10.1371/journal.pone.0114760.

Huang, David, Eric A Swanson, Charles P Lin, Joel S Schuman, William G Stinson, Warren Chang, Michael R Hee, et al. 1991. Optical coherence tomography. *Science* 254: 1178–81.

Huang, Su, Jimin Liu, Looi Chow Lee, Sudhakar K Venkatesh, Lynette L S Teo, Christopher Au, and Wieslaw L Nowinski. 2011. An image-based comprehensive approach for automatic segmentation of left ventricle from cardiac short axis cine MR images. *Journal of Digital Imaging* 24 (4): 598–608.

Iacobellis, Gianluca, and Howard J Willens. 2009. Echocardiographic epicardial fat: A review of research and clinical applications. *Journal of the American Society of Echocardiography* 22 (12): 1311–19.

Ibrahim, El-Sayed H. 2011. Myocardial tagging by cardiovascular magnetic resonance: Evolution of techniques--pulse sequences, analysis algorithms, and applications. *Journal of Cardiovascular Magnetic Resonance: Official Journal of the Society for Cardiovascular Magnetic Resonance* 13 (1): 36.

Isgum, Ivana, Marius Staring, Annemarieke Rutten, Mathias Prokop, Max A. Viergever, and Bran van Ginneken. 2009. Multi-Atlas-Based Segmentation With Local Decision Fusion - Application to Cardiac and Aortic Segmentation in CT Scans. IEEE Transactions on Medical Imaging 28 (7): 1000–1010. doi:10.1109/TMI.2008.2011480.

Isgum, Ivana, Annemarieke Rutten, Mathias Prokop, Marius Staring, Stefan Klein, Josien P W Pluim, Max A Viergever, and Bram van Ginneken. 2010. Automated aortic calcium scoring on low-dose chest computed tomography. *Medical Physics* 37 (2): 714–23.

Jahanzad, Zeinab, Yih Miin Liew, Mehmet Bilgen, Robert A McLaughlin, Chen Onn Leong, Kok Han Chee, Yang F A Aziz, et al. 2015. Regional assessment of LV wall in infarcted heart using tagged MRI and cardiac modelling. *Physics in Medicine and Biology* 60 (10): 4015–31.

Jang, I-K, G Tearney, and B Bouma. 2001. Visualization of tissue prolapse between coronary stent struts by optical coherence tomography: Comparison with intravascular ultrasound. *Circulation* 104 (22): 2754–2754.

Bilmes, Jeff A. 1998. A gentle tutorial of the EM algorithm. doi:10.1016/S0550-3213(97)00753-0.

Jellis, Christine L, and Deborah H Kwon. 2014. Myocardial T1 mapping: Modalities and clinical applications. *Cardiovascular Diagnosis and Therapy* 4 (2): 126–37.

Jiang, Kai, and Xin Yu. 2014. Quantification of regional myocardial wall motion by cardiovascular magnetic resonance. *Quantitative Imaging in Medicine and Surgery* 4 (5): 345–57.

Kang, Doo Kyoung, Nae Jung Im, Soon Mo Park, and Hong Seok Lim. 2011. CT Comparison of Visual and Computerised Quantification of Coronary Stenosis according to Plaque Composition. *European Radiology* 21 (4): 712–21. doi:10.1007/s00330-010-1970-9.

Kang, Dongwoo, Piotr J Slomka, Ryo Nakazato, Reza Arsanjani, Victor Y Cheng, James K Min, Debiao Li, Daniel S Berman, C-C Jay Kuo, and Damini Dey. 2013. Automated knowledge-based detection of nonobstructive and obstructive arterial lesions from coronary CT angiography. *Medical Physics* 40 (2013): 041912.

Kang, Ki-Woon, Hyuk-Jae Chang, Hackjoon Shim, Young-Jin Kim, Byoung-Wook Choi, Woo-In Yang, Jee-Young Shim, Jongwon Ha, and Namsik Chung. 2012. Feasibility of an automatic computer-assisted algorithm for the detection of significant coronary artery disease in patients presenting with acute chest pain. *European Journal of Radiology* 81 (4): e640–46.

Katouzian, Amin, Elsa D Angelini, Stéphane G Carlier, Jasjit S Suri, Nassir Navab, and Andrew F Laine. 2012. A state-of-the-art review on segmentation algorithms in intravascular ultrasound (IVUS) images. *IEEE Transactions on Information Technology in Biomedicine* 16 (5): 823–34.

Kause, Hanne B, Olena G Filatova, Remco Duits, Andrea Fuster , H C Van Assen, L C Mark Bruurmijn, Andrea Fuster, Jos J M Westenberg, Luc M J Florack, and Hans C van Assen. 2013. Direct myocardial strain assessment from frequency estimation in tagging MRI. STACOM 2013: Statistical Atlases and Computational Models of the Heart. *Imaging and Modelling Challenges* pp. 212–219.

Kellman, Peter, and Michael S Hansen. 2014. T1-mapping in the heart: Accuracy and precision. *Journal of Cardiovascular Magnetic Resonance: Official Journal of the Society for Cardiovascular Magnetic Resonance* 16 (1): 2.

Kim, Young Jun, Gong Y Jin, Eun Y Kim, Young M Han, Jei K Chae, Sang R Lee, and Keun S Kwon. 2013. Quantification of coronary artery plaque using 64-slice dual-source CT: Comparison of semi-automatic and automatic computer-aided analysis based on intravascular ultrasonography as the gold standard. *The International Journal of Cardiovascular Imaging* 29 Suppl 2: 93–100.

Klass, Oliver, Susanne Kleinhans, Matthew J Walker, Mark Olszewski, Sebastian Feuerlein, Markus Juchems, and Martin H K Hoffmann. 2010. Coronary plaque imaging with 256-slice multidetector computed tomography: Interobserver variability of volumetric lesion parameters with semiautomatic plaque analysis software. *International Journal of Cardiovascular Imaging* 26 (6): 711–20.

Klingensmith, Jon D., Raj Shekhar, and D Geoffrey Vince. 2000. Evaluation of three-dimensional segmentation algorithms for the identification of luminal and medial-adventitial borders in intravascular ultrasound images. *IEEE Transactions on Medical Imaging* 19 (10): 996–1011.

Kolar, Radim, Radovan Jirik, and Jiří Jan. 2004. Estimator comparison of the nakagami-M parameter and its application in echocardiography. *Radioengineering* 7: 8–12.

König, Andreas, and Volker Klauss. 2007. Virtual histology. *Heart (British Cardiac Society)* 93 (8): 977–82.

Koo, Bon-Kwon, Andrejs Erglis, Joon-Hyung Doh, David V Daniels, Sanda Jegere, Allison M Dunning, Tony Defrance, Alexandra Lansky, Jonathan Leipsic, and James K Min 2011. Diagnosis of ischemia-causing coronary stenoses by noninvasive fractional flow reserve computed from coronary CT angiograms: A prospective multicenter study. *Journal of the American College of Cardiology* 58 (19): 1–23. doi:10.1016/j.jacc.2011.06.066.

Korosoglou, Grigorios, Dirk Mueller, Stephanie Lehrke, Henning Steen, Waldemar Hosch, Tobias Heye, Hans-Ulrich Kauczor, Evangelos Giannitsis, and Hugo A Katus. 2010. Quantitative Assessment of Stenosis Severity and Atherosclerotic Plaque Composition Using 256-Slice Computed Tomography. *European Radiology* 20 (8): 1841–50. doi:10.1007/s00330-010-1753-3.

Kubo, Takashi, Yasushi Ino, Takashi Tanimoto, Hironori Kitabata, Atsushi Tanaka, and Takashi Akasaka. 2011a. Optical coherence tomography imaging in acute coronary syndromes. *Cardiology Research and Practice*: 312978. Doi:10.4061/2011/312978.

Kubo, Takashi, Chenyang Xu, Zhao Wang, Nienke S van Ditzhuijzen, and Hiram G Bezerra. 2011b. Plaque and thrombus evaluation by optical coherence tomography. *The International Journal of Cardiovascular Imaging* 27 (2): 289–98.

Kume, Teruyoshi, Takashi Akasaka, Takahiro Kawamoto, Nozomi Watanabe, Eiji Toyota, Yoji Neishi, Renan Sukmawana, Yoshito Sadahira, and Kiyoshi Yoshida. 2005. Assessment of coronary intima—media thickness by optical coherence tomography. *Circulation Journal* 69: 903–7.

Larsen, Britta A, Gail A Laughlin, Sarah D Saad, Elizabeth Barrett-Connor, Matthew A Allison, and Christina L Wassel. 2015. Pericardial fat is associated with all-cause mortality but not incident CVD: The rancho bernardo study. *Atherosclerosis* 239 (2): 470–75.

Lebenberg, Jessica, Alain Lalande, Patrick Clarysse, Irene Buvat, Christopher Casta, Alexandre Cochet, Constantin Constantinidès, et al. 2015. Improved estimation of cardiac function parameters using a combination of independent automated segmentation results in cardiovascular magnetic resonance imaging. *Plos One* 10 (8): e0135715.

Lesage, David, Elsa D Angelini, Isabelle Bloch, and Gareth Funka-Lea. 2009. A review of 3D vessel lumen segmentation techniques: Models, features and extraction schemes. *Medical Image Analysis* 13 (6): 819–45.

Li, Jiawen, Xiang Li, Dilbahar Mohar, Aidan Raney, Joseph Jing, Jun Zhang, Abbey Johnston, et al. 2014. Integrated IVUS-OCT for real-time imaging of coronary atherosclerosis. *JACC. Cardiovascular Imaging* 7 (1): 101–3.

Liang, Q, I Wendelhag, J Wikstrand, and T Gustavsson. 2000. A multiscale dynamic programming procedure for boundary detection in ultrasonic artery images. *IEEE Transactions on Medical Imaging* 19 (2): 127–42.

Lieback, E, I Hardouin, R Meyer, J Bellach, and R Hetzer. 1996. Clinical value of echocardiographic tissue characterization in the diagnosis of myocarditis. *European Heart Journal* 17: 135–42.

Liu, Hong, Huaifei Hu, Xiangyang Xu, and Enmin Song. 2012. Automatic left ventricle segmentation in cardiac MRI using topological stable-state thresholding and region restricted dynamic programming. *Academic Radiology* 19 (6): 723–31.

Hadhoud M A, Marwa, Mohamed I Eladawy, Ahmed Farag, Franco M Montevecchi, and Umberto Morbiducci. 2012. Left ventricle segmentation in cardiac MRI images. *American Journal of Biomedical Engineering* 2 (3): 131–35.

Macedo, Maysa M G, Welingson V N Guimarães, Micheli Z Galon, Celso K Takimura, Pedro A Lemos, and Marco A Gutierrez. 2015. A bifurcation identifier for IV-OCT using orthogonal least squares and supervised machine learning. *Computerized Medical Imaging and Graphics: The Official Journal of the Computerized Medical Imaging Society* 46 (2): 237–48.

Maddahi, Jamshid, Ernest V Garcia, Daniel S Berman, Alan Waxman, H J Swan, and James Forrester. 1981. Improved noninvasive assessment of coronary artery disease by quantitative analysis of regional stress myocardial distribution and washout of thallium-201. *Circulation* 64 (5): 924–35.

Maejima, N, K Hibi, K Saka, N Nakayama, Y Matsuzawa, M Endo, N Iwahashi, et al. 2014. Morphological features of non-culprit plaques on optical coherence tomography and integrated backscatter intravascular ultrasound in patients with acute coronary syndromes. *European Heart Journal - Cardiovascular Imaging* 16 (2): 190–97.

Mahabadi, Amir A., Joseph M. Massaro, Guido A. Rosito, Daniel Levy, Joanne M. Murabito, Philip A. Wolf, Christopher J. O'Donnell, Caroline S. Fox, and Udo Hoffmann. 2008. Association of Pericardial Fat, Intrathoracic Fat, and Visceral Abdominal Fat with Cardiovascular Disease Burden: The Framingham Heart Study. *European Heart Journal* 30(7): 850–56. doi:10.1093/eurheartj/ehn573.

Marinol, Marco, Federico Veronesil, Giacomo Tarronil, and Cristiana Corsi. 2014. Fully automated assessment of left ventricular volumes, function and mass from cardiac MRI, no. Lv: 109–12.

Markl, Michael, Susanne Schnell, and Alex J Barker. 2014. 4D flow imaging: Current status to future clinical applications. *Current Cardiology Reports* 16 (5): 481.

Markl, Michael, Wolf Wallis, Stefanie Brendecke, Jan Simon, Alex Frydrychowicz, and Andreas Harloff. 2010. Estimation of global aortic pulse wave velocity by flow-sensitive 4D MRI. *Magnetic Resonance in Medicine* 63 (6): 1575–82.

Maurovich-Horvat, Pal, Maros Ferencik, Fabian Bamberg, and Udo Hoffmann. 2009. Methods of laque quantification and characterization by cardiac computed tomography. *Journal of Cardiovascular Computed Tomography* 3 (SUPPL.2). S91–98.

Meade, Robert C, Virinderjit S Bamrah, James D Horgan, Philip P Ruetz, Charles Kronenwetter, and En-Lin Yeh. 1978. Quantitative methods in the evaluation of thallium-201 myocardial perfusion images. *Journal of Nuclear Medicine* 19 (10): 1175–79.

Messroghli, Daniel R, Aleksandra Radjenovic, Sebastian Kozerke, David M Higgins, Mohan U Sivananthan, and John P Ridgway. 2004. Modified look-locker inversion recovery (MOLLI) for high-resolutionT1 mapping of the heart. *Magnetic Resonance in Medicine* 52 (1): 141–46.

Meyer, Mathias, U Joseph Schoepf, Christian Fink, Roman Goldenberg, Paul Apfaltrer, Joachim Gruettner, Diana Vajcs, Stefan O Schoenberg, and Thomas Henzler. 2013. Diagnostic performance evaluation of a computer-aided simple triage system for coronary CT angiography in patients with intermediate risk for acute coronary syndrome. *Academic Radiology* 20 (8): 980–86.

Mignotte, M, and J Meunier. 2001. A multiscale optimization approach for the dynamic contour-based boundary detection issue. *Computerized Medical Imaging and Graphics* 25 (3): 265–75.

Mignotte, Max, Jean Meunier, and Jean-Claude Tardif. 2001. Endocardial boundary E timation and tracking in echocardiographic images using deformable template and markov random fields. *Pattern Analysis & Applications* 4 (4): 256–71.

Min, James K, Jonathon Leipsic, Michael J Pencina, Daniel S Berman, Bon-Kwon Koo, Carlos van Mieghem, Andrejs Erglis, et al. 2012. Diagnostic accuracy of fractional flow reserve from anatomic CT angiography. *JAMA* 308 (12): 1237.

Min, James K, Charles A Taylor, Stephan Achenbach, Bon K Koo, Jonathon Leipsic, Bjarne L Nørgaard, Nico J Pijls, and Bernard De Bruyne. 2015. Noninvasive fractional flow reserve derived from coronary CT angiography. *JACC: Cardiovascular Imaging* 8 (10): 1209–22.

Min, Ji Hye, Sung M Kim, Sunyoung Lee, Jin-Ho Choi, Sung-A Chang, and Yeon H Choe. 2014. Diagnostic performance of algorithm for computer-assisted detection of significant coronary artery disease in patients with acute chest pain: Comparison with invasive coronary angiography. *AJR. American Journal of Roentgenology* 202 (4): 730–37.

Mojsilović, Aleksandra, Miodrag V Popović, Aleksandar N Nešković, and Aleksandar D Popović. 1997. Wavelet image extension for analysis and classification of infarcted myocardial tissue. *IEEE Transactions on Biomedical Engineering* 44 (9): 856–66.

Morikawa, Yoshinobu, Shiro Uemura, Ken-Ichi Ishigami, Tsunenari Soeda, Satoshi Okayama, Yasuhiro Takemoto, Kenji Onoue, et al. 2011. Morphological features of coronary arteries in patients with coronary spastic angina: Assessment with intracoronary optical coherence tomography. *International Journal of Cardiology* 146 (3): 334–40.

Morris, Paul D, Ryan Desmond, Allison C Morton, Richard Lycett, Patricia V Lawford, Rodney D Hose, and Julian P Gunn. 2013. Virtual fractional flow reserve from coronary angiography: Modeling the significance of coronary lesions. *JACC: Cardiovascular Interventions* 6 (2): 149–57. doi:10.1016/j.jcin.2012.08.024.

Nair, Anuja, Barry D Kuban, E Murat Tuzcu, Paul Schoenhagen, Steven E Nissen, and D. Geoffrey Vince. 2002. Coronary plaque classification with intravascular ultrasound radiofrequency data analysis. *Circulation* 106 (17): 2200–2206.

Nakao, Fumiaki, Tooru Ueda, Shigehiko Nishimura, Hitoshi Uchinoumi, Masashi Kanemoto, Nobuaki Tanaka, and Takashi Fujii. 2013. Novel and quick coronary image analysis by instant stent-accentuated three-dimensional optical coherence tomography system in catheterization laboratory. *Cardiovascular Intervention and Therapeutics* 28 (3): 235–41.

Nakatani, Shimpei, Klaudia Proniewska, Elżbieta Pociask, Giulia Paoletti, Sebastiaan de Winter, Takashi Muramatsu, and Nico Bruining. 2013. How clinically effective is intravascular ultrasound in interventional cardiology? Present and future perspectives. *Expert Review of Medical Devices* 10 (6): 735–49.

Neskovic, Aleksandar N, Aleksandra Mojsilovic, Tamislav Jovanovic, Jovan Vasiljevic, Miodrag Popovic, Jelena Marinkovic, Milovan Bojic, and Aleksandar D Popovic. 1998. Infarection with wavelet image decomposition postinfarction period. *Circulation* 8 (7): 634–641.

Nissen, Steven E, John C Gurley, Cindy L Grines, David C Booth, Rick Mcclure, Martin Berk, Charles Fischer, and Anthony N Demaria. 1991. Intravascular ultrasound assessment of lumen size and wall morphology in normal subjects and patients with coronary artery disease. *Circulation* 84 (3): 1087–99.

Norgaard, Bjarne L, Jonathon Leipsic, Sara Gaur, Sujith Seneviratne, Brian S Ko, Hiroshi Ito, Jesper M Jensen, et al. 2014. Diagnostic performance of noninvasive fractional flow reserve derived from coronary computed tomography angiography in suspected coronary artery disease the NXT trial (analysis of coronary blood flow using CT angiography: next steps). *Journal of the American College of Cardiology* 63 (12): 1145–55.

O h-Ici, D, S Jeuthe, N Al-Wakeel, F Berger, T Kuehne, S Kozerke, and D R Messroghli. 2014. T1 mapping in ischaemic heart disease. *European Heart Journal - Cardiovascular Imaging* 15 (6): 597–602.

Osman, Nael, Kerwin, William, McVeigh, Elliot, Prince, Jerry. 1999. Cardiac motion tracking using CINE harmonic phase (HARP) magnetic resonance imaging. *Magnetic Resonance in Medicine* 42 (6): 1048–60.

Oubel, E, M De Craene, AO Hero, A Pourmorteza, M Huguet, G Avegliano, B H Bijnens, and A F Frangi. 2012. Cardiac motion estimation by joint alignment of tagged MRI sequences. *Medical Image Analysis* 16 (1): 339–50.

Pandian, Natesa G, Andrew Weintraub, Andreas Kreis, Steven L Schwartz, Marvin A Konstam, and Deeb N Salem. 1990. Intracardiac, intravascular, two-dimensional, high-frequency ultrasound imaging of pulmonary artery and its branches in humans and animals. *Circulation* 81 (6): 2007–12.

Papafaklis, Michail I, Takashi Muramatsu, Yuki Ishibashi, Lampros S Lakkas, Shimpei Nakatani, Christos V Bourantas, Jurgen Ligthart, et al. 2014. Fast virtual functional assessment of intermediate coronary lesions using routine angiographic data and blood flow simulation in humans: Comparison with pressure wire – fractional flow reserve. *EuroIntervention* 10 (5): 574–83.

Perea, Rosario J, Jose T Ortiz-Perez, Manel Sole, M Teresa Cibeira, Teresa M de Caralt, Susanna Prat-Gonzalez, Xavier Bosch, Antonio Berruezo, Marcelo Sanchez, and Joan Blade. 2015. T1 mapping: Characterisation of myocardial interstitial space. *Insights into Imaging* 6 (2): 189–202.

Petitjean, Caroline, and Jean Nicolas Dacher. 2011. A review of segmentation methods in short axis cardiac MR images. *Medical Image Analysis* 15 (2): 169–84.

Phelps, Michael E, Simon R Cherry. 1998. The changing design of positron imaging systems, *Clinical Positron Imaging*, 1 (1): 31–45.

Piechnik, Stefan K, Vanessa M Ferreira, Erica Dall'Armellina, Lowri E Cochlin, Andreas Greiser, Stefan Neubauer, and Matthew D Robson. 2010. Shortened modified look-locker inversion recovery (ShMOLLI) for clinical myocardial T1-mapping at 1.5 and 3 T within a 9 heartbeat breathhold. *Journal of Cardiovascular Magnetic Resonance* 12 (1): 69.

Prati, Francesco, Evelyn Regar, Gary S Mintz, Eloisa Arbustini, Carlo Di Mario, Ik-Kyung Jang, Takashi Akasaka, et al. 2010. Expert review document on methodology, terminology, and clinical applications of optical coherence tomography: Physical principles, methodology of image acquisition, and clinical application for assessment of coronary arteries and atherosclerosis. *European Heart Journal* 31 (4): 401–15.

Puri, Rishi, E Murat Tuzcu, Steven E Nissen, and Stephen J Nicholls. 2013. Exploring coronary atherosclerosis with intravascular imaging. *International Journal of Cardiology* 168 (2): 670–79.

Puri, Rishi, Matthew I Worthley, and Stephen J Nicholls. 2011. Intravascular imaging of vulnerable coronary plaque: Current and future concepts. *Nature Reviews Cardiology* 8 (3): 131–39.

Qian, Zhen, Qingshan Liu, Dimitris N Metaxas, and Leon Axel. 2011. Identifying regional cardiac abnormalities from myocardial strains using nontracking-based strain estimation and spatio-temporal tensor analysis. *IEEE Transactions on Medical Imaging* 30 (12): 2017–29.

Ritchie, James L, Gene B Trobaugh, Glen W Hamilton, K. Lance Gould, Kenneth A Narahara, John A Murray, and David L Williams. 1977. Myocardial imaging with thallium-201 at rest and during exercise. Comparison with coronary arteriography and resting and stress electrocardiography. *Circulation* 56 (1): 66–71. 862173.

Roujol, Sébastien, Murilo Foppa, Sebastian Weingärtner, Warren J Manning, and Reza Nezafat. 2014. Adaptive registration of varying contrast-weighted images for improved tissue characterization (ARCTIC): Application to T1 mapping. *Magnetic Resonance in Medicine: Official Journal of the Society of Magnetic Resonance in Medicine / Society of Magnetic Resonance in Medicine*: 73(4):1469–1482. doi:10.1002/mrm.25270.

Rubinshtein, Ronen, David A Halon, Tamar Gaspar, Basil S Lewis, and Nathan Peled. 2014. Automatic assessment of coronary artery calcium score from contrast-enhanced 256-row coronary computed tomography angiography. *The American Journal of Cardiology* 113 (1): 7–11.

Rychik, Jack 2005. *Ventricular Function and Blood Flow in Congenital Heart Disease*, edited by Mark Fogel, 1st ed., 101–23. Malden, Massachussets: Blackwell Publishing.

Sakellarios, Antonis I, Kostas Stefanou, Panagiotis Siogkas, Vasilis D Tsakanikas, Christos V Bourantas, Lambros Athanasiou, Themis P Exarchos, et al. 2012. Novel methodology for 3D reconstruction of carotid arteries and plaque characterization based upon magnetic resonance imaging carotid angiography data. *Magnetic Resonance Imaging* 30 (8): 1068–82.

Salerno, Michael and Christopher M Kramer. 2013. Advances in parametric mapping with CMR imaging. *JACC: Cardiovascular Imaging* 6 (7): 806–22.

Santarelli, Maria F, Vincenzo Positano, Nicola Martini, Giuseppe Valvano, and Luigi Landini. 2016. Technological innovations in magnetic resonance for early detection of cardio-vascular diseases, *Current Pharmaceutical Design* 22 (1): 77–89.

Scharf, L, and H Elliott. 1981. Aspects of dynamic programming in signal and image processing. *IEEE Transactions on Automatic Control* 26 (5): 1018–29.

Sethian, J A 1999. *Level Set Methods and Fast Marching Methods*. Cambridge University Press. Cambridge, UK.

Shahzad, Rahil, Daniel Bos, Coert Metz, Alexia Rossi, Hortense Kirişli, Aad van der Lugt, Stefan Klein, et al. 2013a. Automatic quantification of epicardial fat volume on non-enhanced cardiac CT scans using a multi-atlas segmentation approach. *Medical Physics* 40 (9): 091910.

Shahzad, Rahil, Hortense Kirişli, Coert Metz, Hui Tang, Michiel Schaap, Lucas van Vliet, Wiro Niessen, and Theo van Walsum. 2013b. Automatic segmentation, detection and quantification of coronary artery stenoses on CTA. *The International Journal of Cardiovascular Imaging*, 1847–59.

Shahzad, Rahil, Theo van Walsum, Michiel Schaap, Alexia Rossi, Stefan Klein, Annick C Weustink, Pim J de Feyter, Lucas J van Vliet, and Wiro J Niessen. 2013c. Vessel specific coronary artery calcium scoring: An automatic system. *Academic Radiology* 20 (1): 1–9.

Sihan, Kenji, Charl Botha, Frits Post, Sebastiaan de Winter, Nieves Gonzalo, Evelyn Regar, Patrick J W C Serruys, Ronald Hamers, and Nico Bruining. 2009. Fully automatic three-dimensional quantitative analysis of intracoronary optical coherence tomography: Method and validation. *Catheterization and Cardiovascular Interventions* 74 (7): 1058–65.

Singh, Navneet, Alan R Moody, Idan Roifman, David A Bluemke, and Anna E Zavodni. 2016. Advanced MRI for carotid plaque imaging. *International Journal of Cardiovasc Imaging* 32: 83–89.

Stankovic, Zoran, Bradley D Allen, Julio Garcia, Kelly B Jarvis, and Michael Markl. 2014. 4D flow imaging with MRI. *Cardiovascular Diagnosis and Therapy* 4 (2): 173–92.

Sudarshan, Vidya K, E Y K Ng, U Rajendra Acharya, Chou Siaw Meng, Ru San Tan, and Dhanjoo N Ghista. 2015. Computer-aided diagnosis of myocardial infarction using ultrasound images with DWT, GLCM and HOS methods: A comparative study. *Computers in Biology and Medicine* 62: 86–93.

Suinesiaputra, Avan, Patrick J H de Koning, Elena Zudilova-Seinstra, Johan H C Reiber, and Rob J van der Geest. 2012. Automated quantification of carotid artery stenosis on contrast-enhanced MRA data using a deformable vascular tube model. *The International Journal of Cardiovascular Imaging* 28 (6): 1513–24.

Sun, Shanhui, Milan Sonka, and Reinhard R Beichel. 2013. Graph-based IVUS segmentation with efficient computer-aided refinement. *IEEE Transactions on Medical Imaging* 32 (8): 1536–49.

Szilveszter, Bálint, Csilla Celeng, and Pál Maurovich-Horvat. 2016. Plaque Assessment by Coronary CT. *The International Journal of Cardiovascular Imaging* 32: 161. doi:10.1007/s10554-015-0741-8.

Tamaki, N, Y Yonekura, T Mukai, S Kodama, K Kadota, H Kambara, C Kawai, and K Torizuka. 1984. Stress thallium-201 transaxial emission computed tomography: Quantitative versus qualitative analysis for evaluation of coronary artery disease. *Journal of the American College of Cardiology* 4 (6): 1213–21.

Tanami, Yutaka, Masahiro Jinzaki, Satoru Kishi, Matthew Matheson, Andrea L Vavere, Carlos E Rochitte, Marc Dewey, et al. 2015. Lack of association between epicardial fat volume and extent of coronary artery calcification, severity of coronary artery disease, or presence of myocardial perfusion abnormalities in a diverse, symptomatic patient population: Results from the CORE320 Mu. *Circulation. Cardiovascular Imaging* 8 (3): e002676.

Tang, Hui, Theo van Walsum, Robbert S van Onkelen, Reinhard Hameeteman, Stefan Klein, Michiel Schaap, Fufa L Tori, et al. 2012. Semiautomatic carotid lumen segmentation for quantification of lumen geometry in multispectral MRI. *Medical Image Analysis* 16 (6): 1202–15.

Taylor, Charles A, Timothy A Fonte, and James K Min. 2013. Computational fluid dynamics applied to cardiac computed tomography for noninvasive quantification of fractional flow reserve: Scientific basis. *Journal of the American College of Cardiology* 61 (22): 2233–41.

Thilo, Christian, Mulugeta Gebregziabher, Felix G Meinel, Roman Goldenberg, John W Nance, Elisabeth M Arnoldi, Lashonda D Soma, et al. 2015. Computer-aided stenosis detection at coronary CT angiography: Effect on performance of readers with different experience levels. *European Radiology* 25 (3): 694–702.

Trobaugh, Gene B, Fransj Th Wackers, Timothya Derouen, James Ritchie, and Glen W Hamilton. 1978. Thallium-201 myocardial study imaging : An I aterlnstltutional of observer variablIlty for all four-readers. *Journal of Nuclear Medicine* 19 (4): 359–63.

Tsantis, Stavros, George C Kagadis, Konstantinos Katsanos, Dimitris Karnabatidis, George Bourantas, and George C Nikiforidis. 2012. Automatic vessel lumen segmentation and stent strut detection in intravascular optical coherence tomography. *Medical Physics* 39 (1): 503–13.

Tu, Shengxian, Emanuele Barbato, Zsolt Köszegi, Junqing Yang, Zhonghua Sun, Niels R Holm, Balázs Tar, et al. 2014. Fractional flow reserve calculation from 3-dimensional quantitative coronary angiography and TIMI frame count. *JACC: Cardiovascular Interventions* 7 (7): 768–77.

Tu, Shengxian, L Xu, Jurgen Ligthart, Bo Xu, and Karen Witberg. 2012. In vivo comparison of arterial lumen dimensions assessed by co-registered three-dimensional (3D) quantitative coronary angiography, intravascular ultrasound and optical coherence tomography. *The International Journal of Cardiovascular Imaging* 28 (6): 1315–27.

Tung, Kai-Pin, Wen-Zhe Shi, Ranil De Silva, Eddie Edwards, and Daniel Rueckert. 2011. Automatic vessel wall detection in intravascular coronary OCT. In *Proceedings of IEEE International Symposium on Biomedical Imaging: From Nano to Macro*. pp. 610–613. ISBN: 978-1-4244-4128-0.

Ughi, Giovanni J, Tom Adriaenssens, Walter Desmet, and Jan D'hooge. 2012. Fully automatic three-dimensional visualization of intravascular optical coherence tomography images: Methods and feasibility in vivo. *Biomedical Optics Express* 3 (12): 3291–3303.

Ughi, Giovanni J, Tom Adriaenssens, Peter Sinnaeve, Walter Desmet, and Jan D'hooge. 2013. Automated tissue characterization of in vivo atherosclerotic plaques by intravascular optical coherence tomography images. *Biomedical Optics Express* 4 (7): 1014–30.

Ukwatta, Eranga, Jing Yuan, Martin Rajchl, Wu Qiu, David Tessier, and Aaron Fenster. 2013. 3D carotid multi-region MRI segmentation by globally optimal evolution of coupled surfaces. *IEEE Transactions on Medical Imaging* 32 (4): 770–85.

Unal, Gozde, Serhan Gurmeric, and Stéphane Guy Carlier. 2009. Stent implant follow-up in intravascular optical coherence tomography images. *The International Journal of Cardiovascular Imaging* 26 (7): 809–16.

van't Klooster, Ronald, Patrick J H de Koning, Reza Alizadeh Dehnavi, Jouke T Tamsma, Albert de Roos, Johan H C Reiber, and Rob J van der Geest. 2012. Automatic lumen and outer wall segmentation of the carotid artery using deformable three-dimensional models in MR angiography and vessel wall images. *Journal of Magnetic Resonance Imaging: JMRI* 35 (1): 156–65.

van Engelen, Arna, Wiro J Niessen, Stefan Klein, Harald C Groen, Hence J M Verhagen, Jolanda J Wentzel, Aad Van Der Lugt, and Marleen De Bruijne. 2014. Atherosclerotic plaque component segmentation in combined carotid MRI and CTA data incorporating class label uncertainty. *PLoS ONE* 9 (4): 1–14.

van Engelen, Arna, Anouk C van Dijk, Martine T B Truijman, Ronald van T Klooster, Annegreet van Opbroek, Aad van der Lugt, Wiro J Niessen, M Eline Kooi, and Marleen de Bruijne. 2015. Multi-center MRI carotid plaque component segmentation using feature normalization and transfer learning. *IEEE Transactions on Medical Imaging* 34 (6): 1294–1305.

van Ooij, Pim, Wouter V Potters, Jeremy Collins, Maria Carr, James Carr, S Chris Malaisrie, Paul W M Fedak, Patrick M McCarthy, Michael Markl, and Alex J Barker. 2015. Characterization of abnormal wall shear stress using 4D flow MRI in human bicuspid aortopathy. *Annals of Biomedical Engineering* 43 (6): 1385–97.

Van Train, Kenneth F, Jamshid Maddahi, Daniel S Berman, Hosen Kiat, Joseph Areeda, Florence Prigent, and John Friedman. 1990. Quantitative analysis of tomographic stress thallium-201 myocardial scintigrams: A multicenter trial. *Journal of Nuclear Medicine* 31(7): 1168–1179.

von Birgelen, Clemens, Carlo Di Mario, Wenguang Li, Johan C. Schuurbiers, Cornelis J Slager, Pim J de Feyter, Jos R Roelandt and Patrick W Serruys. 1996. Morphometric analysis in three-dimensional intracoronary ultrasound: An in vitro and in vivo study performed with a novel system for the contour detection of lumen and plaque. *American Heart Journal* 132 (3): 516–27.

Vukadinovic, Danijela, Sietske Rozie, Marjon van Gils, Theo van Walsum, Rashindra Manniesing, Aad van der Lugt, and Wiro J Niessen. 2012. Automated versus manual segmentation of atherosclerotic carotid plaque volume and components in CTA: Associations with cardiovascular risk factors. *The International Journal of Cardiovascular Imaging* 28 (4): 877–87.

Vukadinovic, Danijela, Theo van Walsum, Rashindra Manniesing, Sietske Rozie, Reinhard Hameeteman, Thomas T de Weert, Aad van der Lugt, and Wiro J Niessen. 2010. Segmentation of the outer vessel wall of the common carotid artery in CTA. *IEEE Transactions on Medical Imaging* 29 (1): 65–76.

Wahle, Andreas, John J Lopez, Mark E Olszewski, Sarah C Vigmostad, Krishnan B Chandran, James D Rossen, and Milan Sonka. 2006. Plaque development, vessel curvature, and wall shear stress in coronary arteries assessed by X-ray angiography and intravascular ultrasound. *Medical Image Analysis* 10 (4): 615–31.

Wang, Ancong, Jeroen Eggermont, Niels Dekker, Patrick J H de Koning, Johan H C Reiber, and Jouke Dijkstra. 2014. 3D assessment of stent cell size and side branch access in intravascular optical coherence tomographic pullback runs. *Computerized Medical Imaging and Graphics* 38 (2): 113–22.

Wang, Hui, and Amir A. Amini. 2012. Cardiac motion and deformation recovery from MRI: A review. *IEEE Transactions on Medical Imaging* 31 (2): 487–503.

Wang, Lijia, Mengchao Pei, Noel C F Codella, Minisha Kochar, Jonathan W Weinsaft, Jianqi Li, Martin R Prince, and Y Wang. 2015a. Left ventricle: Fully automated segmentation based on spatiotemporal continuity and myocardium information in cine cardiac magnetic resonance imaging (LV-FAST). *BioMed Research International.* Volume 2015 (2015), Article ID 367583, 9 pages. http://dx.doi.org/10.1155/2015/367583.

Wang, Zhao, Michael Jenkins, George Linderman, Hiram Bezerra, Yusuke Fujino, Marco Costa, David Wilson, and Andrew Rollins. 2015b. 3D stent detection in intravascular OCT using a Bayesian network and graph search. *IEEE Transactions on Medical Imaging* 34 (7): 1549–61.

Wang, Zhao, Hiroyuki Kyono, Hiram G Bezerra, Hui Wang, Madhusudhana Gargesha, Chadi Alraies, Chenyang Xu, et al. 2010. Semiautomatic segmentation and quantification of calcified plaques in intracoronary optical coherence tomography images. *Journal of Biomedical Optics* 15 (6): 061711.

Watson, D D, N P Campbell, E K Read, R S Gibson, C D Teates, and G A Beller. 1981. Spatial and temporal quantitationof plane thallium myoardial images. *Journal of Nuclear Medicine* 22 (7): 577–84.

Wei, Jun, Chuan Zhou, Heang-Ping Chan, Aamer Chughtai, Prachi Agarwal, Jean Kuriakose, Lubomir Hadjiiski, Smita Patel, and Ella Kazerooni. 2014. Computerized detection of noncalcified plaques in coronary CT angiography: evaluation of topological soft gradient prescreening method and luminal analysis. *Medical Physics* 41 (8): 081901.

Weingärtner, Sebastian, Mehmet Akçakaya, Sébastien Roujol, Tamer Basha, Cory Tschabrunn, Sophie Berg, Elad Anter, and Reza Nezafat. 2015. Free-breathing combined three-dimensional phase sensitive late gadolinium enhancement and T_1 mapping for myocardial tissue characterization. *Magnetic Resonance in Medicine* 74 (4): 1032–41.

Wenguang, L, W J Gussenhoven, Y Zhong, S H K The, C Di~Mario, S Madretsma, F V Egmond, et al. 1991. Validation of quantitative analysis of intravascular ultrasound images. *International Journal of Cardiac Imaging* 6: 247–53.

Wentland, Andrew L, Thomas M Grist, and Oliver Wieben. 2014. Review of MRI-based measurements of pulse wave velocity: A biomarker of arterial stiffness. *Cardiovascular Diagnosis and Therapy* 4 (2): 193–206.

Wieben, Oliver, Christopher Francois, and Scott B. Reeder. 2008. Cardiac MRI of ischemic heart disease at 3 T: Potential and challenges. *European Journal of Radiology* 65 (1): 15–28.

Wu, LiNa, Tjeerd Germans, Ahmet Güçlü, Martijn W Heymans, Cornelis P Allaart, and Albert C van Rossum. 2014. Feature tracking compared with tissue tagging measurements of segmental strain by cardiovascular magnetic resonance. *Journal of Cardiovascular Magnetic Resonance: Official Journal of the Society for Cardiovascular Magnetic Resonance* 16 (1): 10.

Xu, Chun, James J Pilla, Gamaliel Isaac, J Gorman, A Blom, R Gorman, Zhou Ling, and Lawrence Dougherty. 2010. Deformation analysis of 3D tagged cardiac images using an optical flow method. *J Cardiovasc Magn Reson* 12 (1): 19–25.

Xue, Hui, Andreas Greiser, Sven Zuehlsdorff, Marie-Pierre Jolly, Jens Guehring, Andrew E. Arai, and Peter Kellman. 2013. Phase-sensitive inversion recovery for myocardial T 1 mapping with motion correction and parametric fitting. *Magnetic Resonance in Medicine* 69 (5): 1408–20.

Xue, Hui, Saurabh Shah, Andreas Greiser, Christoph Guetter, Arne Littmann, Marie-Pierre Jolly, Andrew E Arai, Sven Zuehlsdorff, Jens Guehring, and Peter Kellman. 2012. Motion correction for myocardial T1 mapping using image registration with synthetic image estimation. *Magnetic Resonance in Medicine* 67 (6): 1644–55.

Yang, Guanyu, Pieter Kitslaar, Michel Frenay, Alexander Broersen, Mark J Boogers, Jeroen J Bax, Johan H C Reiber, and Jouke Dijkstra. 2012. Automatic centerline extraction of coronary arteries in coronary computed tomographic angiography. *International Journal of Cardiovascular Imaging* 28 (4): 921–33.

Yun Zhu, X Papademetris, A J Sinusas, and J S Duncan. 2010. Segmentation of the left ventricle from cardiac MR images using a subject-specific dynamical model. *IEEE Transactions on Medical Imaging* 29 (3): 669–87.

Zajac, Jakub, Jonatan Eriksson, Petter Dyverfeldt, Ann F. Bolger, Tino Ebbers, and Carl Johan Carlhäll. 2014. Turbulent kinetic energy in normal and myopathic left ventricles. *Journal of Magnetic Resonance Imaging*: 41 (4): 1021–9.

Zerhouni, E A, Parish, D M, Rogers, W J, Yang, A, Shapiro, E P Shapiro. 1988. Human heart: tagging with MR imaging–a method for noninvasive assessment of myocardial motion. *Radiology* 169 (1): 59–63.

Zhang, X, C R McKay, and M Sonka. 1998. Tissue characterization in intravascular ultrasound images. *IEEE* Transactions on Medical *Imaging* 17 (6): 889–99.

8

Realistic Lesion Insertion for Medical Data Augmentation

Aria Pezeshk

Nicholas Petrick

Berkman Sahiner

8.1 Introduction

Development and evaluation of medical image analysis algorithms often require a large number of validated and/or annotated samples (cases). Some cxamples include algorithms for lesion segmentation or volume estimation that can only be evaluated if the true volumes or lesion boundaries are known. Another set of algorithms is computer-aided diagnosis (CAD) systems whose performances rely heavily on the size and representativeness of the training data. Interpreting and labeling large amounts of imaging data is a time-consuming and expensive task that creates a bottleneck in assembling large usable medical image repositories. In many other fields, such as computer vision or speech recognition, it is possible to use crowd-sourcing methods such as Amazon's mechanical Turk to obtain labels or annotations for images and audio in a relatively cheap and timely manner. However, analysis of medical images requires a very high level of expertise, thereby rendering such labeling and annotation methods ineffective. Another bottleneck is caused by the fact that some lesion types or abnormalities have a low prevalence in the imaged patient population. For example, in breast cancer screening in the US, approximately five subjects out of 1,000 have cancer. This means that, if a screening population is used for data collection, one needs to image over 20,000 women to acquire a dataset of 100 cancer cases.

One promising approach to circumvent the aforementioned issues related to unbalanced or small datasets is to increase the available number of labeled or annotated samples by artificially supplementing the dataset. An existing dataset can be expanded to include additional synthetic samples with known properties, pathology, or diagnosis using these methods in order to overcome or reduce the shortcomings of the original set of samples. In this chapter we review some of the existing work in this domain and describe in more detail one specific approach for creating additional samples alongside examples from several modalities and application areas. More specifically, we will use this approach to augment datasets for mammography and chest computed tomography (CT) and to show applications of the augmented datasets in CAD training and evaluation of volumetry algorithms.

8.2 Background

Data augmentation is the artificial boosting of the size of a dataset with samples that are representative of variations observed in the space of real samples but not present in the limited original dataset. In medical image datasets, many factors could limit the utility of a collection of samples. For instance, it is well known that a small training dataset will limit a CAD system's performance (Sahiner et al. 2000, Chan et al. 1999). Furthermore, the number of lesions with a specific condition (e.g., pathology, size, shape, contrast, or location) might be limited due to rarity of the disease that gives rise to that specific condition. Training a CAD system on such a dataset would result in a system that may not perform well for the cases with the specific condition in the true population. Similarly, evaluation of a volumetry algorithm would be incomplete without considering the variations in lesion shape and foreground-background interaction observed in the real sample space. When applied properly, the augmentation procedure can produce a new dataset with the requisite variability of lesion characteristics that can be used in the same fashion as a larger dataset containing real samples.

Data augmentation has been extensively utilized in the context of creating large datasets for classifier training in several fields, such as computer vision, speech recognition, and general pattern recognition. There are three general approaches for this tool. In the first approach, new samples are created directly in the feature space by jittering (e.g., by adding Gaussian noise) or applying other modifications based on feature vectors of the original samples (Melville and Mooney 2003, Chawla et al. 2002, Bishop 1995, Raviv and Intrator 1996, van der Maaten et al. 2013). As the changes are applied in the feature space, this approach might not have a direct physical interpretation in some cases.

The second approach works by applying a simple transformation to each original sample (Papageorgiou and Poggio 2000, Laptev 2007, Krizhevsky and Sutskever 2012). For instance, in the case of images, a sample could be rotated, flipped, translated, or added with noise in order to create new samples that have the same labels as the original samples, but exhibit more diversity. The specific types of alterations applied to the samples are selected such that they reflect the expected variations in the true population. For instance, in a car detection problem, the sample car images can be translated, but not flipped vertically. This technique is particularly popular in deep learning to help the classifier become insensitive to trivial variations in appearance of samples and reduce overtraining.

In the third approach, a generative model is used to create new samples (Pezeshk and Tutwiler 2011, Ho and Baird 1997). The characteristics of the synthetic samples such as their shapes, sizes, degradations, and so on, are determined by model parameters that govern the base algorithm for sample generation, as well as a variety of deterministic and/or random deformations, perturbations, noise levels, etc. The advantage of this approach is that an arbitrary number of new samples with known labels and properties and more diversity, compared to the earlier approaches, can be created. On the other hand, as the new samples are created completely from scratch, greater care is required in order to assure that the generated samples indeed mimic the distribution of the real sample population.

The first approach, feature space augmentation, is only useful for classifier training. However, the two other approaches—trivial transformation of original samples and creating synthetic samples using generative models—are more general and can be used for both classifier training as well as for some of the examples mentioned earlier in the beginning of the chapter, such as evaluation of segmentation and volumetry algorithms. In recent years, several augmentation tools have been developed for medical image datasets following the two latter approaches above and allow for insertion of a lesion into alternate locations on a new image. The lesion being inserted can be one that has been simulated using a generative model, such as a cluster growth model to mimic the appearance of real lesions, or a real lesion extracted from a clinical image. Some examples of methods for generation and insertion of simulated lesions can be found in (Zhang et al. 2007, Li et al. 2009, Shin et al. 2006) for lung nodules, in (Saunders et al. 2006,

Rashidnasab et al. 2013) for mammography, and in (Vaz et al. 2011) for digital breast tomosynthesis. The simulated lesions often follow models that result in near spherical shapes or nearly uniform texture and contrast, or mathematical processes for random growth that produce more variability in shape, but whose biological plausibility cannot be justified. Unless the simulation model has been designed to fit the appearance of true lesions observed in the clinic, it may be difficult to replicate the wide range of characteristics of lesions observed in clinical images using these methods. Methods using real lesions extracted from clinical images, on the other hand (Ambrosini and O'Dell 2010, Peskin and Dima 2010, Madsen et al. 2011), have the advantage that the lesions being inserted are natural by design. However, they cannot introduce variabilities in properties such as lesion shape, size, and contrast, and instead only create new foreground-background interactions by placing lesions in areas with different texture or background compared to the lesions' original location.

For imaging techniques that require image reconstruction from projections, for example, CT or digital breast tomosynthesis (DBT), insertion of a real (e.g., segmented from clinical images), or simulated lesion into a new image, can take place either in the projection space or the image space. When the scanning parameters and the system modulation transfer function are known, a simulated lesion can be passed through an imaging simulation pipeline to create projection data for the lesion. The projections can then be combined with raw projection data of the target, followed by reconstruction into image space. In practice, raw projection data is typically not available, and the imaging simulation pipeline is often only an approximation to physical image acquisition, which may limit the practical application of this insertion method. The alternative, which is lesion insertion in the image space, is often performed using a technique referred to as alpha blending. The image containing the lesion to be inserted, which we refer to as the source image f_s, and the image in which the lesion is to be inserted, which we refer to as the target image f_t, are combined by linearly interpolating their intensity values in a weighted sum whose coefficients (weights) are defined by an alpha matte over each pixel (Porter and Duff 1984):

$$f_{\text{alpha}}(x,y) = \alpha(x,y) f_s(x,y) + (1 - \alpha(x,y)) f_t(x,y) \tag{8.1}$$

In this equation

α is the alpha matte

f_{alpha} is the blended output image

As the alpha matte is defined manually, the regions of interest in the source and target images have to be carefully selected such that salient features in the images do not clash with one another. In addition, extra steps such as feathering (blurring) are needed in order to hide the boundary between the insertion area and its immediate surroundings in the target image. Alpha blending is therefore a time-consuming process in which the perceived quality of the final composite is largely dependent on the skill of the user.

In the next section, we will describe an alternative image space insertion method based on Poisson image editing (Perez et al. 2003) that allows a lesion to be easily extracted from a source image and inserted into a target image using a drag and drop method: the user first draws a casual boundary around a lesion of interest, then selects the insertion point in the target image (Pezeshk et al. 2015b). By minimizing user input and automatically handling a variety of complex scenarios, such as insertion of non-isolated lesions (e.g., those connected to vessels) and insertion into areas with complex texture or salient features, this method vastly simplifies the creation of realistic new images containing lesions. Moreover, we will describe how to use an extracted real lesion as a basis to create a series of new lesions with different properties by applying various transformations to the original lesion prior to its insertion into a new location. When the transformations are selected appropriately, basing the transformed lesions on clinical lesions will result in additional realistic lesions, thereby overcoming the limitation of mathematically modeled lesions that may be too simplistic or difficult to justify biologically. This transformation technique is a generative model for creation of new modified samples based on existing real samples and is, therefore, a hybrid data augmentation method similar to the techniques in (Pishchulin et al. 2011, Heisele and Serre 2007), combining elements from both the second and third augmentation approaches described earlier in the chapter.

8.3 Image Space Lesion Insertion Using Poisson Image Editing

Poisson image editing (Perez et al. 2003) is a popular gradient domain compositing technique for seamless image composition. The underlying principle of this method is that, given the boundary values and the Laplacian of the interior values of a scalar function, it is possible to find a unique solution to the values of the function across the entirety of its domain using the Poisson equation. Similarly, the boundary values along the insertion area of a target image can be interpolated inwards under the guidance of the gradient field of the source image in order to obtain a seamless blending of a source object within the domain of a target image. More formally, let Ω be a region in the image with boundary $\partial\Omega$. Denoting the source and target images as f_s and f_s, respectively, the blended image over the interior of Ω can be found by solving the following equation:

$$\min_f \iint_\Omega \left|\nabla f - \nabla f_s\right|^2 ; \text{with } f\big|_{\partial\Omega} = f_t\big|_{\partial\Omega}, \tag{8.2}$$

where $\nabla . = \left[\partial./\partial x, \partial./\partial y\right]$ is the gradient operator. This notation is illustrated in Figure 8.1. The unique solution to Equation 8.2 can be equivalently found using the Poisson equation with Dirichlet boundary conditions:

$$\Delta f = divf_s \text{ over } \Omega, \text{and } f\big|_{\partial\Omega} = f_t\big|_{\partial\Omega}, \tag{8.3}$$

where $\Delta . = \left(\partial^2./\partial x^2 + \partial^2./\partial y^2\right)$ is the Laplacian operator and $div. = \left(\partial./\partial x + \partial./\partial y\right)$ denotes the divergence operator. This equation can be discretized to obtain a sparse system of linear equations that can be efficiently solved using a sparse solver.

The solution to Equation 8.3 replaces the original contents of f_t within Ω with the contents of f_s. However, if the insertion area Ω within the target image contains salient features, such as areas of high texture or vessels, it may be important to maintain the continuity of these features from outside to the inside of the insertion area. Otherwise, the abrupt disappearance of the features will look unnatural. In such circumstances, the guidance field can be modified by combining the gradient fields of the source and target images so that the composite image will contain both the source object and the salient features in f_t:

$$\Delta f = \begin{cases} divf_t & \text{if } \left|\nabla f_t\right| > \left|\nabla f_s\right| \\ divf_s & \text{otherwise} \end{cases} \tag{8.4}$$

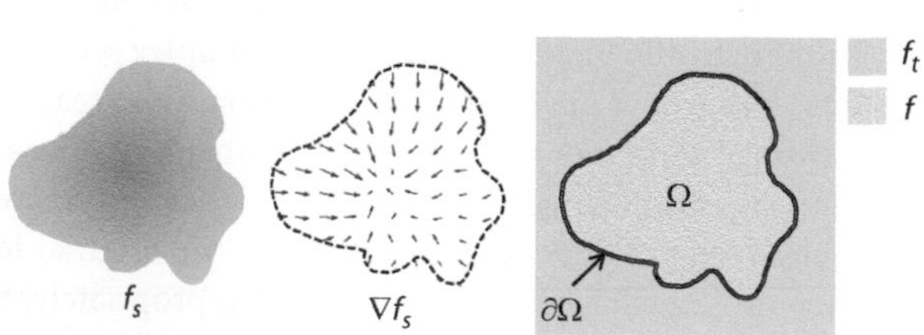

FIGURE 8.1 Poisson blending: f is the inward interpolation of f_t given the boundary conditions in $\partial\Omega$ and the guidance field ∇f_s derived from the source image f_s.

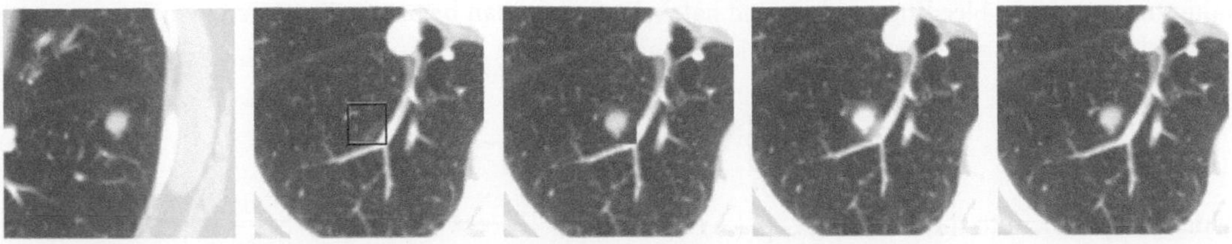

FIGURE 8.2 Gradient mixing. From left to right: Original source image, original target image from a different case, result of direct insertion of source ROI into target, Poisson blend results without gradient mixing, and Poisson blend results with gradient mixing.

Figure 8.2 shows an example of a pulmonary nodule inserted in three different ways into a target image. In the first insertion, a region of interest (ROI) containing the nodule is directly inserted into the target image to demonstrate the large difference between contents of the two images and, more importantly, the obvious inconsistency between the boundaries of the regions. Clearly, it would be very difficult to manually determine the weights of the matte in alpha blending that would resolve the discontinuity across the boundaries. In the second insertion, the original gradient field of the source image was used as the guidance field as in Equation 8.3, thereby resulting in the diffusion of high intensity pixels along the edges of the boundary as well as cutting through the fissure that passes through the insertion area. The third insertion shows the use of the mixing of gradient fields as the guidance fields as in Equation 8.4. In this case, we obtain seamless boundaries and maintain the integrity of the fissure at the same time.

The ROI encompassing the lesion might contain extraneous tissue or anatomical structures. In addition, some lesions may be connected to other structures such as the vasculature. In such cases, additional preprocessing steps are required in order to remove the undesirable elements from the source ROI prior to insertion of the lesion into a new location. Assuming a known segmentation of the lesion Ψ_0, and denoting the area covered by the extraneous object as Ω_{ext} (e.g., found by thresholding parts of selected ROI not covered by Ψ_0), the simplest way to mask the object will be to set the gradient field of f_s over Ω_{ext} to zero, and then to solve for a modified source image f_s^{mod}:

$$\Delta f_s^{\text{mod}} = 0 \text{ over } \Omega_{\text{ext}}, \text{ and } f_s^{\text{mod}}\big|_{\partial\Omega_{\text{ext}}} = f_s\big|_{\partial\Omega_{\text{ext}}} \tag{8.5}$$

However, this solution will lead to the simple unguided inward diffusion of pixels along $\partial\Omega_{\text{ext}}$ and therefore a blurry interpolation result. A better alternative is to replicate the natural background texture of the image within Ω_{ext}. This can be accomplished by replacing the gradient field of f_s over Ω_{ext} with that of a background region, and then solving for f_s^{mod} according to Equation 8.5. To simplify the selection of the background region, we use Poisson reconstruction to seamlessly tile a selected background region in order to obtain a patch with arbitrary size. The gradient field of regions corresponding to Ω_{ext} is then replaced with the gradient field of a region with same size and shape as Ω_{ext} in the tiled background image.

Lesions that have undesirable attachments, such as connected vessels, require an additional processing step, which we refer to as gradient field shaping, so that the tapering of the gradient field along the section of the lesion being detached appears the same as the natural blurring across the edges of the other parts of the lesion. In this step, the intensity of the gradient field in the vicinity of the section being detached is estimated according to the intensity of the gradient field in the other areas surrounding the lesion, while the orientation of the gradient field in these areas is modified based on the orientation of the lesion's border. Full details of these two preprocessing steps can be found in (Pezeshk et al. 2015b).

Poisson editing works best when the boundaries of the source ROI and target insertion area are consistent with one another. When this condition is not satisfied, the diffusion of the inconsistent boundary values may result in bleeding artifacts and halos. This point can be better illustrated by using a change of variables $f' = f - f_s$, and rewriting Equation 8.2 as follows:

$$\min_{f'} \iint_{\Omega} |\nabla f'|^2 ; \text{with } f'|_{\partial} = f_t - f_s|_{\partial\Omega}, \tag{8.6}$$

which is equivalent to the following Laplace equation:

$$\Delta f' = 0 \text{ over } \Omega, \text{and } f'|_{\partial\Omega} = f_t - f_s|_{\partial\Omega} \tag{8.7}$$

Solving for f' is therefore equivalent to the unguided inward interpolation of the difference between the boundary values of f_s and f_t. As the blend result f can be simply computed by adding f_s to f', we conclude that any artifacts in the blended image are a result of differences between the boundaries of the source and target images along the insertion area.

The energy in $\int_{\Omega} |\nabla f'|^2$ can be shown to approach zero if, and only if, the differences between the values along the boundaries of f_s and f_t are constant at every point (Zwillinger 1997). Finding such a boundary manually is clearly very difficult, if not impossible. We, therefore, use the expectation–maximization approach in (Jia et al. 2006) to automatically find an optimal boundary $\partial\Omega_{opt}$ encompassing the lesion boundary $\partial\Psi_0$, and located within the area defined by the original casual boundary $\partial\Omega_0$ drawn by the user for the source lesion. The objective of this approach is to minimize the cost function below through an iterative graph search:

$$E(\partial\Omega, k) = \sum_{p \in \partial\Omega} \left(f_t(p) - f_s(p) - k\right)^2 ; s.t. \Psi_0 \subset \Omega \subset \Omega_0, \tag{8.8}$$

where k is the constant difference between boundaries to be determined

The graph search is conducted using the Dijkstra algorithm (Dijkstra 1959). As the graph's adjacency matrix and thereby its cost matrix form sparse matrices, they can be efficiently stored in memory. More importantly, the optimal boundary can be efficiently computed using implementations of the Dijkstra algorithm that take advantage of this sparsity. Figure 8.3 shows an example of a nodule inserted into an area with a complex background with and without using the optimal boundary option. It can be seen that the discrepancy between the pixel values belonging to source and target images located on the initial user selected boundary causes visible artifacts, whereas the optimal boundary can automatically find the most suitable boundary conditions and thereby produce a seamless insertion.

Once a lesion has been selected by a user, a variety of transformations can be applied to it prior to its insertion into the target image. Modifications to the shape of a lesion can be applied directly in the image domain before computing the gradient field of the source image. Changes to the contrast of the

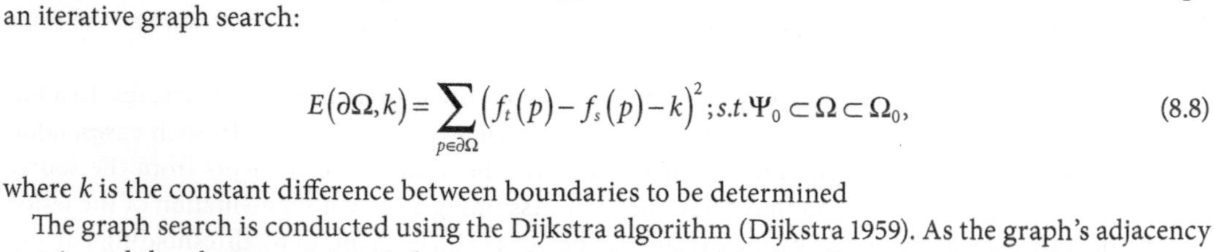

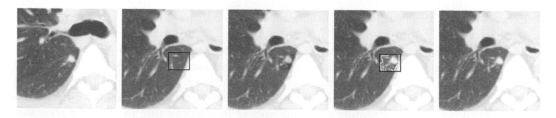

FIGURE 8.3 Effect of optimal boundary. From left to right: Original source image, original target image from a different case, blend results using original boundary (note the darkening artifact around the nodule), blend results using optimal boundary with both boundaries superimposed, and blend results using optimal boundary without boundaries superimposed. Orange and red boxes show the original boundary drawn by the user, and the optimal boundary is shown in yellow.

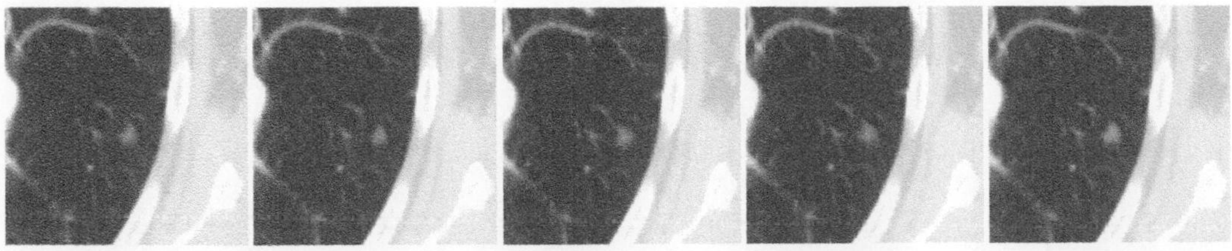

FIGURE 8.4 Augmenting a dataset with transformed copies of a nodule. From left to right: Blending the original, size scaled (0.85), rotated (30°), sheared (0.25 horizontal), and contrast scaled (1.25) copies of a nodule into the same target image.

lesion on the other hand can be obtained by modifying the source gradient field. Figure 8.4 shows an example of blending a nodule modified using single affine transformations of scaling, rotation, and shear, as well as uniform contrast scaling. Other modifications such as combinations of affine transformations, warping, and non-uniform contrast scaling are also possible but not shown here. By selecting different transformation types and parameters, this process can multiply every lesion into many copies that are each different from one another, even though they are all based on the same original lesion. This can be particularly useful for lesions with rare characteristics such as shape, density, and malignancy that are otherwise difficult to obtain from actual patients. Selecting the proper range of parameters that maintain lesion realism largely depends on the specific application and is a topic of further research.

8.4 Applications

8.4.1 Augmentation of Chest CT Images and Reader Study

The largest public CT library of labeled pulmonary nodules is the Lung Image Database Consortium (LIDC) dataset consisting of 1,018 scans from 1,010 patients (Armato III et al. 2011). Each case in this repository was reviewed by four experienced thoracic radiologists, who provided segmentations as well as various characterization scores, such as malignancy, spiculation, lobulation, and subtlety scores, for each detected ≥3 mm nodule. This data is provided in a corresponding XML file for each case. In total, 7,371 lesions were marked by at least one radiologist, of which 2,669 were marked as a ≥3 mm nodule by at least one radiologist. When we look at the number of ≥3 mm nodules detected by all four radiologists, the number of nodules falls to only 928 out of the 2,669. This is a relatively small number of positive samples compared to typical datasets now used in other fields, such as computer vision or speech recognition consisting of tens of thousands to millions of samples (Krizhevsky and Sutskever 2012, Hannun et al. 2014).

We developed a graphical user interface (GUI) for easy extraction and insertion of nodules in the LIDC dataset using our method. The GUI provides the user with general viewing options, including two panels displaying the source and target cases, controls for changing the slice number being displayed and adjusting the window and level values, and ability to zoom and pan. For each source case, the user can draw a rectangular bounding box encompassing the largest cross section of the source nodule in the axial view, a bounding box for a background region devoid of salient objects, and then select the slice number as well as center point of the insertion area in the target case. The GUI subsequently automatically detects which nodule has been selected by the user according to the case number and coordinates of the selected area, extracts the relevant nodule segmentations (i.e., the Ψ_0) by taking the union of all radiologist segmentations for that nodule, and runs the insertion algorithm on a slice-by-slice basis for all slices containing the nodule. The blend results are finally displayed in an auxiliary viewing panel, alongside the original target slices for side by side comparison.

Figure 8.5 shows the results of inserting two consecutive slices of a nodule into a different case. In each slice, the algorithm automatically separates the nodule from any extraneous objects, using a

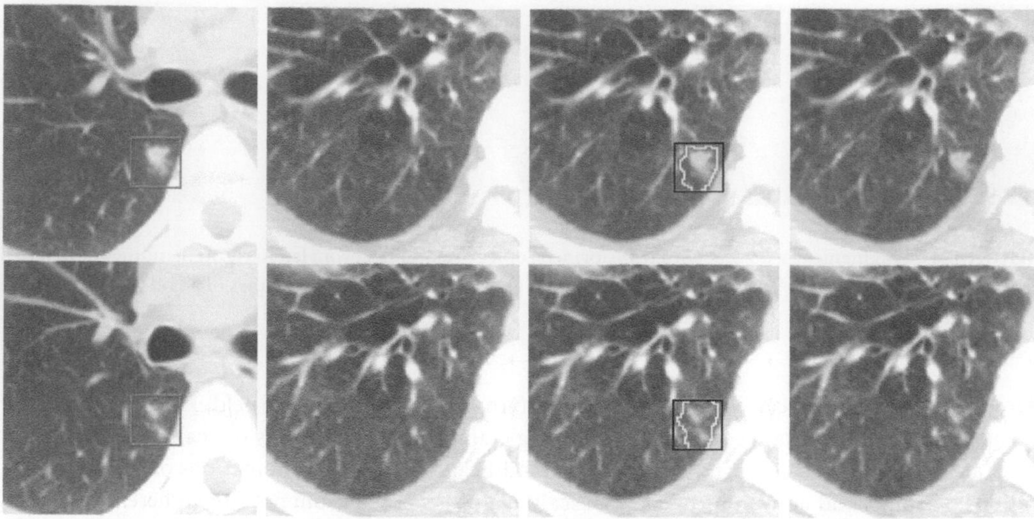

FIGURE 8.5 Blending consecutive slices of a nodule onto a different case. In each row from left to right: Source image, target image, blend results using optimal boundary with both boundaries superimposed, and blend results using optimal boundary without boundaries superimposed. Orange and red boxes show the initial boundary drawn by the user, and the optimal boundary is shown in yellow.

combination of the gradient field shaping process and the optimal boundary, and produces a seamless blend result within a complex target background area while preserving its adjacent vascular structure and texture. Figure 8.6 shows the orthogonal views of an original source nodule and its inserted counterpart, and the zoomed-in views of the same nodules are shown in Figure 8.7. Even though our algorithm blends the nodule slice-by-slice, the inserted nodule is consistent with the original nodule in the coronal and sagittal views. Ground glass and mixed density nodules can similarly be easily inserted into a new location, as the gradient information in the guidance field will simply replicate those textures and variations in the domain of the target images without the need for any additional steps. An example of insertion of a ground glass nodule is shown in Figure 8.8.

We conducted a reader study to assess whether experienced radiologists could reliably distinguish inserted nodules from their original counterparts. All the slices containing an original or inserted nodule were presented to the readers, allowing them to scroll through the slices and adjusting the window level and width, as well as zoom level. The readers were instructed to provide a score between 0 and 100 for each nodule they viewed, with 0 indicating definitely inserted and 100 for definitely original. Three experienced radiologists took part in our reader study, reviewing a total of 110 nodules (55 original and their 55 inserted counterparts) in random order in two separate sessions, and the results were analyzed using receiver operating characteristic (ROC) methodology (for more details see Pezeshk et al. 2015b). The positive and negative samples were defined as original and inserted nodules, respectively, and the score provided by the radiologists for each nodule was used as the decision variable in the ROC analysis. Table 8.1 summarizes the results of this study, showing that the average area under the ROC curve (AUC) for the radiologists was 0.63 ± 0.03. Having defined the binary classification problem in our study as described above, an area under the ROC curve (AUC) close to 0.5 (the chance line) indicates that the inserted nodules are not distinguishable from original nodules.

8.4.2 Augmentation of Mammograms and Reader Study

The blending method discussed in Section 9.3 is a general procedure that does not rely on the physics of the imaging modality, and can therefore be used in different applications without major modifications

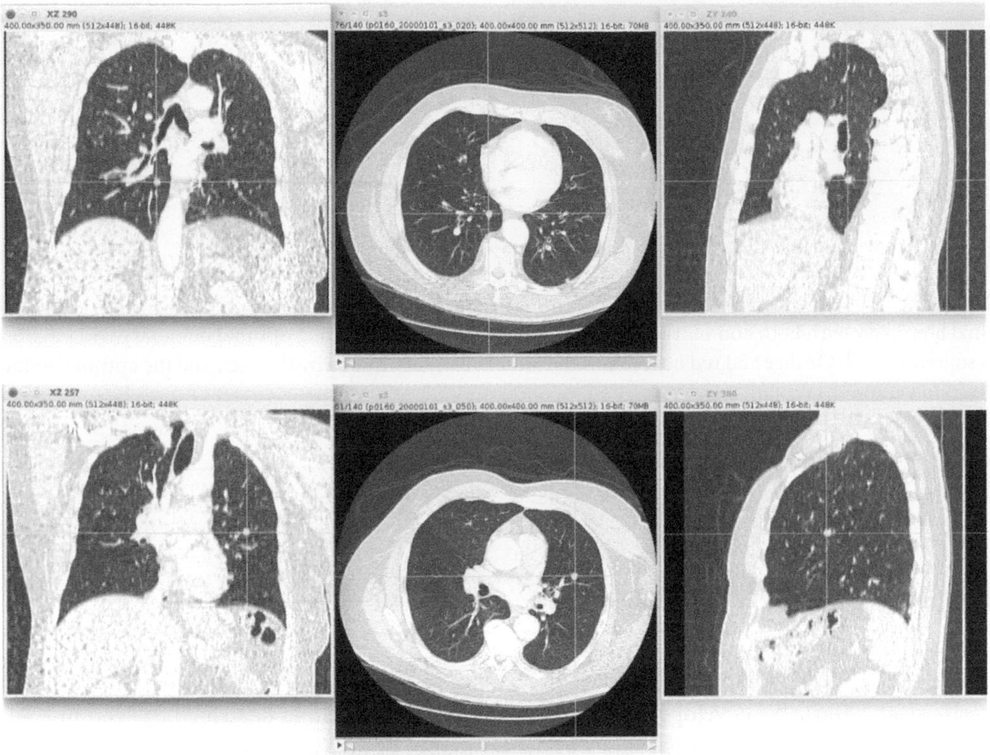

FIGURE 8.6 Comparison of orthogonal views of an original nodule (first row) and its inserted counterpart (second row). In each row from left to right: coronal view, axial view, sagittal view. Crosshairs mark the location of the nodule in each view.

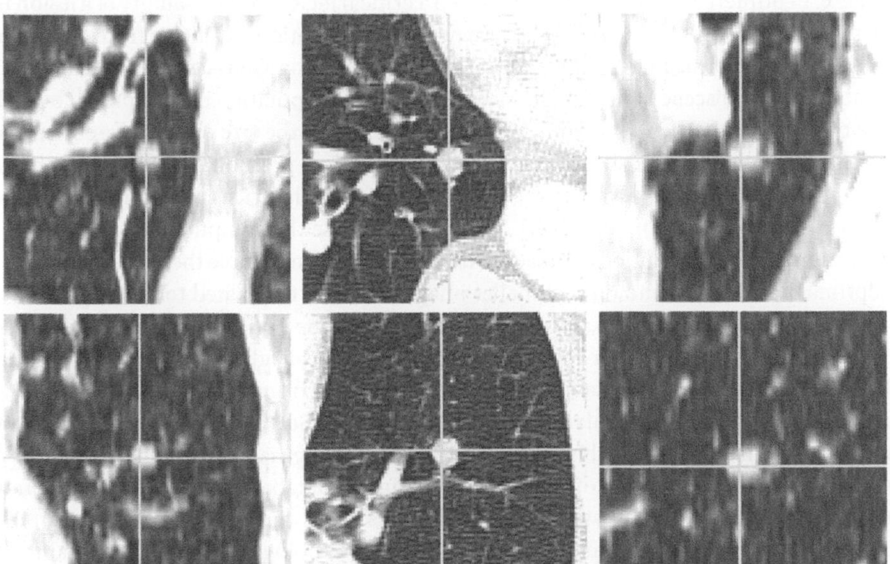

FIGURE 8.7 Zoomed in orthogonal views of the nodules in Figure 8.6. First row shows the original nodule, and second row shows its inserted counterpart.

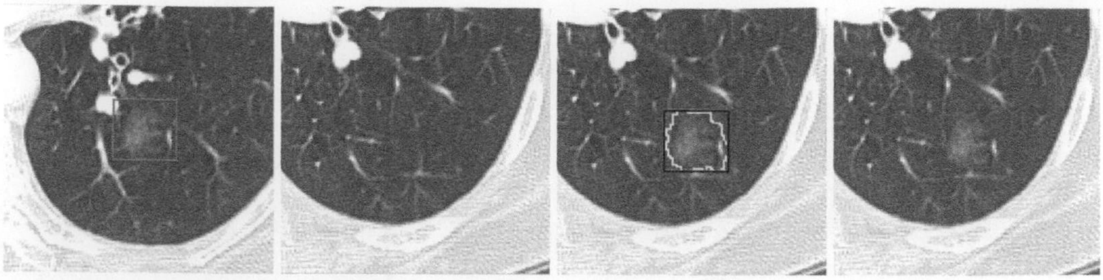

FIGURE 8.8 Blending a ground glass nodule. From left to right: Source image, target image, blend result using optimal boundary with both boundaries superimposed, and blend result using optimal boundary without boundaries superimposed. Orange and red boxes show the initial boundary drawn by the user, and the optimal boundary is shown in yellow.

TABLE 8.1 Summary of Chest CT Reader Study

	Years of Clinical Experience	AUC ± Standard Deviation
Reader 1	>30	0.52 ± 0.06
Reader 2	>35	0.65 ± 0.05
Reader 3	>17	0.71 ± 0.05
Average	>27	0.63 ± 0.03

(Pezeshk et al., 2016). We developed a separate GUI similar to the one described in Section 8.4.1 for simple extraction and insertion of lesions in digital mammography, with the main difference being that the GUI operated on one pair of 2D source and target images instead of the stacks of slices in CT. Figure 8.9 shows an example of a lesion inserted into the contralateral breast of the same patient using our tool.

Digital mammogram images and, consequently, breast lesions have a much higher pixel count compared to their CT counterparts, thereby making speed a critical factor in the usability of a lesion insertion algorithm. Assuming a square $N \times N$ ROI, the system of linear equations resulting from Equation 8.3 will produce a $N^2 \times N^2$ coefficient matrix, and solving this system using Gaussian elimination will require $O(N^6)$ operations. The adjacency and cost matrices used in the computation of the optimal boundary will similarly be of size $N^2 \times N^2$. Even for a 300×300 pixel breast lesion, the size of the different matrices used in creating the final image composite makes the computation cost prohibitive. To address this issue, we can use special solvers that take advantage of the sparsity of these matrices to significantly speed up the insertion process, as discussed in Section 8.3. To further increase the computation speed of the optimal boundary, we used a multi-resolution approach, where we first down-sample the breast image to a smaller size. The optimal boundary found on the smaller image is then extrapolated to the original image size. Compared to using a non-sparse Dijkstra solver for computing the optimal boundary at full resolution, and Gauss-Seidel with successive over-relaxation for the seamless blending of the lesion, these optimizations reduce the insertion time for a sample 300×300 pixel patch from 1,890 s down to 3.47 s.

We conducted a reader study similar to the one described in Section 8.4.1, again, to assess whether experienced radiologists could reliably distinguish between original and inserted lesions. We used mammograms from the University of South Florida Digital Database for Screening Mammography (DDSM) (Heath et al. 2001) for this study. Each original lesion in the study was inserted into the contralateral breast of the same patient using our tool. The two radiologists each read a total of 80 mammograms (40 containing original lesions, and 40 with inserted lesions) over two sessions with a washout period in between the sessions. Twenty-two of the original masses were malignant and 18 were benign. The mass size ranged from 7.6 to 28.4 mm (mean = 15.2 mm, standard deviation = 5.1 mm). The scoring strategy and analysis of results were identical to our previous reader study (see Section 8.4.1 for more

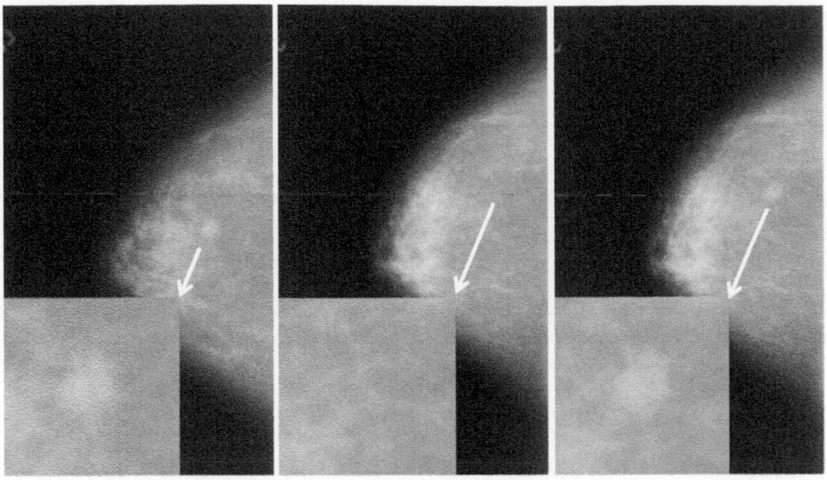

FIGURE 8.9 Example of lesion blending in mammography. From left to right: The source mammogram containing a malignant mass with a zoomed-in view of the original mass in the overlay, the target normal mammogram before blending with a zoomed-in view of the insertion area in the overlay, the target mammogram after blending with a zoomed-in view of the blended mass in the overlay.

TABLE 8.2 Summary of Mammography Reader Study

	Years of Clinical Experience	AUC ± Standard Deviation
Reader 1	>30	0.56 ± 0.07
Reader 2	>17	0.53 ± 0.06
Average	>23.5	0.54 ± 0.05

details). Table 8.2 shows the AUC results of the readers. The readers were found to have an average AUC of 0.54, indicating that their performance in distinguishing the original lesions from their inserted counterparts was close to the chance line.

The efficient MATLAB® code used to create the blend results shown in this chapter has been posted online (Pezeshk 2015).

8.4.3 Effect on CAD Training

We evaluated the effect of creating additional samples using our image blending tool on improving the performance of a CAD system with limited training data (Pezeshk et al. 2015a). This system (Ge et al. 2005, Sahiner et al. 2006) consisted of three main modules: prescreening, feature extraction, and classification. In the prescreening stage, candidate volumes are identified from the interior of the lung. In the next stage, a number of features are extracted from each candidate volume of interest (VOI). The feature vector consists of a total of 30 descriptors based on the statistics and distribution of the 3D gradient field within the VOI, as well as descriptors based on the shape of the object within the VOI, and mean, volume, and standard deviation of voxel intensity values. Finally, a linear discriminant analysis (LDA) classifier is used to produce a discriminant score for each VOI that can be used to classify each VOI into nodule (positive) or non-nodule (negative) classes.

We augmented a small subset of the LIDC dataset by extracting a nodule from its original location and blending it into a new location within the same patient's CT exam. Each inserted nodule therefore provided a new sample located in different slices, lobe, and background compared to its original

TABLE 8.3 AUC Values for the Two CAD Training Scenarios TR1-2 Over Three Experiments E1-3.

	Training # of +'s	Training # of –'s	Test AUC	CI of Test AUC Difference (TR2 – TR1)
E1: TR1	34	398	0.945 ± 0.012	(0.0019, 0.0241)
E1: TR2	87	398	**0.958 ± 0.011**	
E2: TR1	43	498	0.947 ± 0.012	(0.0049, 0.0233)
E2: TR2	112	498	**0.961 ± 0.011**	
E3: TR1	51	598	0.954 ± 0.012	(0.0019, 0.017)
E3: TR2	136	598	**0.963 ± 0.011**	

Note: The Best performance among the three scenarios printed in bold for each experiment. The column titled ci shows the 95% confidence interval for the difference between aucs of TR2 and TR1 in each experiment

counterpart. The effect of augmenting the dataset was assessed by conducting three experiments, each involving two training sets. The first training set, TR1, used only the original nodule VOIs. The second training set, TR2, consisted of all the VOIs in TR1 as well as additional VOIs from the inserted counterparts of the nodules in TR1. The two training sets, therefore, contained the same number of true negatives. The prescreening stage may detect a nodule in multiple VOIs. When multiple overlapping VOIs were detected for a training nodule, all the corresponding feature vectors were used for training the CAD system. The same independent test set containing 107 true positive and 975 true negative VOIs was used for each trained CAD system. For test nodules detected in multiple VOIs, only the VOI with the highest classifier score was used in calculating the test AUC values.

The three experiments, E1, E2, and E3, assessed the difference between using TR1 and TR2 on classifier performance as the numbers of true positives and true negatives were increased. E2 emulated the inclusion of scans from additional patients in training a classifier, compared to E1, and likewise, E3 emulated the inclusion of additional scans compared to E2. The results of these experiments are shown in Table 8.3. In each experiment, TR2 shows a statistically significant improvement in AUC relative to TR1. Moreover, TR1 in E3 contains 50% more true positives than TR1 in E1. However, augmenting TR1 in E1 provides a larger improvement in AUC compared to inclusion of additional real nodules in TR1 in E3. This improvement in performance is essentially free compared to the cost of recruiting and scanning additional patients.

8.4.4 Evaluation of Volumetry Algorithms

We conducted a preliminary study to assess the use of lesion insertion in chest CT for lesion volumetry by comparing the volumes extracted from CT images of physical synthetic nodules with those from lesions inserted using our lesion blending method. An anthropomorphic chest phantom with a vascular insert (N1, Kyoto Kagaku, Kyoto, Japan) was scanned at two different dose levels, once with physical synthetic nodules placed inside the phantom and once without. Each scan was reconstructed using both filtered back projection (FBP) and iterative reconstruction (IR). The synthetic physical nodules consisted of a total of 20 nodules, with various sizes (8, 9, and 10 mm in diameter) and shape characteristics (spherical, ellipsoid, lobulated, and spiculated). The synthetic nodules were placed inside of Styrofoam pellets, to avoid contact with the vascular insert or the lung boundary, and placed in different locations inside of the chest phantom. The scans without the nodules contained empty Styrofoam pellets to provide void volumes that could be later used for insertion of nodules using our image blending tool. The scans were obtained on a Siemens Somatom Definition Flash commercial CT scanner at two dose levels to emulate different image noise characteristics: 1.45 mGy $CTDI_{vol}$ and 22 mGy $CTDI_{vol}$, where CTDI stands for CT dose index.

We used our blending tool to extract each nodule from the scans containing nodules and insert them into a different location on the scans with matching parameters (in terms of dose and reconstruction

TABLE 8.4 Comparison of Performance of Original and Inserted Nodules in Terms of Volumetry

	Average Percent Difference	R^2	p-value
CTDI 1.45 mGy—FBP	−0.33	0.99	>0.05
CTDI 22 mGy—FBP	0.95	0.99	>0.05
CTDI 1.45 mGy—IR	−0.26	0.99	>0.05
CTDI 22 mGy—IR	0.87	0.99	>0.05

filter), but without nodules. A commercial volume estimation tool (iNtuition, TeraRecon Inc.) was then used to compute the volumes of both the physical nodules and their inserted counterparts. We calculated the average percent difference between the volumes extracted from the physical and inserted nodules at a particular dose level and reconstruction filter as follows:

$$APD = 100 \times \frac{1}{M} \sum_{i=1}^{M} \frac{(V_i - V_i')}{V_i'},$$

where:

V_i and V_i' are the estimated volumes of the *i*th inserted and corresponding physical nodules, respectively

M is the total number of synthetic nodules (i.e., 20)

Table 8.4 shows the results of our experiment. The small average percent difference (<1%) of our image blending technique indicates that, on the average, our algorithm introduces minimal volumetric changes compared to the original lesion. Moreover, we conducted a paired t-test to measure the statistical significance of the differences in volume, as well as an R^2 goodness of fit between the volumes estimated from inserted lesions and their physical counterparts. The results of the t-test and the R^2 measurements, also shown in Table 8.4, indicate that in all cases the differences in estimated volumes are not significant (p-values > 0.05), and that the volumes estimated from the inserted nodules closely follow volumes from the physical nodules ($R^2 > 0.99$). These preliminary results are promising in that they indicate the viability of using our image blending tool to create large libraries of lesions with known properties and characteristics in lieu of cumbersome repeated physical measurements.

Conclusion

Artificial augmentation of medical imaging datasets is a powerful tool that can benefit a variety of applications by supplementing and enlarging an existing dataset that is limited in terms of its size or sample characteristics. These benefits are practically free compared to collecting additional patient data, and they can be particularly important for rare diseases or conditions where obtaining real data might be virtually impossible. More work is required to both develop new ways to properly validate these techniques, as well as to investigate new or improved augmentation methods.

References

Ambrosini, R. D. and W. G. O'Dell (2010, March). Realistic simulated lung nodule dataset for testing CAD detection and sizing. In *Proc. SPIE Medical Imaging: Computer-Aided Diagnosis*, Volume 7624, pp. 76242X1–76242X8.

Armato III, S. G. et al. (2011). The lung image database consortium (LIDC) and image database resource initiative (IDRI): A completed reference database of lung nodules on CT scans. *Med. Phys.* 38 (2), 915–931.

Bishop, C. M. (1995, January). Training with noise is equivalent to tikhonov regularization. *Neural Comput.* 7 (1), 108–116.

Chan, H.-P., B. Sahiner, R. F. Wagner, and N. Petrick (1999). Classifier design for computer-aided diagnosis: Effects of finite sample size on the mean performance of classical and neural network classifiers. *Med. Phys.* 26 (12), 2654–2668.

Chawla, N. V., K. W. Bowyer, L. O. Hall, and W. P. Kegelmeyer (2002, June). SMOTE: Synthetic minority over-sampling technique. *J. Artif. Intell. Res.* 16 (1), 321–357.

Dijkstra, E. (1959). A note on two problems in connexion with graphs. *Nu-merische Mathematik* 1 (1), 269–271.

Ge, Z., B. Sahiner, H.-P. Chan, L. M. Hadjiiski, P. N. Cascade, N. Bogot, E. A. Kazerooni, J. Wei, and C. Zhou (2005). Computer-aided detection of lung nodules: False positive reduction using a 3D gradient field method and 3D ellipsoid fitting. *Med. Phys.* 32 (8), 2443–2454.

Hannun, A. Y., C. Case, J. Casper, B. C. Catanzaro, G. Diamos, E. Elsen, R. Prenger, S. Satheesh, S. Sengupta, A. Coates, and A. Y. Ng (2014). Deep speech: Scaling up end-to-end speech recognition. CoRR abs/1412.5567.

Heath, M., K. Bowyer, D. Kopans, R. Moore, and P. Kegelmeyer (2001). The digital database for screening mammography. In M. Yaffe (Ed.), *Proc. Fifth International Workshop on Digital Mammography*, pp. 212–218. Medical Physics Publishing.

Heisele, B., T. Serre, and T. Poggio (2007). A component-based framework for face detection and identification. *Int. J. Comput. Vision* 74 (2), 167–181.

Ho, T. and H. Baird (1997). Large-scale simulation studies in image pattern recognition. *IEEE Trans. on PAMI* 19, 1067–1079.

Jia, J., J. Sun, C.-K. Tang, and H.-Y. Shum (2006). Drag and drop pasting. In *Proc. ACM SIGGRAPH*, pp. 631–636.

Krizhevsky, A., I. Sutskever, and G. E. Hinton (2012). ImageNet classification with deep convolutional neural networks. In P. Bartlett, F. Pereira, C. Burges, L. Bottou, and K. Weinberger (Eds.), *Advances in Neural Information Processing Systems (NIPS)*, Red Hook, NY, pp. 1106–1114. Curran Associates, Inc.

Laptev, I. (2009, April). Improving object detection with boosted histograms. *Image Vision Comput.* 27 (5), 535–544.

Li, X., E. Samei, D. M. Delong, R. P. Jones, A. M. Gaca, C. L. Hollingsworth, C. M. Maxeld, C. W. T. Carrico, and D. P. Frush (2009). Three-dimensional simulation of lung nodules for paediatric multidetector array CT. *BJR* 82 (977), 401–411.

Madsen, M. T., K. S. Berbaum, K. M. Schartz, and R. T. Caldwell (2011, March). Improved implementation of the abnormality manipulation software tools. In *Proc. SPIE Medical Imaging: Image Perception, Observer Performance, and Technology Assessment*, Volume 7966, pp. 7966121–7966127.

Melville, P. and R. Mooney (2003). Constructing diverse classifier ensembles using artificial training examples. In *Proceedings of the 18th International Joint Conference on Artificial Intelligence*, Acapulco, Mexico, pp. 505–510.

Papageorgiou, C. and T. Poggio (2000, June). A trainable system for object detection. *Int. J. Comput. Vision* 38 (1), 15–33.

Perez, P., M. Gangnet, and A. Blake (2003). Poisson image editing. In *Proc. ACM SIGGRAPH*, pp. 313–318.

Peskin, A. P. and A. A. Dima (2010). Modeling clinical tumors to create reference data for tumor volume measurement. In G. Bebis, R. Boyle, B. Parvin, D. Koracin, R. Chung, R. Hammound, M. Hussain, T. Kar-Han, R. Crawfis, D. Thalmann, D. Kao, and L. Avila (Eds.), *Advances in Visual Computing, Volume 6454 of Lecture Notes in Computer Science*, pp. 736–746. Springer Berlin Heidelberg.

Pezeshk, A. (2015). Lesionblender. https://github.com/DIDSR/LesionBlender.

Pezeshk, A., B. Sahiner, W. Chen, and N. Petrick (2015a). Improving CAD performance by seamless insertion of pulmonary nodules in chest CT exams. In *Proc. SPIE Medical Imaging: Computer-Aided Diagnosis*, pp. 94140A-94140A-6.

Pezeshk, A., B. Sahiner, and N. Petrick (2016). Seamless lesion insertion in digital mammography: Methodology and reader study. In *Proc. SPIE Medical Imaging: Computer-Aided Diagnosis*, pp. 97850J- 97850J-6.

Pezeshk, A., B. Sahiner, R. Zeng, A. Wunderlich, W. Chen, and N. Petrick (2015b). Seamless insertion of pulmonary nodules in chest CT images. *IEEE Trans. Biomedical Engineering* 62 (12), 2812–2827.

Pezeshk, A. and R. L. Tutwiler (2011). Automatic feature extraction and text recognition from scanned topographic maps. *IEEE Trans. Geoscience and Remote Sensing* 49 (12), 5047–5063.

Pishchulin, L., A. Jain, C. Wojek, M. Andriluka, T. Thormhlen, and B. Schiele (2011). Learning people detection models from few training samples. In *Proc. IEEE Conference on Computer Vision and Pattern Recognition (CVPR'11)*, pp. 1473–1480.

Porter, T. and T. Duff (1984, January). Compositing digital images. *SIG-GRAPH Comput. Graph.* 18 (3), 253–259.

Rashidnasab, A., P. Elangovan, M. Yip, O. Diaz, D. R. Dance, K. C. Young, and K. Wells (2013). Simulation and assessment of realistic breast lesions using fractal growth models. *Phys. Med. Biol.* 58 (16), 5613–5627.

Raviv, Y. and N. Intrator (1996). Bootstrapping with noise: An effective regularization technique. *Conn. Sci.* 8, 355–372.

Sahiner, B., H.-P. Chan, N. Petrick, R. F. Wagner, and L. M. Hadjiiski (2000). Feature selection and classifier performance in computer-aided diagnosis: The effect of finite sample size. *Med. Phys.* 27 (7), 1509–1522.

Sahiner, B., L. M. Hadjiiski, H.-P. Chan, C. Zhou, and J. Wei (2006). Computerized lung nodule detection on screening CT scans: performance on juxta-pleural and internal nodules. In *Proc. SPIE Medical Imaging: Image Processing*, pp. 61445S-61445S-6.

Saunders, R., E. Samei, J. Baker, and D. Delong (2006). Simulation of mammographic lesions. *Acad. Radiol.* 13 (7), 860–870.

Shin, H.-O., M. Blietz, B. Frericks, S. Baus, D. Savellano, and M. Galanski (2006). Insertion of virtual pulmonary nodules in CT data of the chest: development of a software tool. *Eur. Radiol.* 16 (11), 2567–2574.

van der Maaten, L., M. Chen, S. Tyree, and K. Q. Weinberger (2013). Learning with marginalized corrupted features. In *Proc. ICML*, Volume 28, pp. 410–418.

Vaz, M. S., Q. Besnehard, and C. Marchessoux (2011). 3D lesion insertion in digital breast tomosynthesis images. In *Proc. SPIE Medical Imaging: Physics of Medical Imaging*, Volume 7961, pp. 79615Z1–79615Z10.

Zhang, X., E. Olcott, P. Raffy, N. Yu, and H. Chui (2007, March). Simulating solid lung nodules in MDCT images for CAD evaluation: modeling, validation, and applications. In *Proc. SPIE Medical Imaging: Computer-Aided Diagnosis*, Volume 6514, pp. 65140Z0–65140Z8.

Zwillinger, D. (1997). *Handbook of Differential Equations* (3rd edn.). Boston, MA: Academic Press.

9

Diffuse Lung Diseases (Emphysema, Airway and Interstitial Lung Diseases)

Marcel Koenigkam Santos

Oliver Weinheimer

9.1 Introduction

Diffuse pulmonary diseases may be classified into two large groups of conditions: obstructive and restrictive diseases. This classification is based mainly on clinical features (symptoms and physical examination) and pulmonary function testing. While patients with obstructive disease have airflow limitation/obstruction, patients with restrictive disease have pulmonary expansion limitation/restriction. The most common causes of obstructive disease are chronic obstructive pulmonary diseases (COPD), which includes emphysema and chronic bronchitis, asthma, bronchiectasis and cystic fibrosis (CF). Major conditions causing restrictive lung disease are represented by the interstitial lung diseases (ILD), such as idiopathic pulmonary fibrosis (IPF), pulmonary involvement by autoimmune diseases (systemic sclerosis, rheumatoid arthritis, lupus) and adverse drug reactions. Both groups share the same main symptoms of shortness of breath, cough and exertion. Pulmonary function tests (PFT), such as spirometry, plethysmography and diffusing capacity of the lung for carbon monoxide (DLCO) are used to differentiate obstructive from restrictive lung disease, as well as to assess disease severity and progression. But there are some conditions and situations that may challenge the correct diagnosis and also make adequate assessment of disease severity difficult, even with satisfactory clinical function tests. Some patients may present heterogeneous and mixed patterns of disease, showing restriction associated with obstruction at different degrees. Most clinical tests give only a global picture of the disease and cannot show a regional (right × left lung, basal × apical) or compartmentalized-based (airway, airspace, interstitial, vascular) analysis. For some diseases, PFT has a low sensitivity to detect initial alterations. Instead, pulmonary imaging can present morphological and function information, objectively and in a regional or compartmentalized fashion for the evaluation of diffuse lung diseases, being represented especially by the high resolution computed tomography (HRCT) (Maffessanti and Dalpiaz 2006, Webb and Higgins 2010).

Imaging exams are almost always part of the investigation of restrictive and obstructive lung disease and usually include chest x-ray film and/or HRCT. Routine evaluation of these methods is visual and subjective, most commonly done by a radiologist. X-ray images have lower accuracy for diagnosis of diffuse lung disease. It may help when signs of hyperinflation are identified, probably representing COPD or other obstructive disease, or when interstitial opacities are seen in association with reduced lung volumes, probably representing a fibrotic lung disease or other ILD. On the other hand, HRCT is the imaging modality of choice for evaluation of interstitial and airway diseases. High-resolution tomographic images can identify, with high accuracy, morphological alterations such as emphysema, airway dilation and wall thickening, bronchiectasis, interstitial opacities consistent with inflammation and/or fibrosis and others. Most recently, with the widespread use of multidetector CT (MDCT) technology, volumetric high-resolution images of the entire lung can be obtained within a single breath-hold, also contributing for the method's accuracy increase. Nowadays, it is already well established that the role of HRCT goes beyond diagnosis. For example, in COPD, HRCT is used for disease phenotyping, differentiating a patient with emphysema-predominant disease from airway-predominant disease (Coxson et al. 2014). This is important information for choice of treatment and prognosis (life expectancy, response to treatment, chance of complications). In ILD, HRCT images are essential for disease classification (type of idiopathic interstitial pneumonia) and can even avoid a lung biopsy for diagnosis of IPF (Travis et al. 2013).

So, if HRCT visual assessment already has a very well established role in evaluation of diffuse lung diseases, how could CAD or computed quantification help? Several studies have shown that visual grading of the severity of emphysema showed less agreement with pathology than quantitative CT measures. Size and distribution of emphysema, demonstrated by dedicated quantification software systems, may also influence function and prognosis in patients with COPD (Bankier et al. 1999, Nakano et al. 1999, Aziz et al. 2005, Martinez et al. 2006). In the evaluation of ILD, recent studies have showed that, for example, inter-observer agreement for diagnosis of IPF can be no more than moderate among thoracic radiologists (Walsh et al. 2016), while automated systems could be successfully used in routine clinical practice as a diagnostic aid for identifying disease (Depeursinge et al. 2015). And there are more examples in medical literature showing that computer-aided tools may, indeed, help to more precisely identify and characterize the diffuse lung diseases.

Therefore, in this text, we aim to present an updated review of both clinical and technical aspects related to CAD and quantitative CT (QCT) in the evaluation of diffuse lung diseases, focusing on emphysema and other airway abnormalities, and fibrosing ILD.

9.2 Clinical Applications of CAD for Emphysema and Airway Diseases

CAD and QCT have been most widely used and validated for emphysema in COPD patients, and more recently, for evaluation of bronchi and gas trapping, also in other diseases like asthma and CF. On HRCT, COPD is characterized by low attenuation areas within lung parenchyma, associated with signs of hyperinflation. Muller et al. (1998) were the first to describe and validate the "density mask technique" in correlation with pathology. Using hard-copy images of 10 mm collimation the authors showed that areas with attenuation below a threshold of −910 HU had a good correlation, with the extent of emphysema observed in pathological specimens. Kinsella et al. (1990) were the first to correlate CT measurements of emphysema with PFT, and, approximately 10 years later, a study by Bankier et al. (1999) was the first to demonstrate that QCT correlated better with macroscopic degree of emphysema compared to visual CT scoring. Airway alterations on HRCT are characterized by bronchial dilatation, bronchial wall irregularity and thickening, as well as some opacities related to small airways. Mosaic attenuation and gas trapping on expiratory images can also be related to airway diseases. Quantitative airway analysis is more recent and is strongly related to isometric tridimensional imaging capabilities. The first studies focused on manually measuring few bronchi, sometimes only one (especially the bronchus to right apical segment S1), using zoomed 1 mm images acquired with gaps, in normal patients (Senéterre et al. 1994). From that time until now, many things have changed and dramatically evolved. CT scanners became much faster and robust, enabled better image quality, and, with the development of MDCT, we are now capable of acquiring thin slices of 1 mm or less from the whole lung volume in a single breath-hold. Postprocessing technology has accompanied the process, so that lung and airway segmentation became faster and more precise, with little or no need for manual interference, measurements are faster and more reproducible and more variables can be analyzed. Pulmonary lobes, fissures and vasculature can be automatically detected. Lung volumes and densities are precisely estimated. The whole bronchial tree can be detected and analyzed, until the 7–8th bronchi generations. Different algorithms were developed to measure bronchi diameters and wall thickness. All this evolution changed the focus of radiology studies, and instead of trying to validate measurements obtained with CT in correlation with pathology, which requires an invasive diagnosis, most authors are in search of correlating imaging measurements with function and prognosis, that represent clinical features, are noninvasive and are closer to clinical application and routine.

Kauczor et al. showed that, in patients with COPD, inspiratory and expiratory lung volumes assessed by QCT correlated with measurements obtained in PFT (Kauczor et al. 1998). Later, other authors also described that inspiratory CT lung volume correlates with total lung capacity, showing typical hyperinflation present in COPD patients (Zaporozhan et al. 2005). Emphysema volume (EV) and emphysema index (EI = the relation between emphysema volume and lung volume) correlate with residual volume and forced expiratory volume in first second (FEV1). This parameter, FEV1, obtained with spirometry, is the most important clinical measurement for COPD patients. It is used in disease classification (mild, moderate, severe, very severe) as recommended by the Global Initiative for Chronic Obstructive Lung Disease (GOLD, www.goldcopd.com), the most important worldwide program for COPD management and prevention. It has been demonstrated that emphysema volume, as assessed by QCT, correlates with GOLD categories and can show disease progression in patients with COPD (Pauls et al. 2010, Koenigkam-Santos et al. 2013b). Values of EI higher than 35% have been associated with severe COPD and higher risk for acute exacerbations (Han et al. 2011). Acute exacerbation is an important cause of morbidity (disease progression) and mortality in COPD. For patients with 35% or greater total emphysema, each 5% increase is associated with a 1.18-fold increase in annual exacerbation rate. This same study showed similar results for bronchial wall thickness, so that each 1 mm increase was associated with a 1.84-fold increase in annual exacerbation rate.

Emphysema size and distribution also seem to be important. The extent of emphysema in lower lobes is more closely correlated with PFT. For example, measurements of residual volume and total lung capacity (indicators of hyperinflation) are higher in COPD patients with predominant lower-lobe emphysema than in those with predominant upper-lobe emphysema. The extent of emphysema in the central lung region also has a better correlation with function than that in the peripheral region, so that patients with more central emphysema have lower (worse) values of predicted FEV1 and DLCO (Gurney et al. 1992, Nakano et al. 1999, Saitoh et al. 2000).

Airway measurements and air trapping quantification can also show COPD alterations in relation to pulmonary function (Schroeder et al. 2013). Hasegawa et al. demonstrated that airway luminal area and wall area percent correlate with FEV1, and this correlation becomes more important in distal (smaller) bronchi (Hasegawa et al. 2006). Achenbach et al. showed that wall thickness is significantly higher in smokers than in non-smokers and is even higher in COPD patients. Bronchial wall thickness correlates with the values of FEV1 and predicted FEV1. These correlations can also be higher when only smaller airways are considered (Achenbach et al. 2008). In a lung cancer screening cohort (current and former heavy smokers), Mets et al. showed that air trapping in expiratory images, as assessed by QCT, is independently associated with reduced lung function and accelerated worsening of airflow obstruction (Mets et al. 2013).

QCT has also been used for emphysema treatment control. Patients with severe emphysema may benefit from lung volume reduction therapy, when an emphysematous lobe is resected or collapsed to give room to less-involved lung parenchyma, which contributes more to function. In patients submitted to surgical lung volume reduction of an emphysematous target lobe, volumetric and density changes in the ipsilateral and contralateral lobes, showing expansion of remaining healthier lung, can be detected with CT (Brown et al. 2012). The method can also detect alterations in patients submitted to endobronchial lung volume reduction therapies, as in endobronchial valves (EBV). QCT can detect treated lobe reduced volumes in patients submitted to EBV, in correlation to clinically meaningful improvements in pulmonary function (Coxson et al. 2008, Koenigkam-Santos et al. 2013a)

More recently, QCT has been used to investigate COPD genetics and pathogenesis. Kim et al. showed that some genetic polymorphisms are associated with predominant-emphysema disease while other polymorphisms are associated with predominant-airway COPD, as determined by QCT analysis (Kim et al. 2011). Ostridge et al. showed that some inflammatory marker's concentration obtained with bronchoalveolar lavage (a clinical test performed with bronchoscopy) correlate with the extent of gas trapping, small airways disease and emphysema severity as identified on QCT (Ostridge et al. 2016).

Most clinically significant results of emphysema CT quantification were obtained in patients with α−1 antitrypsin deficiency. For this disease that causes emphysema even in non-smokers, the method is a recognized and validated tool for diagnosis and disease progression control. It has already been used as endpoint in clinical trials and may replace PFT as the gold standard for assessing response to known and novel treatments (Stockley et al. 2010).

Besides COPD, there has also been an increased use of CAD and QCT in the study of other airway diseases, especially asthma and CF. HRCT is capable of identifying bronchial and parenchymal alterations related to inflammation and remodeling in asthmatic patients. QCT has been used to show that asthmatic patients have larger airways, with thicker and denser bronchial walls. These measurements correlate with pathological alterations, disease severity and duration. In patients with persistent asthma, bronchial wall alterations correlate with airflow obstruction as assessed with PFT and response to treatment (Niimi et al. 2000, Montaudon et al. 2009).

Aysola et al. showed that patients with severe asthma have thicker airway walls when compared to patients with mild to moderate disease and healthy subjects. Wall thickness and wall area correlated with pathology findings in biopsy, as well as with FEV1 and response to bronchodilator therapy (Aysola et al. 2008). Lederling et al. showed that airway wall attenuation is higher in patients with asthma, and that this parameter may have higher correlation to clinical airflow obstruction indexes and inflammation when compared to bronchi diameters and wall thickness (Lederlin et al. 2012). Hoshino et al. used QCT to compare two different therapies in patients with asthma. Considering the values of airway wall area and thickness and luminal area, the authors showed that a combination therapy was more effective for reducing airway inflammation when compared to corticosteroid therapy alone. This work is also a great example of the use of CAD and QCT as a therapy outcome parameter in clinical trials (Hoshino and Ohtawa 2012). Brillet et al. also showed a correlation between subsegmental bronchial airway measurements and clinical control of asthma. The authors argued that changes in airways assessed by CT probably represent airway remodeling and structural changes in poorly controlled patients with chronic asthma (Brillet et al. 2013).

Cystic fibrosis is a genetic disorder that especially affects the lungs and pancreas in young Caucasian subjects. In these patients, chronic bacterial infection and inflammation lead to progressive bronchial dilatation, bronchiectasis and parenchymal destruction. There is no cure for CF, but in recent years, treatment has vastly evolved, leading to important increases in life expectancy. Several authors have used QCT to measure airway abnormalities on CF patients. In patients with CF, more bronchi are detected/analyzed when compared to normal subjects, and this parameter in even higher in patients with severe disease, reflecting a higher number of bronchiectasis (Wielpuetz et al. 2013a, Koenigkam-Santos et al. 2016). Airway diameter, wall area and wall thickness also correlate with clinical evaluation scores and function. QCT wall and luminal measures correlate with PFT and disease severity (de Jong et al. 2005, Montaudon et al. 2007). Wielpuetz et al. showed that patients with CF may also have emphysema, as assessed by QCT, even if they are never smokers, and that emphysema volume increases with age and contributes to airflow limitation (Wielpuetz et al. 2013b).

9.3 Clinical Applications of CAD for Interstitial Lung Diseases

Quantitative CT has also been increasingly used in the evaluation of fibrotic interstitial lung diseases (ILD). Different from emphysema and airway diseases, ILD are characterized by the presence of pulmonary opacities on CT, leading to lungs with reduced volume and increased overall attenuation. In daily routine, a radiologist evaluates the presence, distribution and extent of these opacities to classify the ILD, assesses disease severity and performs treatment control. On HRCT, honeycombing (HC) represents pulmonary fibrosis, which means irreversible destruction and is the most important alteration present in patients with the usual interstitial pneumonia (UIP) pattern. UIP is the radiological and pathological manifestation of idiopathic pulmonary fibrosis (IPF), a chronic, progressive and severe disease that has no cure and is associated with very poor prognosis. Even in patients with other known diseases, pulmonary involvement represented by the presence of UIP pattern indicates a worse prognosis, when compared to other pulmonary alterations. Ground glass opacities (GGO), on the other hand, most commonly represent inflammatory process and are associated with other patterns of ILD, like non-specific interstitial pneumonia (NSIP). NSIP most commonly represents a manifestation of pulmonary involvement in some autoimmune diseases or adverse drug reaction, and has a better prognosis when compared to UIP and IPF. Forced vital capacity (FVC) and DLCO are the functional tests mostly recommended in monitoring IPF and other fibrotic lung diseases. However, FVC adequate quantification can be challenging in patients with severe dyspnea and cough and DLCO is affected by other kind of lesions, like pulmonary vasculopathy and emphysema. In its place, many works have shown that CT can be used not only as a diagnostic tool, but also as a reproducible and objective tool to assess disease severity and progression.

Automated identification and quantification of ILD was initially done with basic densitometry or histogram-based methods that could show correlation with physiology and mortality, but there is increasing recognition that a texture-based method is more helpful for classifying and determining the extent of fibrotic abnormalities. Hartley et al. were the first to show that HRCT-derived measures of lung density could be valid, clinically meaningful and objective measures of ILD. The authors studied patients with IPF and exposure to asbestos. In a multivariate analysis, they found that the mean and median gray scale densities were independently associated with presence of moderate-to-severe dyspnea, higher (worse) radiographic categorization, a lower FVC, and a higher concentration of macrophages and eosinophils (inflammation) in the bronchoalveolar lavage fluid (Hartley et al. 1994).

Best et al. showed that quantitative CT indexes based on HRCT histograms in patients with IPF correlate with PFT and could also be predictive of mortality. In a first study, the authors showed that mean lung attenuation, skewness and kurtosis had moderate correlation with function, and that kurtosis could provide predictions of pulmonary function (Best et al. 2003). In a second study, the authors showed this histogram-based analysis could also be predictive of mortality in baseline exams of patients with IPF, and that these measurements could also objectively demonstrate changes over time, reflecting IPF progression (Best et al. 2008). Colombi et al. used a fully automatic histogram-based quantitative evaluation to describe changes over time in extent of fibrosis at MDCT images of patients with IPF. The authors described substantial difference between treated and untreated patients in the 40th and of the 80th percentiles of density histogram, at one-year follow up. They considered that variation of the 40th percentile might reflect changes in overall extent of lung abnormalities, markedly of GGO, while variation of the 80th percentile might reveal the course of reticular opacities (Colombi et al. 2015).

Texture-based quantification tools are more recent and are used to identify and quantify some specific patterns of opacities/alterations on HRCT images, including HC, GGO, reticular opacities (RO) and consolidation. Kauczor et al. were the first to compare a method for automatic identification and quantification of a specific pulmonary lesion pattern (GGO) to traditional density evaluation. In their work, the authors showed that automatic segmentation and quantification of GGO on HRCT by a neural network system was superior to a double-threshold density mask (Kauczor et al. 2000). Since then, several authors demonstrated that texture-based algorithms were computationally efficient and comparable to experienced radiologists evaluation. It is effective in the classification of normal versus abnormal tissue and in distinguishing among typical lesion patterns present in lungs with UIP and other ILD (Zavaletta et al. 2007, Boehm et al. 2008).

Iwasawa et al. used an automatic system to detect and measure fibrosis on HRCT images from patients with IPF. The area of each pattern measured by their system correlated significantly with the area estimated by the radiologists, and the volumes of normal pattern and fibrotic patterns correlated with DLCO. QCT measurements also correlated with survival in a univariate analysis, and multivariate analysis identified HC volume as a significant predictor of survival (Iwasawa et al. 2009). Yoon et al. showed that a texture-based quantification system was comparable to visual assessment for evaluating disease extent in patients with UIP and NSIP, and that it could detect interval changes in correlation with PFT deterioration (Yoon et al. 2013). Iwasawa et al. also used a CAD system on sequential HRCTs of patients with IPF treated with a new anti-fibrotic drug. QCT was able to show that progression of the fibrotic changes (RO and HC) were significantly smaller in the treatment group when compared to control, and that the decline in lung function correlated significantly with the increase in these patterns as assessed by CT. The authors argued that the CAD system could detect even a small progression of pulmonary fibrosis on CT, which visual/subject analysis was not able to identify (Iwasawa et al. 2014). A very recent study by Depeursinge et al. proposed a novel computational approach for the automated classification of UIP. In this study, the system presented by the authors was not only able to identify and quantify the pulmonary lesions (GGO, RO, HC), but also to assess lesions distribution along the lungs (apical × basal, peripheral × central). Their results suggest that an automated system may be useful in routine clinical practice as a diagnostic aid for identifying patients with UIP, so that combining human and computerized evaluations may be synergistic (Depeursinge et al. 2015).

9.4 Computerized Detection and Description of Airways and Lung Parenchyma in CT

The technology of MDCT provides volume datasets with approximately isotropic resolution. Typical resolution values for thin-section MDCT datasets of the thorax in the clinical routine are about 0.4–0.8 mm for in-slice spacing (pixel spacing) and between 0.5 and 1.25 mm for slice thickness and slice spacing. A fast and reliable segmentation and analysis of the lung tissues is of fundamental importance for many clinical applications. In the following text, the methods, detection and description of the airways and the lung parenchyma, as implemented in the software system YACTA ("Yet Another CT Analyzer"), are described. YACTA consists of two main tools: the image processing software and the process manager. The latter is responsible for the organization of the individual images processing tasks on a PC system. YACTA works fully automated—it is possible to send DICOM images from an arbitrary DICOM Viewer via the network to YACTA, or the software can work through any number of MDCT volumes stored on a local device. Due to the complete automation of the software, it is possible to analyze large amounts of data without any user interaction—thus YACTA can be a valuable component for the analysis of large amounts of lung MDCT volumes and an important tool in the context of "Big Data Analysis." The image processing techniques are based on the methods first described by Weinheimer et al. 2003. The software was improved over time and used for many studies, see e.g., Zaporozhan et al. 2005, Achenbach et al. 2008, Wielpütz et al. 2013 or Colombi et al. 2015. The image processing capabilities can be roughly divided in two components: airway analysis and lung parenchyma analysis.

9.4.1 Airway Tree Segmentation

Depending on the quality of the CT data, extraction of the tracheobronchial tree and the lungs can be a very challenging task, especially if it is a fully automated extraction. A CT system should be well calibrated, such that air is at −1000 HU, while water is at 0 HU. In order to verify this, the histogram of the first 10 upper slices of a CT volume is analyzed, and the peak corresponding to air is determined. The difference to −1000 HU is stored as extra corporeal air error (ECAE). The ECAE can be used to adjust HU-dependent parameters in subsequent algorithms to an individual CT volume. An earlier version of the airway segmentation algorithm is described in Weinheimer et al. 2009. The first step is the search of a trachea landmark on the upper slices of a dataset. To achieve this goal, the whole body is detected, so that the search area can be limited to the detected body region. Then, a (dark) region with voxel values <-500 HU (ECAE corrected) is searched; the area must be smaller than $16^2 \times \Pi$ mm^2 and greater than 0 mm^2. The upper limit was empirically determined—it is the highest value we ever saw for a patient. The lower limit was set while performing the study presented in Wielpütz et al. 2014. There, caused by an excessive dynamic collapse of the trachea, the size of the detectable area goes down near to zero. In order to allow the analysis of CT volumes in every respiratory phase, the lower limit of zero is necessary. A region is accepted as a possible trachea region if it is surrounded by airway walls, which is checked by a special edge detection algorithm. The center of gravity of an accepted region is calculated and mapped on the succeeding slice. A similar dark region is searched at the mapped position. If possible trachea points can be found over at least 10 mm of contiguous slices and, additionally, the top of one lung can be found on the last examined slice, then a trachea landmark is found.

The trachea is actually segmented with a 2D region-growing algorithm with threshold −400 HU (ECAE corrected) starting at the previously determined trachea landmark. Again, center of gravity is calculated and mapped on the next slice in basal direction. The procedure is repeated and stops if the bifurcation of the trachea is reached. Figure 9.1a shows the segmented trachea of a COPD patient. The 3D connected trachea region is used to verify the air HU value. To exclude the influence of the trachea wall on tracheal air HU value, the region is eroded, and the difference of the average value within the remaining region to −1000 HU is stored as tracheal air error (TAE).

Voxels at the basal surface of the segmented trachea are used as start points for an iterative, self-adapting, region-growing-based 3D bronchial tree tracer. The tracer uses an N26 neighborhood system. The conditions for growing are selected so restrictively that leaking out of the segmentation into the

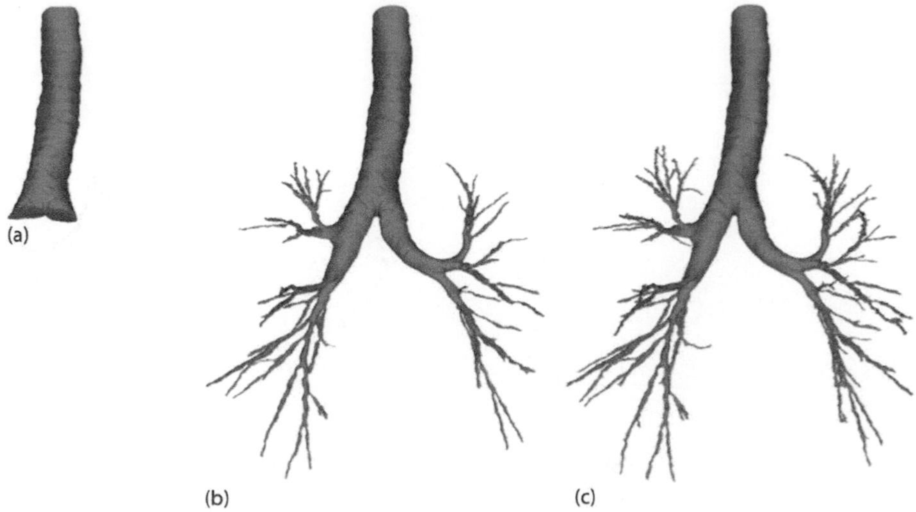

FIGURE 9.1 Result of tracheobronchial tree segmentation in MDCT volume of COPD patient. (a) Initial trachea segmentation. (b) Iterative 3D region-growing-based segmentation after first iteration, and (c) after 95 iterations.

parenchyma is almost impossible. For voxels in a larger bronchus, it is sufficient to consider the HU value of the voxel to be evaluated and the average HU value in its neighborhood. If a voxel is in a smaller bronchus, then it will be cut either by an axial, coronal or sagittal plane in a circular to elliptical way. For each plane, 8 uniformly distributed rays are casted outwards and it is determined if a potential lumen voxel is surrounded by airway wall in one plane in all directions. The HU-value-based default start conditions of this approach are corrected by TAE. After one iteration (see Figure 9.1b), HU-based conditions are changed to allow the segmentation of smaller bronchi. If more than a defined permitted number of voxels are added by a single region-growing process, the marked voxels are reset in order to avoid leaking out into lung parenchyma. Figure 9.1c shows a segmented tracheobronchial tree after 95 iterations.

9.4.2 Airway Tree Labeling

The voxel-based result of the segmentation is skeletonized by an iterative topology-preserving 3D thinning algorithm. This skeletonization algorithm is modeled after Palagyi et al. 2001, and a simple-point definition is used from Klette 2003. A point is called simple if its deletion does not alter the topology of the object. The thinning process is completed when no further simple points could be deleted in an iteration step. Next, the generated skeleton is transformed to an acyclic graph, whereby the branchings of the bronchi correspond to the nodes of the graph (see in Figure 9.2a the graph of the segmented tree from Figure 9.1). Short terminal edges are eliminated before, and possible cycles are eliminated after the graph generation step. Both are usually unwelcome artifacts through image noise (Figure 9.2b).

Actual airway labeling procedure is based on the rule-based method introduced in Gu et al. 2012. In YACTA, the procedure is modified in order to gain more robustness. The trachea, the right main bronchus (RMB) and left main bronchus (LMB) are determined and labeled. The trachea node is defined as root of the graph, a generation number is assigned to each bronchus, representing the hierarchical nature of the bronchial tree. Then, subsequent bronchi in the right lung are assigned to upper lobe (RUL), middle lobe (RML) or lower lobe (RLL). In the left lung, bronchi are assigned to upper lobe (LUL), lingula (LLi) and lower lobe (LLL). Figure 9.2c shows a labeled tracheobronchial tree.

Furthermore, anatomical variations increase the difficulty for labeling of the bronchial tree. Special rules for anatomical variations are implemented. For example, there is a rule for an accessory bronchus originating directly from the trachea (known as pig bronchus, Figure 9.3a), so that this tracheal

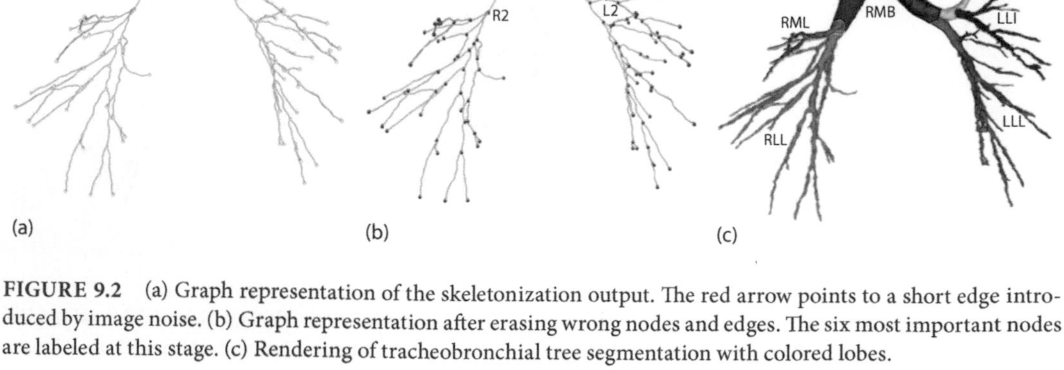

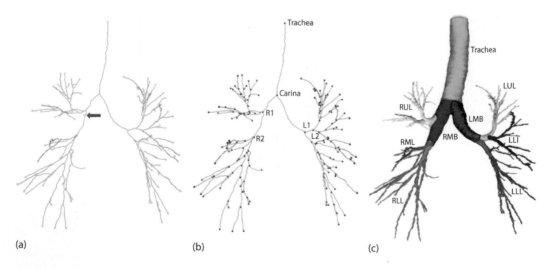

FIGURE 9.2 (a) Graph representation of the skeletonization output. The red arrow points to a short edge introduced by image noise. (b) Graph representation after erasing wrong nodes and edges. The six most important nodes are labeled at this stage. (c) Rendering of tracheobronchial tree segmentation with colored lobes.

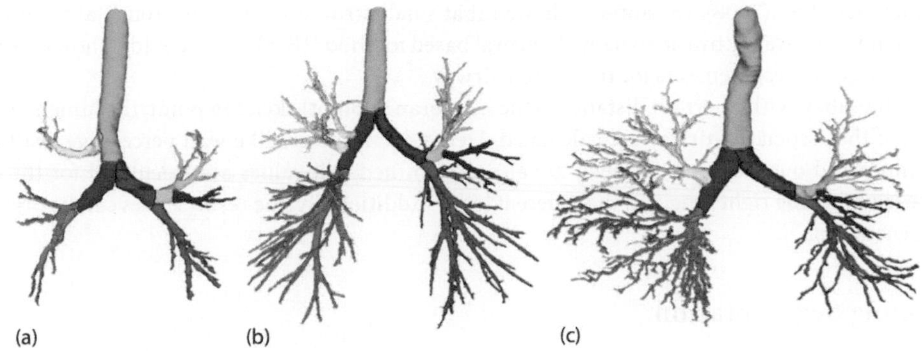

FIGURE 9.3 Tracheobronchial tree variations: (a) Tracheal bronchus (yellow) originating directly from the trachea, cystic fibrosis (CF) patient, (b) accessory bronchus (yellow) originating from the right main bronchus, CF patient. (c) Idiopathic pulmonary fibrosis (IPF) patient.

bronchus will not be assigned to the right upper lobe. The bronchial tree in Figure 9.3b has an accessory bronchus originating from the right main bronchus to the right upper lobe, correctly labeled by the software. Also, different pathological changes increase the difficulty, especially for segmentation and the labeling tasks. Figure 9.3c shows the segmented and labeled tree of an IPF patient.

9.4.3 Airway Tree Analysis

The direction vector of an airway can be determined by the graph representation of the airway tree. Hence, it is possible to calculate an orthogonal plane for every airway and, more precisely, for every skeleton point between two branching. In Figure 9.4a, an orthogonal plane through a subsegmental airway in segment 1 of the RUL is displayed. Virtual rays are radiated from the center of the airway in the calculated orthogonal plane. The airway wall can be determined on the density profiles corresponding to the rays. Figure 9.4b shows the inner (green) and outer (red) airway wall border calculated with the simple full-width-at-half-maximum (FWHM) method for the bronchus chosen in Figure 9.4a. A single profile is plotted in Figure 9.4c, and inner and outer border are marked by vertical lines. The estimation of the airway wall border is hampered by the spread-point function of all CT scanners: CT over-estimates the size of small objects and under-estimates the density (Dougherty and Newman 1999). In the article

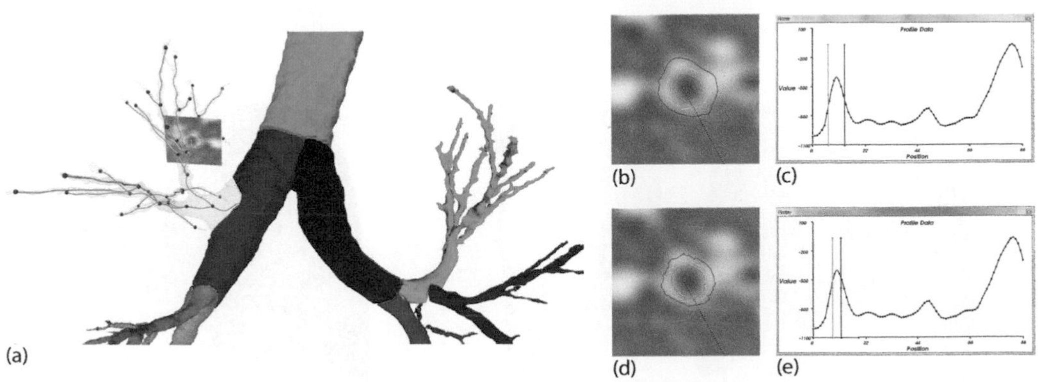

FIGURE 9.4 Determination of the airway geometry. (a) Generated orthogonal plane to subsegmental airway in segment 1 of RUL. (b) Airway measured with FWHM, inner border in green, outer border in red, the HU peaks are marked with yellow dots. (c) The HU profile of the blue dotted line. (d,e) Output of the IBM measurement corresponding (b,c).

by Weinheimer et al. (2008), the authors showed that small structures such as bronchial walls can be measured more accurately by the so named integral based method (IBM). Figure 9.4d,e shows the visual output of the IBM measurements for the chosen airway.

For each usable (with a certain distance to the next branch point) skeleton point, the lumen area and wall area of the associated airway are calculated. Derived parameters like wall percentage, wall thickness or inner and outer airway diameter are also determined. All values are calculated for the entire lung, the left lung, the right lung and the different lobes. Additionally, the result values can be organized by tree generation.

9.4.4 Lung Segmentation

Segmentation and labeling of the tracheobronchial tree are always performed before the segmentation of the lungs. Hence, the position of the main bifurcation of the trachea is known when lung segmentation starts. Body segmentation is performed at the slice position of the main bifurcation. The left and right lungs are searched as low density structures inside the detected body, not marked as bronchial tree previously. If the left or right lungs are not found on this slice, adjacent slices are examined. The method stops if the slice containing the landmark "trachea" is reached in cranial direction. If the search fails in cranial direction, the same number of slices is examined in caudal direction. Then, two independent threshold-based-region-growing algorithms are started, one in the left and one in the right lung, and an optimal thresholding value is defined by Otsu's method. Voxels marked as bronchial tree are excluded. Extra-corporeal voxels are not taken into account when calculating the optimal threshold value. Exclusion of the extra-corporeal air generally leads to a higher optimal threshold and to better segmentation results. A 2D hole-filling algorithm is applied on the region-growing result in order to include high density structures. After this algorithm, mostly all high-density structures surrounded by low-density lung parenchyma are included in the segmentation. Smaller bronchi completely surrounded by lung parenchyma are also marked as lung. If the two region-growing processes deliver just one connected lung region (Figure 9.5a), a fast-marching minimal path extraction method (Mueller et al. 2008) is applied on the axial images to divide the connected lung regions in left and right lung afterward, see Figure 9.5b.

The optional last step in the lung segmentation procedure is performed in order to include fibrotic changes, sub-pleural abnormalities and concavities to the lung. Therefore, the lung segmentation is dilated by a one voxel thick layer and then optionally a morphological closing with a spherical kernel with diameter 10 mm is applied.

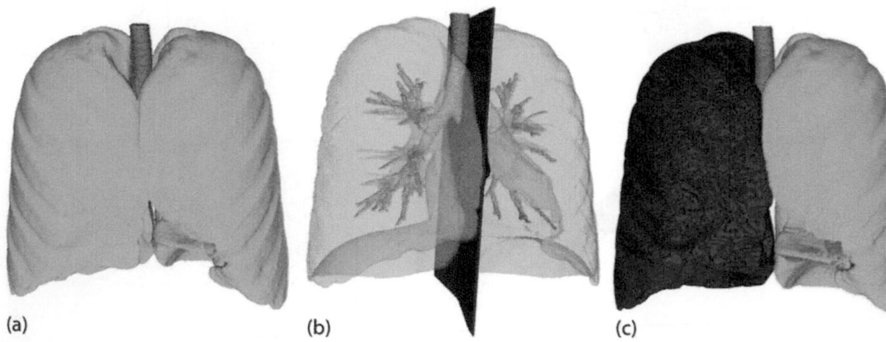

FIGURE 9.5 Lung segmentation steps. (a) Two region-growing processes delivered just one connected lung region. (b) Constructed separating plane between left and right lung. (c) Final result of the lung segmentation procedure: Left and right lung labeled.

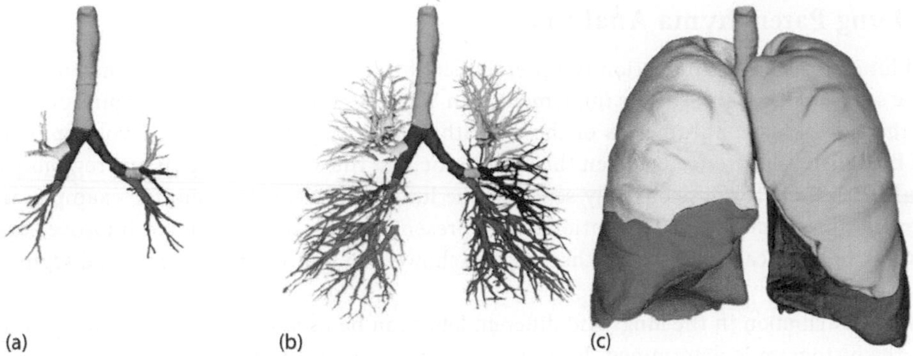

FIGURE 9.6 Steps of lobe segmentation. (a) Labeled tracheobronchial tree. (b) Vessels added to the different lobes, airways and vessels in the same lobe are displayed in the same color. (c) Final result of lobe segmentation algorithm (inspiration scan).

9.4.5 Lung Lobe Segmentation

The lobe segmentation algorithm presupposes a labeled tracheobronchial tree, and the procedure is inspired by the methods described by Lassen et al. 2013.

First, vessels are segmented using a threshold-based algorithm with an adjusted threshold for each CT. Subsequently, 3D-connected vessel objects are generated from the vessel segmentation result. Hereafter, the convex hulls of the bronchi within one lobe are determined and added to the corresponding lobes. Then, vessels are iteratively assigned to the different lobes according to various distance measures and rules (see Figure 9.6b).

Next, the lobe fissures are enhanced in the CT images. This is achieved by calculating the Hessian matrix H followed by the determination of the eigenvalues λ_0, λ_1 and λ_2. The facts that, for locations on the fissure, the most significant eigenvalue $\lambda_0 \ll 0$ and $\lambda_1, \lambda_2 \approx 0$ and that the eigenvector corresponding to λ_0 is orthogonal to the fissure plane are used to enhance the fissures in the images (Wiemker et al. 2005).

The final step uses airway, vessel and fissure information to determine the different lung lobes. Lobe masks are generated by a distance measure to the labeled bronchi and vessels. These initial lobe masks are eroded by some voxel layers, and afterward, the enhanced fissure information is used as input for a watershed transformation. The watershed transformation increases the eroded overlap masks exactly to the lobe fissures, if these are visible on the CT images. Figure 9.6c shows 3D renderings of the lobe segmentation results in an inspiratory scan of a COPD patient. Figure 9.7 displays lobe segmentation results for a CF patient, while Figure 9.7b demonstrates that the algorithm can deliver a high accuracy if the fissures are clearly visible in the CT images.

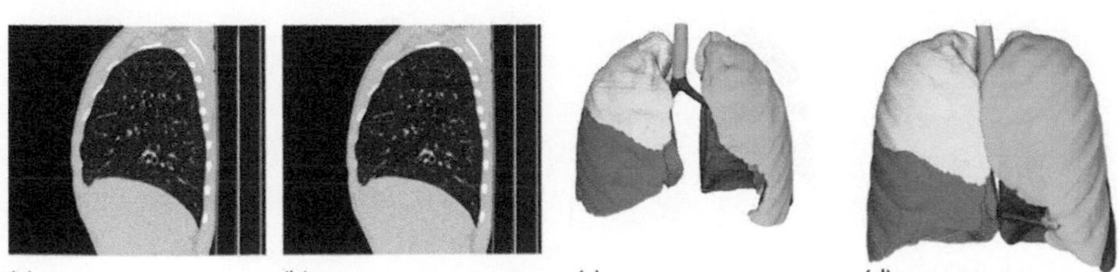

FIGURE 9.7 (a) Sagittal view of a right lung in an inspiratory MDCT scan, in (b) the major and minor fissures are marked by automatic lobe segmentation. (c) Labeled lobes in expiration and (d) labeled lobes in inspiration. All images from the same patient.

9.4.6 Lung Parenchyma Analysis

A good lung and lobe segmentation is a prerequisite for accurate lung parenchyma analysis. Lung and lobe segmentation algorithms implemented in YACTA can be applied on inspiratory and expiratory images, because parameters of the algorithms are adapted to each CT volume. If there is almost no density difference between the lung tissue and the surrounding tissue, region-growing-based algorithms will not accurately segment the lungs. This can happen, for example, in lungs with severe fibrotic changes. The optional step presented in Section 9.5.4 can improve segmentation quality in these cases. However, the software showed, in many studies, very good segmentation results.

Density distribution in the lungs and different lobes can be used for parenchyma analysis. For this reason, the histogram is determined. It measures the frequency with which various gray levels occur in the segmented lung parenchyma. The most important density measurements in inspiratory scans are determined by histogram analysis. The emphysema index is the percentage of voxels with densities lower than a certain threshold (e.g., <−950 HU) and is correlated with the amount of emphysema (see Figure 9.8c,d). The relative position of emphysema regions is determined by measuring the distance to the lobe borders; thereby, the software can distinguish between peripheral and central emphysema. The mean lung density (MLD) is decreased for COPD patients with emphysema and higher for patients with fibrotic changes compared to normal subjects. Percentile values are calculated, the ith percentile is the value below which i percent of the gray values may be found. The 5th percentile is related to the amount of emphysema, 40th and 80th can describe changes in the lung parenchyma of patients with IPF (Colombi et al. 2015).

Expiratory scans can be used for the quantification of air-trapping, when some important measures are calculated. Exp-856 is the percent of the lung voxel <−856. This has been called "percent gas trapping" in some publications. If inspiratory and expiratory images were acquired, air-trapping can be measured by further measures (Hersh et al. 2013): E/I MLA is the expiratory to inspiratory ratio of the mean lung attenuation. The relative volume change, RVC856-950, is defined as the difference between the expiratory and inspiratory values for relative lung volumes, corresponding to lung volume with attenuation between −856 and 950 HU divided by the lung volume without emphysema. Goris et al. 2003 described three thresholds for the definition of air-trapping, the size of the defect areas is expressed as a fraction of the whole analyzed parenchyma region. A1 represents defects on the basis of liberal criteria, while A3 represent defects on the basis of stringent criteria (see Figure 9.8a,b).

All values are calculated for the whole lung, right lung, left lung and for the different lobes. The lobe-based analysis of the airways and the lung parenchyma are important possibilities to describe the severity of airway diseases—at a single time point and over time.

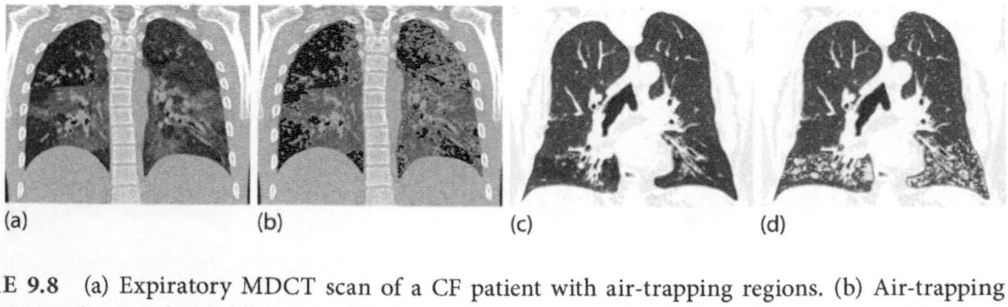

FIGURE 9.8 (a) Expiratory MDCT scan of a CF patient with air-trapping regions. (b) Air-trapping regions marked: A3 red, A2 red and blue, A1 red, blue and green. (c) Inspiratory MDCT scan of a COPD patient with emphysema in the lower parts of both lungs. (d) Emphysema marked in yellow.

Conclusion

In this text, we presented an updated review of both clinical and technical aspects related to CAD and QCT in the evaluation of diffuse lung diseases, focusing on emphysema, airway diseases such as asthma and CF, and also fibrosing ILD.

As discussed, QCT has been mostly used to quantify the features of COPD, specifically emphysema, air trapping and bronchi abnormalities. For quantification of emphysema, the density mask technique is most widely used, with a threshold of −950 HU. On expiratory CT, air trapping may be quantified by evaluating the percentage of lung volume less than a given threshold (mostly <856 HU) and by comparing lung volumes and attenuation on expiration and inspiration. CT parameters correlate well with airway obstruction, PFTs and other clinical aspects of the diseases. 3D model of the airways from volumetric MDCT made it possible to measure dimensions (area, diameters, airway wall thickness and attenuation) of segmental and subsegmental airways. These parameters also correlate with the severity of airflow obstruction, clinical history of COPD exacerbations and may also help in evaluation of other airway diseases such as asthma and CF.

QCT of ILD has been used to classify the types of interstitial abnormalities, help the diagnostic process, determine the extent of disease, and detect subtle changes in volume, density and character of abnormalities in longitudinal evaluation of patients. Novel softwares that use histogram-based or texture analyses have been used to characterize and quantify ILD features. Even if experience with these tools are, so far, not as large as for QCT in evaluation of emphysema, there are several studies showing they can be used in clinical practice to help identify a probable diagnosis, stratify prognosis in early disease, and consistently determine progression of the disease or response to therapy.

References

Achenbach T, Weinheimer O, Biedermann A, et al. 2008 MDCT assessment of airway wall thickness in COPD patients using a new method: correlations with pulmonary function tests. *Eur Radiol* 18:2731–8.

Aysola RS, Hoffman EA, Gierada D, et al. 2008. Airway remodeling measured by multidetector CT is increased in severe asthma and correlates with pathology. *Chest* 134:1183–91.

Aziz ZA, Wells AU, Desai SR, et al. 2005. Functional impairment in emphysema: contribution of airway abnormalities and distribution of parenchymal disease. *AJR Am J Roentgenol* 185:1509–15.

Bankier AA, De Maertelaer V, Keyzer C, Gevenois PA. 1999. Pulmonary emphysema: subjective visual grading versus objective quantification with macroscopic morphometry and thinsection CT densitometry. *Radiology* 211:851–58.

Best AC, Lynch AM, Bozic CM, et al. 2003. Quantitative CT indexes in idiopathic pulmonary fibrosis: relationship with physiologic impairment. *Radiology* 228:407–14.

Best AC, Meng J, Lynch AM, et al. 2008. Idiopathic pulmonary fibrosis: physiologic tests, quantitative CT indexes, and CT visual scores as predictors of mortality. *Radiology* 246:935–40.

Boehm HF, Fink C, Attenberger U, Becker C, Behr J, Reiser M. 2008. Automated classification of normal and pathologic pulmonary tissue by topological texture features extracted from multi-detector CT in 3D. *Eur Radiol* 18:2745–55.

Brillet PY, Grenier PA, Fetita CI, et al. 2013. Relationship between the airway wall area and asthma control score in moderate persistent asthma. *Eur Radiol* 23:1594–602.

Brown MS, Kim HJ, Abtin FG, et al. 2012. Emphysema lung lobe volume reduction: effects on the ipsilateral and contralateral lobes. *Eur Radiol* 22:1547–55.

Colombi D, Dinkel J, Weinheimer O, et al. 2015. Visual vs fully automatic histogram-based assessment of idiopathic pulmonary fibrosis (IPF) progression using sequential multidetector computed tomography (MDCT). *PLoS ONE* 10:e0130653.

Coxson HO, Leipsic J, Parraga G, Sin DD. 2014. Using pulmonary imaging to move chronic obstructive pulmonary disease beyond FEV1. *Am J Respir Crit Care Med* 15;190:135–44.

Coxson HO, Nasute Fauerbach PV, Storness-Bliss C, et al. 2008. Computed tomography assessment of lung volume changes after bronchial valve treatment. *Eur Respir J* 32:1443–50.

Depeursinge A, Chin AS, Leung AN, et al. 2015. Automated classification of usual interstitial pneumonia using regional volumetric texture analysis in high-resolution computed tomography. *Invest Radiol* 50:261–7.

Dougherty G, Newman D. 1999. Measurement of thickness and density of thin structures by computed tomography: a simulation study. *Med Phys* 26:1341–8.

Goris ML., Zhu HJ, Blankenberg F, Chan F, Robinson TE. 2003 An automated approach to quantitative air trapping measurements in mild cystic fibrosis. *Chest* 123;5:1655–63.

Gu S, Wang Z, Siegfried, JM, Wilson D, Bigbee WL, Pu J. 2012. Automated lobe-based airway labeling. *International Journal of Biomedical Imaging* 2012:382806.

Gurney JW, Jones KK, Robbins RA, et al. 1992. Regional distribution of emphysema: correlation of highresolution CT with pulmonary function tests in unselected smokers. *Radiology* 183:457–63.

Han MK, Kazerooni EA, Lynch DA et al. 2011. Chronic obstructive pulmonary disease exacerbations in the COPD Gene study: associated radiologic phenotypes. *Radiology* 261:274–82.

Hartley PG, Galvin JR, Hunninghake GW, et al. 1994. High-resolution CT-derived measures of lung density are valid indexes of interstitial lung disease. *J Appl Physiol (1985)* 76:271–7.

Hasegawa M, Nasuhara Y, Onodera Y, et al. 2006. Airflow limitation and airway dimensions in chronic obstructive pulmonary disease. *Am J Respir Crit Care Med* 15;173:1309–15.

Hersh CP, Washko GR, Estépar RSJ, Lutz S, Friedman PJ, Han MK et al. 2013. Paired inspiratory-expiratory chest CT scans to assess for small airways disease in COPD. *Respiratory Research* 14:42.

Hoshino M, Ohtawa J. 2012. Effects of budesonide/formoterol combination therapy versus budesonide alone on airway dimensions in asthma. *Respirology* 17:639–46.

Iwasawa T, Asakura A, Sakai F, et al. 2009. Assessment of prognosis of patients with idiopathic pulmonary fibrosis by computer-aided analysis of CT images. *J Thorac Imaging* 24:216–22.

Iwasawa T, Ogura T, Sakai F, et al. 2014. CT analysis of the effect of pirfenidone in patients with idiopathic pulmonary fibrosis. *Eur J Radiol* 83:32–8.

de Jong PA, Nakano Y, Hop WC, et al. 2005. Changes in airway dimensions on computed tomography scans of children with cystic fibrosis. *Am J Respir Crit Care Med* 15;172:218–24.

Kauczor HU, Heitmann K, Heussel CP, Marwede D, Uthmann T, Thelen M. 2000. Automatic detection and quantification of ground-glass opacities on high-resolution CT using multiple neural networks: comparison with a density mask. *AJR Am J Roentgenol* 175:1329–34.

Kauczor HU, Heussel CP, Fischer B, Klamm R, Mildenberger P, Thelen M. 1998. Assessment of lung volumes using helical CT at inspiration and expiration: comparison with pulmonary function tests. *AJR Am J Roentgenol* 171:1091–5.

Kim WJ, Hoffman E, Reilly J, et al. 2011. Association of COPD candidate genes with computed tomography emphysema and airway phenotypes in severe COPD. *Eur Respir J* 37:39–43.

Kinsella M, Müller NL, Abboud RT, Morrison NJ, DyBuncio A. 1990. Quantitation of emphysema by computed tomography using a "density mask" program and correlation with pulmonary function tests. *Chest* 97:315–21.

Klette, Gisela. 2003. Simple points in 2D and 3D binary images. *Computer Analysis of Images and Patterns*. Springer Berlin Heidelberg.

Koenigkam-Santos M, Cruvinel DL, Menezes MB, et al. 2016. Quantitative computed tomography airways analysis in patients with cystic fibrosis using automated software: severity evaluation in correlation with spirometry. *Radiol Bras* 49:351–357.

Koenigkam-Santos M, Paula WD, Gompelmann D, et al. 2013a. Endobronchial valves in severe emphysematous patients: CT evaluation of lung fissures completeness, treatment radiological response and quantitative emphysema analysis. *Radiol Bras* 46:15–22.

Koenigkam-Santos M, de Paula WD, Owsijewitsch M, et al. 2013b. Incomplete pulmonary fissures evaluated by volumetric thin-section CT: semi-quantitative evaluation for small fissure gaps identification, description of prevalence and severity of fissural defects. *Eur J Radiol* 82:2365–70.

Lassen B, van Rikxoort EM, Schmidt M, Kerkstra S, van Ginneken B, Kuhnigk JM. 2013. Automatic Segmentation of the Pulmonary Lobes From Chest CT Scans Based on Fissures, Vessels, and Bronchi. *IEEE TMI* 32;2:210–22.

Lederlin M, Laurent F, Portron Y, et al. 2012. CT attenuation of the bronchial wall in patients with asthma: comparison with geometric parameters and correlation with function and histologic characteristics. *AJR Am J Roentgenol* 199:1226–33.

Maffessanti M. and Dalpiaz G. 2006. *Diffuse Lung Diseases: Clinical Features, Pathology, HRCT.* Milan: Springer-Verlag.

Martinez FJ, Foster G, Curtis JL, et al. 2006. Predictors of mortality in patients with emphysema and severe airflow obstruction. *Am J Respir Crit Care Med* 173:1326–34.

Mets OM, de Jong PA, van Ginneken B, et al. 2013. CT air trapping is independently associated with lung function reduction over time. *PLoS One* 16;8:e61783.

Montaudon M, Berger P, Cangini-Sacher A, et al. 2007. Bronchial measurement with three-dimensional quantitative thin-section CT in patients with cystic fibrosis. *Radiology* 242:573–81.

Montaudon M, Lederlin M, Reich S, et al. 2009. Bronchial measurements in patients with asthma: comparison of quantitative thin-section CT findings with those in healthy subjects and correlation with pathologic findings. *Radiology* 253:844–53.

Mueller D. 2008. Fast marching minimal path extraction in ITK. *The Insight Journal.* Chest. October 1988 94(4):782–7.

Muller NL, Staples CA, Miller RR, et al. 1988. "Density mask." An objective method to quantitate emphysema using computed tomography. *Chest* 94:782–7.

Nakano Y, Sakai H, Muro S, et al. 1999. Comparison of low attenuation areas on computed tomographic scans between inner and outer segments of the lung in patients with chronic obstructive pulmonary disease: Incidence and contribution to lung function. *Thorax* 54:384–9.

Niimi A, Matsumoto H, Amitani R, et al. 2000. Airway wall thickness in asthma assessed by computed tomography: Relation to clinical indices. *Am J Respir Crit Care Med* 162:1518–23.

Ostridge K, Williams N, Kim V, et al. 2016. Relationship between pulmonary matrix metalloproteinases and quantitative CT markers of small airways disease and emphysema in COPD. *Thorax* 71:126–32.

Palagyi K, Sorantin E, Balogh E and Kuba A. 2001. A sequential 3D thinning algorithm and its medical applications. *Proc. Inf. Process. Med. Imag. (IPMI 2001)* 2082:409–15.

Pauls S, Gulkin D, Feuerlein S, et al. 2010. Assessment of COPD severity by computed tomography: correlation with lung functional testing. *Clin Imaging* 34:172–8.

Saitoh T, Koba H, Shijubo N, Tanaka H, Sugaya F. 2000. Lobar distribution of emphysema in computed tomographic densitometric analysis. *Invest Radiol* 35:235–43.

Schroeder JD, McKenzie AS, Zach JA, et al. 2013. Relationships between airflow obstruction and quantitative CT measurements of emphysema, air trapping, and airways in subjects with and without chronic obstructive pulmonary disease. *AJR Am J Roentgenol* 201:W460–70.

Senéterre E, Paganin F, Bruel JM, Michel FB, Bousquet J. 1994. Measurement of the internal size of bronchi using high resolution computed tomography (HRCT). *Eur Respir J* 7;3:596–600.

Stockley RA, Parr DG, Piitulainen E, Stolk J, Stoel BC, Dirksen A. 2010. Therapeutic efficacy of α-1 antitrypsin augmentation therapy on the loss of lung tissue: An integrated analysis of 2 randomised clinical trials using computed tomography densitometry. *Respir Res* 5;11:136.

Travis WD, Costabel U, Hansell DM, et al. 2013. An official American Thoracic Society/European Respiratory Society statement: Update of the international multidisciplinary classification of the idiopathic interstitial pneumonias. *Am J Respir Crit Care Med* 15;188:733–48.

Walsh SL, Calandriello L, Sverzellati N, Wells AU, Hansell DM, UIP Observer Consort. 2016. Interobserver agreement for the ATS/ERS/JRS/ALAT criteria for a UIP pattern on CT. *Thorax* 71:45–51.

Webb R. and Higgins CB. 2010. *Thoracic Imaging: Pulmonary and Cardiovascular Radiology*. Philadelphia, PA: LWW.

Weinheimer O, Achenbach T, Bletz C, Düber C, Kauczor HU, Heussel CP. 2008. About objective 3-d analysis of airway geometry in computerized tomography. *IEEE Trans. Med. Imaging* 27;1:64–74

Weinheimer O, Achenbach T, Buschsiewke C, Heussel CP, Uthmann T, Kauczor HU. 2003. Quantification and characterization of pulmonary emphysema in multislice-CT: A fully automated approach. *Medical Data Analysis, Lecture Notes in Computer Science*, Vol. 2868, Springer-Verlag.

Weinheimer O, Achenbach T, Düber C.. 2009. Fully automated extraction of airways from CT scans based on self-adapting region growing. In *Proc. of the Second International Workshop on Pulmonary Image Analysis*. London, UK, September 20, 2009. MICCAI, pp. 315–321.

Wielpütz MO, Eberhardt R, Puderbach M, Weinheimer O, Kauczor HU, Heussel CP. 2014. Simultaneous assessment of airway instability and respiratory dynamics with low-dose 4D-CT in chronic obstructive pulmonary disease: A technical note. *Respiration; International Review of Thoracic Diseases*, 87;4:294–300.

Wielpütz MO, Eichinger M, Weinheimer O, et al. 2013a. Automatic airway analysis on multidetector computed tomography in cystic fibrosis: correlation with pulmonary function testing. *J Thorac Imaging* 28:104–13.

Wielpütz MO, Weinheimer O, Eichinger M, et al. 2013b. Pulmonary emphysema in cystic fibrosis detected by densitometry on chest multidetector computed tomography. *PLoS One* 21;8:e73142.

Wiemker R, Bülow T, Blaffert T. 2005. Unsupervised extraction of the pulmonary interlobar fissures from high resolution thoracic CT data. *International Congress Series*, 1281;0:1121–1126.

Yoon RG, Seo JB, Kim N, et al. 2013. Quantitative assessment of change in regional disease patterns on serial HRCT of fibrotic interstitial pneumonia with texture-based automated quantification system. *Eur Radiol* 23:692–701.

Zaporozhan J, Ley S, Eberhardt R, et al. 2005. Paired inspiratory/expiratory volumetric thin-slice CT scan for emphysema analysis: comparison of different quantitative evaluations and pulmonary function test. *Chest* 128:3212–20.

Zavaletta VA, Bartholmai BJ, Robb RA. 2007. High resolution multidetector CT aided tissue analysis and quantification of lung fibrosis. *Acad Radiol* 14:772–87.

10

Computerized Detection of Bilateral Asymmetry

Arianna Mencattini

Paola Casti

Marcello Salmeri

Rangaraj M. Rangayyan

In this chapter, the challenge of detecting bilateral asymmetry in mammograms is discussed, along with the related difficulties encountered by radiologists. Previous work dealing with automatic detection of asymmetric findings between the left and right breasts of a patient is described, stressing the reason for searching new algorithmic solutions. The fundamental notion of bilateral masking, based on the paradigm by Tabár, will be recalled and framed in the protocol described to improve the detection of asymmetry between the two breasts of a patient. Novel methods, which have been recently published in the literature, based on the analysis of semi-variance and on the quantification of structural similarity, will be described and applied to publicly available mammography databases. An overview of the results achieved by the presented methodology and of the comparative analysis performed will be illustrated and discussed. Future research directions inspired by the presented results will be debated at the end of the chapter. In detail, Section 10.1 will introduce the reader to bilateral asymmetry and breast cancer. In Section 10.2 we will give an overview of the work found in the literature on this topic. In Section 10.3 we will describe the two publicly available databases that we used for validating the approach. Section 10.4 will give an insight of the methods developed. Finally, Section 10.5 will present the results of classification and validation procedures and Section 10.6 will provide a discussion on a few challenging aspects.

10.1 Introduction

The routine assessment of a mammographic examination includes diligent comparison of each mammogram with the contralateral mammogram of the same view, displayed side by side. As a consequence, pairing of anatomical structures in the left and the right breasts is fundamental for accurate detection of asymmetric regions and planning a suitable follow-up. The importance of detecting bilateral asymmetry as an early sign of breast cancer has grown since 2007, when the mammographic descriptor *developing asymmetry* was proposed [1,2] and later included in the fifth edition of the Breast Imaging Reporting and Data System (BI-RADS) atlas [3]. A mammographic developing asymmetry is defined as a focal asymmetry that either is new or has increased in size or conspicuity compared with images from previous examinations [1]. Therefore, to understand the developing asymmetry, one must first understand bilateral asymmetry, which is still a vague concept for Computer-Assisted Detection (CADe) systems [4]. Automated methods for quantification of asymmetry as part of a CADe system can improve the accuracy of interpretation of mammograms and assist radiologists in the reporting process. The immediate consequence is that the efficacy of breast cancer screening programs can be improved. When masses, microcalcifications, and architectural distortion are not visible in the breast, the identification of asymmetric findings in a given pair of mammograms may be the only clue to breast disease that is detectable on standard mammographic projections [5]. Bilateral asymmetry has proved to be an indicator of increased risk of developing breast cancer [6–8], stressing the importance of special surveillance and follow-up observations of the patients to establish the nature of the verified asymmetry.

During the customary comparison of the left and right breasts, a greater area of tissue with fibro-glandular density found in one of the two breasts is reported as an asymmetric finding, either local or global. Asymmetric findings on mammograms may indicate a developing or underlying mass. They can be subtle in presentation and, hence, overlooked or misinterpreted by radiologists.

The difficulty with the detection of asymmetry arises because the bilateral anomalies caused by a developing or underlying pathological process need to be differentiated from the physiological differences between the two breasts and distortions due to projection artifacts. These confounding factors worsened by the fact that no mass, calcifications, or architectural distortion may be present and can cause overlooking or misinterpretation, even by experienced radiologists [9]. Clinical studies have reported that asymmetry accounts for 3%–9% of breast cancer cases incorrectly reported by radiologists, showing no evidence of a tumor [10]. Evidence also suggests that asymmetric distribution of fibro-glandular density is a common source of false-positive diagnosis [11].

The development of accurate methods for detection of bilateral asymmetry as part of a CADe system for mammography may lead to the achievement of two different, yet equally valuable objectives. First, it assists radiologists in a more accurate reporting process [12]. Second, in conjunction with direct risk factors such as aging and family history, it can be used to optimize screening programs and prevention plans by assessing the near-term individualized risk for having or developing breast cancer [13,14].

10.2 Related Work

Over the last two decades, many researchers have been addressing detection of signs of breast cancer via computerized analysis of mammograms [15–17]. Some of the reported work used asymmetry between the right and the left mammograms as a strategy to improve the accuracy of CADe systems in detecting masses or microcalcifications [18–22]. Only a limited number of studies have addressed the detection of asymmetry as a distinctive sign for diagnosis [23–29].

The existing approaches can be divided into two categories: direct and indirect methods. Indirect ones stage the paradigmatic situation of quantifying asymmetry as the difference of extracted descriptors. On the other hand, only a few approaches try to compare the left and right breast tissue, directly producing an index of asymmetry. Concerning indirect approaches, the better the descriptors represent the difference between the two mammographic densities, the better their difference will be representative of the level of asymmetry. However, the complexity of breast tissue makes it difficult to quantify such degree of asymmetry in terms of simple textural or pixel-intensity descriptors. On the other hand, the main difficulty in direct approaches is to design methods for comparative analysis of the two breasts that reach a sufficiently high sensitivity while being robust to artifact, ill-positioning, and physiological disparity.

In the context of indirect approaches, the earliest work on detection of bilateral asymmetry was performed by Lau and Bischof [24] who defined an asymmetry measure combining descriptors of roughness, brightness, and directionality. The scope of the work was to localize 13 suspicious asymmetric areas in a set of 10 asymmetric pairs of mammograms.

Miller and Astley [25] proposed a semiautomated method to classify pairs of mammograms as normal or abnormal using shape, texture, and density features. The procedure consisted of segmentation of mammograms into fat and non-fat regions, extraction of asymmetry measurements from the two regions, and classification. In order to obtain a proper match between corresponding areas of the two breasts of a subject, radiologists' annotations were used to segment the non-fat regions of the mammograms. With a set of 47 normal and 28 asymmetric pairs of mammograms, an accuracy of 76% was obtained using linear discriminant analysis (LDA) and leave-one-out (LOO) cross-validation.

The study conducted by Ferrari et al. [23] introduced the application of Gabor wavelets and rose diagrams to quantify differences in oriented textural patterns of mammograms. In a subsequent work [26], the same authors improved the accuracy of their method to 84.4% by including morphological and density features. Pairs of mediolateral oblique (MLO) views from the mini Mammographic Image Analysis Society (mini-MIAS) database [30] were used for validation of their results, of which 22 were normal, 14 asymmetric, and eight contained architectural distortion.

In the work of Tzikopoulos et al. [27], 114 differential features were tested on the whole mini-MIAS database, which includes 15 asymmetric pairs of mammograms; the remaining 146 pairs were treated as being symmetric, despite the presence of masses, microcalcifications, and architectural distortion. The reported accuracy was 84.5%.

The studies of Wang and colleagues [28,29] demonstrated the feasibility of applying a computerized scheme, based on the use of asymmetric density features, for the assessment of early signs of breast cancer. Their method was tested on 100 pairs of normal mammograms and 100 pairs of prior mammograms, and achieved an area under the receiver operating characteristic (ROC) curve (A_z) of 0.78 ± 0.02.

In 2011, the work by Mencattini et al. [31] was based on density analysis and detection of fibroglandular disc through adaptive clustering techniques, on implementation of algorithms for detection of bilateral asymmetries based on Gabor filters, and on the use of a Bayes classifier with the LOO method to assess the degree of asymmetry of the two breasts. Mammograms from the mini-MIAS database and from the digital database for screening mammography (DDSM) were used to test the methods.

More recently, a separate category of approaches has been proposed with the aim to improve automatic detection of bilateral asymmetry. Inspired by the work by Tabár [32], a preliminary masking procedure was applied in order to select paired restricted regions of the left and right breasts [33]. Such a scenario facilitated and optimized the comparison step, by focusing the attention of the algorithm on confined regions. The approach was demonstrated to reduce false positives and to improve sensitivity in detecting actual structural asymmetry. Moreover, the strategy provides a promising back-end procedure for the subsequent task of localizing focal asymmetric regions.

In the work of Casti et al. [34], the diversity between corresponding directional structures was quantified by applying measures of similarity to paired masked strips. A novel application of Moran's index [35] was designed to measure the angular covariance between rose diagrams related to the phase and magnitude responses of multidirectional Gabor filters. Moran's index is a measure of spatial autocorrelation based on the computation of the correlation in a signal among nearby locations in space. A set of 128 mammograms from the DDSM database, including 32 normal and 32 asymmetric pairs, was used to validate the procedure.

Sequentially, the studies of Casti et al. [36] presented a novel method to detect asymmetry in mammograms based upon bilateral analysis of the spatial distribution of density within paired mammographic strips. Various differential measures of spatial correlation of gray-scale values were computed with reference to the position of the nipple.

In the method developed by Casti et al. in [36], changes in structural information of the extracted paired regions were investigated using spherical semivariogram descriptors and correlation-based

TABLE 10.1 Summary of Indirect Methods and Performance Statistics for the Analysis of Bi-Lateral Mammographic Asymmetry

Authors	Dataset	Methods and Results
Miller and Astley [25]	75 MLO and mediolateral (ML) mammograms including 47 normal and 28 asymmetric pairs.	Six shape, brightness, and topology features extracted from manually marked regions of fibroglandular components and selected using the F-test; LDA classifier and LOO cross-validation: Accuracy = 76%.
Ferrari et al. [23]	80 MLO mammograms from mini MIAS [30] including 20 normal, 14 were asymmetric, and six architectural distortion cases.	Gabor wavelets, Karhunen-Loève transform, Otsu's method, rose diagrams, and three statistical features selected using exhaustive combination; QDA classifier and LOO cross-validation: Accuracy = 74.4%.
Rangayyan et al. [26]	88 MLO mammograms including 22 normal cases, 14 asymmetric cases, and eight architectural distortion cases from mini-MIAS [30].	Alignment of the phase responses of Gabor wavelets with reference to the corresponding pectoral muscle edges, four directional features selected using exhaustive combination; quadratic Bayesian classifier and LOO cross-validation: Accuracy = 84.4%.
Tzikopoulos et al. [27]	All of the 322 MLO mammograms from mini-MIAS [30] including 15 asymmetric cases; the remaining cases treated as symmetric.	Minimum cross-entropy thresholding, 18 differential first-order statistical features selected using the t-test; SVM classifier and LOO cross-validation: Accuracy = 85.7%.
Wang et al. [29]	800 randomly selected full-field digital mammograms, including CC and MLO views of 100 normal cases and 100 verified positive cases for having high-risk of developing breast cancer.	20 selected features including statistical, textural, and density features from automatically extracted regions of interest of CC views and the entire segmented breast areas of MLO views; genetic algorithms, ANN classifier, and leave-one-patient-out cross-validation: Area under the ROC curve of 0.78.
Tan et al. [56]	Full-field digital prior mammograms of 645 cases, including CC and MLO views of 362 normal cases and 283 verified positive cases for having high-risk of developing breast cancer.	9 differential density features in addition to woman's age selected using the sequential forward floating algorithm; SVM classifier and 10-fold cross-validation: Area under the ROC curve of 0.716.
Mencattini et al. [31]	22 normal and 22 abnormal pairs of mammograms including asymmetry and architectural distortions.	21 differential directional features computed using two rose diagrams; PCA and Bayes classifier: Sensitivity 0.82, Specificity 0.82.

structural similarity indices in the spatial and complex wavelet domains. The spatial distribution of gray-scale values, as well as of the magnitude and phase responses of multidirectional Gabor filters, were used to represent the structure of mammographic density and of the directional components of breast tissue patterns, respectively. All the available mammograms containing a reported bilateral asymmetry were used for test.

Tables 10.1 and 10.2 summarize methods (indirect and direct, respectively) for the analysis of bilateral asymmetry and achieved performance reported in previously published work. The state-of-the-art approaches suggest that automatic detection of asymmetry in mammograms can be achieved, but more effort is needed to devise new methods to improve performance levels and to progress toward clinical application.

The last study reported in Table 10.2 presents innovation with respect to the state-of-the-art approaches listed in Tables 10.1 and 10.2. Moreover, it is the result of fusion and optimization of previous results and, at present, represents an overview of how detection of bilateral asymmetry may be addressed by means of a CADe system. For this reason, the rest of this chapter will provide a detailed insight to the approach described by Casti et al. in [36] and of the results obtained through its application to available asymmetric mammograms when compared to normal cases.

TABLE 10.2 Summary of Direct Methods and Performance Statistics for the Analysis of Bilateral Mammographic Asymmetry

Authors	Dataset	Methods and Results
Casti et al. [34]	128 mammograms from DDSM, 32 pairs with asymmetric finding and 32 normal pairs randomly selected.	12 angular similarity indices extracted by extending the Moran's index to the comparison of phase and magnitude Gabor filter response; Fisher LDA classifier and Bayes QDA with LOPO cross-validation: $A_z = 0.84$ and $A_z = 0.82$, respectively.
Casti et al. [36]	188 mammograms (94 pairs), of which 128 (64 pairs) are from DDSM including CC and MLO projections of the two breasts of each subject, and 60 MLO mammograms (30 pairs) are from mini-MIAS (15 asymmetric and 15 normal pairs).	12 differential spherical semivariogram descriptors, nine correlation-based similarity indices, and three correlation-based descriptors in the wavelet domain; LDA, Bayes QDA, and ANN-RBF classifiers compared using LOO per patient, LOO per pair, two-fold per patient, two-fold per pair: Accuracy up to 94%.

Note: Linear discriminant analysis (LDA), quadratic discriminant analysis (QDA), support vector machine (SVM), artificial neural network with radial basis function (ANN-RBF), leave one out (LOO), leave one patient out (LOPO), craniocaudal (CC), principal component analysis (PCA).

10.3 Dataset of Mammograms with Asymmetry

The procedures for detection of bilateral asymmetry were tested by selecting all of the available cases with reported asymmetry. For this reason, images were taken from the two public database of screening mammography, the mini-MIAS [30] and the DDSM [37]. In addition, to allow future exhaustive comparative analysis, and with the need to include control cases, an equal number of normal cases in descending order starting from the first available normal case (i.e., images mdb003 and mdb004 for mini-MIAS and case 0002 of volume normal_01 for DDSM) were selected. The dataset of images thus obtained is composed of a total of 188 mammograms (94 pairs), of which 60 MLO mammograms (30 pairs) are from the mini-MIAS database and 128 (64 pairs) are from the DDSM database including CC and MLO projections of the two breasts of each subject. The asymmetric pairs, 47 in total, include 15 cases of asymmetry from the mini-MIAS database and 16 pairs of focal asymmetry and 16 pairs of global asymmetry from the DDSM database. The images have spatial resolution of 42.5, 43, 50, or 200 μm, and pixel depth of 8, 12, or 16 bits/pixel (bpp). Proven ground truth of the asymmetric findings was available for all of the cases. The approach was tested and validated on the whole dataset of mammograms (188 images), by implementing different cross-validation methodologies. All images from the two datasets were first down-sampled using bicubic interpolation to a spatial resolution of 300 μm to reduce the computational burden of the procedure and also to unify the resolution of images from different databases. This setting is related to the size of parenchymal dense tissue that is relevant in the analysis of asymmetry. Finer details could be related to noise and cannot improve the overall sensitivity of the approach; coarser pixel size would lead to loss of details.

10.4 A CADe System for the Detection of Bilateral Asymmetry

As introduced in the previous sections, the system developed by Casti et al. in [38] integrated an indirect approach based on the quantification of absolute differences between semivariogram descriptors, with a direct procedure based on the extraction of structural similarity indices via spatial cross-correlation metrics. Both the strategies were applied to paired regions of the left and right mammograms, thus providing a local characterization of bilateral asymmetry. In addition, a high-level comparison of the whole breast region was performed to tackle the complexity and heterogeneity of the appearance of bilateral asymmetry. The overall strategy can be viewed as a synergy of direct and indirect comparisons which

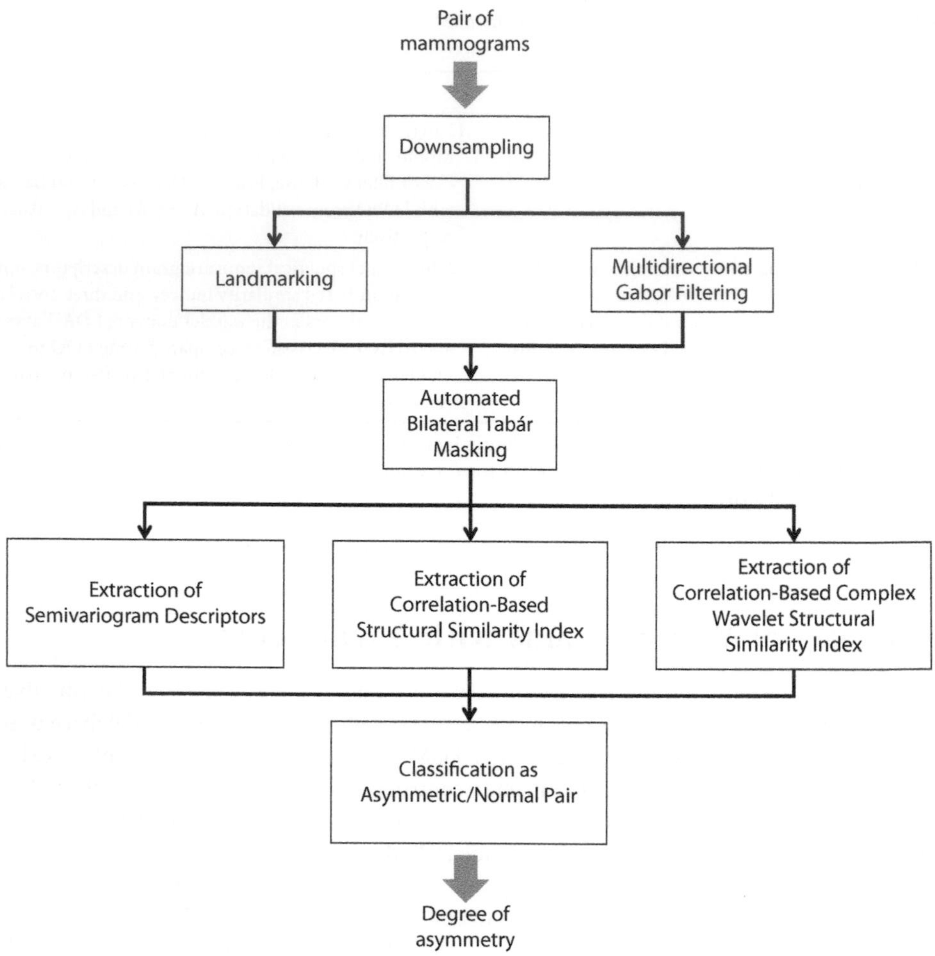

FIGURE 10.1 A flowchart of the approach for detection of bilateral asymmetry.

were performed at various levels of analysis with the aim of improving the capability of the system to extract the structural differences between paired mammograms.

More formally, the overall system was based on three main steps: computerized masking of the left and right mammograms of a patient via a landmarking algorithm, quantification of the differences between paired mammographic regions by means of measures of similarity or dissimilarity, and classification of pairs of mammograms as normal or asymmetric by means of the extracted features. Figure 10.1 provides a schematic illustration of the architecture of the method. Each step will be detailed in the following sections.

10.4.1 Tabár Masking Procedures

Tabár introduced guidelines for the use of bilateral masking procedures by radiologists [32,39]. The studies conducted by Tabár reinforce the underlying assumption that performing effective comparison between the two mammograms of a patient requires a localized analysis of anatomically matched regions. In this work, it was assumed that the structural information to be quantified is space-variant, thus requiring accurate matching of the areas under investigation. Following Tabár's criteria for interpretation of mammograms automated implementations of Tabár masking procedures have been presented in [33,36,38,40,41].

The procedure for the analysis of breast asymmetry requires effective matching between the regions of the two breasts that will be compared at each step. For this task, paired matching points, or landmarks, on the mammograms are required. The landmarks, which include the breast skin-line, the pectoral muscle (in MLO views), and the nipple, were detected by using previously developed algorithms [42–44]. The approaches were essentially based on the application of Gabor filters, phase propagation, and analysis of the gradient vector field. Such anatomical reference structures are fundamental for the implementation of Tabár-like bilateral masking, since they are also included in the radiological protocol for viewing.

As an example, a malignant asymmetric case from DDSM and a normal case from mini-MIAS are given in Figure 10.2a,d, respectively. The focal asymmetric region is outlined in red for the DDSM case. The detected landmarks along with the binary masks of the obtained breast regions and linear approximations of pectoral muscle edges are superimposed on the images. The green squares correspond to the positions of the nipple. For each pair of images, segmentation of paired strips was performed to derive corresponding mammographic regions, as follows:

Masking #1: Horizontal (for CC views) and annular (for MLO views) strips were segmented by dividing the line from the topmost to the lowest pixel inside the breast region (for CC views) and the perpendicular line from the nipple to the pectoral muscle (for MLO views) into eight equally spaced segments and annuli, as shown in Figure 10.2b,e, respectively.

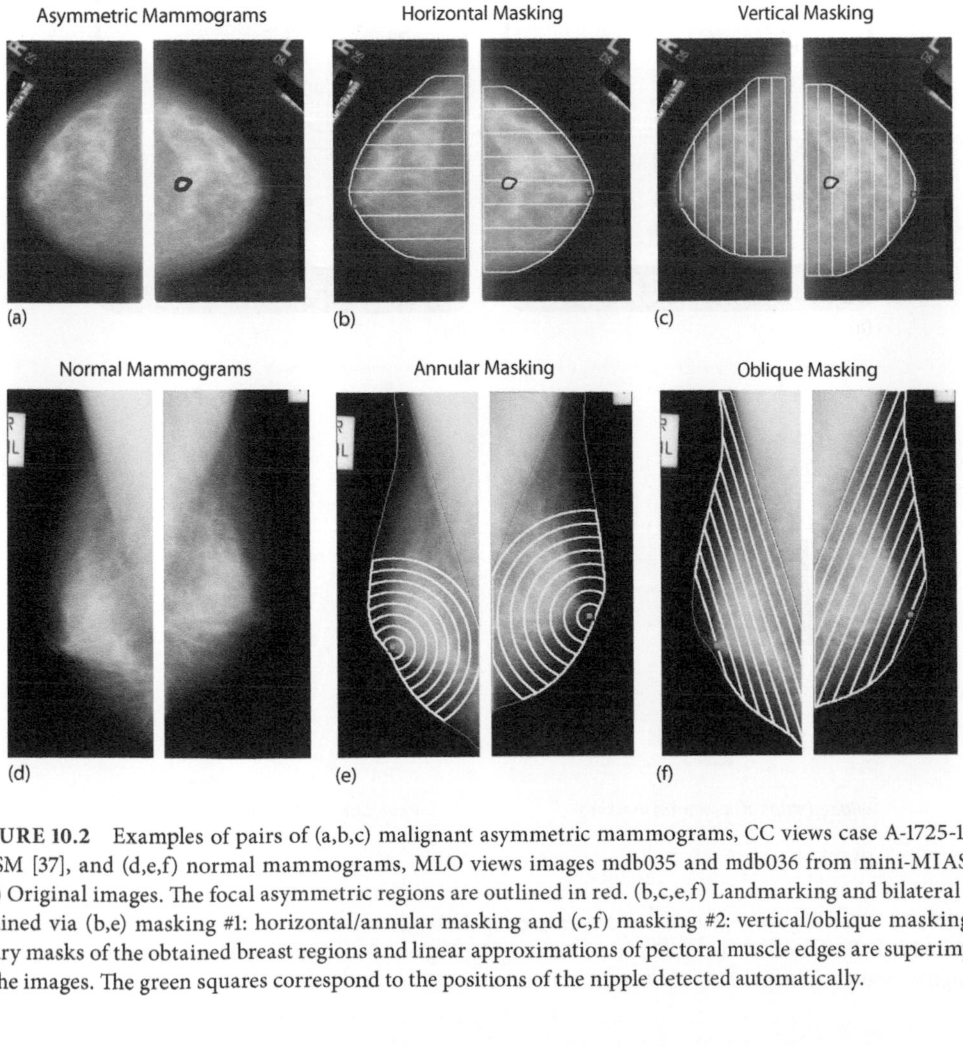

FIGURE 10.2 Examples of pairs of (a,b,c) malignant asymmetric mammograms, CC views case A-1725-1 from DDSM [37], and (d,e,f) normal mammograms, MLO views images mdb035 and mdb036 from mini-MIAS [30]. (a,d) Original images. The focal asymmetric regions are outlined in red. (b,c,e,f) Landmarking and bilateral strips obtained via (b,e) masking #1: horizontal/annular masking and (c,f) masking #2: vertical/oblique masking. The binary masks of the obtained breast regions and linear approximations of pectoral muscle edges are superimposed on the images. The green squares correspond to the positions of the nipple detected automatically.

Masking #2: Vertical (for CC views) and oblique (for MLO views) strips were segmented parallel to the chest wall and the pectoral muscle, respectively, by dividing the perpendicular line from the nipple to the chest-wall (for CC views) and the perpendicular line from the nipple to the pectoral muscle line (for MLO views) into eight equally spaced segments, as shown in Figure 10.2c,f.

In order to make computationally effective the extraction of similarity indices and facilitate the interpretation of the results for the user, rectangular regions were derived from the paired regions by extracting the largest rectangles enclosed in each segmented strip, as illustrated in Figure 10.3. Additionally, to maintain a radiologist-like approach to the bilateral matching, the method also included the whole breast region from which a rectangular central region was derived. The annular strips obtained via masking #1 for MLO views in Figure 10.2e were substituted by horizontal strips, as shown in Figure 10.3c. Each segmented strip extracted from a left mammogram was flipped left to right and paired with the corresponding strip in the contralateral mammogram, and used, in addition to the whole breast regions, for feature extraction.

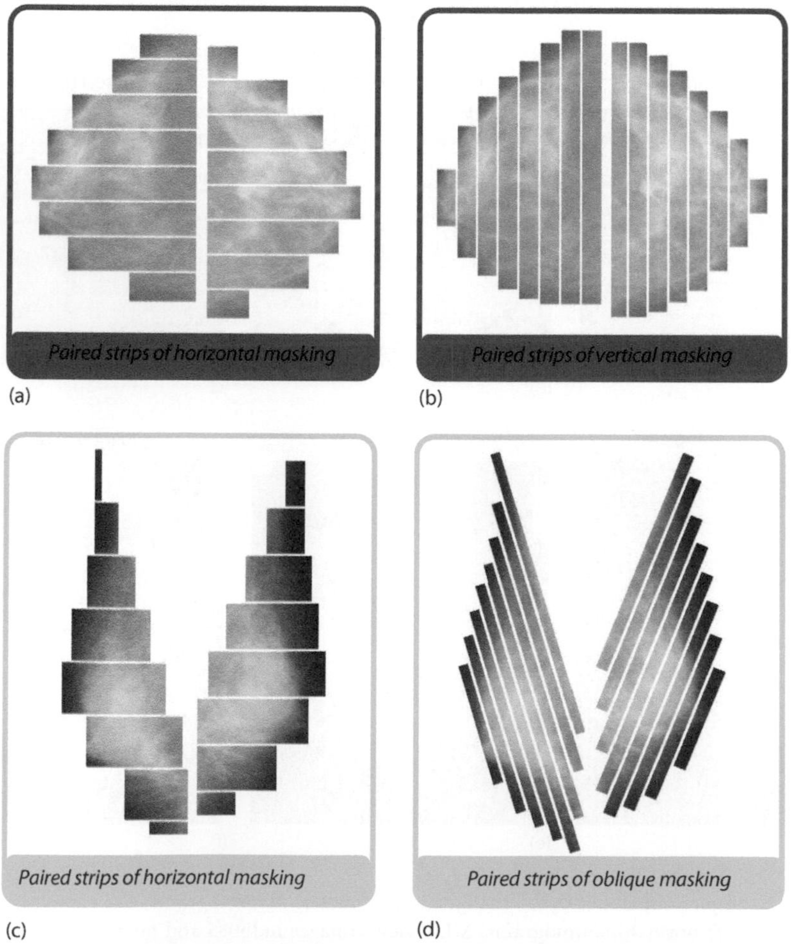

FIGURE 10.3 Rectangular regions derived by extracting the largest rectangles enclosed in each strip of the masking procedures illustrated in Figure 10.2. (a,c) Rectangular regions from masking #1: horizontal masking. (b,d) Rectangular regions from masking #2: vertical/oblique masking.

10.4.2 Characterization of Bilateral Asymmetry

As mentioned before, the procedure considered in this chapter can be seen as a hybrid approach halfway between a direct comparison of left and right breast tissue, with or without the application of a preliminary masking step, and an indirect evaluation through the computation of differences of features extracted separately from each mammogram. For the task of describing details of the method, we first provide a description of indirect evaluation, and will then get into the details of the direct procedure.

10.4.2.1 Indirect Methods

Analysis of the variogram [45] allows quantification of the degree of spatial dependence between samples. Ericeira et al. [18] used the empirical values of variogram and cross-variogram functions computed from pairs of windows of size 32×32 pixels to detect masses in mammograms. We hypothesize that characterization of the difference in the spatial and statistical characteristics of pixels between a given mammographic region and the contralateral matched one may provide measures of asymmetry.

The semivariogram, $\gamma(h)$, is defined for a given set of $N(h)$ pairs of pixels separated by a distance h as:

$$\gamma(h) = \frac{1}{2N(h)} \sum_{m=1}^{N(h)} \left[f(\mathbf{u}_{m,0}) - f(\mathbf{u}_{m,h}) \right]^2, \tag{10.1}$$

where $\mathbf{u}_{m,0}$ and $\mathbf{u}_{m,h}$ $m = 1,2,\ldots,N(h)$, are the vectors of spatial coordinates (x, y) of the $N(h)$ pairs of pixels, and f is the gray-scale level at the given spatial locations.

Semivariogram analysis was used to derive four spherical semivariogram descriptors as follows. First, the images were further down-sampled by a factor of five to a resolution of 1.5 mm/pixel, to reduce the computational cost. The maximum value of h to be investigated, h_{max}, was set equal to one-half of the maximum distance between pairs of pixels in each region. The range of distances from 0 to h_{max} was divided into 20 equally spaced bins, and pixel pairs were aggregated accordingly to estimate the empirical semivariogram γ_n for representative distances of each aggregate. Least-squares fitting of each empirical semivariogram was performed using a spherical structure function, as:

$$\hat{\gamma}(h) = \begin{cases} a + s\left(\dfrac{3h}{2r} - \dfrac{h^3}{2r^3}\right), & \text{if } h \leq r, \\ a + s, & \text{if } h > r, \end{cases} \tag{10.2}$$

where:

- *a*, the *nugget*, represents the discontinuity at the origin due to small-scale variations
- *s*, the *sill*, gives an estimate of the variance of pixels
- *r*, the *range* of influence of the spatial structure, corresponds to the distance at which the semivariogram stops increasing [46].

The spherical semivariogram functions obtained using the whole breast regions are illustrated for the asymmetric malignant case A-1725-1 of DDSM [37] and the normal case mdb035/mdb036 of mini-MIAS [30] in Figure10.4a,b, respectively; the nugget, the sill, and the range are also indicated. Separate anisotropic semivariograms, $\gamma(h_\alpha)$, were also computed for the $N(h_\alpha)$ pairs of gray-scale values separated by the lag distance vector, $\gamma(h_\alpha)$, oriented at the angle α, with $\alpha = 0°, 30°,\ldots, 180°$, to quantify the behavior of the autocorrelation structures in different directions of analysis. When the structural variations among pixel values are dependent on the direction of analysis, the behavior is referred to as geometric anisotropy, g, and can be quantified by the anisotropy ratio, as the range of the anisotropic semivariogram in the direction producing the longest range, divided by the range in the direction with the smallest range.

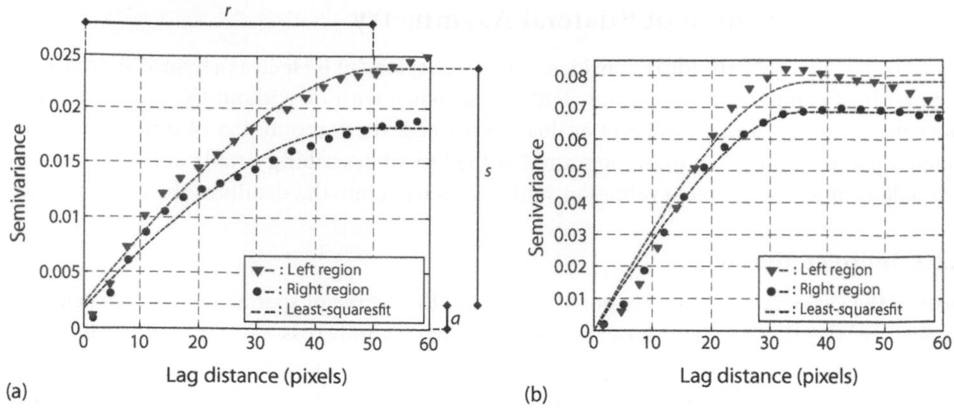

FIGURE 10.4 Examples of semivariograms computed for (a) the asymmetric malignant case A-1725-1 of DDSM [37] and (b) the normal case mdb035/mdb036 of mini-MIAS [30]. The empirical values and the corresponding spherical least-squares fits are shown for the left and right breast regions in red and blue, respectively. The nugget, a, the sill, s, and the range, r, are indicated.

The absolute differences between the four parameters derived from each region of the right mammogram and the corresponding parameters derived from the matching region of the left mammogram, $\Delta V_a = |a_R - a_L|$, $\Delta V_S = |s_R - s_L|$, $\Delta V_R = |r_R\, r_L|$, and $\Delta V_g = |g_R\, g_L|$, were used as spherical semivariogram descriptors. Higher values of the descriptors indicate lower similarity between the paired regions analyzed and are expected to represent asymmetric cases. The differential spherical semivariogram descriptor obtained for the asymmetric and normal case examples are: $\Delta V_a = 0.4 \times 10^{-3}$, $\Delta V_s = 4.2 \times 10^{-3}$, and $\Delta V_R = 4.2$ pixels, for the malignant asymmetric case and $\Delta V_a = 0.2 \times 10^{-10}$, $\Delta V_s = 9.5 \times 10^{-3}$, and $\Delta V_R = 1.0$ pixels, for the normal case. For the case of the features extracted via bilateral masking, the differential values obtained from the eight strips of each masking procedure were summed together [41].

10.4.2.2 Direct Methods

With the development of the structural similarity index (SSIM) and complex wavelet SSIM (CW-SSIM) index for image quality assessment [47,48], the key role of structural information in human image perception has been pointed out. Analysis of the structural content within images consists of quantifying patterns of distortion apart from the mean intensity and contrast of the image. The same approach can be effectively applied to quantify structural similarity between paired mammographic regions.

In order to perform comparisons at different levels of information, as a preliminary step, the two images were filtered using a set of $N = 18$ real Gabor filters, with $\tau = 6$ pixels and $l = 8$, oriented at different angles [49]. The filtering step allowed extraction of the directional components of the breast tissue patterns and subsequent quantification of the differences in the orientation of the structures of the breast parenchyma. The resulting images of magnitude and phase were used, in addition to the original gray-scale images, to derive the paired bilateral strips, as described in Section 10.4.1.

Due to the differences in size between the regions to analyze, we introduce a Correlation-Based Structural SIMilarity (CB-SSIM) index. Given the pair of right and left rectangular regions, x_R and y_L, of size $M \times N$ and $P \times Q$ pixels, respectively, the CB-SSIM index is defined as follows:

$$S(\mathbf{x}_R, \mathbf{y}_L) = \frac{\left(2\mu_R\mu_L + K_1\right)\left\{2\max\left[\mathrm{corr}(\mathbf{x}_R, \mathbf{y}_L)\right] + K_2\right\}}{\left(\mu_R^2 + \mu_L^2 + K_1\right)\left\{\max\left[\mathrm{corr}(\mathbf{x}_R, \mathbf{x}_R)\right] + \max\left[\mathrm{corr}(\mathbf{y}_L, \mathbf{y}_L)\right]\right\}}, \quad (10.3)$$

where μ_R and μ_L are the mean values of pixels within the right and left regions, respectively, and:

$$\text{corr}(\mathbf{x}_R,\mathbf{y}_L)= \sum_{m=1}^{M}\sum_{n=1}^{N}\{[\mathbf{x}_R(m,n)-\mu_R][\mathbf{y}_L(m+p,n+q)-\mu_L]\}, \tag{10.4}$$

with $-P+1 \le p \le M-1$ and $-Q+1 \le q \le N-1$, is the cross-correlation in the two-dimensional space between the right and left regions; $corr(\mathbf{x}_R, \mathbf{x}_R)$ and $corr(\mathbf{y}_L, \mathbf{y}_L)$ are the corresponding autocorrelation functions. As indicated in the work by Wang et al. [48], two small positive constants, K_1 and K_2, equal to 0.01 and 0.03, respectively, were added to the formulation of the index to improve its robustness. The CB-SSIM index is equal to the standard SSIM index, when the regions $\mathbf{x}_R$ and $\mathbf{y}_L$ have the same size, that is, if $N = M$ and $P = Q$, and the value of unity is achieved when the two regions are identical.

The concept of spatial structural similarity was extended to the complex wavelet domain by Sampat et al. [47] to achieve insensitivity to scale and image distortions that are not related to the actual differences in the structure of the images being compared. In particular, the rotation-invariance properties of steerable pyramid filters [50,51] have been shown to be effective in the computation of CW-SSIM [47]. This aspect is crucial in applications for mammography, and especially in bilateral mammographic comparison, where the distortions caused by compression and relative translation of the two breasts during the imaging procedure may cause false-positive results. Similarly to what we did in the spatial domain, we define a Correlation-Based Complex Wavelet SIMilarity index, CB-CW-SSIM, as follows:

$$\tilde{S}(\mathbf{c}_R,\mathbf{c}_L)=\frac{2\max_{p,q}\left\{\left|\sum_{i,j}\left[\mathbf{c}_R(i,j)\mathbf{c}_L^*(i+p,j+q)\right]\right|\right\}}{\sum_{i,j}|\mathbf{c}_R(i,j)|^2+\sum_{s,t}|\mathbf{c}_L(s,t)|^2}, \tag{10.5}$$

where $\mathbf{c}_R$ and $\mathbf{c}_L$ are the complex wavelet coefficients obtained, respectively, by decomposing the regions $\mathbf{x}_R$ and $\mathbf{y}_L$ with a 3-scale, 16-orientation steerable pyramid decomposition procedure [47,50,51]. The asterisk denotes the complex conjugate. Examples related to the computation of the correlation-based structural similarity indices, S and $\tilde{S}$, are shown in Figures 10.5 and 10.6 for the Gabor magnitude and phase responses of the central regions extracted from the malignant asymmetric and normal cases in Figure 10.2. The relative scale between the left and right regions has been preserved to illustrate size differences related to corresponding paired areas of the two breasts of a patient. The corresponding feature values obtained for the two cases are also provided. The corresponding structural similarity descriptor values are: $S_m = 2.2\times10^{-13}$ and $\tilde{S}_\Phi = 3.2\times10^{-6}$ in the spatial domain, $\tilde{S}_M = 2.2\times10^{-6}$ and $\tilde{S}_\Phi = 1.3\times10^{-6}$ in the complex wavelet domain, respectively, for the Gabor magnitude and phase responses of the asymmetric case. For the normal case, the corresponding structural similarity descriptor values are: $S_m = 3.9\ 10^{-12}$ and $\tilde{S}_\Phi = 1.4\times10^{-4}$ in the spatial domain, $\tilde{S}_M = 7.1\times10^{-6}$ and $\tilde{S}_\Phi = 7.1\times10^{-6}$ in the complex wavelet domain, respectively, for the Gabor magnitude and phase responses. Analysis of the normalized cross-correlation functions shown in Figures 10.5 and 10.6 and the related feature values indicate relatively low values of similarity for both cases. This is due to the inherent differences between the two breasts of a patient and to the additional dissimilarity introduced by compression and positioning of the breast during the mammographic examination. The magnitude and phase responses of the normal case in Figures 10.5 and 10.6, however, show more diffuse areas of higher cross-correlation and, as expected, higher values of the corresponding structural similarity features than the magnitude and phase responses of the asymmetric case.

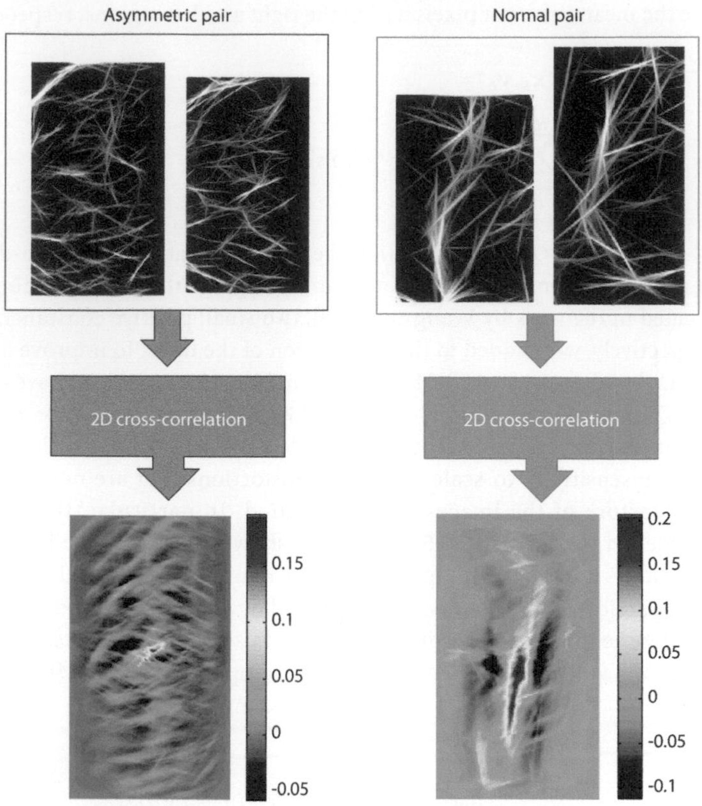

FIGURE 10.5 Analysis of cross-correlation of Gabor magnitude responses for the central regions extracted from the asymmetric case A-1725-1 of DDSM [37] and the normal case mdb035/mdb036 of mini-MIAS [30]. The left and right regions are illustrated, preserving the relative scale, together with the obtained normalized cross-correlation functions.

10.5 Results of Pattern Classification

In this section, we report results of the application of the described method for bilateral asymmetry detection. Features extracted are first evaluated in terms of their effectiveness in identifying the disparity (Section 10.5.1) between left and right mammograms in asymmetric pairs. Then, machine-learning approaches are trained and tested on the available datasets, and results of recognition are illustrated and compared (Section 10.5.2).

10.5.1 Performance of Individual Features

The individual classification performance of the features extracted from the bilateral regions was first analyzed in terms of area under the ROC curve, A_z, computed through the ROCKIT software package [52] for the two datasets of mammograms, mini-MIAS and DDSM. Performance measures were computed by implementing cross-validation within each dataset and also combining the two sets in a unique validation session. Automatic selection of the features in the training set of each experiment was implemented by stepwise logistic regression (SWR) [53] based on the F-statistic. In the feature selection step, different p-values of the F-statistic were selected, in correspondence to which diverse combinations of features were obtained. Only selected results among those achieved are reported in this work, together with the sets of the most frequently selected features. Standard machine-learning

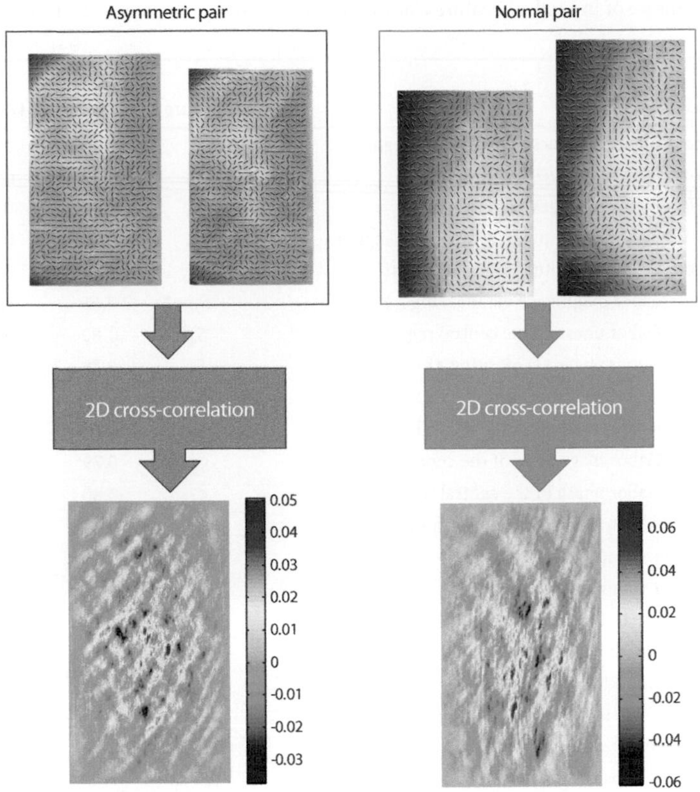

FIGURE 10.6 Analysis of cross-correlation of Gabor phase responses for the central regions extracted from the asymmetric case A-1725-1 of DDSM [37] and the normal case mdb035/mdb036 of mini-MIAS [30]. The left and right regions are illustrated, preserving the relative scale, together with the obtained normalized cross-correlation functions.

approaches were applied: LDA, the Bayes QDA classifier [54], and a two-layer ANN-RBF [55]. First, each pair of mammograms was analyzed individually; second, combination of the features extracted from the CC and MLO projections of the same patient was performed for the DDSM dataset to explore the performance of two-view analysis. The LOO per patient method was used for cross-validation of results. In addition, two-fold cross-validation was applied to the combined set of mammograms, DDSM + MIAS, including 47 asymmetric and 47 normal cases, to test the robustness of the described approach to increasing heterogeneity in the training set; the procedure was repeated 100 times and the results averaged over the repetitions. The A_z (and standard error, SE) and 95% confidence interval, $I_{95\%}$, values were also obtained. Sensitivity, specificity, and accuracy rates were computed at the operating point on the experimental ROC curve closest to the vertex (0, 1).

The classification performance of individual features in discriminating between normal and asymmetric pairs of mammograms is reported in terms of A_z in Table 10.3 for various datasets of images. Values higher than 0.5 indicate behavior according to expectation, that is in the case of the differential spherical semivariogram descriptors, ΔV, lower values for normal pairs, and higher values for asymmetric pairs. The opposite trend is expected, instead, in the case of the similarity indices, S and $\tilde{S}$. The obtained values of A_z indicate that all of the similarity indices follow the expected trend. The highest value of 0.88 was obtained by the CB-CW-SSIM index applied on the intensity values of the central regions of the two mammograms, $\tilde{S}_I$, for the DDSM dataset. Analysis of the results indicates that

TABLE 10.3 Performance of Individual Features for the Classification of Normal vs. Asymmetric Pairs of Mammograms.

	Description	Feature	MIAS	DDSM	DDSM + MIAS
CB-SSIM	From intensity values of the central region	S_{I1}	0.65	0.63	0.59
	From intensity values via masking #1	S_{I2}	**0.80**	0.64	0.69
	From intensity values via masking #2	S_{I3}	0.65	0.61	0.57
	From Gabor magnitude of the central region	S_{M1}	0.72	0.75	0.72
	From Gabor magnitude via masking #1	S_{M2}	0.75	0.58	0.63
	from Gabor magnitude via masking #2	S_{M3}	0.65	0.53	0.57
	From Gabor phase of the central region	S_{Φ}	**0.82**	**0.85**	0.84
	From Gabor phase via masking #1	S_{Φ}	0.78	0.70	0.72
	From Gabor phase via masking #2	S_{Φ}	0.77	0.79	0.79
CB-CW-SSIM	From intensity values of the central region	$\tilde{S}_I$	0.80	**0.88**	**0.85**
	From Gabor magnitude of the central region	$\tilde{S}_M$	0.79	**0.81**	**0.81**
	From Gabor phase of the central region	$\tilde{S}_{\Phi}$	0.80	0.84	**0.81**
Semi-vanogram	Nugget from the whole breast region	ΔV_{a1}	0.44	0.59	0.54
	Nugget via masking #1	ΔV_{a2}	0.43	0.69	0.60
	Nugget via masking #2	ΔV_{a3}	0.40	0.61	0.55
	Differential sill from the whole breast region	ΔV_{s1}	0.40	0.51	0.47
	Sill via masking #1	ΔV_{s2}	0.49	0.47	0.46
	Sill via masking #2	ΔV_{s3}	0.35	0.55	0.51
	Range from the whole breast region	ΔV_{r1}	0.56	0.57	0.56
	Range via masking #1	ΔV_{r2}	0.49	0.65	0.60
	Range via masking #2	ΔV_{r3}	0.72	0.64	0.65
	Geometrical anisotropy from the whole breast region	ΔV_{g1}	0.65	0.68	0.67
	Geometrical anisotropy via masking #2	ΔV_{g2}	0.29	0.54	0.47
	Geometrical anisotropy via masking #3	ΔV_{g3}	0.48	0.47	0.47

S and $\tilde{S}$: structural similarity descriptors. ΔV: differential semivariogram descriptors.
CB-SSIM: Correlation-Based Structural SIMilarity index
CB-CW-SSIM: Correlation-Based Complex Wavelet Structural SIMilarity index.
Masking #1: horizontal/annular masking. *Masking #2:* vertical/oblique masking.
Note: Results are given in terms of A_z for various datasets of images. Cases with $A_z > 0.8$ are in bold.

CW-SSIM and CB-CW-SSIM exhibit stronger discriminating ability than the spherical semivariogram descriptors.

The correlation-based similarity approach facilitates comparison between images that are inherently different in origin and size, while sharing some degree of similarity that needs to be quantified. In addition, such an approach avoids point-by-point comparison of the breast tissue; the drawbacks of which have been already discussed. In order to demonstrate quantitatively the improvement achieved by the correlation-based extension of the structural similarity index, we compared the performance obtained by the correlation-based descriptors with respect to the better-known structural similarity indices, SSIM [48] and CW-SSIM [47], applied to the various mammographic regions extracted as described in Section 10.4.1. Since point-by-point correspondence is required for the standard similarity analysis, regions of equal size were derived by removing the extra pixels along the x and y directions and then used for computation of the features. Table 10.4 summarizes the A_z values achieved by the SSIM and CW-SSIM indices in the classification of normal versus asymmetric pairs of mammograms. The results indicate a general decrease in the performance of the individual features, SSIM and CW-SSIM, with respect to the performance achieved with the CB-SSIM and CB-CW-SSIM indices, as summarized in

TABLE 10.4 Performance of the SSIM and CW-SSIM Indices in the Classification of Normal vs. Asymmetric Pairs of Mammograms

	Description	MIAS	DDSM	DDSM + MIAS
CW-SSIM	From intensity values of the central region	0.63	**0.81**	0.75
	From Gabor magnitude of the central region	0.57	0.63	0.60
	From Gabor phase of the central region	0.62	0.56	0.58
SSIM	From intensity values of the central region	0.55	0.71	0.62
	From intensity values via masking #1	0.38	0.73	0.60
	From intensity values via masking #2	0.56	0.66	0.59
	From Gabor magnitude of the central region	0.58	0.51	0.53
	From Gabor magnitude via masking #1	0.67	0.53	0.58
	From Gabor magnitude via masking #2	0.48	0.57	0.55
	From Gabor phase of the central region	0.65	0.71	0.69
	From Gabor phase via masking #1	0.64	0.72	0.70
	From Gabor phase via masking #2	0.55	0.65	0.60

Note: Results are given in terms of A_z for the various datasets of images. Cases with $A_z > 0.8$ are in bold.
CW-SSIM: Complex Wavelet Structural SIMilarity index.

Table 10.3. Poorer results were observed, in particular, for the mini-MIAS dataset, for which a more effective matching between the compared regions appears to be relevant. The results also point out that the use of more sophisticated descriptors is needed for comparative analysis of the directional components of pairs of mammograms, as indicated by the lower A_z values obtained by the SSIM and CW-SSIM indices derived from the magnitude and phase responses of Gabor filters.

10.5.2 Performance of Classification and Cross-Validation

The results of pattern classification using the LDA, Bayes QDA, and ANN-RBF classifiers with the LOO per patient cross-validation method are reported in Table 10.5. The sets of features selected more than 50% of the time in the training dataset via SWR are listed for each experiment. The best A_z (SE) values obtained on a per-pair-of-mammogram basis for the MIAS and DDSM datasets individually were, respectively, 0.88 (0.04) and 0.90 (0.04). Analysis of the corresponding ROC curves indicated accuracies up to 0.87 and 0.91, respectively, for the MIAS and DDSM datasets. The combination of the two datasets on a per-pair-of-mammograms basis, DDSM + MIAS, led to A_z of 0.83 (0.04), 0.77 (0.05), and 0.87 (0.04), respectively, with the LDA, BQDA, and ANN-RBF classifiers. The best accuracy was achieved with the BQDA classifier when two of the similarity indices were often selected.

Two-view analysis was performed only for the DDSM dataset, for which both CC and MLO views were available, by combining the features extracted from the two different views on a per-patient basis. The overall best performance was achieved using the ANN-RBF classifier, with the A_z value of 0.93 (0.04), with the corresponding sensitivity, specificity, and accuracy of 1, 0.88, and 0.94, respectively, calculated on a per-patient basis. Figure 10.7 displays the binormal ROC curves estimated by ROCKIT and related to two-view analysis for the three classifiers used. Results for the normal pair versus asymmetric pair classification for the 94 pairs of mammograms of the combined dataset (DDSM + MIAS) using the features selected via SWR and the LDA classifier for several cross-validation methods are summarized in Table 10.6. As expected, higher values of A_z, up to 0.86, were obtained when the features were selected using the entire dataset of mammograms.

TABLE 10.5 Results of Pattern Classification Using the Features Selected via stepwise Logistic Regression and the Leave-one-patient-out cross-validation Procedure.

Classifier	Selected features	A_z(SE)	$I_{95\%}$	TPR	TNR	Acc.
(a) DDSM Dataset (single view) 64 cases, 32 patients						
LDA	S_{M3}, $\tilde{S}_I$	0.90 (0.04)	[0.79–0.95]	0.81	0.91	0.86
BQDA	S_{M3}, $\tilde{S}_I$	0.83 (0.05)	[0.72–0.91]	0.84	0.84	0.84
ANN-RBF	ΔV_{a2}, S_{M3}, $\tilde{S}_I$	0.90(0.04)	[0.78–0.96]	0.84	0.97	0.91
(b) M1AS Dataset (single view) 30 cases, 30 patients						
Classifier	Selected features	A_z(SE)	$I_{95\%}$	TPR	TNR	Acc.
LDA	ΔV_{a2}, ΔV_{r3}, ΔV_{g2}, $\tilde{S}_{\Phi}$	0.87 (0.07)	[0.69–0.96]	0.80	0.93	0.87
BQDA	ΔV_{a2}, ΔV_{s1}, ΔV_{s3}, ΔV_{r3}, ΔV_{g1}, ΔV_{g2}, S_{I1}, S_1, S_2, $\tilde{S}_{\Phi}$	0.84 (0.08)	[0.64–0.95]	0.80	0.93	0.87
ANN-RBF	$\tilde{S}_{\Phi}$	0.88 (0.06)	[0.71–0.96]	0.87	0.80	0.83
(c) DDSM + MIAS Datasets (single view) 94 cases, 62 patients						
Classifier	Selected features	A_z(SE)	$I_{95\%}$	TPR	TNR	Acc.
LDA	$\tilde{S}_I$	0.83 (0.04)	[0.73–0.90]	0.68	0.89	0.79
BQDA	S_{M3}, $\tilde{S}_I$	0.77 (0.05)	[0.65–0.85]	0.84	0.80	0.83
ANN-RBF	S_{M3}, $\tilde{S}_I$	0.87 (0.04)	[0.79–0.93]	0.74	0.87	0.81
(d) DDSM Dataset (two-view) 32 cases, 32 patients						
Classifier	Selected features	A_z(SE)	$I_{95\%}$	TPR	TNR	Acc.
LDA	CC: ΔV_{s1}, ΔV_{r2}, ΔV_{g1}, MLO: ΔV_{s1}, $\tilde{S}_I$	0.85(0.07)	[0.68–0.95]	0.81	0.75	0.78
BQDA	CC: ΔV_{s1}, ΔV_{r2}, ΔV_{g1}, MLO: ΔV_{s1}, $\tilde{S}_I$	0.78(0.09)	[0.58–0.91]	0.75	0.75	0.75
ANN-RBF	CC: ΔV_{s2}, ΔV_{r2}, ΔV_{g1}, S_{M3} MLO: ΔV_{s1}, $\tilde{S}_I$	0.93(0.06)	[0.73–0.99]	1	0.88	0.94

Note: Results are provided on a per-pair-of-mammograms basis (single view) for the various datasets of images and on a per-patient basis (two-view) for the DDSM. Features selected more than the 50% of the time during the cross-validation process are listed. Sensitivity, specificity, and accuracy rates were computed at the operating point on the experimental ROC curve closest to the vertex (0,1). True Positive Rate (TPR), True Negative Rate (TNR), Accuracy (Acc)

10.6 Discussion and Challenging Aspects

In the study presented herein, a strategy for the analysis of structural similarity has been presented. The approach consists of sequential steps focused on landmarking of the two mammograms, on automatic bilateral masking of the overall breast regions, on the application of multidirectional Gabor filtering, and on the extraction of spherical semivariogram descriptors and of structural similarity features. The last step was performed by introducing correlation-based structural similarity indices in both spatial (CB-SSIM) and complex wavelet (CB-CW-SSIM) domains, which extended the SSIM and CW-SSIM indices, previously proposed, to facilitate quantitative comparison of regions of different sizes for which, due to the non-stationary nature of the tissue patterns under observation, point-by-point comparisons are inherently not meaningful.

The effectiveness of the whole strategy is due to the contribution of the performance of each of the steps highlighted above. The landmarking procedures have already demonstrated high performance in previous studies in detecting the correct nipple position and in delineating the breast skin line contour and the pectoral muscle. The small residual inaccuracies of the landmarking procedure are rendered inconsequential by the fact that point-by-point comparisons are not made. The non-stationary nature of the tissue patterns under examination requires a region-based approach. Moreover, registration of left and right regions would be needed in order to perform point-wise correspondence, which could cause registration

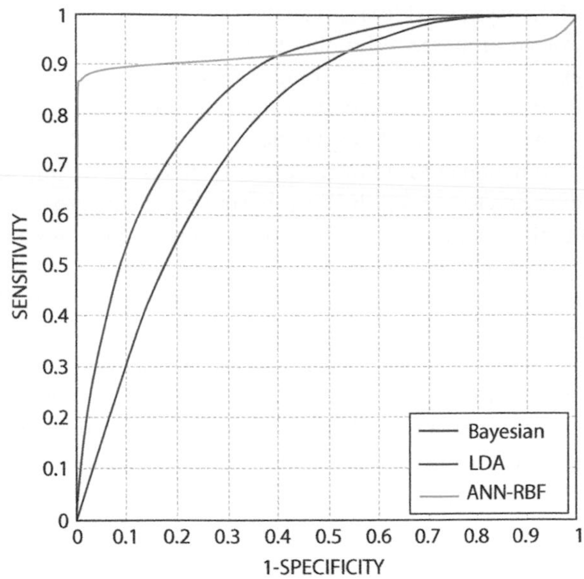

FIGURE 10.7 Binormal ROC curves estimated using ROCKIT [52 for the dataset of 32 cases of the DDSM database (two-view analysis) using leave-one-patient-out cross-validation and stepwise logistic regression for feature selection. The curves represent the performance of classification of normal versus asymmetric cases on a per-patient basis.

TABLE 10.6 Results of ROC Analysis for the 94 Cases of the Combined Dataset of Mammograms (DDSM + MI AS) Using the Features Selected via SWR and the LDA Classifier for Several Cross-Validation Methods

	Feature Selection	
Cross-Validation Method	Training Set	Entire Dataset
2-fold, pair of images	0.81 ±0.03	0.86 ±0.02
2-fold, patient	0.80 ±0.04	0.82 ±0.03
LOO, pair of images	0.83	0.86
LOO, patient	0.83	0.85

errors and alteration of the patterns of breast parenchyma. In this regard, the correlation-based structural similarity approach facilitates comparison between images of different size and origin, avoiding enforcement of an ideal or perfect match. The performance of the individual features is, by itself, effective in detecting asymmetric signs in the distribution of the fibroglandular components of the two mammograms of a patient. CB-CW-SSIM, in particular, showed the best discriminatory power, especially for the case of MLO views of the MIAS dataset, where the differential distortions due to positioning and compression of the breast are more evident. Results of pattern classification showed that the features were also complementary to one another, as demonstrated by the improvement achieved via feature selection procedures.

Given the undoubted valence of considering two distinct datasets for validating the approach, additional studies will be required to consider larger sets of mammograms, including full-field digital mammograms and database-independent validation procedures, considering one database for training and a different one for testing. Such further investigations will be helpful to determine a stable optimal set of features and the most effective pattern classification model. Surely, more extensive testing of the proposed procedures on larger datasets could confirm the presented results and help in proceeding toward clinical application. In the present study, which is entirely focused on the detection of bilateral asymmetry by using all of the asymmetric cases available in public databases of mammograms, we have demonstrated the ability of the proposed techniques to detect pathological differences in the fibroglandular components of breast tissue patterns in mammograms with accuracy up to 0.94. An accuracy of 0.87, with the corresponding sensitivity and specificity of 0.80 and 0.93, respectively, was achieved using all of the asymmetric cases of the mini-MIAS database. The results obtained with the DDSM cases also point out that the combination of the information extracted from CC and MLO views can provide higher accuracy than the single-view approach. The robustness of our approach is indicated by the good results obtained by combining mammograms from two different databases and acquired at multiple hospitals by customized device settings.

Continuing in this direction, we identify as challenging a further study on the localization of the sites of asymmetry in the breast with the aim to focus the radiologist's attention on a smaller portion of the tissue. Such results will provide assistance in focusing further examinations. In addition, the malignant nature of asymmetry could be recognized with respect to the benign/normal nature of asymmetric portions of the breast. Such indication could help radiologists in assigning appropriate priority to the patient examined and planning the most effective follow-up procedures.

References

1. J.W.T. Leung and E.A. Sickles. Developing asymmetry identified on mammography: Correlation with imaging outcome and pathologic findings. *American Journal of Roentgenology,* 188(3):667–675, 2007.
2. E.A. Sickles. The spectrum of breast asymmetries: Imaging features, work-up, management. *Radiologic Clinics of North America,* 45(5):765–771, 2007. cited By 9.
3. C.J. D'Orsi, E.A. Sickles, and E.B. Mendelson. *BI-RADS: Mammography.* American College of Radiology, 5th edition, 2013.
4. E.R. Price, B.N. Joe, and E.A. Sickles The developing asymmetry: Revisiting a perceptual and diagnostic challenge. *Radiology,* 274(3):642–651, 2015.
5. E.A. Sickles. Mammography: Asymmetries, masses, and architectural distortion. In J. Hodler, G.K. von Schulthess, and C.L. Zollikofer, editors, *Diseases of the Heart and Chest, Including Breast 2011–2014,* 255–258. Springer, 2011.
6. J.A. Harvey, L.L. Fajardo, and C.A. Innis. Previous mammograms in patients with impalpable breast carcinoma: retrospective vs blinded interpretation. *Am. J. Roentgenol,* 161(6):1167–1172, 1993.
7. D. Scutt, G.A. Lancaster, and J.T. Manning. Breast asymmetry and predisposition to breast cancer. *Breast Cancer Research,* 8:R14, 2006.
8. B. Zheng, J.H. Sumkin, M.L. Zuley, X. Wang, A.H. Klym, and D. Gur. Bilateral mammographic density asymmetry and breast cancer risk: a preliminary assessment. *Eur. J. Radiol,* 81(11):3222–3228, 2012.
9. A.S. Majid, E.S. de Paredes, R.D. Doherty, N.R. Sharma, and X. Salvador. Missed breast carcinoma: pitfalls and pearls. *Radiographics,* 23:881–895, 2003.
10. H.C. Burrell, A.J. Evans, A.R.M. Wilson, and S.E. Pinder. False-negative breast screening assessment. What lessons can we learn? *Clinical Radiology,* 56(5):385–388, 2001.
11. A. Venkatesan, P. Chu, K. Kerlikowske, E.A. Sickles, and R. Smith-Bindman. Positive predictive value of specific mammographic findings according to reader and patient variables. *Radiology,* 250(3):648–657, 2009.
12. F.J. Gilbert, S.M. Astley, M.G.C. Gillan, O.F. Agbaje, M.G. Wallis, J. James, C.R.M. Boggis, and S.W. Duffy. Single reading with computer-aided detection for screening mammography. *N. Engl. J. Med.,* 359(16):1675–1684, 2008.
13. N.F. Boyd, H. Guo, L.J. Martin, L. Sun, J. Stone, E. Fishell, R.A. Jong, A. Chiarelli, S. Minkin, and M.J. Yaffe. Mammographic density and the risk and detection of breast cancer. *N. Engl. J. Med.,* 356(3):227–236, 2007.

14. R.A. Smith, D. Manassaram-Baptiste, D. Brooks, V. Cokkinides, M. Doroshenk, D. Saslow, R.C. Wender, and O.W. Brawley. Cancer screening in the United States, 2014: A review of current American Cancer Society guidelines and issues in cancer screening. *CA-Cancer J. Clin.*, 64(1):30–51, 2014.
15. K. Doi. Computer-aided diagnosis in medical imaging: historical review, current status and future potential. *J. Comp. Med. Imaging Graphics*, 31 (4–5): 198–211, 2007.
16. R.M. Rangayyan. *Biomedical Image Analysis.* CRC Press, Taylor & Francis Group, 6000 Broken Sound Parkway, NW, Suite 300 Boca Raton, FL 33487-2742, 2004.
17. J. Tang, R.M. Rangayyan, J. Xu, I. El Naqa, and Y. Yang. Computer-aided detection and diagnosis of breast cancer with mammography: recent advances. *IEEE Trans. Inf. Technol. Biomed.*, 13(2):236–251, 2009.
18. D.R. Ericeira, A.C. Silva, A.C. de Paiva, and M. Gattass. Detection of masses based on asymmetric regions of digital bilateral mammograms using spatial description with variogram and cross-variogram functions. *Comput. Biol. Med.*, 43(8):987–999, 2013.
19 M. Karnana and K. Thangavel. Automatic detection of the breast border and nipple position on digital mammograms using genetic algorithm for asymmetry approach to detection of microcalcifications. *Comput. Meth. Prog. Bio.*, 87(1):12–20, 2007.
20. N. Karssemeijer. Automated classification of parenchymal patterns in mammograms. *Phys. Med. Biol*, 43(2):365–378, 1998.
21. X. Wang, L. Li, W. Xu, W. Liu, D. Lederman, and B. Zheng. Improving performance of computer-aided detection of masses by incorporating bilateral mammographic density asymmetry: an assessment. *Acad. Radiol*, 19(3):303–310, 2012.
22. F.F. Yin, M.L. Giger, K. Doi, C.J. Vyborny, and R.A. Schmidt. Computerized detection of masses in digital mammograms: automated alignment of breast images and its effect on bilateral subtraction technique. *Med. Phys.*, 21(3):445–452, 1994.
23. R.J. Ferrari, R.M. Rangayyan, J.E.L. Desautels, and A.F. Frère. Analysis of asymmetry in mammograms via directional filtering with Gabor wavelets. *IEEE Trans. Med. Imag.*, 20(9):953–964, 2001.
24. T.K. Lau and W.F. Bischof. Automated detection of breast tumors using the asymmetry approach. *Comput. Biomed. Res.*, 24:273–295, 1991.
25. P. Miller and S. Astley. Automated detection of mammographic asymmetry using anatomical features. *Int. J. Pattern Recogn. Artif. Intell*, 7(6):1461–1476, 1993.
26. R.M. Rangayyan, R.J. Ferrari, and A.F. Frère Analysis of bilateral asymmetry in mammograms using directional, morphological, and density features. *J. Electron. Imaging*, 16(1): 013003-013003-12, 2007.
27. S.D. Tzikopoulos, M.E. Mavroforakis, H.V. Georgiou, N. Dimitropoulos, and S. Theodoridis. A fully automated scheme for mammographic segmentation and classification based on breast density and asymmetry. *Comput. Meth. Prog. Bio.*, 102(l):47–63, 2011.
28. X. Wang, D. Lederman, J. Tan, X.H. Wang, and B. Zheng. Computerized detection of breast tissue asymmetry depicted on bilateral mammograms: a preliminary study of breast risk stratification. *Acad. Radiol.*, 17(10):1234–1241, 2010.
29. X. Wang, D. Lederman, J. Tan, X.H. Wang, and B. Zheng. Computerized prediction of risk for developing breast cancer based on bilateral mammographic breast tissue asymmetry. *Med. Eng. Phys.*, 33(8):934–942, 2011.
30. J. Suckling, J. Parker, D.R. Dance, S. Astley, I. Hutt, C.R.M. Boggis, I. Ricketts, E. Sta-makis, N. Cerneaz, S.L. Kok, P. Taylor, D. Betal, and J. Savage. The Mammographic Image Analysis Society digital mammogram database. *Excerpta Medica, International Congress Series* 1069: 242–248, 1994.
31. A. Mencattini, M. Salmeri, and P. Casti. Bilateral asymmetry identification for the early detection of breast cancer. In *Medical Measurements and Applications Proceedings (MeMeA), 2011 IEEE International Workshop on*, pages 613–618, 2011.

32. L. Tabár, T. Tot, and P.B. Dean. Breast cancer, the art and science of early detection with mammography: perception, interpretation, histopathologic correlation. George Thieme Verlag, New York, NY, Verlage, Germany, Thieme, 2008.
33. P. Casti, A. Mencattini, and M. Salmeri. Characterization of the breast region for computer assisted Tabár masking of paired mammographic images. In *IEEE Computer-Based Medical Systems (CBMS), 2012 25th International Symposium on*, pages 1–6, 2012.
34. P. Casti, A. Mencattini, M. Salmeri, A. Ancona, F. Mangieri, and R.M. Rangayyan. Masking procedures and measures of angular similarity for detection of bilateral asymmetry in mammograms. In *Conf. Proc. IEEE e-Health and Bioeng. (EHB)*, Iasi, Romania, 21–23, Nov. 2013.
35. P.A.P. Moran. Notes on continuous stochastic phenomena. *Biometrika*, 37(1/2):17–23, 1950.
36. P. Casti, A. Mencattini, M. Salmeri, and R.M. Rangayyan. Spatial correlation analysis of mammograms for detection of asymmetric findings. In *Breast Imaging (IWDM 2014) in Lecture Notes in Computer Science*, pages 558–564. Springer, 2014.
37. M. Heath, K. Bowyer, D. Kopans, R. Moore, and W.P. Kegelmeyer. The Digital Database for Screening Mammography. In *Proc. 5th International Workshop on Digital Mammography*, pages 212–218. Medical Physics Publishing, 2001.
38. P. Casti, A. Mencattini, M. Salmeri, and R.M. Rangayyan. Analysis of structural similarity in mammograms for detection of bilateral asymmetry. *IEEE Trans. Med. Imag.*, 34(2):662–671, 2015. DOI:10.1109/TMI.2014.2365436.
39. H. Zhang, P. Heffernan, and L. Tabár. User interface and viewing workflow for mammography workstation. US Patent 2009/0185732 A1, 2009.
40. P. Casti, A. Mencattini, M. Salmeri, A. Ancona, F. Mangieri, and R.M. Rangayyan. Design and analysis of contour-independent features for classification of mammographic lesions. In *Conf. Proc. IEEE e-Health and Bioeng. (EHB)*, Iasi, Romania, 21–23 Nov. 2013.
41. P. Casti, A. Mencattini, M. Salmeri, and R.M. Rangayyan. Semivariogram analysis and spherical modeling to detect structural bilateral asymmetry in mammograms. *Int. J. CARS*, 9(Suppl. 1):S231–S234, 2014.
42. P. Casti, A. Mencattini, M. Salmeri, A. Ancona, F.F. Mangieri, M.L. Pepe, and R.M. Rangayyan. Automatic detection of the nipple in screen-film and full-field digital mammograms using a novel Hessian-based method. *J. Digit. Imaging*, 26(5):948–957, 2013.
43. P. Casti, A. Mencattini, M. Salmeri, A. Ancona, F.F. Mangieri, M.L. Pepe, and R.M. Rangayyan. Estimation of the breast skin-line in mammograms using multidirectional Gabor filters. *Comput. Biol. Med.*, 43(11):1870–1881, 2013.
44. R.J. Ferrari, R.M. Rangayyan, J.E.L. Desautels, R.A. Borges, and A.F. Frère. Automatic identification of the pectoral muscle in mammograms. *IEEE Trans. Med. Imag.*, 23(2):232–245, 2004.
45. E.H. Isaaks and R.M. Srivastava. *An Introduction to Applied Geo statistics.* Oxford University Press, 1989.
46. M. A. Oliver. Determining the spatial scale of variation in environmental properties using the variogram. In N. Tate and P.M. Atkinson, editors, *Modelling Scale in Geographical Information Science*, pages 193–219. John Wiley & Sons, 2001.
47. M.P. Sampat, Z. Wang, S. Gupta, A.C. Bovik, and M.K. Markey. Complex wavelet structural similarity: a new image similarity index. *IEEE Trans. Image Processing*, 18(11):2385–2401, 2009.
48. Z. Wang, A.C. Bovik, H.R. Sheikh, and E.P. Simoncelli. Image quality assessment: from error visibility to structural similarity. *IEEE Trans. Image Processing*, 13(4):1–14, 2004.
49. F.J. Ayres and R.M. Rangayyan. Design and performance analysis of oriented feature detectors. *J. Electron. Imaging*, 16(2): 023007-023007-12, April 2007.
50. J. Portilla and E.P. Simoncelli. A parametric texture model based on joint statistics of complex wavelet coefficients. *Int. J. Comput. Vision*, 40(1):49–71, 2000.
51. E.P. Simoncelli, W.T. Freeman, E.H. Adelson, and D.J. Heeger. Shiftable multi-scale transforms. *IEEE Trans. Inf. Theory*, 38(2):587–607, 1992.

52. ROCKIT software. http://www.radiology.uchicago.edu/.
53. N.R. Draper and H. Smith. *Applied Regression Analysis.* Wiley-Interscience, 1998.
54. R.O. Duda, P.E. Hart, and D.G. Stork. *Pattern Classification.* Wiley-Interscience, 2nd edition, 2001.
55. S. Haykin. *Neural Networks: A Comprehensive Foundation.* Prenctice Hall, 1999.
56. M. Tan, B. Zheng, P. Ramalingam, and D. Gur. Prediction of near-term breast cancer risk based on bilateral mammographic feature asymmetry. *Acad. Radiol.*, 20(12):1542–1550, 2013.

11

Computer-Aided Diagnosis of Breast Cancer with Tomosynthesis Imaging

Heang-Ping Chan

Ravi K. Samala

Lubomir M. Hadjiiski

Jun Wei

11.1 Introduction

Breast cancer is the most prevalent cancer in women worldwide, and the second-most common cause of cancer death in many countries, including the United States [1,2]. Mammography has been found to be effective in reducing breast cancer mortality by a number of cohort and case-control studies [1,3], although the cost of over-diagnosis has been a topic of controversy and study in recent years. A major limitation of screening mammography is the low sensitivity in dense breasts [4,5] due to the reduced conspicuity of lesions obscured by overlapping dense fibroglandular tissue. Another limitation is the high recall rate. Many of these recalls are caused by overlapping tissue that resembles a lesion and requires diagnostic workup. Finally, many malignant and benign lesions have similar mammographic appearance and cannot be distinguished even by further diagnostic workup. The positive predictive value of recommended biopsies ranges from only about 15%–30% [6]. Recalls and benign biopsies not only cause patient anxiety, but also increased healthcare costs.

Digital breast tomosynthesis (DBT) is a new imaging modality that has been introduced into clinical use in the past few years. In the United States, three commercial systems have been approved by the Food and Drug Administration since 2011. DBT is a limited-angle tomographic technique in which a small number (e.g., 9–25) of projection images of the compressed breast are acquired over a small angular range (e.g., 11°–60°). With proper reconstruction, a stack of reconstructed image slices covering the breast volume can be obtained. DBT provides high spatial resolution on slices reconstructed parallel (or at small angles) to the detector plane but with low resolution in the depth direction. DBT reduces the overlap of fibroglandular tissue that can obscure cancerous lesions on mammograms, thereby alleviating a major problem that limits the sensitivity of breast cancer detection in mammography.

A number of clinical trials have been conducted to evaluate the addition of DBT as an adjunct to the digital mammogram (DM), that is, combining DBT and DM (DBT + DM), in comparison to DM alone in screening settings [7–10]. Other investigators compared the cancer detection and recall rates in screening populations before and after the DBT + DM mode was introduced into their clinical practice [11,12]. All these studies found significant improvement in cancer detection and reduction in recalls. Lang et al. [13] compared two-view DM alone and one-view DBT alone in a screening population and found that one-view DBT significantly improved the cancer detection rate, but increased the recall rate while maintaining the same positive predictive value. Although the DBT + DM mode could achieve increased cancer detection rate and reduced recall rate compared to DM alone, it doubles the radiation dose to the screening population. Recently efforts are being made to synthesize a mammogram-like image (SM) from the DBT to obviate the need for the DM. Skaane et al. [14] showed that a newer version of SM (C-view) combined with DBT was not significantly different from the DBT+DM mode in a large screening study and concluded that DBT + SM was acceptable for routine clinical use. Gilbert et al. [15] compared the DBT + DM mode and DBT + SM with DM alone in a screen-recalled population and observed significant increase in specificity and sensitivity for invasive cancers, but marginal increase in sensitivity for all cancers using the DBT + DM mode; however, DBT + SM increased specificity significantly but no significant increase in sensitivity for all cancers.

Although the studies found that DBT increased the detectability of breast cancer and reduces recall rates compared to DM, most studies did not analyze the detection of non-calcified lesions and the detection of microcalcifications separately. In a few studies that reported the performance of DBT in the detection of microcalcifications, the results were not as consistent. In an early study with 98 subjects, Poplack et al. [16] found that the recall rate could be reduced by 40% with the addition of DBT to DM, but the conspicuity of microcalcifications were inferior in 8 of the 14 cases. Gur et al. [17] compared DM alone to DBT alone and DM + DBT. They found that DM+DBT could reduce recall rate by 30%; however, three benign microcalcification clusters that were seen in DM were not visible in DBT, whereas six benign masses not seen in DM were seen in DBT. Wallis et al. [18] found that two-view DBT provided significantly higher detection for both masses and microcalcifications than DM. Kopans et al. [19] also reported that the clarity of calcifications in DBT acquired with a GE prototype system was better than or comparable to that in DM in 92% of 119 cases with relevant calcifications. Andersson et al. [20] found that the visibility of calcifications in DBT were comparable to that in DM for the 13 cancer cases with calcifications in their study. However, Spangler et al. [21] found that the sensitivity and specificity of calcification detection in DM were higher than those in DBT in a dataset with 20 malignant and 40 benign calcification cases.

Various methods have been studied to improve the detection of microcalcifications in DBT, including the use of DM in combination with one-view or two-view DBT [18,22–28], the use of a synthesized DM-like image from DBT to replace the directly acquired DM [14,15,29], development of computer-aided detection methods for DBT [30–41], and the enhancement of the visibility of microcalcifications by improving reconstruction and image processing methods [42–46].

Regardless of the method of implementing DBT (combo DBT + DM, DBT + SM, replacing one or both DM views with DBT), one of the major concerns of integrating DBT into clinical practice is the change in workflow. A DBT volume contains a large number of reconstructed slices that need to be read by radiologist. Even at 1-mm slice thickness, the number of slices per view of the breast will range from about 30 to over 80. Although the correlation between adjacent slices and the less-complex background make it much more efficient in reading each slice than reading a regular mammogram, studies showed that the time required for interpretation of a DBT + DM examination was about 50%–100% longer than that for reading DM alone [7,18,47,48]. If the caseload for a radiologist has to be maintained at essentially the same level as DM due to the limited resources available for screening, radiologists inevitably will tend to speed up the reading. The DBT + DM or DBT + SM approach allows radiologists to search for microcalcifications in the two-dimensional (2D) DM or

SM, but the search for subtle microcalcifications even in 2D DM is known to be a challenging task; the additional blur and noise in the SM synthesized from DBT may make it more challenging. Soft-tissue lesions such as masses and architectural distortion will be more visible in DBT slices, but it requires scrolling through the hundreds of slices in the 4-view screening examination. The chance for oversight of subtle lesions in the large search space may not be negligible under the time constraint. Detection of microcalcifications in DBT is especially important if DBT would replace DM for screening because no other imaging modalities can detect calcification as effectively as DM and calcification is an important sign of early stage breast cancer. Computer-aided detection (CAD), therefore, is expected to play a similarly important, if not more important, role for DBT as for DM in clinical practice.

11.2 Imaging Characteristics of Breast Tomosynthesis

To design effective computer vision techniques for CAD, it is important to understand the imaging characteristics of DBT. The image acquisition geometry of a typical DBT system is shown in Figure 11.1. The x-ray system is basically a digital mammography system, except that the x-ray source is rotated along an arc or moved linearly over a limited angle range and takes a small number of low-dose mammograms along the way. However, different DBT manufacturers may have different designs for the image acquisition process. For example, the detector can be stationary or may be rotated around the pivot point in the opposite direction while the x-ray source is moved to different locations for acquisition of the projection views (PVs). The x-ray source can be moved continuously while the PVs are taken with short x-ray pulses to minimize the blurring by the source motion, or moved in a velocity mode or a step-and-shoot mode such that the x-ray source is stationary during acquisition of the projections. Another design that

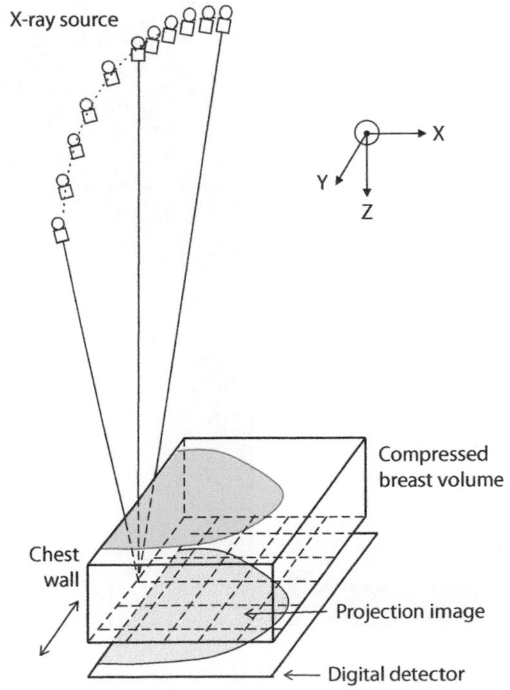

FIGURE 11.1 Geometry and image acquisition of a typical breast tomosynthesis system. The legend shows a coordinate system being referred to in the other figures in this chapter.

uses an array of stationary x-ray sources placed at proper locations to take PVs at different angles is also being developed. DBT is not a true three-dimensional (3D) imaging modality because image acquisition is limited to a small angular range (e.g., 11°–60°) compared to computed tomography (CT) of over 180°. The information of the object in the depth direction that is void of projections is insufficient to permit accurate reconstruction of the details. The depth resolution is mainly determined by the tomographic scan angle: the larger the angle, the higher the depth resolution. In addition, the small number of projections taken within the imaging arc with relatively large sampling intervals and the distribution of the PVs (e.g., uniform and non-uniform) also affect the reconstructed image quality [49,50]. However, the spatial resolution on the reconstructed DBT planes parallel to the detector is almost as high as that of digital mammograms, which allows DBT to maintain the spatial details of subtle breast lesions such as microcalcifications and small spiculated lesions similar to mammograms, while gaining the advantage of separating the images of the overlapping tissue into thinner layers. Therefore, a tomosynthesis volume is different from a CT volume that can provide nearly isotropic spatial resolution and can be viewed at any cross-sectional planes.

An example of a DBT and a mammogram of the same breast in mediolateral oblique (MLO) view is shown in Figure 11.2. The DBT was imaged with an experimental system that acquires 21 projections over a 60° tomographic angle at 3° angular intervals. The system uses a CsI/a:Si flat panel detector with a 0.1 mm × 0.1 mm pixel pitch. The simultaneous algebraic reconstruction technique (SART) was used for the reconstruction at 0.1 mm × 0.1 mm pixel size and 1 mm slice interval. The breast contains an invasive ductal carcinoma manifested as a mass with calcifications. The DBT slice shows the irregular-shaped mass and its extended spiculations clearly, while the same mass on the mammogram appears as an ill-defined density, similar to the adjacent normal breast tissue. Figure 11.3 shows the same DBT volume in three perpendicular planes. The DBT slice parallel to the detector plane (x-y plane) has high spatial

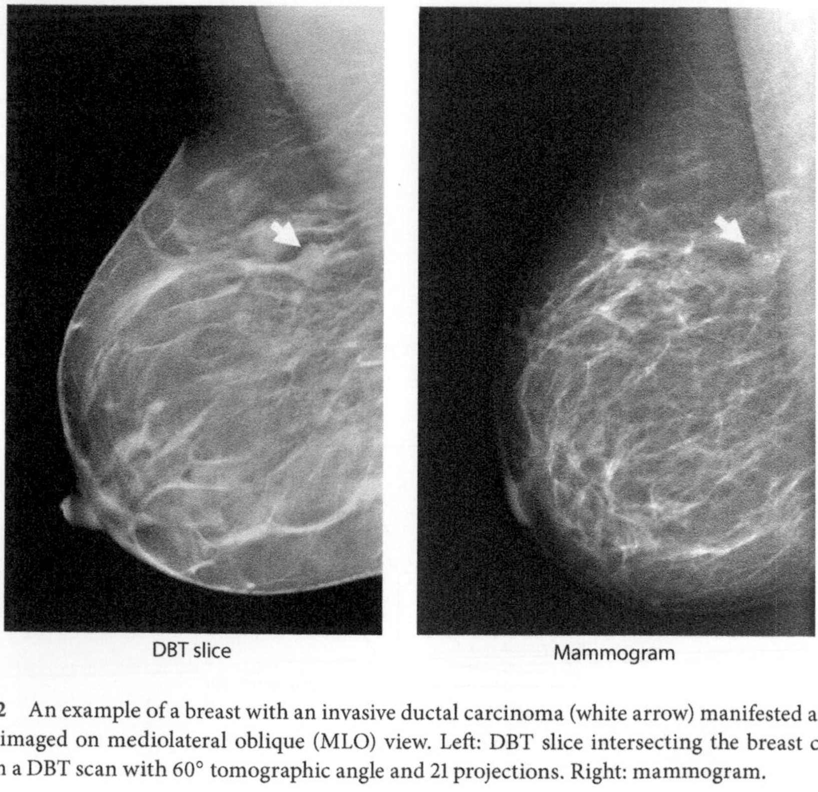

FIGURE 11.2 An example of a breast with an invasive ductal carcinoma (white arrow) manifested as a mass with calcification imaged on mediolateral oblique (MLO) view. Left: DBT slice intersecting the breast cancer reconstructed from a DBT scan with 60° tomographic angle and 21 projections. Right: mammogram.

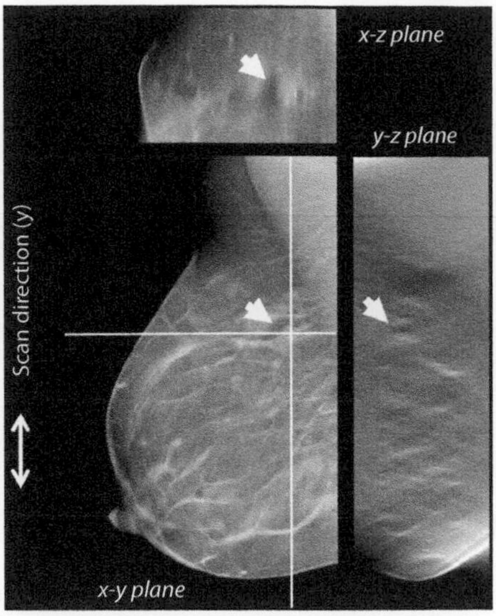

FIGURE 11.3 Reconstructed DBT volume of the same breast shown in Figure 11.2 displayed in three perpendicular planes. The simultaneous algebraic reconstruction technique (SART) was used for the reconstruction at 1-mm slice interval. The invasive ductal carcinoma is indicated by a white arrow. The image plane parallel to the detector plane (*x*-*y* plane) has high spatial resolution, similar to that of a mammogram. The other two planes (*x*-*z* and *y*-*z* planes) that are perpendicular to the detector plane have very low resolution. The horizontal and vertical lines on the DBT slice indicate the locations where the *x*-*y* plane and the *y*-*z* plane are relative to the *x*-*y* plane.

resolution, similar to that of a mammogram, whereas the other two perpendicular planes are dominated by the angular patterns of the x-ray paths without clear structures that resemble breast tissue. Figure 11.4 shows an example of a DBT volume with a cluster of microcalcifications from a high nuclear grade ductal carcinoma in situ (DCIS). The inter-plane artifacts can be seen clearly as the extension of the long bright shadows of a dense calcification along the x-ray paths on the *x*-*z* plane and the *y*-*z* plane. The shape of an object in DBT is, therefore, distorted along the depth direction and casts a shadow on the adjacent slices. It is important to take into consideration these imaging properties during feature extraction for image analysis in DBT.

It is known that the image quality of DBT depends on the image acquisition parameters, including the tomographic scan angle, the angular increment, and the number of projections, in addition to other factors that affect the image quality of x-ray imaging systems. The visibility of breast lesions also depends on the physical properties such as size and contrast of the lesions, as well as the structured noise in the images. The best combination of the DBT acquisition parameters for each type of lesions has been an area of interest for research and development in DBT. A number of simulation and modeling studies [51–55] or experimental evaluations [49,50,56–59] have been conducted to examine the dependence of image quality measures on DBT acquisition parameters.

In the studies by Zhang et al. [49] and Lu et al. [50], DBT scans of phantoms acquired at 60° angle and 3° increments with a total of 21 PVs were used. They selected six subsets of 11 PVs from the original DBT scans to simulate DBT acquired with different tomographic angles and uniform or non-uniform angular increments. The contrast-to-noise ratio (CNR), the full-width-at-half-maximum (FWHM), and the artifact spread function (ASF) of calcification-like and mass-like objects in the reconstructed DBT volumes were calculated to estimate the visibility of the objects on the DBT slices, the spatial blur on the *x*-*y* plane and along the *z*-direction, respectively. The results showed that DBT acquired with a wide scan angle or, for a fixed scan angle, having a large fraction of PVs at large angles was superior to those acquired with a

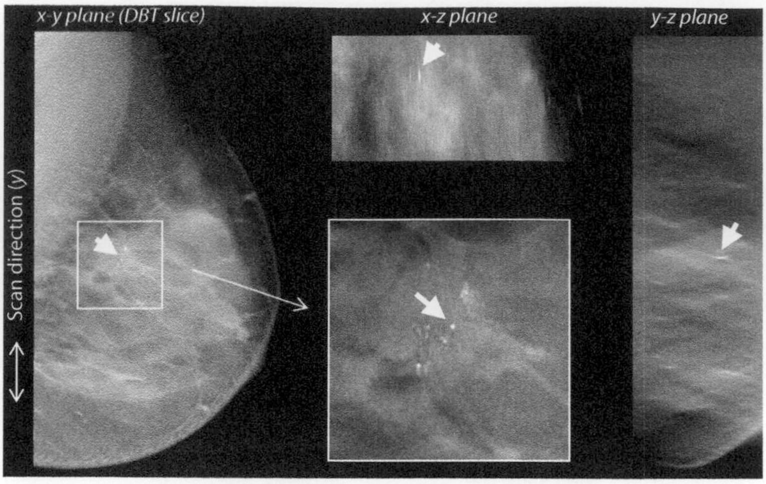

FIGURE 11.4 Reconstructed DBT volume of a breast with a cluster of microcalcifications in a high grade ductal carcinoma in situ (DCIS) displayed in three perpendicular planes. SART was used for the reconstruction at 1-mm-thick slice interval. The x-ray source moved in the y-direction. The image plane parallel to the detector plane (*x-y* plane) has high spatial resolution, similar to that of a digital mammogram. The other two planes (*x-z* and *y-z* planes) that are perpendicular to the detector plane have very low resolution. The white arrow points to the same dense calcification that causes inter-plane artifacts extending several mm along the depth (z) direction on the *x-z* plane and the *y-z* plane. (Reprinted from Chan H-P., *Computer Aided Detection and Diagnosis in Medical Imaging*, CRC Press, Boca Raton, FL, 2015.)

narrower scan angle, as measured by the ASF in the *z*-direction. On the *x-y* planes, the effect of PV distributions on spatial blur depended on the directions. In the x-ray source scan direction, the PV distributions with a narrow scan angle or a large fraction of PVs at small angles had smaller FWHM, that is, less spatial blur. In the direction perpendicular to the scan direction, the difference in the spatial blur among the different PV distributions was negligibly small. In addition, for small objects such as subtle microcalcifications, PV distributions with a narrow scan angle or a large fraction of PVs at small angles yielded higher CNR than those with a wide scan angle. Recently, Park et al. [60] conducted experimental studies to evaluate the effects of variable PV distribution and variable angular dose distributions in DBT acquisition on the reconstructed image quality of microcalcifications in breast phantom and observed similar results.

Chan et al. [58] and Goodsitt et al. [59] further investigated the impact of the imaging parameters on the image quality of signals in DBT by observer performance studies using an experimental DBT system that allows acquisition of projections at variable scan angles, angular increments and number of PVs. One observer performance study [58] evaluated the detectability of simulated microcalcifications in DBT of heterogeneous breast phantoms acquired at seven acquisition geometries, using different combinations of scan angle and uniform or non-uniform angular intervals. Another observer preference study [59] compared the visual quality of low-contrast objects for 12 different acquisition geometries. These studies showed that a large tomographic angle was better for reducing overlapping tissue and improving the detectability of low-contrast objects such as soft-tissue lesions, whereas narrow tomographic angles provided higher detectability of microcalcifications.

Figure 11.5 shows DBT volumes of the same breast shown in Figures 11.2 and 11.3 reconstructed at three combinations of tomographic angle and angular intervals. The original DBT was acquired with 60°, 3° angular increments and 21 projection views (PVs). Two other geometries, wide angle (60°, 6°, 11 PVs) and narrow angle (30°, 3°, 11 PVs), were simulated by reconstruction with a subset of 11 PVs. Although the x-ray dose was reduced by about half, and the noise was higher for the reconstructions with the subsets of 11 PVs, the main effects of acquisition geometry on the appearance of the mass, microcalcifications and tissue

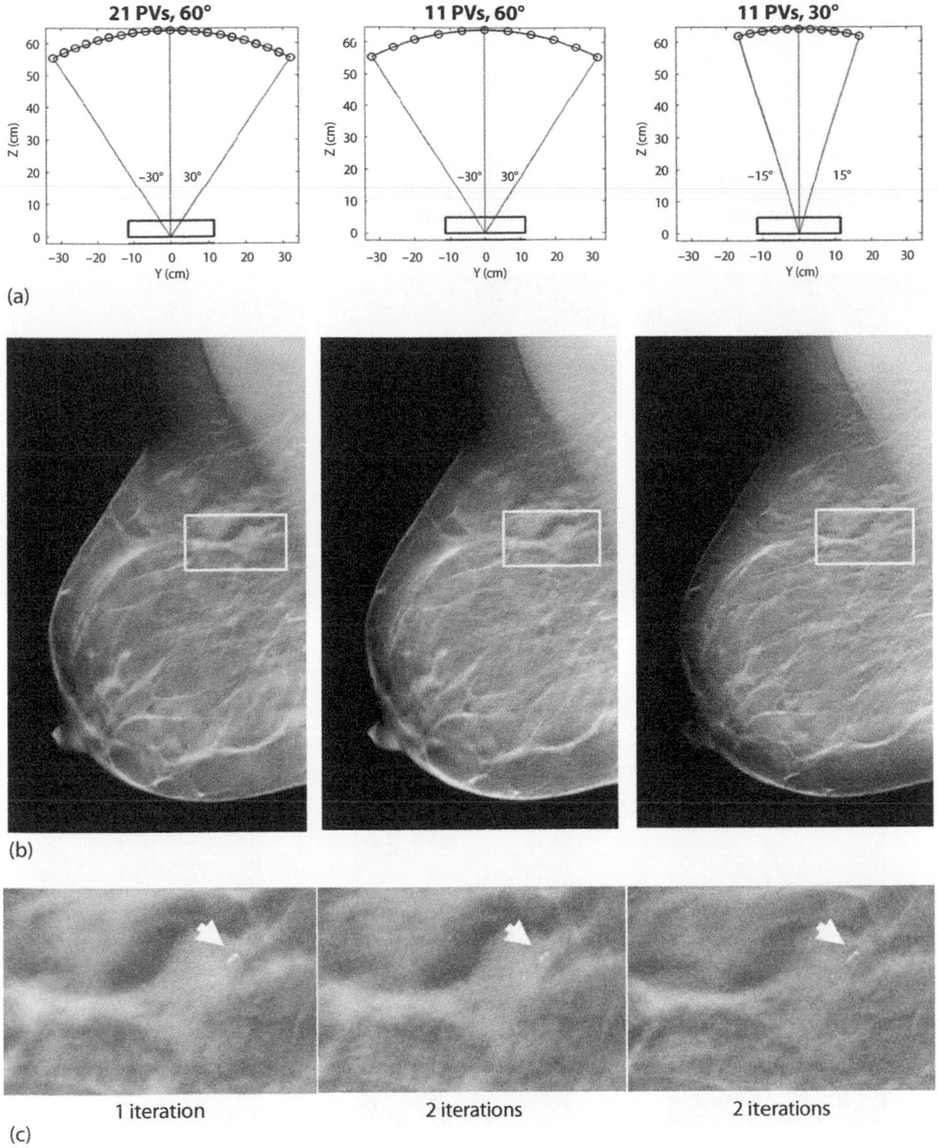

FIGURE 11.5 Comparison of reconstructed DBT images for three acquisition geometries using SART. The original DBT was acquired with 60°, 3° angular increments and 21 projection views (PVs). The other two geometries, middle column: wide angle (60°, 6°, 11 PVs), and right column: narrow angle (30°, 3°, 11 PVs), were simulated by reconstruction with a subset of 11 PVs. Although the x-ray dose is reduced by about half and the noise will be higher for the reconstructions with the subsets, the effects of acquisition geometry on the appearance of the mass, microcalcifications, and tissue texture patterns are demonstrated. The number of iterations for the DBT by the 11PV-reconstructions was doubled so that the number of PV updates is approximately equal to that of the 21PV-reconstruction. (a) PV distributions of three geometries. (b) DBT slice (x-y plane) intersecting an invasive ductal carcinoma (white box). (c) and (d) The enlarged region of interest showing the spiculated mass with calcifications. The mass shows higher contrast in the wide-angle DBT, whereas the calcifications are sharper and higher contrast in the narrow-angle DBT. Both the signal and noise increase as the number of iterations increases.

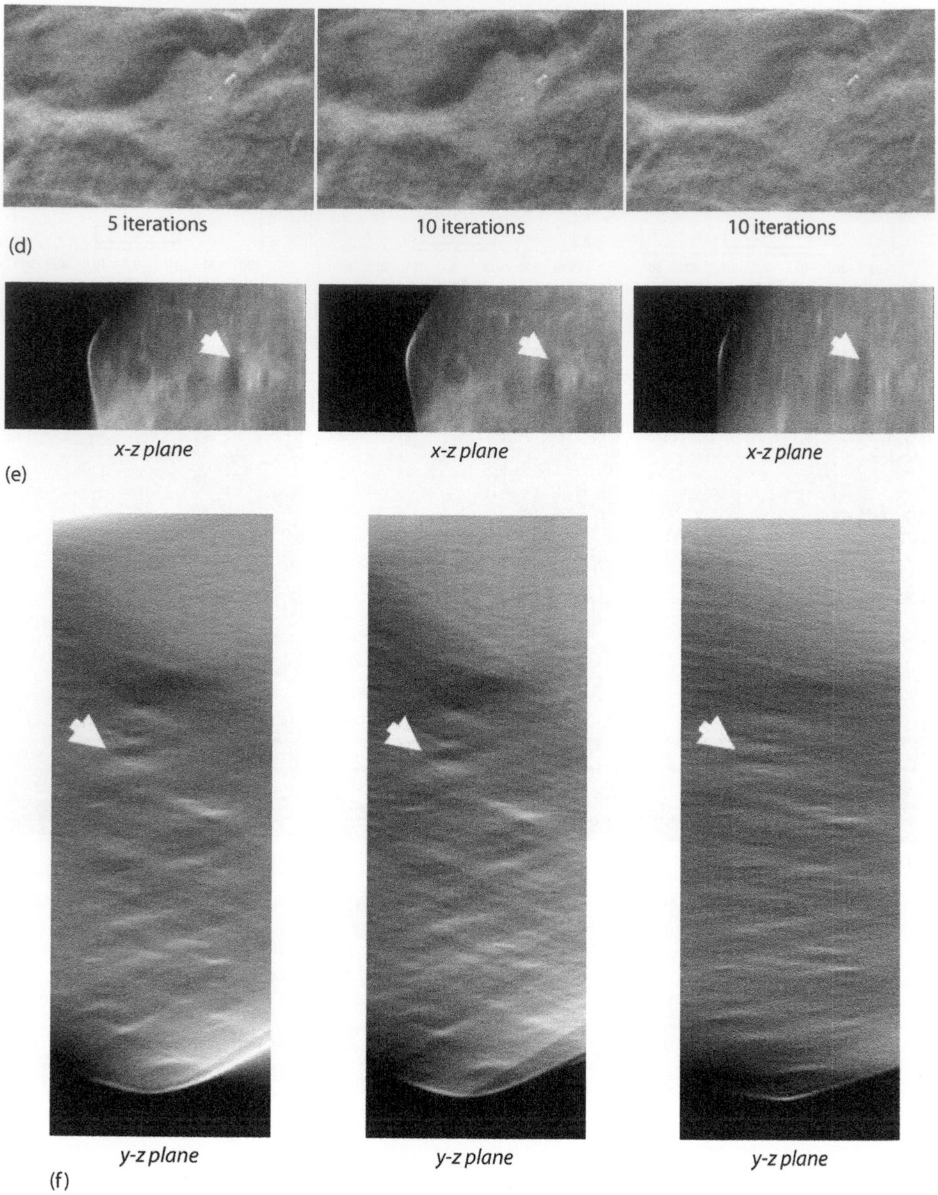

FIGURE 11.5 (CONTINUED) (e) and (f) The inter-plane artifacts of the narrow-angle DBT extend longer than those of the wide-angle DBT, indicating that the wide-angle DBT has better depth resolution and less overlapping tissue shadows than the narrow-angle DBT.

texture patterns can be seen by comparison of the images. The DBT was reconstructed with SART; the number of iterations for the DBT by the 11 PV-reconstructions was doubled so that the number of PV updates was approximately equal to that of the 21 PV-reconstruction to reduce the impact of fewer updates on the subset reconstruction. It is shown that the spiculated mass and the fibroglandular tissue have higher contrast in the wide-angle (60°) DBT than those in the narrow-angle (30°) DBT; however, the calcifications are sharper in the narrow-angle DBT. Both the signal and noise increase as the number of iterations increases.

On the cross-sectional images perpendicular to the detection plane, the image texture is dominated by the patterns of x-ray paths and the inter-plane artifacts of the narrow-angle DBT extend longer than those of the wide-angle DBT. These examples illustrate that wide-angle DBT has better depth resolution and less overlapping tissue shadows than the narrow-angle DBT, which results in DBT slices having a background with less fibrous textures and reduced tissue overlap, as evident by comparing the DBT slices in the second row.

In addition to the image characteristics of various types of lesions and their different dependences on the DBT acquisition geometry, the design of a DBT system often has to take into consideration the trade-offs among many other factors, such as the detector efficiency, the x-ray source output, the readout speed and lag of the detector, the scanning and breast compression time, and the mechanical stability and precision, while under the constraint of maintaining low radiation dose to the patient. The optimal design of a DBT system that can balance the image quality requirements of various types of lesions at the lowest possible radiation dose is still a topic of continued investigation.

A number of reconstruction methods have been applied to DBT reconstruction, including shift-and-add, tuned aperture computed tomography (TACT), maximum likelihood-convex (ML-convex) algorithm, matrix inversion (MITS), filtered back projection (FBP) and simultaneous algebraic reconstruction technique (SART) [61–65]. Reconstruction methods have a strong impact on image quality of DBT. Studies to improve the reconstruction methods and artifact reduction techniques are on-going [42,44,66,67]. Reconstruction methods specifically designed to enhance microcalcifications and reduce noise are also under investigation [44, 68–70]. DBT images are usually reconstructed in slices parallel to the detector plane. The spatial resolution on the reconstructed DBT slices can approach that of the digital detector if the geometry of the scanning system is accurately known and patient motion is kept at a minimum. However, some degree of blurring is inevitable due to the reconstruction from multiple PVs with different x-ray incident angles and oblique incidence of the x-ray beam to the detector, especially at large projection angles [71]. Super-resolution has been observed in DBT when reconstruction is performed with finer grids [46,72,73]. Because of the lack of PVs at large projection angles, the spatial resolution in the direction perpendicular to the detector plane (the depth or z-direction) is poor. The depth resolution is mainly determined by the tomographic angle: the larger the angle, the higher the depth resolution and the less the inter-plane artifact but with a trade-off of greater blurring on the DBT plane due to oblique intersection of the x-ray paths with a reconstructed slice of finite thickness. This blurring may be reduced by reconstruction with an adaptive grid approach along the depth direction for small objects such as microcalcifications [74]. Regardless of the reconstruction methods, tomosynthesis cannot provide true 3D information due to the lack of sampling over a wide angular range.

11.3 Computer-Aided Detection in DBT

DBT is composed of a number of low-dose DMs taken at slightly different projection angles. The PVs, together with the acquisition geometry, contain all the available information for signal detection in DBT. However, the individual PVs are noisy due to the low-dose acquisition. A DBT volume can be reconstructed from the PVs by an appropriate technique, which can enhance the signal and reduce noise by combining the information from the multiple projections. If both the set of PVs and the reconstructed DBT are available, CAD methods can be developed by combining the information from both in many different ways. One approach is to use the set of PVs as input and combine the information from the PVs in the process. Another approach is to use the reconstructed DBT volume (slices) as input and analyze images as a 3D volume or 2D slices. A third approach is to use both sets of images as input and combines the information at different stages of detection. Although the PVs and DBT volume basically contain the same information, the computer-vision techniques designed for the different sets of images may utilize the information differently. Combining the information extracted from the different forms of images derived from the original PV images may improve signal detection or characterization.

The recent development of methods for generating a 2D synthetic mammogram from the DBT images leads to an additional option of lesion detection, namely, detection in the synthetic mammogram, which may be combined with the approaches described above. However, it should be noted that some synthetic mammogram generation methods rely on detecting potential lesions with CAD to enhance the conspicuity of the lesions on the synthetic mammograms [75]. The sensitivity of detecting lesions in this type of synthetic mammograms will depend on the sensitivity and false-positive rate of the CAD methods used in the generation of the synthetic mammogram. On the other hand, if the synthetic mammogram is generated from the DBT without using CAD, the image quality and lesion detectability is most likely poorer than a DM because all overlapping tissue remains in the synthetic mammogram and additional blurring may result from the multiple-projection reconstruction and the limited depth resolution of the reconstructed volume.

11.3.1 Computer-Aided Detection of Microcalcifications

Detection of subtle microcalcifications in DBT by human or computer vision is challenging because of the large search space and the noisy background. CAD methods for detection of microcalcifications in the projection views (PVs), the reconstructed slices or the reconstructed volume have been studied. Peters et al. [76] detected calcifications on a small set of DBT. A band-pass, filter-based, wavelet kernel was used to separate the potential calcification candidates from the background on the PVs. A feature map was generated for each PV image, and the correspondence between 2D and 3D locations determined by the DBT acquisition geometry was used as a criterion to identify the calcifications. Park et al. [77] applied a 2D CAD algorithm developed for digitized screen-film mammograms (SFM) to the PV and the reconstructed DBT slices. Reiser et al. [36] developed an algorithm to detect microcalcifications in PV images to avoid the dependence of the CAD performance on the reconstruction algorithm. van Schie et al. [37] estimated a non-uniform noise model from each individual DBT-reconstructed volume which was used for normalization of the local contrast feature. Potential microcalcifications were detected by thresholding the local contrast feature, and the microcalcification candidates within a 5 mm radius were grouped to form microcalcification clusters. The detection strategies developed by our research laboratory are described below.

11.3.1.1 Microcalcification Detection in DBT Volume

A CAD system generally consists of several major stages: preprocessing for signal enhancement, prescreening for candidate signals, feature extraction and analysis for false positive reduction and final decision for identifying detected signals. A number of preprocessing methods have been investigated to improve the detectability of microcalcifications in DBT. Sahiner et al. [31] developed a CAD system for detection of microcalcifications, as shown in Figure 11.6. Two parallel processing methods are designed to identify microcalcification candidates and cluster seed candidates. For identifying microcalcification candidates, a 2D contrast-to-noise ratio (CNR) enhancement filter is applied to the DBT slices to enhance potential microcalcifications and reduce the low frequency background. Adaptive thresholding and region growing are then applied to the CNR-enhanced volume to segment the individual microcalcification candidates. For identifying cluster seed candidates, 3D multiscale filtering is applied to the DBT volume, and the eigenvalues of Hessian matrices are calculated at each voxel. Multiscale calcification response representing the intensity, size and shape information are then derived from the Hessian eigenvalues, which is further weighted by the CNR-enhanced volume voxel by voxel, resulting in an enhancement-modulated calcification response (EMCR) volume. With adaptive thresholding and region growing, potential calcifications are segmented from the EMCR volume and a set of top-ranked candidates are used as cluster seeds. A dynamic clustering process then groups the individual microcalcifications into clusters using the cluster seeds as the starting point and a distance criterion to determine cluster membership. The cluster candidates identified in the clustering process will undergo feature analysis and the clusters that do not satisfy the criteria are excluded as false positives (FPs). The

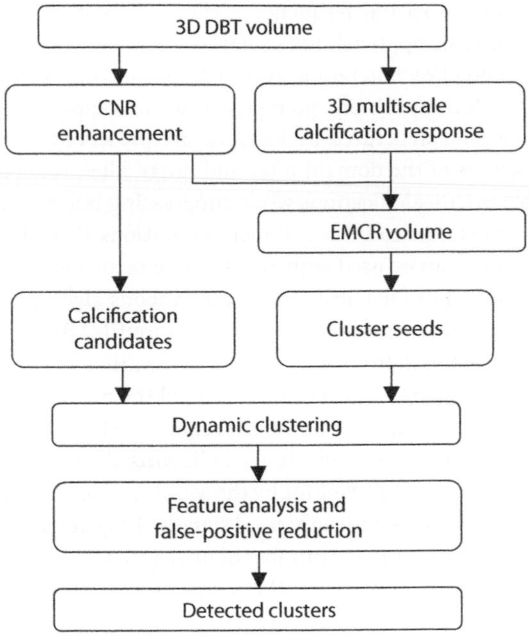

FIGURE 11.6 CAD system for detection of microcalcifications in DBT.

likelihood of a remaining cluster being a true cluster is determined as the highest CNR value among the cluster members. The sensitivity and specificity (or the number of FPs per DBT volume) can then be adjusted by applying a decision threshold to the cluster likelihood value. The overall performance of the CAD system can be described by a free-response receiver operating characteristic (FROC) curve that plots the relationship between the sensitivity and the number of FPs per DBT volume as the decision threshold is varied.

11.3.1.2 Enhancement of Microcalcification by Regularized Reconstruction

The microcalcifications in DBT can be enhanced by regularized reconstruction. Sidky et al. demonstrated that non-convex total p-variation regularization method (TpV) [44] with properly chosen regularization can reduce noise and increase the conspicuity of microcalcifications and masses on the reconstructed slices. However, the TpV method can cause *staircasing* (i.e., contouring) artifacts in the soft tissue background. Lu et al. [42,43,68,78] investigated methods to enhance the CNR of subtle microcalcifications while preserving the texture of the breast parenchyma in DBT. They found that incorporation of multiscale bilateral filtering (MSBF) into iterative DBT reconstruction is a promising approach [68].

Bilateral filtering [79] is a nonlinear filter that exploits both the geometric uniformity in the spatial background and the intensity difference of the signals in an image to selectively smooth the noise and enhance the sharpness of the signals. At each pixel of an image, bilateral filtering applies the product of two Gaussian filters, referred to as the domain filter and the range filter, to a neighborhood centered at the pixel. The domain and the range filters weight the intensity value of a neighboring pixel based on its distance and its intensity difference, respectively, from the central pixel, and the sum of all the weighted intensity value from the neighborhood yields the bilateral filtered value of the pixel on the image. The degrees of denoising and the signal enhancement are determined by the selection of the standard deviations of the Gaussian domain filter and range filter.

For the application to DBT, it is important to smooth the noise while preserving both the tissue structures, such as the spiculations and mass margins, and the small signals, such as microcalcifications that have large differences in the spatial frequency contents. Based on these image characteristics, Lu et al. [68] designed a multiscale approach to regularize noise between iterations of iterative reconstruction techniques. At the end of each iteration, every DBT slice is decomposed into several frequency bands via Laplacian pyramid decomposition. No regularization is applied to the low-frequency bands, so that subtle edges of masses and structured background are preserved. Bilateral filtering, with properly selected standard deviations of the domain filter and range filter, is applied to the high-frequency bands to selectively enhance microcalcifications while suppressing noise. The regularized DBT images are used for updating in the next iteration. The number of iterations also affects the overall image quality of DBT. MSBF regularization can be used with any iterative reconstruction techniques. Examples of MSBF-regularized reconstruction of DBT using the simultaneous algebraic reconstruction technique (SART), in comparison to TpV reconstruction and SART without MSBF, are shown in Figure 11.7. The MSBF method not only achieved higher CNR of microcalcifications than SART alone or TpV reconstruction, but also reduced contouring artifacts and preserved the mass margin and the parenchyma.

The microcalcification enhancement by the MSBF-regularized SART offers new opportunity to improve the detection accuracy of microcalcifications. Following the framework of the CAD system in Figure 11.6, Samala et al. [38] used the DBT volume by the MSBF-regularized SART as input and adapted the processing techniques to the images with enhanced signals. They designed new criteria to reduce FP clusters based on the size, CNR values and the number of microcalcifications in the cluster, cluster shape and cluster-based maximum intensity projection. They demonstrated that, with the MSBF enhancement

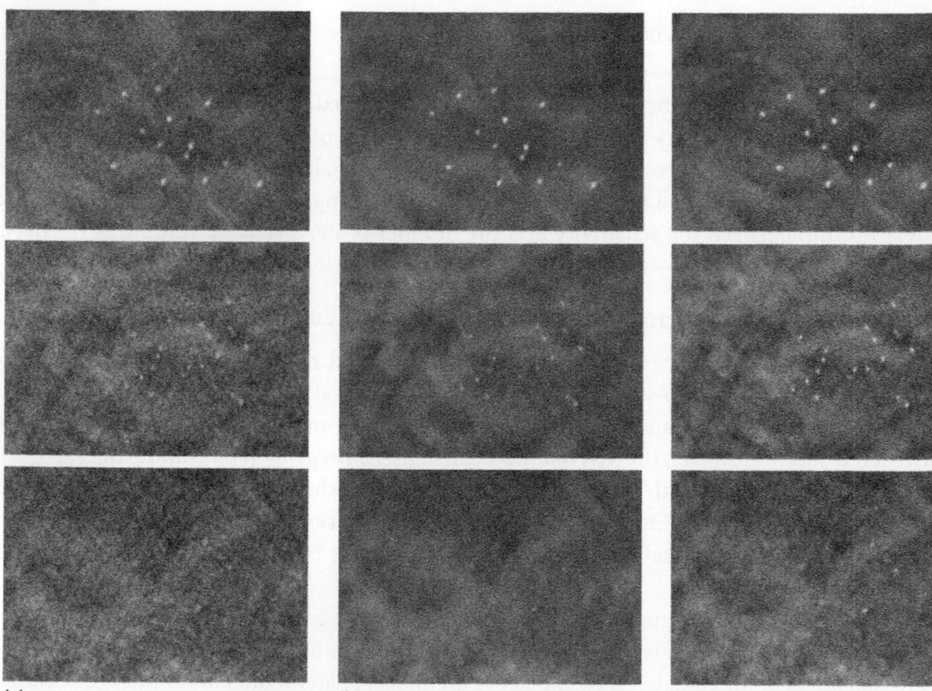

FIGURE 11.7 Examples of microcalcifications reconstructed with three methods. (a) SART, (b) non-convex total p-variation with $p = 0.8$, (c) SART with multiscale bilateral filtering, where the standard deviation of the domain filter was 2 and the standard deviation of the range filter was adaptively calculated from noise patches in the DBT volume being reconstructed. The focus slices after five iterations are shown. Simulated microcalcification clusters of three contrast groups in the breast phantoms: (upper row) high contrast, (middle row) median contrast, (lower row) low contrast. The same window and level settings were applied to images in the same row. (Reprinted from Lu Y, et al. *Medical Physics*, 42(1), 182–195, 2015.)

in combination with properly designed adaptive threshold criteria, effective microcalcification feature analysis and FP reduction techniques, the CAD system achieved a significant improvement in the detection of clustered microcalcifications in DBT compared to without MSBF enhancement.

The improved microcalcification detection in DBT with MSBF indicates the promise of denoising without blurring the high-frequency signals and edges. Inspired by this approach, Zheng et al. [80] proposed a new regularization method for iterative reconstruction, referred to as the spatially weighted non-convex (SWNC) regularization method. Similar to the MSBF, the SWNC method considers the spatial and intensity differences between pixels within a small neighborhood centered at each pixel to estimate the presence of signal or noise. However, the SWNC regularizer is incorporated into the formulation of the image reconstruction cost function, which can then be minimized with any suitable iterative algorithm. It is shown that, with proper selection of the parameters, the SWNC regularization method can further increase the CNR of microcalcifications while preserving the appearance of the spiculations and the breast parenchyma. The effectiveness of the SWNC method in improving microcalcification detection is yet to be investigated.

11.3.1.3 Microcalcification Detection in 2D Planar Projection (PPJ) Image

The decomposition of the high-frequency and the low-frequency information in the DBT slices suggests a new approach to detect the microcalcifications. Because the microcalcifications are mainly contained in the high-frequency band and already separated from the low frequency structured background, the detection of microcalcifications by the CAD system may focus on the stack of high-frequency slices. Moreover, the 3D spatial distribution of the individual microcalcifications in a cluster is sparse and is more difficult to be differentiated from noise. Samala et al. [40] proposed to generate a planar projection (PPJ) image and perform the detection in 2D. The PPJ image is obtained by a maximum intensity projection of the high-frequency DBT volume in the direction perpendicular to the detector plane at the final iteration of the SART reconstruction. The corresponding locations of the clusters, if any, can be mapped back to the DBT volume. The example of a biopsy-proven case of ductal carcinoma in situ (DCIS) shown in Figure 11.8 demonstrates the improvement in conspicuity of the microcalcifications on the PPJ image, compared to that in the DBT slices.

Samala et al. [40] designed a 2D approach for detection of microcalcifications in DBT, taking advantage of the PPJ image, as shown in Figure 11.9. With the PPJ image as input, no structured background removal is needed. For prescreening of microcalcification candidates, iterative gray-level thresholding is performed by applying a threshold that automatically steps from high to low value by analysis of the histogram of the PPJ image. At each threshold, the pixels exceeding the threshold value are subjected to region-growing using 8-connectivity to form individual objects. The threshold is reduced and the above process is repeated until the number of segmented objects reaches a desired value. Each object is further refined by a second region-growing segmentation using a gray-level threshold adaptive to the local statistic. The size and CNR features of the microcalcification candidates after the refined segmentation are

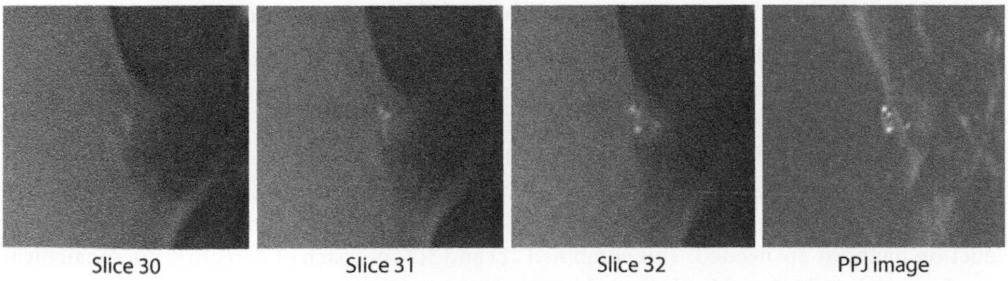

FIGURE 11.8 Planar projection (PPJ) image of a breast with biopsy-proven ductal carcinoma in situ. A 3D region of interest from the DBT volume enclosing the cluster is shown as slices 30–32. The region shown is 15 × 15 mm (150 × 150 pixels). (Reprinted from Samala RK, et al. *Physics in Medicine and Biology*, 59(23), 7457–7477, 2014.)

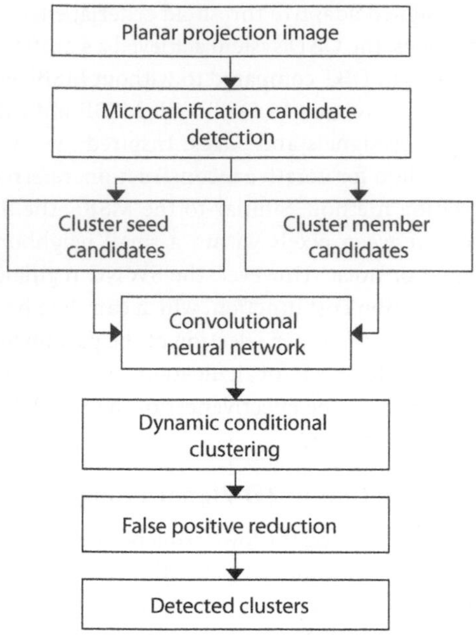

FIGURE 11.9 CAD system for detection of clustered microcalcifications in DBT by applying a 2D approach to the PPJ image.

used for FP reduction in the subsequent stages. The objects are ranked according to their CNR value. N top ranked objects are considered to be indicators of potential cluster locations, referred to as cluster seeds. Another set of N_s high ranking objects are kept as potential cluster members. A trained convolution neural network (CNN) is then used to differentiate true microcalcifications from FP objects, most of which come from edges of fibrous tissue and ducts, as well as artifacts such as the high contrast edges of metal clips from previous biopsy and inter-plane artifacts from the clips. Based on automatic analysis of the CNR histogram of the remaining candidate objects in the input PPJ image, CNR threshold criteria adaptive to this image are estimated for microcalcifications of different degrees of subtlety as stratified by their size and CNR values. Following the ranking of the cluster seeds and the cluster member candidates, a dynamic conditional clustering process then hierarchically forms cluster candidates by a distance criterion. Clusters that do not satisfy the adaptive CNR threshold criteria in combination with the size and the number of microcalcifications in the cluster are eliminated in the clustering process. FROC analysis indicates that the performance of the 2D approach in detection of microcalcifications was significantly higher than that of the detection in the DBT volume.

11.3.1.4 Microcalcification Detection by Joint DBT-PPJ Approach

Both the DBT volume and the PPJ image are reconstructed from the same set of projection-view images that provide the same information content. However, the DBT and PPJ images utilize the information in different ways and result in different detectability of the microcalcifications. Samala et al. [41] exploited the potential of improving detection by combining the two approaches based on the assumptions that a cluster detected in both 2D and 3D is more likely to be a true cluster and that each of the approaches may detect some clusters that the other misses. The trade-off is that the FPs may double so that more effective FP reduction methods are needed. This combined 2D and 3D approach of detecting microcalcifications is referred to as joint-CAD and is illustrated in Figure 11.10.

The joint-CAD system incorporates the individual steps from each of the CAD_{DBT} and CAD_{PPJ} systems within one framework using task-specific strategies to take advantage of the combined information. The joint-CAD system can be broadly divided into several stages: (1) preprocessing of the

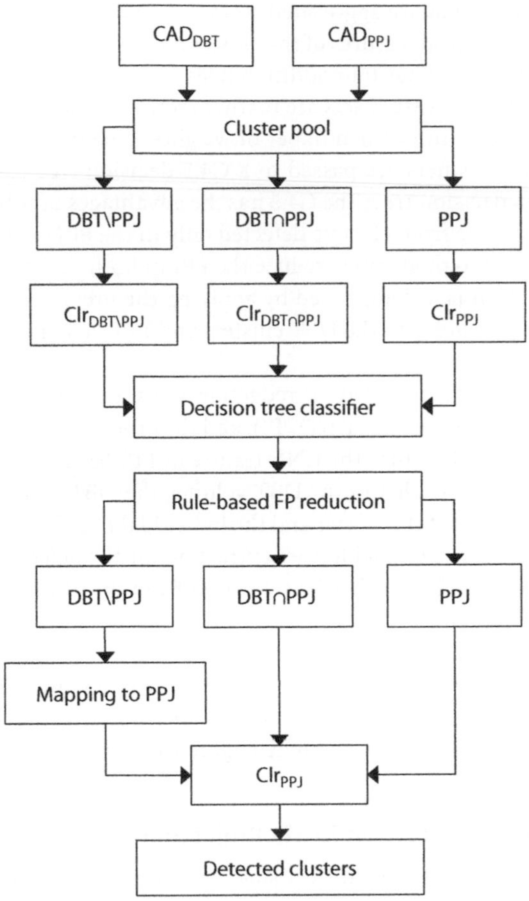

FIGURE 11.10 Joint-CAD system: cluster candidates detected by the individual CAD$_{DBT}$ and CAD$_{PPJ}$ systems form a cluster pool, which are grouped into three types: clusters detected by the CAD$_{DBT}$ system alone (DBT\PPJ), by the CAD$_{PPJ}$ system alone (PPJ), by both systems (DBT∩PPJ). Each group will undergo FP reduction by a classifier designed for that type, and finally by a decision tree classifier and a rule-based FP classification to further reduce the FP clusters. The remaining DBT\PPJ clusters are mapped to the PPJ space and features are extracted from the PPJ image. All clusters are then screened by the classifier trained for the PPJ clusters for a final FP reduction. (Clr = classifier).

DBT volume, (2) generation of microcalcification candidates and cluster candidates in DBT and PPJ, (3) FP reduction of clusters, (4) mapping of clusters from the DBT domain to the PPJ domain and (5) final FP reduction. The CAD$_{DBT}$ and CAD$_{PPJ}$ systems are basically the same as the individual systems, described above. After the dynamic clustering stage in each system, the clusters are combined into a cluster pool and grouped into three types: clusters detected by DBT alone (DBT\PPJ), clusters detected by both systems (DBT∩PPJ), and clusters detected by PPJ alone (Figure 11.10). To determine if a cluster detected in the DBT path overlaps with a cluster detected on the PPJ image, the DBT cluster is projected onto the PPJ image. If the centroid of the DBT cluster falls within the bounding box of a PPJ cluster or vice versa, the two clusters are considered overlapped. If more than one DBT clusters overlap with the PPJ cluster, the DBT cluster with the highest CNR score is kept while the other DBT clusters will be considered FPs. This will reduce the number of low score clusters that are more likely to be FPs to be passed from the DBT volume to the cluster pool. Three different classifiers are trained; Clr$_{DBT}$ and Clr$_{PPJ}$ are trained on the set of DBT and PPJ cluster features, respectively, and Clr$_{DBT}$∩$_{PPJ}$ is trained on the cluster features that are aggregated from the (DBT∩PPJ) clusters. The cluster features are derived from the morphological features of the cluster members [81] for the DBT (or PPJ) clusters. For the PPJ clusters, the feature set has four additional features derived from the CNN output scores for the cluster members [82]. The feature sets are further reduced by feature selection, and linear discriminant classifiers are used to limit the number of weights to be trained [81,83]. The discriminant scores generated from the classifiers are passed to a C4.5 decision tree classifier [84] with reduced error pruning to generate a decision tree. The C4.5 has the advantages of inherently handling missing values, which in this study came from clusters detected only in the PPJ or the DBT path. The purpose of the classifier at this stage is to moderately reduce the FPs in both paths while keeping the sensitivity high. Further FP reduction is accomplished by applying the previously developed decision rules for the CAD_{DBT} and CAD_{PPJ} systems to the DBT clusters and PPJ clusters, respectively, in the second FP reduction step.

The resulting clusters from the cascaded FP reduction steps are categorized into two groups; one contains those detected in DBT alone (i.e., DBT\PPJ), and the other contains clusters detected either in PPJ alone or in both DBT and PPJ. Since the CNR feature in PPJ has better discriminatory power for TP and FP candidates compared to that in the DBT volume, the DBT\PPJ clusters are mapped to the PPJ domain, that is, the locations of the cluster and the individual members in the cluster are projected onto the PPJ image, and segmentation and feature extraction for the objects are performed on the PPJ image. All clusters at this stage will therefore have features from the PPJ image, which will be input to the trained classifier for PPJ, Clr_{PPJ}, for final FP classification.

The joint CAD system was evaluated in a set of DBT test cases reconstructed by MSBF regularized-SART, from which the PPJ image was also generated for each DBT volume. The FROC analysis showed that the performance of the joint-CAD system is significantly higher than those of the CAD_{DBT} and CAD_{PPJ} systems alone, indicating that the combined approach is useful for improving microcalcification detection in DBT.

11.3.1.5 Microcalcification Detection in 2D Projection Views

If the 2D projection views for a DBT scan are available, an alternative approach is to detect the microcalcifications in the 2D projection views before reconstruction, making the CAD system independent of the reconstruction technique. Wei et al. [39] has developed a multi-channel response (MCR) approach for detection of microcalcifications in DBT using the set of projection view images as input. The main processing steps of the 2D CAD system are shown in Figure 11.11.

The projection-view images in a DBT scan are basically low-dose digital mammograms. Some of the CAD methods developed for mammograms can be applied to the projection views but the methods have to be adapted to the noisy images. The input images are preprocessed with the difference-image technique [85,86] to enhance microcalcification candidates and remove the low-frequency background on the individual projections. An iterative global thresholding technique then extracts a desired number of microcalcification candidates having top ranking of filter response on each projection view. Each candidate is segmented with a region-growing method based on a gray-level threshold adaptive to the local CNR of the candidate pixels and an 8-connectivity criterion. The shape, size, CNR and a refined centroid location of each candidate are determined after the local segmentation and used for preliminary reduction of FPs.

The microcalcification candidates are then subjected to the multi-channel response analysis. The Channelized Hotelling Observer (CHO) [87] that has been used to model signal detection tasks is adapted to extract signal response from the individual microcalcification candidates on the projection views. With the CHO model, a region of interest (ROI) centered at a microcalcification on a projection view image can be represented by a multi-channel filter bank using a set of orthonormal basis functions $\{b_1, \dots b_N\}$. The channelized basis functions are a set of optimal templates, each of which can characterize an image in specific frequency bands. A given ROI image can be decomposed into a linear combination of multiple channels:

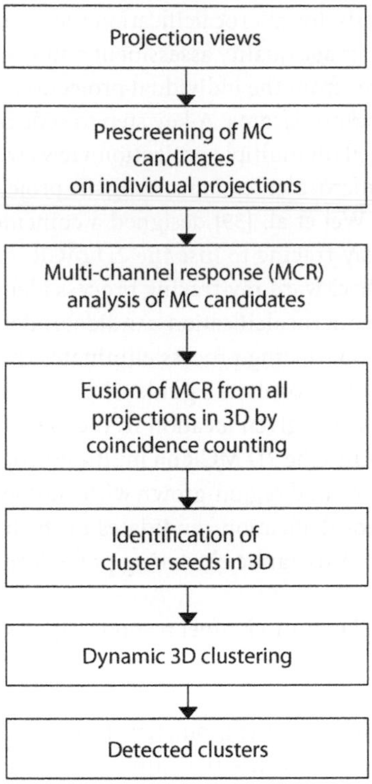

FIGURE 11.11 CAD system for detection of microcalcifications in DBT by applying a 2D approach to the projection view images. Multi-channel response (MCR) and coincidence counting methods are designed to extract signal response and fuse the response in 3D for FP reduction.

$$ROI(x,y) = \sum_{i}^{N} f_i b_i(x,y)$$

The ROI is therefore characterized by a set of channelized responses $\{f_1, \ldots, f_N\}$ that can be treated as a vector $\mathbf{f} = \{f_1, \ldots, f_N\}$ in the space spanned by the set of orthonormal basis functions $\{b_1, \ldots b_N\}$. With this representation, the task of differentiating the true and false microcalcification candidates can be formulated as a linear classification model to classify a given vector **f** into one of the two classes:

$$MCR(f) = (\overline{m_2} - \overline{m_1})^{\mathrm{T}} \Sigma^{-1} f$$

where:

$\overline{m_k}$ is the mean vector for class k, $k = 1,2$

Σ is an $N \times N$ covariance matrix for both classes

Therefore, the multi-channel response, MCR(f), is a weighted sum of the individual channel response $\{f_1, \ldots, f_N\}$ that can be used as a decision variable for ROC analysis. From the analysis of the channel responses of a set of training ROI samples containing true microcalcifications and FPs, the type of multi-channel basis functions, the number of channels N that are effective for characterizing microcalcifications can be determined by maximizing the separation between the two classes. Wei et al. [39] investigated two types of orthogonal basis functions: the Laguerre-Gaussian (LG) polynomials and the

Hermite-Gaussian (HG) polynomials, for microcalcification representation. The LG functions are commonly used in the CHO model for image quality assessment and simple signal detection tasks.

The microcalcification candidates from the individual projection views include many FPs because of the noisy nature of the low dose mammograms. A key step to reduce FPs is to utilize the fact that true microcalcifications are likely imaged on multiple projection views whereas FPs from noise are random. The location correspondence of a microcalcification among the projection views is encoded in the imaging geometry of the x-ray system. Wei et al. [39] designed a coincidence counting method based on a two-stage backward and forward ray-tracing to fuse the 2D MCR into a 3D response using the known geometry of the DBT system. The backward ray-tracing process identifies all possible locations (voxels) in the 3D breast volume where the microcalcification candidates detected on the projection views may originate from, while the forward ray-tracing process eliminates the redundant locations by using the coincidence counts accumulated in the voxels as a guide. Higher coincidence counts, and concurrently higher 3D MCR value, accumulated at a given location in the 3D volumes indicate a higher likelihood that an object at that location generates the 2D MCR on multiple projection views. The remaining voxels of high 3D MCR values are segmented and region-grown with 26-connectivity, and the resulting objects correspond to the location of microcalcification candidates in the breast volume. The locations of low MCR values are eliminated as FPs. A dynamic clustering procedure is then used to identify microcalcification clusters.

Wei et al. [39] showed that this 2D multi-channel response approach achieved high accuracy for the detection of microcalcifications in DBT, and the performances are comparable using either the LG or the HG polynomials as basis functions.

The set of projection views together with the geometry of the DBT system contain the same information as the reconstructed DBT volume because the latter is derived from the former. A well-designed reconstruction technique can combine the information from the projection views effectively by enhancing the signal and reducing noise. However, the performance of a CAD system using the reconstructed volume as input may depend on the reconstruction technique and parameters used because they affect the resulting image quality. On the other hand, detection of signals on the individual projection views may be more difficult because of the low exposure and thus noisy images but an advantage of this approach is that the CAD system will be independent of reconstruction. Fusion of the detected candidates on projection views in 3D using the DBT system geometry is, therefore, a crucial step to differentiate true and false signals. The coincidence counting method was found to increase the classification accuracy significantly. Further improvement of the fusion methods that may better utilize the information from multiple projection views in DBT warrants more extensive investigations if a CAD system that does not depend on a specific reconstruction technique is preferred.

11.3.2 Computer-Aided Detection of Masses

Non-calcified lesions, including masses, architectural distortion and focal asymmetry, are other important signs of breast cancer. For detection of masses, Chan et al. developed gradient field analysis and feature extraction methods in the DBT volume [88,89]. They also developed a 2D approach in which mass detection was performed on individual projection views, and the mass likelihood score estimated for the candidate lesions from the projection views were back-projected to the 3D volume to merge into a 3D mass likelihood score. Comparison of the combined 2D and 3D approaches to that by the 2D approach or 3D approach alone showed that the detection accuracy was improved significantly by the combined approach [90–92]. Chan et al. also compared the mass detection accuracy for different reconstruction methods with different number of iterations [93], evaluated the effect of different number of projection views and dose on detection [94] and compared mass detection in DBT and conventional mammograms [95]. Wei et al. evaluated false positive reduction method to further improve the mass detection performance [96]. Reiser et al. applied 2D and 3D radial gradient index segmentation methods to mass detection in DBT and compared the detection performance in the reconstructed DBT slices or 3D volume [97]. They also applied a mass detection

CAD algorithm originally developed for mammograms directly to the projection view images and merged the mass candidates from the projection views for feature analysis in DBT [98]. Peters et al. preprocessed the projection views with a wavelet filter and extracted the mass contours using several segmentation methods [99,100]. Jerebko et al. applied CAD algorithms developed for mammograms to projection views [101]. Singh et al. prescreened for mass candidates on projection view images, shifted and added to generate the suspicious locations in the 3D volume, and reduced false positives by an information-theoretic approach using a knowledge database from reconstructed DBT slices [102–105]. Mazurowski et al. [106] applied a mass CAD system developed for screen-film mammography to a single slice per DBT volume (i.e., central slice intersecting a mass for abnormal cases or a random slice for normal cases), and showed that their template-matching approach by mutual information trained with a knowledge database of mammographic masses can be transferred to DBT. Van Schie et al. applied a CAD system trained for mass detection on mammograms to DBT slabs of various thicknesses and compared three methods to merge the detected objects and determine their locations in 3D [107]. Palma et al. [108] developed a two-channel method that detected masses with a fuzzy approach. The second detection channel models the convergence characteristics of architectural distortions using an *a contrario* approach. Kim et al. [109] proposed a boosting framework to combine features extracted from the DBT volume with those from the projection views and showed that the sensitivity of the combined approach was significantly higher than those of the individual approaches. Using the same dataset, Kim et al. [110] compared the performance of mass detection in the DBT volume alone by combining detection in the DBT volume and in a simulated projection image. However, the detection performances using the DBT volume alone reported in the two studies were very different and no comparison of the two studies was provided so that it is unknown which approach might be more effective. Recently, Morra et al. [111] evaluated a commercial CAD system for DBT that performed detection in the reconstructed volume using proprietary methods. Review of CAD methods for computerized detection of masses in DBT can be found in the literature [112], and the details will not be discussed here.

Architectural distortion (AD) of the breast is similar to spiculated mass except that it does not have a central dense region like a mass. It appears as distortion of normal breast parenchyma or spiculations radiating from the central region. AD is associated with high positive predictive value for breast cancer, but it has a high false negative rate on mammograms [113,114], likely due to its low contrast and masking by overlapping tissue. It is therefore important to develop CAD methods to assist in AD detection. Some ADs may be detectable by CAD methods developed for masses, but in general it will require specifically designed detection techniques to achieve high sensitivity due to the lack of central density in AD. CAD techniques for detection of AD in mammograms can be found in the literature [115]. Partyka et al. [116] showed that DBT is superior to DM in detection of AD. Some of the CAD techniques developed for AD in mammograms may be applicable to detection in DBT slices or can be generalized to detection in the DBT volume. To date, there has not been a CAD system designed for AD in DBT. One reason may be due to the much lower prevalence of AD than that of mass, so it will take more time to collect a large dataset for CAD development.

Focal asymmetry is a concentration of density in a local region but does not fit the criteria of a mass; it appears on two views of the same breast but not in the corresponding region in the contralateral breast. The focal density will likely be detected as mass candidate at the prescreening stage of a mass CAD system but it may be dismissed as false positives during feature classification because their extracted features may not satisfy the criteria designed for masses. To distinguish focal asymmetry from normal dense tissue, it will require the analysis of corresponding regions in the two views of the same breast and in the views of the contralateral breast, similar to radiologists' strategy in reading screening mammograms, which were also found to be useful for improving sensitivity and specificity in CAD of masses on mammograms [117–121]. Some computer vision techniques developed for detecting focal asymmetry on bilateral mammograms were reviewed in [115]. These two-view and four-view CAD approaches will provide the framework for guiding the future development of similar techniques in DBT.

11.4 Summary

DBT is a new breast imaging technology that has been introduced into clinical use in recent years. Studies have shown that DBT can improve sensitivity of breast cancer detection and reduce recall rates. However, the detection of breast cancer manifested as subtle clustered microcalcifications in DBT is not as promising, and the impact of increase in reading time and how the radiologists adapt to the increased workload is still uncertain. CAD is expected to be a useful adjunct to DBT. The discussion in this chapter has focused in CAD for microcalcifications. However, computer-assisted visualization and interpretation of DBT will be equally important for all types of lesions. CAD methods have already been incorporated into technologies for generating synthetic mammograms for radiologists' preview of DBT. The preview synthetic mammogram may evolve to be a fast prescreening tool for many radiologists, and lesions not apparent in synthetic mammogram may be more likely to be overlooked. This application of CAD essentially uses CAD as a first reader, and, as such, it is crucial that the CAD system has very high sensitivity for detecting suspicious lesions so that they can be enhanced on the synthetic mammograms. Continued development of CAD methods to improve the sensitivity and specificity of lesion detection and characterization will therefore be an important component of DBT. An accurate and efficient CAD system, regardless of whether it is used as a second look or visualization aid, will accelerate the adaptation of DBT into routine clinical use and alleviate the need for digital mammograms. Nevertheless, the impact of CAD as a second reader on radiologists' interpretation and especially as a tool for generating synthetic mammograms with enhanced lesions to assist radiologists in prescreening should be investigated rigorously.

The development of CAD methods for DBT is similar to that for mammography in many aspects. However, the possibilities of developing computer vision techniques using the 2D projection views before reconstruction, the 2D reconstructed slices, the 3D reconstructed volume, the derivatives of the reconstructed images, such as the planar projection and synthetic mammogram, or a combination of the available images in different ways offers great opportunities for exploiting the information for lesion detection and diagnosis in DBT, but it is much more challenging to select the best approach to the problem. In addition, many alternative methods may be implemented for each approach, and the information from the different approaches may be combined at different stages of the algorithms. The development of an effective CAD system for DBT is a very high dimensional optimization problem and requires extensive efforts. The best combination of the techniques and parameters may depend on the quality of the DBT images, which, in turn, depends on the imaging system and acquisition geometries (e.g., the tomographic angle, the angular increments, and the number and distribution of the projection views), as well as the reconstruction and regularization techniques. Retraining is most likely necessary to translate a CAD system developed for one acquisition geometry or a specific reconstruction technique to another as the DBT system and reconstruction technique evolve over time. Even if a CAD system is designed with the goal to be adaptive to these variables, rigorous validation is still essential to assure the system can perform within expectation [122,123].

Common to any machine-learning problems is the fact that there are a number of parameters in the various image processing and decision-making steps in the CAD system. The optimization of these parameters depends on learning the image characteristics from the training set. In general, a more complex algorithm with a larger number of parameters requires a larger number of training samples, and overtraining with a limited training set will generalize poorly to independent cases [83,124–129]. The flexibility of combining the various 2D and 3D processing approaches for lesion detection or diagnosis in DBT increases the dimensionality of the parameter space and the complexity of the CAD system. The "test" set used in validating the performance during the optimization process becomes a part of the training set after numerous repeated uses so that it cannot guard against overfitting to the training or test cases. It will be imperative to use a large-enough set to train and validate the CAD systems, but ultimately the robustness of the developed systems can only be assured by testing with sequestered, independent cases representative of the intended population [123].

Acknowledgments

The work in digital breast tomosynthesis conducted at the University of Michigan was supported by National Institutes of Health awards R01 CA151443 and R21/R33 CA120234 (PI: Heang-Ping Chan), and from the efforts of many current and former members of the CAD Research Laboratory and many of our clinical colleagues in the Department of Radiology.

References

1. Lauby-Secretan B, Scoccianti C, Loomis D, Benbrahim-Tallaa L, Bouvard V, Bianchini F, Straif K. Int Agcy Res Canc Handbook W. Breast-Cancer Screening: Viewpoint of the IARC Working Group. *New England Journal of Medicine* 2015;372(24): 2353–8.
2. Siegel RL, Miller KD, Jemal A. Cancer Statistics, 2015. *Ca-a Cancer Journal for Clinicians* 2015;65(1): 5–29.
3. Hendrick RE, Helvie MA, Hardesty LA. Implications of CISNET modeling on number needed to screen and mortality reduction with digital mammography in women 40–49 years old. *American Journal of Roentgenology* 2014;203(6): 1379–81.
4. Mandelson MT, Oestreicher N, Porter PL, White D, Finder CA, Taplin SH, White E. Breast density as a predictor of mammographic detection: Comparison of interval- and screen-detected cancers. *Journal of the National Cancer Institute* 2000;92(13): 1081–7.
5. Pediconi F, Catalano C, Roselli A, Dominelli V, Cagioli S, Karatasiou A, Pronio A, Kirchin MA, Passariello R. The challenge of imaging dense breast parenchyma is magnetic resonance mammography the technique of choice? A comparative study with X-ray mammography and whole-breast ultrasound. *Investigative Radiology* 2009;44(7): 412–21.
6. Rosenberg RD, Yankaskas BC, Abraham LA, Sickles EA, Lehman CD, Geller BM, Carney PA, Kerlikowske K, Buist DSM, Weaver DL, Barlow WE, Ballard-Barbash R. Performance benchmarks for screening mammography. *Radiology* 2006;241(1): 55–66.
7. Skaane P, Bandos AI, Gullien R, Eben EB, Ekseth U, Haakenaasen U, Izadi M, Jebsen IN, Jahr G, Krager M, Niklason LT, Hofvind S, Gur D. Comparison of digital mammography alone and digital mammography plus tomosynthesis in a population-based screening program. *Radiology* 2013;267(1): 47–56.
8. Skaane P, Bandos AI, Gullien R, Eben EB, Ekseth U, Haakenaasen U, Izadi M, Jebsen IN, Jahr G, Krager M, Hofvind S. Prospective trial comparing full-field digital mammography (FFDM) versus combined FFDM and tomosynthesis in a population-based screening programme using independent double reading with arbitration. *European Radiology* 2013;23(8): 2061–71.
9. Bernardi D, Caumo F, Macaskill P, Ciatto S, Pellegrini M, Brunelli S, Tuttobene P, Bricolo P, Fanto C, Valentini M, Montemezzi S, Houssami N. Effect of integrating 3D-mammography (digital breast tomosynthesis) with 2D-mammography on radiologists' true-positive and false-positive detection in a population breast screening trial. *European Journal of Cancer* 2014;50(7): 1232–8.
10. Caumo F, Bernardi D, Ciatto S, Macaskill P, Pellegrini M, Brunelli S, Tuttobene P, Bricolo P, Fanto C, Valentini M, Montemezzi S, Houssami N. Incremental effect from integrating 3D-mammography (tomosynthesis) with 2D-mammography: Increased breast cancer detection evident for screening centres in a population-based trial. *Breast* 2014;23(1): 76–80.
11. Rose SL, Tidwell AL, Bujnoch LJ, Kushwaha AC, Nordmann AS, Sexton R. Implementation of breast tomosynthesis in a routine screening practice: An observational study. *American Journal of Roentgenology* 2013;200(6): 1401–8.
12. Friedewald SM, Rafferty EA, Rose SL, Durand MA, Plecha DM, Greenberg JS, Hayes MK, Copit DS, Carlson KL, Cink TM, Barke LD, Greer LN, Miller DP, Conant EF. Breast cancer screening using tomosynthesis in combination with digital mammography. *JAMA-Journal of the American Medical Association* 2014;311(24): 2499–507.

13. Lang K, Andersson I, Rosso A, Tingberg A, Timberg P, Zackrisson S. Performance of one-view breast tomosynthesis as a stand-alone breast cancer screening modality: Results from the Malmo Breast Tomosynthesis Screening Trial, a population-based study. *European Radiology* 2016;26(1): 184–90.
14. Skaane P, Bandos AI, Eben EB, Jebsen IN, Krager M, Haakenaasen U, Ekseth U, Izadi M, Hofvind S, Gullien R. Two-view digital breast tomosynthesis screening with synthetically reconstructed projection images: Comparison with digital breast tomosynthesis with full-field digital mammographic images. *Radiology* 2014;271(3): 655–63.
15. Gilbert FJ, Tucker L, Gillan MG, Willsher P, Cooke J, Duncan KA, Michell MJ, Dobson HM, Lim YY, Purushothaman H, Strudley C, Astley SM, Morrish O, Young KC, Duffy SW. The TOMMY trial: A comparison of TOMosynthesis with digital Mammography in the UK NHS Breast Screening Programme—A multicentre retrospective reading study comparing the diagnostic performance of digital breast tomosynthesis and digital mammography with digital mammography alone. *Health Technology Assessment* 2015;19(4): 1–136.
16. Poplack SP, Tosteson TD, Kogel CA, Nagy HM. Digital breast tomosynthesis: Initial experience in 98 women with abnormal digital screening mammography. *American Journal of Roentgenology* 2007;189(3): 616–23.
17. Gur D, Abrams GS, Chough DM, Ganott MA, Hakim CM, Perrin RL, Rathfon GY, Sumkin JH, Zuley ML, Bandos AI. Digital breast tomosynthesis: Observer performance study. *American Journal of Roentgenology* 2009;193(2): 586–91.
18. Wallis MG, Moa E, Zanca F, Leifland K, Danielsson M. Two-view and single-view tomosynthesis versus full-field digital mammography: High-resolution x-ray imaging observer study. *Radiology* 2012;262:788–96.
19. Kopans D, Gavenonis S, Halpern E, Moore R. Calcifications in the breast and digital breast tomosynthesis. *Breast Journal* 2011;17(6): 638–44.
20. Andersson I, Ikeda DM, Zackrisson S, Ruschin M, Svahn T, Timberg P, Tingberg A. Breast tomosynthesis and digital mammography: A comparison of breast cancer visibility and BIRADS classification in a population of cancers with subtle mammographic findings. *European Radiology* 2008;18:2817–25.
21. Spangler ML, Zuley ML, Sumkin JH, Abrams G, Ganott MA, Hakim C, Perrin R, Chough DM, Shah R, Gur D. Detection and classification of calcifications on digital breast tomosynthesis and 2D digital mammography: A comparison. *American Journal of Roentgenology* 2011;196(2): 320–4
22. Gennaro G, Toledano A, di Maggio C, Baldan E, Bezzon E, La Grassa M, Pescarini L, Polico I, Proietti A, Toffoli A, Muzzio PC. Digital breast tomosynthesis versus digital mammography: A clinical performance study. *European Radiology* 2010;20(7): 1545–53.
23. Teertstra HJ, Loo CE, van den Bosch MAAJ, van Tinteren H, Rutgers EJT, Muller SH, Gilhuijs KGA. Breast tomosynthesis in clinical practice: initial results. *European Radiology* 2010;20(1): 16–24.
24. Svahn TM, Chakraborty DP, Ikeda D, Zackrisson S, Do Y, Mattsson S, Andersson I. Breast tomosynthesis and digital mammography: a comparison of diagnostic accuracy. *British Journal of Radiology* 2012;85(1019): E1074–E82.
25. Gennaro G, Hendrick RE, Toledano A, Paquelet JR, Bezzon E, Chersevani R, di Maggio C, La Grassa M, Pescarini L, Polico I, Proietti A, Baldan E, Pomerri F, Muzzio PC. Combination of one-view digital breast tomosynthesis with one-view digital mammography versus standard two-view digital mammography: per lesion analysis. *European Radiology* 2013;23(8): 2087–94.
26. Thibault F, Dromain C, Breucq C, Balleyguier CS, Malhaire C, Steyaert L, Tardivon A, Baldan E, Drevon H. Digital breast tomosynthesis versus mammography and breast ultrasound: a multireader performance study. *European Radiology* 2013;23(9): 2441–9.

27. Waldherr C, Cerny P, Altermatt HJ, Berclaz G, Ciriolo M, Buser K, Sonnenschein MJ. Value of One-View Breast Tomosynthesis Versus Two-View Mammography in Diagnostic Workup of Women With Clinical Signs and Symptoms and in Women Recalled From Screening. *American Journal of Roentgenology* 2013;200(1): 226–31.
28. Shin SU, Chang JM, Bae MS, Lee SH, Cho N, Seo M, Kim WH, Moon WK. Comparative evaluation of average glandular dose and breast cancer detection between single-view digital breast tomosynthesis (DBT) plus single-view digital mammography (DM) and two-view DM: correlation with breast thickness and density. *European Radiology* 2015;25(1): 1–8.
29. Gur D, Zuley ML, Anello MI, Rathfon GY, Chough DM, Ganott MA, Hakim CM, Wallace L, Lu A, Bandos AI. Dose Reduction in Digital Breast Tomosynthesis (DBT) Screening using Synthetically Reconstructed Projection Images: An Observer Performance Study. *Academic Radiology* 2012;19(2): 166–71.
30. Chan H-P, Sahiner B, Wei J, Hadjiiski LM, Zhou C, Helvie MA. Digital breast tomosynthesis: Computerized detection of microcalcifications in reconstructed breast volume using a 3D approach. *Proceedings of SPIE* 2010;7624: 76241D.
31. Sahiner B, Chan H-P, Hadjiiski LM, Helvie MA, Wei J, Zhou C, Lu Y. Computer-aided detection of clustered microcalcifications in digital breast tomosynthesis: A 3D approach. *Medical Physics* 2012;39(1): 28–39.
32. Hadjiiski L, Chan H-P, Wei J, Sahiner B, Zhou CA, Helvie MA. Digital breast tomosynthesis: Feasibility of automated detection of microcalcification clusters on projections views. *Proceedings of SPIE* 2010;7624: 76241F.
33. Wei J, Chan H-P, Hadjiiski L, Helvie MA, Zhou C, Lu Y. Computer-aided detection of microcalcifications in digital breast tomosynthesis (DBT): A multichannel signal detection approach on projection views. *Proceedings of SPIE* 2012;8315: 83152D.
34. Samala RK, Chan H-P, Lu Y, Hadjiiski L, Wei J, Sahiner B, Helvie M. Detection of microcalcifications in breast tomosynthesis reconstructed with multiscale bilateral filtering regularization. *Proceedings of SPIE* 2013;8670: 86701L.
35. Bernard S, Muller S, Onativia J. Computer-aided microcalcification detection on digital breast tomosynthesis data: A preliminary evaluation. *IWDM 2008: Lecture Notes in Computer Science* 2008;5116:151–7.
36. Reiser I, Nishikawa RM, Edwards AV, Kopans DB, Schmidt RA, Papaioannou J, Moore RH. Automated detection of microcalcification clusters for digital breast tomosynthesis using projection data only: A preliminary study. *Medical Physics* 2008;35(4): 1486–93.
37. van Schie G, Karssemeijer N. Noise model for microcalcification detection in reconstructed tomosynthesis slices. *Proceedings of SPIE* 2009;7260: 72600M.
38. Samala RK, Chan H-P, Lu Y, Hadjiiski L, Wei J, Sahiner B, Helvie MA. Computer-aided detection of clustered microcalcifications in multiscale bilateral filtering regularized reconstructed digital breast tomosynthesis volume. *Medical Physics* 2014;41(2): 021901-1 (14 pages).
39. Wei J, Chan H-P, Hadjiiski LM, Helvie MA, Lu Y, Zhou C, Samala R. Multichannel response analysis on 2D projection views for detection of clustered microcalcifications in digital breast tomosynthesis. *Medical Physics* 2014;41(4): 041913-1 (11 pages).
40. Samala RK, Chan H-P, Lu Y, Hadjiiski LM, Wei J, Helvie MA. Digital breast tomosynthesis: Computer-aided detection of clustered microcalcifications on planar projection images. *Physics in Medicine and Biology* 2014;59(23): 7457–77.
41. Samala RK, Chan H-P, Lu Y, Hadjiiski LM, Wei J, Helvie MA. Computer-aided detection system for clustered microcalcifications in digital breast tomosynthesis using joint information from volumetric and planar projection images. *Physics in Medicine and Biology* 2015;60(21): 8457–79.
42. Lu Y, Chan H-P, Wei J, Hadjiiski LM. Selective-diffusion regularization for enhancement of microcalcifications in digital breast tomosynthesis reconstruction. *Medical Physics* 2010;37:6003–14.

43. Lu Y, Chan H-P, Wei J, Hadjiiski L, Zhou C. Multiscale regularized reconstruction for enhancing microcalcification in digital breast tomosynthesis. *Proceedings of SPIE* 2012;8313: 831322.
44. Sidky EY, Pan X, Reiser I, Nishikawa RM, Moore RH, Kopans DB. Enhanced imaging of microcalcifications in digital breast tomosynthesis through improved image-reconstruction algorithms. *Medical Physics* 2009;36(11): 4920–32.
45. Das M, Gifford HC, O'Connor JM, Glick SJ. Penalized maximum likelihood reconstruction for improved microcalcification detection in breast tomosynthesis. *IEEE Transactions on Medical Imaging* 2011;30(4): 904–14.
46. Acciavatti RJ, Maidment ADA. Observation of super-resolution in digital breast tomosynthesis. *Medical Physics* 2012;39(12): 7518–39.
47. Bernardi D, Ciatto S, Pellegrini M, Anesi V, Burlon S, Cauli E, Depaoli M et al. Application of breast tomosynthesis in screening: Incremental effect on mammography acquisition and reading time. *British Journal of Radiology* 2012;85(1020): E1174–E8.
48. Dang PA, Freer PE, Humphrey KL, Halpern EF, Rafferty EA. Addition of tomosynthesis to conventional digital mammography: Effect on image interpretation time of screening examinations. *Radiology* 2014;270(1): 49–56.
49. Zhang Y, Chan H-P, Goodsitt MM, Schmitz A, Eberhard J, Claus B. Investigation of different PV distributions in digital tomosynthesis mammography (DTM). *Proceedings of 9th International Workshop on Digital Mammography* 2008;LNCS 5116: 593–600.
50. Lu Y, Chan H-P, Wei J, Goodsitt MM, Carson PL, Hadjiiski L, Schmitz A, Eberhard JW, Claus BEH. Image quality of microcalcifications in digital breast tomosynthesis: Effects of projection-view distributions. *Medical Physics* 2011;38(10): 5703–12.
51. Hu YH, Zhao B, Zhao W. Image artifacts in digital breast tomosynthesis: Investigation of the effects of system geometry and reconstruction parameters using a linear system approach. *Medical Physics* 2008;35(12): 5242–52.
52. Gifford HC, Didier CS, Das M, Glick SJ. Optimizing breast-tomosynthesis acquisition parameters with scanning model observers. *Proceedings of SPIE* 2008;6917: 69170S.
53. Chawla AS, Lo JY, Baker JA, Samei E. Optimized image acquisition for breast tomosynthesis in projection and reconstruction space. *Medical Physics* 2009;36:4859–69.
54. Sechopoulos I, Ghetti C. Optimization of the acquisition geometry in digital tomosynthesis of the breast. *Medical Physics* 2009;36(4): 1199–207.
55. Reiser I, Nishikawa RM. Task-based assessment of breast tomosynthesis: Effect of acquisition parameters and quantum noise. *Medical Physics* 2010;37(4): 1591–600.
56. Mertelmeier T, Ludwig J, Zhao B, Zhao W. Optimization of tomosynthesis acquisition parameters: Angular range and number of projections. *Proceedings of 9th International Workshop on Digital Mammography* 2008;LNCS 5116: 220–7.
57. Tucker AW, Lu JP, Zhou O. Dependency of image quality on system configuration parameters in a stationary digital breast tomosynthesis system. *Medical Physics* 2013;40(3): 031917.
58. Chan H-P, Goodsitt MM, Helvie MA, Zelakiewicz S, Schmitz A, Noroozian M, Paramagul C et al. Digital breast tomosynthesis: Observer performance of clustered microcalcification detection on breast phantom images acquired with an experimental system using variable scan angles, angular increments, and number of projection views. *Radiology* 2014;273(3): 675–85.
59. Goodsitt MM, Chan H-P, Schmitz A, Zelakiewicz S, Telang S, Hadjiiski L, Watcharotone K et al. Digital breast tomosynthesis: Studies of the effects of acquisition geometry on contrast-to-noise ratio and observer preference of low-contrast objects in breast phantom images. *Physics in Medicine and Biology* 2014;59(19): 5883–902.
60. Park HS, Kim YS, Kim HJ, Choi YW, Choi JG. Optimization of configuration parameters in a newly developed digital breast tomosynthesis system. *Journal of Radiation Research* 2014;55(3): 589–99.

61. Niklason LT, Christian BT, Niklason LE, Kopans DB, Castleberry DE, Opsahl-Ong BH, Landberg CE et al. Digital tomosynthesis in breast imaging. *Radiology* 1997;205:399–406.
62. Wu T, Stewart A, Stanton M, McCauley T, Phillips W, Kopans DB, Moore RH et al. Tomographic mammography using a limited number of low-dose cone-beam projection images. *Medical Physics* 2003;30:365–80.
63. Suryanarayanan S, Karellas A, Vedantham S, Baker SP, Glick SJ, D'Orsi CJ, Webber RL. Evaluation of linear and nonlinear tomosynthetic reconstruction methods in digital mammography. *Academic Radiology* 2001;8:219–24.
64. Dobbins JT, Godfrey DJ. Digital x-ray tomosynthesis: Current state of the art and clinical potential. *Physics in Medicine and Biology* 2003;48:R65–R106.
65. Zhang Y, Chan H-P, Sahiner B, Wei J, Goodsitt MM, Hadjiiski LM, Ge J, Zhou C. A comparative study of limited-angle cone-beam reconstruction methods for breast tomosynthesis. *Medical Physics* 2006;33(10): 3781–95.
66. Zhang Y, Chan H-P, Sahiner B, Wei J, Zhou C, Hadjiiski LM. Artifact reduction methods for truncated projections in iterative breast tomosynthesis reconstruction. *Journal of Computer Assisted Tomography* 2009;33:426–35.
67. Zhang Y, Chan H-P, Sahiner B, Wu Y-T, Zhou C, Ge J, Wei J, Hadjiiski LM. Application of boundary detection information in breast tomosynthesis reconstruction. *Medical Physics* 2007;34(9): 3603–13.
68. Lu Y, Chan H-P, Wei J, Hadjiiski L, Samala R. Multiscale bilateral filtering for improving image quality in digital breast tomosynthesis. *Medical Physics* 2015;42(1): 182–95.
69. Vieira MAC, de Oliveira HCR, Nunes PF, Borges LR, Bakic PR, Barufaldi B, Acciavatti RJ, Maidment ADA. Feasibility study of dose reduction in digital breast tomosynthesis using non-local denoising algorithms. *Medical Imaging 2015: Physics of Medical Imaging*; 2015.
70. Zheng J, Fessler JA, Chan H-P. Digital breast tomosynthesis reconstruction with detector blur and correlated noise. *Proceedings of 4th International Conference on Image Formation in X-Ray Computed Tomography* 2016;CT-Meeting 2016: 21–4.
71. Mainprize JG, Bloomquist AK, Kempston MP, Yaffe MJ. Resolution at oblique incidence angles of a flat panel imager for breast tomosynthesis. *Medical Physics* 2006;33(9): 3159–64.
72. Acciavatti RJ, Maidment ADA. Oblique reconstructions in tomosynthesis. II. Super-resolution. *Medical Physics* 2013;40(11): 111912–1 (19 pages).
73. Zheng J, Fessler JA, Chan HP. Segmented Separable Footprint Projector for Digital Breast Tomosynthesis and Its application for Subpixel Reconstruction. *Medical Physics* 2017;44(3): 986-1001.
74. Claus B, Chan H-P. Digital breast tomosynthesis reconstruction with an adaptive voxel grid. *Proceedings of SPIE* 2014;9033: 90335A.
75. van Schie G, Mann R, Imhof-Tas M, Karssemeijer N. Generating synthetic mammograms from reconstructed tomosynthesis volumes. *IEEE Transactions on Medical Imaging* 2013;32(12): 2322–31.
76. Peters G, Muller S, Bernard S, Iordache R, Wheeler F, Bloch I. Reconstruction-independent 3D CAD for calcification detection in digital breast tomosynthesis using fuzzy particles. In *Proceedings of Conference 10th Iberoamerican Congress on Pattern Recognition, CIARP 2005: Progress in Pattern Recognition, Image Analysis and Applications*. November 15–18; Havana, Cuba: Springer-Verlag Berlin, 2005.
77. Park SC, Zheng B, Wang XH, Gur D. Applying a 2D based CAD scheme for detecting micro-calcification clusters using digital breast tomosynthesis images: An assessment—art. no. 691507. *Proceedings of SPIE* 2008;6915:6915071–8.
78. Lu Y, Chan H-P, Fessler JA, Hadjiiski L, Wei J, Goodsitt MM. Adaptive diffusion regularization for enhancement of microcalcifications in digital breast tomosynthesis (DBT) reconstruction. *SPIE Medical Imaigng Conference: Physics of Medical Imaging*, 2011;7961: 796117(1–9).

79. Zhang B, Allebach JP. Adaptive bilateral filter for sharpness enhancement and noise removal., *IEEE Transactions on Image Processing* 2008;17(5): 664–78.
80. Zheng J, Fessler JA, Chan H-P. Digital breast tomosynthesis reconstruction using spatially weighted non-convex regularization. *Proceedings of SPIE* 2016;9783: 978369.
81. Chan H-P, Sahner B, Wagner RF, Petrick N. Effects of sample size on classifier design for computer-aided diagnosis. *Proceedings of SPIE* 1998;3338:845–58.
82. Ge J, Hadjiiski LM, Sahiner B, Wei J, Helvie MA, Zhou C, Chan H-P. Computer-aided detection system for clustered microcalcifications: Comparison of performance on full-field digital mammograms and digitized screen-film mammograms. *Physics in Medicine and Biology* 2007;52(4): 981–1000.
83. Way T, Sahiner B, Hadjiiski L, Chan H-P. Effect of finite sample size on feature selection and classification: A simulation study. *Medical Physics* 2010;37(2): 907–20.
84. Quinlan JR. *C4.5: Programs for Machine Learning*. San Mateo, CA: Morgan Kaufmann; 1993.
85. Chan H-P, Doi K, Galhotra S, Vyborny CJ, MacMahon H, Jokich PM. Image feature analysis and computer-aided diagnosis in digital radiography. 1. Automated detection of microcalcifications in mammography. *Medical Physics* 1987;14:538–48.
86. Ge J, Sahiner B, Hadjiiski LM, Chan H-P, Wei J, Helvie MA, Zhou C. Computer aided detection of clusters of microcalcifications on full field digital mammograms. *Medical Physics* 2006;33(8): 2975–88.
87. Barrett HH, Yao J, Rolland JP, Myerss KJ. Model observers for assessment of image quality. *Proceedings of National Academy of Sciences* 1993;90:9758–65.
88. Chan H-P, Wei J, Sahiner B, Rafferty EA, Wu T, Roubidoux MA, Moore RH, Kopans DB, Hadjiiski LM, Helvie MA. Computerized detection of masses on digital tomosynthesis mammograms: a preliminary study. *Proceedings of 7th International Workshop on Digital Mammography* 2004;IWDM-2004: 199–202.
89. Chan H-P, Wei J, Sahiner B, Rafferty EA, Wu T, Roubidoux MA, Moore RH, Kopans DB, Hadjiiski LM, Helvie MA. Computer-aided detection system for breast masses on digital tomosynthesis mammograms: preliminary experience. *Radiology* 2005;237:1075–80.
90. Chan H-P, Wei J, Sahiner B, Rafferty EA, Wu T, Ge J, Roubidoux MA et al. Computer-aided detection on digital breast tomosynthesis (DBT) mammograms: comparison of two approaches. *RSNA Program Book* 2004;2004: 447.
91. Chan H-P, Wei J, Zhang Y, Moore RH, Kopans DB, Hadjiiski LM, Sahiner B, Roubidoux MA, Helvie MA. Computer-aided detection of masses in digital tomosynthesis mammography: Combination of 3D and 2D detection information. *Proceedings of SPIE* 2007;6514:161–6.
92. Chan H-P, Wei J, Zhang YH, Helvie MA, Moore RH, Sahiner B, Hadjiiski L, Kopans DB. Computer-aided detection of masses in digital tomosynthesis mammography: Comparison of three approaches. *Medical Physics* 2008;35(9): 4087–95.
93. Chan H-P, Wei J, Wu T, Sahiner B, Rafferty EA, Hadjiiski LM, Helvie MA, Roubidoux MA, Moore RH, Kopans DB. Computer-aided detection on digital breast tomosynthesis (DBT) mammograms: Dependence on image quality of reconstruction. *RSNA Program Book* 2005;2005: 269.
94. Chan H-P, Wei J, Zhang YH, Sahiner B, Hadjiiski L, Helvie MA. Detection of masses in digital breast tomosynthesis mammography: Effects of the number of projection views and dose. *Proceedings of 9th International Workshop on Digital Mammography* 2008;IWDM-2008: 279–85.
95. Chan H-P, Wei J, Sahiner B, Hadjiiski LM, Helvie MA. Comparison of computerized mass detection in digital breast tomosynthesis (DBT) mammograms and conventional mammograms. *Proceedings of SPIE* 2009;7260: 0S1–7.
96. Wei J, Chan H-P, Zhang Y, Sahiner B, Zhou C, Ge J, Wu YT, Hadjiiski LM. Classification of breast masses and normal tissues in digital tomosynthesis mammography. *Proceedings of SPIE* 2008;6915: 6915081–6.

97. Reiser I, Nishikawa RM, Giger ML, Wu T, Rafferty E, Moore RH, Kopans DB. Computerized detection of mass lesions in digital breast tomosynthesis images using two- and three dimensional radial gradient index segmentation. *Technology in Cancer Research and Treatment* 2004;3(5): 437–41.
98. Reiser I, Nishikawa RM, Giger ML, Wu T, Rafferty EA, Moore RH, Kopans DB. Computerized mass detection for digital breast tomosynthesis directly from the projection images. *Medical Physics* 2006;33:482–91.
99. Peters G, Muller S, Bernard S, Iordache R, Bloch I. Reconstruction-independent 3D CAD for mass detection in digital breast tomosynthesis using fuzzy particles. *Proceedings of SPIE* 2006;6144: 61441Z.
100. Peters G, Muller S, Grosjean B, Bernard S, Bloch I. A hybrid active contour model for mass detection in digital breast tomosynthesis: Art. no. 65141V. *Proceedings of SPIE* 2007;6514: 65141V.
101. Jerebko A, Quan Y, Merlet N, Ratner E, Singh S, Lo JY, Krishnan A. Feasibility study of breast tomosynthesis CAD system. *Proceedings of SPIE* 2007;6514:141–8.
102. Singh S, Tourassi GD, Lo JY. Breast mass detection in tomosynthesis projection images using information-theoretic similarity measures. *Proceedings of SPIE* 2007;6514:6514151–8.
103. Singh S, Tourassi GD, Baker JA, Samei E, Lo JY. Automated breast mass detection in 3D reconstructed tomosynthesis volumes: A featureless approach. *Medical Physics* 2008;35(8): 3626–36.
104. Singh S, Tourassi GD, Lo JY. Effect of similarity metrics and ROI sizes in featureless computer aided detection of breast masses in tomosynthesis. *IWDM 2008: Lecture Notes in Computer Science* 2008;5116:286–91.
105. Singh S, Tourassi GD, Chawla AS, Saunders RS, Samei E, Lo JY. Computer aided detection of breast masses in tomosynthesis reconstructed volumes using information-theoretic similarity measures. *Proceedings of SPIE* 2008;6915:6915051–8.
106. Mazurowski MA, Lo JY, Harrawood BP, Tourassi GD. Mutual information-based template matching scheme for detection of breast masses: From mammography to digital breast tomosynthesis. *Journal of Biomedical Informatics* 2011;44(5): 815–23.
107. van Schie G, Wallis MG, Leifland K, Danielsson M, Karssemeijer N. Mass detection in reconstructed digital breast tomosynthesis volumes with a computer-aided detection system trained on 2D mammograms. *Medical Physics* 2013;40(4): 041902–1 (11 pages).
108. Palma G, Bloch I, Muller S. Detection of masses and architectural distortions in digital breast tomosynthesis images using fuzzy and a contrario approaches. *Pattern Recognition* 2014;47(7): 2467–80.
109. Kim DH, Kim ST, Ro YM. Improving mass detection using combined feature representations from projection views and reconstructed volume of DBT and boosting based classification with feature selection. *Physics in Medicine and Biology* 2015;60(22): 8809–32.
110. Kim ST, Kim DH, Ro YM. Detection of masses in digital breast tomosynthesis using complementary information of simulated projection. *Medical Physics* 2015;42(12): 7043–58.
111. Morra L, Sacchetto D, Durando M, Agliozzo S, Carbonaro LA, Delsanto S, Pesce B et al. Breast cancer: Computer-aided detection with digital breast tomosynthesis. *Radiology* 2015;277(1): 56–63.
112. Chan H-P. Chapter 4. Detection and diagnosis of breast mass in digital tomosynthesis. In: Li Q, Nishikawa RM, editors. *Computer Aided Detection and Diagnosis in Medical Imaging*. Boca Raton, FL: CRC Press; 2015. 57–74.
113. Burrell HC, Sibbering DM, Wilson RM, Pinder SE, Evans AJ, Yeoman LJ, Elston CW, Ellis IO, Blamey RW, Robertson J. Screening interval breast cancers: Mammographic features and prognostic factors. *Radiology* 1996;199:811–7.
114. Warren RML, Young JR, McLean L, Lyons K, Wilson ARM, Evans A, Duffy SW, Warsi IM. Radiology review of the UKCCCR breast screening frequency trial: Potential improvements in sensitivity and lead time of radiological signs. *Clinical Radiology* 2003;58(2): 128–32.

115. Rangayyan RM, Ayres FJ, Desautels JEL. A review of computer-aided diagnosis of breast cancer: Toward the detection of subtle signs. *Journal of the Franklin Institute-Engineering and Applied Mathematics* 2007; 344 (3–4): 312–48.
116. Partyka L, Lourenco AP, Mainiero MB. Detection of mammographically occult architectural distortion on digital breast tomosynthesis screening: Initial clinical experience. *American Journal of Roentgenology* 2014;203(1): 216–22.
117. Wu Y-T, Wei J, Hadjiiski LM, Sahiner B, Zhou C, Ge J, Shi J, Zhang Y, Chan H-P. Bilateral analysis based false positive reduction for computer-aided mass detection. *Medical Physics* 2007;34:3334–44.
118. Wei J, Chan H-P, Sahiner B, Zhou C, Hadjiiski LM, Roubidoux MA, Helvie MA. Computer-aided detection of breast masses on mammograms: Dual system approach with two-view analysis. *Medical Physics* 2009;36:4451–60.
119. Wei J, Chan H-P, Zhou C, Wu Y-T, Sahiner B, Hadjiiski LM, Roubidoux MA, Helvie MA. Computer-aided detection of breast masses: Four-view strategy for screening mammography. *Medical Physics* 2011;38:1867–76.
120. Zheng B, Leader JK, Abrams GS, Lu AH, Wallace LP, Maitz GS, Gur D. Multiview-based computer-aided detection scheme for breast masses. *Medical Physics* 2006;33(9): 3135–43.
121. van Engeland S, Karssemeijer N. Combining two mammographic projections in a computer aided mass detection method. *Medical Physics* 2007;34(3): 898–905.
122. Huo ZM, Summers RM, Paquerault S, Lo J, Hoffmeister J, Armato SG, Freedman MT et al. Quality assurance and training procedures for computer-aided detection and diagnosis systems in clinical use. *Medical Physics* 2013;40(7): 077001-1 (13 pages).
123. Petrick N, Sahiner B, Armato SG, Bert A, Correale L, Delsanto S, Freedman MT et al. Evaluation of computer-aided detection and diagnosis systems. *Medical Physics* 2013;40(8): 087001-1 (17 pages).
124. Fukunaga K. *Introduction to Statistical Pattern Recognition*. 2nd edn New York: Academic Press; 1990.
125. Chan H-P, Sahiner B, Wagner RF, Petrick N. Classifier design for computer-aided diagnosis in mammography: Effects of finite sample size. *Medical Physics* 1997;24:1034–5.
126. Sahiner B, Chan H-P, Petrick N, Wagner RF, Hadjiiski LM. Feature selection and classifier performance in computer-aided diagnosis: The effect of finite sample size. *Medical Physics* 2000;27:1509–22.
127. Sahiner B, Chan H-P, Hadjiiski LM, Zhou C, Wei J, Petrick N. Comparison of resampling techniques for classifier performance estimation: The effects of feature selection, feature space dimensionality, and design sample size. *Medical Image Perception Conference X*, September 11–14, Durham, NC, 2003.
128. Sahiner B, Chan H-P, Hadjiiski L. Classifier performance prediction for computer-aided diagnosis using a limited data set. *Medical Physics* 2008;35:1559–70.
129. Sahiner B, Chan H-P, Hadjiiski LM. Classifier performance estimation under the constraint of a finite sample size: Resampling schemes applied to neural network classifiers. *Neural Networks* 2008;21:476–83.

12

Computer-Aided Diagnosis of Spinal Abnormalities

Marcello H. Nogueira-Barbosa

Paulo Mazzoncini de Azevedo-Marques

12.1 Introduction

There has been increasing interest in computerized methods for analysis and diagnosis of spinal diseases in the last decade (Rak and Tönnies 2016). Most published papers evaluated the spine via magnetic resonance imaging (MRI) due to its clinical importance, though radiography and computed tomography (CT) have also been the subject of many studies. MRI is the reference standard for diagnostic imaging for spinal diseases due to its high contrast resolution for soft tissues and for bone tissue. MRI provides different acquisition protocols dedicated to specific requirements based on the clinical presentation. Additionally, MRI does not use ionizing radiation for image acquisition, an important issue considering patients' safety (Semelka et al. 2007, Semelka et al. 2015).

The spine may be subdivided into its cervical, thoracic, lumbar, sacral, and coccygeal regions located, in order, from the superior to the inferior of the patient. Each vertebra, except for the first cervical vertebra, is composed of an anterior part, named the vertebral body, and a posterior vertebral arch (Figure 12.1).

Most published papers have focused on analysis of the vertebrae or intervertebral discs in the lumbar or lumbosacral region, solely or combined with the lower thoracic region. Cervical and cervicothoracic regions have been preferred in spinal canal and spinal cord studies (Rak and Tönnies 2016). Regarding assessment of the intervertebral discs and vertebral abnormalities, it could be interesting in the future to shift from a regional approach to evaluation of the whole spine. This trend greatly depends on the availability of and advances in MRI equipment and computer analysis capacity.

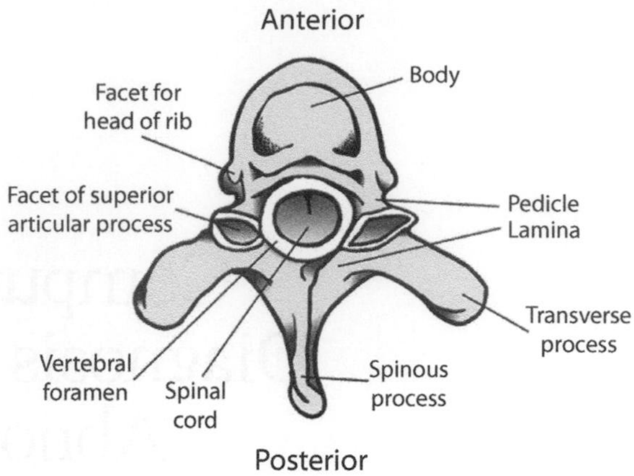

FIGURE 12.1 Schematic drawing of a thoracic vertebra showing the vertebral anatomical elements. The vertebral body corresponds to an anterior bone mass. The posterior vertebral arch is composed by the pedicle, the lamina and the articular and transverse processes, structures that appear both on the right and left side of the vertebra. The spinous process closes the posterior vertebral arch at the midline. The vertebral foramen is delimited anteriorly by the vertebral body and posteriorly by the posterior arch, and it accommodates the spinal cord inside.

In this chapter, we present an up-to-date review of clinical and technical aspects related to computer-aided diagnosis (CAD) in the evaluation of spinal diseases, with specific sections dedicated to the vertebrae and intervertebral discs. We have not found CAD systems developed for diagnosis of spinal cord diseases. Considering the importance of MRI in the clinical routine, we will primarily, but not exclusively, focus on this technique, and therefore, a specific section regarding MRI aspects related to image processing is included.

12.2 Magnetic Resonance Imaging

The extraction of quantitative information from routine MRI clinical sequences is an increasing and challenging tendency. Though most routine clinical sequences were not initially designed for quantitative evaluation, MRI pixel signals have data that could be used for quantitative analysis. Quantitative MRI techniques, on the other hand, are increasingly available for the assessment of spinal diseases, such as MRI relaxometry (Blumenkrantz et al. 2010), magnetic resonance spectroscopy—MRS (Fayad et al. 2010, Lee et al. 2010), and diffusion-weighted imaging—DWI (Khoo et al. 2011). These quantitative MRI sequences are not universally implemented in the clinical routine, especially for the investigation of degenerative spinal disease.

Signal intensity ranges are measurable from the routine clinical MRI sequences, but these measurements are not necessarily reproducible between regions of the same patient positioned differently relative to the receiver coil, or between different acquisition parameters, different equipment, and different clinical facilities. Image texture can vary from one scanner or manufacturer to the next. The gray levels representing the signal intensities measured with routine MRI sequences are dimensionless quantities, measured on an arbitrary scale, and are not specific physical measurable quantities. There are also unavoidable imperfections in the image acquisition process mainly related to magnetic field inhomogeneity.

As in any sectional imaging method, we also must consider the partial volume effect arising across tissue boundaries when assessing spine MRI. Across the boundary of two different tissues, the measured signal intensity is a combination of the tissues' signal intensities, weighted by their volumetric contribution to the image voxel.

MRI voxels in clinical routine acquisitions are often anisotropic, with the voxel dimension proportionally enlarged in a particular direction. This may influence the image segmentation quality. The use of anisotropic voxels is justified in the clinical routine due to the related reduction in image acquisition time and increase in efficiency, but it creates drawbacks in image processing.

The localization and labeling of intervertebral discs and vertebrae are crucial before segmentation and any automated diagnosis, since this information is necessary to guide clinical follow up and eventual treatment.

12.3 CAD Dedicated to Vertebral Abnormalities

The spine comprises a relatively constant number of irregular-shaped bones named vertebrae, assembled one on top of the following one. Each vertebra articulates superiorly (cranially) and inferiorly (caudally) with the adjacent vertebrae. The anterior region of each vertebra is named the vertebral body, which has its superior and inferior surfaces, referred to as the vertebral plateaus or vertebral end plates; each end plate is in contact with an intervertebral disc (Figure 12.2).

Computer-assisted methods applied to the investigation of vertebral bodies have focused on detection or classification of abnormalities. The normal vertebral body typically presents a nearly rectangular shape in lateral radiography and in sagittal planes of CT and MRI (Figure 12.3). Aging and degeneration may lead to small marginal bone outgrowths, known as osteophytes that may cause subtle shape abnormalities (Figure 12.4). Small focal depressions with rounded contours are frequently seen in vertebral end plates, usually idiopathic and representing small intravertebral herniation of the intervertebral disc (Figure 12.4). Such a focal vertebral contour abnormality is commonly named a Schmörl node. Schmörl nodes and marginal osteophytes are commonly present in vertebral bodies, as seen in radiography, CT and MRI, often without clinical significance.

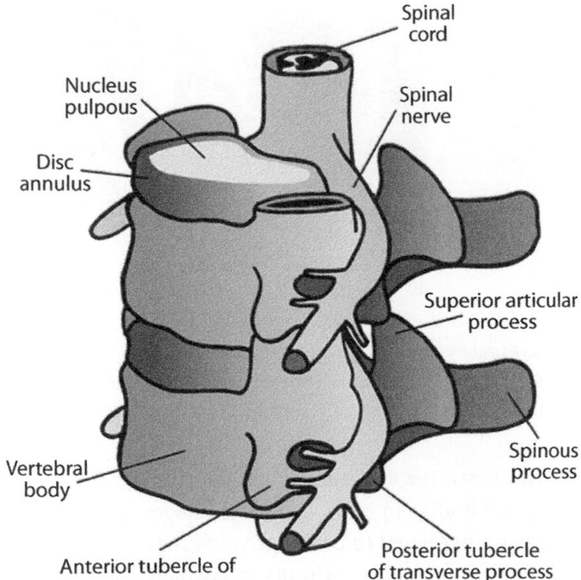

FIGURE 12.2 Schematic drawing of a spine segment or unit composed by two adjacent vertebrae and the corresponding intervertebral discs. The intervertebral disc at the top of the figure is depicted with its central portion, the nucleus pulposus, and its peripheral region, the anullus fibrosus, composed by concentric fibrous rings.

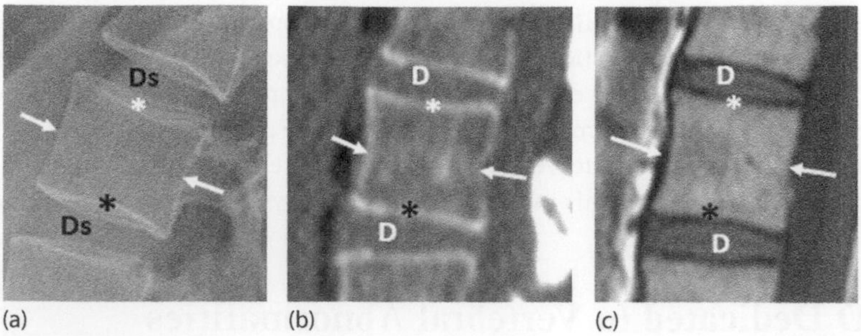

FIGURE 12.3 Localized zoomed-in images illustrating typical vertebral body aspect in different diagnostic imaging modalities. (a) Shows the vertebral body in radiography. (b) Shows the vertebral body in CT sagittal reconstruction. (c) Demonstrates the vertebral body in a sagittal plane of T1-weighted MRI. Black asterisks indicate the inferior vertebral plateau or inferior end plate in each imaging modality. White asterisks indicate the superior vertebral plateau or superior end plate in each case. White arrows indicate the anterior and posterior walls of the vertebral bodies. Ds = disc space; D = intervertebral disc.

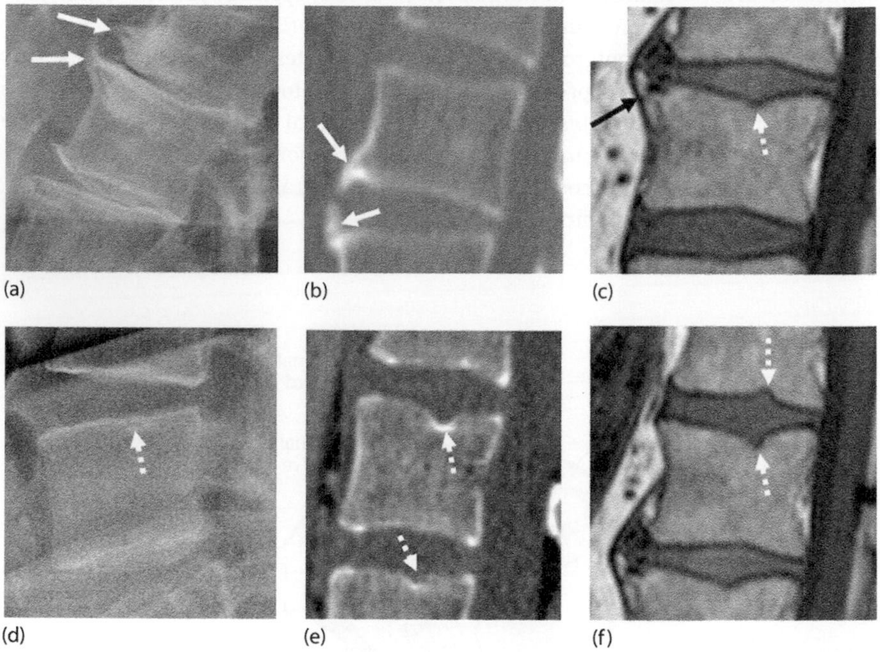

FIGURE 12.4 Examples of marginal osseous outgrowth in vertebral bodies, known as marginal osteophytes, indicated by solid arrows. Dashed arrows indicate small intravertebral disc material herniations, also known as Schmörl nodes. (a) and (d) Correspond to radiographic images, (b) and (e) to CT images, and (c) and (f) to T1-weighted MRI.

Normal vascular marks and foramina are important anatomical landmarks that may cause difficulties in segmentation of vertebral bodies in cross-sectional images obtained with CT and MRI. When present, vascular marks are usually related to drainage veins that merge in the posterior aspect of the vertebral body. Typically, there is a focal discontinuity of cortical bone at the middle third of the posterior vertebral body wall (Figure 12.5). Some important anatomical landmarks and characteristics in spine imaging are summarized on Table 12.1.

Previous literature on computer-assisted methods focused on detection of vertebral compression fracture (VCF), usually related to osteoporosis. More recently, computed-assisted methods have

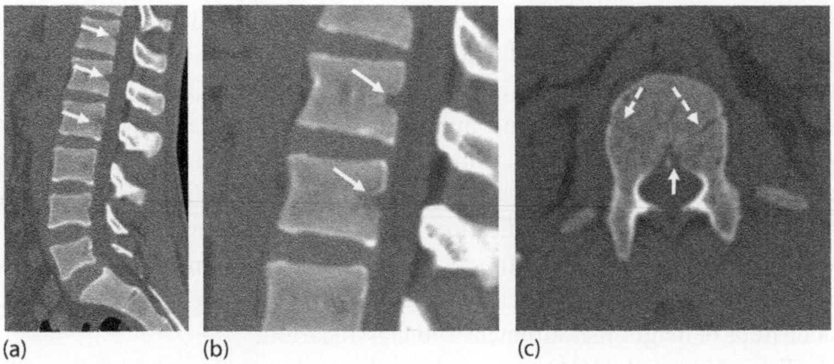

FIGURE 12.5 CT images demonstrating vascular channels or marks in vertebral bodies. (a)and (b) Correspond to a sagittal CT reconstruction of the lumbar spine and a detail of the same image. (c) Is an axial sectional CT image to show the vascular channels (dashed arrows). The solid-white arrows indicate the basivertebral foramen in the sagittal plane and the axial plane.

been used to assess the differentiation between benign (osteoporotic) and malignant (metastatic) VCFs.

12.3.1 CAD of Vertebral Body Fracture Based on Radiography

The application of computerized methods to spine radiographs is mainly dedicated to detection and diagnosis of vertebral body deformities related to osteoporosis. Osteoporosis affects one-third of women and one-fifth of men above the age of 50 years, according to the International Osteoporosis Foundation Annual Report (2004). The most important and most common outcome of osteoporosis is VCF (Vogt et al. 2000). Patients with VCF have a five-fold increased risk for subsequent fractures (Black et al. 1999). Furthermore, osteoporotic VCFs are associated with increased morbidity (Scane et al. 1999, Adachi et al. 2001, Tosteson et al. 2001, Oleksik et al. 2005) and mortality (Cooper et al. 1993, Hasserius et al. 2003, Kanis et al. 2004). After osteoporotic VCF diagnosis, pharmacologic treatment may reduce the risk of new fractures in the future (Liberman et al. 1995, Black et al. 1996, Chesnut III et al. 2000).

In clinical practice, osteoporosis-related VCF secondary to bone fragility is identified as a partial vertebral body collapse using diagnostic imaging techniques. Despite clinical importance, osteoporotic VCFs are frequently missed (Gehlbach et al. 2000, Kim et al. 2004, Bartalena et al. 2009). Therefore, methods to improve recognition of VCF are desirable.

Kasai et al. (2006) developed an automatic CAD system for VCF based on chest x-ray images. They extracted a vertebral area automatically using the posterior skin line. This area of interest was straightened so that vertebral end plates were oriented horizontally. Edge candidates were enhanced by a horizontal line enhancement filter, and multiple thresholding techniques, followed by feature analysis, were used for identification of vertebral end plates. After end-plate detection, the height of each vertebra was determined and fractured vertebrae were detected by comparison of the measured vertebral height with the expected height. The accuracy for the detection of vertebral end plates ranged between 70.9% and 76.6%, compared to the manual marking performed by radiologists. The sensitivity in the detection of VCFs varied from 75% to 95%.

Ribeiro et al. (2012) proposed a CAD method for the detection of VCFs using lateral radiographs of the lumbar spine. Gabor filters and an artificial neural network were applied to extract the superior and the inferior end plates of each vertebral body. Next, the anterior (Ha), posterior (Hp), and middle (Hm) heights of the vertebral bodies were derived automatically. Measured vertebral body heights were used to obtain VCF classification of each vertebral body as proposed by Genant et al. (1993). According to the criteria of Genant et al., a difference of more than 20% between the three measures of vertebral height represents a deformity indicative of VCF, considering the highest value as the reference. Grading of VCF was performed using the method of Genant et al.: normal (grade 0) if the percentage difference between vertebral heights is <20%, grade 1 deformity if the difference is 20%–25%, grade 2 deformity if the difference is 25%–40%, and grade 3 if the difference is >40%. Considering the classification of vertebral bodies as normal or fractured (abnormal), the algorithm's sensitivity, specificity, and accuracy were, respectively, 84.4%, 92.1%, and 87.3%, using the diagnosis of the radiologist as the gold standard. The CAD results achieved sensitivity of 78% and specificity of 95% using the same classification rules with manual measurements and the classification performed by a senior radiologist as the gold standard. Nevertheless, the proposed CAD methodology showed limitations in the steps of semiautomatic segmentation and identification of vertebral end plates, resulting in the loss of 36% of vertebral bodies for the subsequent steps of height measurement and classification.

More recently, Franchini et al. (2016) proposed another automatic method for vertebral morphometry measurements. The authors analyzed lateral spine radiographs by combining three different techniques, phase symmetry, Harris corner, and active shape models (ASM). The algorithm starts by searching for vertebrae through local measures of phase symmetry; the borders of vertebrae are highlighted in this step. Directional Log-Gabor wavelets are used next, resulting in filters with radial and angular components. A threshold is then applied, resulting in a black-and-white image. A Harris corner detector is used to detect the corners of each vertebral body. The ASM method employs a statistical model of vertebral shape repeatedly deformed to adapt its borders to the shape of the vertebra being analyzed. After identification of the vertebrae, the subsequent step is calculation of vertebral morphometry. The purpose is to identify the edges of each vertebra and to fix six morphometric points to calculate Hp, Hm, and Ha. Franchini et al. (2016) used the Melton approach to define a morphometric VCF, deciding on a VCF when the ratio of the heights in the same vertebra or between adjacent vertebrae are <0.85 (Melton et al. 1989). The algorithm for detection of vertebrae reached 89.1% sensitivity and 100% specificity compared to manually segmented vertebrae. Detection was missed in cases of vertebrae located at borders of an image or in high brightness areas. For each automatically detected vertebral body, the algorithm classified the vertebra as normal, biconcave deformity, crushing deformity, and wedge deformity. The accuracy of such classification was 92.8% using the classification provided by an experienced operator as the reference standard. No false-positive diagnosis of vertebral deformities was documented.

12.3.2 CAD of Vertebral Body Fracture Based on CT

Ghosh et al. (2011) proposed a fully automated CAD system for detection of lumbar VCFs in CT images using three height deviation measures of vertebral bodies as features. They obtained sensitivity, specificity, and accuracy of 91.67%, 98.41%, and 97.33%, respectively, using support vector machine (SVM), quadratic discriminant analysis (QDA), and naive Bayes classifiers together, considering the majority vote approach.

Al-Helo et al. (2013) developed an automated CAD system for detection of lumbar vertebral body fractures in CT images. The authors performed localization, labeling, and segmentation of vertebrae, and achieved the diagnosis for each vertebra. They performed segmentation via a coordinated system with an ASM and then refined the segmentation with a gradient vector flow active contour (GVF-Snake). The authors provide two machine-learning solutions, including a supervised learner (neural networks (NN)) and an unsupervised learner (k-means). They obtained a diagnostic accuracy of 93.2% on average using NN, and 98% on average accuracy using k-means. K-means resulted in sensitivity over 99% and specificity of 87.5%.

CT CAD methods for VCF showed to be accurate but there are some important points to consider before such methods may be implemented in the clinical scenario. CT is one of the largest contributors to radiation doses in patients (Semelka et al. 2007, Semelka et al. 2015), and therefore, it is not justifiable to use CT in screening for osteoporosis-related VCF. While CT is an important modality for spine trauma evaluation, MRI is the most adequate technique for the majority of clinical scenarios of spinal disease investigation. Despite cost and limited availability, MRI has better resolution contrast for bone marrow and soft tissue, and does not use ionizing radiation.

12.3.3 CAD Methods for Classification of Benign and Malignant Vertebral Body Fractures Based on MRI

Non-traumatic VCFs characteristically occur with deformity and partial collapse of the affected vertebral body. They are especially common in the elderly population due to bone failure and may be secondary to osteoporotic bone fragility or vertebral metastasis. Vertebral fragility fractures secondary to osteoporosis are an increasingly important health issue (Kondo 2008, Oei et al. 2013). The elderly population also has a high incidence of VCFs related to metastatic cancer in bone (Tehranzadeh and Tao 2004).

MRI is effective in early detection of VCFs (Uetani et al. 2004, Prasad and Schiff 2005). The distribution of signal intensity throughout the vertebral body is an important criterion to discriminate between benign and malignant VCFs using MRI (Cuenod et al. 1996). A malignant vertebral collapse typically exhibits diffuse low-signal intensity throughout the vertebral body in T1-weighted MRI (Cuenod et al. 1996). Less commonly, a malignant VCF may show a focal nodular low-signal lesion (Figure 12.6). Osteoporotic VCFs, on the other hand, characteristically demonstrate partial preservation of the normal bone marrow adipose tissue in T1-weighted MRI, with the low-signal abnormality distributed in an approximate band-like shape close to the impacted vertebral plateau (Figure 12.6). In clinical practice, the radiologist also compares the signal intensity in the vertebral body's bone marrow with the signal intensity in the intervertebral discs. Such a comparison helps the radiologist to decide if bone marrow is infiltrated by malignant disease. Usually, normal vertebral bodies exhibit higher signal intensity than the intervertebral discs in T1-weighted MRI because of the high-intensity bone marrow adiposity. A vertebral body with bone marrow signal intensity equal to or lower than the signal intensity of a neighboring disc may be suspicious for malignancy in adequate clinical context (Kaplan et al. 2001).

Analysis of the shape of a vertebral body can also help in the differentiation between benign and malignant VCFs (Cuenod et al. 1996). Non-traumatic VCFs may cause multiple types of changes in the contours of vertebrae. Malignant VCFs could result in a bulge or convexity of the posterior vertebral body wall, though this may occur together with a concave deformation of the vertebral end plates. Malignant processes may cause the contours of vertebrae to be relatively rounded or smoothened due to bulging neoplastic tissue. In the case of benign VCFs, the vertebral plateaus may acquire a more accentuated concave shape, and subchondral bone impaction may result in rough contours with indentations.

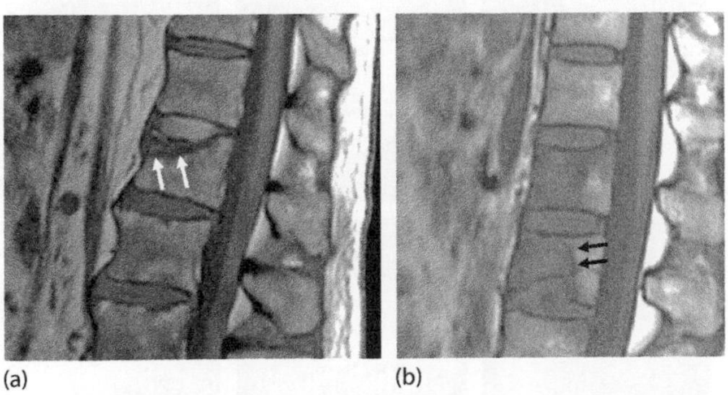

(a) (b)

FIGURE 12.6 T1-weighted sagittal plane MRI of two different patients. (a) Exemplifies a benign fracture related to osteoporotic vertebral compression fracture. White arrows indicate the low signal line or band related to bone trabecular impactation. (b) Illustrates a malignant vertebral compression fracture secondary to bone marrow neoplasm infiltration with a focal substitution lesion (black arrows).

Benign VCFs may cause posterior wall fragment retropulsion with angulated or irregular contours (Cuenod et al. 1996, Jung et al. 2003).

Azevedo-Marques et al. (2015) developed a CAD system to help in the differentiation between malignant and benign VCFs using T1-weighted MRI of the lumbar spine. They used sagittal plane MRI from 47 patients, including 19 malignant and 54 benign VCFs. Spectral and fractal features were extracted from manually segmented images of 73 fractured vertebral bodies. The classification of malignant versus benign VCFs was performed using a k-nearest-neighbor (kNN) classifier with the Euclidean distance. Features derived from Fourier and wavelet transforms, together with the fractal dimension, were able to obtain correct classification rate up to 94.7% with area under the receiver operating characteristic curve (AUC) up to 0.95.

Frighetto-Pereira et al. (2016) developed an algorithm using midsagittal images of T1-weighted MRI of the lumbar spine to assist in VCF diagnosis. They used features of gray levels, texture, and shape of vertebral bodies to detect VCFs, and further to classify VCFs as benign or malignant. The authors described a Fourier-descriptor-based feature (*FDF*) and analyzed its performance in conjunction with nine shape factors including convex deficiency (*CD*), compactness (C_o), and the seven measures based on central invariant moments as proposed by Hu (1962). Furthermore, they analyzed three statistical measures of gray levels, including the coefficient of variation (*CV*), skewness (*Skew*), and kurtosis (*Kurt*) (Joanes and Gill 1998), and the 14 texture features proposed by Haralick et al. (1973). The lumbar vertebral bodies were manually segmented and statistical features of gray levels were computed from their histograms. Extraction of texture and shape features was performed to analyze the signal variations in and contours of the vertebral bodies (Figures 12.7 and 12.8). The kNN method, a neural network with radial basis functions, and a naïve Bayes classifier were used with feature selection. They compared the classification obtained by these classifiers with the final diagnosis of each case, including biopsy for the malignant fractures and clinical and laboratory follow up for the benign fractures. Results reported show AUC of 0.97 in distinguishing between normal and fractured vertebral bodies, and 0.92 in discriminating between benign and malignant fractures.

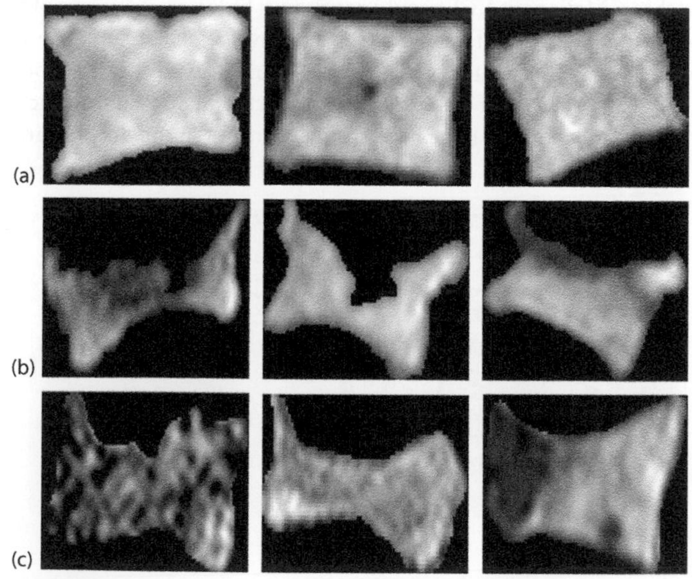

FIGURE 12.7 Nine examples of vertebral body ROIs from T1-weighted MRI of the lumbar spine. Row (a): Normal vertebral bodies. Row (b): Benign VCFs. Row (c): Malignant VCFs. See also Figure 12.8.

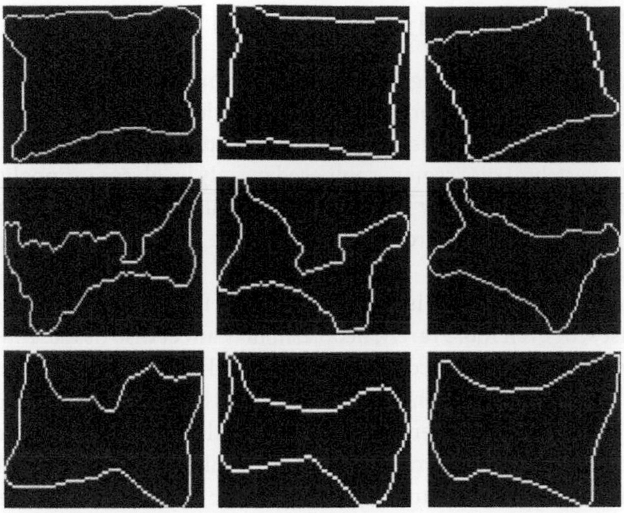

FIGURE 12.8 Nine examples of vertebral body contours ROIs. Row (a): Contours of normal vertebral bodies. Row (b): Contours of benign VCFs. Row (c): Contours of malignant VCFs. See also Figure 12.7.

12.3.4 CAD Methods for Subchondral Bone Marrow Abnormalities Related to Intervertebral Disc Degeneration

Signal-intensity abnormalities in vertebral body bone marrow adjacent to degenerated discs are commonly observed on MRI and have been described as taking three main forms (Modic and Ross 2007). Such bone marrow abnormalities are considered reactive or reparative changes secondary to abnormal mechanical stresses acting on the vertebral body due to intervertebral disc degeneration. Type I changes show decreased signal intensity in T1-weighted MRI and increased signal intensity in T2-weighted MRI, and represent vascularized fibrous tissue within the bone marrow. Type II changes show increased signal intensity in T1-weighted MRI and isointense or slightly hyperintense signal in T2-weighted MRI, findings related histologically to fat marrow replacement in the vertebral body adjacent to the degenerated disc. Type III changes are characterized by decreased signal intensity in both T1- and T2-weighted MRI, and usually correlate with extensive reparative bony sclerosis. The clinical importance of marrow changes associated with degenerative disc disease remains unclear. Type I changes seem to be correlated with a higher prevalence of active low back pain (Modic and Ross 2007).

Vivas et al. (2015) describe an approach for early detection of degenerative changes in lumbar intervertebral discs in MRI using a semiautomatic classifier. They used 115 discs extracted from MRIs of 23 patients with diagnosis of degenerative disc disease or back pain. The sagittal T2 MRI images were first evaluated by a specialist physician in training and then analyzed by the software to identify Modic changes and intervertebral disc hernia (protrusion or extrusion), which produced the results "normal or Modic" and "normal or abnormal," respectively. Modic is a semi-quantitative classification with three degrees based on signal intensity changes in the superior and inferior vertebral plateaus adjacent to each disc. Several classifiers were tested to determine the best parameter for defining disc disease. The authors conclude that the semiautomatic classifier could be a useful tool for early diagnosis or established disease. However, at the reported stage of development, software performance was limited, and the results showed 65%–60% certainty for Modic rating and 61%–58% for disc herniation, when compared with clinical evaluations.

12.4 CAD Dedicated to Intervertebral Disc Disease

The intervertebral disc is a fibrocartilaginous structure connecting and articulating two adjacent vertebral bodies, being also responsible for significant absorption of mechanical load. Intervertebral disc histology comprises two distinct regions named nucleus pulposus and annulus fibrosus. Nucleus pulposus is the central intervertebral disc region, composed mainly by glycosaminoglycans and collagen type II fibers, and typically with a strongly hydrated extracellular matrix. Annulus fibrosus is composed by several concentric ring layers of fibrocartilage with type I and type II collagen, and represents the peripheral intervertebral disc region.

The most common intervertebral disc disease is degeneration, although less frequently infection, inflammatory diseases, or even more rarely, neoplasm may affect disc tissue. Intervertebral disc degeneration usually occurs with loss of glycosaminoglycans and water content, with concomitant increase in type I collagen. Such biochemical compositional changes lead to structural failure. Annulus fibrosus fibers may tear and subsequently nucleus pulposus material may be displaced to peripheral regions of the disc. Focal disc material displacement beyond the limits of the intervertebral disc space delimited by the vertebral bodies gives rise to intervertebral disc herniation. Intervertebral disc herniation may cause or may contribute to spinal cord or nerve root compression when the disc displacement occurs toward neural tissue. Intervertebral disc displacement may contribute to spinal canal obliteration, a situation that may characterize spinal canal stenosis. Spinal cord and nerve root compression symptoms vary according to the compressed structure and depending on the level of compression. Intervertebral disc degeneration alone, without disc herniation or nerve compression, may cause localized back pain. Nerve root compression secondary to abnormal disc morphology is usually accompanied by pain radiating to the upper or the lower limb, respectively, in the case of nerve compression in the cervical or lumbar regions.

Intervertebral disc degeneration, disc herniation, and degenerative spinal stenosis are common diagnoses in the clinical routine and affect millions of people. Though any of these entities may be incidentally found in asymptomatic subjects, they may cause pain and disability and are among the biggest causes of labor incapacity and increasing healthcare costs. Spinal degenerative disease may include structural or morphological changes in other tissues than the intervertebral disc, such as the subchondral bone of the vertebral plateaus, the interapophyseal joints, and ligaments.

MRI is the modality of choice for evaluation of degenerative spinal disease. Many different intervertebral segments may be affected and several different anatomical structures may be involved in several degrees and combinations. Therefore, MRI diagnosis of spinal degenerative disease usually depends on a detailed and time consuming evaluation by the radiologist. Interrater agreement on MRI spinal degenerative abnormalities is often reported as moderate, at best. Therefore, CAD methods may offer a reliable way to quantify and classify degenerative spinal disease. Different algorithms have been developed for this purpose using several computational techniques and with variable success rates.

Michopoulou et al. (2009) proposed a texture-based pattern recognition method for cervical spine intervertebral disc degeneration. They used T2-weighted midsagittal 1.5 T cervical spine MRI and developed a classification system based on least squares minimum distance. The classification employed two classes, normal or degenerated intervertebral disc. Statistical analysis revealed statistically significant differences ($p < 0.05$) between normal and degenerated intervertebral discs for the texture feature values. The sensitivity, specificity, and accuracy achieved were, respectively, 96%, 92%, and 94%. The results have a potential value for use in a decision support tool for intervertebral disc degeneration assessment, but the system was tested on a relatively limited set of intervertebral discs, 25 normal and 25 degenerated. The proposed computerized approach is dependent on intervertebral disc ROIs manually delineated by a specialist, and this may be a limiting factor in use of the system in clinical practice.

Ghosh et al. (2011) described a robust CAD system for intervertebral disc herniation using lumbar spine T2-weighted MRI. They extracted a combination of features: raw, local binary patterns, Gabor, gray-level co-occurrence matrix, shape, and intensity features. The authors applied a probabilistic model

TABLE 12.1 Summary of the Common Appearance of Specific Anatomical Structures of the Spine in Different Imaging Modalities

Anatomical Structure	Imaging Technique		
	Radiography	CT	MRI
Vertebral body	Normal shape: nearly rectangular, anteroposterior diameter > height. Right and left borders of vertebral plateaus may project separately, with cylindrical appearance	Normal shape: nearly rectangular, anteroposterior diameter > height	Normal shape: nearly rectangular, anteroposterior diameter > height
Cortical or compact bone (bone surface)	High attenuation of x-rays, therefore depicted as bright pixels	High attenuation of x-rays, therefore depicted as bright pixels	Relaxation is fast in conventional sequences, therefore depicted as black or dark pixels
Cancelous or trabecular bone (inside vertebral body)	Discrete to moderate heterogeneity, with the higher density trabeculae intermixed with the low-attenuation porous regions	Discrete to moderate heterogeneity, with the higher density trabeculae intermixed with the low-attenuation porous regions	Bone trabeculae are frequently not well demonstrated in conventional T1-weighted and T2-weighted MRI spin echo or fast spin echo sequences. Spin echo and fast spin echo T1-weighted images have high contrast resolution to show substitution of normal adipose marrow tissue by other cell types with higher water content in the porous bone
Intervertebral disc	Not directly identified; disc space height gives indirect information about disc collapse	Normal disc tissue appears like homogeneous soft tissue, with intermediate gray level. Disc herniations and internal fissures may be identified	Normal disc tissue appears homogeneous with intermediate signal in T1-weighted images. T2-weighted images usually shows the annulus fibrosus with very low signal intensity. The normal nucleus pulposus is richly hydrated and demonstrate high signal on T2-weighted images. Degenerated disc is desiccated and shows dark signal on T2-weighted MRI
Spinal cord	Not identified	Soft tissue density, intermediate gray-level pixels. CT has low sensitivity for lesions and low contrast resolution between normal tissue and disease	Intermediate gray-level pixels. MRI has the highest sensitivity for lesions

for automatic localization and labeling of intervertebral discs from midsagittal MRI, searching for initial starting points inside each intervertebral disc for active shape model-based segmentation. They used dimensionality reduction techniques, principal component analysis (PCA) and linear discriminant analysis (LDA), and tested SVM, kNN, and naïve Bayes classifiers. LDA with naïve Bayes and LDA with kNN (k=5) obtained the best results. For PCA32+LDA+kNN (k=5), the sensitivity, specificity, and accuracy were 98.11%, 95.08%, and 96.0%, respectively. For PCA32 + LDA + Bayes, the sensitivity, specificity, and accuracy were 96.23%, 99.18%, and 98.29%, respectively.

In the next step, the same group developed a CAD framework for lumbar disc herniation in MRI, using a two-level classifier (Koh et al. 2012). In the first level, each classifier makes its own diagnosis, normal or abnormal, for lumbar disc herniation. The authors used four classifiers at the first level, a perceptron classifier, a least mean square (LMS) classifier, an SVM classifier, and a k-means classifier. In the second level, an ensemble classifier achieved a weighted sum of the first-level classifiers' scores, resulting in the final classification. The method was validated using a set of 70 clinical 3-T lumbar spine MRI, employing both T1-weighted and T2-weighted sagittal images for each patient. With this limited patient data, the proposed CAD obtained sensitivity, specificity and F-measure of 99.0%, 100.0%, and 98.9%, respectively. The CAD system achieved a high accuracy for lumbar intervertebral disc herniation, and the authors argued that their system could speed up the diagnosis by a factor of 30 times compared to a radiologist's diagnosis. The last consideration about the speed-up factor deserves comment, since it was based on an average of 15 min for the radiologist's diagnosis in the clinical environment; radiologists could be faster than that, on average, for lumbar spine MRI reading in other institutions. However, most important to note in this case, is that the average time used in the clinical scenario certainly includes more sophisticated and complex analysis, not restricted to decision on the presence or absence of lumbar intervertebral disc herniation.

Oktay et al. (2014) presented a new CAD method for the diagnosis of degenerative disc disease in midsagittal lumbar MRI. The system was tested and validated using 1.5 T midsagittal T2-weigthed and T1-weighted MRI from 102 subjects. After intervertebral disc detection and labeling, the discs were segmented using active appearance models (AAMs). Intensity, shape, context, and texture features were extracted with various techniques. An SVM classifier was applied to classify the discs as normal or degenerated. Oktay et al. (2014) implemented and evaluated the two previously mentioned systems (Ghosh et al. 2011, Koh et al. 2012) in order to compare the performance of those systems with their own system. Oktay et al. (2014) obtained lower accuracies for the methods of Ghosh et al. (2011) and Koh et al. (2012) than the previously reported accuracies by the same authors in their original analysis. Otkay and collaborators also encountered in their experiments higher sensitivity, specificity and accuracy when compared to the previously reported methods, with their own system results being 94.6%, 89.8%, and 92.81% in terms of sensitivity, specificity and accuracy, respectively. They mentioned that the previous systems were proposed for lumbar intervertebral disc herniation diagnosis, but they considered that those systems were also expected to diagnose degenerative disc disease, and that disc herniation is also a type of disc degeneration. It is not clear how they managed the criteria for disc degeneration that they used for the previous systems to compare them with their own CAD system. They did not use a statistical approach to compare the results of their experiments using the three systems, but instead used the average values of sensitivity, specificity and accuracy for the comparison.

Pfirrmann et al. (2001) described a semi-quantitative classification to provide a standardized and reliable assessment of MRI disc degeneration, allowing comparison of data from different investigations. This classification is not used in the clinical practice, since it is not related to prognosis and clinical symptoms. On the other hand, this classification is important in the research scenario, because it allows statistical comparison between different groups of patients and permits analysis of the evolution of disc disease longitudinally, using clinical routine MRI sequences.

Barreiro et al. (2014) developed a CAD system for intervertebral disc degeneration based on the classification of disc degeneration proposed by Pfirrmann et al. (2001), a semi-quantitative scale with five degrees of degeneration. The dataset used consists of T2-weighted midsagittal lumbar spine MRI of 210 discs obtained from 42 volunteers. Binary masks of manually segmented discs (Figure 12.9) were used to compute the centroids of the regions, estimate the curvature of the spine by polynomial fitting, normalize intensities, and extract ROIs. Texture analysis was performed using the texture features of Haralick et al. (1973), and moments were computed for each disc. The classification of discs was performed using the full attribute vectors with 15 features, with a multilayer perceptron (MLP) ANN. This

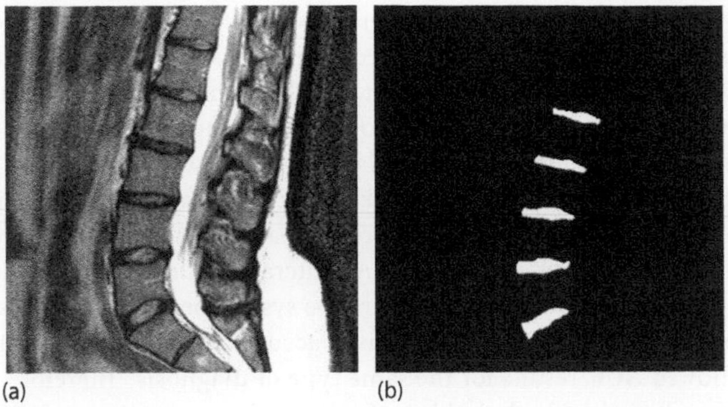

FIGURE 12.9 (a) T2-weighted MRI of the lumbar spine in the sagittal plane. (b) Binary masks of manually segmented intervertebral discs.

system obtained an average true-positive rate of 75.2% and an average AUC of 0.78 in the diagnosis of disc degeneration according to the classification of Pfirrmann et al. (2001).

Ruiz-España et al. (2015) also described a computer-aided approach to classify degenerative lumbar spinal disease based on the classification of Pfirrmann et al. (2001). They used a set of 52 lumbar intervertebral discs from 14 patients for training and a set of 243 lumbar intervertebral discs from 53 patients for testing. The authors used T2-weighted sagittal and axial 1.5T lumbar spine MRI. The same radiologist performed classification according to Pfirrmann et al. (2001) for disc degeneration and also classified each intervertebral disc morphology as normal, bulging disc or herniated disc (protrusion or extrusion). Semiautomatic analyses were performed by the same engineer blinded to the radiologist's classifications. The manual tasks were limited to placing a seed point in the intervertebral disc. A signal intensity method was used for segmentation and to detect spinal stenosis. They used the gradient vector flow algorithm to extract disc shape features and to detect contour abnormalities. The authors reported high reproducibility in the diagnosis of intervertebral disc herniation and spinal canal stenosis. The method was not adequate to identify correctly disc bulging. The CAD system presented was able to detect and also quantify spinal canal narrowing.

Conclusion

The literature shows significant advances in the development of computer-aided systems for the diagnosis of spinal diseases in the last 10 years. Robust CAD systems have been proposed for the diagnosis of vertebral body compression fractures with radiography, CT and MRI, and for the diagnosis of intervertebral disc degeneration and disc herniation using lumbar spinal MRI. Remarkable progress has been achieved in semiautomatic classification of benign osteoporotic and malignant vertebral compression fractures from routine lumbar spinal MRI. Some general observations need to be highlighted, without losing sight of the importance of these systems developed by different groups and the potential future application of the proposed methods in clinical practice.

First, we would like to point out the necessity of future work to validate the mentioned CAD systems in a practical and more generalizable environment. Most of the CAD systems were validated by the same group that developed it, using a relatively limited set of patients and usually without external validation from a different institution or hospital. We found only one example of an attempt to reproduce the results from a different research group (Oktay et al. 2014). For practical reasons, most studies were based on midsagittal MRI, but in real clinical practice, disc herniations, vertebral fractures, and malignant spinal bone infiltration may occur in focal and non-uniform patterns, and the midsagittal image may be insufficient for the diagnosis of a small, but not negligible portion of patients. The published CAD systems also need generalization to be applicable to the whole spine and not only to specific segments, such as the cervical or lumbar spine.

Second, because the published CAD systems were validated with limited numbers of patients compared to the general population, it seems to us necessary to use a systematic statistical approach to analyze the diagnostic performance of CAD system. The published papers usually give results in terms of average sensitivity, specificity, accuracy, or AUC, but no confidence intervals are calculated or published. In the same way, confidence intervals would be desirable when one intends to compare the performance of different systems. We do not know if global average accuracies of 92% and 94% calculated for two different systems using a sample of patients have statistically significant differences, unless we evaluate this. Our analysis of CAD literature for spinal disorders also shows that different authors may elect distinct ways to measure the system performance. It is not an easy task to compare results from the literature if one group used accuracy, a second group used the F-measure, and a third paper showed AUC results for the same type of diagnosis. Therefore, standardization of CAD performance results is highly desirable to facilitate the comparison of different systems presented in the literature.

Finally, it would be desirable to test if the already developed spinal CAD systems could influence the diagnosis of radiologists, clinicians, or surgeons, and if this influence would lead to positive effects in terms of the final diagnosis.

In summary, important spinal CAD systems, especially MRI-based CAD systems, have been developed and published, but further validation in a clinical scenario with more generalization is necessary before the CAD systems get approved and used in the clinical routine.

Acknowledgments

We thank Lucas Frighetto-Pereira and Marcelo Barrero for the preparation of Figures 12.7 through 12.9.

References

Adachi JD, Loannidis G, Berger C et al. 2001. The influence of osteoporotic fractures on health-related quality of life in community-dwelling men and women across Canada. *Osteoporos Int* 12: 903–908.

Al-Helo S, Alomari RS, Ghosh S, Chaudhary V, Dhillon G, Al-Zoubi MB, Hiary H, Hamtini TM. 2013. Compression fracture diagnosis in lumbar: A clinical CAD system. *Int J CARS* 8: 461–469.

Azevedo-Marques PM, Spagnoli HF, Frighetto-Pereira L, Menezes-Reis R, Metzner GA, Rangayyan RM, Nogueira-Barbosa MH. 2015. Classification of vertebral compression fractures in magnetic resonance images using spectral and fractal analysis. *Conf Proc IEEE Eng Med Biol Soc*: 723–726. doi: 10.1109/EMBC.2015.7318464.

Barreiro MS, Nogueira-Barbosa MH, Rangayyan RM, Reis RM, Pereyra LC, Azevedo-Marques PM. 2014. Semiautomatic classification of intervertebral disc degeneration in magnetic resonance images of the spine. *Biosignals and Biorobotics Conference (2014): Biosignals and Robotics for Better and Safer Living (BRC), 5th ISSNIP-IEEE*. doi: 10.1109/BRC.2014.6880984.

Bartalena T, Giannelli G, Rinaldi MF, Rimondi E, Rinaldi G, Sverzellati N, Gavelli G. 2009. Prevalence of thoracolumbar vertebral fractures on multidetector CT: Underreporting by radiologists. *Eur J Radiol* 69: 555–559.

Black DM, Arden NK, Palermo L, Pearson J, Cummings SR. 1999. Prevalent vertebral deformities predict hip fractures and new vertebral deformities but not wrist fractures. *J Bone Miner Res* 14: 821–828.

Black DM, Cummings SR, Karpf DB, Cauley JA, Thompson DE, Nevitt MC, Bauer DC, Genant HK, Haskell WL, Marcus R, Ott SM, Torner JC, Quandt SA, Reiss TF, Ensrud KE. 1996. Randomised trial of effect of alendronate on risk of fracture in women with existing vertebral fractures. Fracture intervention trial research group. *Lancet* 348: 1535–1541.

Blumenkrantz G, Zuo J, Li X, Kornak J, Link TM, Majumdar S. 2010. In vivo 3.0-tesla magnetic resonance T1ρ and T2 relaxation mapping in subjects with intervertebral disc degeneration and clinical symptoms. *Magn Reson Med* 6: 1193–1200.

Chesnut CH 3rd, Silverman S, Andriano K, Genant H, Gimona A, Harris S, Kiel D, LeBoff M, Maricic M, Miller P, Moniz C, Peacock M, Richardson P, Watts N, Baylink D. 2000. A randomized trial of nasal spray salmon calcitonin in postmenopausal women with established osteoporosis: the prevent recurrence of osteoporotic fractures study. PROOF study group. *Am J Med* 109: 267–276.

Cooper C, Atkinson EJ, Jacobsen SJ, O'Fallon WM, Melton LJ III. 1993. Population-based study of survival after osteoporotic fractures. *Am J Epidemiol* 137: 1001–1005.

Cuenod CA, Laredo JD, Chevret S et al. 1996. Acute vertebral collapse due to osteoporosis or malignancy: Appearance on unenhanced and gadolinium-enhanced MR images. *Radiology*, 199: 541–550.

Fayad LM, Wang X, Salibi N, Barker PB, Jacobs MA, Machado AJ, Weber KL, Bluemke DA. 2010. A feasibility study of quantitative molecular characterization of musculoskeletal lesions by proton MR spectroscopy at 3 T. *AJR Am J Roentgenol* 195: W69–W75.

Franchini R, Conversano F, Pisani P, Casciaro E, Peccarisi M, Quarta E, Grimaldi A, Muratore M, Casciaro S. 2016. Automatic method for vertebral morphometry measurements. *IET Sci Meas Technol* 10: 1–8.

Frighetto-Pereira L, Menezes-Reis R, Metzner GA, Rangayyan RM, Nogueira-Barbosa MH, Azevedo-Marques PM. 2015. Classification of vertebral compression fractures in magnetic resonance images using shape analysis. *Proceeding of IEEE E-Health and Bioengineering (EHB)*, Iasi, Romania, 1–4.

Frighetto-Pereira L, Rangayyan RM, Metzner GA, Azevedo-Marques PM, Nogueira-Barbosa MH. 2016. Shape, texture, and statistical features for classification of benign and malignant vertebral compression fractures in magnetic resonance images. *Comput Biol Med* 73: 147–156.

Gehlbach SH, Bigelow C, Heimisdottir M, May S, Walker M, Kirkwood JR. 2000. Recognition of vertebral fracture in a clinical setting. *Osteoporos Int* 11: 577–582.

Genant HK, Wu CY, van Kuijk C, Nevitt MC. 1993. Vertebral fracture assessment using a semiquantitative technique. *J. Bone Miner Res*, 8: 1137–1149.

Ghosh S, Alomari RS, Chaudhary V, Dhillon G. 2011a. Automatic lumbar vertebra segmentation from clinical CT for wedge compression fracture diagnosis. *Int J Comput Assist Radiol Surg* 8: 461–469.

Ghosh S, Alomari RS, Chaudhary V, Dhillon G. 2011b. Composite features for automatic diagnosis of intervertebral disc herniation from lumbar MRI. *33rd Annual International Conference of the IEEE EMBS Boston*, Massachusetts, MA, August 30–September 3.

Haralick RM, Shanmugam K, Dinstein IH. 1973. Textural features for image classification. *IEEE Trans Syst Man Cybern*, SMC-3: 610–632.

Hasserius R, Karlsson MK, Nilsson BE, Redlund-Johnell I, Johnell O. 2003. Prevalent vertebral deformities predict increased mortality and increased fracture rate in both men and women: A 10-year population-based study of 598 individuals from the Swedish cohort in the European Vertebral Osteoporosis Study. *Osteoporos Int* 14: 61–68.

Hu MK. 1962. Visual pattern recognition by moment invariant. *IRE Trans Inf Theory*, 8: 179–188.

Joanes DN and Gill CA. 1998. Comparing measures of sample skewness and kurtosis. *J R Stat Soc.* 47: 183–190.

Jung HS, Jee WH, McCauley TR, Ha KY, Choi KH. 2003. Discrimination of metastatic from acute osteoporotic compression spinal fractures with MR imaging. *Radiographics* 23: 179–188.

Kanis JA, Oden A, Johnell O, De Laet C, Jonsson B. 2004. Excess mortality after hospitalization for vertebral fracture. *Osteoporos Int* 15:108–112.

Kaplan PA, Helms CA, Dussault R, Anderson MW. 2001. *Musculoskeletal MRI*, 1st edn: Philadelphia, PA: Saunders Elsevier.

Kasai S, Li F, Shiraishi J, Li Q, Doi K. 2006. Computerized detection of vertebral compression fractures on lateral chest radiographs: preliminary results with a tool for early detection of osteoporosis. *Med Phys* 33: 4664–4675.

Khoo MMY, Tyler PA, Saifuddin A, Padhani AR. 2011. Diffusion-weighted imaging (DWI) in musculoskeletal MRI: A critical review. *Skeletal Radiol* 40: 665–681.

Kim N, Rowe BH, Raymond G, Jen H, Colman I, Jackson SA, Siminoski KG, Chahal AM, Folk D, Majumdar SR. 2004. Underreporting of vertebral fractures on routine chest radiography. *Am J Roentgenol* 182: 297–300.

Koh J, Chaudhary V, Dhillon G. 2012. Disc herniation diagnosis in MRI using a CAD frameworkand a two-level classifier. *Int J CARS* 7: 861–869.

Kondo KL. 2008. Osteoporotic vertebral compression fractures and vertebral augmentation. *Semin Interventional Radiol* 25:413–425.

Lee CW, Lee JH, Kim DH, Min HS, Park BK, Cho HS, Kang HG, Suh JS, Ehara S. 2010. Proton magnetic resonance spectroscopy of musculoskeletal lesions at 3 T with metabolite quantification. *Clin Imaging* 34: 47–52.

Liberman UA, Weiss SR, Bröll J, Minne HW, Quan H, Bell NH, Rodriguez-Portales J, Downs RW Jr, Dequeker J, Favus. 1995. Effect of oral alendronate on bone mineral density and the incidence of fractures in postmenopausal osteoporosis. The alendronate phase III osteoporosis treatment study group. *M. N Engl J Med* 333: 1437–1443.

Melton LJ, Kan SH, Frye MA et al. 1989. Epidemiology of vertebral fractures in women. *Am J Epidemiol* 129: 1000–1011.

Michopoulou S, Boniatis I, Costaridou L, Cavouras D, Panagiotopoulos E, Panayiotakis G. 2009. Computer assisted characterization of cervical intervertebral disc degeneration in MRI. *J Instrum* 4. doi: 10.1088/1748-0221/4/05/P05022.

Modic MT, Ross JS. 2007. Lumbar degenerative disk disease. *Radiology* 245: 43–61.

Oei L, Rivadeneira F, Breda FLSJ, et al. 2013. Review of radiological scoring methods of osteoporotic vertebral fractures for clinical and research settings. *Eur Radiol* 23: 476–487.

Oktay AB, Albayrak NB, Akgul YS. 2014. Computer aided diagnosis of degenerative intervertebral disc diseases from lumbar MR images. *Comput Med Imaging Graphics* 38: 613–619.

Oleksik AM, Ewing S, Shen W, van Schoor NM, Lips P. 2005. Impact of incident vertebral fractures on health related quality of life (HRQOL) in postmenopausal women with prevalent vertebral fractures. *Osteoporos Int* 16: 861–870.

Pfirrmann CWA, Metzdorf A, Zanetti M, Hodler J, Boos N. 2001. Magnetic resonance classification of lumbar intervertebral disc degeneration. *Spine* 26:1873–1878.

Prasad D, Schiff D. 2005. Malignant spinal-cord compression. *Lancet Oncol*, 6: 15–25.

Rak M, Tönnies KD. 2016. On computerized methods for spine analysis in MRI: A systematic review. *Int J Comput Assist Radiol Surg* 11: 1445–1465.

Ribeiro EA, Nogueira-Barbosa MH, Rangayyan RM, Azevedo-Marques PM. 2012. Detection of vertebral compression fractures in lateral lumbar X-ray images. *XXIII Congresso Brasileiro em Engenharia Biomédica (CBEB), Porto de Galinhas, PE, Brazil*, 1–4.

Ruiz-España S, Arana E, Moratal D. 2015. Semiautomatic computer-aided classification of degenerative lumbar spine disease in magnetic resonance imaging. *Comput Biol Med* 62: 196–205.

Scane AC, Francis RM, Sutcliffe AM, Francis MJ, Rawlings DJ, Chapple CL. 1999. Case-control study of the pathogenesis and sequelae of symptomatic vertebral fractures in men. *Osteoporos Int* 9: 91–97.

Semelka RC, Armao DM, Elias J Jr, Huda W. 2007. Imaging strategies to reduce the risk of radiation in CT studies, including selective substitution with MRI. *J Magn Reson Imaging* 25: 900–909.

Semelka RC, Armao DM, Elias Junior J, Huda W. 2015. Imaging strategies to reduce the risk of radiation in CT studies, including selective substitution with MRI. *JMag Res Imaging* 25: 900–909.

Tehranzadeh J, Tao C. 2004. Advances in MR imaging of vertebral colapse. *Semin Ultrasound, CT, MR* 25: 440–461.

Tosteson AN, Jönsson B, Grima DT, O'Brien BJ, Black DM, Adachi JD. 2001. Challenges for model-based economic evaluations of postmenopausal osteoporosis interventions. *Osteoporos Int* 12: 849–-857.

Uetani M, Hashmi R, Hayashi K. 2004. Malignant and benign compression fractures: Differentiation and diagnostic pitfalls on MRI. *Clin Radiol*, 59: 124–132.

Vivas ELA, Oliva FJC, Aguilar RQF, González AG, Cruz JG, Ávila JMJ. 2015. Application of semiautomatic classifier for MODIC and disk hernia changes in magnetic resonance. *Coluna/Columna* 14:18–22.

Vogt TM, Ross PD, Palermo L, Musliner T, Genant HK, Black D, Thompson DE. 2000. Vertebral fracture prevalence among women screened for the fracture intervention trial and a simple clinical tool to screen for undiagnosed vertebral fractures. *Mayo Clin Proc* 75: 888–896.

13

CAD of GI Diseases with Capsule Endoscopy

Yixuan Yuan

Max Q.-H. Meng

13.1 Introduction

Gastrointestinal (GI) diseases such as bleeding, ulcer and tumor can be cured or controlled in their early stages, otherwise they will deteriorate into cancers or some other fatal diseases [1]. The best way to diagnose these diseases in their early stages is to directly view the GI tract. Thus, endoscopy is widely used as a direct and effective diagnostic technology. Visualization of the upper and lower part of the GI tract (esophagus, stomach, duodenum, terminal ileum and colon) can be achieved by the use of conventional endoscopic techniques [2,3], such as colonoscopy and upper endoscopy. But these techniques require the insertion of the endoscope into the body of the patient through the mouth or anus, thus they are deeply inconvenient and painful to undergo, even dangerous for patients with cardiovascular issues. Moreover, the 7 m length, and the numerous windings required to reach the GI tract, make traditional endoscopies incapable of visualizing the entire small bowel due to its physical inaccessibility.

Wireless capsule endoscopy (Figure 13.1a) is a revolutionary device that provides direct, painless and noninvasive inspection of the GI tract [4]. It was developed in 2001 by Given Imaging. As shown in Figure 13.1b, it is a disposable, pill-shaped device that consists of one lens, Light Emitting Diodes (LEDs), a Complementary Metal Oxide Semiconductor (CMOS) camera, two batteries and a radio transmitter. The patient swallows the capsule, which captures images of the GI tract at a speed of two frames per second (fps). These images are compressed and transmitted wirelessly to a special recorder attached to the patient's waist. The entire inspective process lasts approximately 8 h until the batteries are exhausted. Finally, the images are downloaded from the recorder to a computer; thus, the clinicians, with the aid of the special software, can review the images and analyze potential sources of various GI diseases [5,6].

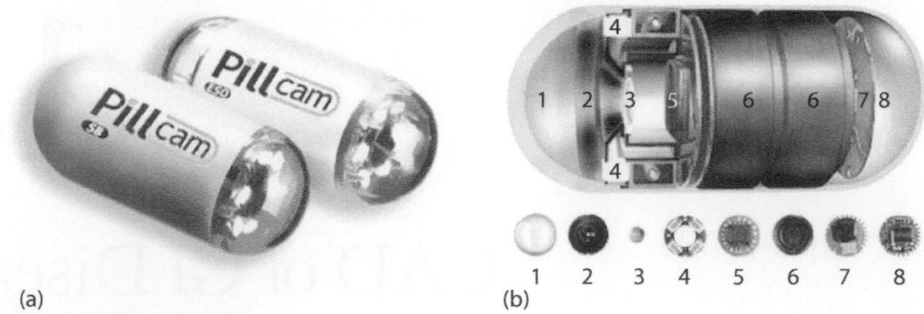

FIGURE 13.1 Wireless capsule endoscopy. (a) A typical WCE device. (b) Detailed components of a WCE device: (1) optical dome. (2) lens holder, (3) lens, (4) illuminating LEDS, (5) CMOS, (6) battery, (7) application specific integrated circuit transmitter, (8) antenna.

Since the capsule travels along the digestive tract with the physical peristalsis, the examination of the entire small intestine has become comfortable and painless to the patient.

No matter how advantageous and revolutionary WCE is over traditional techniques, there are challenging issues to deal with. One of the most crucial drawbacks that prevent WCE from being widely used is the time-consuming task of inspecting the images produced. The endoscopic capsule captures approximately 57,000 images (2 fps for 8 h approximately) during the whole diagnosis process and it takes more than 2 h for the clinician to review and analyze the data [7]. Moreover, abnormal findings are often visible in only one or two frames, thus they are easily missed [8]. In addition, the spatial characteristics of abnormal images such as shape, texture, size and the contrast with their surroundings vary, and it can be difficult for clinicians to reliably detect abnormality in all circumstances [9, 10]. Therefore, it is crucial to design an automatic computer-aided system to assist clinicians to analyze abnormal images.

Polyp is one of the most common intestinal diseases among adults, occurring in 30%–50% of adults under 50 years old. For people more than 50 years old, this percentage goes up to 90%. As shown in Figure 13.2a–c, polyp is defined as growing protrusions of mucosa inside the intestine due to excessive proliferation of tissue and inflammation or deep-seated malformations [11]. Although sometimes polyps are benign, virtually all colon and rectal cancers start from them [12]. Therefore, it is crucial for clinicians to detect polyps in their early stage automatically and cure them before they deteriorate to cancer cells.

In this chapter, we propose a novel automatic polyp-recognition method. The contribution of our work is summarized as follows:

1. Instead of using the existing Bag of Words (BoW) method to do the polyp-recognition task as in previous studies [10,13,14], we propose a novel strategy using improved Locality-Constrained Linear Coding (LLC) method combined with the Probabilistic Latent Semantic Analysis (PLSA) model to describe WCE images. In our method, each patch feature is represented by hybrid features: SIFT and CLBP. Then, the improved LLC method utilizes the locality constraints to project each integrated descriptor of images into its local-coordinate system, while the PLSA method explores the relationship among the visual words and the images by introducing high-level semantic features.
2. The intensive experiments are performed to analyze influences of the parameters in WCE polyp-recognition tasks. The experimental results show that the proposed approach outperforms the state-of-the-art approaches for polyp recognition.

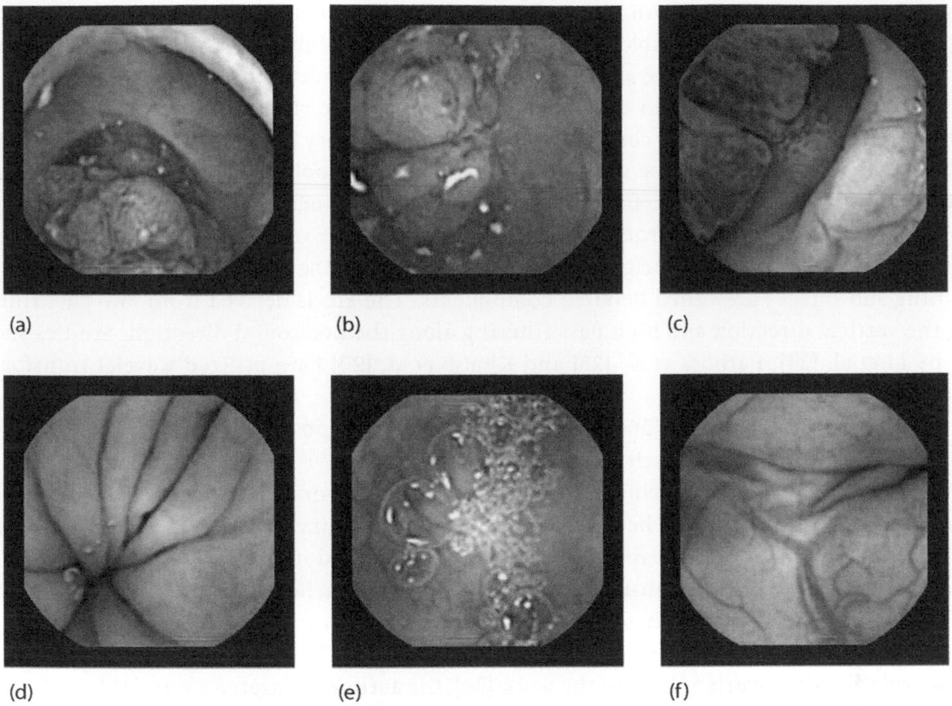

FIGURE 13.2 Examples of WCE images. (a–c) are the polyp images and (d–f) are the normal images.

13.2 Related Works

13.2.1 Review of Existing Polyp Recognition Methods

Many efforts and computational approaches toward polyp recognition in WCE images have been reported in literature. These methods extracted image features and then utilized classification methods to recognize the polyp frames in the WCE images.

The common feature used to distinguish the polyp from the normal images is color [15]. Examples of normal images and polyp images are shown in Figure 13.2. The WCE images are originally obtained in RGB color spaces, but the color images are transformed into different color spaces for the property of color invariance. Zhao et al. [16] extracted features from opponent color spaces. The images in the methods [11,17] are converted in to HSI/HSV color spaces before extracting texture features. The color statistic information and color histogram are also usually used to describe polyp images.

Besides color features, texture is another kind of frequently employed approach in polyp recognition, which includes filter-based features and local binary pattern features.

The filter-based feature is widely used in polyp recognition, since the filter cans describe images in different spaces and good filter banks can separate the polyp region from the normal region. A Gabor filter [18] is defined as a product of a sinusoidal function and a Gaussian kernel. By manipulating the frequency of the sinusoidal function and the standard deviation of the Gaussian kernel, we can obtain different band-pass filters. The methods in [19–21] used Gabor filters to preprocess images. In [11,22], Log-Gabor filters are applied on images to obtain polyp region candidates based on two advantages compared with Gabor features. The first advantage is there is no direct current (DC) component, which enables filtering out the absolute intensity of the original image and retaining meaningful details. The second advantage is an extended tail at the high-frequency end, thus Log-Gabor filters are able to encode natural images more efficiently than Gabor functions, which suppress higher frequency components. The wavelet theory [23] has been also widely applied to analyze texture feature because wavelet provides a capable tool for multi-resolution analysis of images. It transforms images into four sub-images which are generally denoted as LL, LH, HL and HH. The LL sub-image comes from low-pass filtering in both vertical direction and horizontal direction; thus, it looks like the original image, and it is called the approximation component. The remaining sub-images are called detailed components. The HL is derived from low-pass filtering along the vertical direction and high pass filtering along the horizontal direction. Studies such as those by Li et al. [24], Barbosa et al. [25] and Khatib et al. [20] have utilized wavelet transform to extract texture features.

The traditional LBP operator [26] is a texture descriptor proposed by Ojala et al. in 1996. It has been widely used in texture classification and shows impressive discriminative power. The LBP descriptor is obtained by comparing each pixel with its neighbors and recording the corresponding result as a binary number. Then the histogram of the binary numbers is used as the texture descriptor. LBP and its extended approaches [27,28] are adopted in analysis of WCE images as the methods of extracting texture feature. Yuan et al. [19] proposed a novel feature in which Monogenic LBP [27] is used after the image has been processed by Gabor transforms. And in [20], the texture features of GI tract images are represented by 144-dimensional feature vectors, which combine the LBP and rotationally invariant LBP. In the work [24], the author integrated a wavelet transform and a uniform local binary pattern together to characterize images. Our previous work [29] utilized the complete LBP (CLBP) [28] to characterize images since the magnitude part of the CLBP contains complementary features that may be of usage for texture classification.

As illustrated in Figure 13.2, the polyps in the WCE images also show different shape characteristics. Shape is another primary low-level image feature exploited by clinicians. A histogram of oriented gradients (HOG) [30] is a common shape feature extraction method based on local gradient orientation content. The methods in [31,32] extracted this feature for polyp recognition. Our previous study in [10] used the HOG to represent the local patch features. The method in [17] utilized Zernike moments to extract the shape feature based on its invariant to rotation, scale and translation.

13.2.2 Review of Feature Coding Methods

Although many methods have already been proposed to diagnose polyps automatically, the majority of published papers extracted features from a whole WCE image, and the features extracted from the non-polyp region may bring some noisy and redundant information to the classification. Along with rapid progress in the application of local descriptors, the Bag of Word (BoW) (Figure 13.3) approach has appeared as a promising alternative method for WCE image classification [10,13,14,33]. It detects points of interest in the image, extracts the corresponding features and then applies the clustering methods on the obtained features to get visual words. Then the images are represented by histograms of visual words through image coding methods.

Given an image, a set of M-dimensional local features $X = [x_1, x_2, \ldots, x_P] \in \mathbb{R}^{M \times P}$ can be extracted, where P is the number of image patches. Let B be the visual words, and $Z = [z_1, z_2, \ldots, z_P]$ be the final feature. We illustrate the BoW method and its variants as follows.

Traditional BoW method utilizes the Vector Quantization (VQ) [33] to represent local features, and each local feature is encoded by their nearest visual word. It is calculated by solving the following constrained least-square representation problem:

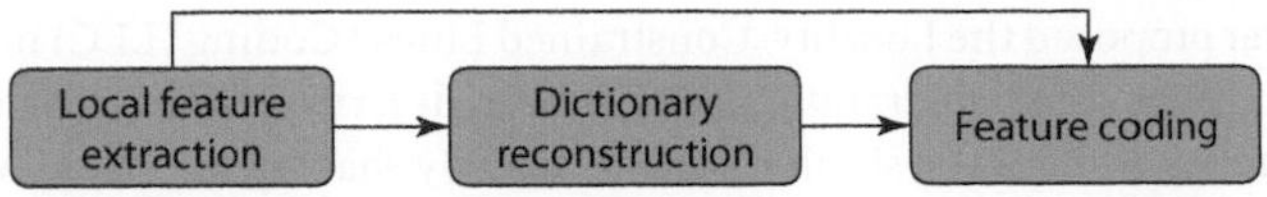

FIGURE 13.3 Illustration of BoW based methods.

$$\min_{Z} \sum_{i=1}^{P} \left\| x_i - Bz_i \right\|^2$$
$$\text{subject to } \left\| z_i \right\| = 1, z_i \geq 0 \forall i, \tag{13.1}$$

where the constraints $\|\mathbf{z}_i\| = 1$ and $\mathbf{z}_i \geq 0$ illustrate that only one nonzero element in z_i is assigned to 1 to characterize x_i. The illustration of VQ coding method is shown in Figure 13.4a. The final feature of the image is then represented by the histogram of visual words.

Our previous paper [10] proposed an improved BoW method to assist polyp recognition in the WCE images. Instead of utilizing a single Scale-Invariant Feature Transform (SIFT) feature [34,35] in the traditional BoW method, we extracted different textural features from the neighborhoods of the key points and integrated them together as synthetic descriptors to carry out recognition tasks. However, since each feature is only encoded by its nearest visual word, this hard-coding assignment of VQ usually leads to quantization loss.

To further improve the feature coding method, Yang et al. [36] proposed the sparse coding with spatial pyramid matching (ScSPM) method where sparse coding (SC) was used instead of VQ to obtain nonlinear codes (Figure 13.4b). It uses l_1 norm of z_i as the sparsity regularization term and the objective function is as follows:

$$\min_{Z} \sum_{i=1}^{P} \left\| x_i - Bz_i \right\|^2 + \lambda \left\| z_i \right\|$$
$$\text{subject to } 1^T z_i = 1, \forall i. \tag{13.2}$$

This cost function of the sparse coding can achieve good balances in reconstruction error and sparsity of the code vector. However, the sparsity constraint does not necessarily guarantee similar coding vectors for near local descriptors. Instead, it may lead completely different feature vectors even for similar images.

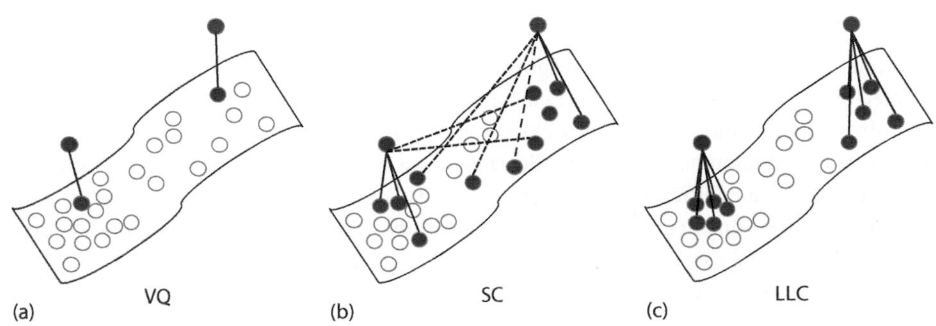

FIGURE 13.4 Illustration of coding methods. The red and deep blue points represent the patch features while the purple points show the selected visual words for coding bases.

Wang et al. [37] later proposed the Locality-Constrained Linear Coding (LLC) method (Figure 13.4c). This method utilizes the locality constraints to project each descriptor into its local-coordinate system and captures the correlations between similar local features by sharing the visual words.

13.2.3 Review of PLSA Model

Although the BoW model works quite well in some classification tasks, there are some disadvantages associated with the BoW model since it is usually affected by the synonyms of visual words and fails to reveal the semantics among words. To overcome this shortcoming, a generative model named Probabilistic Latent Semantic Analysis (PLSA) [38] has been proposed to disambiguate visual words by introducing a latent topic layer between the image and visual words. PLSA [39] was introduced by Hofmann originally for information retrieval and later extended for unsupervised learning and document classification [38]. In image analysis, it is assumed that each image consists of multiple topics, and the occurrences of visual words in images are results of topic mixture. In this way, the PLSA model could reduce the dimension of images since the number of topics is smaller than the number of visual words in images. The PLSA models have already been applied in medical image analysis [40–44] and yielded good performance. The method [40] used the PLSA to analyze image features and then classified images into different organs. Reza et al. [41] utilized the PLSA on the extracted features to provide more stable representation of the x-ray images.

13.3 Methods

In this chapter, we propose a novel strategy to recognize polyp images from WCE video clips. The framework of the proposed method is illustrated in Figure 13.5. For each WCE image, we first apply SIFT detector to extract key points, as shown in Figure 13.5a, and the extract an 8*8 patch size around the key points to calculate local texture features (Figure 13.5b). Since the polyp in the WCE image shows different texture features compared with surrounding mucosa, we extract the traditional SIFT feature and the CLBP feature and fuse them together to characterize the patch features. Then, the k-means clustering method (Figure 13.5e) is applied on the integrated features to obtain visual words, which we represent the WCE images as LLC histograms (Figure 13.5f). To further discriminate WCE images, we apply the PLSA model on the LLC histograms to discover latent topics and obtain the final features. Finally, the SVM method (Figure 13.5h) is applied on histograms of topics to recognize polyps in the WCE images. Specifically, there are two parameters in our method: the cluster numbers and the topic numbers. We carry out a detailed study on the influence of these parameters in terms of the recognition performance.

13.3.1 Key Points Extraction

Different from the feature describing an image in a holistic way, in our proposed method, a WCE image is represented by the combination of features around the key points. Key points in an image are defined as salient points that contain rich local information, thus the features around the neighborhood of these points could represent an image well. Actually, there exist many popular key point detectors, including Laplacian of Gaussian, Difference of Gaussian, Harris Laplace, Hessian Laplace, Harris Affine and scale-invariant feature transform (SIFT) [45]. In our study, we apply the SIFT method to detect key points due to its scale and rotation invariance properties. The SIFT is proposed by Lowe et al. in 1999 [34] and was further improved and developed by him [35]. The key points can be obtained through the following steps.

First, the scale space of an image is defined as a convolution of a variable-scale Gaussian $G(x, y, k\sigma)$ with input image $I(x, y)$. This is

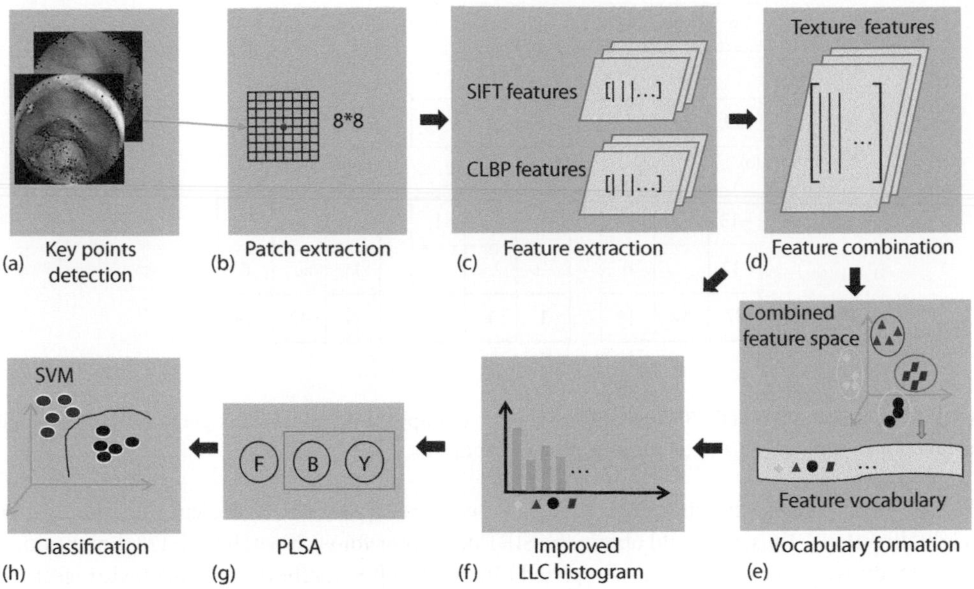

FIGURE 13.5 Workflow of the proposed method.

$$L(x,y,\sigma)=G(x,y,k\sigma)\otimes I(x,y), \tag{13.3}$$

where:

$$G(x,y,k\sigma)=\frac{1}{2\pi\sigma^2}e^{-\frac{(x^2+y^2)}{2k\sigma^2}},\ k=1,2,\ldots,n$$

After the convolution, the difference of the Gaussians (DOGs) that occurs at multiple scales is given by

$$L(x,y,\sigma)=\big(G(x,y,k_i\sigma)-G(x,y,k_j\sigma)\big)\otimes I(x,y)=L(x,y,k_i\sigma)-L(x,y,k_j\sigma). \tag{13.4}$$

Then, the key point is identified as local extrema of the DoG images across scales by comparing each pixel in the DoG images to its neighbors at the same and neighboring scales. If the pixel value is smaller or greater than all of the compared pixels, it is selected as a SIFT key point candidate.

Different patch sizes around the key points present different features of the WCE images. Previous experiments show that the patch size 8×8 gives best classification result [10]. Thus we fixed the patch size as 8×8 to carry out the experiments.

13.3.2 Feature Extraction

As shown in Figure 13.2, the WCE images with or without polyps exhibit different characteristics on their mucosa surfaces. In order to capture diversity characteristics of images, we extract different features: SIFT and CLBP in the neighborhoods of the key points and then combine them together to represent image local features.

When calculating SIFT feature, we determine 16×16 neighborhood areas of the key points and calculate the gradient magnitude and direction of each pixel in the neighborhood region. After that, we compute

9	12	34
11	22	28
99	64	56

(a)

0	0	1
0		1
1	1	1

(b)

Original LBP:
(1,1,1,1,0,0,0,1)

−13	−10	12
−11		6
77	42	34

(c)

−1	−1	1
−1		1
1	1	1

(d)

13	10	12
11		6
77	42	34

(e)

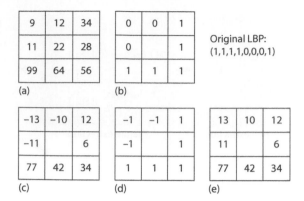

FIGURE 13.6 Illustration of CLBP. (a) 3×3 image patch example; (b) original LBP operator; (c) the local differences; (d) the sign component; and (e) the magnitude components.

the histogram of gradient direction in each 4×4 subregion, in which the gradient direction is divided into eight directions. Thus, we could obtain the SIFT descriptor for each patch with 128 dimension.

In this study, we propose to use the completed LBP (CLBP) [28] method to extract texture features in which each local region is represented by the center pixel and a local difference sign-magnitude transform (LDSMT). CLBP-Center (CLBP_C) represents the image gray level and is coded into a binary code by a global threshold. After computing LDSMT, the signs (CLBP_S) and the magnitudes (CLBP_M) of the local region can be calculated, respectively. Since these two vectors are complementary, they can perfectly recover the original difference vectors. We illustrated the calculation step of CLBP using an example given in Figure 13.6. Figure 13.6a shows the original 3×3 local structure of the WCE image with CLBP_C 22. The difference vector (Figure 13.6c) is [6,34,42,77,−11,−13,−10,12]. After applying LDSMT, the sign vector of this local region (Figure 13.6d) is [1,1,1,1,−1,−1,−1,1] and the magnitude vector (Figure 13.6e) is [6,34,42,77,11,13,10,12]. Finally, we obtain the rotation invariant texture features by combining CLBP_C, CLBP_S, and CLBP_M into the hybrid final CLBP histograms with 200 dimensions. It is clearly that the original LBP only uses the sign vector to code the local pattern as an 8-bit string [1,1,1,1,0,0,0,1], where −1 is coded as 0. Equipped with more complementary information, the CLBP method has more accurate performance to describe the images.

Then we integrate these two features together to describe local information of polyp images. Let $SIFT_i$ and $CLBP_i$ be the SIFT feature and CLBP feature of the interest point i. The final feature (x_i) will be

$$x_i = \left[SIFT_i, CLBP_i \right] \tag{13.5}$$

where the dimension of this integrated feature is 128 + 200

13.3.3 Calculation of Visual Words

After representing images as the features of the interest points, we apply the k-means algorithm on the integrated features extracted from training datasets to generate the visual vocabulary. The resultant cluster centers serve as a vocabulary of visual words. In the experiments, considering the trade-off between discrimination and generalization, we need to choose an appropriate vocabulary size. For example, if the size of visual words is too small, some features of key points will be assigned to the same cluster, although they are not similar, which leads the discrimination ability to decrease. On the other hand, the vocabulary may be difficult to generalize with extra processing time. In an attempt to address this problem, we change the vocabulary size from 100 to 400 and evaluate the corresponding classification performance.

13.3.4 Formation of LLC Histogram

Once obtaining the codebooks, we utilize the LLC method [37] to obtain the image representation. It applies locality constraint to select similar basis of local image descriptors from the codebook and captures the correlations between similar local features by sharing visual words.

Let $\mathbf{X} = [\mathbf{x}_1, \mathbf{x}_2, \ldots, \mathbf{x}_P] \in \mathbb{R}^{328\ P}$ be a set of 328 dimensional descriptor vectors for P patches in an image, $\mathbf{B} = [\mathbf{b}_1, \mathbf{b}_2, \ldots, \mathbf{b}_K] \in \mathbb{R}^{328\ K}$ be the visual words and $\mathbf{Z} = [\mathbf{z}_1, \mathbf{z}_2, \ldots, \mathbf{z}_P] \in \mathbb{R}^{K \times P}$ be the code for the image with LLC method. Then the objective function of LLC is specified as

$$\min_{\mathbf{Z}} \sum_{i=1}^{P} \left\| \mathbf{x}_i - \mathbf{B}z_i \right\|^2 + \lambda \left\| \mathbf{d}_i \odot \mathbf{z}_i \right\|^2 \qquad (13.6)$$
$$\text{subject to } \mathbf{1}^T \mathbf{z}_i = 1, \forall i,$$

where:

ψ is a weight parameter for adjusting the weights of locality

$\odot$ denotes element-wise multiplication

$\mathbf{d}_i \in \mathbb{R}^K$ is locality adapter that gives different freedom for each basis vector proportional to its similarity to the input descriptor x_i

$$\mathbf{d}_i = exp\left(\frac{dist(\mathbf{x}_i, \mathbf{B})}{\sigma}\right),$$
$$dist(\mathbf{x}_i, \mathbf{B}) = \left[dist(\mathbf{x}_i, \mathbf{b}_1), \cdots, dist(\mathbf{x}_i, \mathbf{b}_K)\right], \qquad (13.7)$$

where:

$dist(\mathbf{x}_i, \mathbf{b}.)$ is the Euclidean distance between $\mathbf{x}_i$ and each element of the codebook $\mathbf{B}$.

σ is used for adjusting the weight decay speed for locality adapter

13.3.5 Max Pooling Strategy

With the LLC method, the original patch descriptors X are encoded as Z. Then, these features are pooled together by max-pooling method to generate a WCE image representation f.

Suppose an image contains P patches and the codebook size is K, then after the max pooling, the image feature $f = [f_1, f_2, \ldots f_j, \ldots, f_K]$ will be represented by a K dimensional vector, and f_j is calculated by

$$f_j = max\left\{z_{1j}, z_{2j}, \ldots z_{Pj}\right\}. \qquad (13.8)$$

Since we extract multiple textural features around the neighborhood of interest points, the final feature of the improved LLC method could characterize images more accurately.

13.3.6 PLSA

Although the LLC method achieved good performance for polyp recognition in WCE images [10], it is usually affected by the synonyms of visual words and, thus, fails to reveal the latent semantics among words. It has been reported in literature [38] that Probabilistic Latent Semantic Analysis (PLSA) can disambiguate visual words, discover latent relations and result in the discovery of objects. Compared with the LLC model, the PLSA model assumes that the latent topics exist among visual words and each image can be represented by topics rather than words. Consequently, PLSA can discover topical similarities among words and simultaneously avoid the polysemy of words.

PLSA was introduced by Hofmann originally for information retrieval and later extended for unsupervised learning and image classification. The underlying idea of PLSA is that each image can be considered as a mixture of topics, where the topic is a probability distribution of visual words. Suppose we have a collection of M images $F = [f_1, f_2, \ldots f_j, \ldots, f_M]$ with visual words from a vocabulary $B = \{b_1, b_2, , b_k, b_K\}$. Let the probability of an image be $P(f_j)$, the probability of the latent topic y_i to the image f_j be $P(y_i|f_j)$ and the probability of the word b_k to the topic y_i be $P(b_k|y_i)$. Then the joint probability between a word b_k and the image f_j can be defined as following

$$P(f_j, b_k) = \sum_{i=1}^{N} P(f_j)P(b_k|y_i)P(y_i|f_j). \tag{13.9}$$

The goal is to determine a model that gives high probability to the words that appear in the images, then the parameters $P(f_j)$, $P(b_k|y_i)$, $P(y_i|f_j)$ can be estimated by maximizing the following log-likelihood function,

$$\begin{aligned} L &= \sum_{j=1}^{M}\sum_{k=1}^{K} n(f_j, b_k)\, logP(f_j, b_k) \\ &\propto \sum_{j=1}^{M}\sum_{k=1}^{K} n(f_j, b_k) log \sum_{i=1}^{N} P(b_k \mid y_i)P(y_i \mid f_j), \end{aligned} \tag{13.10}$$

where $n(f_j, b_k)$ is the number of occurrence of the term b_k in the image f_j.

Since the log-likelihood defined by Equation 13.10 is non-convex, the global optimality of this function is difficult to obtain. Thus, Expectation–Maximum (EM) algorithm [46] is applied to solve the above optimization problem to obtain the image representation based on topics $(P(y_i|f_j))$.

In general, the EM algorithm seeks to find out, by an iterative two-step process, the maximum likelihood estimate (MLE) of likelihood function of the complete data, including both observable data and latent aspects. The first step is called the expectation step (E-step), where posterior probabilities are computed for the latent variables, based on the current estimates of the parameters. The second step is the maximization step (M-step), where parameters are updated based on maximizing the so-called expected, complete, data log-likelihood, which depends on the posterior probabilities computed in the E-step.

In the E-step of the EM algorithm, the conditional probability distribution of the latent topic y_i is computed as follows:

$$\mathbf{P}(y_i \mid b_k, f) = \frac{P(b_k \mid y_i)P(y_i \mid f_j)}{\sum_{l=1}^{N} P(b_k \mid y_l)P(y_l \mid f_j)}, \tag{13.11}$$

In the M-step, the word probability $P(b_k|y_i)$ conditioned on the topic and the image probability $P(y_i|f_j)$ conditioned on the topic are updated as follows based on the new expected value $\mathbf{P}(y_i|b_k, f_j)$.

$$\mathbf{P}(b_k \mid y_i) = \frac{\sum_{j=1}^{M} n(b_k, f_j)P(y_i \mid b_k, f_j)}{\sum_{t=1}^{K}\sum_{j=1}^{M} n(b_t, f_j)P(y_i \mid b_t, f_j)}, \tag{13.12}$$

$$\mathbf{P}(y_i \mid f_j) = \frac{\sum_{k=1}^{K} n(b_k, f_j) P(y_i \mid b_k, f_j)}{n(f_j)}. \tag{13.13}$$

The EM algorithm starts with random values for $P(y_i)$, $P(b_k|y_i)$, and $P(y_i|f_j)$. After each EM iterations defined by Equations 13.11 through 13.13, we could compute the likelihood value L defined by Equation 13.10. If it is higher than the previously computed maximum value, the new value of L replaces the old one. The whole problem configuration, particularly the set of $P(f_j|y_i)$ values, is saved corresponding to the maximum likelihood value. After a number of iterations, when the algorithm converges and no further improvements in L could be obtained, the most dominant topic of each image is ascertained to be y_i where $P(f_j|y_i)$ is maximum for $i = l$. Thus, $P(yi|fj)$ is obtained as the final image feature with the latent topics. Then, we apply LibSVM [47] with Gaussian radial basis function kernel to carry out the polyp image recognition task.

13.4 Experiment Results

13.4.1 Image Acquisition and Experimental Setup

The images used for the development and evaluation of the proposed approach were extracted from 10 patients' WCE videos. The examinations were conducted using the by Pillcam SB WCE system, and Rapid Reader 6.0 software (Both from Given Imaging Ltd) was employed to export the images from the video sequence. In this chapter, we composed a dataset that consists of 500 polyp and 500 normal images. The 500 polyp images were obtained from different polyp regions to achieve the lowest possible similarity. Furthermore, the normal images include both simple and confusing healthy tissue (folds, villus and bubbles) to simulate the actual discrimination process.

13.4.2 Experiment Results of the Polyp Recognition

In this chapter, we proposed a novel strategy using LLC and PLSA to recognize polyps in the WCE images. In the proposed method, there are two free parameters: the number of visual words *(K)* and the number of topics (0). The classification performance of our model by using different sets of these parameters was tested. *K*-values between 100 and 400 and the number of Q between 10 and 60 were analyzed, with increments of 50 and 10, respectively. Figure 13.7 illustrates the classification performance for each parameter combination. The best polyp classification performance is obtained with the number of visual words, 250, and the number of topics, 50. The corresponding accuracy, sensitivity and specificity are 90.53%, 88.13% and 92.96%, respectively.

To further show the effectiveness of our features, we plotted six examples of the feature histograms with the above parameter sets (the number of visual words set at 250, and the number of topics at 50) in Figure 13.8. The six histograms correspond to the six example images in Figure 13.2. We can see that the histograms of the different types of images show different characteristics. The histograms of polyps demonstrate extremely high value for specific topics, while the histograms of the normal images show uniformity distribution on the different topics. Thus, the polyp images and the normal WCE images could be further classified.

13.4.3 Comparison Results with the Traditional LLC Method

To validate the efficiency of the PLSA, we compared our method with the one without the PLSA, which just utilizes the LLC to encode the images. We set the number of topics as 50 and plotted the corresponding classification accuracy and sensitivity under different sizes of visual words in Figure 13.9. We found

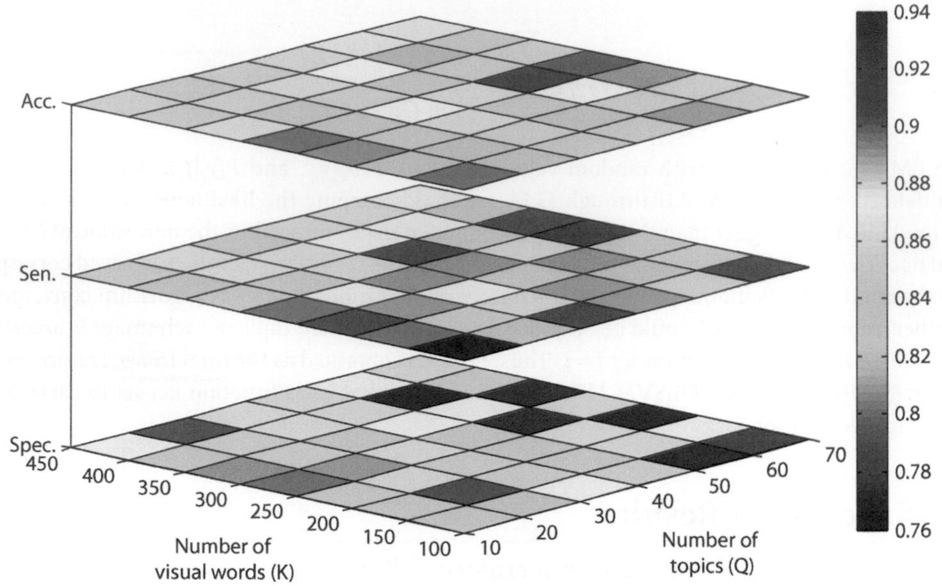

FIGURE 13.7 Polyp classification performance in the WCE image.

that our method shows higher accuracy and sensitivity compared with the original LLC method. This result validates the importance of the PLSA.

13.4.4 Comparison Results with Existing Polyp Recognition Methods

To further evaluate the performance of the proposed method, we compared it with state-of-the-art polyp diagnosis methods [13,19,24,]. The method in [13] introduced a BoW method with the local feature of Gabor filters and color histograms to recognize the polyps, while Li et al. [24] proposed a new texture feature that combines the advantages of the wavelet transform and the uniform local binary pattern descriptor to discriminate between normal images and polyp images.

Table 13.1 shows the accuracy, sensitivity and specificity by the existing polyp recognition methods and ours, respectively. It is clear that our proposed method shows better classification performance with higher accuracy, sensitivity and specificity. The proposed method demonstrates an improvement of 7.26%, 4.81%, 3.02% in accuracy and 6.22%, 4.81%, 1.94% in sensitivity for polyp image detection compared with the methods [13,19,24,], respectively. This result validates that the proposed method possesses superior ability to characterize WCE images and demonstrates good discriminative capability for polyp recognition.

In [19,24], the authors extracted features from the whole image to represent the image information, but the features extracted from the non-polyp region may bring noisy and redundant information to the classification. Although the method [13] utilized the local feature to encode the images, it didn't consider the latent relationship among image and visual words. In our method, we first utilized the improved LLC method to encode the image local features and then applied the PLSA model to analyze the relationship among image and visual words by introducing the latent topics. Thus, our proposed method shows superior recognition performance compared with the existing methods.

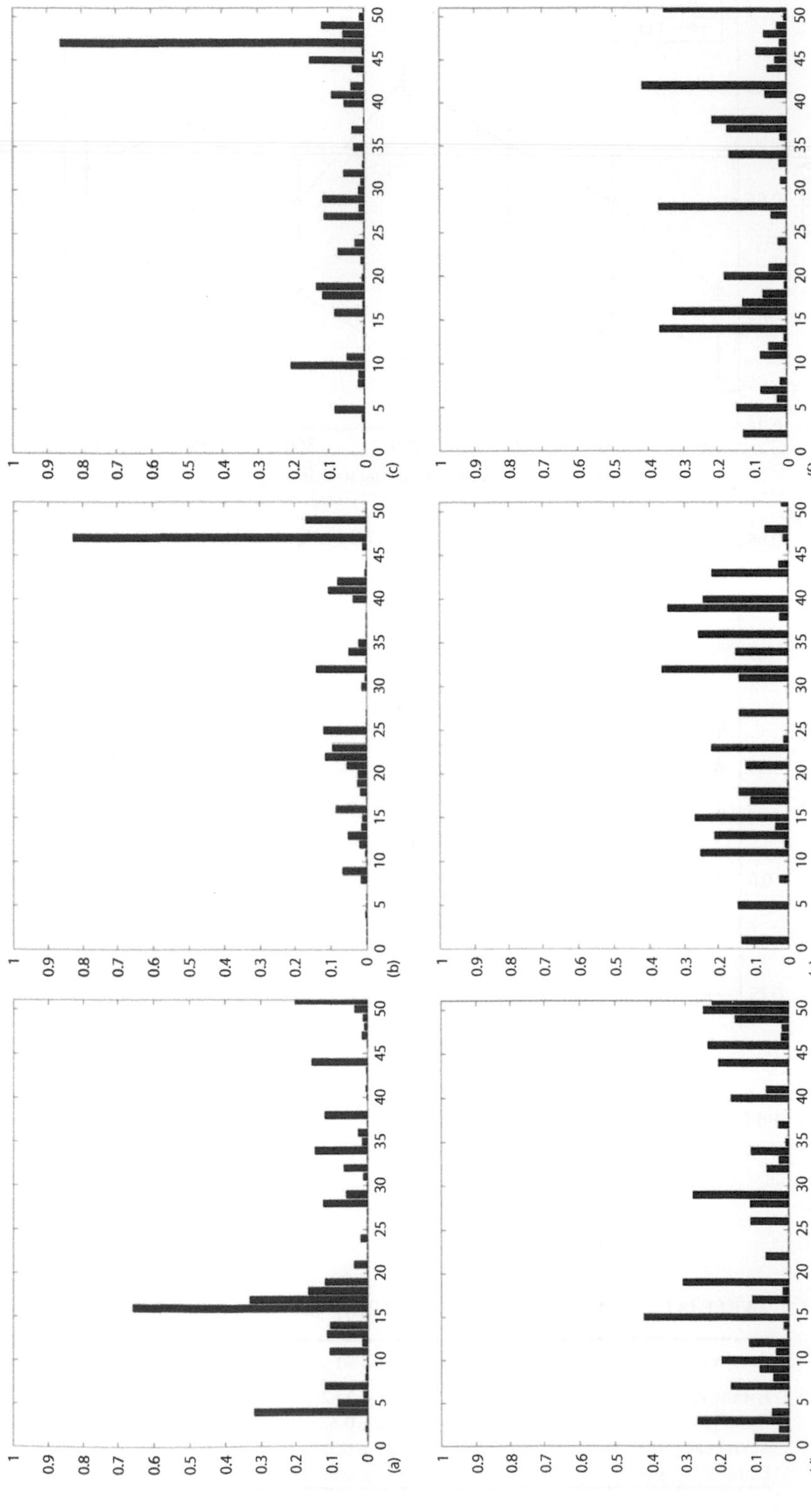

FIGURE 13.8 Illustration of feature histograms for Figure 13.2.

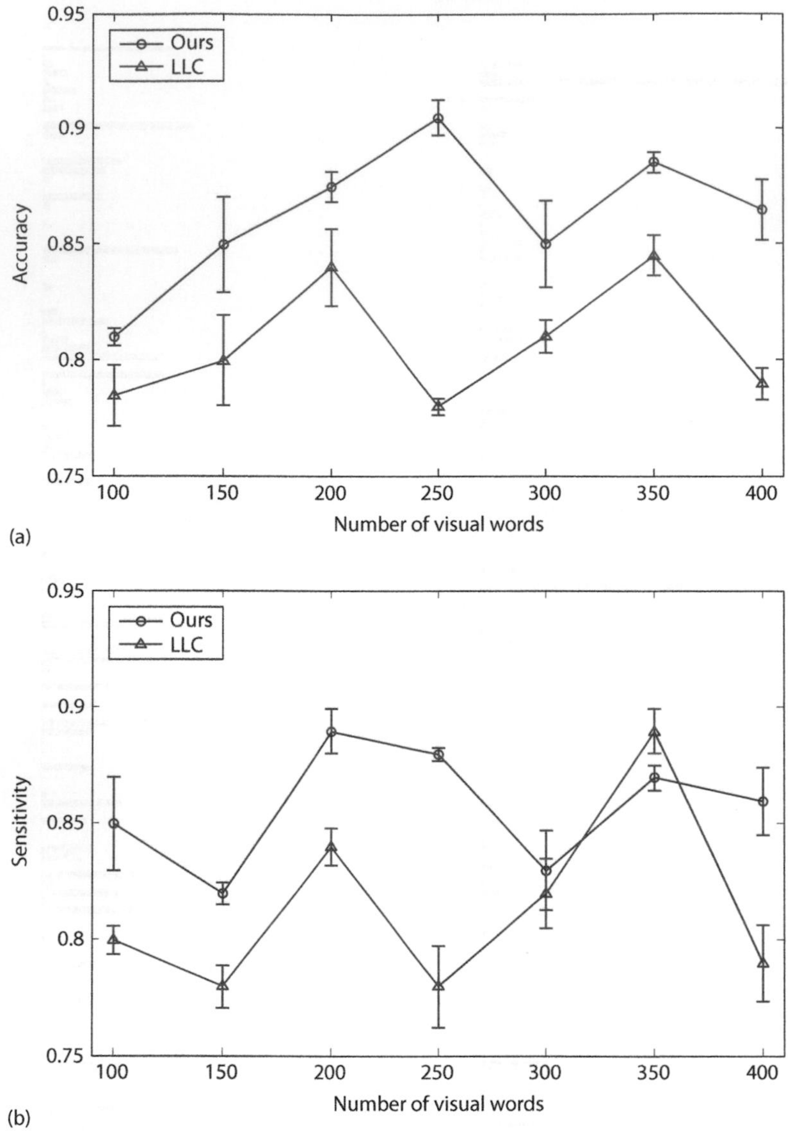

FIGURE 13.9 Comparison of classification performance. (a) Accuracy, (b) sensitivity.

TABLE 13.1 Performance Comparison of Polyp Classification Methods

Acc.(%)	Sen.(%)	Spe.(%)
Hwang et al. [13] 83.27 ± 1.22	81.91 ± 0.82	84.52 ± 0.94
Lietal [24] 85.72 ± 1.42	83.32 ± 0.62	88.17 ± 0.74
Yuan et al. [19] 87.51 ± 0.42	86.19 ± 0.31	88.96 ± 1.38
Ours* 90.53 ± 0.77	88.13 ± 0.31	92.96 ± 0.82

Conclusion

This chapter has introduced a novel method for polyp recognition in WCE images based on improved LLC method and PLSA model. Instead of extracting features from a complete WCE image, we proposed an improved LLC method to represent WCE images. Since the features are extracted directly from the neighborhoods of the key points, they provide accurate descriptions of the WCE images. The PLSA model explores the latent semantic topics among the images and the visual words; thus, it could provide better image representation.

Furthermore, we evaluated the relationships between the different visual words, number of topics and the classification performance. Based on the comprehensive experimental studies, the best performance of polyp classification is obtained with the number of visual words at 250 and topic size at 50. The obtained results reported a promising polyp detection accuracy of 90.53%, demonstrating that the effectiveness of our proposed method. Results show that the proposed method outperforms existing polyp recognition methods in terms of classification criteria with our current datasets.

References

1. Caffrey, Colm Mc, Chevalerias, Olivier, Mathuna, Cian O., and Twomey, Karen, "Swallowable-capsule technology," *IEEE Trans. Pervasive Comput.*, 7, 1, 23–29, 2008.
2. Upchurch, Bennie R. and Vargo, John J., "Small bowel enteroscopy." *Rev. Gastroenterol. Disord.*, 3, 8, 169–177, 2008.
3. Manno, Mauro, Manta, Raffaele, and Conigliaro, Rita, "*Single-Balloon Enteroscopy*," Springer-Verlag Inc., Springer, Milan 2012.
4. Iddan, Gavriel, Meron, Gavriel, Glukhovsky, Arkady, and Swain, Paul, "Wireless capsule endoscopy," *Nature*, 405, 417, 2000.
5. Lee, Noel M. and Eisen, Glenn M., "10 years of capsule endoscopy: An update," *Expert Rev. Gastroenterol. Hepatol.*, 4, 4, 503–512, 2010.
6. Yuan, Yixuan, Wang, Jiaole, Li, Baopu, and Meng, Max, "Saliency based ulcer detection for wireless capsule endoscopy diagnosis," *IEEE Trans. Med. Imag.*, 34, 10, 2046–2057, 2015.
7. Segui, Santi, Drozdzal, Michal, Vilarino, Fernando, Malagelada, Carolina, Azpiroz, Fernando, Radeva, Petia, and Vitria, Jordi, "Ategorization and segmentation of intestinal content frames for wireless capsule endoscopy," *IEEE Trans. Inf. Technol. Biomed.*, 16, 6, 1341–1352, 2012.
8. Pan, Guobing and Wang, Litong, "Swallowable wireless capsule endoscopy: Progress and technical challenges," *Gastroenterol. Res. Pract.*, 1012, 1–9, 2012.
9. Yu, Marcia, "M2A capsule endoscopy: A breakthrough diagnostic tool for small intestine imaging," *Gastroenterol. Nurs.*,25, 1, 24–27, 2002.
10. Yuan, Yixuan, Li, Baopu, and Meng, Max Q.-H., "Improved bag of feature for automatic polyp detection in wireless capsule endoscopy images," *IEEE Trans. Autom. Sci. Eng.*, 13, 2, 529–535, 2016.
11. Karargyris, Alexandras and Beourbakis, Nikolaos, "Detection of small bowel polyps and ulcers in wireless capsule endoscopy videos," *IEEE Trans. Biomed. Eng.*, 58, 10, 2777–2786, In Conference Las Vegas, NV 2011.
12. Cappell, Mitchell S, "The pathophysiology, clinical presentation, and diagnosis of colon cancer and adenomatous polyps," *Med. Clin. North Am.*, 89, 1, 1–42, 2005.
13. Hwang, Sae, "Bag-of-visual-words approach to abnormal image detection in wireless capsule endoscopy videos," *Advances in Visual Computing.* Springer, Berlin Heidelberg, 320–327, 2011.
14. Gueye, Limamou, Yildirim-Yayilgan, Sule, Alaya-Cheikh, Faouzi, and Balas-ingham, Ilangko, "Automatic detection of colonoscopic anomalies using capsule endoscopy," *Proc. IEEE Conf. Image Process. (ICIP)*, 1061–1064, 2015.

15. Fisher, Mark, and Mackiewicz, Michal, "Colour image analysis of wireless capsule endoscopy video: A review," *Color Med. Image Anal*, 129–144, Springer, the Netherlands 2013.
16. Zhao, Qian and Meng, Max Q.-H., "Polyp detection in wireless capsule endoscopy images using novel color texture features," *WCICA 2011*, 948–952, Taipei, Taiwan 2011.
17. Li, Baopu, Fan, Yichen, Meng, Max Q.H., and Qi, Lin, "Intestinal polyp recognition in capsule endoscopy images using color and shape features," *2009 IEEE International Conference on Robotics and Biomimetics (ROBIO)*, 1490–1494, Guilin, China 2009.
18. Choi, Wing-Pong, Tse, Siu-Hong, Wong, Kwok-Wai, and Lam, Kin-Man, "Simplified gabor wavelets for human face recognition,," *Pattern Recognit.*, 41, 3, 1186–1199, 2008.
19. Yuan, Yixuan and Meng, Max Q-H, "A novel feature for polyp detection in wireless capsule endoscopy images," *The 2014 IEEE/RSJ International Conference on Intelligent Robots and Systems*, 5010–5015, Chicago, IL 2014.
20. El Khatib, Alaa, Werghi, Naoufel, and Al-Ahmad, Hussain, "Automatic polyp detection: A comparative study," *2015 37th Annual International Conference of the IEEE Engineering in Medicine and Biology Society (EMBS)*, 2669–2672, Milan, Italy 2015.
21. Hwang, Sea and Celebi, M. E., "Polyp detection in wireless capsule endoscopy videos based on image segmentation and geometric feature," *2010 IEEE International Conference on Acoustics, Speech and Signal Processing*, 678–681, Dallas, TX 2010.
22. Karargyris, Alexandras and Bourbakis, Nikolaos, "Identification of polyps in wireless capsule endoscopy videos using log gabor filters," 2009 *IEEE/NIH Life Science Systems and Applications Workshop (LiSSA)*, 143–147, Bethesda, MD 2009.
23. Mallat, Stephane G., "A theory for multiresolution signal decomposition: The wavelet representation," *IEEE Trans. Pattern Anal. Mach. Intell.*, 11, 7, 674–693, 1989.
24. Li, Baopu and Meng, Max Q.-H., "Automatic polyp detection for wireless capsule endoscopy images," *Expert Syst. Appl.*, 39, 12, 10952–10958, 2012.
25. Barbosa, Daniel C., Roupar, Dalila B., Ramos, Jaime C, Tavares, A. and Lima, Carlos S., "Automatic small bowel tumor diagnosis by using multi-scale wavelet-based analysis in wireless capsule endoscopy images," *Biomed. Eng. Online*, 11:3, 2012.
26. Ojala, Timo, Pietikainen, Matti, and Maenpaa, Topi, "Multiresolution gray-scale and rotation invariant texture classification with local binary patterns," *IEEE Trans. Pattern Anal. Mach. Intell.*, 24, 7, 971–987, 2002.
27. Zhang, Lin, Zhang, Lei, Guo, Zhenhua, and Zhang, David, "Monogenic-LBP: A new approach for rotation invariant texture classification," *ICIP 2010*, 24, 7, 2677–2680, 2010.
28. Guo, Zhenhua, Zhang, Lei, and Zhang, David, "A completed modeling of local binary pattern operator for texture classification," *IEEE Trans. Image Process.*, 19, 6, 1657–1663, 2010.
29. Li, Baopu, Meng, Max Q.-H., and Hu, Chao, "A comparative study of endoscopic polyp detection by textural features," *IEEE 2012 10th World Congress on Intelligent Control and Automation (WCICA 2012)*, 4671–4675, Beijing, China 2012.
30. Dalai, Navneet and Triggs, Bill, "Histograms of oriented gradients for human detection," *CVPR 2005*, 1, 886–893, 2005.
31. Iwahori, Yuji, Hattori, Akira, Adachi, Yoshinori, Bhuyan, Manas K., Woodham, Robert J., and Kasugai, Kunio, "Automatic detection of polyp using hessian filter and HOG features," *Procedia Comput. Sci.*, 60, 730–739, 2015.
32. Bae, Seung H., and Yoon, Kuk-Jin, "Polyp detection via imbalanced learning and discriminative feature learning," *IEEE Trans. Pattern Anal. Mach. Inteli.*, 24, 7, 971–987, 2002.
33. Csurka, Gabriella and Dance, Christopher and Fan, Lixin and Willamowski, Jutta and Bray, Cédric, "Visual categorization with bags of keypoints," *ECCV Workshop 2004*, 1, 1–1, 2004.
34. Lowe, David G., "Object recognition from local scale-invariant features," *IEEE Trans. Pattern Anal. Mach. Inteli.*, 24, 7, 971–987, 2002.

35. Lowe, David G., "Distinctive image features from scale-invariant keypoints," *ICCV 1999*, 2, 1150–1157, 1999.
36. Yang, Jianchao and Yu, Kai and Gong, Yihong and Huang, Tingwen, "Linear spatial pyramid matching using sparse coding for image classification," *CVPR 2009*, pp. 1794 –1801, Miami, FL 2009.
37. Wang, Jinjun, Yang, Jianchao, Yu, Kai and Lv, Fengjun and Huang, Thomas and Gong, Yihong, "Locality-constrained linear coding for image classification," *CVPR 2010*, 3360–3367, San Francisco, CA 2010.
38. Hofmann, Thomas, "Unsupervised learning by probabilistic latent semantic analysis," *Machine Learning,* 42, 1–2, 177–196, 2001.
39. Hofmann, Thomas, "Probabilistic latent semantic indexing," *ACM SIGIR 1999*, 50–57, Berkeley, CA 1999.
40. Shen, Yao, Guturu, Parthasarathy, and Buckles, Bill P., "Wireless capsule endoscopy video segmentation using an unsupervised learning approach based on probabilistic latent semantic analysis with scale invariant features," *IEEE Trans. Inf. Technol. Biomed.*, 16, 1, 98–105, 2012.
41. Reza, Zare M., Mueen, Abdullah, Awedh, Mohammad, and Chaw, Seng W., "Automatic classification of medical X-ray images: hybrid generative-discriminative approach," *IET Image Process.*, 7, 5, 523–532, 2013.
42. Cruz-Roa, Angel, González, Fabio, Galaro, Joseph, Judkins, Alexander R., Ellison, David, Baccon, Jennifer, Madabhushi, Anant, and Romero, Eduardo, "A visual latent semantic approach for automatic analysis and interpretation of anaplastic medulloblastoma virtual slides," *MICCAI 2012*, 157–164, Nice, France 2012.
43. Zhang, Fan, Song, Yang, Cai, Weidong, Hauptmann, Alexander G., Liu, Sidong, Pujol, Sonia, Kikinis, Ron, Fulham, Michael J., Feng, David D., and Chen, Mei, "Dictionary pruning with visual word significance for medical image retrieval," *Neurocomputing*, 177, 75–88, 2016.
44. Zheng, Yushan, Jiang, Zhiguo, Shi, Jun, and Ma, Yibing, "Pathology image retrieval by block LBP based pLSA model with low-rank and sparse matrix decomposition," *Advances in Image and Graphics Technologies,* 327–335, Beijing, China 2014.
45. Jiang, Yu-Gang, Ngo, Chong-Wah, and Yang, Jun, "Towards optimal bag-of-features for object categorization and semantic video retrieval," *ACM CIVR 2007*, 494–501, Amsterdam, The Netherlands 2007.
46. Dempster, Arthur P., Laird, Nan M., and Rubin, Donald B., "Maximum likelihood from incomplete data via the EM algorithm," *J. R. Stat. Soc. Series B Stat. Methodol.*, 1–38, 1977.
47. Chang, Chih-Chung and Lin, Chih-Jen, "Libsvm: A library for support vector machines," *ACM Trans. Intell. Syst. Technol.*, 2, 3, 27, 2011.

14

Texture-Based Computer-Aided Classification of Focal Liver Diseases using Ultrasound Images

Jitendra Virmani

Vinod Kumar

Contents

14.1 Introduction

Chronic liver diseases are very prevalent not only in India but in many parts of the world, and these liver diseases can either affect the whole liver (i.e., diffuse liver disease) or certain localized parts of the liver parenchyma (focal liver disease). Several characteristics of liver sonography have been used to evaluate the degree of severity of liver disease, including changes of the liver surface, inferior edge, echotexture, echogenicity, and diameters of hepatic and splenic veins. However, the measurements of these characteristics are always subjectively based on clinician observation. Several studies have shown that characterization accuracy of diffused diseases using only simple visual interpretation by physicians is around 70%. For confirmative diagnosis of patients with diffuse parenchyma liver disease and malignant focal liver lesions, needle biopsy of the liver has long been the primary practice. However, the pathological measurement of these diseases may be severely biased due to sampling error in the biopsy specimen. Furthermore, apart from discomfort to the patient, liver biopsy is an invasive procedure that may result in morbidity or even mortality. Therefore, a reliable noninvasive way for measuring liver histology would be a better choice in diagnosis and monitoring of liver disease.

It has also been established by many researchers that noninvasive, computer-assisted diagnosis yields more dependable results than through pathological investigations. Ultrasound measurements that describe the structure of soft tissue are potentially very useful tissue signatures, since one feature of diffuse and focal disease processes of the liver is the disruption of the normal tissue architecture. Ultrasound echoes from soft tissues, when displayed as a B-scan image, form a texture pattern.

Echotexture has been reported to be a good correlate of liver pathology and has been used in several studies to evaluate diffuse, as well as focal, parenchymal changes.

In the present work, a computerized tissue classification system to detect the presence of focal liver disease based on texture analysis has been proposed.

14.2 Texture-Based Computer-Aided Classification of Focal Liver Disease Using US Images

The differential diagnosis between atypical focal liver lesions (FLLs) from B-Mode ultrasound (US) images is a difficult and confusing task for radiologists due to the existence of a wide variety of sonographic appearances even within individual classes of FLLs [1–7]. Even then, B-Mode US is considered as a primary choice for characterization of FLLs, mainly due to its nonionizing, noninvasive, inexpensive nature and real time imaging capabilities [2,7]. A brief description of sonographic appearances of different liver image classes considered in the present study are reported in Table 14.1.

The disadvantages associated with use of conventional gray scale US for characterization of FLLs include: (1) limited sensitivity for detection of small FLLs (< 2 cm) developed on cirrhotic liver which is already nodular and coarse-textured [1–3,5,12], (2) sonographic appearance of HCC and MET lesions which are highly overlapping [1–3,5,14,18], (3) sonographic appearances of cystic metastasis and atypical cyst which is often overlapping [1,3], (4) sonographic appearances of atypical HEM which sometimes mimic atypical MET and HCC, [5,13,14,17,20,21], and (5) difficulty characterizing isoechoic lesions with very slim difference in contrast between regions inside the lesion and the surrounding liver parenchyma in some cases [13,14].

It is very much desirable to reduce these limitations and to build an efficient Computer-Aided Classification (CAC) system for characterization of FLLs using B-Mode US images. Typical FLLs can be easily diagnosed, even by an inexperienced radiologist, from their classic sonographic appearances. But differential diagnosis between atypical FLLs and HCC lesions developed on cirrhotic liver from B-Mode US images is considered a difficult task for radiologists in routine practice [3,5,12–14,16–21].

The experienced, participating radiologists opined that the textural characteristics of the liver parenchyma surrounding the lesion should contribute to the effective characterization of FLLs. The present work investigates the contribution of texture information extracted from inside lesion ROIs (IROIs) and surrounding lesion ROIs (SROIs) in characterization of FLLs using conventional gray scale B-Mode US images. The participating radiologists were of the view that the training dataset used in the

TABLE 14.1 Brief Description of Sonographic Appearances of Different Liver Image Classes

Image Class	Sonographic Appearance (typical)	Sonographic Appearance (Atypical)
Normal	Appear homogeneous with slightly increased echogenicity as compared to the right kidney [8–12].	No atypical appearance.
Cyst	Appear as round, anechoic lesion with posterior acoustic enhancement and well defined thin imperceptible wall [2,3,5,13,14].	Appear with internal echoes and thickened irregular walls.
HEM	Appear as well circumscribed uniformly hyperechoic lesion [2,3,5,13,14].	May appear as isoechoic or even hypoechoic mimicking the sonographic appearance of certain atypical MET and HCC lesions [2,13,15–17].
MET	Appear with "target" or "bull's-eye" appearance, that is, hypoechoic center surrounded by a hyperechoic rim [2,13,14,18–20].	Appear with extremely variable sonographic appearances ranging from anechoic, hypoechoic, isoechoic, hyperechoic and even with mixed echogenicity [1–3,5,14,18].
HCC	There is no typical sonographic appearance for HCCs. The sonographic appearances of SHCC vary from hypoechoic to hyperechoic. LHCC appear frequently with mixed echogenicity [3,5,12,14].	

present study is a comprehensive and diversified set consisting of representative images from various subclasses.

The study in [2], used statistical features based on first order statistics (FOS); second order statistics, that is, gray-level co-occurrence matrix (GLCM); higher order statistics, that is, gray level run length matrix (GLRLM); spectral features, that is, Gabor wavelet-transform-based (GWT) features; and Laws' texture features for classification of NOR, Cyst, HEM, HCC, and MET liver classes. The study in [22], reported classification between NOR, benign, malignant liver classes by using FOS, and GLRLM statistical features with linear discriminant analysis and neural network (NN) classifier. The study in [23] used GLCM, autocorrelation, Laws' and edge frequency based texture features, and an NN classifier for classification of NOR, Cyst, HEM, and malignant liver classes. In research studies [22,23], malignant lesions are considered as a single class; however, diagnosis of malignant lesions such as HCC or MET is clinically significant for effective treatment and management of liver malignancies [2,3,24]. A recent study carried out by the authors of the present work [24] reported the effect of texture analysis of regions surrounding the lesions for designing an exclusive computer-aided diagnostic (CAD) system for diagnosis of HCC and MET liver malignancies using GLCM, GLRLM, FPS (Fourier power spectrum), and Laws' texture features, along with SVM classifier. In another related study [25], multiscale, wavelet, packet texture descriptors are used with NN classifier for binary classification tasks, that is HEM versus HCC, HEM versus MET, and HCC versus MET.

The studies [13,14,26,27] experimented with classification between five liver image classes, that is Cyst, HEM, HCC, MET, and NOR liver, by using a large feature vector consisting of texture features extracted by using FOS, GLCM, GLRLM, GWT, and Laws' feature extraction methods.

In present study, the performance of CAC systems based on single PCA-NN based multi-class classifier design has been compared with CAC system based on PCA-NN based hierarchical classifier design.

14.2.1 Dataset Description

The design of a CAC system for the characterization of FLLs was carried out with 108 B-Mode liver US images, that is, 21 NOR, 12 Cyst, 15 HEM, 28 HCC, and 32 MET images. As per the radiologists, there is wide variability of sonographic appearances even within SHCCs and LHCCs; therefore; no sonographic appearance can be considered typical for HCC. The size of SHCCs and LHCCs in the acquired image database varied from 1.5 to 1.9 cm and 2.1 to 5.6 cm, respectively. The dataset used in this study is a comprehensive and representative dataset offering a high degree of variability encountered during subjective analysis of (1) typical and atypical cases of Cyst, HEM, and MET lesions; (2) SHCC and LHCC lesions; and (3) NOR liver cases.

The sample images for (1) NOR liver tissue; (2) typical case of Cyst, HEM, and MET lesions; and (3) SHCC and LHCC lesions are shown in Figure 14.1.

The sample images for atypical cases of Cyst, HEM, and MET lesion are shown in Figure 14.2.

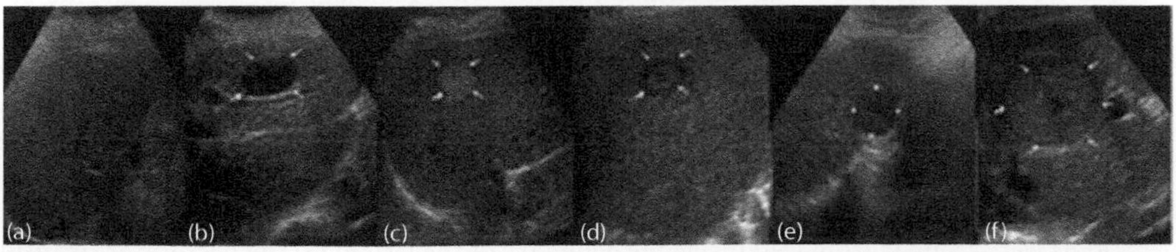

FIGURE 14.1 B-Mode ultrasound liver images with appearance of:(a) Normal liver (homogeneous echotexture with medium echogenicity); (b) Typical cyst (thin walled anechoic lesion with posterior acoustic enhancement); (c) Typical HEM (well circumscribed uniformly hyperechoic appearance); (d) Typical MET ("target" or "bull's-eye" appearance, i.e., hypoechoic center surrounded by a hyperechoic rim); (e) Hypoechoic SHCC; (f) Heterogeneous echotexture represents complex and chaotic structure exhibited by LHCC due to coexistence of areas of necrosis, fibrosis and active growth areas.

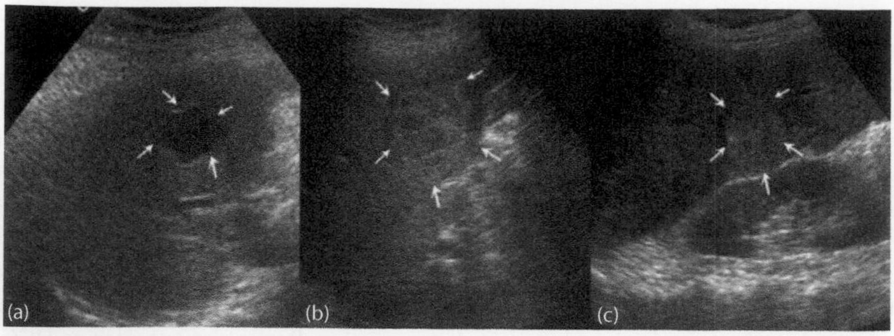

FIGURE 14.2 Sample image variants of atypical cases: (a) Atypical cyst with internal echoes and irregular walls; (b) Atypical HEM with heterogeneous echotexture; (c) Hyperechoic atypical MET with heterogeneous echotexture.

14.2.2 Data Collection Protocols

The following protocols were followed for data collection:

1. Only HCC lesions developed on cirrhotic liver were included as the existence of HCC on normal liver is very rare.
2. Labeling of HCC lesions as SHCC (< 2 cm) or LHCC was done by observing the size of the lesion in transverse and longitudinal views.
3. Labeling of HCC as SHCC or LHCC and Cyst, HEM, and MET as typical or atypical lesion was carried out during data collection in order to ensure that the training dataset used for classifier design should consist of representative cases from all image subclasses.

14.2.3 ROI Extraction Protocols

The protocols followed for selection of ROIs in different experiments carried out in the present research work, as well as the protocols followed for selection of ROI size, are discussed in this section.

14.2.3.1 Selection of ROIs

The following protocols were followed for cropping the regions of interest (ROIs) from the image database:

1. The ROIs were cropped by an experienced, participating radiologist by using a graphic user interface-based (GUI) ROI manager software developed in the Biomedical Instrumentation Laboratory, Indian Institute of Technology, Roorkee. The ROI manager software provided the radiologist the flexibility to load the image, choose the ROI size and shape, move the ROI to any desired location over the image, freeze the ROI at any location, and crop the ROIs together after the position of all the ROIs for an image is frozen.
2. (1) two types of ROIs (i.e., IROIs and SROIs) were extracted; (2) maximum non-overlapping IROIs were cropped from well inside the boundary of the lesion; (3) the necrotic areas inside the lesions were avoided while cropping IROIs; (4) for each lesion, a single SROI was cropped from region surrounding the lesion approximately at the same depth as that of the center of the lesion; (5) due care was taken while cropping SROI, such that the SROI contain only liver parenchyma without any inhomogeneous structures like hepatic ducts, blood vessels, etc.; and (6) for each NOR image, maximum ROIs were cropped at the same depth, a single

extreme ROI was considered as SROI, and all other ROIs at the same depth were considered as IROIs.

14.2.3.2 Selection of ROI Size

The selection of ROI size plays an important role as texture measurements are sensitive to ROI size. In other studies, it has been demonstrated that ROI size must be at least 800 pixels to provide good sampling distribution for estimating reliable statistics [28–30], whereas, in a few other related researches, a sample size of at least 1000 pixels is suggested to estimate reliable statistics [15,31]. However, for characterization of diffuse liver diseases different ROI sizes ranging from 32 × 32 pixels [32–34], 40 × 40 pixels [35,36] to 64 × 64 pixels [25,37] have been considered. Similarly, for characterization of FLLs, the study reported in [13] used ROI size of 25 × 25 pixels for computing texture features, and in studies [22,23], the use of ROI size of 10 × 10 pixels is reported. The use of 10 × 10 pixels, and even 25 × 25 pixels, as ROI size yields smaller number of pixels in comparison to minimum 800 pixels required to estimate reliable statistics [12,14,35,36]. The study reported in [20] used 64 × 64 pixels as ROI size extracted from high-resolution scanned images, instead of real US images. It is otherwise difficult to select such a large ROI size, keeping in view the size of small lesions and resolution of images obtained from US machines. After interaction with the radiologists, ROI size of 32 × 32 pixels was considered appropriate for the present study considering the facts such as:

1. There is sufficient evidence in the literature that ROI size must be at least 800 pixels to provide good sampling distribution for estimating reliable statistics [14,36]; as ROI size of 32 × 32 gives 1024 pixels, it can be assumed that the computed texture parameters are reliable estimates.
2. During initial discussions with the participating radiologists, an attempt was made to mark larger ROI sizes, but few practical difficulties were faced. Certain lesions had necrotic area; radiologists opined that the necrotic area inside lesions must be avoided while extracting IROIs, and it was not possible to consider large ROI size for these lesions. Also, participating radiologists were of the view that SROI for each lesion, and ROIs for each normal and cirrhotic liver image must be selected by avoiding the inhomogeneous structures like hepatic ducts, blood vessels, etc., which was practically difficult by considering larger ROI size.
3. For real-time implementation, small ROI size is always favorable, as time taken for feature extraction and classification is obviously less in comparison to large ROI size. Also, with small ROI size, more samples are available for classifier design.

The entire image database was stored on a personal computer with a Pentium Core-2-Duo, 2.67 GHz processor, and 1.97 GB RAM.

The sample images of NOR, Cyst, HEM, HCC and MET cases from the acquired image database with ROIs marked are shown in Figure 14.3a–e.

In the present work, two types of features are considered for analysis, that is texture features computed from IROIs and texture ratio features computed by taking the ratio of texture feature computed from IROI, and texture feature computed from corresponding SROI. It can be noted that NOR liver image in Figure 14.3a contains 4 IROIs and a corresponding SROI. Thus, 4 instances of texture feature sets and 4 instances of texture ratio feature sets are obtained. Similarly, from Cyst, HEM, HCC, and MET lesions shown in Figure 14.3b–e, 6, 4, 4, and 2 instances of texture feature sets and 6, 4, 4 and 2 instances of texture ratio feature sets are obtained. The distribution of clinically acquired database of 108 B-Mode liver US images (including NOR, Cyst, HCC, HEM, and MET images) among various liver image subclasses is shown in Figure 14.4.

The bifurcation of total FLLs, among typical, atypical, SHCC, and LHCC lesions and bifurcation of total ROIs, among IROIs and SROIs belonging to typical, atypical, SHCC, and LHCC lesions is shown in Figure 14.5.

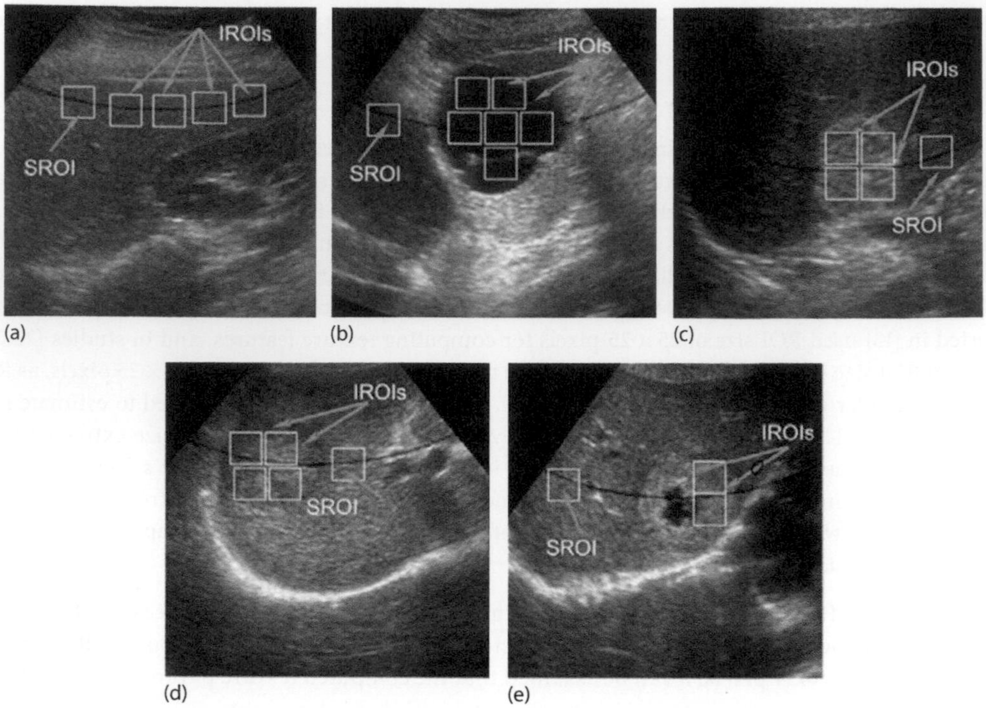

FIGURE 14.3 (a–e) NOR, Cyst, HEM, HCC, and MET images with IROIs and SROI marked. **Note:** As shown in (e), necrotic area within the lesions is avoided while cropping IROIs. For NOR images, all the IROIs and SROI are taken at same depth, and a single extreme ROI is considered as SROI, as shown in (a). For liver images with lesions, maximum non-overlapping IROIs are taken from well within the boundary of each lesion, and a single SROI is extracted from liver parenchyma surrounding the lesion at approximately the same depth as that of the center of the lesion, by avoiding inhomogeneous areas like blood vessels, hepatic ducts, etc., as shown in (b) to (e). (From Virmani et al. Journal of Medical Engineering and Technology, 37(4) 292–306, 2013, © Taylor and Francis.)

14.2.4 Proposed CAC System Designs

In the present study, exhaustive experimentation, (described in Table 14.2), has been carried out for designing an efficient CAC system for characterization of FLLs using B-Mode US images.

14.2.4.1 Experiment No 1: To Design and Evaluate the Performance of CAC System for Characterization of FLLs Using Single PCA-NN Based Multi-Class Classifier Design

The bifurcation of the acquired dataset into the training dataset and testing dataset for experiment no. 1 is shown in Figure 14.6.

To ensure generality, the training data was chosen carefully in consultation with experienced participating radiologists so as to include typical and atypical cases of Cyst, HEM, and MET lesions, as well as cases of SHCC and LHCC lesions, for designing a robust classifier with representative cases for all image subclasses. Two sets of images were created for each image class: ROIs from one set of images were used for training, and ROIs from the other set were used for testing to avoid any biasing.

The final dataset consisting of total 380 IROIs and 111 SROIs was stored on a PC (Pentium Core-2-Duo, 2.67 GHz with 1.97 GB RAM).

The block diagram for the design of the CAC system for characterization of FLLs using single, multi-class classifier design is shown in Figure 14.7.

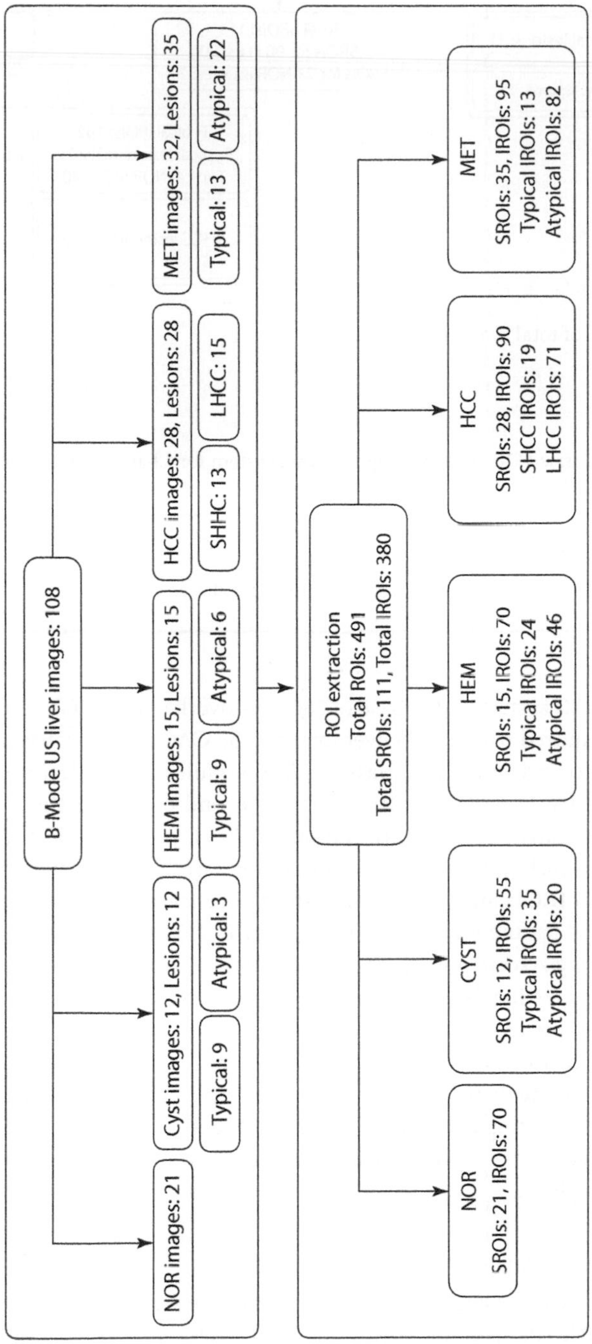

FIGURE 14.4 Dataset description: CAC system for FLLs.

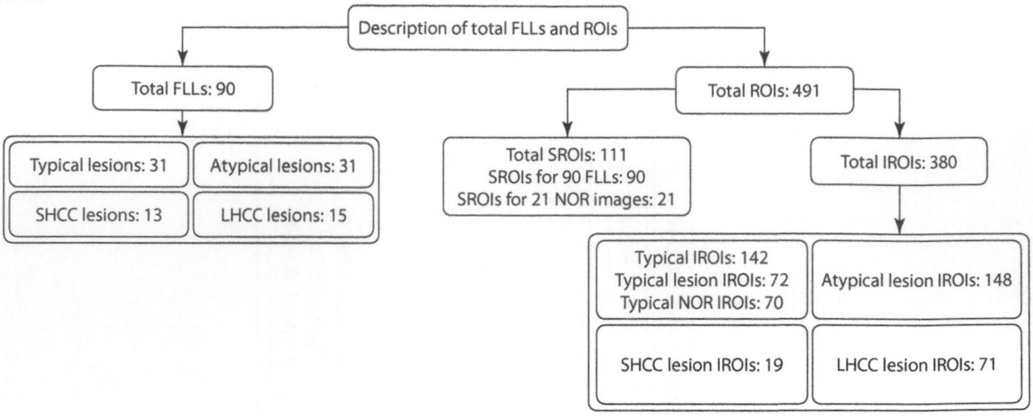

FIGURE 14.5 Description of total FLLs and ROIs.

TABLE 14.2 Experiments Carried Out for the Design of CAC System for Characterization of FLLs Using B: -Mode US Images

Experiment 1: To design and evaluate the performance of CAC system for characterization of FLLs using single PCA-NN based multi-classifier design.
Experiment 2: To design and evaluate the performance of CAC system for characterization of FLLs using PCA-NN based hierarchical classifier design

For the design of the CAC system, a database of 380 non-overlapping IROIs and 111 SROIs was created from 108 clinically acquired B-Mode US liver images. The CAC system consisted of feature extraction, feature dimensionality reduction, and classification modules. In the feature extraction module, statistical texture features based on first order statistics, second order statistics (i.e., GLCM features [25,27,38,39]), higher order statistics (i.e., GLRLM features [39–42]), spectral features (i.e., FPS features [39,42], and GWT features [40,43]), as well as spatial filtering-based Laws' texture features [27,39,44,45], are computed from all 380 IROIs and 111 SROIs. Both texture features and texture ratio features are considered for analysis. The feature set of 208 texture features (104 texture features + 104 texture ratio features) is normalized using z-score normalization. The normalized feature set is partitioned into the training data feature set and testing data feature set. The bifurcation of instances of individual classes in the training data feature set and testing data feature set is described in Figure 14.6. In the dimensionality reduction module, PCA is carried out on the training data feature set and a reduced training dataset of PCA derived principal components (PCs) is obtained. The reduced testing dataset is obtained by projecting the data points of the testing data feature set in the direction of PCs of the training data feature set. In the classification module, the classifier neural network is trained and tested with reduced feature sets obtained after applying PCA.

14.2.4.1.1 Feature Extraction Module

The general idea of feature extraction is to obtain mathematical descriptors for both visually extractable and visually non-extractable sonographic features of FLLs. These mathematical descriptors are either morphological (based on shape or contour of the lesion) or textural features (based on intensity distribution) [38]. Both these morphological as well as textural features are significant for developing CAD systems for breast lesions from B-Mode US images [46–49]. Radiologists opined that morphological features of FLLs do not give any significant information about their characterization, as is also evident

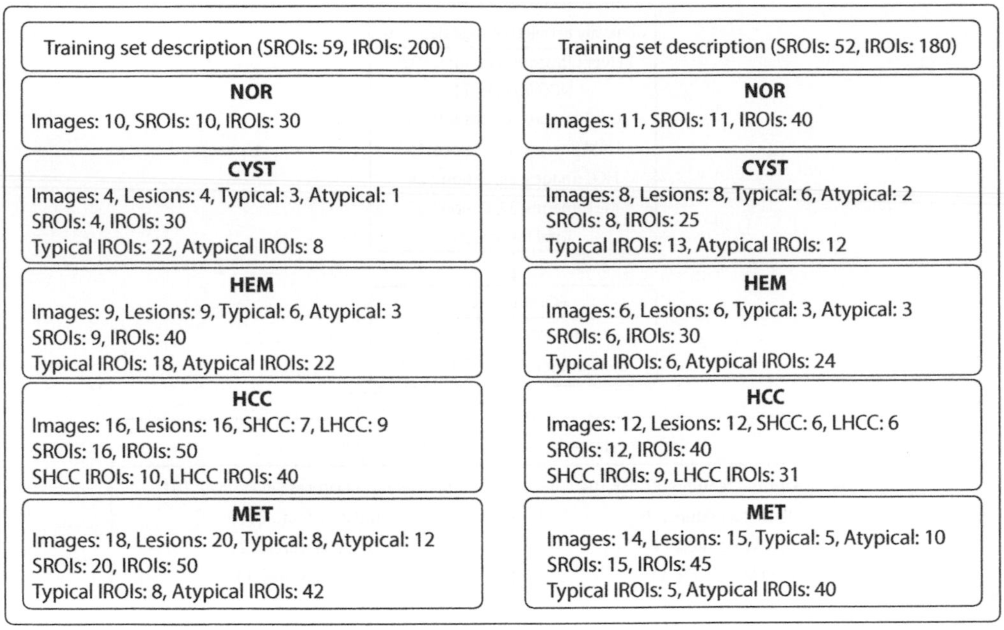

Training set description (SROIs: 59, IROIs: 200)
NOR
Images: 10, SROIs: 10, IROIs: 30
CYST
Images: 4, Lesions: 4, Typical: 3, Atypical: 1
SROIs: 4, IROIs: 30
Typical IROIs: 22, Atypical IROIs: 8
HEM
Images: 9, Lesions: 9, Typical: 6, Atypical: 3
SROIs: 9, IROIs: 40
Typical IROIs: 18, Atypical IROIs: 22
HCC
Images: 16, Lesions: 16, SHCC: 7, LHCC: 9
SROIs: 16, IROIs: 50
SHCC IROIs: 10, LHCC IROIs: 40
MET
Images: 18, Lesions: 20, Typical: 8, Atypical: 12
SROIs: 20, IROIs: 50
Typical IROIs: 8, Atypical IROIs: 42
Training set description (SROIs: 52, IROIs: 180)
NOR
Images: 11, SROIs: 11, IROIs: 40
CYST
Images: 8, Lesions: 8, Typical: 6, Atypical: 2
SROIs: 8, IROIs: 25
Typical IROIs: 13, Atypical IROIs: 12
HEM
Images: 6, Lesions: 6, Typical: 3, Atypical: 3
SROIs: 6, IROIs: 30
Typical IROIs: 6, Atypical IROIs: 24
HCC
Images: 12, Lesions: 12, SHCC: 6, LHCC: 6
SROIs: 12, IROIs: 40
SHCC IROIs: 9, LHCC IROIs: 31
MET
Images: 14, Lesions: 15, Typical: 5, Atypical: 10
SROIs: 15, IROIs: 45
Typical IROIs: 5, Atypical IROIs: 40

FIGURE 14.6 Description of training dataset and testing dataset. **Note:** The size of SHCC lesion varied from 1.5 to 1.9 cm and size of LHCC lesion varied from 2.1 to 5.6 cm.

from other related researches. The proposed CAD systems for characterization of FLLs from B-Mode US have relied on textural features only [13,20,22,23,39]. In the present work, initially, a wide variety of visual and non-visual echotexture features are extracted by using statistical, spectral, and spatial filtering-based feature extraction methods.

FOS Features: Six total FOS texture features, that is, average gray level, standard deviation, smoothness, third moment, uniformity and entropy FOS, are computed for each ROI [50].

GLCM Features: A total of 13 GLCM mean features, that is, angular second moment, contrast, correlation, variance, inverse difference moment, sum average, sum variance, sum entropy, entropy GLCM, difference variance, difference entropy, information measures of correlation-1, and information measures of correlation-2, are computed for each ROI [27,39].

GLRLM Features: A total of 11 GLRLM features, that is, short-run emphasis; long-run emphasis; low gray level-run emphasis; high gray level-run emphasis; short-run, low gray level emphasis; short-run, high gray level emphasis; long-run, low gray level emphasis; long-run, high gray level emphasis; gray level non-uniformity; run-length, non-uniformity; and run percentage; are computed for each ROI [39–41,49].

FPS Features: Two spectral features, that is, radial sum and angular sum of the discrete Fourier transform, are computed for each ROI [39,42].

GWT Features: Multiscale features estimated at various frequency and orientations of Gabor filters provide useful description of texture [40,51–54]. Two statistical features, mean and standard deviation, are computed by using a set of Gabor wavelets at three scales and seven orientations, resulting in a set of $(2 \times 3 \times 7 = 42)$ texture features for each ROI.

Laws' Features: Spatial filtering-based Laws' texture features determine texture properties by performing local averaging, edge detection, spot detection, wave detection, and ripple detection in texture [28]. Law's texture features can be computed by using special 1D filters of length 3, 5, 7 and 9 [40,39]. Different filter lengths correspond to different resolutions for extraction of texture

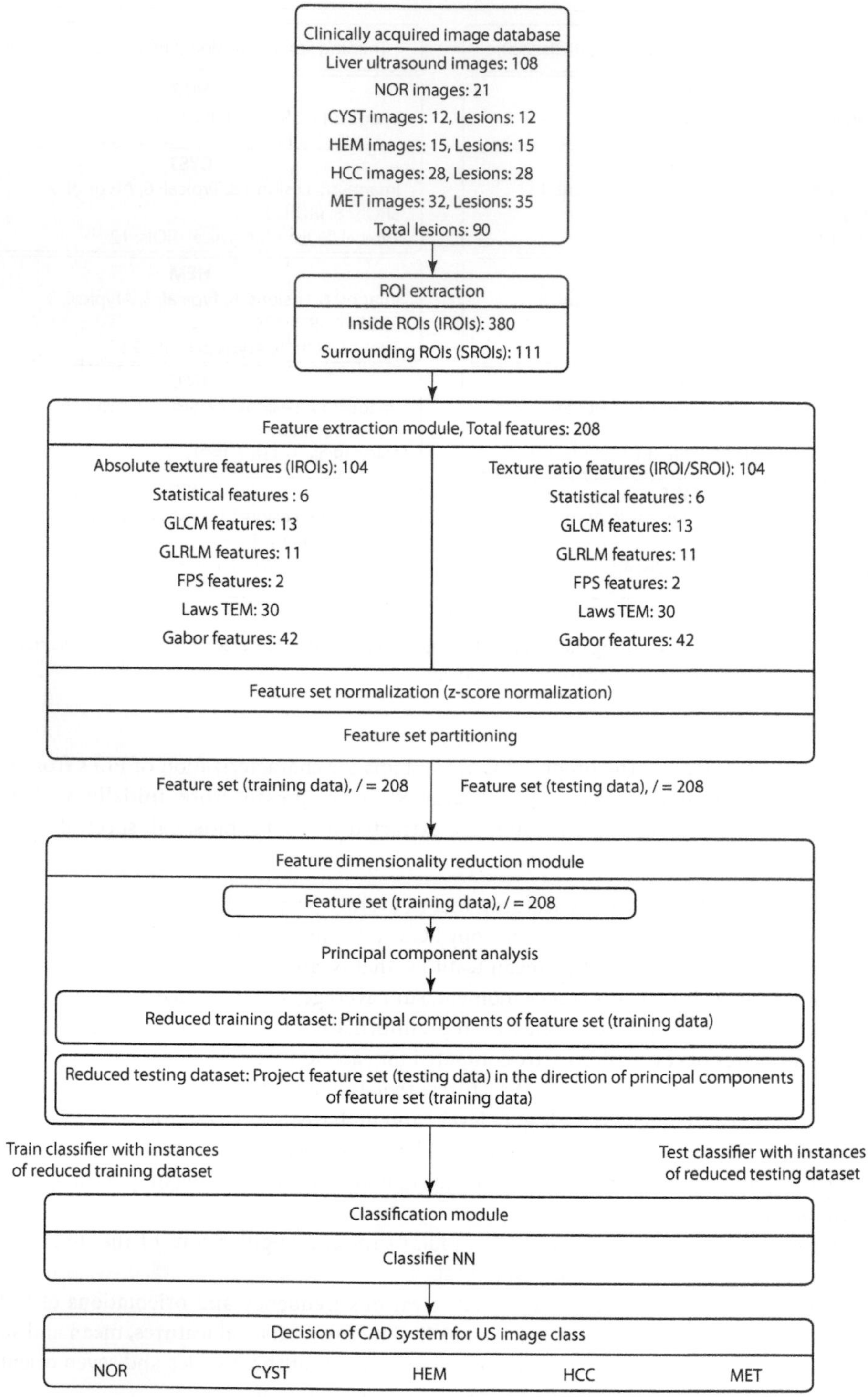

FIGURE 14.7 CAC systems for FLLs using single multi-class classifier based design.

features from a ROI. In the present work, 1D filters of length 5, that is, L5 = [1, 4, 6, 4, 1], E5 = [−1, −2, 0, 2, 1], S5 = [−1, 0, 2, 0, −1], W5 = [−1, 2, 0, −2, 1], and R5 = [1, −4, 6, −4, 1], are used. A total of twenty-five 2D filters are generated by combining these 1D filters. These 2D filters are convolved with the ROI images. A 15 × 15 square window is applied to the resulting convolved images in order to compute texture energy images. Out of 25 2D filters, 10 filters are identical to each other if they are rotated by 90°. Texture energy images computed from these pairs of identical filters can be combined to obtain rotation invariant image. Thus, a total of 15 rotational invariant texture energy images are obtained for each ROI. Two statistics, that is, mean and standard deviation, computed from each rotational invariant texture energy image result in (15 × 2 = 30) texture features for each ROI.

For detection and characterization of FLLs initially, three texture feature vectors (TFVs) are computed using FOS, GLCM, GLRLM, FPS, GWT, and Laws' texture feature extraction methods. The brief description of these TFVs is tabulated in Table 14.3.

14.2.4.1.2 Feature Space Dimensionality Reduction Module

The combined TFV of length 208 computed for the present classification task, TFV3, may contain redundant and correlated features. If these are used for classifier design, the performance of CAC system in terms of accuracy and reliability can be degraded. PCA reduces the dimensionality of the feature space by retaining most of the original variability in the data. PCA is widely used for finding essential attributes/variables, that is, PCs needed for the classification task [28,55–57].

In the present work, PCA is used as a preprocessing step for reducing the dimensionality of the feature space before classification. The main steps in PCA algorithm are (1) standardization of the values of all the features in the dataset to zero mean and unit variance, (2) obtaining the covariance matrix form the training dataset, (3) obtaining the eigenvalues and eigenvectors from the covariance matrix (eigenvectors are the directions of the PCs), and (4) projecting the data points in the testing dataset in the direction of PCs of the training dataset [55–57]. As the computed PCs are uncorrelated to each other and are arranged in decreasing order of variance explained by each PC, it is reasonable to step through the first few PCs for building the classification model [55–58]. In the present work, the optimal number of PCs to be retained for classification task is determined empirically by repeated experimentations carried out by stepping through the first 15 PCs to build the classification models, (i.e., by first considering only the first PC, and then the first 2 PCs, first 3 PCs and so on up to the first 15 PCs and verifying the performance of resulting classification models). The number of PCs yielding the maximum classification accuracy are used as features of reduced TFV, as described in Table 14.3.

14.2.4.1.3 Classification Module

NN classifier: The NN architecture consists of an input layer, hidden layer, and output layer. The optimum values for number of hidden layer neurons and the number of PCs to be retained in order to design

TABLE 14.3 Description of TFVs

Texture Feature Vectors (TFVs)	(l)
TFV1: TFV consisting of 104 texture features (6 FOS, 13 GLCM, 11 GLRLM, 2 FPS, 42 Gabor and 30 Laws' features), computed from IROIs.	104
TFV2: TFV consisting of 104 texture ratio features (6 FOS, 13 GLCM, 11 GLRLM, 2 FPS, 42 Gabor and 30 Laws' features), computed by taking the ratio of texture feature computed from IROI and the corresponding SROI.	104
TFV3: Combined TFV consisting of 104 texture features (TFV1) and 104 texture ratio features (TFV 2)	208
TFV4: Reduced TFV consisting of first few PCs, obtained by subjecting combined TFV (TFV3) to PCA	6

Note: RTFV: Reduced texture feature vector. The length of RTFV, that is, length of TFV4 implies optimal number of PCs retained for the classification task, obtained by repeated experimentations carried out by stepping through first 15 PCs to build the classification models.

an NN model is determined empirically by repeated experimentation for the number of hidden layer neurons {4,5,…,9,10} and the number of PCs {1,2,…,14,15}. The PCA-NN model for the present classification task consist of 6 neurons (corresponding to first 6 PCs), one hidden layer with 10 neurons and one output layer with 5 neurons (corresponding to NOR, Cyst, HEM, HCC, and MET classes). The output of the neuron corresponding to the labeled class is set to 1, and the output of other neurons is set to 0. The learning of the network is supervised and back-propagation algorithm with adaptive learning rate and momentum is used to obtain the desired input-output relationship [59,60].

*14.2.4.1.4 **Results- Experiment I: CAC System Based on Single PCA-NN Based Five-Class Classifier***

Rigorous experimentation has been carried out for performance analysis of the proposed CAC system for characterization of FLLs. Initial experiments were carried out using original features vectors (TFV1, TFV2, and TFV3), primarily for the validation of the fact that texture information from the surrounding liver parenchyma significantly contributes toward classification of FLLs. It was observed that the combined TFV, TFV3 (consisting of both texture features and texture ratio features), yielded better classification performance in comparison with TFV1 (consisting of texture features computed from IROIs) and TFV2 (consisting of texture ratio features computed from IROIs and corresponding SROI). These results indicate that texture information from inside lesion ROIs (IROIs) and surrounding lesion ROIs (SROIs), collectively, yield significant improvement in classification accuracy. Thus, only combined TFV, that is, TFV3, was considered for dimensionality reduction through PCA, resulting in a reduced TFV, or TFV4. The classification performance of the feature set consisting of instances of TFV1, TFV2, TFV3, and TFV4 is tested by using single, five-class, NN classification models. The results obtained are reported in Table 14.4.

TABLE 14.4 Classification Performance: CAC Design Based on Single Five-Class NN Classification Models with TFV1, TFV2, TFV3, and TFV4.

TFV (*l*)		CM					OCA (%)	ICA_{Class} (%)
TFV1 (104)		NOR	CYST	HEM	HCC	MET	66.1	
	NOR	31	0	0	5	4		ICA_{NOR} : 77.5
	CYST	0	23	2	0	0		ICA_{CYST}: 92.0
	HEM	0	0	15	10	5		ICA_{HEM}: 50.0
	HCC	0	1	8	15	16		ICA_{HCC}: 37.5
	MET	1	0	2	7	35		ICA_{MET}: 77.7
TFV2 (104)	NOR	35	0	0	2	3	75.0	ICA_{NOR} : 87.5
	CYST	1	19	5	0	0		ICA_{CYST}: 76.0
	HEM	0	0	23	0	7		ICA_{HEM}: 76.6
	HCC	7	1	0	24	8		ICA_{HCC}: 60.0
	MET	1	0	8	2	34		ICA_{MET}: 75.5
TFV3 (208)	NOR	34	0	0	4	2	82.7	ICA_{NOR} : 85.0
	CYST	0	23	2	0	0		ICA_{CYST}: 92.0
	HEM	0	0	25	1	4		ICA_{HEM}: 83.3
	HCC	1	0	0	35	4		ICA_{HCC}: 87.5
	MET	3	0	6	4	32		ICA_{MET}: 71.1
TFV4 (6)	NOR	33	0	2	4	1	87.7	ICA_{NOR} : 82.5
	CYST	0	24	0	0	1		ICA_{CYST}: 96.0
	HEM	1	0	28	0	1		ICA_{HEM}: 93.3
	HCC	1	2	0	36	1		ICA_{HCC}: 90.0
	MET	1	0	4	3	37		ICA_{MET}: 82.2

Source: Virmani et al. *Journal of Medical Engineering and Technology*, 37(4) 292–306, 2013, © Taylor & Francis.

Note: TFV: Texture feature vector, *l*: Length of TFV, CM: Confusion matrix, OCA: Overall classification accuracy, ICA: Individual class accuracy, Here, ICA of class A, denoted as ICAA refers to (Number of cases correctly classified as class A/Total number of cases in class A).

It can be observed from Table 14.4, that TFV1 (consisting of texture features computed from IROIs) yielded maximum OCA value of 66.1% with NN classification model consisting of an input layer (104 neurons), one hidden layer (10 neurons), and an output layer (5 neurons). TFV2 (consisting of texture ratio features computed from IROIs and corresponding SROI) yielded maximum OCA value of 75% with NN classification model consisting of an input layer (104 neurons), one hidden layer (10 neurons), and an output layer (5 neurons). Combined TFV, or TFV3 (consisting of texture features and texture ratio features) yielded maximum OCA value of 82.7% with NN classification model consisting of an input layer (208 neurons), one hidden layer (10 neurons), and an output layer (5 neurons). It can be noted that maximum OCA value of 82.7% is achieved by using TFV3, or combined TFV, consisting of 104 texture features and 104 texture ratio features. Thus, TFV3 is considered for feature-space dimensionality reduction by using PCA to design an efficient CAC system using NN classifier. It can be observed that reduced TFV, or TFV4 consisting of the first six PCs yields maximum OCA value of 87.7% with NN classification model consisting of an input layer (6 neurons), one hidden layer (10 neurons), and an output layer (5 neurons).

From Table 14.4, it can be observed that texture features computed from IROIs yield OCA value of 66.1%, with the ICA values of 77.5%, 92%, 50%, 37.5%, and 77.7% for NOR, Cyst, HEM, HCC, and MET classes, respectively. The texture ratio features yield OCA value of 75%, with the ICA values of 87.5%, 76%, 76.6%, 60%, and 75.5% for NOR, Cyst, HEM, HCC, and MET classes, respectively. It can be concluded that texture-ratio features have more discrimination ability than texture features for classification of FLLs from B-Mode US. It can be observed that combined TFV, that is, TFV3 (consisting of texture features and texture ratio features), yields the OCA value of 82.7%, with the ICA values of 85%, 92%, 83.3%, 87.5%, and 71.1% for NOR, Cyst, HEM, HCC, and MET classes, respectively. In comparison to ICA values obtained by using texture-ratio features only, the values obtained by using texture features and texture-ratio features collectively show improvement by 16%, 6.7%, and 27.5%, for Cyst, HEM, and HCC classes, respectively. However, there is little fall in the ICA values for NOR and MET cases by 2.5% and 4.4%, respectively. These results indicate that both texture features and texture-ratio features contribute toward effective classification of FLLs from B-Mode US. The results are in agreement with the premise that radiologists visualize the textural characteristics of regions inside and outside the lesions for differential diagnosis between FLLs. Finally, it is observed that optimal reduced TFV with the first six PCs yields the maximum OCA value of 87.2%, with the ICA values of 82.5%, 96%, 93.3%, 90%, and 82.2% for NOR, Cyst, HEM, HCC, and MET classes, respectively.

14.2.4.2 Experiment No. II: To Design and Evaluate the Performance of CAC System for Characterization of FLLs Using PCA-NN Based Hierarchical Classifier Design

It is worth mentioning that the hierarchical CAC (HCAC) design provides the possibility to go stepwise from the general classification problem, that is, normal versus abnormal liver tissue, to the more particular classification problem, which is the identification of exact liver abnormality with hierarchical framework of classifiers. In the present experiment, an efficient hierarchical CAC (HCAC) system is designed with texture features computed from IROIs and SROIs by using a comprehensive and representative image database with (1) typical and atypical cases of liver Cyst, primary benign lesion (PBL, i.e., HEM) and secondary malignant lesion (SML, i.e., MET); (2) small and large cases of primary malignant lesion (PML, i.e., HCC); and (3) normal (NOR) liver tissue. The proposed CAC system for characterization of FLLs using hierarchical classifier design consists of feature extraction, dimensionality reduction, and classification module, as shown in Figure 14.8.

14.2.4.2.1 Feature Extraction Module

In the present work, initially, a wide variety of texture features are extracted by using statistical, spectral, and spatial filtering-based feature extraction methods. As described earlier, a feature vector of 104 texture features, that is, 6 FOS features, 13 GLCM features [14,36], 11 GLRLM features [41,44,49], 2 FPS features [14], 42 Gabor features [13,43], and 30 Laws' texture features [35] is computed for each IROI. Similarly, a feature vector of 104 texture ratio features is computed for each IROI and a corresponding

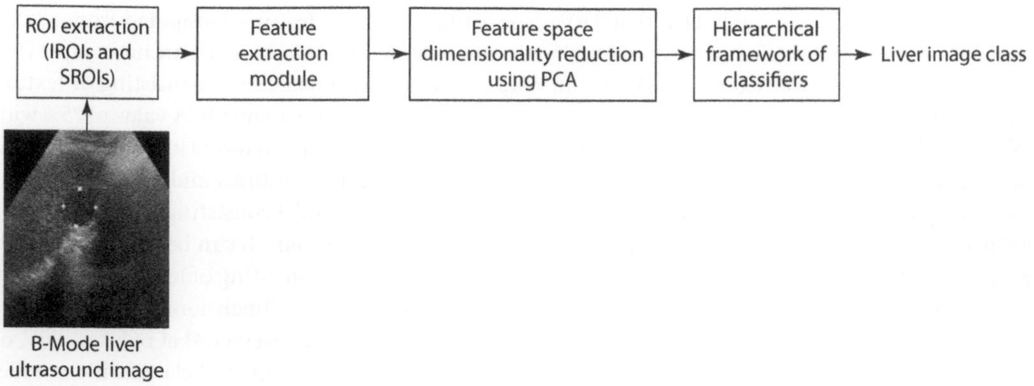

FIGURE 14.8 Generalized block diagram of a hierarchical computer-aided diagnostic (HCAC) system.

SROI. Finally, a combined feature set consisting of 208 features (104 texture features + 104 texture ratio features) is inputted to the dimensionality reduction module.

14.2.4.2.2 Feature-Space Dimensionality Reduction Module

In the present work, feature-space dimensionality reduction has been carried out using principal component analysis (PCA). The optimal number of principal components (PCs) to be retained for the classification task is determined empirically by repeated experiments carried out by stepping through the first 15 PCs to build the classification model [26,41].

14.2.4.2.3 Classification Module

The classification module consists of four binary classifiers arranged in a hierarchical framework. These four classifiers provide stepwise classification for the generalized five class classification problem. The first classifier is used to classify an unknown test case into normal (NOR) or abnormal (ABNOR) case. If the test case is predicted as abnormal (ABNOR), the second classifier is activated for classification into Cyst or "Other Lesion" (OL) class. If the test case is predicted as belonging to OL class, the third classifier is activated for classification into a PBL or Malignant Lesion (ML) class. If the prediction of the third classifier is for a ML class, the fourth classifier is activated for further classification into PML or SML classes. The block diagram of the classification module of the proposed HCAC system is shown in Figure 14.9.

Feature space dimensionality reduction using PCA algorithm is applied individually before designing each binary classifier.

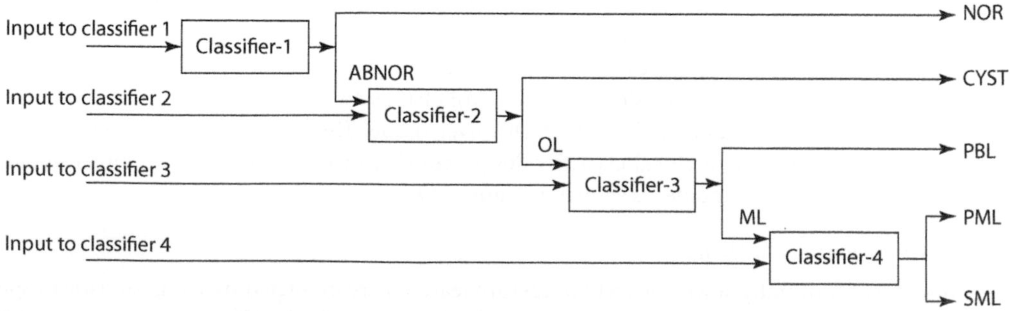

FIGURE 14.9 Block diagram of the classification module of proposed HCAC system. **Note:** OL: Other lesion; ML: Malignant lesion; PBL: Primary benign lesion (HEM); PML: Primary malignant lesion (HCC); SML: Secondary malignant lesion (MET).

NN Classifier: The architecture of NN classifier consists of three layers (i.e., an input layer, hidden layer, and output layer). Instances of feature vectors consisting of an optimal number of PCs obtained for the binary classification tasks (i.e., NOR/ABNOR, CYST/OL, PBL/ML, and PML/SML) are fed to the input layer of corresponding binary NN (BNN) classifiers. The number of neurons in the hidden layer is obtained by trial-and-error procedure. Experiments were carried out with different numbers of hidden neurons, and it was observed that, with 4 neurons in the hidden layer, a reasonable trade-off between accuracy and convergence is obtained. The output layer of each NN has two neurons corresponding to the considered classes. For designing each NN classifier, the output of the neuron corresponding to the labeled class is set to 1 and the output of other neuron is set to 0, that is, the learning of each NN classifier is supervised, and a back-propagation algorithm with adaptive learning rate and momentum is used to obtain the desired input-output relationship [13,26].

The block diagram of classification module for PCA-NN based HCAC system for classification of FLLs is shown in Figure 14.10.

14.2.4.2.4 *Results- Experiment 2: PCA-NN based HCAC system*

Four classification tasks were considered, corresponding to four classifiers arranged in a hierarchical framework. Each binary classifier was trained independently. The design of HCAC system for characterization of FLLs was carried out with 108 B-Mode liver US images, that is, 21 NOR, 12 Cyst, 15 PBL, 28 PML, and 32 SML images. The dataset description and the bifurcation of the dataset into disjoint training and testing datasets for each binary classifier is shown in Table 14.5.

In this experiment the performance of binary classifiers of PCA-NN based HCAC system is evaluated, and the results obtained are given in Table 14.6.

By visualizing the performance of individual binary classifiers of PCA-NN-based HCAC system (shown in Table 14.6), some interesting facts are observed:

1. For classification between NOR/ABNOR classes, the accuracy of 95.5% as obtained by using PCA-NN1 classifier.
2. For classification of ABNOR cases between CYST/OL classes, the accuracy of 97.1% is obtained by using PCA-NN2 classifier.
3. For classification of OL cases between PBL/ML classes, the accuracy of 93% is obtained by using PCA-NN3 classifier.
4. For classification of ML cases between PML/SML classes, all the four classifiers (i.e., PCA-NN1, PCA-NN2, PCA-NN3, and PCA-NN4) yield the same accuracy of 98.8%. It may be noted that PCA-NN4 is considered as best classifier for PML/SML classes because it is designed with only 3 PCs (least number of PCs) for the classification task.

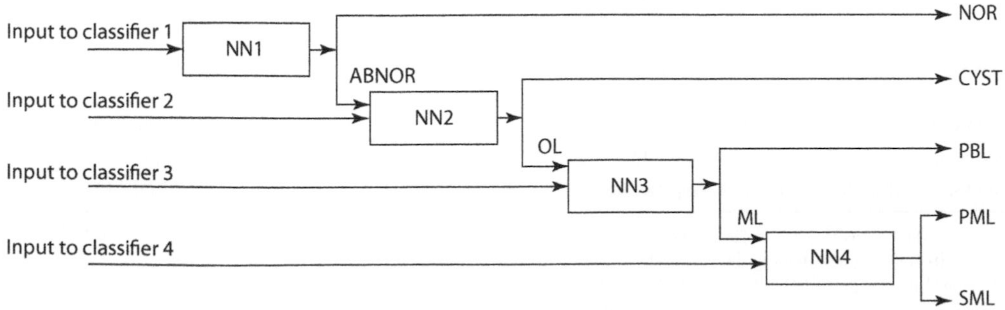

FIGURE 14.10 Block diagram of classification module for PCA-NN based HCAC system. **Note:** NOR: Normal; ABNOR: Abnormal; OL: Other lesion; ML: Malignant Lesion; PML: Primary malignant lesion; SML: Secondary malignant lesion.

TABLE 14.5 Dataset Description: HCAC System for FLLs

Dataset Description: 108 *B-Mode Liver US Images*									
NOR Images: 21		Cyst Images: 12		PBL Images: 15		PML Images: 28		SML Images: 32	
		Cyst Lesions: 12		PBL Lesions: 15		PML Lesions: 28		SML Lesions: 35	
		typ: 9	atyp: 3	typ: 9	atyp: 6	Small: 13	Large: 15	typ: 13	atyp: 22
Description: 491 *ROIs* (*SROIs*: 111, *IROIs*: 380) *Description*: 111 *SROIs*									
NOR: 21		Cyst: 12		PBL: 15		PML: 28		SML: 35	
Description: 380 IROIs									
NOR: 70		Cyst: 55		PBL: 70		PML: 90		SML: 95	
typ: 70	atyp: Nil	typ: 35	atyp: 20	typ: 24	atyp: 46	Small: 19	Large: 71	typ: 13	atyp: 82

Dataset Description: Binary Classifier NOR/ABNOR	
Training Set Description	*Testing Set Description*
Images: 57, SROIs: 59. IROIs: 200	Images: 51, SROIs: 52, IROIs: 180
NOR: Images: 10, SROIs: 10, *IROIs*: 30	NOR: Images: 11, SROIs: 11, *IROIs*: 40
ABNOR: Images: 47, SROIs: 49, *IROIs*: 170	ABNOR: Images: 40, SROIs: 41, *IROIs*: 140
Cyst: Images: 4, Lesions: 4, typ.: 3, atyp.: 1	Cyst: Images: 8, Lesions: 8, typ.: 6, atyp.: 2
SROIs: 4, *IROIs:* 30, typ. IROIs: 22, atyp. IROIs: 8	SROIs: 8, *IROIs:* 25, typ. IROIs: 13, atyp.IROIs: 12
PBL: Images: 9, Lesions: 9, typ.: 6, atyp.: 3	PBL: Images: 6, Lesions: 6, typ.: 3, atyp.: 3
SROIs: 9, *IROIs:* 40, typ. IROIs: 18, atyp. IROIs: 22	SROIs: 6, *IROIs:* 30, typ. IROIs: 6, atyp. IROIs: 24
PML: Images: 16, Lesions: 16, small : 7, large: 9	PML: Images: 18, Lesions: 20, small: 8, large: 12
SROIs: 16, *IROIs:* 50, small IROIs: 10, large IROIs: 40	SROIs: 12, *IROIs:* 40, small IROIs: 9, large IROIs: 31
SML: Images: 18, Lesions: 20, typ.: 8, atyp.: 12	SML: Images: 14, Lesions: 15, typ.: 5, atyp.: 10
SROIs: 20, *IROIs:* 50, typ. IROIs: 8, atyp. IROIs: 52	SROIs: 15, *IROIs:* 45, typ. IROIs: 5, atyp.IROIs: 40

Dataset Description: Binary Classifier CYST/OL	
Training Set Description	*Testing Set Description*
Images: 47, SROIs: 49, *IROIs:* 170	Images: 40, SROIs: 41, *IROIs:* 140
Cyst: Images: 4, Lesions: 4, typ.: 3, atyp.: 1	Cyst: Images: 8, Lesions: 8, typ.: 6, atyp.: 2
SROIs: 4, *IROIs:* 30, typ. IROIs: 22, atyp. IROIs: 8	SROIs: 8, *IROIs:* 25, typ. IROIs: 13, atyp. IROIs: 12
OL*: Images: 43, SROIs: 45, *IROIs:* 140	OL*: Images: 32, SROIs: 33, *IROIs:* 115

Dataset Description: Binary Classifier PBL/ML	
Training Set Description	*Testing Set Description*
Images: 43, SROIs: 45, IROIs: 140	Images: 32, SROIs: 33, *IROIs:* 115
PBL: Images: 9, Lesions: 9, typ.: 6, atyp.: 3	PBL: Images: 6, Lesions: 6, typ.: 3, atyp.: 3
SROIs: 9, IROIs: 40, typ. IROIs: 18, atyp. IROIs: 22	SROIs: 6, *IROIs:* 30, typ. IROIs: 6, atyp. IROIs: 24
ML*: Images: 34, SROIs: 36, IROIs: 100	ML*: Images: 26, SROIs: 27, *IROIs:* 85

Dataset Description: Binary Classifier PML/SML	
Training Set Description	*Testing Set Description*
Images: 34, SROIs: 36, *IROIs:* 100	Images: 26, SROIs: 27, *IROIs:* 85
PML: Images: 16, Lesions: 16, small.: 7, large.: 9	PML: Images: 12, Lesions: 12, typ.: 6, atyp.: 6
SROIs: 16, *IROIs: 50*, small IROIs: 10, large IROIs: 40	SROIs: 12, *IROIs: 40*, small IROIs: 9, large IROIs: 31
SML: Images: 18, Lesions: 20, typ.: 8, atyp.: 12	SML: Images: 14, Lesions: 15, typ.: 5, atyp.: 10
SROIs: 20, *IROIs:* 50, typ. IROIs: 8, atyp. IROIs: 42	SROIs: 15, *IROIs:* 45, typ. IROIs: 5, atyp. IROIs: 40

Note: typ.: Typical; atyp.: Atypical; OL: Other lesion (Further bifurcation of OL* class into PBL, PML and SML subclasses in training and testing dataset is shown in dataset bifurcation for ABNOR image class); ML: Malignant lesion (Further bifurcation of ML* class into PML and SML subclasses in training and testing dataset is shown in dataset bifurcation for ABNOR image class). The size of small PML lesion varied from 1.5 to 1.9 cm and size of large PML lesion varied from 2.1 to 5.6 cm; (a) Total lesions = 90 [(typ. lesions (31) + atyp. lesions (31) + small PML lesions (13) + large PML lesions (15)]; (b) Total SROIs = 111 [90 SROIs corresponding to 90 lesions + 21 SROIs corresponding to 21 NOR images]; (c) Total typ. IROIs = 142 [typ. lesion IROIs (72) + typ. NOR IROIs (70)]; (d) Total atyp. IROIs = 148; (e) Total small PML lesion IROIs = 19; (f) Total large PML lesion IROIs = 71; (g) Total IROIs = (c) + (d) + (e) + (f) = 142 + 148 + 19 + 71 = 380.

TABLE 14.6 Performance Obtained by PCA-NN Based HCAC System

Classifier	PCs		CM		Acc_Bin_Class (%)	OCA (%)	ICA (%)
			NOR	ABNOR			
PCA-NN1	11	NOR	37	3	95.5 (172/180)	91.6 (165/180)	92.5 (37/40)
		ABNOR	5	135			96.4 (135/140)
			Cyst	OL			
PCA-NN2	15	Cyst	22	3	97.1 (136/140)		88.0 (22/25)
		OL	1	114			99.1 (114/115)
			PBL	ML			
PCA-NN3	11	PBL	30	0	93.0 (107/115)		100 (30/30)
		ML	8	77			90.5 (77/85)
			PML	SML			
PCA-NN4	3	PML	40	0	98.8 (84/85)		100 (40/40)
		SML	1	44			97.7 (44/45)

Note: Acc_Bin_Class: Accuracy of binary classifier, OCA: Overall classification accuracy, ICA: Individual class accuracy, NOR: Normal liver, ABNOR: Abnormal liver, OL: Other lesion, PBL: Primary benign lesion, ML: Malignant lesion, PML: Primary malignant lesion, SML: Secondary malignant lesion, PCA: Principal component analysis, NN: Neural Network, PCs: Principal components.

Conclusion

The overall classification accuracy achieved from single, multi-class, PCA-NN-based classifier design is 87.2%, and the OCA achieved from PCA-NN-based hierarchical classifier design is 91.6%. Hierarchical design offers the possibility of stepwise characterization of focal liver diseases. The main conclusion of the present research work is that CAC system designs using texture features (computed from IROIs) and texture-ratio features (computed from IROIs and corresponding SROI) collectively enhance the performance of the system for characterization of FLLs. Thus, it can be concluded that the texture of the region surrounding the lesion contributes effectively for differential diagnosis between different FLLs.

Acknowledgments

The authors acknowledge the support provided by Dr. Niranjan Khandelwal and Dr. Naveen Kalra of the Department of Radio diagnosis and Imaging, PGIMER, Chandigarh, for the collection and labeling of liver US images with FLLs.

References

1. Baert, A.L., and Sartor, K., *Focal Liver Lesions-Detection, Characterization, Ablation*, Springer Berlin Heidelberg, New York, pp. 167–77, 2005.
2. Bates, J., *Abdominal Ultrasound How Why and When* 2nd Ed., Churchill Livingstone, Oxford, pp. 80–107, 2004.
3. Harding, J., and Callaway, M., Ultrasound of focal liver lesions, *RAD Magazine*, Vol. 36, No. 424, pp. 33–34, 2010.
4. Huang, Y.L., Chen, J.H., Shen, W.C., Diagnosis of hepatic tumors with texture analysis in non-enhanced computed tomography images, *Academic Radiology*, Vol. 13, No. 6, pp. 713–720, 2006.
5. Jeffery, R.B., Ralls, P.W. *Sonography of Abdomen*, Raven, New York, 1995.
6. Namasivayam, S., Salman, K., Mittal, P.K., Martin, D. Small W.C., Hypervascular hepatic focal lesions: Spectrum of imaging features, *Current Problems in Diagnostic Radiology*, Vol. 36, No. 3, pp. 107–123, 2007.

7. Soye, J.A., Mullan, C.P., Porter, S., Beattie, H., Barltrop, A.H., and Nelson, W.M., The use of contrast-enhanced ultrasound in the characterization of focal liver lesions, *Ulster Medical Journal*, Vol. 76, No. 1, pp. 22–25, 2007.
8. Minhas, F., Sabih, D., and Hussain, M., Automated classification of liver disorders using ultrasound images, *Journal of Medical Systems*, Vol. 36, No. 5, pp. 3163–3172, 2012.
9. Virmani, J., Kumar, V., Kalra, N., and Khandelwal, N., A rapid approach for prediction of liver cirrhosis based on first order statistics, In *Proceedings: IEEE International Conference on Multimedia, Signal Processing and Communication Technologies*, IMPACT-2011, pp. 212–215, 2011.
10. Virmani, J., Kumar, V., Kalra, N., and Khandelwal, N., Prediction of liver cirrhosis based on multiresolution texture descriptors from B-mode ultrasound, *International Journal of Convergence Computing*, Vol. 1, No.1, pp. 19–37, 2013.
11. Virmani, J., Kumar, V., Kalra, N., and Khandelwal, N., SVM-based characterization of liver cirrhosis by singular value decomposition of GLCM matrix, *International Journal of Artificial Intelligence and Soft Computing*, Vol. 3, No. 3, pp. 276–296, 2013.
12. Virmani, J., Kumar, V., Kalra, N., and Khandelwal, N., SVM-Based characterization of liver ultrasound images using wavelet packet texture descriptors, *Journal of Digital Imaging*, Vol. 26, No. 3, pp. 530–543, 2013.
13. Mittal, D., Kumar, V., Saxena, S.C., Khandelwal, N., and Kalra, N., Neural network based focal liver lesion diagnosis using ultrasound images, *Computerized Medical Imaging and Graphics*, Vol. 35, No. 4, pp. 315–323, 2011.
14. Virmani, J., Kumar, V., Kalra, N., and Khandelwal, N., Characterization of primary and secondary malignant liver lesions from B-mode ultrasound, *Journal of Digital Imaging*, Vol. 26, No. 6, pp. 1058–1070, 2013.
15. Beussink, L.S., Atypical hepatic hemangioma, *Journal of Diagnostic Medical Sonography*, Vol. 25, No. 1, pp. 67–70, 2009.
16. Marsh, J.I., Gibney, R.G., and David, K.B., Hepatic hemangioma in the presence of fatty infiltration: An atypical sonographic appearance, *Gastrointestinal Radiology*, Vol. 14, No. 3, pp. 262–264, 1989.
17. Tsurusaki, M., Kawasaki, R., Yamaguchi, M., Sugimoto, K., Fukumoto, T., Ku, Y., and Sugimura, K. Atypical hemangioma mimicking hepatocellular carcinoma with a special note on radiological and pathological findings, *Japanese Journal of Radiology*, Vol. 27, No. 3, pp. 156–160, 2009.
18. Scheible, W., Gosink, B.B., and Leopold, G.R., Gray scale echographic patterns of hepatic metastatic disease, *American Journal of Roentgenology*, Vol. 129, No. 6, pp. 983–987, 1977.
19. Tiferes, D.A., and Ippolito, G.D., Liver neoplasms: Imaging characterization, *Radiologia Brasileira*, Vol. 41, No. 2, pp. 119–127, 2008.
20. Yoshida, H., Casalino, D.D., Keserci, B., Coskun, A., Ozturk, O., and Savranlar, A., Wavelet packet based texture analysis for differentiation between benign and malignant liver tumors in ultrasound images, *Physics in Medicine and Biology*, Vol. 48, No. 22, pp. 3735–3753, 2003.
21. Sekiguchi, R., Kuwajima, A., Nagamoto, M., Ohno, H., and Tamura, M., Hepatocellular carcionoma: The diagnostic difficulties of ultrasonography and analysis of risk factors in MHTS, *Journal of Medical Systems*, Vol. 17, No. 3–4, pp. 133–137, 1993.
22. Burges, C.J.C., A tutorial on support vector machines for pattern recognition, *Journal of Data Mining and Knowledge Discovery*, Vol. 2, No. 2, pp. 1–43, 1998.
23. Cao, G.T., Shi, P.F., and Hu, B., Liver fibrosis identification based on ultrasound images captured under varied imaging protocols, *Journal of Zhejiang University Science*, Vol. 6B, No. 2, pp. 1107–1114, 2005.
24. Bass, T., Gravenhorst, F., Fischer, R., Khawaja, A., and Dossel, O., Comparison of three t-wave delineation algorithms based on wavelet filter bank, correlation and PCA, *Computing in Cardiology*, Vol. 37, pp. 361–364, 2010.
25. Haralick, R.M., Shanmugam, K. and Dinstein, I., Textural features for image classification, *IEEE Transactions on Systems, Man and Cybernetics*, Vol. SMC-3, No. 6, pp. 610–621, 1973.

26. Virmani, J., Kumar, V., Kalra, N., and Khandelwal, N., PCA-SVM based CAD system for focal liver lesions from B-Mode ultrasound, *Defence Science Journal*, Vol. 63, No. 5, pp. 478–486, 2013.
27. Virmani, J., Kumar V., Kalra, N., and Khandelwal, N., A comparative study of computer-aided classification systems for focal hepatic lesions from B-mode ultrasound. *Journal of Medical* Engineering *and* Technology, Vol. 37 , No. 4 pp. 292–306, 2013
28. Du, C., Linker, R., and Shaviv, A., Identification of agricultural mediterranean soils using mid-infrared photoacoustic spectroscopy, *Geoderma*, Vol. 143, No. 1, pp. 85–90, 2008.
29. Kadah, Y.M., Farag, A.A., Zurada, J.M., Badawi, A.M., and Youssef, A.M., Classification algorithms for quantitative tissue characterization of diffuse liver disease from ultrasound images, *IEEE Transactions in Medical Imaging*, Vol. 15, No. 4, pp. 466–478, 1996.
30. Singh, M., Singh, S., and Gupta, S., A new measure of echogenicity of ultrasound images for liver classification, In *Proceedings: 24th IEEE Canadian Conference on Electrical and Computer Engineering*, CCECE' 11, pp. 317–320, 2011.
31. Padhye, N.S., Brazdeikis, A., and Verklan, M.T., Monitoring fetal development with magnetocardiography, In *Proceedings: IEEE Engineering in Medicine and Biology Society*, Vol. 26, pp. 3609–3610, 2004.
32. Lee, W.L., and Hsieh, K.S., Computer assisted characterization for ultrasonic liver tissue by fusion of classifiers, In *Proceedings: 2nd International Conference on Innovative Computing Information and Control*, ICICIC'2007, Kumamoto, pp. 100–103, 2007.
33. Pavlopoulos, S., Kyriakou, E., Koutsouris, D., Blekas, K., Stafylopatis, A., and Zoumpoulis, P.,Fuzzy neural network based texture analysis of ultrasonic images, *IEEE Engineering in Medicine and Biology Magazine*, Vol. 19, No. 1, pp. 39–47, 2000.
34. Wu, C.M., Chen, Y.C., and Hsiesh, K.S., Texture features for classification of ultrasonic liver images, *IEEE Transactions on Medical Imaging*, Vol. 11, No. 2, pp. 141–152, 1992.
35. Virmani, J., Kumar, V., Kalra, N., and Khandelwal, N., Prediction of cirrhosis from liver ultrasound B-mode images based on laws' masks analysis, In *Proceedings: IEEE International Conference on Image Information Processing*, ICIIP-2011, Wakhnaghat, Shimla, India, pp. 1–5, 2011.
36. Virmani, J., Kumar, V., Kalra, N., and Khandelwal, N., Prediction of cirrhosis based on SVD decomposition of GLCM matrix and a neural network classifier, In *Proceedings: IEEE International Conference on Developments in E-systems Engineering*, DeSe'11, Dubai, pp. 146–151, 2011.
37. Mojsilovic, A., Popovic, M., and Sevic, D., Classification of the ultrasound liver images with the 2N×l-D wavelet transform, In *Proceedings: IEEE International Conference on ImageProcessing*, Lausanne, Switzerland, pp. 367–370, 1996.
38. Nawaz, S., and Dar, A.H., Hepatic lesions classification by ensemble of SVMs using statistical features based on co-occurrence matrix, In *Proceedings: 4th IEEE International Conference on Emerging Technologies*, ICET' 2008, pp. 21–26, 2008.
39. Badawi, A.M., Derbala, A.S., and Youssef, A.B.M.,Fuzzy logic algorithm for quantitative tissue characterization of diffuse liver diseases from ultrasound images, *International Journal of Medical Informatics*, Vol. 55, No. 2, pp. 135–147, 1999.
40. Chu, A., Sehgal, C.M., and Greenleaf, J.F., Use of gray value distribution of run lengths for texture analysis, *Pattern Recognition Letters*, Vol. 11, No. 6, pp. 415–420, 1990.
41. Dasarathy, B.V., and Holder, E.B., Image characterizations based on joint gray level-run length distributions, *Pattern Recognition Letters*, Vol. 12, No. 8, pp. 497–502, 1991.
42. Galloway, R.M.M., Texture analysis using gray level run lengths, *Computerized Graphics and Image Processing*, Vol. 4, No. 2, pp. 172–179, 1975.
43. Weszka, J.S., Dyer, C.R., and Rosenfeld, A., A comparative study of texture measures for terrain classification, *IEEE Transactions on Systems Man and Cybernetics*, Vol. SMC-6, No. 4, pp. 269–285, 1976.
44. Lee, C.C., and Chen, S.H., Gabor wavelets and SVM classifier for liver diseases classification from CT images, In *Proceedings: IEEE International Conference on Systems, Man and Cybernetics*, Taipei, Taiwan, pp. 548–552, 2006.

45. Backes, A.R., Gonclaves, W.N., Martinez, A.S., and Bruno, O.M., Texture analysis and classification using deterministic tourist walk, *Pattern Recognition*, Vol. 43, No. 3, pp. 685–694, 2010.
46. Rachidi, M., Marchadier, A., Gadois, C., Lespessailles, E., Chappard, C., and Benhamou, C.L., Laws' masks descriptors applied to bone texture analysis: An innovative and discriminant tool in osteoporosis, *Skeletal Radiology*, Vol. 37, No. 6, pp. 541–548, 2008.
47. Kim, S.H., Lee, J.M., Kim, K.G., Kim, J.H., Lee, J.Y., Han, J.K., and Choi, B.I., Computer-aided image analysis of focal hepatic lesions in ultrasonography: Preliminary results, *Abdominal Imaging*, Vol. 34, No. 2, pp. 183–91, 2009.
48. Diao, X.F., Zhang, X.Y., Wang, T.F., Chen, S.P., Yang, Y., and Zhong, L., Highly sensitive computer aided diagnosis system for breast tumor based on color doppler flow images, *Journal of Medical Systems*, Vol. 35, No. 5, pp. 801–809, 2011.
49. Huang, Y.L., Wang, K.L., and Chen, D.R., Diagnosis of breast tumors with ultrasonic texture analysis using support vector machines, *Neural Computing and Applications*, Vol. 15, No. 2, pp. 164–169, 2006.
50. Moayedi, F., Azimifar, Z., Boostani, R., and Katebi, S., Contourlet based mammography mass classification, In *Proceedings: ICIAR, LNCS*, Vol. 4633, pp. 923–934, 2007.
51. Nandi, R.J., Nandi, A.K., Rangayyan, R.M., and Scutt, D., Classification of breast masses in mammograms using genetic programming and feature selection, *Medical and Biological Engineering and Computing*, Vol. 44, No. 8, pp. 683–94, 2006.
52. Golemati, S., Tegos, T.J., Sassano, A., Nikita, K.S., and Nicolaides, A.N., Echogenicity of B-Mode sonographic images of the carotid artery: Work in progress, *Journal of Ultrasound in Medicine*, Vol. 23, No. 5, pp. 659–669, 2004.
53. Bianconi, F., and Fernandez, A., Evaluation of the effects of Gabor filter parameters on texture classification, *Pattern Recognition*, Vol. 40, No. 12, pp. 3325–3335, 2007.
54. Clausi, D.A., and Jernigan, M.E., Designing Gabor filters for optimal texture seperability, *Pattern Recognition*, Vol. 33, No.11, pp. 1835–1849, 2000.
55. Kadir, A., Nugroho, L.E., Susanto, A., and Santosa, P.I., Performance improvement of leaf identification system using principal component analysis, *International Journal of Advanced Science and Technology*, Vol. 44, pp. 113–124, 2012.
56. Sachdeva, J., Kumar, V., Gupta, I., and Khandelwal, N., A dual neural network ensemble approach for multiclass brain, *International Journal for Numerical Methods in Biomedical Engineering*, Vol. 28, No. 11, pp. 1107–1120, 2012.
57. Frank, M.W., David, U.J.K., Stefan, B., Gunnar, S., Cristian, L., and Dossel, O., Predicting tissue conductivity influences on body surface potentials: An efficient approach based on principal component analysis, *IEEE Transactions on Biomedical Engineering*, Vol. 58, No. 2, pp. 265–273, 2011.
58. Suganthy, M., and Ramamoorthy, P., Principal component analysis based feature extraction, morphological edge detection and localization for fast iris recognition, *Journal of Computer Science*, Vol. 8, No. 9, pp. 1428–1433, 2012.
59. Chong, C.C., Jia, J.C., and Mital, D.P., Classification of multi-spectral images using BP neural networks classifiers: Input coding assignment, In *Proceedings: IEEE Region 10's Ninth Annual International Conference on Frontiers of Computer Technology*, TENCON'94, Singapore, Vol. 2, pp. 867–871, 1994.
60. Daponte, J.S., and Sherman, P., Classification of ultrasonic image texture by statistical discriminant analysis and neural networks, *Computerized Medical Imaging and Graphics*, Vol. No.1, pp. 3–9, 1991.

15

CAD of Dermatological Ulcers (Computational Aspects of CAD for Image Analysis of Foot and Leg Dermatological Lesions)

Marco Andrey Cipriani Frade

Guilherme Ferreira Caetano

Éderson Dorileo

15.1 Wound Healing

Skin wounds are characterized by the disruption of anatomical continuity of living tissue, caused by an injury, compromising the functionality [1–3]. In order to restore tissue loss, a complex process involving migration, biomolecular signaling network, proliferation, differentiation of multiple cell types—mainly, epidermal, fibroblasts and infiltrating inflammatory cells—and synthesis of matrix components occur in a well-regulated process through overlapping phases like inflammation, proliferation and remodeling [4–7].

It is expected that a wound will follow the physiological healing process. However, because of vascular impairment, repetitive tissue damage and infections, some wounds are trapped in an uncoordinated and long inflammatory phase, clinically known as chronic wounds. The most common reasons include pathophysiological causes, such as venous, arterial, neuropathic and pressure ulcers and burns [8,9].

The time for a chronic wound epithelialization is significantly greater than usual wounds, and, in some cases, the wound has increased its area, mainly due to the highly rich environment in proteases, which remain in a persistent inflammatory state. The persistent inflammatory stimulus is related to complications that include tissue hypoxia, necrosis, exudate and bacterial infections. The continuous inflammatory state triggers numerous tissue responses, leading to impaired wound healing [10,11]

The development of new tools to monitor the healing process and determine the real state of wounds may represent a possible optimization of the treatment efficacy [12]. Monitoring the wound on a regular basis by computer-aided analysis, associated to its etiology, can provide insights on which phase of healing the wound is in, which will help practitioners provide better health care with the choice of an appropriate intervention to likely enhance that particular phase of the wound repair [8,11,12].

15.2 Measurements of Wounds and Tissue Analyses by Images

Diagnosis of skin lesions in dermatology are based mainly on visual assessment of pathological regions and the evaluation of macroscopic features. This fact indicates that correct diagnosis is highly dependent on the observer's experience and visual perception [13]. In order to help clinicians making decisions on diagnostics, some tools have been used to assist them to treat different wound types and also to work on clinical images.

15.2.1 Wound Healing Rate

In clinical practice, doctors are able to observe wound closure according to the rate of change in wound surface area, also known as wound healing rate. This is important information about how the wound is responding to the treatment, since the most established clinical marker during a wound follow-up is its size. A regular evaluation is critical in assessing the efficacy of treatments, procedures and/or estimated healing time [14].

The gold standard for measuring wound size and comparisons during follow-up is tracing the wound area using a transparent film. Keeping data from different days allows clinical staff to monitor the healing progress. Even though this method is relatively noninvasive and useful, there is a risk of wound contamination and secondary damages while the measurements have been taken. Furthermore, the tracing samples should be stored in a safe place to avoid losing data [15].

With the possibility of acquiring images from the wounds and saving them on a computer disc, wound healing rate (WHR) or ulcer healing rate (UHR) has become one of the best ways to quantify the healing progress (reepithelialization) and can be calculated following the equation: $\left[A_i - A_f / A_i\right]$, where A_i represents the initial wound area and A_f represents the final area. The results usually range from −1.0 to 1.0; they can also be shown as percentages (when multiplied by 100). WHR equal to 1 or 100% means complete reepithelialization; greater than 0 or 0% means a decrease in the wound area; lower than 0 or 0% represents an increase in wound area; while equal to 0 or 0% means no changes. All images must be acquired at the same distance from the wounds, assisted by a scale bar or ruler [16–22].

Figure 15.1 shows some images of ulcers taken to be used for the wound healing rate (Figure 15.1a–c). Furthermore, an example of images for computer-aided assessment using ImageJ software to measure the wound area during follow-up is shown. Briefly, the image is uploaded to the ImageJ interface, and, using a ruler or bar scale, a known distance is traced to set up the software (Figure 15.1d). Then, by using the polygon selections, the wound is traced all around the edges, and finally, based on the beginning configuration (Figure 15.1e), the wound area is calculated by the software.

15.2.2 Wound Tissues Analysis by Image

Instead of using tracing to measure the wound area, photography has been a valuable tool in medicine and biomedical research, since an image can provide enough information about wound progression, contamination, inflammation, variations and so forth, leading to a possible diagnosis. It is considered as a nonnvasive technique and may help keep reliable data for follow up [23]. Moreover, digital photographs

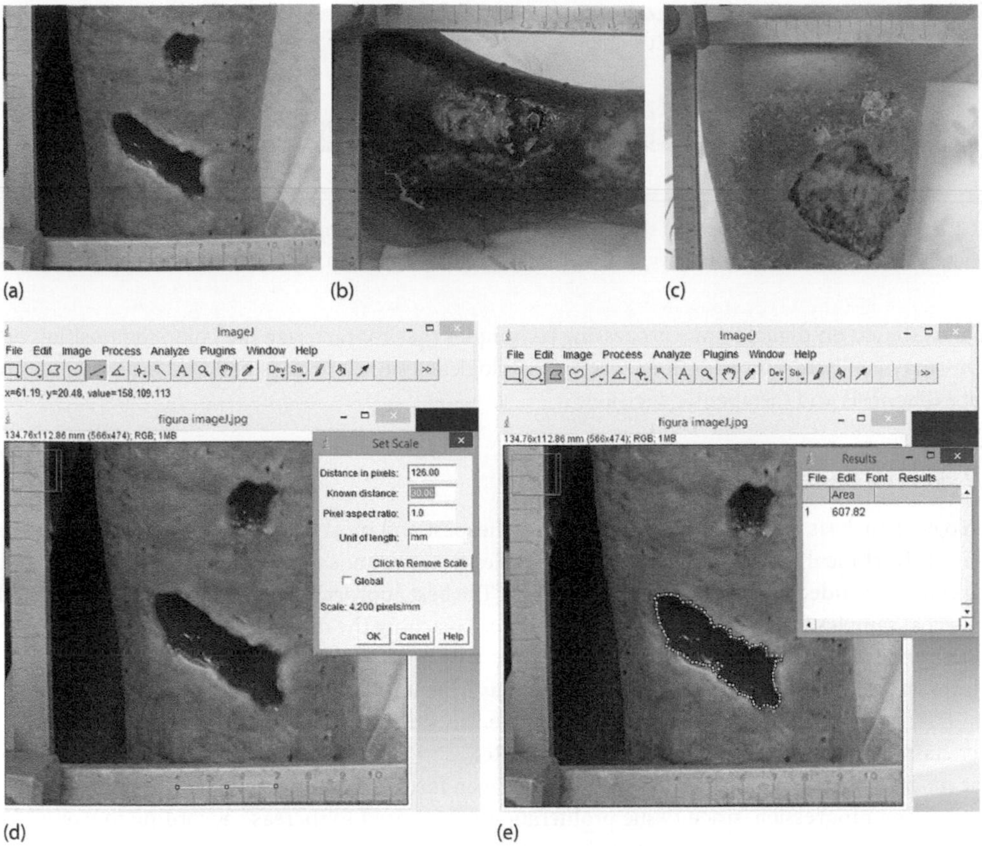

FIGURE 15.1 Example images of skin ulcers for wound healing rate monitoring: (a–c) skin ulcers in different aspects; (d,e) skin ulcer in ImageJ software for wound healing rate calculation during follow up.

(especially with high resolution to get as much detail as possible without flash) can be shared instantly between medical staff for interdisciplinary analyses [17–20, 24].

The use of digital cameras and computer-aided assessment has been considered affordable and practical for clinical or *in vivo* analyses. Several software are available to measure wound dimensions, surface area, volume (if wound depth is provided), colorimetric analyses for monitoring changes in surface area by tissue type, granulation tissue, fibrin/sphacelus, necrosis, cell counting, fluorescent specific staining, and so on. [20,25,26]. Some of the most-used software include ImageJ (National Institutes of Health), AutoCAD® 14 software (Autodesk®, Inc., San Rafael, CA, USA), DigiSkin™ (Wharton, NJ, USA), Image Pro Plus, Visiopharm and LEICA QWin Image.

A clinical observation of wound skin may provide clues that can help with the diagnosis, prognosis of healing or determination of severity, mainly because wounds (or ulcers) may appear due to patients' pathologies associated [27]. In some cases, a small wound may become a chronic wound, which could include a mixture of black necrotic eschar (scar), yellow fibrin (slough) and red granulation tissue, (Figure 15.2). Some lesions also exhibit white areas corresponding to hyperkeratotic tissue, or callus, around the border. The Red-Yellow-Black-White (RYKW) model, as indicated by such categorization of the tissue composition of an ulcer, is used by clinicians as a descriptive tool [14,27,28].

Quantification of the color distribution of skin lesions by image processing techniques may help in the analysis of the dynamics of the pathological process, as well as the progress of healing [29]. In a clinical scenario, the possibility of retrieving images from an established database that are similar to the case

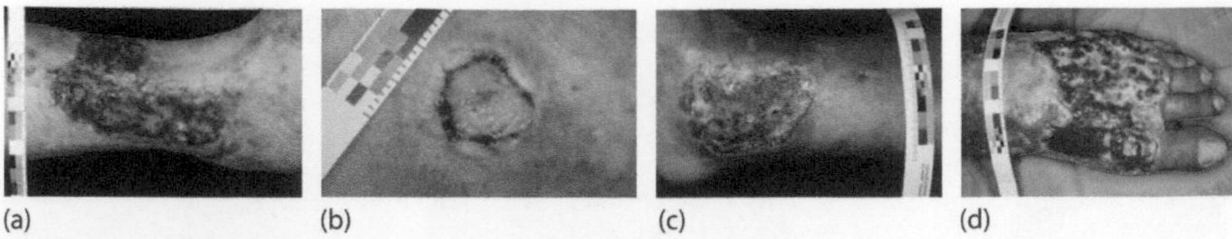

FIGURE 15.2 Examples of images of skin ulcers with different tissue composition: (a) mainly composed of granulation; (b) mainly fibrin; (c) mainly necrotic tissue; (d) mixed composition.

on hand—based on digital image processing techniques that characterize the color composition of the lesion—may facilitate understanding how dermatological lesions are classified and may assist in arriving at a diagnosis and therapeutic decision.

15.2.3 Histopathological Analysis

Histological analysis is still considered one of the cheapest and most useful techniques for wound measurement. In clinical and biomedical studies, samples from wounds are able to be biopsied and studied using computer-aided histomorphometric analysis. The best approach to get the study done is to collect histological samples from the center of the wound until reaching the uninjured edge, in order to compare the ulcerated area and the surrounding uninjured skin by microscopy after histological staining [12].

The most widely used stain in wound pathology, as in general dermatopathology, is Hematoxylin and Eosin (H&E): hematoxylin stains nuclei blue, and eosin stains cytoplasm and collagen fibers red [30]. Other stains, such as trichromes and Picrosirius Red, can be used in order to measure collagen, one of the most important skin proteins. Studying collagen formation allows medical staff to have an idea about healing progression, since tissue proliferation is supposed to increase according to wound healing. Immunohistochemical markers can also be applied to highlight tissue components, specific protein content, microorganisms or foreign materials [12].

Furthermore, as recently reported, computer-aided histomorphometric analysis of histological sections by ImageJ software was performed on extracellular matrix quantification compared to biochemical assay during an *in vivo* wound healing follow up. Even though there is a very important standardization process during image quantification, the method has been proven to be simple, rapid and low-cost [31].

15.3 CBIR (Computer-Based Image Retrieval) System

Another situation using images in dermatology is presented. This work presents color image processing methods for the analysis of dermatological images in the context of a CBIR (Computer-Based Image Retrieval) system. CBIR refers to the use of techniques that allow clinicians to compare the similarities between images in a database or their similarities with a given external query image as input. CBIR measures similarity based on quantitative information, which is derived from the images themselves, rather than relying on the accompanying text or annotation.

CBIR feasibility is related to the characterization of the image contents in terms of quantitative features [32]. The features of the query image are compared with the features of each image in the database. Then, the images having high similarity with respect to the query image are retrieved and displayed. CBIR techniques applied to the context of medical images is a useful tool and could provide physicians with assistance in the form of a display of relevant past cases with proven pathology, along with the associated clinical, diagnostic and other information.

Originally, the use of CBIR systems under the medical domain were mostly applied to gray level radiology images [33]. CBIR applications in the domain of dermatology allow the exploration of additional color information for image description and indexing. In general, techniques can differ in terms of feature extraction and classification, segmentation of the region of interest and results analysis [22,27,28,31].

Rahman et al. [34] explored features such as color and textures for matching by similarity: 358 images of pigmented skin lesions were classified as benign, dysplastic nevi and melanoma. Quantitative evaluation was only based on precision curve and shows an average precision of around 60%. A CBIR approach was proposed for skin lesion images, classified as actinic keratosis, basal cell carcinoma, melanocytic nevus, squamous cell carcinoma and seborrhoeic keratosis [35]. The system relies on color and texture features and the classification is using genetic algorithms. Precision values were observed between 67% and 82%.

The use of k-nearest neighbor (KNN) classifier and Euclidean distance were used before [36]. Images were evaluated for 10 different color model representations and classified in four categories. Best performance results of 75% were obtained using CIE-Lab color representation for K = 11. While the precision values of the presented works seem to be very representative, the comparison between the works is a challenging task, as the classes of the images in the database are usually not comparable. Moreover, results are often not presented in terms of precision x recall (ROC curves). Most of the time, only the best results are presented. As it is well known, the results can vary significantly, depending on the number of representative values present on the database or the number of K parameters when using KNN classifiers.

In this work, qualifying tests of the tissues were carried with predominant lesions that were separated according to the RYKW model: necrotic tissue (black), tissues with fibrin (yellow), granulation tissue (red), hyperkeratosis tissue (white). The images indexing and retrieval were carried out based on features of color and texture, which were obtained by RGB (Red, Green and Blue) and HSI (Hue, Saturation and Intensity) components of the dermatological lesions regions.

In the next topic we will discuss how color and texture features can be extracted, using RGB and HIS models, combined with GLCM (Gray-level Co-occurrence Matrix).

15.3.1 Features Extraction for Images Indexing

Indexing of images is performed using image descriptors based on color and texture features extraction. This is a crucial step in the development of CBIR systems. However, to make sure that only relevant features are used, images have to be segmented in such a way that only relevant information can be used by the descriptor. Manual segmentation allowed the images be represented in two subregions: foreground (including the lesion) and background. For each foreground subregion, images of basic color components, such as RGB, were generated. In addition to that, two other images were generated, one based on the hue component and the other based on the saturation component.

A total of 50 color and texture descriptors were extracted from the RGB, hue, and saturation (HS) components to index the images. They approach is described in detail below.

15.3.1.1 Color Features

Four color (first order) features were calculated directly from the histogram of the five work images: mean, standard deviation, skewness and kurtosis.

Skewness is a measure of dispersion that describes the "flattening" of the distribution function curve. Skewness is the measure of data symmetry of the probability distribution of a variable.

15.3.1.2 Texture Features

According to the authors, it was assumed that texture information is stored in the co-occurrence matrix (GLCM—Gray-level Co-occurrence Matrix), from which texture features can be calculated [37]. Therefore, five of the 14 Haralick's features were derived from an averaged GLCM for a distance of one pixel at 0, 45, 90 and 135 degrees. The second order features chosen were homogeneity, contrast, correlation, entropy and local homogeneity, as suggested by [38]. They are shown in Table 15.1.

TABLE 15.1 Haralick's Features Calculated as GLCM-Based Texture Descriptors.

Descriptor	Equation	
Homogeneity	$\sum_i \sum_j M(i,j)^2$	(1)
Contrast	$\sum_i \sum_j (i-j)^2 M(i,j)$	(2)
Correlation	$\frac{1}{\sigma_1 \sigma_1} \sum_i \sum_j ijM(i,j) - \mu_i \mu_j$	(3)
Entropy	$\sum_i \sum_j M(i,j) logM(i,j)$	(4)
Local Homogeneity	$\sum_i \sum_j \frac{M(i,j)}{1+(i-j)^2}$	(5)

Where μ_i and σ_i are the horizontal mean and variance, while μ_j and σ_j are the vertical ones. The five texture features mentioned above were derived from six multispectral co-occurrence matrices, obtained from the RGB components using the multispectral fusion model, as proposed elsewhere [39]. The method is an extension of BLCM-based Haralick's method and was developed to take into account the existing correlation between the color components (Figure 15.3).

The CBIR web system was developed using Java EE platform, Java 2 SDK Standard Edition, Apache server, MySQL, Eclipse SDK and classes of ImageJ software. The retrieval engine used a K-nearest-neighbor (KNN) classifier based on the Euclidean distance. Conducted experiments selected 50 images, including 15 randomly selected images for each class of fibrin, granulation and mixed composition, and 5 images of necrotic tissue. The tests considered K-values of the KNN classifier, as varying from 5 to 215 in steps of 5. For every K-value, the 50 images selected as above were used, one by one, as the query image.

The performance of the CBIR system was measured in terms of precision and recall, and it had a visual classification made by the medical specialist doctors as a ground truth reference to analyze the predominant kind of tissue in the lesions. For every retrieval experiment, mean precision and recall values were computed. The precision of retrieval is computed as a fraction of the number of relevant images

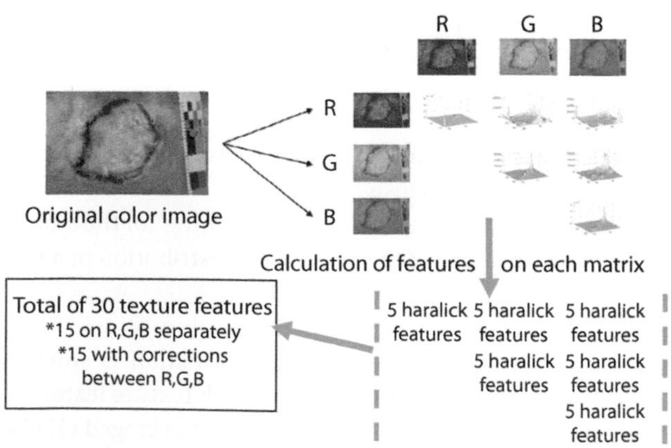

FIGURE 15.3 Multispectral fusion model to texture features extraction.

retrieved by the search to the total number of images retrieved by the search. In computing precision, each retrieved image receives a binary weight of unity or zero, representing a relevant image or not. In order to compute recall, the number of relevant documents retrieved by the search was divided by the total number of existing relevant images. In such classification context, precision and recall can also be defined as follows:

$$Precision = \frac{tp}{tp + fp} \tag{15.1}$$

$$Recall = \frac{tp}{tp + fn} \tag{15.2}$$

where:

tp is the true positive
fp is the false positive
fn is the false negative

The performance of the CBIR system was evaluated based on measurements of both precision and recall.

15.4 Protocol for CAD in Dermatological Ulcers

15.4.1 Images Acquisition

A database consisting of 215 dermatologic images was prepared, based on 63 consecutive medical examinations of outpatients at the Neurovascular Ulcers Outpatient Clinic of the Clinical Hospital of Ribeirao Preto at the University of Sao Paulo. It was composed of 79 images of granulation, 43 images of fibrin, 8 images of necrotic tissue and 85 images of mixed tissue.

Images were obtained based on a specific protocol that was determined after initial tests. The protocol assumed that the photographic clinical images are acquired through clinical routine*: performed by different people, in different rooms under non-specified illumination. Therefore, the protocol was concerned to improve the quality and uniformity of the photographic images during acquisition process. The protocol was defined as follows below.

15.4.1.1 Clinical Photography Equipment

Images are acquired using the same equipment configuration (Canon EOS 5DO), with a spatial resolution of two mega pixels, using circular flash and a 50 mm macro lens, as well as a polarization filter. The circular flash allows the objective lens to be closer to the lesion and assure uniformity to the reflected light. The polarization filter reduces the effect of flash noising in the image. Finally, a 50 mm macro lens is able to make pictures for more than 90% of dermatology cases [33].

15.4.1.2 Image Resolution

The typical size of the color images is around 1747 × 1165 pixels. Such resolution is considered a good compromise between pixel information for image processing analysis and photographic quality for clinical use.

* Project developed under approval of the Ethics and Research Committee of the Clinical Hospital of the School of Medicine of the University of Sao Paulo, Ribeirao Preto. Letter N. 415/2008. Process HCRP n. 9646/2007.

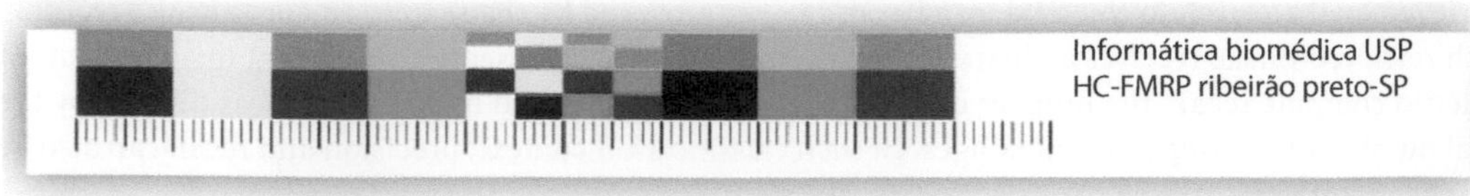

FIGURE 15.4 Disposable ruler with color patches and millimetric indicators used for *image normalization.*

15.4.1.3 Surgical Drape

A blue surgical drape is used to create an image background. The blue color was chosen because it is a color not expected within ulcer lesions or on the part of the body being imaged.

15.4.1.4 Image Subject

The image's subject (lesion) is centralized without intersecting the borders of the image.

15.4.1.5 Disposable Rulers

Disposable rulers with color patches and millimetric graduation (Figure 15.4) are included in the images to facilitate color normalization and dimension quantification of the lesions.

The segments with RGB and CMYK (Cyan, Magenta, Yellow and Black) color models are useful tools to image normalization. They minimize the effect of natural variation in lighting between the different clinical rooms.

15.5 Results

We will present in this section the experimental results obtained with the approach described above. A graphical representation of the variation of the mean precision value (P) versus recall (R) is shown in Figure 15.5. Note that the area under the curve indicates the performance of the CBIR system.

Figure 15.5b shows a graph of mean precision versus recall using only 15 images of lesions with mixed tissue composition as the query images. It is possible to see an improvement in the results when compared with Figure 15.5a, in which all the images of the base are used as query in the search. Figures 15.5c,d show similar results for images of lesions with granulation and fibrin, respectively. This result is promising because images of lesions with mixed tissue composition tend to be more difficult to classify in terms of diagnosis and therapeutic outcome.

Overall, it is seen that as K increases, the precision drops rapidly. However, it is important to note that even for the higher values of K, there are some relevant images retrieved. Considering the use under the proposed CAD scenario, the user will be able to choose the images about which more information is desired. The other images may be easily disregarded or rejected from further analysis. The worst result was observed in Figure 15.5e, which concerned lesions with necrotic tissue. This result was likely associated with the availability of only a few samples of this class of image in the database.

15.6 Conclusion

Image processing methods for wound healing rate, histological analysis and content-based image retrieval techniques for color and texture analysis of dermatological lesions were presented in this chapter. They were presented under the context of CAD-based applications to assist medical staff monitoring and diagnosing skin ulcers.

Furthermore, a case-study was presented, and the results obtained in this work support the potential of using CBIR systems to assist in the diagnosis of a lesion, based on analysis of previously classified cases. For the cases in which query results bring back images with different tissue

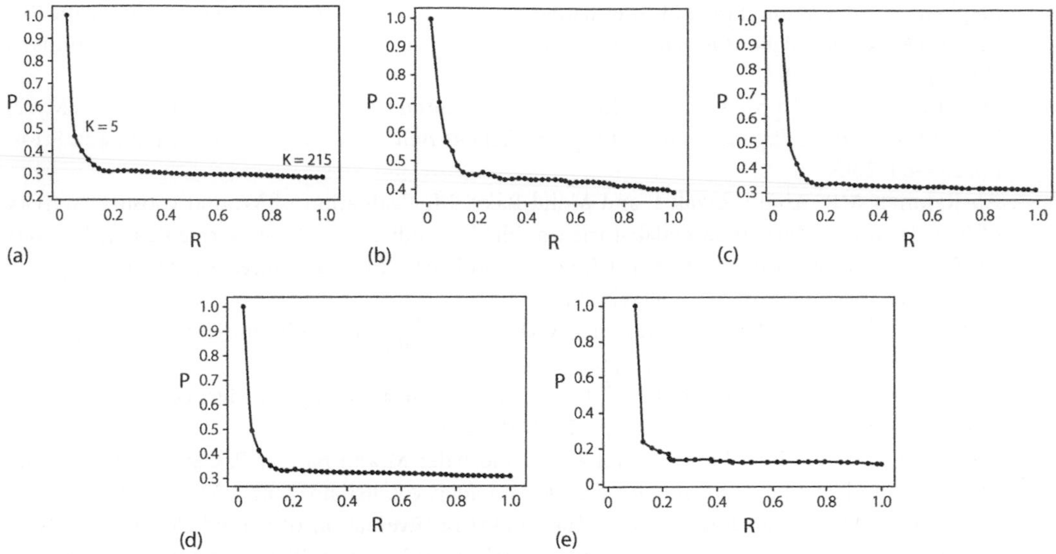

FIGURE 15.5 Graphs of mean precision and recall (PxR) based on the variation of K in KNN. The area under the curve indicates the performance of the CBIR system: the larger the area, the better the performance. (a) Mean PxR graph having all images used in the experiment as query in the search. (b) P x R results using only 15 images of lesions with mixed tissue composition as query. (c) P x R results using only 15 images of lesions with granulation as the query. (d) P x R results using only 15 images with fibrin as the query. (e) Mean P x R graph for only 5 images with necrotic tissue as query images.

composition, for example, the clinician would be able to analyze them, determine which images are similar to the query image, evaluate the classification of the retrieved images and decide on the classification of the image on hand. The clinician could also gain an understanding on why some of the retrieved images have been classified into different classes. In this manner, the CBIR system can assist in arriving at a classification decision when the image on hand is at the boundaries between different categories.

The obtained results demonstrate the potential of the proposed methodology to recover similar images under the context of color images of dermatological ulcers. Although the initial results obtained are not ideal, we believe that it is possible to improve the methodology by including additional features related to the multispectral color composition of regions within lesions, as well as features related to the edges of the lesions that include the interaction with the surrounding normal skin.

References

1. Velnar T, Bailey T and Smrkolj V. The wound healing process: an overview of the cellular and molecular mechanisms. *Journal of International Medical Research*, vol. 37, no. 5, pp. 1528–42, 2009.
2. Shrimanker M, Patel N, Modi H and Dave R. A review: Screening models for wound healing activity in animals. *American Journal of PharmTech Research*, vol. 3, no. 3, pp. 237–51, 2013.
3. Mulkalwar S, Behera L, Golande P, Manjare R and Patil H. Evaluation of wound healing activity of topical phenytoin in an excision wound model in rats. *International Journal of Basic and Clinical Pharmacology*, vol. 4, no. 1, pp. 139–43, 2015.
4. Koschwanez HE and Broadbent E. The use of wound healing assessment methods in psychological studies: a review and recommendations. *British Journal of Health Psychology*, vol. 16, no. Pt 1, pp. 1–32, 2011.

5. Stephens P, Caley M and Peake M. Alternatives for animal wound model systems. In: Gourdie RG, Myers TA, editors. *Wound Regeneration and Repair Methods and Protocols*. New York: Humana Press, pp. 177–201, 2013.
6. Stojadinovic O and Tomic-Canic M. Human ex vivo wound healing model. In: Gourdie RG, Myers TA, editors. *Wound Regeneration and Repair Methods and Protocols*. New York: Humana Press, pp. 255–64, 2013.
7. Ab-Rahman MR, Abdul-Razak F and Mohd-Bakri M. Evaluation of Wound Closure Activity of Nigella sativa, Melastoma malabathricum, Pluchea indica, and Piper sarmentosum Extracts on Scratched Monolayer of Human Gingival Fibroblasts. *Evidence-Based Complementary and Alternative Medicine*, vol. 2014, 190342, 2014.
8. Menke NB, Ward KR, Witten TM, Bonchev DG and Diegelmann RF. Impaired wound healing. *Clinics in Dermatology*, vol. 25, no. 1, pp. 19–25, 2007.
9. Stojadinovic A, Carlson JW, Schultz GS, Davis TA, and Elster EA. Topical advances in wound care. *Gynecologic Oncology*, vol. 111, no. 2, pp. S70–S80, 2008.
10. Schreml S, Szeimies RM, Prantl L, Karrer S, Landthaler M and Babilas P. Oxygen in acute and chronic wound healing. *British Journal of Dermatology*, vol. 163, no. 2, pp. 257–68, 2010.
11. Ab-Rahman MR, Abdul-Razak F and Mohd-Bakri M. Evaluation of wound closure activity of Nigella sativa, Melastoma malabathricum, Pluchea indica, and Piper sarmentosum extracts on Scratched Monolayer of Human Gingival Fibroblasts. *Evidence-Based Complementary and Alternative Medicine*, vol. 2014, 190342, 2014.
12. Romanelli M, Miteva M, Romanelli P, Barbanera S and Dini V. Use of diagnostics in wound management. *Current Opinion in Supportive and Palliative Care*, vol. 7, no. 1 pp. 106–10, 2013.
13. Oduncu H, Hoppe A, Clarck M, Williams RJ, and Harding KG. Analysis of skin wound images using digital color image processing: A preliminary communication. *The International Journal of Lower Extremity Wounds*, vol. 3, no. 3, pp. 151–6, 2004.
14. Papazoglou ES, Zubkov L, Mao X, Neidrauer M, Rannou N and Weingarten MS. Image analysis of chronic wounds for determining the surface area. *Wound Repair Regen*, vol. 18, no. 4, pp. 349–58, 2010.
15. Most D, Efron DT, Shi HP, Tantry US and Barbul A. Characterization of incisional wound healing in inducible nitric oxide synthase knockout mice. *Surgery*, vol. 132, no. 5, pp. 866–76, 2002.
16. Okuma CH et al. Development of lamellar gel phase emulsion containing marigold oil (Calendula officinalis) as a potential modern wound dressing. *European Journal of Pharmaceutical Sciences*, vol. 71, pp. 62–72, 2015.
17. Masson-Meyers DS, Andrade TAM, Leite SN and Frade MAC. Cytotoxicity and wound healing properties of Copaifera langsdorffii oleoresin in rabbits. *International Journal of Natural Product Science*, vol. 3, no. 3, pp. 10–20, 2013.
18. Fronza M, Caetano GF, Leite MN, Bitencourt CS, Paula-Silva FW, Andrade TA, et al. Hyaluronidase modulates inflammatory response and accelerates the cutaneous wound healing. *PLoS One*, vol. 9, no. 11, e112297, 2014.
19. Leite SN, Leite MN, Caetano GF, Ovidio PP, Jordao Junior AA and Frade MA. Phototherapy improves wound healing in rats subjected to high-fat diet. *Lasers in Medical Science*, vol. 30, no. 5, pp. 1481–8, 2015.
20. Caetano GF, Frade MA, Andrade TA, Leite MN, Bueno CZ, Moraes AM, et al. Chitosan-alginate membranes accelerate wound healing. *Journal of Biomedical Materials Research Part B: Applied Biomaterials*, vol. 103, no. 5, pp. 1013–22, 2015.
21. Minatel DG, Frade MA, Franca SC and Enwemeka CS. Phototherapy promotes healing of chronic diabetic leg ulcers that failed to respond to other therapies. *Lasers in Surgery and Medicine*, vol. 41, no. 6, pp. 433–41, 2009.
22. de-Avila-Santana L, Alves JM, Andrade TA, Kajiwara JK, Garcia SB, Gomes FG, et al. Clinical and immunohistopathological aspects of venous ulcers treatment by Low-Intensity Pulsed Ultrasound (LIPUS). *Ultrasonics*, vol. 53, no. 4, pp. 870–9, 2013.

23. Goldman RJ and Salcido R. More than one way to measure a wound: An overview of tools and techniques. *Advances in Skin and Wound Care*, vol. 15, no. 5, pp. 236–43, 2002.
24. Pereira SM, Frade MA, Rangayyan RM and Azevedo Marques PM. Classification of color images of dermatological ulcers. *IEEE Journal of Biomedical and Health Informatics*, vol. 17, no. 1, pp. 136–42, 2013.
25. Frade MA, Andrade TA, Aguiar AF, Guedes FA, Leite MN and Passos WR, et al. Prolonged viability of human organotypic skin explant in culture method (hOSEC). *Anais Brasileiros de Dermatologia*, vol. 90, no. 3, pp. 347–50, 2015.
26. Andrade TA, Iyer A, Das PK, Foss NT, Garcia SB, Coutinho-Netto J, et al. The inflammatory stimulus of a natural latex biomembrane improves healing in mice. *Brazilian Journal of Medical and Biological Research*, vol. 44, no. 10, pp. 1036–47, 2011.
27. Maglogiannis I, Pavlopoulos S and Koutsouris D. An integrated computer supported acquisition and charcterization system for pigmented skin lesions in dermatological images. *IEEE Transactions on Information Technology in Biomedicine*, 2005.
28. Dorileo EAG, Frade MA, Roselino AM, Rangayyan RM and Azevedo-Marques PM. Color image processing and content-based image retrieval techniques for the analysis of dermatological lesions. *IEEE Engineering in Medicine and Biology Society*, 2008.
29. Wannous H, Treuillet S and Lucas Y. Supervised tissue classification from color images for a complete wound assesment tool. *IEEE, 29th Annual International Conference of the EMBS*, Lyon, 2007.
30. Bancroft JD and Layton C. The hematoxylins and eosin. In: Suvarna SK, Layton C, Bancroft JD, editors. *Bancroft's Theory and Practice of Histological Techniques.* China: Churchill Livingstone Elsevier, pp. 173–86, 2013.
31. Caetano GF, Fronza M, Leite MN, Gomes A and Frade MA. Comparison of collagen content in skin wounds evaluated by biochemical assay and by computer-aided histomorphometric analysis. *Pharmaceutical Biology*, vol. 54, no. 11, pp. 2555–9, 2016.
32. Tarallo AS, Gonzaga A and Frade MA. Artificial neural networks applied to the segmentation and classification of images of cutaneous ulcers. *IEEE 7th International Conference on Bioinformatics and Bioengineering*, Boston, MA, 2007.
33. Mûller H, Michoux N, Bandon D and Geissbuhler A. A review of content-based image retrieval systems in medical applications, clinical benefits and future directions. *International Journal of Mediclal Informatics*, vol. 73, pp. 1–23, 2004.
34. Rahman MM, Desai BC and Bhattacharya P. Image retrieval-based decision support system for dermatoscopic images. In: *IEEE Symposium on Computer-Based Medical Systems*, Los Alamitos, CA, USA, IEEE Computer Society, pp. 285–290, 2006.
35. Ballerini L, Li X, Fisher RB, Aldridge B and Rees J. Content-Based Image Retrieval of Skin Lesions by Evolutionary Feature Synthesis. In: MICCAI-09 workshop MCBR-CDS: Medical content-based retrieval for clinical decision support. *LNCS*, vol 5853. Berlin: Springer, pp. 31–8, 2009.
36. Bosman HH, Petkov N and Jonkman MF. Comparison of color representations for content-based image retrieval in dermatology. *Skin Research and Technology*, vol. 16, no. 1, pp. 109–13, 2010.
37. Nayler J. Clinical photography: A guide for the clinician. *Journal of Postgraduate Medicin*, vol. 49, pp. 256–262, 2003.
38. Haralick R and Shanmugam K. Textural features for image classification. *IEEE Transactions on Systems, Man and Cybernetics*, vol. 3, pp. 610–21, 1973
39. Conners RW and Harlow CA. A theoretical comparison of texture algorithms. *IEEE Transactions on Pattern Analysis and Machine Intelligence*, vol. 2, pp. 204–222, 1980.

16

In Vivo Bone Imaging with Micro-Computed Tomography

Steven K. Boyd

Pierre-Yves Lagacé

16.1 Introduction

16.1.1 Bone Structure

Bone is a living tissue composed mostly of collagen (type 1), minerals (hydroxyapatite) and water. Bone tissue has mechanical and physiological functions within the body, including support, protection, hematopoiesis (blood cell formation, in the bone marrow) and mineral storage.

At the macroscopic level, two compartments can be distinguished in most bones: cortical and trabecular. Cortical bone is dense and compact with a porosity ranging from 5% to 30%, and it makes up approximately 80% of total bone mass in the body (Langton and Njeh 2004). It is found at the external surface of bones and in the diaphyseal walls of long bones. Trabecular bone, also called cancellous or spongy bone, is less dense, with a porosity up to 90%. It is mostly found in the epiphysis of long bones and in vertebrae. Figure 16.1 illustrates the cortical and trabecular bone compartments at the proximal femur.

The basic structural element of cortical bone is the osteon, a cylindrical structure (~250 μm diameter) that is composed of concentric bony lamellae, with a hollow canal (Haversian canal) in its center through which pass capillaries and nervous fibers (Martin and Burr 1989). They are connected by Volkmann's canals, which run perpendicularly to osteons and allow for fluid flow and nutrient transport. Trabecular bone is organized much differently than cortical bone at the macroscopic level. It is much less densely packed and is composed of a network of interconnecting plates and rods called trabeculae (Martin and Burr 1989). Trabeculae range in size from 75 to 200 μm. As for osteons, trabeculae are composed of lamellae, which are, however, organized longitudinally

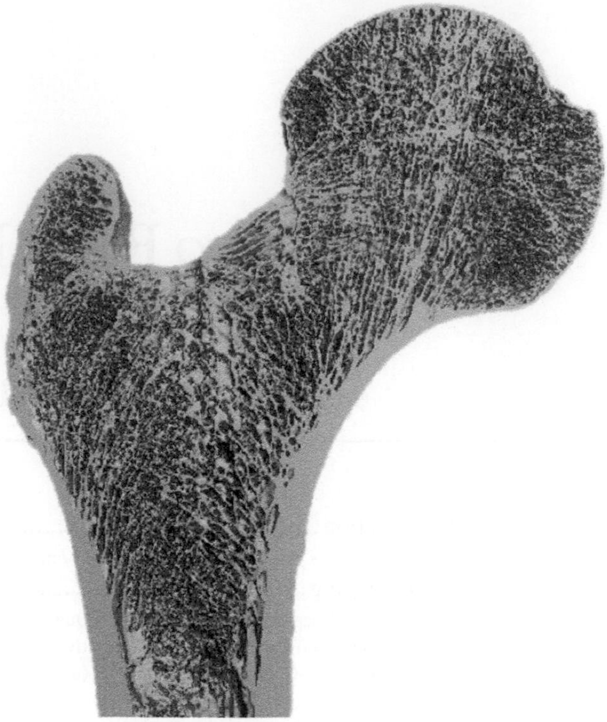

FIGURE 16.1 Coronal slice of the proximal femur illustrating the cortical and trabecular bone compartments.

(Kragstrup et al. 1983). Both bone compartments contribute to the mechanical properties of whole bones, and alterations to either of them will impact the ability of a bone to fulfill its functions, mechanical and physiological.

16.1.2 Mechanical Properties of Bone

Bones make up our skeleton and play an important role in support and protection of our body and of organs. However, disease and extreme loading conditions may compromise the integrity of bones and lead to altered mechanical properties and fractures. Understanding the mechanical properties of bones is essential to understand, predict and prevent fractures in healthy and diseased bone. As such, determinants of bone strength which are accessible *in vivo* have been and are still studied extensively.

The most important and most often considered mechanical properties of bone tissue are the material properties, such as Young's modulus (modulus of elasticity), or the structural properties, such as strength (compressive and tensile). The Young's modulus of a material is determined from mechanical tensile testing of bone specimens and corresponds to the ratio of the applied stress (applied force/area, F/A) to the relative deformation sustained by the material (strain, length change/initial length, $\Delta L/L$), parallel to the applied stress. Young's modulus is expressed in units of stress (Gigapascals, GPa). Strength represents the maximum force sustained by a material before it fails and is expressed in N (Newtons).

The elastic properties and strength of a whole bone depend on the mechanical properties of its material constituents, their amounts, and their organization at the macroscopic and microscopic level (Liu et al. 2010; MacNeil and Boyd 2007b; Sornay-Rendu et al. 2007). As such, the properties of cortical and trabecular bone vary between individuals and sites, depending on the local density and organization of both compartments.

16.1.2.1 Trabecular Bone Properties

Mechanical properties of trabecular bone are influenced by three main factors: apparent density (bone volume fraction), microarchitecture and mineral-collagen ratio (Goulet et al. 1994; Gibson 1985; Carter and Hayes 1976; Carter et al. 1980; Carter and Hayes 1977; Hodgskinson et al. 1989; Van Der Linden et al. 2001). The number, thickness and arrangement of the trabeculae are especially important as determinants of trabecular bone strength (Seeman 1999).

Bone mineral density (BMD) has been studied the most extensively as a predictor of bone strength. Apparent BMD refers to the structural density of bone and is directly related to its porosity and the tissue material properties. Apparent density is used to express the amount of mineralized tissue present in bone and is directly related to areal BMD (aBMD) measured *in vivo* in clinical routine using dual-energy x-ray absorptiometry (DXA) (Goldstein 1987). Correlating bone mechanical properties to apparent density can therefore be very convenient for in vivo assessment of bone strength and fracture risk. Linear and power functions have been proposed to relate mechanical properties to apparent density with varying levels of success, with power functions being generally more successful than linear functions (Goldstein 1987; Weaver and Chalmers 1966; Galante et al. 1970; McElhaney et al. 1970; Lindahl 1976; Carter and Hayes 1977; Carter et al. 1980; Martens et al. 1983; Stone et al. 1983; Kaplan et al. 1985).

The mechanical properties of highly porous materials depend on pore configuration and the effect of loading on the cell walls (tension, compression, bending), which varies considerably in trabecular bone, between sites and individuals. Spatial arrangement of the trabeculae (trabecular microstructure) has an impact on the mechanical properties (stiffness and strength) of trabecular bone and whole bones (Hodgskinson and Currey 1990a,b; Ciarelli et al. 2000), beyond apparent density alone. For example, a study by Bell et al. (1967) showed that transverse trabeculae (perpendicular to the direction of the principal stresses) were lost faster than longitudinal trabeculae. This would make longitudinal trabeculae more likely to buckle due to a lack of lateral support, explaining why trabecular bone loses strength at a faster rate than it loses tissue. It has also been shown that the strength of trabecular bone samples was highly anisotropic, with measured strength in the longitudinal direction (superior-inferior, aligned with the principal stress direction) being twice that measured in the transverse direction (medial-lateral, perpendicular to principal stress direction).

16.1.2.2 Cortical Bone Properties

The main determinants of cortical bone strength are cross-sectional area, thickness and porosity (MacNeil and Boyd 2007b; Spadaro et al. 1994). Porosity of cortical bone is mostly attributed to Haversian canals, absorption cavities and Volkmann channels. Mechanical properties of cortical bone are highly influenced by porosity, with the modulus of elasticity of cortical bone being related to the latter by a power function (Schaffler and Burr 1988).

Mechanical properties of cortical bone also depend on the direction of the applied stresses. For example, compressive strength and Young's modulus of up to 215 MPa and 34 GPa have been reported (Burstein et al 1976; Reilly et al. 1974), while tensile values are slightly lower (170 MPa and 29 GPa) (Burstein et al. 1976). As mentioned earlier, osteons, the basic structural element of cortical bone, are oriented longitudinally, which results in a much higher modulus of elasticity and strength in the longitudinal direction than in the transverse direction. Property differences between the longitudinal and transverse directions have been reported to be as much as twofold (Reilly and Burstein 1975; Yoon and Katz 1976; Pope and Outwater 1974). Properties of cortical bone also vary significantly between sites and individuals due to structural and mineralization variations (Ashman et al. 1984; Ashman et al. 1989).

16.1.2.3 Whole Bone Properties

The mechanical properties of a whole bone are the result of the properties and organization of its cortical and trabecular compartments. Stiffness and strength are determined by the size of the bone, thickness and density of the cortex, and density and microstructure of the trabecular component. Mechanical structural properties of bone tissue (trabecular and cortical) can be measured on bone samples with classical mechanical tests, such as compression, tension, torsion and bending. In these mechanical tests, properties of whole bones are assessed on bone segments to which are applied site-specific loads meant to simulate forces occurring *in vivo* in various situations, such as standing, walking or falling (Langton and Njeh 2004; Eckstein et al. 2004; Lochmüller et al. 2002; MacNeil and Boyd 2008b; Muller et al. 2003; Pistoia et al. 2002; Keyak 2000; Keyak 2001). However, these procedures are time consuming, expensive and, most of all, cannot be applied to individuals. Finite element (FE) analysis, which will be discussed in Section 16.3.5, provides a valuable alternative to estimate mechanical properties of bone *in vivo*.

16.1.3 Bone Physiology

During development, the majority of bone mass is accumulated during childhood and adolescence (Rizzoli et al. 2001) and peaks at puberty (Bailey et al. 1999). Bone modeling also occurs through periosteal apposition and endosteal resorption to define the shape and size of bones (Duan et al. 2003; Kontulainen et al. 2006). A remodeling process then occurs throughout life, which consists, in normal cases, of a continuous and mostly balanced bone resorption and formation processes. Continuous remodeling allows bone to adapt to its mechanical demands and repair itself (Parfitt 2002; Frost 1987). Bone formation typically dominates until an individual reaches 20–30 years of age, when bone mass peaks. Following this, remodeling skews itself toward more resorption, resulting in slow bone loss (del Rio Barquero et al. 1992). Trabecular bone is more responsive to metabolic stimuli than cortical bone (up to 8 times) (Burr and Gallant 2012), resulting in faster loss of trabecular bone, especially in women during and after menopause (Langton and Njeh 2004) or astronauts in space flight (Smith et al. 2014; Sibonga et al. 2015). Such alterations of the remodeling process can lead to bone deficiencies, which lead to decreased modulus of elasticity and strength, and increased fracture risk (Ammann and Rizzoli 2003; Wang et al. 2010).

16.1.3.1 Bone Loss and Osteoporosis

Bone loss occurs through different mechanisms depending on the type of bone tissue. Loss of trabecular bone occurs through perforation of horizontal (supporting) trabeculae, which irreversibly disrupts the trabecular network and compromises its structural integrity (Mosekilde 1993). Cortical bone loss typically occurs by expansion of the bone marrow cavity concurrently with the external diameter, which results in a net thinning of the cortical bone layer, again resulting in altered mechanical properties (Einhorn 1992). Amounts of these changes vary greatly between individuals and skeletal sites, but have been reported to be of greater magnitude in the lower extremities (Langton and Njeh 2004).

16.1.3.2 Osteoporosis

Osteoporosis is one of the most common metabolic bone diseases and is associated with low bone density and skeletal fragility. It is defined by the National Institute of Health (NIH) as "a disease characterized by low bone mass and structural deterioration of bone tissue, leading to bone fragility and an increased risk of fractures of the hip, spine, and wrist" (Diseases 2015). Osteoporosis influences bone turnover and remodeling cycle (Langton and Njeh 2004), with higher rates of bone resorption than formation, which results in decreased bone mass and altered microstructure, as illustrated in Figure 16.2.

In the United States, it is estimated that more than 53 million people, men and women, either have osteoporosis or have lower than normal bone mass (osteopenia) (Diseases 2015). In England and Wales, it has reported that one in two women and one in five men are affected by osteoporosis after the age of 50 (van Staa et al. 2001). Prevalence of osteoporosis at the hip, as reported by the third National Health and Nutrition Survey (Looker et al. 1997), is as high as 21% in Caucasian women. The World Health Organization (WHO) estimates that, after the age of 50, 4–6 million

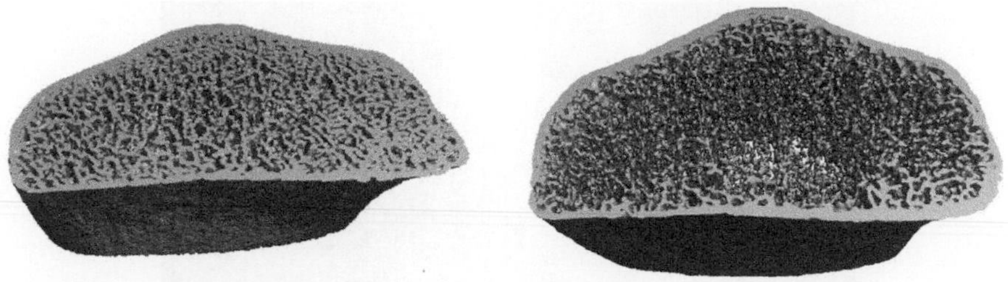

FIGURE 16.2 Normal (left) and osteoporotic (right) bone microstructure.

women are osteoporotic at the hip, and 13–17 million are osteopenic (Looker et al. 1997). The prevalence of osteoporosis is high in postmenopausal women because estrogen deficiency associated with menopause results in an imbalance between bone resorption and formation, leading to bone loss. Alterations in bone density and microstructure associated with osteoporosis lead to reduced mechanical properties and an increased fracture risk, notably in the spine, hip and wrist (Diseases 2015; Canada 2008).

Osteoporosis and its associated bone property changes remain asymptomatic until a fragility fracture occurs, caused by a minor impact or fall (Langton and Njeh 2004). Globally, at least 30% of women and 20% of men will suffer a fragility fracture associated with osteoporosis in their lifetime (Cummings et al. 1989). The healthcare costs associated with osteoporosis reached close to $2 billion in Canada in the mid-1990s (Wiktorowicz et al. 2001) and close to $14 billion in the US (Langton and Njeh 2004). Therefore, screening and diagnosis of osteoporosis is important in preventing fragility fractures from occurring. The most common osteoporotic fractures are vertebral fractures, which can remain asymptomatic, being found only when performing imaging for unrelated clinical reasons. Other common fragility fracture sites are the proximal femur and distal radius (Cummings et al. 1989), as a result of bone fragility and forces sustained during a fall or other impact. Bone fragility is due to alterations in trabecular and cortical bone. The strength (and fragility) of vertebrae is mostly influenced by the trabecular compartment, that of the radius is mostly impacted by cortical porosity, and fragility of the proximal femur is mostly a result of decreased trabecular bone volume and increased cortical porosity (Chen and Kubo 2014).

Causes of osteoporosis are varied and include estrogen deficiency in older women, which initiates a biological response that decreases bone turnover (Järvinen et al. 2003) and results in reduced bone density and altered microstructure. Risk factors also include: gender, age, body size, ethnicity, family history, sex hormones, calcium and vitamin D intake, medication, lifestyle, smoking and alcohol consumption (Diseases 2015; Canada 2008). A few of these risk factors can obviously not be eliminated, such as gender, age and ethnicity. However, others can be altered in an attempt to reduce one's risk of developing osteoporosis. For example, osteoporosis can be prevented and treated with an adequate intake of calcium and vitamin D, and exercising, which will stimulate bone formation, and by avoiding tobacco, alcohol and medications that cause bone loss (Diseases 2015; Canada 2008). In individuals who have osteoporosis, fall prevention (to avoid impacts causing fractures) and therapeutic medications (to balance the bone remodeling process) are also generally recommended and useful in preventing fragility fractures associated with osteoporosis.

16.1.3.3 Diagnosis

Diagnosis of osteoporosis aims at identifying individuals who are at risk of fragility fractures, in order to help prevent those fractures from occurring through targeted treatment and intervention. Currently, osteoporosis is diagnosed using a bone density index, most commonly derived from DXA examination, which is recommended following a medical assessment that includes consideration of risk factors. The index resulting from a DXA exam is the aBMD (mg/cm^2), which is a 2D measurement of BMD obtained

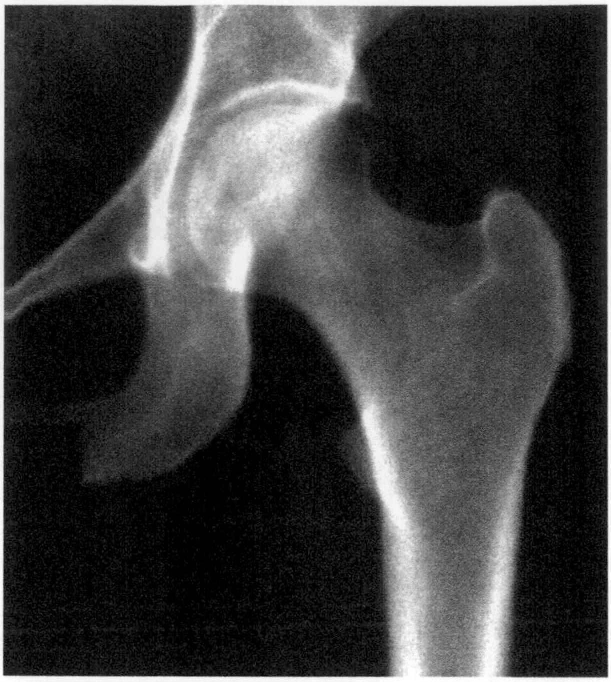

FIGURE 16.3 DXA image of a proximal femur.

at predefined regions of interest of the proximal femur, lumbar spine or distal radius. Figure 16.3 illustrates a DXA evaluation of the proximal femur, which is used clinically as a predictor of bone strength (Bouxsein et al. 1999) and to classify an individual as normal, osteopenic (slightly low aBMD) or osteoporotic (low aBMD with increased fracture risk) (Diseases 2015; Canada 2008) by comparison with a young, healthy population.

The literature contains significant evidence that the current DXA-derived aBMD index is limited in its prediction of fracture risk. For example, over 80% of fragility fractures occur in women not classified as osteoporotic (Stone et al. 2003). It has also been shown that BMD accounts for only 75%–85% of the variation in bone strength in normal individuals (McCalden et al. 1993), and that other factors, including bone size, shape and microstructure, play an important role determining bone strength (Suominen 1993). The inclusion of these parameters in the diagnosis of osteoporosis could undoubtedly improve the accuracy of bone strength estimates and help establish a realistic fracture risk for individuals. Unfortunately, these cannot be accounted for with current DXA-based diagnosis because it is a two-dimensional projectional technique.

16.2 Bone Imaging

Assessment and measurement of bone is most often performed using x-ray based modalities such as radiography and computed tomography (CT). In these modalities, x-rays are emitted from an x-ray tube and then attenuated as they pass through the body. The exiting flux is detected to create a gray-scale image, where gray level values depend on the amount of attenuation sustained by the x-rays over each path of propagation through the body. The amount of x-ray attenuation is dependent on the thickness of matter and its linear attenuation coefficient:

$$I_{out} = I_{in} e^{-\mu x} \tag{16.1}$$

where:

I_{out} is the intensity of the exiting x-ray flux
I_{in} is the intensity of the flux entering the body
μ is the linear attenuation coefficient, and x is the thickness of matter

It is important to note that the coefficient of absorption, μ, varies with material type, material density and x-ray photon energy. The variation of absorption properties with x-ray photon energy, and, for different tissue types, is the driving principle behind all x-ray techniques, which allows clinicians to gain some knowledge about material density.

16.2.1 Bone Density Imaging: Absorptiometry

Absorptiometry techniques have been used since the 1960s as a means of estimating bone density. The first techniques to be implemented used a mono-energetic and narrow photon beam and a constant, soft-tissue thickness assumption to estimate bone density based on the absorption characteristics revealed in the resulting image. Single photon absorptiometry (SPA) evolved into single x-ray absorptiometry (SXA), which used a single polychromatic x-ray beam from an x-ray tube instead of a monochromatic beam from a radionuclide for convenience, improved spatial resolution and reduced examination time (Kelly et al. 1994). These techniques were based on the assumption that overall thickness was constant, which required the use of a water bath in which the imaged structure was immersed. SXA has shown to be useful for the diagnosis of osteoporosis, but the constant thickness assumption limited its precision and applicability to central skeletal sites such as the spine and hip.

The different absorption properties of matter at different photon energies have later been exploited in dual photon absorptiometry (DPA) in order to overcome the limitation of SPA and SXA linked to the uniform thickness assumption. DPA uses a narrow photon beam emitted by a radionuclide at two energies. Simultaneous measurement of both energies of radiation allowed accounting for varying thickness of the diagnosis site without the need for a water bath, and therefore extended the application of absorptiometry to the torso.

16.2.1.1 Dual-Energy X-Ray Absorptiometry

DXA absorptiometry was developed in the 1970s and succeeded DPA as an aBMD measurement tool in the late 1980s (Cullum et al. 1989). It is used today as the clinical standard for the evaluation of aBMD. DXA measures the transmission of x-rays at high and low energy to then correct for varying overall thickness of the imaged segment. Compared to DPA, DXA significantly improves upon examination time and precision/accuracy by using an x-ray tube instead of a radionuclide. Early generations of DXA scanners used a pencil beam and required scanning of the entire region of interest. The current generation uses a cone beam coupled with a 2D array of detectors, which allows for considerably faster acquisition times, resulting in reduced motion artifacts. Peripheral DXA systems (pDXA) have also been introduced that allow precise measurement of aBMD of the appendicular skeleton (specifically radius and calcaneus) with a compact system and at lower cost than full-size DXA systems.

Each DXA system uses its own specific algorithms to extract aBMD information, but the principle is generally the same:

1. Dual-energy images are acquired
2. Bone contours are detected using an edge detection algorithm
3. Area of bone is calculated from the contour
4. aBMD value is reported as the average of all bone pixels
5. Bone mineral content (BMC) is determined by multiplying aBMD by the area of bone

DXA is currently the clinical standard for the evaluation and follow up of BMD in osteoporosis. Sites at which trabecular bone is present in large amounts are generally preferred, since trabecular bone is more responsive to metabolic stimuli than cortical bone. Therefore, the lumbar spine (L1-L4), the proximal femur (femoral neck, trochanter, total hip), and peripheral sites (distal radius, calcaneus) are usually where DXA evaluation of aBMD is performed. Total body composition exams are also performed to estimate global aBMD (Kelly et al. 1998) and evaluate lean mass across the whole body.

Fracture risk is estimated from DXA exams based on the distribution of aBMD at different locations (e.g., between L1 and L4 at the lumbar spine) or from absolute measurements (e.g., at the proximal femur, and peripheral sites). As mentioned earlier, however, DXA evaluation of aBMD is unable to accurately predict fracture risk for all individuals, and other parameters in addition to aBMD should be considered when assessing fracture risk.

Relatively recently, the trabecular bone score (TBS) has been developed as an index of bone microarchitecture and strength, derived from DXA images of the lumbar spine (Pothuaud et al. 2008; Bousson et al. 2015). The TBS relies on texture analysis of a DXA image to derive a single value index which reflects the quality of bone microarchitecture, a high value representing high-quality microarchitecture. A recent review (Bousson et al. 2015) of the TBS concludes that, while this quantitative index is reproducible and easy to measure, it does not actually reflect trabecular microarchitecture. TBS does however discriminate and predict fracture as well or better than lumbar spine aBMD, and its addition to aBMD may improve fracture prediction. However, even though work is in progress to integrate the TBS in fracture prediction algorithms, there is currently not sufficient evidence to justify the use of TBS in patient follow-up and diagnosis (Bousson et al. 2015).

16.2.1.2 Radiation Dose in DXA

Ionizing radiation is known to increase the stochastic risk of developing a cancer over the lifetime. Therefore, although there are undeniably situations where x-ray imaging is beneficial, the radiation doses should be understood and minimized when possible.

Effective radiation dose to an individual, reported in Sv (Sievert), is a function of the amount of radiation emitted by the system that is absorbed by the body, and the sensitivity of the organs that receive radiation. Effective radiation doses involved vary with system configuration, scan parameters and individuals, and they need to be measured using radiological phantoms for specific imaging protocols to get accurate estimates. Doses involved in DXA remain low when compared to other x-ray exams. For example, radiation doses of the order of 5 μSv have been reported for a total body DXA scan, which is optimized for measuring lean mass (not bone) (Scafoglieri et al. 2011). Typical values for bone scans are on the order of 0.1 μSv for the forearm, 2 μSv for the femur, and 10 μSv the spine (Njeh et al. 1996; Bezakova et al. 1997; RadiologyInfo.org 2015). As a reference, the radiation dose associated with a chest x-ray is of the order of 100 μSv, and natural background radiation on Earth is of the order of 3000 μSv/year (~8 μSv/day) (RadiologyInfo.org 2015).

16.2.1.3 Measurement Error in DXA

Measurement errors in DXA are affected by region of interest definition, segmentation, artifacts, patient positioning, motion, and weight loss. Phantom studies have shown that some systems could measure aBMD of a spine phantom with an accuracy within 2% (Blake 1996). aBMD measurements made with DXA systems from different manufacturers are generally not directly comparable, however, and should not be used to evaluate aBMD changes in a longitudinal study (Blake et al. 1996; Heilmann et al. 1998).

Measurement precision (reproducibility) can be very good (1%–2% depending on the site), especially with well-trained and experienced staff (Adams and Bishop 2008). The precision of DXA measurements is highly influenced by the ability to reposition the patient adequately for repeated exams and the sensitivity of the measurement to patient positioning (short-term precision). Changes in the soft tissue composition of a subject between visits over a longer period of time, for example, when gaining or losing weight, can also influence bone density measurement (long-term precision).

The major limitation of DXA derived aBMD comes from the fact that it is a 2D measurement, which is affected by bone size and bone orientation. Because of this, significantly different aBMD values can be obtained when assessing individuals with actual bone densities that are similar. DXA is therefore best adapted to the evaluation of bone density changes in a single individual (with minimal weight gain or loss), when using the same system for repeated measurements. It is also useful for measuring lean mass of an individual through whole body scans. Since it relies on 2D projection images, DXA is also unable to distinguish between trabecular and cortical bone compartments, or to resolve bone microarchitecture, both of which are important contributors to bone strength. As it will be discussed in the coming sections, 3D modalities such as CT and QCT provide the ability to perform 3D measurements of BMD and bone architecture, which can significantly improve the prediction of bone strength from imaging data.

16.2.1.4 Clinical Data

The main objective of assessing BMD (and bone quality) in osteoporosis is to assess the level of bone fragility, in order to intervene at the appropriate time to prevent fragility fractures. As discussed previously, BMD is measured at key regions of interest (lumbar spine, proximal femur and distal radius). DXA measures areal aBMD in units of mg/cm^2, whereas a 3D technique such as quantitative CT (QCT) measures volumetric BMD (vBMD) with units of mg/cm^3. The general term BMD is ubiquitous and often interchangeably used for either 2D or 3D measurements. Strictly, BMD should only be used when a 3D acquisition technique is used because, by default, density is a volumetric measure—we use that default definition throughout and refer to areal BMD explicitly as aBMD. It is clear that a volumetric measure of BMD is technically correct and tends to provide more robust measurements than the areal measurement from DXA because they are not influenced by bone and patient size. Furthermore, 3D techniques can provide distinct measurements for the cortical and trabecular bone compartments.

In clinical practice, aBMD of an individual is often compared to that of a matching reference population (sex, age, ethnicity). Data for different age, sex and ethnic groups are collated by health organizations around the world, for example, in the National Health and Nutrition Examination Survey (Looker et al. 1998), and used as a basis for comparison.

This comparison is often made with the use of the T-score, which compares an individual's aBMD value ($aBMD_{individual}$) to a population's peak value ($aBMD_{peak,\ population}$), normalized by the standard deviation of the population ($Std\_dev_{peak,\ population}$). This score gives an indication of how many standard deviations away from the population mean the individual is located.

$$T\_score_{SD} = \frac{aBMD_{individual} - aBMD_{peak,\ population}}{Std\_dev_{peak,\ population}} \tag{16.2}$$

The T-score can also express an individual's aBMD ($aBMD_{individual}$) as a percentage of the peak aBMD of the population ($aBMD_{peak,\ population}$).

$$T\_score_{\%} = \frac{aBMD_{individual}}{aBMD_{peak,\ population}} \times 100 \tag{16.3}$$

Z-scores, which compare an individual's aBMD to that of an age-matched reference group, are also used in clinical practice. Similar to T-scores, Z-scores can be presented as $Z\text{-}score_{SD}$ or $Z\_score_{\%}$. It is well documented that aBMD decreases with age, and that the prevalence of osteoporosis is higher in older age groups. Therefore, Z-scores are not used to define osteoporosis, but for comparing an individual's relative skeletal health.

Individuals are classified as normal, osteopenic or osteoporotic based on their bone mineral content (BMC) or aBMD (WHO Study Group 1994). An individual is classified as normal if his/her aBMD is within one standard deviation (SD) of the population's normal peak bone mass ($T\_score_{SD} > -1$). aBMD's between 1 and 2.5 SD's below normal peak aBMD ($-2.5 < T\_score_{SD} \leq -1$), and more than 2.5 SD's below ($T\_score_{SD} < -2.5$) are classified as osteopenic and osteoporotic, respectively (Langton and Njeh 2004).

This approach is simple, but has some significant drawbacks. For example, the normal peak population aBMD data comes from young healthy individuals, so this type of comparison cannot account for different patterns of bone loss (Langton and Njeh 2004). Additionally, these simple thresholds merely classify individuals' aBMD and don't realistically assess levels of fracture risk over the lifetime associated with decreased aBMD (Langton and Njeh 2004). However, the timing of preventive treatment can be adjusted based on T-scores in order to reduce the risk of developing severe osteoporosis and suffering fragility fractures. Classification of individuals based on T-scores alone also fails to account for differences in the timing and amount of bone loss at different sites. It has been shown that the timing of bone loss changes significantly between the proximal femur (in the 20s), the spine (in the 30s) and the distal radius (at menopause) (Slemenda et al. 1996; Compston 2001).

By failing to account for other parameters that influence bone strength and fracture risk, T-score thresholds, therefore, lead to significant misclassification of individuals. For example, at the lumbar spine, an aBMD that is 2 SD lower than the normal peak value has a sensitivity and specificity for fracture of only 60% (Langton and Njeh 2004). As a result, many individuals not classified as osteoporotic suffer from fragility fractures, and many who are classified as osteoporotic never do. This conundrum clearly has important implications for osteoporosis treatment decisions. The inclusion of bone microstructure parameters in addition to aBMD as descriptors of *in vivo* bone quality may provide the additional information required to more reliably assess disease progression, treatment effects and fracture risk (Muller and Ruegsegger 1995). However, until recently, assessment of bone microstructure *in vivo* has remained impossible.

The driving objective of bone mineral density assessment and diagnosis of osteoporosis is the identification of individuals who are at risk of suffering a fragility fracture, in order to prevent these fractures from occurring. Fracture risk prediction algorithms have been developed, which integrate various parameters such as aBMD, T- or Z-scores, age and life expectancy, secondary causes of osteoporosis and epidemiological data (Cummings et al. 1993; Grampp et al. 1997) to establish fracture risk and a treatment plan. As discussed earlier, timing and magnitude of bone loss will vary from site to site, and as such, assessment of aBMD has a better predictive power of fracture risk at the site where it is assessed. Some studies have shown that combining aBMD measurements from different sites could significantly improve the prediction of overall relative fracture risk (Lu et al. 2001).

16.2.2 Quantitative Computed Tomography (QCT)

CT is a 3D x-ray based imaging modality that is particularly well suited to the imaging of bone. In CT, a source-detector array spins around a body to acquire a series of profiles or views, which are then used to reconstruct 2D axial slices of the imaged body. Successive slices of 0.5–3 mm in thickness and with an in-plane resolution of 200–500 μm are generally acquired (Glüer 2013) by advancing the body in the gantry to obtain volume data of the region of interest. The slices are 2D gray-level images with gray values expressing the attenuation coefficient in Hounsfield Units (HU).

A density calibration phantom can be included in the field of view of the scan to gain density information from the CT scan's gray-level images. This technique is called quantitative CT (QCT) and is used to assess volumetric BMD (vBMD) *in vivo*. Gray-level values (HU) of the known densities of the phantom are used to calculate the density associated with all voxels. Local vBMD and average vBMD values can then be calculated for different regions of interest such as different vertebral levels, the hip, etc. (Kalender et al. 1987; Steiger et al. 1990; Lang et al. 1999).

QCT has a major advantage over DXA in that it allows for the measurement of 3D volumetric BMD (vBMD), as opposed to the 2D areal BMD (aBMD) measured by DXA. It is, therefore, not affected by bone size, which makes it more robust and reliable for measuring BMD. QCT is, therefore, much more suited for use in population studies and to evaluate growing children. Since it is a 3D modality, QCT also allows for the evaluation of the trabecular and cortical bone compartments separately. The access to 3D vBMD information for each compartment of bone (trabecular and cortical) is of great interest, as it allows for patient-specific FE analysis, which opens the door to the development of personalized mechanical bone models based on the geometry and density information obtained from QCT measurements. These can then be used to simulate various loading patterns and determine the distribution of stresses and strains in the bone, and to estimate fracture load (Cody et al. 2000; Cody et al. 1999). However, this comes at the cost of increased radiation dose, with one vertebral exam involving doses of the order of 80 μSv (Kalender 1992) and a higher cost, when compared to DXA.

Sites commonly imaged by QCT include the spine and hip, and the measurements include BMD, bone mineral content and bone volume for the trabecular, cortical and total bone regions. This is of particular interest since it has been shown, in an *in vitro* study, that trabecular vBMD of the trochanter was highly correlated to fracture load in a side-fall loading configuration (Lang et al. 1997).

16.2.2.1 Dose in QCT

Radiation doses in QCT and CT can vary greatly depending on system configuration, scan parameters and patient size. As opposed to CT, QCT exams do not cover the entire body or entire body segments, and generally focus on a few narrow bands of a few millimeters. Radiation doses for a QCT exam, including a scout image, have been reported to be in the range of 60–1400 μSv, depending on the system and estimation method used (Kalender 1992; Huda and Morin 1996). This indicates that the radiation dose from QCT is of the same order of magnitude as other radiological examinations (1500 μSv for a full spine exam), and much lower than for CT exams (10,000 μSv for a chest & pelvis exam) (RadiologyInfo.org 2015) It should also be noted that radiation dose is very dependent on a number or parameters, and that new reconstruction techniques (iterative reconstruction) has greatly reduced QCT dose without degrading image quality.

16.2.2.2 Measurement Error

The main source of error in QCT measurement is caused by the partial volume effect, which results from the contributions of various tissues (bone, fat, red marrow) to the gray value of one voxel. The presence of fat and fatty marrow in a QCT volume has been reported to result in accuracy errors of 5%–15% (Langton and Njeh 2004). The presence of various tissues in QCT volumes tends to lower the HU values and therefore the density measurement. This source of error can be corrected for by using dual-energy acquisitions (Goodsitt et al. 1994; Goodsitt et al. 1987; Glüer et al. 1988), which can be performed with dose-neutral protocols when compared to standard CT (van Elmpt et al. 2016; Omoumi et al. 2015). Precision errors in QCT are caused by discrepancies in slice and region of interest placement. This can be minimized by automatic or careful manual positioning of the region of interest based on the scout view acquired prior to scanning.

By providing slice images of bone density, QCT allows for the precise and accurate measurement of bone density and geometry, and the discrimination between trabecular and cortical bone. This allows for the development of personalized mechanical models of bones for use in FE models, which allow for the estimation of the mechanical properties of bones and fracture risk. The ability to account for micro-architecture of trabecular bone would further improve the quality and reliability of these estimates. Over the past 20 years, great progress has been made in the development of imaging technologies that are able to resolve trabecular microarchitecture *in vivo* and *in vitro*.

16.2.3 Peripheral Quantitative Computed Tomography (pQCT)

Full-body QCT systems are adapted to the assessment of vBMD at central and peripheral sites. However, the cost of these full size CT systems is somewhat prohibitive, especially when imaging peripheral sites, where a full-size system is not required. pQCT systems have been developed and used in research since the 1970s to reduce the cost of assessment of vBMD at peripheral sites, and to allow for lower radiation doses (Hangartner et al. 1985).

pQCT systems function on the same principles as full-size QCT systems, but with smaller systems, allowing for quicker acquisition times and lower radiation doses. These systems can be used for imaging small animals or specimens, as well as the human appendicular skeleton, *in vivo*. Resolution of pQCT scanners range from 70 to 500 μm for small animal systems to 200–800 μm for larger clinical systems (Langton and Njeh 2004). With their smaller systems and higher attainable resolution, pQCT systems are able to separate cortical from trabecular bone compartments and to identify small changes in volume and mineral content (Ferretti et al. 1995). Moments of inertia, volume and thickness of the cortical shell can also be measured with such systems.

16.2.3.1 Measurement Errors

Accuracy of pQCT has been reported to be of the order of 2% *in vitro* (Louis et al. 1995; Takada et al. 1996; Louis et al. 1996). Precision of the measurement strongly depends on appropriate positioning of the scan site, in order to measure vBMD at the same location in the volume. This is especially important in the forearm, where the proportion of trabecular bone changes significantly with distance from the ulnar styloid process (Schlenker and VonSeggen 1976). Short-term precision of pQCT systems used in clinical settings has been reported to be of the order of 1%–2% *in vivo*, depending on the compartment of interest (trabecular, cortical, total bone region) (Butz et al. 1994). These values could be improved by using multiple slice scanners that would allow for the evaluation of larger volumes, which would, in turn, reduce the sensitivity of the measurement to volume position (Langton and Njeh 2004).

16.2.3.2 Clinical Utility of pQCT

The low radiation dose and robust measurement afforded by pQCT systems make them an effective clinical tool (Langton and Njeh 2004). Bone changes associated with aging, disease and drug use, can be studied with pQCT, which can, in turn, be useful for the classification of osteoporotic/osteopenic individuals and the prediction of fracture risk. Additionally, since it focuses of peripheral sites, which are not very sensitive to radiation, effective doses associated with pQCT are low (~2 μSv). This is of the same order of magnitude as radiation doses associated with spine DXA and much lower than that of spine QCT. In addition to the low radiation dose of the system, the simplicity and short duration of the exam make it a very useful tool. This type of tool has opened the door to studies of cortical and trabecular bone differences/changes in populations of individuals (Butz et al. 1994; Grampp et al. 1995; Takagi et al. 1995). It also allows for discriminating between osteoporotic and non-osteoporotic individuals (Butz et al. 1994; Takagi et al. 1995), monitoring individuals during treatment (Schneider et al. 1999; Ruegsegger et al. 1995) and predicting fracture load *in vitro* (Augat et al. 1996; Spadaro et al. 1994). The ability to measure cortical bone separately from trabecular bone has also revealed the importance of the cortical shell in bone properties and the importance of its role in the weakening of bone in osteoporosis (Langton and Njeh 2004). pQCT systems are also used in pre-clinical animal studies for the evaluation and validation of drugs that have an impact on bone density.

16.2.4 Micro-Computed Tomography (μCT)

pQCT allows for studying the density of the trabecular and cortical compartments of bones separately. However, its resolution is typically not sufficient to resolve details of the trabecular microstructure. μCT has been recently developed in an attempt to resolve 3D trabecular structure and its response to drugs or diseases. Resolution of μCT systems can reach less than 10 μm, with fields of view up to 100 mm and more. At these high resolutions, the measurements provided by these systems compare to parameters measured by histomorphometry, in 3D, and without the need for destructive techniques (Müller et al. 1996).

Micro-CT systems function on the same principle as a classical CT, but with a micro-focused x-ray source and extremely precise motion control of the system and source. For radiation safety, μCT units, similar to x-ray microscopy systems shown in Figure 16.4, are typically enclosed in a lead-shielded box that protects the users from any radiation, without requiring a lead-shielded room and separate area for the users.

With μCT, it is possible to image bone specimens and small animals at a resolution sufficient to resolve individual trabeculae and to extract relevant histomorphometric parameters from the resulting 3D microstructure model (Odgaard and Gundersen 1993; Parfitt et al. 1983). These parameters include trabecular thickness, trabecular separation, trabecular bone volume, connectivity and anisotropy (Müller et al. 1996). Additionally, 3D models resulting from μCT imaging can be used in FE modeling

FIGURE 16.4 Xradia 3D x-ray microscopy system (Carl Zeiss Microscopy, Jena, Germany).

studies that can correlate various bone microstructure parameters to the mechanical properties of bone, as measured *in vitro* (Müller and Rüegsegger 1996). Unfortunately, dedicated μCT scanners, due to the constraints of their small physical systems, are not adapted to the imaging of bone *in vivo* in humans; however, in recent years, high resolution pQCT (HR-pQCT) systems have been developed, which now allow for the imaging of 3D bone microstructure in humans *in vivo* at peripheral sites. This new technology is presented in Section 16.3.

16.2.5 CT Compared to DXA

CT-based methods have clear advantages over DXA. They allow for the measurement of trabecular bone separate from cortical bone, provide true volumetric BMD as opposed to aBMD, characterize 3D geometric parameters (moment of inertia, cortical thickness) and are not affected by bone size. These advantages make CT-based methods more accurate and precise for the measurement of *BMD in vivo*, which opens up the possibility for longitudinal studies, for example, the influence of drugs, growing and aging on bone quality. Additionally, with QCT, it is possible to separate muscle from bone and fat, which allows for the study of muscle cross-section areas and opens the door to the study of correlations between muscle and bone parameters.

CT-based methods also have some limitations, however. CT scanners cost much more, have limited accessibility and higher radiation doses when compared to DXA systems. However, QCT techniques that focus on small and peripheral (pQCT) regions of interest can perform measurements with effective radiation doses at acceptably safe levels. In the last decade or so, the appearance of HR-pQCT has opened new doors for the study of trabecular microstructure and mechanical strength from the FE method *in vivo*.

Magnetic resonance imaging (MRI) is also being explored for the assessment of bone structure (Krug et al. 2008). One major advantage of MRI is that it does not involve ionizing radiations, but its drawbacks are associated with an increased cost, lower resolution and the inability to provide density information.

16.3 HR-pQCT for Bone and Joint Imaging

16.3.1 Scanner Hardware and Specifications

HR-pQCT has been commercially available since the mid-2000s as a clinical research tool to assess bone microarchitecture and volumetric BMD (vBMD) *in vivo*, noninvasively and with low radiation doses.

The first generation of HR-pQCT scanner, the XtremeCT (Scanco Medical, Brüttisellen, Switzerland) was introduced in 2005 for dedicated assessment of bone microstructure at the distal radius and tibia (Boutroy et al. 2005). This first system allowed for the acquisition of a 9 mm thick region of interest at an isotropic resolution of 82 μm in under 3 min, and with an effective radiation dose of less than 3 μSv (Boutroy et al. 2005; MacNeil and Boyd 2007b). Sample scanner specifications and parameters are presented in Table 16.1. By providing new possibilities and insight in the study of bone quality *in vivo*, noninvasively and non-destructively, this HR-pQCT system (Figure 16.5) has been used extensively in osteoporosis research: over 200 peer-reviewed journal articles using this technology have been published since 2004 (Cheung et al. 2013).

In 2014, a second-generation system was developed that addresses the challenge of motion artifacts by performing scans faster, resulting in improved reproducibility compared to the first-generation system. Hardware components were updated to reduce the scan time, increase the resolution and field of view, without affecting image quality or radiation dose. A new forearm cast (Figure 16.6) was designed to help prevent any motion during scanning. The improvements implemented in the second-generation scanner improve the reliability of the measurements, which proves to be particularly useful in the context of longitudinal studies of bone fragility and treatment efficacy.

TABLE 16.1 HR-pQCT Scanner Specifications

	Xtreme CT I	Xtreme CT II	
	XT-I	Low Resolution	High Resolution
X-Ray tube			
Voltage (kV)	60	68	68
Intensity (μA)	900	1460	1460
Scan protocol			
Integration time (ms)	100	36	43
Number of projections	750	750	900
Radiation (μSv)	<3	<3	<5
Scan time (min)	2.8	1.4	2
Resolution			
Voxel size (μm)	82	91	61
MTF at 10% (μm)	130	120	95
Field of view			
Diameter (mm)	126	140	140
Stack height (mm)	9	10	10
Number of slices	110	110	164
Max scan length	150	200	200

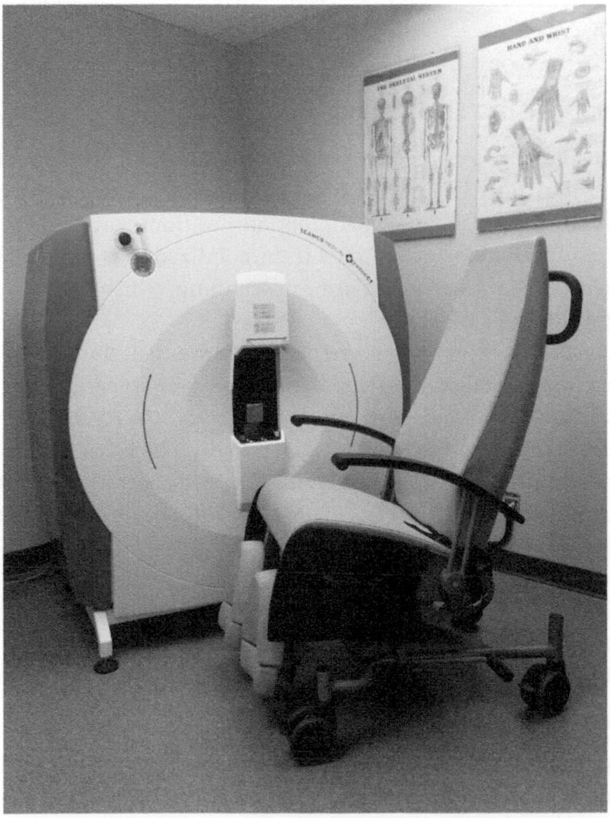

FIGURE 16.5 HR-pQCT system (XtremeCT and XtremeCTII, Scanco Medical, Brüttisellen, Switzerland).

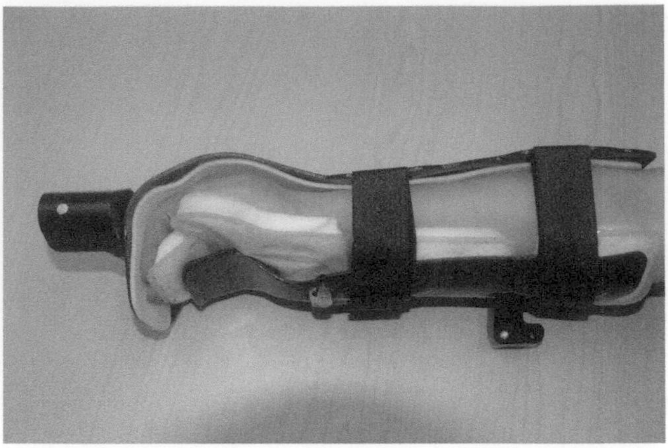

FIGURE 16.6 Forearm cast for the XtremeCTII system (Scanco Medical, Brüttisellen, Switzerland).

As with the first generation HR-pQCT systems, scans are graded for motion (Pauchard et al. 2012), and subjects are rescanned if the amount of motion is determined to be unacceptable for the desired application. This second-generation system reduces the scan time to less than 2 min, while maintaining image quality and low radiation dose. A higher resolution of 61 μm is also attainable *in vivo* with a slightly increased scan time (2 min) and radiation dose (<5 μSv), or of 30 μm (in vitro) with a much longer scan time and increased radiation dose. Sample scanner specifications and parameters of the system are presented in Table 16.3.

16.3.2 Patient Positioning and Image Acquisition

Standard imaging protocols have been developed for HR-pQCT, which involve scanning the non-dominant radius and tibia. As illustrated in Figure 16.5, the gantry of the system is narrow and shallow, which only allows distal sites to be imaged. To ensure appropriate positioning within the gantry and to prevent excessive subject motion, the limb being scanned is immobilized in a carbon fiber cast (Figure 16.6). As with classical CT, a scout view is acquired prior to scanning, which allows for identification of the proper region of interest (ROI) for the scan.

Because HR-pQCT acquires images at high resolution, movement during the scan results in image artifacts that will affect image quality and the accuracy and reproducibility of the measurements. As noted earlier, a carbon fiber cast is used to prevent motion during the scan. Generation of the sinogram is monitored during the scan for signs of excessive motion, in which case the scan can be stopped (to avoid any unnecessary radiation) and repeated. With the first generation of scanner, up to 22.7% of scans were reported to require a repeated scan (Pialat et al. 2012). Improvements made to the hardware in the second generation scanner have allowed for the reduction of acquisition times, which should result in reduced re-scans due to motion.

16.3.3 Image Processing and Analysis

16.3.3.1 Segmentation

Assessment of bone quality with HR-pQCT involves segmentation of the trabecular and cortical compartments, measurement of BMD and calculation of morphometric indices (Boutroy et al. 2008; Buie et al. 2007; MacNeil and Boyd 2007b). From this analysis, various material and structural parameters and properties can be measured. Various methods, automatic and semiautomatic, have been proposed for the segmentation of cortical and trabecular bone. Semiautomatic methods rely on the operator to initialize the solution by manually drawing contour lines around the region of interest, after which algorithms fit them to gradient boundaries in the image (Kass et al. 1988; Laib et al. 1998). Even if contours are projected between slices to reduce user interaction, the initialization of the solution by manually drawing contours is time consuming and can be the source of errors. A fully automated algorithm is integrated within the software package for all users of the system (Burghardt et al. 2010). This algorithm uses a series of image dilations and erosions to generate a mask for the trabecular and cortical bone compartments in the images (Buie et al. 2007). Other segmentation algorithms are currently being researched and developed to improve the speed and reliability of trabecular and cortical bone segmentation.

16.3.3.2 Registration

One of the interests of the HR-pQCT technology is the ability to perform longitudinal studies to assess bone microarchitecture changes at different time points. To do so accurately, one must ensure that the exact same region is assessed at each time point. Various similarity-based registration techniques have been developed to register 3D volumes to be compared (MacNeil and Boyd 2008a).

16.3.3.3 Image Analysis and Measures

BMD is obtained by comparing the gray-level values of an image (in HU) to those of a calibration phantom consisting of a series of materials of known densities. In the case of HR-pQCT the calibration phantom is scanned separately from individuals during a daily and weekly calibration process (Laib et al. 1998). Following the scan of an individual, density values are obtained for each voxel by plotting the HU values against the known density values of the phantom and interpolating between known densities. This technique has been demonstrated to show very high correlations between sites and scanner manufacturers (Ito et al. 2010).

One of the main advantages of HR-pQCT is that it allows characterization of bone microstructure and measurement of a set of morphometric indices, in addition to vBMD of cortical and trabecular bone. The resolution of HR-pQCT (61–82 μm) is at the limit to resolve bone microstructure, since it is close to the size of the trabeculae. In the 82 μm system microstructure measurements are derived from parameters that are less affected by resolution (Laib et al. 1998), but in the 61 μm system the resolution is now sufficient to directly measure those parameters without the need to derive parameters. Microstructure is evaluated as follows:

- Trabecular bone volume fraction (BV/TV): derived from trabecular vBMD (Laib et al. 1998), which is measured accurately by HR-pQCT.
- Trabecular number (Tb.N, number of trabeculae per unit length): calculated from distance between the longitudinal axes of trabeculae.
- Trabecular thickness (Tb.Th): derived from BV/TV and Tb.N. (Boutroy et al. 2005)
- Trabecular separation (Tb.Sp): derived from BV/TV and Tb.N. (Boutroy et al. 2005)
- Cortical thickness: direct measurement (Laib et al. 1998)
- Cortical porosity: derived similarly to BV/TV, from the fraction of void space to total cortical region (Buie et al. 2007).

These measurements obtained from HR-pQCT imaging have been shown to be highly correlated to μCT measurements made on cadavers (MacNeil and Boyd 2007a; Liu et al. 2010). Precision of the HR-pQCT measurement of density and morphological parameters has been shown to be smaller than 1% and 4.5%, respectively (MacNeil and Boyd 2008a).

In addition to the standard measurements, direct and derived, some secondary measurements can be made from HR-pQCT scans. For example, the structure model index (SMI) (Hildebrand and Rüegsegger 1997b), which quantifies the plate- and rod-like structure of trabecular bone, connectivity density

(Odgaard and Gundersen 1993) and individual trabecular segmentation (Liu et al. 2010), can be calculated. Additional measurements specific to some applications and disease are also being developed, such as measurement of bony erosions in rheumatoid arthritis (Barnabe et al. 2013), which will add value and applications to HR-pQCT measurements.

Even though cortical bone significantly contributes to whole bone strength (MacNeil and Boyd 2007b; Spadaro et al. 1994), early HR-pQCT studies have mostly focused on the study of trabecular microstructure *in vivo* (Boutroy et al. 2008; Liu et al. 2010; Hildebrand and Rüegsegger 1997a). Characterizing the quality and strength of cortical bone also plays a key role in the prediction of fracture risk. Cortical bone has a higher density than trabecular bone, but contains pores, which provide space for vessels and nerves. Cortical porosity tends to increase with age and disease (Bousson et al. 2001; Stein et al. 1999) and has been shown to be a predictor of bone strength (McCalden et al. 1993; Wachter et al. 2001; Yeni et al. 1998). Cortical porosity could account for as much as 76% of the age-related decrease in bone strength at the proximal femur (McCalden et al. 1993). Unfortunately, clinical DXA assessment of bone is not able to resolve microarchitecture (Snyder and Schneider 1991), and as such, cortical porosity cannot be used to predict bone strength *in vivo* with traditional clinical tools. As will be discussed in Section 16.3.5, FE analysis can be applied to models obtained from HR-pQCT scans to predict bone strength.

16.3.4 Accuracy and Precision

The HR-pQCT system has been tested for both accuracy and precision. Accuracy is generally tested on *ex vivo* bones using μCT as the gold standard and aims at evaluating the ability of the system to match gold-standard measurements. Precision testing can be done *in vivo* and *ex vivo* and aims at evaluating the ability of the system to yield the same measurements from repeated scans. Precision measurements have been done in single- and multi-center studies (Burghardt et al. 2013).

Accuracy tests have shown moderate to good agreement between HR-pQCT and μCT for: morphological parameters ($r^2 = 0.59–0.98$), connectivity density ($r^2 = 0.90$), mechanical stiffness ($r^2 = 0.73$) and cortical porosity ($r^2 = 0.80$) (Cheung et al. 2013).

Precision tests performed in single-center studies have demonstrated that HR-pQCT measurements were reproducible, with precision errors generally under 1% for vBMD, 2.5%–6.3% for morphological measurements, and < 3.5% for FE strength estimates (Cheung et al. 2013). Movement artifacts are typically higher at the radius, resulting in higher errors than for the tibia and for those measured *ex vivo*. A recent multi-center precision study has shown that multi-center precision errors on a realistic forearm phantom were smaller than 3% for density measures and 4%–5% for measurements of trabecular structure (Burghardt et al. 2013). This shows that the variability between measurements made with different scanners is comparable to the error introduced by mild patient motion. In other words, precision data obtained from single-center studies could lead to an underestimation of error in multi-center studies, and therefore to an underestimation of the number of subjects required to achieve adequate statistical power.

16.3.5 Finite Element Analysis

FE models derived from QCT images have shown to be valid in their ability to estimate bone strength (Bessho et al. 2007; Dragomir-Daescu et al. 2011; Keyak et al. 1998; Koivumäki et al. 2012; Viceconti et al. 2004). However, even though the FE technique is able to assess bone strength and structure at peripheral and central sites, the time and computational power required to solve FE models have been a major obstacle to its implementation in clinical practice so far.

The FE method is a numerical method that is routinely used in engineering to resolve the behavior of a body under a predefined set of loading conditions. In FE analysis, a volume of interest is divided into a finite number of sub-volumes of simple and known shape (generally tetrahedrons or hexahedrons), called elements, which are connected to each other at their corners (nodes). By discretizing a volume in this way, the behavior of a body can be expressed with a set of coupled differential equations, which can then be solved to determine unknown parameters at each node. When coupled with 3D imaging of bone and known values of bone tissue properties (as estimated from mechanical tests), the FE method can estimate, for example, stresses and strains within a bone segment under known loading conditions.

FE analysis can be very complex and computationally intensive, resulting in excessively long computing times for large high-resolution and mechanically realistic models. Simplifying assumptions are often used to accelerate computations, while yielding realistic estimates of bone strength and behavior. For example, under normal loading conditions, the deformation of bones is relatively small (strains $\leq 0.5\%$), and FE analysis using the under-formed bone shape is usually adequate. In this physiological range of deformations, the behavior of bone tissue is linear elastic, so linear FE analysis is typically performed. It has been shown that properties from the linear analysis can be used to estimate tensile strength of bone (MacNeil and Boyd 2008b). When large forces are considered, however, strains that are larger than physiological values and behaviour outside of the linear elastic zone occur. In this case, changes in geometry and properties, due to the deformation, must be accounted for by performing a nonlinear FE analysis. Resolution of the model can be reduced to simplify and speed up the calculations (Koivumäki et al. 2012). Another simplifying assumption that is commonly used is that of isotropy of bone properties. In reality bone tissue is anisotropic, so care should be taken when interpreting FE results obtained using isotropic bone properties. Living bone contains blood, marrow and fat, which influence its mechanical behavior, but are not always considered in FE, which may lead to errors even though their contribution to strength is small. Poroelastic FE can account for these components by modeling the interaction of fluids and elastic solids in a porous volume. Parameters such as total stress, pore pressure, strain in the solid phase and fluid content are used to express the poro-visco-elastic behavior of bone. In order to do so, additional parameters must be measured and used as input variables, such as the porosity and permeability of the solid phase. Often, poro-elastic models are too complex to solve when complex geometries (e.g., from μCT) are used.

Definition of a FE model usually requires the following steps:

- Definition of the geometry of interest. This is generally realized with 3D imaging modalities such as CT or MR.
- Division of the volume in finite elements (meshing) and attribution of mechanical properties. The geometry of interest can first be segmented from the 3D dataset acquired in the previous step, and then postprocessed in a meshing software.
- Attribution of mechanical properties to each volume region. Alternatively, in the case of pQCT, density values are assigned to each voxel based on its gray level, to which mechanical properties can be directly attributed (Muller and Ruegsegger 1995; van Rietbergen et al. 1995).
- Application of the boundary values (loads).
- Solution of the system of equations.

The generation of the mesh often requires significant user input and can be very time consuming. Automated methods for the generation of FE models from 3D datasets have been developed over the last two decades to allow for faster and more robust FE analysis of large datasets (Muller and Ruegsegger 1995; van Rietbergen et al. 1995). For example, based on a predefined threshold, software can identify bone and convert it into a set of 3D elements with the same dimensions. Fast and special-purpose solvers are then used to solve the system (Keyak et al. 1990; Van Rietbergen et al. 1996; Van Rietbergen et al. 2003). Despite being considerably faster than classical FE techniques, this method can lead to oscillating stress values at the surface of bones, due to the presence of the 3D elements, which lead to sharp "edges" (Guldberg et al. 1998). This problem can be solved by a second automatic method that uses marching cube algorithms to generate an FE model with smooth surface (Frey et al. 1994; Muller and Ruegsegger 1995). This second technique does not, however, have purpose solvers, so computation times remain higher, sometimes prohibitively so.

These methods have been successfully applied to datasets obtained from QCT (Cody et al. 1999; Keaveny et al. 2010; Keyak et al. 1998) and HR-pQCT (Boutroy et al. 2008; Burghardt et al. 2010; MacNeil and Boyd 2008b).

FE analysis is performed on high-performance workstations using specialized software packages such as ANSYS, Abaqus and others. General solvers are usually adapted for whole bone FE models, which comprise up to 10^5 elements. Larger models, such as those including the microstructure of bone as obtained from μCT imaging, can require special purpose FE packages because models can easily have more than 25M elements.

16.3.5.1 Whole Bone Level

At the whole bone level, a bone is generally considered to be a continuous material with a cortical and trabecular phase. 3D geometry of bones in these "whole-bone" models are generally obtained from CT or MR, and both phases are modeled as continuous materials; porosity of trabecular bone is not modeled explicitly. Cortical bone is characterized with its average properties and is generally modeled as a transversally isotropic material, with a longitudinal modulus of 15–20 GPa and smaller values of transverse modulus of elasticity (Langton and Njeh 2004). Trabecular bone is characterized with mechanical properties depending on its density (Carter and Hayes 1977; Rice et al. 1988), and as such, its modulus of elasticity can vary considerably (0.1–2000 MPa). Density information must therefore be obtained from imaging (CT, DXA), in addition to the geometrical information. Some techniques have also been employed to estimate trabecular parameters such as spacing, directionality and anisotropy (Odgaard 1997).

Whole bone FE models, as illustrated in Figure 16.7, are used in tissue mechanics and implant design studies to estimate stresses and strains in bones under different load configurations, and failure load. They can also be used to study stress increases due to altered bone density (osteoporosis) (Pinilla et al. 1996). These studies are useful to predict the fracture risk associated with disease or falls. FE models used in these studies are generally validated by creating an FE model of a cadaveric specimen and comparing the results to measurements performed during mechanical testing of the specimen. Good agreement of FE-estimated failure load and mechanical testing-measured loads has been reported (Lotz et al. 1991a,b). FE analysis at the whole bone model can also be useful in a clinical setting to estimate bone strength. For example, CT-derived FE has been shown to be more reliable than QCT or DXA measurements to predict fracture load of the femur (Cody et al. 1999). FE modeling at the whole bone level is also used for the study of the stress-shielding effect resulting from implantation of stiff joint replacement prosthesis, which often results in significant bone resorption in areas of low stress (Weinans et al. 1993; Skinner et al. 1994; Pressel et al. 2000; Stone et al. 1999). Further study of bone remodeling is also possible using more complex FE methods with variable structure (geometry, bone density) (Huiskes et al. 1987; Cowin and Hegedus 1976; Carter 1987). This allows for quick and relatively inexpensive evaluation of implant design on the mechanical environment of bone surrounding the implants. Modeling of bone microstructure is not critical in this case, as the objective is simply to compare stresses in the normal and operated bones.

16.3.5.2 Trabecular Level

In order to get more accurate estimates of stresses and strains, or to consider samples of small size, bone should be modeled at the trabecular microstructure model. At this level, trabecular bone is no longer considered to be a continuous material, and the microarchitecture of the trabecular bone is accounted for. In this case, bone tissue, which is composed of collagen and minerals, is considered to be a continuous material, which is characterized with property data obtained from mechanical testing. The elastic modulus of bone tissue has been reported to range from 1 to 15 GPa, depending on the sample and the test procedure (Rho et al. 1993). Trabecular bone tissue is usually modeled as an isotropic and homogeneous material; the use of an effective isotropic modulus has been shown to be sufficient.

FIGURE 16.7 Finite element models of bone. Blue, green, yellow and red coloring represent increasing stresses (scales are different for both images). Left: Coronal slice of an FE model of the proximal femur at the whole bone level, derived from CT. Right: Axial volume cut of an FE model of the distal tibia at the trabecular level, derived from HR-pQCT.

Over the last two decades, FE analysis at the microscopic, trabecular level (μFE), as seen in Figure 16.7, has allowed great advancements in the characterization of the mechanical properties of trabecular bone as a function of its morphology, the quantification of normal loading conditions of bone, and the analysis of trabecular bone properties *in vivo*. At this level, FE models are built almost exclusively with automated techniques based directly on image data (usually μCT or μMR), such as the voxel conversion or marching cube methods discussed earlier (Van Rietbergen et al. 1996; Muller and Ruegsegger 1995; Ulrich et al. 1998). These techniques are required because the complexity of the trabecular structure and large number of elements (10^5–10^6/cm^3) cannot efficiently be modeled with traditional meshing tools. Combined with the continuously improving performance of computer, these automatic techniques allow large and complex FE models to be solved in a reasonable amount of time (Van Rietbergen et al. 1996).

Depending on the resolution of the imaging modality, two approaches can be used integrate bone tissue properties in the FE model. At lower resolutions, such as with QCT, partial volume effects are significant, so tissue density values are converted to tissue modulus with the appropriate density-modulus relationships. When resolution is sufficiently high, such as with HR-pQCT and μCT, the dataset can be binarized and all bone elements assigned a homogeneous bone-tissue modulus, or, where mineralization differences are expected, in-homogeneities can be incorporated based on gray-level differences (Bourne and van der Meulen 2004).

Characterization of the elastic mechanical properties of trabecular bone as a function of its morphology (Hollister et al. 1994; Van Rietbergen et al. 1996) offers major advantages over mechanical testing: characterization of anisotropy in every direction, modification of microstructure, modification of bone tissue properties, quick characterization of large sets of specimens (Kabel et al. 1999; Ulrich et al. 1999), and more.

Over the years, the use of μFE has shown that:

- Microstructure alone is responsible for the anisotropic behavior of trabecular bone (Kabel et al. 1999)
- The mechanical properties of trabecular bone are orthotropic (Yang et al. 1998)
- There is a strong relationship of trabecular bone mechanical properties and fabric density (Kabel et al. 1999)
- There is a strong relationship between trabecular bone mechanical properties and bone volume fraction (Yang et al. 1998)

When studying the post failure behavior of bone, large strains should be expected, which usually requires nonlinear FE models. These models are complex and computationally intensive, but have been shown to be feasible (Niebur et al. 2000) and to provide valuable insight into the yield behavior of bone.

16.3.6 Applications

Since its introduction, HR-pQCT has led to key findings of bone microarchitecture differences with gender, age, disease and therapies, and of bone strength and fracture risk (Cheung et al. 2013).

16.3.6.1 Fracture Risk Assessment

One of the driving objectives for the imaging of bone microstructure is to estimate the risk of fragility fracture of individuals. The analysis of vBMD and bone microstructure parameters derived from HR-pQCT can provide insight into bone quality and mechanical properties, including strength and fracture outcomes. Trabecular and cortical parameters have been found to be significantly different in individuals with and without hip fractures, and in individuals with hip fractures compared to wrist fractures (Vico et al. 2008). Bone microstructure parameters, both cortical and trabecular, were also shown to contribute to fracture risk, independent of aBMD (Sornay-Rendu et al. 2007, Boutroy et al. 2008, Vilayphiou et al. 2010).

Estimates of bone strength from HR-pQCT derived FE models can be used to better predict fracture risk in specific individuals. For example, principal component analyses of different parameters associated with fragility fractures has revealed that FE-derived stiffness was one of the main contributors to fractures (Vilayphiou et al. 2010; Boutroy et al. 2008). Other studies have found significant associations of fractures with FE-derived bone strength (Melton et al. 2007), beyond aBMD or microstructure alone (Boutroy et al. 2008). Associations such as those reported above are useful at the population level for indicating potential interventions or treatment of a disease (Feng 2010). However, in order for microarchitecture measurements and FE bone strength estimates to have clinical relevance, we must go beyond general associations with fractures and use these to develop classification models with which to predict fracture risk and help with clinical decision making. For classification applications, association models are useful and required, but not sufficient on their own (Feng 2010; Pepe et al. 2004). Classification models based on several machine-learning methods, such as artificial neural networks, gradient boosting machines and support vector machines (SVM) have been developed, each having their own advantages and disadvantages. Such methods have shown to be able to classify individuals with fractures in several studies (Atkinson et al. 2012; Chiu et al. 2006) (Lee et al. 2008; Burges 1998; Schnackenburg et al. 2011; Sode et al. 2010).

16.3.6.2 Clinical Data

16.3.6.2.1 Cortical Porosity

A study by Nishimaya (Nishiyama et al. 2010) showed very good agreement of Ct.Po and Ct.Th measurements between HR-pQCT and μCT in 10 cadaveric forearms. In this same study, the assessment of cortical bone at the distal radius and tibia of 280 women demonstrated that Ct.Po was higher at both sites in postmenopausal women than premenopausal. Cortical bone was also shown to be thinner and more porous in osteopenic and osteoporotic women than in normal women. This study demonstrates the value and relevance of HR-pQCT as a tool for assessing cortical bone parameters *in vivo*, and investigates variations and associations of these parameters with age and disease.

In a more recent study (Nishiyama et al. 2012) using HR-pQCT in a population of 398 children and adolescents, Nishimaya et al. showed that peri- and post-pubertal girls had higher cortical density (Ct.BMD) and lower cortical porosity than boys of the same groups. These same boys had higher trabecular bone volume ratios (BV/TV) and cortical cross-sectional area than the girls, and higher FE-estimated failure load/fall force ratio at every stage except during early puberty, leading to lower risk of fracture. Within sex, the most mature boys and girls had the highest cortical density and lower porosity at both the radius and tibia. Peri- and post-pubertal subjects had higher FE-estimated bone strength than younger (pre-pubertal) individuals. This clearly demonstrates the potential of HR-pQCT for the evaluation of age- and gender-related differences in bone microstructure and strength.

16.3.6.2.2 Age-Related Bone Changes

Research on age-related bone changes (DXA, HR-pQCT and others) has, to date, mostly consisted of cross-sectional studies of populations of different age groups. Unfortunately, even though these studies provide valuable knowledge about differences in bone density and structure between age groups, they provide limited information as to how bone microstructure evolves during aging because secular trends are undoubtedly affecting the collected results.

Large longitudinal studies are needed to gather normative data and better understand changes in bone density and structure with aging. Large multi-site prospective studies would also be of great interest for fracture prediction. Such studies are underway in Canada and worldwide (e.g., the Canadian Multicentre Osteoporosis Study: CaMos), involving HR-pQCT, DXA and other clinical tools, which are currently available at different locations. These studies, involving HR-pQCT, will allow the development of age- and sex-specific reference data of bone microstructure and its evolution. This reference data will in turn allow comparison of individuals to population-level data, which will improve our understanding of bone structure and changes and add to the clinical relevance of HR-pQCT measurements. As an example, a recently published study (Burt et al. 2014) evaluated bone density and strength changes in a youth cohort of male and females (16–29 years). This study highlighted bone quality differences between sex and skeletal sites and variations in bone density and strength with age at the tibia with females, with a peak at 16–19 years and subsequent decrease by 29 years. Boys show significant changes in many bone quality parameters (Tb.Th. BV/TV, and others, periosteal surface) by late puberty, compared to girls, who show much more subtle changes (Andrews 2012).

16.3.6.2.3 Other Applications

The ability of HR-pQCT to image bones at high resolutions *in vivo* opens new possibilities for the study of various bone and joint diseases other than osteoporosis, which were not possible before, due to dose and size limitations associated with high resolution scanners.

For example, HR-pQCT has recently been evaluated for the study of metacarpophalangeal (MCP) joint space (Barnabe et al. 2013). The method was demonstrated to have good reproducibility, and HR-pQCT was shown to be a valuable tool for the assessment of the activity and progression of rheumatoid arthritis.

The use of μCT *in vivo* in humans could also provide valuable information on the development of osteoarthritis. Subchondral bone sclerosis and subchondral bone plate thickening are visible on x-ray in individuals with severe knee osteoarthritis (OA). It has also been shown, using DXA and QCT, that OA is associated with subchondral and periarticular bone density changes, both in idiopathic and post-traumatic OA (Brandt et al. 1991). The sequence of events leading to OA is not well understood and still the subject of debate within the scientific community. HR-pQCT offers unique new opportunities to study the evolution of bone microarchitecture linked to injury and OA at peripheral sites including the knee. Hardware upgrades made to the second generation HR-pQCT scanner include a slighter wider and deeper gantry. This opens the door to the evaluation of bone microstructure *in vivo* at the knee, as seen in Figure 16.8.

In conclusion, we are on the cusp of major improvements in computer-aided diagnosis of bone disease such as OP and joint diseases involving bone changes. Newly developed *in vivo* techniques based on CT (HR-pQCT) now allow for the analysis of bone structure (cortical thickness and porosity, and trabecular number, spacing, thickness, etc.). This newly available data, which characterizes the structure of cortical and trabecular far beyond BMD, can be combined with advanced computational analyses, such as the FE method. Together, this provides a basis for conducting noninvasive assessments of bone strength, and these techniques are providing important insight into how bone adapts with the normal aging process, in the progression of diseases and the effects of treatments. At the time of writing this chapter, computer-aided diagnosis and therapy planning for osteoporosis is still based on DXA-derived aBMD measurement. However, awareness is growing that bone strength

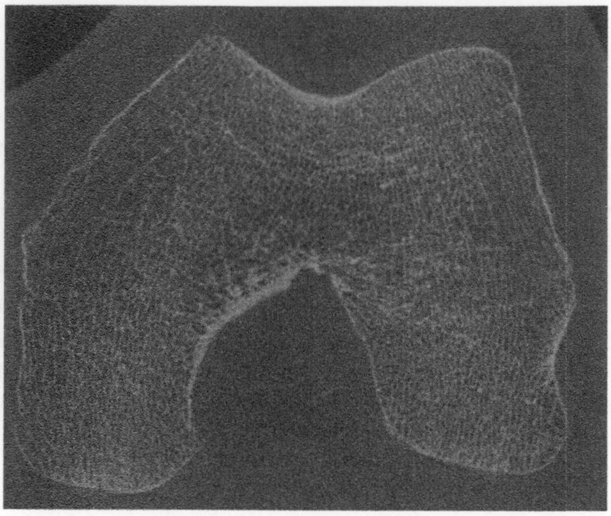

FIGURE 16.8 HR-pQCT scan of the knee (axial slice). At the time of writing this chapter, such an image is not used for diagnosis, but offers new opportunities in the study, early diagnosis and follow-up of joint injury and OA.

and fracture risk are highly dependent on bone microstructure. Hopefully, with the availability of new HR-pQCT based measurements of bone structure *in vivo*, it will be possible in the near future to improve our diagnostics beyond the simple measure of areal BMD from a DXA measurement to include these advanced technologies.

References

Adams, J.E., and N. Bishop. 2008. Dual energy X-ray absorptiometry (DXA) in adults and children. In *Pediatric Bone: Biology and Diseases*, ed. F. H. Glorieux, M. Pettifor and H. Juppner, 151–8. American Society for Bone and Mineral Research.

Ammann, P., and R. Rizzoli. 2003. Bone strength and its determinants. *Osteoporosis International* 14 Suppl 3:S13–18.

Andrews, N. A. 2012. Noninvasive evaluation of bone microarchitecture and strength: Better than DXA? *IBMS BoneKEy* 9:1–3.

Ashman, R. B., S. C. Cowin, W. C. Van Buskirk, and J. C. Rice. 1984. A continuous wave technique for the measurement of the elastic properties of cortical bone. *Journal of Biomechanics* 17 (5):349–61.

Ashman, R. B., J. Y. Rho, and C. H. Turner. 1989. Anatomical variation of orthotropic elastic moduli of the proximal human tibia. *Journal of Biomechanics* 22 (8–9):895–900.

Atkinson, E. J., T. M. Therneau, L. Joseph Melton, J. J. Camp, S. J. Achenbach, S. Amin, and S. Khosla. 2012. Assessing fracture risk using gradient boosting machine (GBM) models. *Journal of Bone and Mineral Research* 27 (6):1397–404.

Augat, P., H. Reeb, and L. E. Claes. 1996. Prediction of fracture load at different skeletal sites by geometric properties of the cortical shell. *Journal of Bone and Mineral Research* 11 (9): 1356–63.

Bailey, D. A., H. A. McKay, R. L. Mirwald, P. R. Crocker, and R. A. Faulkner. 1999. A six-year longitudinal study of the relationship of physical activity to bone mineral accrual in growing children: The university of Saskatchewan bone mineral accrual study. *Journal of Bone and Mineral Research* 14 (10):1672–79.

Barnabe, C., H. Buie, M. Kan, E. Szabo, S. G. Barr, L. Martin, and S. K. Boyd. 2013. Reproducible metacarpal joint space width measurements using 3D analysis of images acquired with high-resolution peripheral quantitative computed tomography. *Medical Engineering and Physics* 35 (10):1540–4.

Bell, G. H., O. Dunbar, J. S. Beck, and A. Gibb. 1967. Variations in strength of vertebrae with age and their relation to osteoporosis. *Calcified Tissue Research* 1:75–86.

Bessho, Masahiko, I. Ohnishi, J. Matsuyama, T. Matsumoto, K. Imai, and K. Nakamura. 2007. Prediction of strength and strain of the proximal femur by a CT-based finite element method. *Journal of Biomechanics* 40 (8):1745–53.

Bezakova, E., P. J. Collins, and A. H. Beddoe. 1997. Absorbed dose measurements in dual energy X-ray absorptiometry (DXA). *The British Journal of Radiology* 70 (February):172–179.

Blake, G. M. 1996. Replacing DXA scanners: Cross-calibration with phantoms may be misleading. *Calcified Tissue International* 59 (1):1–5.

Blake, G. M., R. J. Herd, and I. Fogelman. 1996. A longitudinal study of supine lateral DXA of the lumbar spine: A comparison with posteroanterior spine, hip and total-body DXA. *Osteoporosis International* 6 (6):462–470.

Bourne, B. C., and M. C. van der Meulen. 2004. Finite element models predict cancellous apparent modulus when tissue modulus is scaled from specimen CT-attenuation. *Journal of Biomechanics.* 37 (0021-9290 (Print)): 613–21.

Bousson, V., C. Bergot, B. Sutter, T. Thomas, S. Bendavid, C. L. Benhamou, H. Blain, et al., . 2015. Trabecular bone score: Where are we now? *Joint, Bone, Spine: Revue du Rhumatisme* 82 (5):320–5.

Bousson, V., A. Meunier, C. Bergot, E. Vicaut, M. A Rocha, M. H. Morais, A M. Laval-Jeantet, and J. D. Laredo. 2001. Distribution of intracortical porosity in human midfemoral cortex by age and gender. *Journal of Bone and Mineral Research* 16 (7):1308–17.

Boutroy, S., M. L. Bouxsein, F. Munoz, and P. D. Delmas. 2005. In vivo assessment of trabecular bone microarchitecture by high-resolution peripheral quantitative computed tomography. *The Journal of Clinical Endocrinology and Metabolism* 90 (12):6508–15.

Boutroy, S., Bert Van Rietbergen, E. Sornay-Rendu, F. Munoz, M. L. Bouxsein, and P. D. Delmas. 2008. Finite element analysis based on in vivo HR-pQCT images of the distal radius is associated with wrist fracture in postmenopausal women. *Journal of Bone and Mineral Research* 23 (3):392–9.

Bouxsein, M. L., B. S. Coan, and S. C. Lee. 1999. Prediction of the strength of the elderly proximal femur by bone mineral density and quantitative ultrasound measurements of the heel and tibia. *Bone* 25 (1):49–54.

Brandt, K. D., S. L. Myers, D. Burr, and M. Albrecht. 1991. Osteoarthritic changes in canine articular cartilage, subchondral bone, and synovium fifty-four months after transection of the anterior cruciate ligament. *Arthritis and Rheumatism* 34 (12):1560–70.

Buie, H. R., G. M. Campbell, R. Joshua Klinck, J. A. MacNeil, and S. K. Boyd. 2007. Automatic segmentation of cortical and trabecular compartments based on a dual threshold technique for in vivo micro-CT bone analysis. *Bone* 41 (4):505–15.

Burges, Christopher J. C. 1998. A tutorial on support vector machines for pattern recognition. *Data Mining and Knowledge Discovery* 2 (2):121–67.

Burghardt, A. J., H. R. Buie, A. Laib, S. Majumdar, and S. K. Boyd. 2010. Reproducibility of direct quantitative measures of cortical bone microarchitecture of the distal radius and tibia by HR-pQCT. *Bone* 47 (3):519–28.

Burghardt, A. J., J. B. Pialat, G. J. Kazakia, S. Boutroy, K. Engelke, J. M. Patsch, A. Valentinitsch, et al. 2013. Multicenter precision of cortical and trabecular bone quality measures assessed by high-resolution peripheral quantitative computed tomography. *Journal of Bone and Mineral Research* 28 (3):524–36.

Burghardt, A. J., G. J. Kazakia, S. Ramachandran, T. M. Link,and S. Majumdar. 2010. Age- and gender-related differences in the geometric properties and biomechanical significance of intracortical porosity in the distal radius and tibia. *Journal of Bone and Mineral Research* 25 (5):983–93.

Burr, David B., and Maxime A. Gallant. 2012. Bone remodelling in osteoarthritis. *Nature Reviews Rheumatology* 8 (11):665–73.

Burstein, A. H., D. T. Reilly, and M. Martens. 1976. Aging of bone tissue: mechanical properties. *The Journal of Bone and Joint Surgery. American* 58 (1):82–6.

Burt, L. A., H. M. Macdonald, D. A. Hanley, and S. K. Boyd. 2014. Bone microarchitecture and strength of the radius and tibia in a reference population of young adults: An HR-pQCT study. *Archives of Osteoporosis* 9 (1):183.

Butz, S., C. Wüster, C. Scheidt-Nave, M. Götz, and R. Ziegler. 1994. Forearm BMD as measured by peripheral quantitative computed tomography (pQCT) in a German reference population. *Osteoporosis International* 4 (4):179–84.

Canada, Health. 2008. Seniors and Aging - Osteoporosis. Last Modified 2008-07-18, accessed December. http://hc-sc.gc.ca/hl-vs/iyh-vsv/diseases-maladies/seniors-aines-ost-eng.php.

Carter, D. R. 1987. Mechanical loading history and skeletal biology. *Journal of Biomechanics* 20 (11–12):1095–109.

Carter, D. R., and W. C. Hayes. 1976. Bone compressive strength: The influence of density and strain rate. *Science* 194 (4270): 1174–6.

Carter, D. R., and W. C. Hayes. 1977. The compressive behavior of bone as a two-phase porous structure. *The Journal of Bone and Joint Surgery. American* 59 (7):954–62.

Carter, D. R., G. H. Schwab, and D. M. Spengler. 1980. Tensile fracture of cancellous bone. *Acta Orthopaedica Scandinavica* 51 (5):733–41.

Chen, H., and K. Y. Kubo. 2014. Bone three-dimensional microstructural features of the common osteoporotic fracture sites. *World Journal of Orthopaedics* 5 (4):486–95.

Cheung, A. M., J. D. Adachi, D. A. Hanley, D. L. Kendler, K. S. Davison, R. Josse, J. P. Brown, et al., 2013. High-resolution peripheral quantitative computed tomography for the assessment of bone strength and structure: A review by the Canadian Bone Strength Working Group. *Current Osteoporosis Report* 11 (2):136–46.

Chiu, J.-S, Y.-C. Li, F.-C. Yu, and Y.-F. Wang. 2006. Applying an artificial neural network to predict osteoporosis in the elderly. *Studies in Health Technology and Informatics* 124:609–14.

Ciarelli, T. E., D. P. Fyhrie, M. B. Schaffler, and S. A Goldstein. 2000. Variations in three-dimensional cancellous bone architecture of the proximal femur in female hip fractures and in controls. *Journal of Bone and Mineral Research* 15 (1):32–40.

Cody, D. D., G. W. Divine, K. Nahigian, and M. Kleerekoper. 2000. Bone density distribution and gender dominate femoral neck fracture risk predictors. *Skeletal Radiology* 29 (3):151–61.

Cody, D. D., G. J. Gross, F. J. Hou, H. J. Spencer, S. A. Goldstein, and D. P. Fyhrie. 1999. Femoral strength is better predicted by finite element models than QCT and DXA. *Journal of Biomechanics* 32 (10):1013–20.

Compston, J. E. 2001. Sex steroids and bone.

Cowin, S. C., and D. H. Hegedus. 1976. Bone remodeling I: Theory of adaptive elasticity. *Journal of Elasticity* 6 (3):313–26.

Cullum, I. D., P. J. Ell, and J. P. Ryder. 1989. X-ray dual-photon absorptiometry: A new method for the measurement of bone density. *The British Journal of Radiology* 62 (739):587–92.

Cummings, S. R., D. M. Black, M. C. Nevitt, W. Browner, J. Cauley, K. Ensrud, H. K. Genant, L. Palermo, J. Scott, and T. M. Vogt. 1993. Bone density at various sites for prediction of hip fractures *The Study of Osteoporotic Fractures Research Group*. *Lancet* 341 (8837):72–5.

Cummings, S. R., D. M. Black, and S. M. Rubin. 1989. Lifetime risks of hip, Colles', or vertebral fracture and coronary heart disease among white postmenopausal women. *Archives of Internal Medicine* 149:2445–8.

del Rio Barquero, L., M. Romera Baures, J. Pavia Segura, J. Setoain Quinquer, L. Serra Majem, P. Garces Ruiz, C. Lafuente Navarro, and F. M. Domenech Torné. 1992. Bone mineral density in two different socio-economic population groups. *Journal of Bone and Mineral Research* 18:159–168.

Diseases, National Institute for Arthritis and Musculoakeletal and Skin. 2015. Osteoporosis Overview. National Institute for Health, Last Modified June 2015, accessed December. http://www.niams.nih.gov/Health_Info/Bone/Osteoporosis/overview.asp.

Dragomir-Daescu, D., J. Op Den Buijs, S. McEligot, Y. Dai, R. C. Entwistle, C. Salas, L. J. Melton III, K. E. Bennet, S. Khosla, and S. Amin. 2011. Robust QCT/FEA models of proximal femur stiffness and fracture load during a sideways fall on the hip. *Annals of Biomedical Engineering* 39 (2):742–55.

Duan, Y., T. J. Beck, X.-F. Wang, and E. Seeman. 2003. Structural and biomechanical basis of sexual dimorphism in femoral neck fragility has its origins in growth and aging. *Journal of Bone and Mineral Research : The Official Journal of the American Society for Bone and Mineral Research* 18 (10):1766–74.

Eckstein, F., V. Kuhn, and E. M. Lochmüller. 2004. Strength prediction of the distal radius by bone densitometry - Evaluation using biomechanical tests. *Annals of Biomedical Engineer*ing 32 (3):487–503.

Einhorn, T. A. 1992. Bone strength: the bottom line. *Calcified Tissue International* 51 (5):333–9.

Feng, Ziding. 2010. Classification versus association models: Should the same methods apply? *Scandinavian Journal of Clincal and Laboratory Invistigation* 242;53–8

Ferretti, J. L., O. Gaffuri, R. Capozza, G. Cointry, C. Bozzini, M. Olivera, J. R. Zanchetta, and C. E. Bozzini. 1995. Dexamethasone effects on mechanical, geometric and densitometric properties of rat femur diaphyses as described by peripheral quantitative computerized tomography and bending tests. *Bone* 16 (1):119–124.

Frey, P., B. Sarter, and M. Gautherie. 1994. Fully automatic mesh generation for 3-D domains based upon voxel sets. *International Journal for Numerical Methods in Engineering* 37 (16):2735–53.

Frost, H. M. 1987. Bone "mass" and the "mechanostat": A proposal. *The Anatomical rec*ord 219 (1):1–9.

Galante, J., W. Rostoker, and R. D. Ray. 1970. Physical properties of trabecular bone. *Calcified Tissue Research* 5 (1):236–46.

Gibson, L. J. 1985. The mechanical behaviour of cancellous bone. *Journal of Biomechanics* 18 (5):317–28.

Glüer, C. C. 2013. Quantitative computed tomography in children and adults. In *Primer on the Metabolic Bone Diseases and Disorders of Mineral Metabolism*, 264–276. Wiley.

Glüer, C. C., U. J. Reiser, C. A. Davis, B. K. Rutt, and H. K. Genant. 1988. Vertebral mineral determination by quantitative computed tomography (QCT): accuracy of single and dual energy measurements. *Journal of Computer Assisted Tomography* 12 (2):242–58.

Goldstein, S. a. 1987. The mechanical properties of trabecular bone: dependence on anatomic location and function. *Journal of Biomechanics* 20 (11–12):1055–61.

Goodsitt, M. M., P. Hoover, M. S. Veldee, and S. L. Hsueh. 1994. The composition of bone marrow for a dual-energy quantitative computed tomography technique. A cadaver and computer simulation study. *Investigative Radiology* 29 (7):695–704.

Goodsitt, M. M., D. I. Rosenthal, W. R. Reinus, and J. Coumas. 1987. Two postprocessing CT techniques for determining the composition of trabecular bone. *Investigative Radiology* 22 (3):209–15.

Goulet, R. W., S. A Goldstein, M. J. Ciarelli, J. L. Kuhn, M. B. Brown, and L. A Feldkamp. 1994. The relationship between the structural and orthogonal compressive properties of trabecular bone. *Journal of Biomechanics* 27 (4):375–89.

Grampp, S., H. K. Genant, A Mathur, P. Lang, M. Jergas, M. Takada, C. C. Glüer, Y. Lu, and M. Chavez. 1997. Comparisons of noninvasive bone mineral measurements in assessing age-related loss, fracture discrimination, and diagnostic classification. *Journal of Bone and Mineral Research* 12 (5):697–711.

Grampp, S., P. Lang, M. Jergas, C. C. Glüer, A. Mathur, K. Engelke, and H. K. Genant. 1995. Assessment of the skeletal status by peripheral quantitative computed tomography of the forearm: Short-term precision in vivo and comparison to dual X-ray absorptiometry. *Journal of Bone and Mineral Research* 10 (10):1566–76.

Guldberg, R. E., S. J. Hollister, and G. T. Charras. 1998. The accuracy of digital image-based finite element models. *Journal of Biomechanical Engineering* 120 (2):289–95.

Hangartner, T. N., T. R. Overton, C. H. Harley, L. van den Berg, and P. M. Crockford. 1985. Skeletal challenge: An experimental study of pharmacologically induced changes in bone density in the distal radius, using gamma-ray computed tomography. *Calcified Tissue International* 37 (1):19–24.

Heilmann, P., Chr Wüster, Chr Prolingheuer, M. Götz, and R. Ziegler. 1998. Measurement of forearm bone mineral density: Comparison of precision of five different instruments. *Calcified Tissue International* 62 (5):383–7.

Hildebrand, T., and P. Rüegsegger. 1997a. A new method for the model-independent assessment of thickness in three-dimensional images. *Journal of Microscopy* 185 (1):67–75.

Hildebrand, T., and P. Rüegsegger. 1997b. Quantification of bone microarchitecture with the structure model index. *Computer Methods in Biomechanics and Biomedical Engineering* 1 (1):15–23.

Hodgskinson, R., and J. D. Currey. 1990a. The effect of variation in structure on the Young's modulus of cancellous bone: A comparison of human and non-human material. *Proceedings of the Institution of Mechanical Engineers. Part H, Journal of Engineering in Medicine* 204 (2):115–21.

Hodgskinson, R., and J. D. Currey. 1990b. Effects of structural variation on Young's modulus of non-human cancellous bone. *Proceedings of the Institute of Mechanical Engineers* [H.] 204 (0954-4119 LA - eng PT - Journal Article SB - IM):43–52.

Hodgskinson, R., J. D. Currey, and G. P. Evans. 1989. Hardness, an indicator of the mechanical competence of cancellous bone. *Journal of Orthopaedic Research* 7 (5):754–8.

Hollister, S. J., J. M. Brennan, and N. Kikuchi. 1994. A homogenization sampling procedure for calculating trabecular bone effective stiffness and tissue level stress. *Journal of Biomechan*ics 27 (4):433–44.

Huda, W., and R. L. Morin. 1996. Patient doses in bone mineral densitometry. *The British Journal of Radiology* 69 (821):422–5.

Huiskes, R., H. Weinans, H. J. Grootenboer, M. Dalstra, B. Fudala, and T. J. Slooff. 1987. Adaptive bone-remodeling theory applied to prosthetic-design analysis. *Journal of Biomechanics* 20 (11–12):1135–50.

Ito, Masako, N. Wakao, T. Hida, Y. Matsui, Y. Abe, K. Aoyagi, M. Uetani, and A. Harada. 2010. Analysis of hip geometry by clinical CT for the assessment of hip fracture risk in elderly Japanese women. *Bone* 46 (2):453–7.

Järvinen, T. L. N., Pekka Kannus, and H. Sievänen. 2003. Estrogen and bone--a reproductive and locomotive perspective. *Journal of Bone and Mineral Research* 18 (11):1921–31.

Kabel, J., B. Van Rietbergen, A Odgaard, and R. Huiskes. 1999. Constitutive relationships of fabric, density, and elastic properties in cancellous bone architecture. *Bone* 25 (4):481–486.

Kalender, W. A. 1992. Effective dose values in bone mineral measurements by photon absorptiometry and computed tomography. *Osteoporosis International* 2 (2):82–7.

Kalender, W. A., E. Klotz, and C. Suess. 1987. Vertebral bone mineral analysis: An integrated approach with CT. *Radiology* 164 (2):419–23.

Kaplan, S. J., W. C. Hayes, J. L. Stone, and G. S. Beaupre. 1985. Tensile strength of bovine trabecular bone. *Journal of Biomechanics* 18 (9):723–7.

Kass, Michael, A. Witkin, and D. Terzopoulos. 1988. Snakes: Active contour models. *International Journal of Computer Vis*ion 1 (4):321–31.

Keaveny, T. M., D. L. Kopperdahl, L. Joseph Melton, P. F. Hoffmann, B. Shreyasee Amin, L. Riggs, and S. Khosla. 2010. Age-dependence of femoral strength in white women and men. *Journal of Bone and Mineral Research* 25 (5):994–1001.

Kelly, T. L., N. Berger, and T. L. Richardson. 1998. DXA body composition: Theory and practice. *Applied Radiation and Isotopes* 49 (5–6):511–3.

Kelly, T. L., G. Crane, and D. T. Baran. 1994. Single X-ray Absorptiometry of the Forearm: Precision, Correlation, and Reference Data. *Calcified Tissue International* 54 (3):212–8.

Keyak, J. H. 2000. Relationships between femoral fracture loads for two load configurations. *Journal of Biomechanics* 33 (4):499–502.

Keyak, J. H., S. A Rossi, K. A. Jones, and H. B. Skinner. 1998. Prediction of femoral fracture load using automated finite element modeling. *Journal of Biomechanics* 31:125–33.

Keyak, Joyce H. 2001. Improved prediction of proximal femoral fracture load using nonlinear finite element models. *Medical Engineering and Physics* 23 (3):165–73.

Keyak, J. H., J. M. Meagher, H. B. Skinner, and C. D. Mote. 1990. Automated three-dimensional finite element modelling of bone: A new method. *Journal of Biomedical Engineering* 12 (5):389–97.

Koivumäki, J. E. M., J. Thevenot, P. Pulkkinen, V. Kuhn, T. M. Link, F. Eckstein, and T. Jämsä. 2012. Ct-based finite element models can be used to estimate experimentally measured failure loads in the proximal femur. *Bone* 50 (4):824–9.

Kontulainen, S. A, H. M. Macdonald, and H. A McKay. 2006. Change in cortical bone density and its distribution differs between boys and girls during puberty. *Journal of Clinical Endocrinology and Metabolism* 91:2555–61.

Kragstrup, J., F. Melsen, and L. Mosekilde. 1983. Thickness of lamellae in normal human iliac trabecular bone. *Metabolic Bone Disease and Related Research* 4 (5):291–5.

Krug, R., J. Carballido-Gamio, A. J. Burghardt, G. Kazakia, B. H. Hyun, B. Jobke, S. Banerjee, M. Huber, T. M. Link, and S. Majumdar. 2008. Assessment of trabecular bone structure comparing magnetic resonance imaging at 3 Tesla with high-resolution peripheral quantitative computed tomography ex vivo and in vivo. *Osteoporosis International* 19 (5):653–61.

Laib, A., H. J. Häuselmann, and P. Rüegsegger. 1998. In vivo high resolution 3D-QCT of the human forearm. *Technology and Health Care* 6:329–37.

Lang, T. F., J. Li, S. T. Harris, and H. K. Genant. 1999. Assessment of vertebral bone mineral density using volumetric quantitative CT. *Journal of Computer Assisted Tomography* 23 (1):130–7.

Lang, T. F., J. H. Keyak, M. W. Heitz, P. Augat, Y. Lu, A. Mathur, and H. K. Genant. 1997. Volumetric quantitative computed tomography of the proximal femur: Precision and relation to bone strength. *Bone* 21 (1):101–8.

Langton, C. M., and C. F. Njeh. 2004. *The Physical Measurement of Bone*. Edited by C. G. Orton, J. A. Spaan and J. G. Webster, *Series in Medical Physics and Biomedical Engineering*. London: Institute of Physics Publishing.

Lee, S., J. W. Lee, J.-W. Jeong, D.-S. Yoo, and S. Kim. 2008. A preliminary study on discrimination of osteoporotic fractured group from nonfractured group using support vector machine. *Conference Proceedings: Annual International Conference of the IEEE Engineering in Medicine and Biology Society*. 2008:474–7.

Lindahl, O. 1976. Mechanical properties of dried defatted spongy bone. *Acta Orthopaedica* 47 (1):11–9.

Liu, X. S., A. Cohen, E. Shane, E. Stein, R. Halley, S. L. Kokolus, P. T. Yin, D. J. McMahon, J. M. Lappe, R. R. Recker, and X. E. Guo. 2010. Individual trabeculae segmentation (ITS)-based morphological analysis of high-resolution peripheral quantitative computed tomography images detects abnormal trabecular plate and rod microarchitecture in premenopausal women with idiopathic osteoporosis. *Journal of Bone and Mineral Research* 25 (7):1496–505.

Liu, X. S., X. Henry Zhang, K. K. Sekhon, M. F. Adams, D. J. McMahon, J. P. Bilezikian, E. Shane, and X. Edward Guo. 2010. High-resolution peripheral quantitative computed tomography can assess microstructural and mechanical properties of human distal tibial bone. *Journal of Bone and Mineral Research* 25 (4):746–56.

Lochmüller, E. M. M., O. Groll, V. Kuhn, and F. Eckstein. 2002. Mechanical strength of the proximal femur as predicted from geometric and densitometric bone properties at the lower limb versus the distal radius. *Bone* 30 (1):207–16.

Looker, A. C., E. S. Orwoll, C. C. Johnston, R. L. Lindsay, H. W. Wahner, W. L. Dunn, M. S. Calvo, T. B. Harris, and S. P. Heyse. 1997. Prevalence of low femoral bone density in older U.S. adults from NHANES III. *Journal of Bone and Mineral Research* 12 (11):1761–8.

Looker, A C., H. W. Wahner, W. L. Dunn, M. S. Calvo, T. B. Harris, S. P. Heyse, C. C. Johnston, and R. Lindsay. 1998. Updated data on proximal femur bone mineral levels of US adults. *Osteoporosis International* 8 (5):468–89.

Lotz, J. C., E. J. Cheal, and W. C. Hayes. 1991a. Fracture prediction for the proximal femur using finite element models: Part I--linear analysis. *Journal of Biomechanical Engineering* 113 (4):353–60.

Lotz, J. C., E. J. Cheal, and W. C. Hayes. 1991b. Fracture prediction for the proximal femur using finite element models: Part II--nonlinear analysis. *Journal of Biomechanical Engineering* 113 (4):361–5.

Louis, O., S. Soykens, J. Willnecker, P. Den Van Winkel, and M. Osteaux. 1996. Cortical and total bone mineral content of the radius: Accuracy of peripheral computed tomography. *Bone* 18 (5):467–72.

Louis, O., J. Willnecker, S. Soykens, P. Van den Winkel, and M. Osteaux. 1995. Cortical thickness assessed by peripheral quantitative computed tomography: accuracy evaluated on radius specimens. *Osteoporosis International* 5 (6):446–9.

Lu, Y., H. K. Genant, J. Shepherd, S. Zhao, A. Mathur, T. P. Fuerst, and S. R. Cummings. 2001. Classification of osteoporosis based on bone mineral densities. *Journal of Bone and Mineral Research* 16 (5):901–10.

MacNeil, J. A., and S. K. Boyd. 2007a. Accuracy of high-resolution peripheral quantitative computed tomography for measurement of bone quality. *Medical Engineering and Physics* 29 (10):1096–105.

MacNeil, J. A., and S. K. Boyd. 2008a. Improved reproducibility of high-resolution peripheral quantitative computed tomography for measurement of bone quality. *Medical Engineering and Physics* 30 (6):792–9.

MacNeil, Joshua A, and Steven K. Boyd. 2007b. Load distribution and the predictive power of morphological indices in the distal radius and tibia by high resolution peripheral quantitative computed tomography. *Bone* 41 (1):129–37.

MacNeil, J. A, and S. K. Boyd. 2008b. Bone strength at the distal radius can be estimated from high-resolution peripheral quantitative computed tomography and the finite element method. *Bone* 42 (6):1203–13.

Martens, M., R. van Audekercke, P. Delport, P. De Meester, and J. C. Mulier. 1983. The mechanical characteristics of cancellous bone at the upper femoral region. *Journal of Biomechanics* 16 (12):971–83.

Martin, R. B., and D. B. Burr. 1989. *Structure Function, and Adaptation of Compact Bone*. Raven Press: New York, NY :275–6.

McCalden, R. W., J. A. McGeough, M. B. Barker, and C. M. Court-Brown. 1993. Age-related changes in the tensile properties of cortical bone. The relative importance of changes in porosity, mineralization, and microstructure.*Journal of Bone and Joint Surgery (American Volume)* 75 (8):1193–205.

McElhaney, J. H., J. L. Fogle, J. W. Melvin, R. R. Haynes, V. L. Roberts, and N. M. Alem. 1970. Mechanical properties on cranial bone. *Journal of Biomechanics* 3 (5):495–511.

Melton III, L. J., B. L. Riggs, G. H. van Lenthe, S. J. Achenbach, R. Muller, M. L. Bouxsein, S. Amin, E. J. Atkinson, and S. Khosla. 2007. Contribution of in vivo structural measurements and load/strength ratios to the determination of forearm fracture risk in postmenopausal women. *Journal of Bone and Mineral Research* 22 (9):1442–8.

Mosekilde, L. 1993. Vertebral structure and strength in vivo and in vitro. *Calcified Tissue International* 53 Suppl 1:S121-5;discussion S125–6.

Muller, M.E., C. E. Webber, and M. L. Bouxsein. 2003. Predicting the failure load of the distal radius. *Osteoporosis International* 14 (4):345–52.

Müller, R., T. Hildebrand, H. J. Häuselmann, and P. Rüegsegger. 1996. In vivo reproducibility of three-dimensional structural properties of noninvasive bone biopsies using 3D-pQCT. *Journal of Bone and Mineral Research* 11 (11):1745–50.

Muller, R., and P. Ruegsegger. 1995. Three-dimensional finite element modelling of non-invasively assessed trabecular bone structures. *Medical Engineering and Physics* 17 (2):126–33.

Müller, Ralph M. Hahn, M. Vogel, G. Delling, and P. Rüegsegger. 1996. Morphometric analysis of non-invasively assessed bone biopsies: Comparison of high-resolution computed tomography and histologic sections. *Bone* 18 (3):215–20.

Müller, R., and P. Rüegsegger. 1996. Analysis of mechanical properties of cancellous bone under conditions of simulated bone atrophy. *Journal of Biomechanics* 29 (8):1053–60.

Niebur, G. L., M. J. Feldstein, J. C. Yuen, T. J. Chen, and T. M. Keaveny. 2000. High-resolution finite element models with tissue strength asymmetry accurately predict failure of trabecular bone. *Journal of Biomechanics* 33 (12):1575–83.

Nishiyama, K. K., H. M. Macdonald, H. R. Buie, D. A. Hanley, and S. K. Boyd. 2010. Postmenopausal women with osteopenia have higher cortical porosity and thinner cortices at the distal radius and tibia than women with normal aBMD: An in vivo HR-pQCT study. *Journal of Bone and Mineral Research* 25 (4):882–90.

Nishiyama, K. K., H. M. Macdonald, S. A. Moore, T. Fung, S. K. Boyd, and H. A. McKay. 2012. Cortical porosity is higher in boys compared with girls at the distal radius and distal tibia during pubertal growth: an HR-pQCT study. *Journal of Bone and Mineral Research* 27 (2):273–82.

Njeh, C. F., K. Apple, D. H. Temperton, and C. M. Boivin. 1996. Radiological assessment of a new bone densitometer--the Lunar EXPERT. *The British Journal of Radiology* 69 (820):335–0.

Odgaard, A. 1997. Three-dimensional methods for quantification of cancellous bone architecture. *Bone* 20 (4):315–28.

Odgaard, A., and H. J. Gundersen. 1993. Quantification of connectivity in cancellous bone, with special emphasis on 3-D reconstructions. *Bone* 14:173–82.

Omoumi, P., F. Becce, D. Racine, J. G. Ott, G. Andreisek, and F. R. Verdun. 2015. Dual-energy CT: Basic principles, technical approaches, and applications in musculoskeletal imaging (part 1). *Semin Musculoskelet Radiology* 19 (5):431–7.

Parfitt, A. M. 2002. Targeted and nontargeted bone remodeling: Relationship to basic multicellular unit origination and progression. *Bone* 30 (1):5–7.

Parfitt, A. M., C. H. Mathews, A. R. Villanueva, M. Kleerekoper, B. Frame, and D. S. Rao. 1983. Relationships between surface, volume, and thickness of iliac trabecular bone in aging and in osteoporosis. Implications for the microanatomic and cellular mechanisms of bone loss. *The Journal of Clinical Investigation* 72 (4):1396–409.

Pauchard, Y., A. M. Liphardt, H. M. Macdonald, D. A. Hanley, and S. K. Boyd. 2012. Quality control for bone quality parameters affected by subject motion in high-resolution peripheral quantitative computed tomography. *Bone* 50 (6):1304–10.

Pepe, Margaret Sullivan, Holly Janes, Gary Longton, Wendy Leisenring, and Polly Newcomb. 2004. Limitations of the odds ratio in gauging the performance of a diagnostic, prognostic, or screening marker. *American Journal of Epidemiology* 159 (9):882–90.

Pialat, J. B., A. J. Burghardt, M. Sode, T. M. Link, and S. Majumdar. 2012. Visual grading of motion induced image degradation in high resolution peripheral computed tomography: Impact of image quality on measures of bone density and micro-architecture. *Bone* 50 (1):111–8.

Pinilla, T. P., K. C. Boardman, M. L. Bouxsein, E. R. Myers, and W. C. Hayes. 1996. Impact direction from a fall influences the failure load of the proximal femur as much as age-related bone loss. *Calcified Tissue International* 58 (4):231–235.

Pistoia, W., B. van Rietbergen, E. M. Lochmüller, C. A. Lill, F. Eckstein, and P. Rüegsegger. 2002. Estimation of distal radius failure load with micro-finite element analysis models based on three-dimensional peripheral quantitative computed tomography images. *Bone* 30:842–8.

Pope, M. H., and J. O. Outwater. 1974. Mechanical properties of bone as a function of position and orientation. *Journal of Biomechanics* 7:61–66.

Pothuaud, L., P. Carceller, and D. Hans. 2008. Correlations between grey-level variations in 2D projection images (TBS) and 3D microarchitecture: applications in the study of human trabecular bone microarchitecture. *Bone* 42 (4):775–87.

Pressel, T., M. Lengsfeld, R. Leppek, and J. Schmitt. 2000. Bone remodelling in humeral arthroplasty: Follow-up using CT imaging and finite element modeling an in vivo case study. *Archives of Orthopaedic and Trauma Surgery* 120:333–5.

RadiologyInfo.org. 2015. Radiation Dose in X-Ray and CT Exams. Last Modified 2015-06-24, accessed December. http://www.radiologyinfo.org/en/info.cfm?pg=safety-xray.

Reilly, D. T., and a H. Burstein. 1975. The elastic and ultimate properties of compact bone tissue. *Journal of Biomechanics* 8 (6):393–405.

Reilly, D. T., A. H. Burstein, and V. H. Frankel. 1974. The elastic modulus for bone. *Journal of Biomechanics* 7 (3):271–5.

Rho, J. Y., R. B. Ashman, and H. Turner. 1993. Young's modulus of trabecular and cortical bone material: Ultrasonic and microtensile measurements. *Journal of Biomechanics* 26 (2):111–9.

Rice, J. C., S. C. Cowin, and J. A. Bowman. 1988. On the dependence of the elasticity and strength of cancellous bone on apparent density. *Journal of Biomechanics* 21 (2):155–68.

Rizzoli, R., J. P. Bonjour, and S. L. Ferrari. 2001. Osteoporosis, genetics and hormones. *Journal of Molecular Endocrinology* 26 (2):79–94.

Ruegsegger, P., A. Keller, and M. A. Dambacher. 1995. Comparison of the treatment effects of ossein-hydroxyapatite compound and calcium-carbonate in osteoporotic females. *Osteoporosis International* 5 (1):30–4.

Scafoglieri, A, S Provyn, J Wallace, O Louis, J Tresignie, and I Bautmans. 2011. Whole body composition by Hologic QDR 4500/A DXA system: System reliability versus user accuracy and precision. In *Applications and Experiences of Quality Control*, edited by O Ivanov, 45–62. InTech.

Schaffler, M. B., and D. B. Burr. 1988. Stiffness of compact bone: effects of porosity and density. *Journal of Biomechan*ics 21 (1):13–16.

Schlenker, R. A, and W. W. VonSeggen. 1976. The distribution of cortical and trabecular bone mass along the lengths of the radius and ulna and the implications for in vivo bone mass measurements. *Calcified Tissue Research* 20 (1):41–52.

Schnackenburg, K. E., H. M. Macdonald, R. Ferber, J. P. Wiley, and S. K. Boyd. 2011. Bone quality and muscle strength in female athletes with lower limb stress fractures. *Medicine and Science in Sports and Exercise* 43 (11):2110–9.

Schneider, P. F., M. Fischer, B. Allolio, D. Felsenberg, U. Schroder, J. Semler, and J. R. Ittner. 1999. Alendronate increases bone density and bone strength at the distal radius in postmenopausal women. *Journal of Bone and Mineral Research* 14 (8):1387–93.

Seeman, E. 1999. The structural basis of bone fragility in men. *Bone* 25 (1):143–7.

Sibonga, J. D., E. R. Spector, S. L. Johnston, and W. J. Tarver. 2015. Evaluating bone loss in ISS astronauts. *Aerosp Med Hum Perform* 86 (12 Suppl):A38–44.

Skinner, H. B., A S. Kim, J. H. Keyak, and C. D. Mote. 1994. Femoral prosthesis implantation induces changes in bone stress that depend on the extent of porous coating. *Journal of Orthopaedic Research* 12 (4):553–563.

Slemenda, C., C. Longcope, M. Peacock, S. Hui, and C. C. Johnston. 1996. Sex steroids, bone mass, and bone loss: A prospective study of pre-, peri-, and postmenopausal women. *The Journal of Clinical Investigation* 97:14–21.

Smith, S. M., S. A. Abrams, J. E. Davis-Street, M. Heer, K. O. O'Brien, M. E. Wastney, and S. R. Zwart. 2014. Fifty years of human space travel: Implications for bone and calcium research. *Annual Review of Nutrition* 34:377–400.

Snyder, S. M., and E. Schneider. 1991. Estimation of mechanical properties of cortical bone by computed tomography. *Journal of Orthopaedic Research* 9 (3):422–31.

Sode, M., A. J. Burghardt, G. J. Kazakia, T. M. Link, and S. Majumdar. 2010. Regional variations of gender-specific and age-related differences in trabecular bone structure of the distal radius and tibia. *Bone* 46 (6):1652–60.

Sornay-Rendu, E., S. Boutroy, F. Munoz, and P. D. Delmas. 2007. Alterations of cortical and trabecular architecture are associated with fractures in postmenopausal women, partially independent of decreased BMD measured by DXA: the OFELY study. *Journal of Bone and Mineral Research* 22 (3):425–33.

Spadaro, J. A, F. W. Werner, R. A Brenner, M. D. Fortino, L. A Fay, and W. T. Edwards. 1994. Cortical and trabecular bone contribute strength to the osteopenic distal radius. *Journal of Orthopaedic Research* 12 (2):211–8.

Steiger, P., J. E. Block, S. Steiger, A. F. Heuck, A. Friedlander, B. Ettinger, S. T. Harris, C. C. Glüer, and H. K. Genant. 1990. Spinal bone mineral density measured with quantitative CT: Effect of region of interest, vertebral level, and technique. *Radiology* 175 (2):537–43.

Stein, M. S., S. A Feik, C. D. Thomas, J. G. Clement, and J. D. Wark. 1999. An automated analysis of intracortical porosity in human femoral bone across age. *Journal of Bone and Mineral Research* 14 (4):624–32.

Stone, J. L., G. S. Beaupre, and W. C. Hayes. 1983. Multiaxial strength characteristics of trabecular bone. *Journal of Biomechanics* 16 (9):743–52.

Stone, K. L., D. G. Seeley, L. Lui, J. A Cauley, K. E. Ensrud, W. S. Browner, M. C. Nevitt, and S. R. Cummings. 2003. BMD at multiple sites and risk of fracture of multiple types: Long-term results from the Study of Osteoporotic Fractures. *Journal of Bone and Mineral Research* 18 (11):1947–54.

Stone, K. D., J. J. Grabowski, R. H. Cofield, B. F. Morrey, and K. N. An. 1999. Stress analyses of glenoid components in total shoulder arthroplasty. *Journal of Shoulder and Elbow Surgery* 8 (2):151–8.

Suominen, H. 1993. Bone mineral density and long term exercise. An overview of cross-sectional athlete studies. *Sports Medicine (Auckland, N.Z.)* 16 (5):316–30.

Takada, M., K. Engelke, S. Hagiwara, S. Grampp, and H. K. Genant. 1996. Accuracy and precision study in vitro for peripheral quantitative computed tomography. *Osteoporosis International* 6 (3):207–12.

Takagi, Y., Y. Fujii, A. Miyauchi, B. Goto, K. Takahashi, and T. Fujita. 1995. Transmenopausal change of trabecular bone density and structural pattern assessed by peripheral quantitative computed tomography in Japanese women. *Journal of Bone and Mineral Research* 10 (0884-0431 LA - eng PT - Journal Article SB - IM):1830-1834.

Ulrich, D., B. Van Rietbergen, H. Weinans, P. Ru, B. Van Rietbergen, H. Weinans, and P. Rüegsegger. 1998. Finite element analysis of trabecular bone structure: A comparison of image-based meshing techniques. *Journal of Biomechanics* 31 (12):1187–92.

Ulrich, D., B. van Rietbergen, A. Laib, and P. Ruegsegger. 1999. The ability of three-dimensional structural indices to reflect mechanical aspects of trabecular bone. *Bone* 25 (1):55–60.

Van Der Linden, J. C., D. H. Birkenhäger-Frenkel, J. A N. Verhaar , and H. Weinans. 2001. Trabecular bone's mechanical properties are affected by its non-uniform mineral distribution. *Journal of Biomechanics* 34 (12):1573–80.

van Elmpt, W., G. Landry, M. Das, and F. Verhaegen. 2016. Dual energy CT in radiotherapy: Current applications and future outlook. *Radiotherapy and Oncology* 119 (1):137–44.

Van Rietbergen, B., R. Huiskes, F. Eckstein, and P. Rüegsegger. 2003. Trabecular bone tissue strains in the healthy and osteoporotic human femur. *Journal of Bone and Mineral Research* 18 (10):1781–8.

Van Rietbergen, B., A. Odgaard, J. Kabel, and R. Huiskes. 1996. Direct mechanics assessment of elastic symmetries and properties of trabecular bone architecture. *Journal of Biomechanics* 29 (12):1653–7.

van Rietbergen, B., H. Weinans, R. Huiskes, and A. Odgaard. 1995. A new method to determine trabecular bone elastic properties and loading using micromechanical finite-element models. *Journal of Biomechanics* 28 (1):69–81.

Van Rietbergen, B., H. Weinans, R. Huiskes, and B. J. W. Polman. 1996. Computational strategies for iterative solutions of large fem applications employing voxel data. *International Journal for Numerical Methods in Engineering* 39 (16):2743–67.

van Staa, T. P., E. M. Dennison, H. G. Leufkens, and C. Cooper. 2001. Epidemiology of fractures in England and Wales. *Bone* 29 (6):517–22.

Viceconti, M., M. Davinelli, F. Taddei, and A. Cappello. 2004. Automatic generation of accurate subject-specific bone finite element models to be used in clinical studies. *Journal of Biomechanics* 37 (10):1597–605.

Vico, L., M. Zouch, A. Amirouche, D. Frère, N. Laroche, B. Koller, A. Laib, T. Thomas, and C. Alexandre. 2008. High-resolution pQCT analysis at the distal radius and tibia discriminates patients with recent wrist and femoral neck fractures. *Journal of Bone and Mineral Research* 23 (11):1741–50.

Vilayphiou, N., S. Boutroy, E. Sornay-Rendu, B. Van Rietbergen, F. Munoz, P. D. Delmas, and R. Chapurlat. 2010. Finite element analysis performed on radius and tibia HR-pQCT images and fragility fractures at all sites in postmenopausal women. *Bone* 46 (4):1030–7.

Wachter, N. J., P. Augat, G. D. Krischak, M. Mentzel, L. Kinzl, and L. Claes. 2001. Prediction of cortical bone porosity in vitro by microcomputed tomography. *Calcified Tissue International* 68 (1):38–42.

Wang, Q., X-F. Wang, S. Iuliano-Burns, A. Ghasem-Zadeh, R. Zebaze, and E. Seeman. 2010. Rapid growth produces transient cortical weakness: a risk factor for metaphyseal fractures during puberty. *Journal of Bone and Mineral Research* 25 (7):1521–6.

Weaver, J. K., and J. Chalmers. 1966. Cancellous bone: Its strength and changes with aging and an evaluation of some methods for measuring its mineral content. *The Journal of Bone and Joint Surgery. American vol*ume 48 (2):289–98.

Weinans, H., R. Huiskes, B. van Rietbergen, D. R. Sumner, T. M. Turner, and J. O. Galante. 1993. Adaptive bone remodeling around bonded noncemented total hip arthroplasty: A comparison between animal experiments and computer simulation. *Journal of Orthopaedic Research* 11 (4):500–13.

WHO Study Group. 1994. Assessment of fracture risk and its application to screening for postmenopausal osteoporosis. *World Health Organization Technical Report Series* (843):1–129.

Wiktorowicz, M. E., R. Goeree, A. Papaioannou, J. D. Adachi, and E. Papadimitropoulos. 2001. Economic implications of hip fracture: health service use, institutional care and cost in Canada. *Osteoporosis International* 12 (4):271–78.

Yang, G., J. Kabel, B. Van Rietbergen, A. Odgaard, R. Huiskes, and S. C. Cowin. 1998. Anisotropic Hooke's law for cancellous bone and wood. *Journal of Elasticity* 53:125–146.

Yeni, Y. N., C. U. Brown, and T. L. Norman. 1998. Influence of bone composition and apparent density on fracture toughness of the human Femur and Tibia. *Bone* 22 (1):79–84.

Yoon, H. S., and J. L. Katz. 1976. Ultrasonic wave propagation in human cortical bone-I. Theoretical Considerations for Hexagonal Symmetry. *Journal of Biomechanics* 9 (6):407–12.

17

Augmented Statistical Shape Modeling for Orthopedic Surgery and Rehabilitation

Bhushan Borotikar

Tinashe Mutsvangwa

Valérie Burdin

Enjie Ghorbel

Mathieu Lempereur

Sylvain Brochard

Eric Stindel

Christian Roux

17.1 Introduction and Overview

As surgeons rely more and more on computers for pre-surgical and in-surgical parameters, the onus is on researchers and engineers to make the computer diagnosis reliable within clinical limits. Researchers have long targeted this single approach in a multifaceted way. Statistical shape modeling is one of the emerging techniques in the computer vision, image analysis and machine-learning domains, which, if applied effectively, could change the future course of how medical diagnosis and subsequent treatment is imparted to the patient. Statistical shape models (SSMs) have been extensively used in computer vision pipelines and in medical image processing fields [1]. SSMs describe shape distribution within a population [2,3] from which one can analyze shape variations and construe shape characterization.

The use of statistical shape models of bony structures prospers in the medical imaging literature, with applications from orthopedic surgical planning [4,5] and biomechanics research, [6] to three dimensional (3D) shape reconstruction from single [7] and bi-planar images [8,9]. Their ubiquity lies in their ability to model biological shapes, which, on top of being highly variable between individuals, can also be highly complex. For example, when registered with new image sets, SSMs can facilitate evaluation, diagnosis and treatment of musculoskeletal diseases [2,10]. Such patient-specific models have medical applications such as evaluating bone motion tracking or computer-aided orthopedic surgeries [1,11,12]. Furthermore, complex shapes in the human body include bony structures such as vertebrae, pelvis and scapulae. There have been several reports on the development of SSMs for the pelvis [13–15], vertebrae [16–18] and shoulder bones [19]. After many reports on the lower limbs, some clinicians and engineers are beginning to report on studies of the more difficult upper limbs [20–23].

The basic approach used in making an SSM is to establish the pattern of "anatomical" variation in the shape and spatial relationships of the structures in a given sample of object instances. Shape is an attribute that remains after all the effects of location, size (scale) and orientations are removed. A statistical shape is a statistical representation of such shape that can be used to either study a generalized morphology of that shape or conform to a subject-specific shape through known and valid correspondences. In order to be able to extract statistical information from several instances of the object, the instances have to be brought into correspondence. This is central to the generation of SSMs for surfaces as it directly affects the model quality. According to van Kaick et al. [24] a definition of meaningful correspondence depends on the task at hand. This ranges from a simpler case of identifying portions of shapes that are geometrically similar, to more complex problem of relating elements that represent the same parts or serve the same function on shapes. Thus the matching parts may differ significantly in their geometry, structure within the context of the whole shape, or even topology. This makes traditional landmarking for establishing correspondence impractical, tedious and error-prone. Thus, the majority of work in SSM development is centered on establishing reliable methods for automated, dense correspondence of objects [25]. Registration, thus, becomes the most viable methodology to obtain such dense correspondence. Statistical analysis via Principal Component Analysis (PCA) is used to give a parameterization of this variability, providing an appropriate representation of shape and allowing shape constraints to be applied [26]. This type of modeling allows a 3D morphometric analysis to reveal the important shape parameters and also how multiple parameters change together, rather than focusing on one parameter at a time [27].

It is common to assess the computed correspondences in terms of the quality of the SSMs that they generate in the computational anatomy community [24]. The models are evaluated on criteria such as generality, specificity, or compactness [24,26,28,29] with the assumption that accurate correspondences lead to accurate models. Briefly, generality functions are focused on measuring a model's capability to represent unseen instances of the species of the object modeled. Specificity is an evaluation of the model's ability to generate instances of the object that are close to those in the training set. Finally, compactness is the ability to use a small number of parameters to cover more shape instances in the training sets [30]. While these criteria are generally reported in the literature, a comparative platform is needed. This would facilitate the exchange of models for validation of methodology by other research groups and for collaborative efforts. In this regard, a recent development has been the Statismo framework for PCA-based statistical modeling [31]. Despite its capability and wide usage in medical image segmentation, application of SSMs in surgical planning, biomechanics modeling, and clinical treatments has not been extensively explored. This could partially be due to the complexity involved in building the SSMs. It could also be attributed to our inability to validate these models with anatomical data that has clinical relevance. Thus to use SSMs in biomechanical analysis as a clinical diagnostic tool, one would first require a necessary validation of their statistical integrity, computational stability and prediction ability for clinically relevant areas and is termed as "clinical validity" for the remainder of this chapter.

This work reports on an integrated pipeline for building an automated, unbiased global SSM of the scapula and humerus for morphometric as well as segmentation purposes. The chapter also explains the methodology to build augmented SSMs to be used for clinical diagnosis. Validity of such augmented SSMs is also determined using a global, as well as local, validation methodology. Scapula bone is especially focused on, as attempts to date to build models of this bone have evaded automation, as in [32], which required manual landmarking to initialize the B-spline-based non-rigid registration. Another reported attempt [33] did not present a global model, which required watershed segmentation in order to constraint the anatomical correspondence. This chapter first briefs the development of an SSM-building pipeline for anatomical bone shapes in detail. This is done to thoroughly understand the techniques used for various steps involved in the SSM building. The remainder of this chapter is arranged as follows. The second section explains how bone models are constructed from computed tomography (CT) data by using semiautomatic segmentation approach. The third section explains different registration methods and gives a detailed account of the rigid and non-rigid registration techniques used in this chapter. The fourth section explains construction of SSMs using registered sample sets that are in correspondence. The fifth section evaluates the robustness of the methods employed to build the SSM using the measures of generality, specificity, and compactness. The sixth section explains the development of augmented SSM using anatomical landmarks in order to use the statistical shape model as a clinical and surgical tool. The seventh section discusses the proposed application of such augmented SSMs in computer-aided surgery and pre-surgical planning.

17.2 Building Image-Based Bone Models

Image-based model building has been in use for a long time now. Manual or automatic segmentation of image slices is the most commonly used method to build 3D bone models. Bone models can be represented in any geometric data type including: (1) point sets, (2) line segment sets (polylines), (3) implicit curves $\left(\vec{g}(x,y,z)=0\right)$, (4) parametric curves ($x(\mu)$, $y(\mu)$, $z(\mu)$), (5) triangle sets (faceted surfaces), (6) implicit surfaces ($g(x,y,z) = 0$), (7) parametric surfaces ($x(\mu,\upsilon)$, $y(\mu,\upsilon)$, $z(\mu,\upsilon)$), and so on. The most typically used data representation is, however, triangle sets or faceted surfaces. In this study, triangular representation of bone data has been adopted. This section explains how data was collected for shape modeling pipelines and how bone models were acquired/developed using the data.

17.2.1 Data Acquisition through Radiography

17.2.1.1 Scapula and Humerus Data

Data from dry bones was collected for both scapula and humerus. Scapula data was acquired via CT scans and consisted of a total of 84 scapulae. Initially, 27 scapulae (13 left-side) were acquired from the anatomy laboratory at CHRU Brest (Centre Hospitalier Regional et Universitaire, Brest, France). Humerus data consisted of 28 right-side humeri, and the 3D data of the humeri was obtained from a previously developed dataset from dry bone scans. Each scapula and humerus was inspected for signs of trauma, and only those exhibiting no trauma were included in the imaged sample. No other criterion was used in the selection, and no other descriptive information such as sex and age were available. Both of the bones were imaged using the SIEMENS SOMATOM Definition AS (Siemens Healthcare Solutions, Forchheim, Germany) scanner at the CHRU Brest. The protocol used for the imaging was developed by an expert technician at CHRU Brest. It was chosen to maximize the quality of the images and is presented in Table 17.1. The left and right scapulae were imaged separately, as shown in Figure 17.1. A single acquisition per side proved to be the most economical and time-efficient method for acquisitions. For each side, the rows of scapulae were placed as separately as possible from the next row. The same was done for each column of scapulae. This made the semiautomated segmentation and separation of the scapula easier. In addition, a flat transmissive platform was placed on the resting bed so the scapulae were rested on three points (the superior angle, the inferior angle and the acromion processes). The glenoid region is a clinically important part of the interface between the scapula and the humerus, and, thus, it was important to image this region with the best image quality. Care was taken to place the scapula with the glenoid surface parallel to the XY-plane.

TABLE 17.1 CT Imaging Protocol for Imaging Dry Scapulae

Scan Parameter	Value
Width	500 mm–512 kpixels
Height	500 mm–512 kpixels
Resolution	1.024 pixels per mm
Pixel size	0.98 × 0.98 mm
Manufacturer's model name	SOMATOM definition AS
Slice thickness	0.6 mm
kVp	70
Data collection diameter	500 mm
Device serial number	11,067
Software versions(s)	Syngo MI.PET/CT 2012A
Protocol name	Extremity
Reconstruction diameter	500mm
Distance source to detector	1085.6 mm
Distance source to patient	595 mm
Gantry/detector tilt	0°
Exposure time	1000 ms
X-ray tube current	75 mA
Collimation	19.2
Collimation	0.6
Pitch factor	0.8
Reconstruction diameter	488 mm

Note: Important parameters for scanning are explained as: (1) *Pixel Size:* Square element (area) in image mapping average μ in corresponding voxel. thus pixel size here represents 0.98 mm width and 0.98 mm height. (2) *Slice Thickness:* Slice thickness determined by width of x-ray beam determined by collimator adjustment. (3) *KVp:* Voltage between filament cathode and anode. higher potential accelerates electrons more giving them and the resultant x-rays more energy. (4) *X-ray Tube Current:* Current flowing through tube filament. larger current produces great x-ray intensity. (5) *Pitch Factor:* The pitch is the ratio of the patient table increment to the total nominal beam width for the CT scan. The pitch factor relates the volume coverage speed to the thinnest sections that can be reconstructed.

Additionally, imaging data for 57 scapulae, including pathological samples, were obtained from a database collected by Imascap (http://www.imascap.com/).

17.2.1.2 Hippocampi Data

To facilitate a comparison of the proposed methodology and other state-of-the-art methods, publicly available, 42-segmented, hippocampi data from Styner and colleagues [34] was obtained as in Rasoulian et al. [28]. The hippocampi CT data had voxel sizes between $20 \times 35 \times 17$ mm^3 and $26 \times 41 \times 24$ mm^3.

17.2.2 Image Processing: Segmentation and Model Creation

Segmentation was performed in a 3D modeling software called Amira v5.4.3 (Visage Imaging: http://www.vsg3d.com/). Each scapula was segmented and, using the "extract image" feature in Amira, a 3D surface mesh was extracted from the labeled images. The meshes were also smoothed in Amira. The Amira smoothing module smooths a surface by shifting each vertex toward the average position of its neighbors. One problem encountered during the segmentation was the identification and delineation of the thinnest part of the scapula, the area of the Supraspinous and Infraspinous fossae. The thickness of

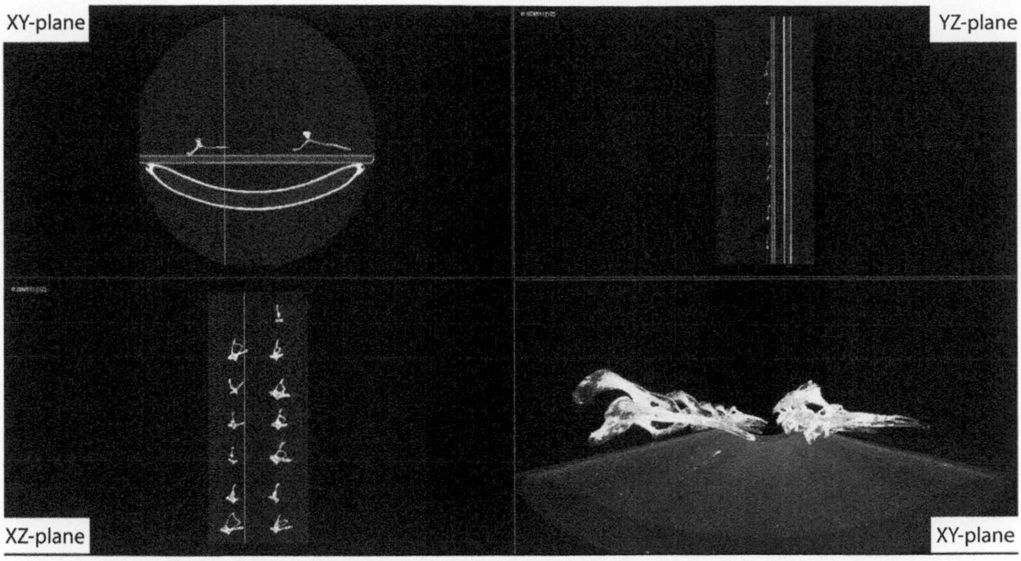

FIGURE 17.1 Scapula CT resultant images in 4 views. The glenoid plane was kept parallel to the XY-plane to get the best resolution for this part. Due to the high resolution data acquisition (0.98 × 0.98×0.6 mm^3), the transmissive resting platform became visible at high threshold values.

this area was often less than 1.0 mm and, given a Y-plane pixel size of 0.98 mm, made it difficult to identify except at a high gray-scale threshold of the images. However, the high threshold made the normally transmissive platform visible in the images. In turn, this made it difficult to select the boundary of the resting regions of the scapula. The authors, thus, had to carefully select the boundary of the scapulae and the platform in each slice, manually negating the semiautomation of the segmentation and making it a very lengthy process to segment the scapulae. A solution to this problem would be to image each scapula separately; although, this would be more costly.

Further postprocessing involved remeshing the smoothed surface to a lower point-cloud density more suitable as input data for the modeling process. The typical vertex count after segmentation was 250,000, which was too large for the registration process that followed. This decimation was performed in Amira using the Remesh Module. The Amira Remesh Module approach used is based on Lloyd relaxation (this determines the number of passes used for relaxing the points) and attempts an isotropic vertex placement in order to achieve a high triangle quality. Since vertex placement is modulated by either the curvature or a density field, isotropy of vertex placement will only hold with respect to the modulated vertex distribution [35]. The remeshing module in Amira provides a parameter to determine the density contrast of vertices. The density contrast determines the influence of the density field or the surface curvature on the triangle area in certain regions of the surface [35]. The difference of the remeshing with two different density contrast parameters is shown in Figure 17.2.

A higher density contrast results in higher density of vertices around regions of high curvature. A zero value for the density contrast results in evenly spread vertices. To assess the effect of different vertex density contrasts on the registration algorithm presented in the following sections, the scapulae data was remeshed using two density contrast values, that is, 0 and 5.

The number of vertices chosen in each remesh procedure was 15,000. This was a compromise between the number of vertices to adequately incorporate the surface geometry and the computational cost of the registration as will be explained in the next sections. The left scapulae were mirrored so all the scapulae now represented the right side. This was done to increase the sample size of the dataset, and, thus, we worked with 27 right-side scapulae. Similar to the scapulae data, further postprocessing of humerus

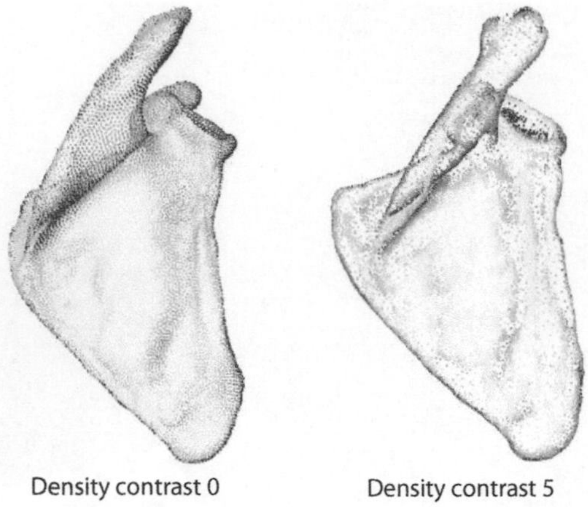

FIGURE 17.2 Two instances of the same scapula with different vertex density contrasts.

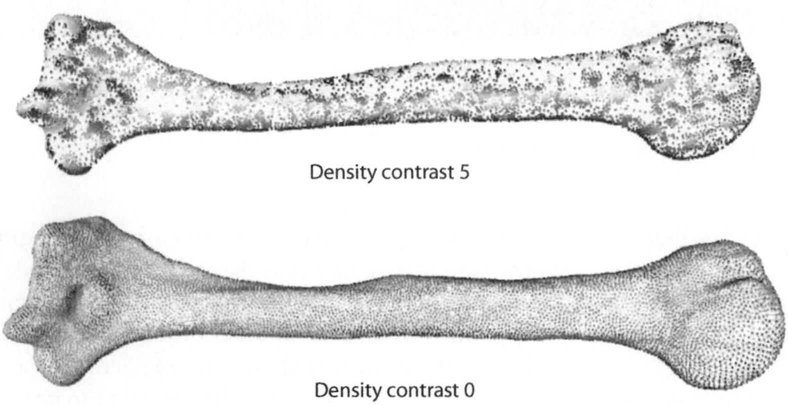

FIGURE 17.3 The same humerus with two different density contrast values and with 15,000 vertices each.

data involved remeshing the smoothed surface to a point cloud density more suitable as input data for the modeling process. This decimation was performed in Amira using the Remesh Module, and two datasets were processed: one set with a vertex density contrast 0 and another with 5, as illustrated for one example in Figure 17.3.

The postprocessing of these meshes involved remeshing to match the vertex count of the 27 scapulae dataset described above, which is 15,000 vertices. In addition, all meshes were configured to be right-sided. These data were not included in the density contrast analysis, registration quality analysis and model. However, they were used in a larger scapulae model reported later in this chapter. The segmentation and postprocessing was performed for the scapula dataset, although the final vertex count for these meshes after remeshing was 6,000. This lower number of vertices was chosen to limit the computational expense of the model validation procedure.

17.3 Registration

17.3.1 Basics of Registration

Registration is the process of finding the spatial transform that maps points from one image or shape to the homologous points in another image or shape [36]. An important distinction in medical image registration is the difference between intra-subject and inter-subject registration. In intra-object registration, repeated image scans of the same object are used to capture the effect of disease development, treatment progress and also to mitigate against the effects of pose differences and other imaging artifacts. The repeated imaging of the same object provides redundancy of information, which may facilitate a better reconstruction of the image in both 2D and especially in 3D, where pose is important. On the other hand, inter-object registration is useful for objective and quantitative assessment of abnormalities. In addition, the computed deformation field maybe used to encode pattern of anatomic variability within a population. This chapter focuses on the latter, that is, inter-object registration.

Furthermore, two sub-problems exist in the registration process. The first is the estimation of the transformation from one instance to another, and this transformation is based on a prior identification of primitive counterparts on both instances called correspondence. The combination of these primitives is the problem of matching, the second sub-problem [37–39]. In a two-object situation, one of them is taken as reference and the other as mobile. The transformation is estimated to be the move from the mobile instance to the reference. If more than two instances are to be registered, then this is called multiple registration or multiple object registration. In this case, different strategies are possible, with the aim wherever possible to share available information on all instances to eliminate the sources of error and make the process more robust [37]. Closely related to the correspondence problem is the issue of the most representative shape of the collection of objects. This is important for selecting a reference topology during the establishment of correspondence. Selecting one specific instance from the sample leads to the obvious problem of bias induced by the choice of reference shape, even if the procedure is repeated a second time with the mean shape as reference in a bootstrapping framework [25]. Only by examining a whole set of shapes of a class of objects can one learn which are the important features [40], a process which is not possible through the sequential, pairwise registration necessary in this kind of correspondence establishment. The obvious solution is a groupwise registration framework for establishing correspondences [28,41,42]. In addition, because of this dependency on registration, the quality of the registration plays a critical role in the SSM developmental pipeline.

17.3.1.1 Rigid versus Non-Rigid

Surface registration may consider rigid or non-rigid shapes. Rigid registration assumes that two (or more) surfaces are related by a rigid transformation. Non-rigid registration allows deformation (e.g., morphing, articulation) between them. Rigid registration is a challenging problem. First, the data itself poses many difficulties, which may include noise, outliers and limited amounts of overlap. Noise may take the form of perturbations of points, or unwanted points close to a 3D surface. Outliers are unwanted points far from the surface, which can seriously affect results if not discarded. Limited overlap arises due to different parts of the object being in view in each scan; typically the number of scans is kept low for efficiency, with few points in common between successive scans. Further problems may arise due to self-occlusion when the object is scanned from certain viewing angles. While such problems can be mitigated by careful scanning, they are hard to avoid completely. Second, variations in initial positions and orientations (and what is known about them), as well as resolutions of data, can also affect algorithm performance and must be taken into account when comparing rates of convergence, methods of correspondence determination and approaches to optimization.

Non-rigid registration is even more difficult, as it not only faces the above challenges but also needs to account for deformation, so the solution space is much larger. Unlike the rigid case, where a few correspondences are sufficient to define one candidate-rigid transformation for hypothesis testing, both deformation and alignment in the non-rigid case, without strong prior assumptions, often require a lot more reliable correspondences to define. Establishing meaningful and natural correspondences, however, is a challenging problem in its own right. Choice of appropriate representation for the deformation and suitable tools for the evaluation of non-rigid registration methods are two different problems. Recent success in rigid surface registration, coupled with the development of scanning devices that can capture time varying surfaces, have brought non-rigid registration into focus.

17.3.2 Rigid Registration

17.3.2.1 Iterative Closest Point (ICP)

Popular rigid registration algorithms include the ICP algorithm by [43] and the Softassign Procrustes by [44]. Both accept two surfaces with potentially different numbers of vertices as input and deliver the optimal similarity transformation from one surface to the other as a result. The ICP is a registration algorithm combining both the steps of mapping and the estimation of the transformation. The ICP algorithm first updates point correspondences with respect to the current estimation of the pose estimation transform $\boldsymbol{T}$; thus, an initial guess of $\boldsymbol{T}_0$ is required. The second step is an update of the current estimation transform $\boldsymbol{T}$ through a pose estimation algorithm. An important point to note with ICP is that if the initial solution is too far from the optimal solution, the algorithm may converge to a local minimum [37]. The mapping in the ICP is based on finding the nearest neighbor, and, to be accurate, a simple and effective initialization is done by way of aligning all the objects by their main axes of inertia. Technical details of implementation of ICP used in the project, including the methodology of nearest neighbor search, the K-d tree, are provided in [37] and are adopted here.

17.3.2.2 Robust Iterative Closest Point (ICPR)

A variation of ICP [45] is called ICPr [46,47]. The main drawback of the ICP is its lack of robustness to outliers or missing data on the surfaces of objects being registered. Indeed, in the ICP, all pairings/matchings participate equally in the estimation of the transformation. The quality of the calculated transformation strongly depends on the quality of the original data. The registration could be more robust through computing the nearest neighbor of each point of the moving object, but keeping only the supposed good matches. A selection of the Euclidean distance between matched points is performed to select only reliable matches. The central modification of ICPr from the traditional ICP is the inclusion of a weighting of the matching or the "interim corresponding points" during each ICP iterative cycle. This weight reflects the reliability of a matched pair [37]. Again, a complete treatment of the implementation of ICPr is provided in [37,45]. Suffice it to say, however, that the chosen m-estimator used is the Tukey biweight estimator [48]. The introduction of a robust estimator in the ICP algorithm eliminates unlikely pairings and, thus, reduces the influence of disturbances on the calculation of the registration. The use of an adaptive point of discharge into the estimator allows the algorithm to evolve during the iterations and, so, become more selective in the refinement of registration. The discharge point is greater than the median value of residuals, and this process ensures that there are always at least 50% of valid matches preventing too many zero-weight matches.

17.3.2.3 Iterative Median Closest Point (IMCP)

As mentioned earlier in Section 17.3.2.1, the transformation T is estimated to be the move from a mobile instance to the reference. In the multiple instance case, that is, two or more mobile objects, the difficulty is to perform a simultaneous registration of all objects. A further variation of the ICP incorporating ICPr has been previously developed at Telecom, Bretagne [46,47]. The algorithm, called the IMCP, facilitates mitigation against the commonly used practice of selecting a specific reference from the sample as a reference object, a practice which necessarily introduces a bias in the registration. Instead of using a reference point cloud, the algorithm iteratively creates a reference virtual cloud of points using statistically derived information from all the instance point clouds. For a particular mobile point in a particular instance, the associated virtual point is computed along a linearly interpolated line between a robustly calculated centroid of matching neighbor points and that particular mobile point. The robustness of the centroid is guaranteed by the weighted median of the matching neighbor points to the mobile point being investigated.

In [37], the IMCP was previously used to remove artifacts from multiple images in different poses of the same object, relying on redundancy of information to reconstruct the rigid object from the multiple images. The method is illustrated in the example shown in Figure 17.4. Here, artificially synthesized data

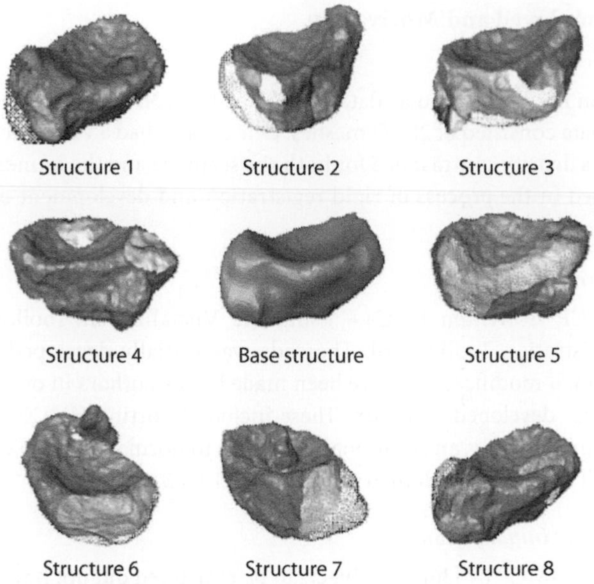

FIGURE 17.4 The testing of IMCP as performed by [37]. Artificially created modifications (structures 1–8) of the base structured shown in the middle were generated based on a synthetic navicular (a bone in the human foot). The IMCP was able to correctly co-register all the structures and obtain the original base structure.

was created through cloning of the basic structure in eight copies, simulation of the various causes of distortion using Boolean operators (protrusions, extrusions and perforations), introduction of Gaussian noise and decorrelation of the topology of the media. The originality of the IMCP relies, in part, on the registration of the instances of the movement on the virtual forms representing the consensus of the information in the dataset acquired. By using the information of the weight of the whole of the clouds of points, it is possible to reconstruct the rigid object that emerged as the consensus of the instances, as shown in Figure 17.4. The reconstruction of the rigid form is the culmination of a process of fusion of the instances of the movement, taking into account the weighting of each of the vertices.

In this project, the very same concept of information "redundancy" expected in multiple objects of the same biological "shape" species and the consensus of the dataset with weights of reliability can be used to provide a true rigid form of the shape objects being investigated. The hypothesis is that this inter-object registration, using only very reliable points, can provide an "intrinsic consensus shape" of the object. All variation from this true object can be viewed in the same way as noise artifacts and difference in pose in the work of [37].

The following is adopted from [37]. In order to get the intrinsic consensus shape, the first step of this process is to merge the whole of the clouds of points in a single cloud (which is, therefore, non-mesh). Each point of this cloud has a weight w which reflects the reliability of its positioning, and more specifically, the weight w can be interpreted as the probability that the point is representative of Gaussian disturbance of the virtual form outcome of the consensus of all the objects registered. The goal of the iterative reconstruction process is, therefore, to move toward a cloud of points for which the average of the weight would be as close as possible to 1.0. The cloud obtained would, therefore, be equivalent to the form of the real object. To achieve this goal, we must, therefore, in general, move the low-weight points to the areas of higher reliability. The Laplacian is one of the most-favored techniques employed for the elimination of noise in a mesh. When filtered with Laplacian, a point P converges to the barycenter of its direct neighbors. By using the weight of the clouds, the Laplacian filter can resist movement of points with high weights and increase that of points with low weight. Further details of the mesh fusion used in IMCP are provided in [37].

17.3.2.4 Methods Employed and Motivation

17.3.2.4.1 Data

As mentioned in Section 17.2.1, the scapulae data were in the form 3D meshes of the 27 scapulae obtained from CT. The humeri data consisted of 28 3D meshes. One dataset had a vertex density contrast of 0 and the same number with a density contrast of 5 for both the scapulae and the humeri. Both sets of data for each structure were used in the process of rigid registration and development of virtual shapes using IMCP.

17.3.2.4.2 IMCP Software

The algorithm for IMCP is written in C++ using the Visualization Toolkit (VTK) and Insight Segmentation and Registration Toolkit [36]. The code was initially developed by J.J. Jacq (LaTIM - INSERM UMR 1101). Small modifications have been made by the authors in order for the code to work in Windows, since it was developed in Linux. These included porting UNIX-based C++ methods to Windows. In terms of functionality, an additional algorithm to normalize all objects with respect to size was added, and this will be further explained in Section 17.3.2.4.3.

17.3.2.4.3 Centroid Size and Initialization

In the previous version of IMCP by [36], the objects to be registered did not have size differences, since it was developed for intra-object registration. In the current SSM development pipeline, there was a big variation on the sizes of different scapulae and humeri. To remove the effect of size in the registration, the centroid size of each object was used to scale each object to unit size. Centroid size is the square root of the sum of squared distances of a set of landmarks, in this case vertices, from their centroid. Centroid size is used in geometric morphometrics because it is approximately uncorrelated with every shape variable, when landmarks are distributed around mean positions by independent noise of the same small variance at every vertex and in every direction. This ties in well with the concept of statistical shape analysis, where shape is the geometric information that is left after removing the effects of size, position and orientation. The Procrustes processes to follow remove the other effects of position and orientation/pose. The dataset is also initialized by aligning all instances using their main axes of inertia, as described earlier.

17.3.2.4.4 Intrinsic Consensus Shape

17.3.2.4.4.1 Scapula

Comparison of virtual shapes with different density contrast Figure 17.5 shows the developed scapulae consensus shapes. The results of the scapulae with density contrast of 0 are better than those with density contrast of 5. The rest of the analyses of the scapulae were done using data with density contrast of 0.

Postprocessing virtual shapes The point-wrap module in Amira was used to postprocess the intrinsic consensus shape. This algorithm performs a surface reconstruction from a set of unorganized points. It models a probe sphere, which is being "dropped" onto and then "rolled over" the set of points. Every three points the sphere rests on during this tour become a triangle in the resulting surface. The result is (almost) guaranteed to be an oriented manifold [49]. Further postprocessing included smoothing and remeshing to get an isotropic mesh of 7,500 vertices (Figure 17.6), exactly half the number of the dataset, for reasons to be explained in the coming section on non-rigid registration (Section 17.3.3).

While the point-wrap algorithm in Amira performs well in extracting the exterior surface from an unordered point cloud, there are still areas of the mesh which show some small geometric indentations which are not typical in the scapula. For such indentations (Figure 17.7), further processing was required to get a smooth virtual shape, which here will be called the *mean-virtual* (MV) shape. This postprocessing was done using non-rigid registration described in the next section. Since there is no ground truth for the median consensus shape, this assessment of the quality of the mesh is under the assumption that the median shape should be smooth.

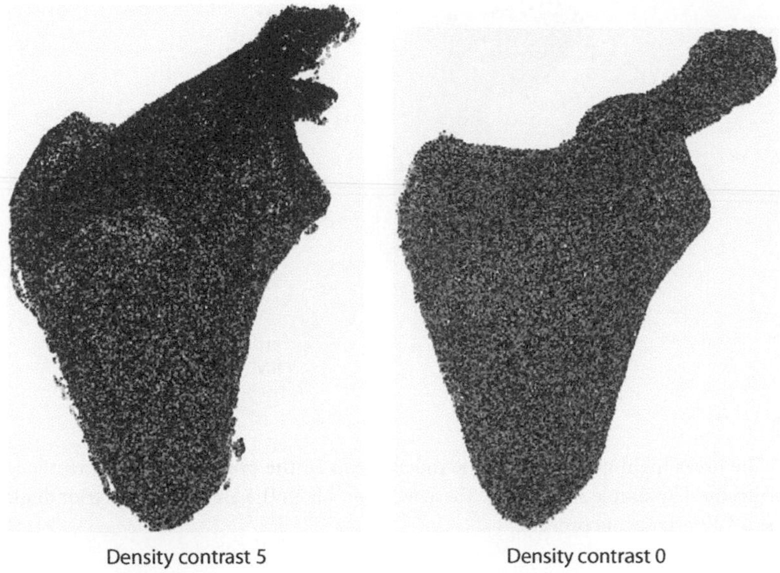

FIGURE 17.5 The intrinsic consensus shape point cloud results for two different vertices' density contrast. It is clear that the mesh fusion process in IMCP works better with density contrast of 0.

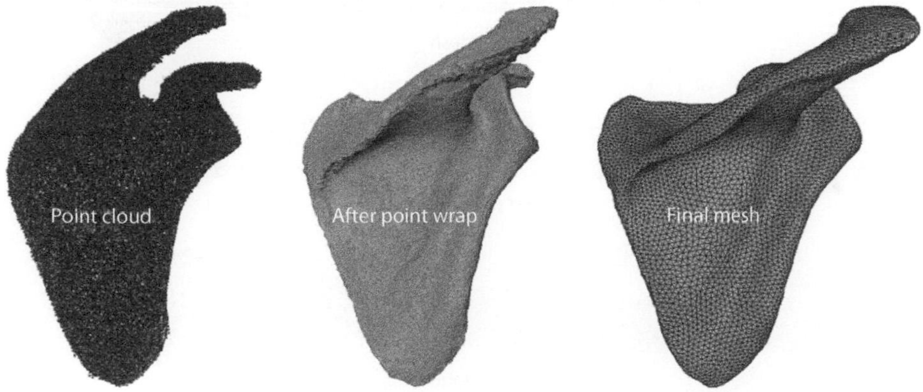

FIGURE 17.6 The results of the postprocessing step. The final mesh represents an intrinsic scapula shape formed from the median consensus points from the dataset with points of high reliability weight. The final mesh here is outlined to show the vertex quality.

17.3.2.4.4.2 ***Humerus***

Comparison of virtual shapes with different density contrast The results of the density contrast test for the humerus mirrored those of the scapulae data and are not presented here. Suffice it to say that all the remaining analyses of the humeri data were done on data with a density contrast of 0.

Postprocessing humerus virtual shape Figure 17.8 shows the results of the postprocessing of the virtual shape using the example of humeri data. Postprocessing included smoothing and remeshing to get an isotropic mesh of 7,500 vertices, exactly half the number of the dataset, for reasons to be explained in Section 17.3.3. The same indentations from the mesh fusion process as in the scapulae data were found and are shown in Figure 17.9. The humerus virtual mesh, thus, also required postprocessing using non-rigid registration.

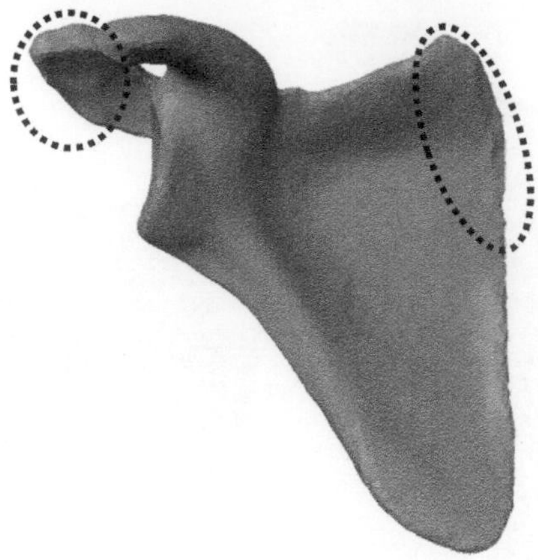

FIGURE 17.7 The areas highlighted show some indentations in the point-wrapped intrinsic consensus shape which cannot be removed by simple smoothing. These were mainly in the area of the superior angle of the scapula and the inferior side of the coracoid process.

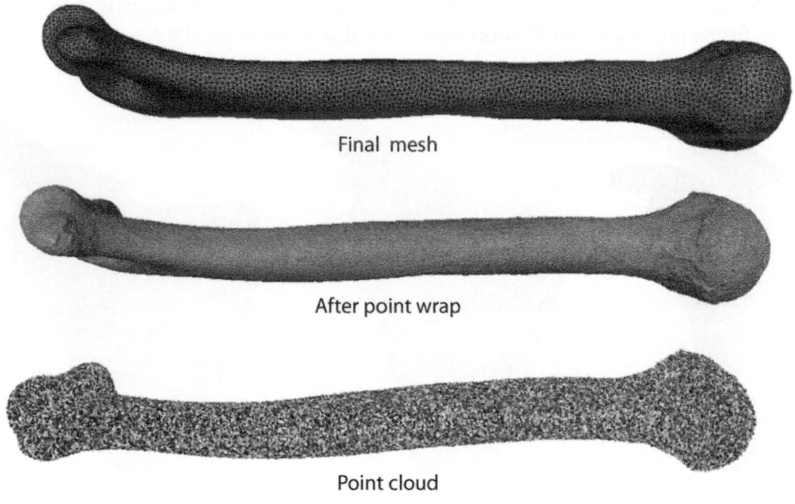

FIGURE 17.8 The results of the postprocessing step. The final mesh represents an intrinsic humerus shape formed from the median consensus points from the dataset with points of reliability weight greater than 0.99999.

17.3.3 Non-Rigid Registration

17.3.3.1 Coherence Point Drift

Nonparametric image registration estimates the transformation as an unknown function without explicit parameterization. Such a problem is ill-posed, because the solution is not unique. To constrain the problem, one can use a regularization term that penalizes some undesirable transformation properties, for example, oscillations [50]. Myronenko [50] uses the example of estimation of a non-rigid transformation over the image domain given only a set of corresponding landmarks. The solution to such a problem is not unique: "there are an infinite number of transformations that will match the corresponding landmarks, but have totally different behavior in the remaining image parts" [50]. An imposition

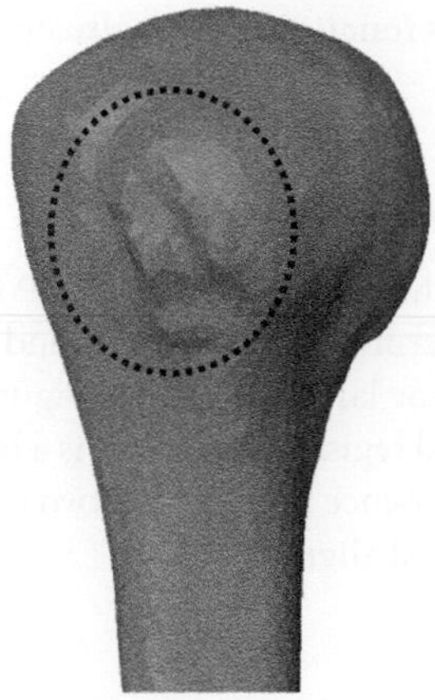

FIGURE 17.9 The highlighted area shows indentations in the virtual shape which cannot be removed by simple smoothing. These areas were around the anatomical neck and the lesser tubercle.

of an additional requirement of using the smoothest transformation, as defined by the corresponding regularization term can provide a unique solution. The coherence point drift (CPD) algorithm considers the alignment of two point sets as a probability density estimation problem. Basically, the Gaussian mixture model (GMM) centroids, represented by the points of a mobile object's point cloud, are fit to the reference object's point cloud, by maximizing the likelihood. This optimization is performed with the expectation-maximization algorithm. During the optimization process, the GMM centroids are forced to move coherently as a group, to ensure preservation of the topological structure of the point set. A concise mathematical description provided by [51] and adopted here is given below:

Let a mesh be considered as a point cloud S consisting of N points $x_i \in \mathbb{R}^3$

$$S \equiv s = \left[\boldsymbol{x}_1^T, \ldots, \boldsymbol{x}_N^T\right]^T \in \mathbb{R}^{3N} \tag{17.1}$$

Let the reference mesh, in our case the virtual shape s_{ref} be the reference to which we register all the instances $\{s_t\}_{t=1}^T$ to, where $T = 27$ for the scapulae and 28 for the humeri. The displacement function v for the mobile mesh is defined as:

$$\hat{\boldsymbol{s}}_t = \boldsymbol{s}_t + v\left(\boldsymbol{s}_t\right) \tag{17.2}$$

with s_t as the initial centroid positions, $\hat{s}_t$ and v, respectively, are obtained by minimizing the following energy function:

$$E\left(\hat{\boldsymbol{s}}_t\right) = -\sum_{s=1}^{S} \log\left\{\sum_{n=1}^{N} e^{-\frac{1}{2}\left\|\frac{x_s - y_n}{\sigma}\right\|^2}\right\} + \frac{\lambda}{2}\phi\left(v\right) \tag{17.3}$$

where:

$\phi(v)$	is a regularization term to ensure that the displacement field is smooth,
x_s	denotes a point on the reference mesh s_{ref} and y_n a point of the transformed mesh $\hat{\boldsymbol{s}}_t$, respectively,
S and N	refer to the number of points within the respective mesh.

The parameter λ determines the trade-off between data fitting and smoothness of the deformation field, which is to be decided empirically. Practically, there is another parameter β related to the standard deviation of the Gaussian filter (Green's function) ,which also had to be obtained empirically. The details of this are described further on.

17.3.3.2 Methods Employed and Motivation

The problem with standard point-matching algorithms like ICP for landmark creation is the restriction to similarity transformations [25]. Determination of corresponding points by proximity alone can lead to obviously wrong correspondences for large variations. Figure 17.10 shows the virtual shape and a single scapula instance after IMCP rigid registration. There is a lack of exact anatomical alignment and, thus, is not possible to assign correspondence between the two objects. This necessitates the use of non-rigid registration to get better anatomical alignment.

17.3.3.2.1 Coherence Point Drift Software

The CPD software toolbox was developed by Myronenko [50]. The MATLAB-based toolbox is for rigid, affine and non-rigid point set registration and matching and allows aligning two N-D point sets and recovering the correspondences. It is freely available for academic use. Several parameters have to be set depending on the size of the dataset. Recommendations are provided in the help file on how to choose the appropriate parameters. The regularization parameter λ and another value β (explained below), however, have to be found empirically. The parameters in the help file are listed below in Table 17.2 with the recommended values and those used by the authors together with the justification of the choices.

17.3.3.2.2 Testing for Values of β and λ

One instance of the scapulae and one instance of the humeri were non-rigidly registered to their respective virtual shapes to test for various combinations of β and λ parameters. In the interests of time, the maximum iteration value for these tests was 150. Only integer values of the extremes of the range were tested for each parameter while the other was held at the recommended value, and then the process was reversed. The assessment of each parameter that affects the registration was judged qualitatively by looking at the registration results. As a rule, a good registration should show a heterogeneous interspersion of the two surfaces (Figure 17.11).

The final values found empirically to perform good registration are in Table 17.3.

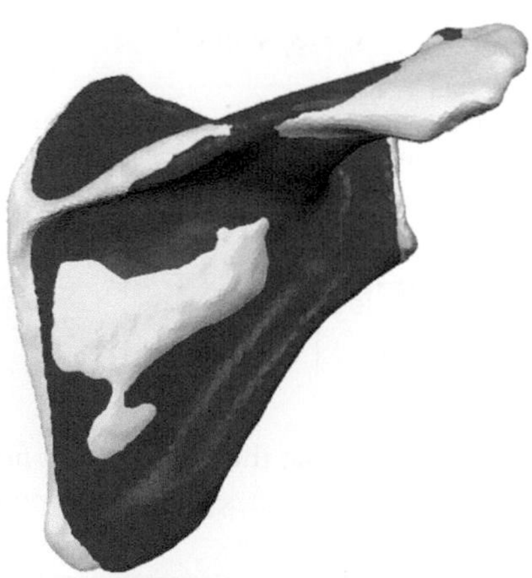

FIGURE 17.10 After the IMCP rigid registration the virtual shape (in red) and a scapula instance are not perfectly aligned and thus it is not possible to assign correspondence. This was the case of both the scapulae data and the humeri data.

TABLE 17.2 Parameters in the CPD Toolbox

Parameter	Function	Recommended	Value Used	Justification
B	This is the standard deviation of the Gaussian filter (Green's function) and gives the flexibility to control the range of filtered frequencies and thus the amount of spatial smoothness	3 (range is 1–5)	See Table 3.2	
λ	This is the regularization weight which determines the trade-off between data fitting and smoothness of the deformation field	2 (range is 1–5)	See Table 3.2	
Maximum iterations	The maximum number of iterations allowed	150	300	There is no need to restrict the time taken to convergence. Ideally the algorithm only needs to run once in the model creation process
Outliers	The weight of noise and outliers	0–1	0	None of the data was considered as noise as point to point correspondence of all the vertices was the aim
FGT	Use a Fast Gauss transform (FGT) for the matrix vectors commutations or to find the largest eigenvectors in case of large data problems	0 or 1	1	The datasets were very large with 15,000 vertices for each instance
Number of eigenvectors:	The number of largest eigenvectors to use allows utilization of Woodbury formula for fast linear system solution	$\sqrt{\text{ref}}\_\text{length}$	86	As per recommendation
Normalize	Normalization to unit variance and zero mean before registering	Yes	Yes	As per recommendation
Tolerance	The stopping criteria although its use did not seem to be useful			

17.3.3.2.3 Getting the Mean-Virtual Shape

As mentioned in the previous sections, the processed virtual shape exhibited some indentations which are attributed to mesh fusion process. The first use of non-rigid registration was to register each of the instances for both the scapula and the humerus on their respective virtual shapes. After this registration the correspondence across the whole dataset as determined by the CPD algorithm was obtained. The instance data used are the rigidly registered instances after IMCP. The results of this process for the humerus are shown in Figure 17.12. The effect of this procedure is removing the small indentations from the point wrap process and smoothing out the virtual shapes. The resultant shape will be called

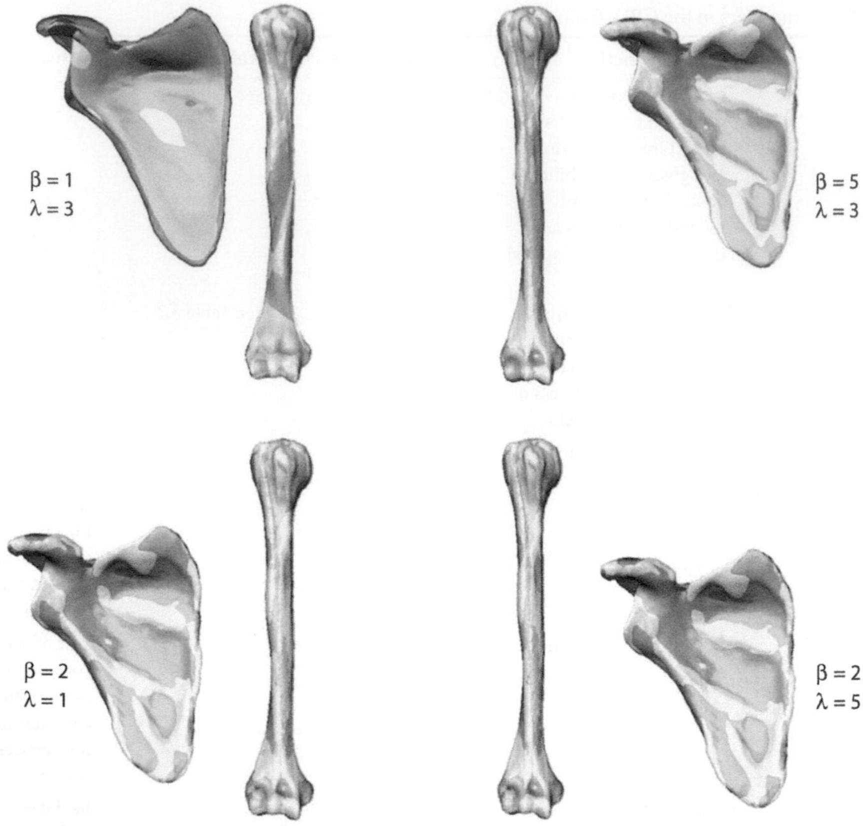

FIGURE 17.11 A qualitative assessment of the non-rigid registration for different values of λ and β for both the scapula and the humerus. The red surface in all cases is the respective virtual shape and the yellow surfaces are the same instances registered with different values of λ and β. In each case the transparency of the red surfaces was increased to clearly show the results of the rigid registration. The results show that λ value does not seem to affect the results of the registration to a great degree. The β value however, greatly affects the results of the registration. At a value of $\beta = 1$, the effect is to perform an over smoothing to the point that folds start occurring in the humerus and an unrealistically thin scapula emerges. At the other extreme, that of $\beta = 5$, there is very little deformation of the object toward the respective virtual shape meaning the original shapes virtually remain in the same position.

TABLE 17.3 λ and β Parameters Used in the CPD Toolbox for Non-Rigid Registration

Scapula	Humerus
$\beta = 2$	$\beta = 2$
$\lambda = 3$	$\lambda = 3$

the mean-virtual shape in each case. The process of finding correspondence across the whole sample is described in more detail in the next section.

17.3.3.2.4 Finding Correspondence Across the Sample

The non-rigid registration of each instance to the virtual mean shape is not perfect using CPD. After CPD with the virtual shape and all the instances having the same number of vertices, there was often

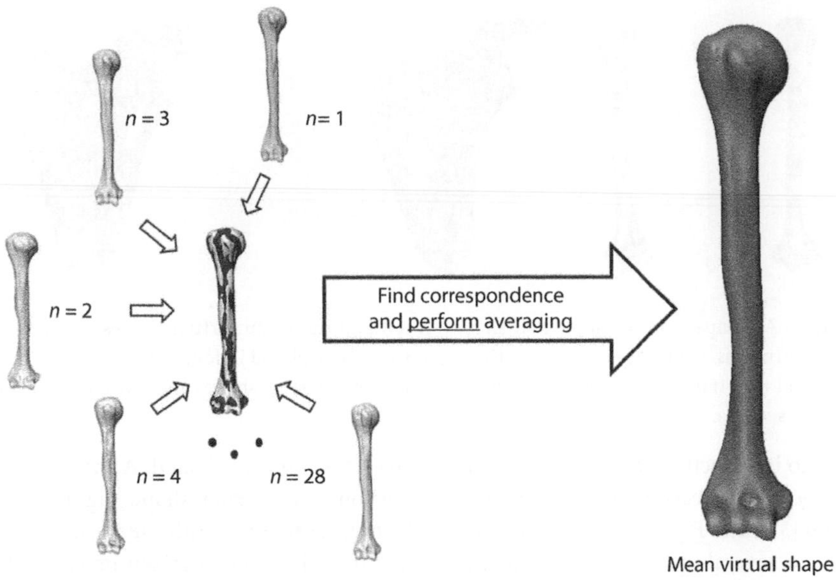

FIGURE 17.12 Obtaining the mean-virtual shape for the humerus. The original instances are non-rigidly registered onto the virtual shape and an average shape is calculated. The result is a smoothed out version of the mean virtual shape. The same procedure was followed for the scapula.

multiple assignment of correspondence of vertices in an instance to those in the virtual shapes. That is to say, it often occurred that two or more vertices on an instance corresponded to one vertex on the virtual shape. This was the justification for using half the number vertices on the virtual shape as on each of the instances. While this reduced the number of multiple assignments of correspondence, occasionally this still occurred. To further refine the process, in a situation where there was more than one corresponding value from the vertices of an instance to the vertices of the virtual shape, the multiple corresponding values were averaged. The net effect of this was to have a one-to-one correspondence of all the instance vertices and the virtual shape. However, the final number of corresponding vertices was often slightly less than the number of vertices of the virtual shape. In fact, Table 17.4 shows the actual number of corresponding vertices across the sample after non-rigid registration for both the scapula and the humerus.

As mentioned earlier, the results of the process for establishing MV shape are smoother virtual shapes for both the scapula and the humerus (Figure 17.13). It is important to note that the mean-virtual shapes are not very different in shape to their virtual shape counterparts. This smoothing process, however, does have a slight overall contraction component similar to that produced by a Laplacian filter. This is more evident in the scapula than for the humerus.

Based on a visual qualitative inspection, the mean-virtual shapes are considered to represent a smooth "intrinsic shape" for both the scapula and the humerus. They form an excellent reference shape for the next component of non-rigid registration in which they are deformed to each of the instances to in fact, sample the instances. Going back to the example of Schwartz's PhD thesis [37], the mean-virtual shape

TABLE 17.4 Number Vertices of the Original Virtual Shape Compared to Corresponding Points After Non-Rigid Registration

	Number of Vertices of Virtual Shape ($N_{virtual}$)	Number of Corresponding Vertices Across Sample ($N_{after\_non\_rigid}$)	%
Humerus	7500	7390	98.5
Scapula	7500	7057	94.1

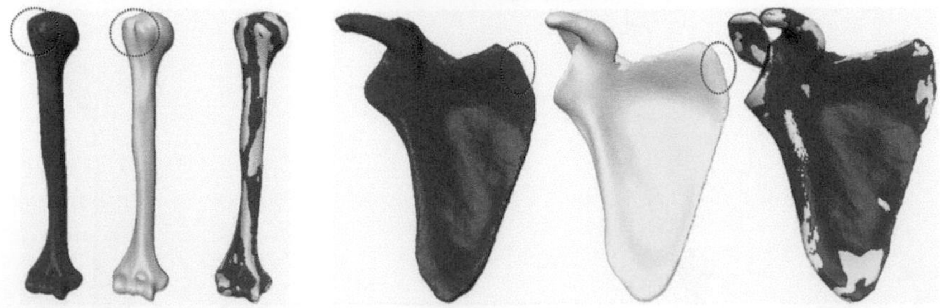

FIGURE 17.13 A comparison of the virtual shapes in red and the mean virtual shapes in green for both the scapula and the humerus. Highlighted areas are the regions which displayed indentations in the virtual shapes and after the non-rigid registration have been smoothed out. The mean virtual shapes also exhibit a slight contraction especially for the scapula.

is considered to be structural base for the species of objects being investigated. All the variation exhibited by the species of objects is variation from this mean consensus virtual shape. Figure 17.14 shows the example from [37], and Figure 17.15 shows an equivalent representation with the mean-virtual shape of the scapula and several instances. The difference is that in [37], the synthesized images of the navicular bone were of the same object (intra-object registration) but with artifacts and differences in pose, whereas in the present study (Figure 17.15), the objects are different scapulae (inter-object registration).

17.3.3.2.5 Deforming Mean-Virtual Shape to Each of the Instances

The next step taken was to deform the mean-virtual shape using CPD to sample each of the instances. Because the shape is the same and the mean-virtual shape is deformed for each instance, the non-rigidly registered samples are guaranteed to be in correspondence. In addition, since the mean-virtual shapes are constructed from the median consensus of all the objects, they represent an unbiased reference with points of the highest reliability across the sample. The mean-virtual shapes were remeshed to the same

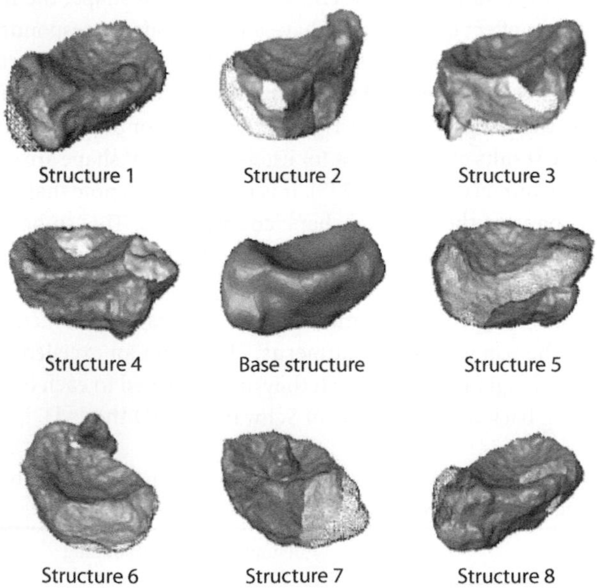

FIGURE 17.14 The previously shown example by [37]. Artificially created modifications (structures 1–8) of the base structured shown in the middle were generated based on a synthetic navicular (a bone in the human foot). The IMCP was able to correctly co-register all the structures and obtain the original base structure.

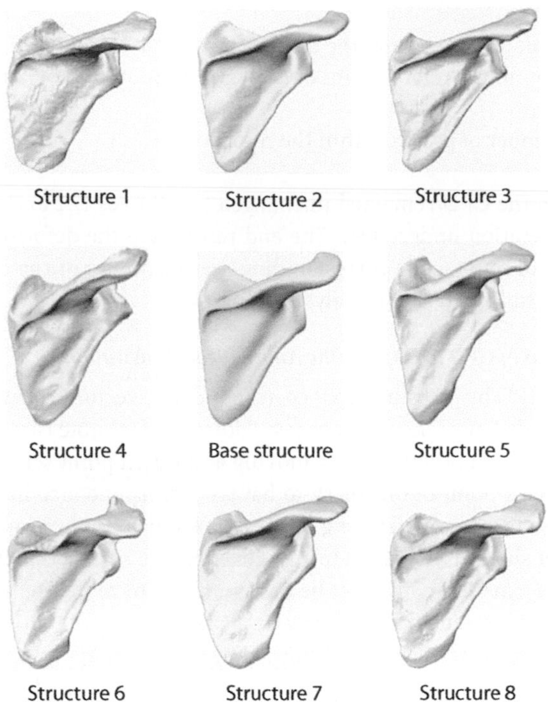

FIGURE 17.15 The scapula mean-virtual and some instances of scapulae. The same notation is used here to describe the objects as structures to show the multi-object equivalence of the work by Schwartz [52].

number of vertices as the original objects, that is, 15,000 vertices. To justify this, we confirmed that there is no longer any dependence on the posterior correspondence values calculated by the CPD algorithm to establish correspondence across the samples. This was the case in the previous non-rigid registration, since the same objects were being deformed non-rigidly to the instances. Another important point to note is the instances used in each case were the original post IMCP rigidly registered objects. That is to say, the sampling by the mean-virtual shape was on the original data.

A reformulation of the mathematical operation may serve to clarify the procedure. It is the reverse of the previous non-registration procedure. Let a mesh be considered as a point cloud S consisting of N points $x_i \in \mathbb{R}^3$

$$S \equiv s = \left[x_1^T, \ldots, x_N^T\right]^T \in \mathbb{R}^{3N} \tag{17.4}$$

Let the reference mesh, in this case, the mean-virtual shape be $s_{mean\_virtual}$ to which we register to the entire sample $\{s_t\}_{t=1}^{T}$ to, where $T = 27$ for the scapulae case and 28 for the humeri. The displacement function v for the mean-virtual shape is defined as:

$$\hat{s}_{mean\_virtual} = s_{mean\_virtual} + v\left(s_{tmean\_virtual}\right) \tag{17.5}$$

with $s_{mean\_virtual}$ as the initial centroid positions, $\hat{s}_{mean\_virtual}$ and v, respectively, are obtained by minimizing the following energy function:

$$E\left(\hat{s}_{mean\_virtual}\right) = -\sum_{s=1}^{S} \log\left\{\sum_{n=1}^{N} e^{-\frac{1}{2}\left\|\frac{x_s - y_n}{\sigma}\right\|^2}\right\} + \frac{\lambda}{2}\phi(v) \tag{17.6}$$

where:

$\phi(v)$ is regularization term to ensure the displacement field is smooth

x_s denotes a point on the reference mesh s_t and y_n a point of the transformed mesh $\hat{s}_{mean\_virtual}$, respectively

S and N refer to the number of points within the respective mesh.

The parameters used in the CPD remained unchanged (see Tables 17.2 and 17.3). For further clarity, Figure 17.16 shows the operation undertaken. The end results are the deformed mean-virtual shapes' vertices which form a set of one-to-one corresponding mesh samples of the original instances. These then form the data for the statistical shape analyses to follow.

17.3.3.2.6 Multiple Iterations of CPD to Improve the Anatomical Registration

In the implementation of CPD by [53], an output correspondence vector C for the corresponding points in the target object is provided after the CPD process. However, [54] note that CPD suffers from many-to-one correspondences between points of these moving and target point sets. In particular, there is no direct constraint to enforce a point of one mesh to have a correspondence in another. This makes the algorithm unable to achieve a proper matching in some specific configurations. Indeed, an analysis of the correspondence vector showed this many-to-one effect on the correspondence values. However, the correspondence vector information could still be used in order to refine the correspondence transference procedure. In the work presented here, a greater surface-to-surface intimacy between the mean-virtual shape and the original instances was established by iteratively applying CPD while also utilizing the correspondence vector information. After an iteration of CPD, intermediate positions of the current mean-virtual estimates' vertices were obtained by averaging the positions just after that deformation with the positions of corresponding vertices on the original instances obtained in the correspondence vector. These intermediate positions were used as the moving points in the next iteration of CPD. After several iterations, a better anatomical estimation of the original instance by the mean-virtual estimate was achieved while still guaranteeing topological correspondence. Figure 17.17 illustrates the process using the example of four vertices each for both surfaces.

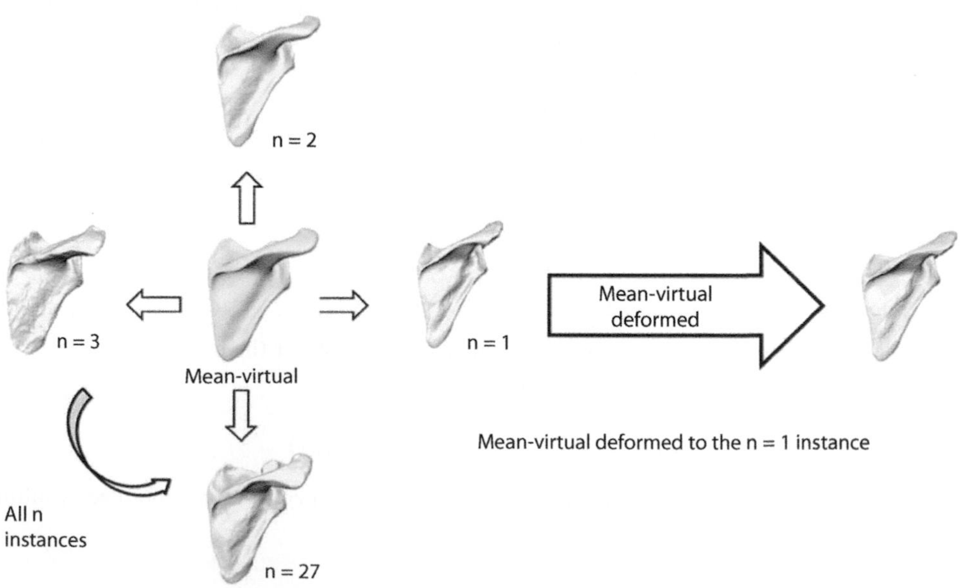

FIGURE 17.16 A schematic of the scapula mean-virtual shape deformed to the original scapula objects. On the right is an example of the mean-virtual shape deformed to instance 17.

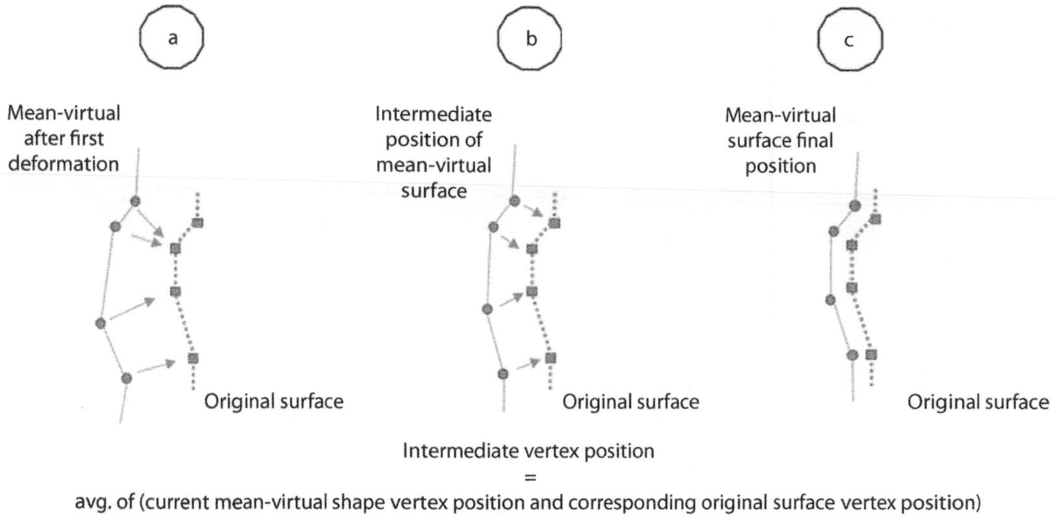

FIGURE 17.17 A four vertex illustration of the effect of iterative application of CPD on the mean-virtual shape toward the original shape. (a) The many to one correspondence often occurs with CPD, but in (b), an averaging of the vertex positions of the mean-virtual current estimate and the current "corresponding" vertex positions in the original surface occurs. This generally has the effect of creating a better intermediate surface for the next iteration of correspondence transference and results in (c), a more anatomically aligned mean-virtual estimation of the original surface.

Typical cycles of CPD application are shown in Figure 17.18 for a scapula and a humerus. The blue-dotted box in Figure 17.18b shows mean-virtual estimates after one iteration of CPD, and the red box shows the results after the final iteration (four in the work reported here).

17.3.3.2.7 Assessing Registration Quality

In general, CPD performs well, but a quantitative assessment of the registration results was required. This was because the described pipeline was very sensitive to how well the mean-virtual shape estimated the original instances. The point-to-surface distribution of distances between the mean-virtual estimate of an instance and the original instance was performed using several different distance measures between two triangulated surfaces. For each vertex of one surface, the closest point on the other surface was computed, and the following measures were computed: (1) Mean distance; (2) Standard deviation from the mean distance; (3) Root mean square (RMS) distance; (4) Hausdorff distance [55]. The original instance surface was always used as the reference surface as these calculations of surface-distance measures are asymmetric.

Figure 17.19 shows the results of distance measures computed for the scapula after one iteration of CPD (top left figure) and after four iterations of CPD (bottom left figure) on a logarithmic scale for the distances. The Hausdorff distance measure was halved on average by repeated iterations of CPD (Table 17.5). The rest of the distance measures showed, on average, a five-fold reduction in magnitude between one and four iterations of CPD. After multiple iterations of CPD the distance measures were all in the sub-millimeter range except for the Hausdorff distance.

17.3.3.3 Correspondence Quality

The objective of this test was to evaluate the correspondence transference. Only humerus data was used in the assessment of correspondence quality. This was purely because there were anatomical landmarks in the humerus sample that could be more consistently identified as extreme points of curvature than on

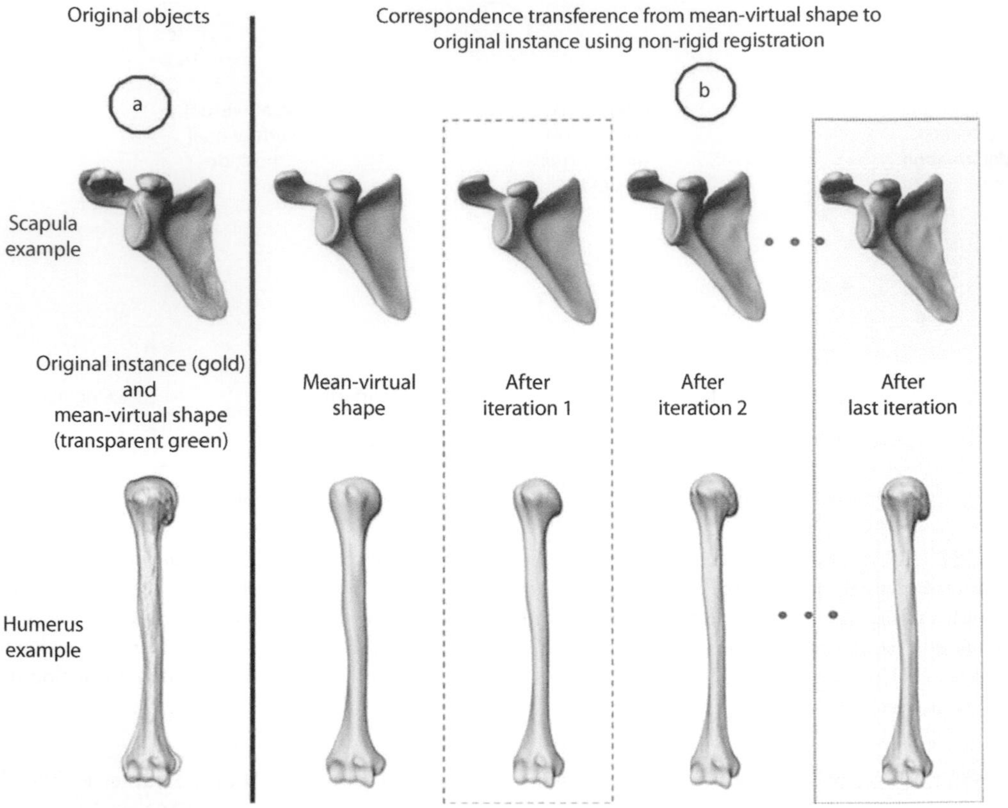

FIGURE 17.18 (a) Example original instances in gold with the mean-virtual shapes superimposed on top in transparent green. (b) This shows the transference of correspondence using the mean-virtual shape over several iterations for both the example scapula and humerus. The end results are mean-virtual estimates of the original instances. The blue-dotted box shows the result of one iteration of CPD (see above), and the red box shows the result of four iterations using the methodology developed.

the scapula data. Three easily identifiable landmarks, the medial and lateral condyles and lesser tubercle, were selected on a decimated mean-virtual shape of the humerus with 1,000 vertices (see Figure 17.20). The mean-virtual shape was deformed in the manner described previously toward the original humerus instances with 15,000 vertices. The indices of the vertices describing the three landmarks were then compared between the mean-virtual humerus and the mean-virtual estimations of each of the original instances. To obtain the exact vertex, indices of the landmarks in their general areas of location the following procedures were performed: (1) Obtain the vertex normals for each of the vertices; (2) Obtain the Gauss curvature of the vertices; (3) The landmarks are those with the highest gauss curvature and normals most perpendicular to the long axis of the humerus. The results of the correspondence assessment of three humerus landmarks are shown in Table 17.6.

17.3.3.4 Mean-Virtual Shape Stability

The importance of the mean-virtual shape was in obtaining an unbiased, underlying reference structure from which real world instances can be considered noisy perturbations of the mean-virtual shape. In light of this, the stability of the mean-virtual shape becomes critical, and an exhaustive test of how stable this shape is with varying number of instances will have to be conducted in future as part of assessing model quality. However, a preliminary test was performed using the dry bone scapula data as

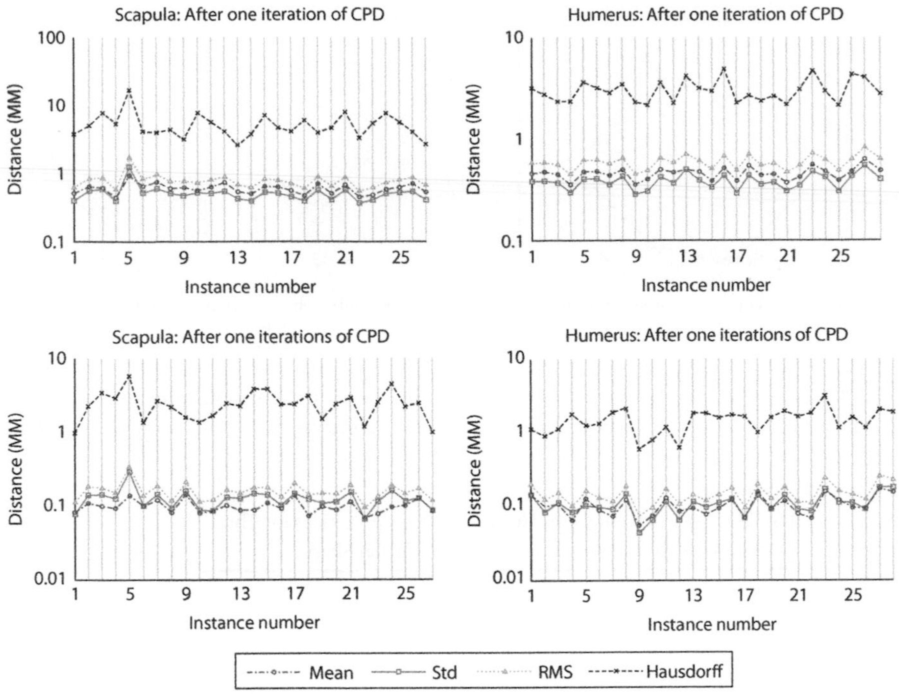

FIGURE 17.19 The point-to-surface distances distribution for the scapula; Top left: After a single CPD iteration, and bottom left: After 4 iterations of CPD using the method described. The point-to-surface distances distribution for the humerus; top right: After a single CPD iteration and bottom right: After 4 iterations of CPD using the method described.

TABLE 17.5 Summary of Registration Error Statistics, Average of All Measurements Across Samples

	Scapula (mm)		Humerus (mm)	
	1 Iteration	4 Iterations	1 Iteration	4Iterations
Mean	0.58	0.10	0.45	0.10
STD	0.51	0.13	0.38	0.11
RMS	0.78	0.16	0.59	0.15
Max	5.22	2.46	3.00	1.41
Average edge length	1.51		1.62	

scapulae from both sides of the human shoulder were available. The mean point-to-surface distances between a model created with 14 left-sided scapulae, and all 27 scapulae were calculated. The same was done for 13 right-sided scapulae, and the color-coded distance results for both sets are shown in Figure 17.21. The acromion process was the only anatomical region with a high level of variability. This is expected, as there have been reports of the variability of this structure across the population. In the work presented here, this is, in fact, one of the major modes of variation in the scapula model presented here (Figure 17.24). Thus, from this preliminary test, the all-instance mean-virtual shape was considered to be relatively stable.

The CPD performs well, considering the great variability in anatomy especially for the scapula. The CPD seems to be sensitive to the geometry of the reference shape. Since the algorithm has the constraint of coherent movement of points, some of the deformities on instance surfaces challenge this constraint. An instant observation is that features which do not appear on the virtual shape, such as

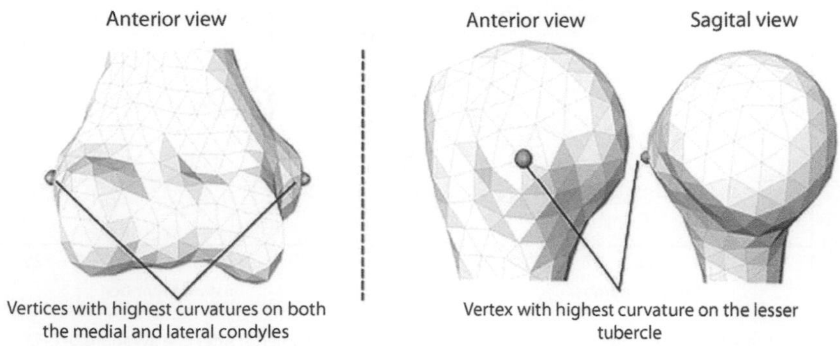

FIGURE 17.20 The landmarks chosen on the humerus whose vertex indices were compared across the sample. The landmarks were selected on the criteria that they are easy to identify across all samples as real anatomical landmarks.

TABLE 17.6 Percentage of Humerus Instance Vertices for Three Landmarks with the Same Index as Mean-Virtual Shape

Landmark	% Agreement
Lateral condyle	89
Medial condyle	100
Anterior lesser tubercle	100

Note: Only one landmark had less than 100% agreement in indices across the humerus samples and this was the lateral condyle with two instances which did not have the same index for this landmark.

some ossifications or areas of trauma, are not estimated well by the mean-virtual shape. A new correspondence test was introduced, in which the index of vertices of known extreme landmarks on the humerus was tracked. By using curvature and normals to determine the landmarks, the error associated with manual landmarking was eliminated allowing for a confident assessment of the tracking of landmark index across the humerus sample. In this regard, the modeling pipeline performed reasonably well in the ability to assign correspondence to the right landmarks across the sample. The humerus example used here was an ideal case, and such kind of testing would be more difficult for a structure like the scapula. However, the determination of correspondences between unstructured point sets is especially difficult when one shape features a certain structural detail and the other does not [56]. This is particularly true of the scapula, which exhibits, at least in our dry bone dataset, structural details in some areas on some sample instances and totally lacking in others, for example, the suprascapula notch. This often resulted in fairly low mean and RMS point-to-surface distances, but some considerably larger values for the Hausdorff distances (Figure 17.20) during the non-rigid registration process to establish correspondence. However, the multiple applications of CPD resulted in an improved registration and, therefore, more localized correspondence across the samples. The tracking of landmarks identified by the International Society of Biomechanics (ISB) could be advantageously used for a systematic assessment of the scapula correspondence.

17.4 Statistical Shape Models

17.4.1 Shape Model Treatment

The previous section dealt with the establishment of correspondence across all the sample population of scapulae and humeri. The MV "estimates" of each of the instances were now considered the primary

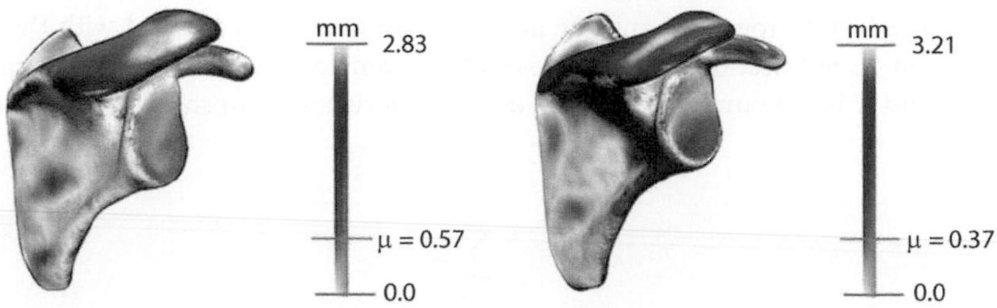

FIGURE 17.21 The left figure is a color-coded distance map of the difference between the mean-virtual shape created with left-sided scapula only and the one made using all the scapulae available. The right figure is the same for the right-sided scapula, although mirrored to the same side to facilitate a direct comparison.

data for the statistical shape analysis in this section. By modeling the variation across corresponding vertices, shape models of the bone structures under investigation could be observed. However, the data had to be rigidly aligned using Procrustes registration before building the model to establish true shape representations of the data.

17.4.1.1 Procrustes Superposition

Procrustes superposition (Figure 17.22) is a least-squares method that estimates the parameters for location and orientation that minimize the sum of squared distances between corresponding points on two configurations [57]. A mathematical description of an n-point/landmark shape in k-dimensions is a concatenation of each dimension into a $(k \times n)$ vector. Establishing a common coordinate reference with respect to scale, position and rotation, aligns all the objects in a given set to obtain true shape representation. The Procrustes method for comparing shapes is a method used to all n-corresponding points for N objects, with optimal superimposing of landmarks. The method minimizes the distances between corresponding landmarks by removing all registration effects using some constraints. Various minimization constraints exist, but the most popular is that which minimizes the sum of the squared distances between corresponding points [58].

In a multi-sample situation, the sample members are superimposed to a mean. However, there is usually no meaningful mean prior to the superposition. A solution is to use an iterative process to align shapes to a changing mean shape. Any of the shapes in the sample can be used as an initial mean and the others aligned to it. A new mean from the aligned configurations may then be calculated as the arithmetic average location of individual vertices in the sample and scaled to unit centroid size. This process is guaranteed to produce monotonically decreasing sum-of-squared deviations of the sample configurations around the estimated mean [57]. Convergence is declared when the mean shape does not change significantly between iterations [60].

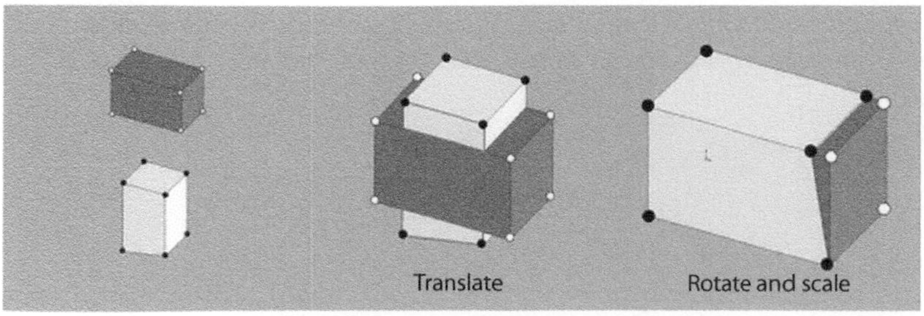

FIGURE 17.22 Shape alignment using Procrustes superposition. (From Mutsvangwa, T., *Human Biology*, University of Cape Town: Cape Town, 2009.)

The following is adopted from [59], and the notation should not be confused with the notation in the previous section on registration. Mathematically, for a sample of corresponding configurations S_i, where $i = 1,\ldots,N$ and N is the number of objects, $\bar{\alpha}$, the Procrustes mean shape after convergence can be estimated by:

$$\bar{\alpha} = \frac{1}{N}\sum_{i=1}^{N}\alpha_i \tag{17.7}$$

and gives the Procrustes mean vertices ($\overline{\alpha_{jx}}, \overline{\alpha_{jy}}, \overline{\alpha_{jz}}$), where $j = 1,\ldots,,n$ (n is the number of corresponding points). The full Procrustes fit vertices S_i^P are found by fitting each shape S_i to the Procrustes mean $\bar{\alpha}$ using, for example, a least squares method [58], and, thus, each S_i^P has vertices: $\left(S_{ijx}^P, S_{ijy}^P, S_{ijz}^P\right)$. Procrustes residuals S_i^{PR} [61] are the difference between the full Procrustes fit vertices and the Procrustes mean vertices and are here represented as: $\left(S_{ijx}^{PR}, S_{ijy}^{PR}, S_{ijz}^{PR}\right)$. Procrustes residuals have statistically useful properties [62] and can be used in principal component analysis to explore shape variability [63]. However, a tangent space projection serves to facilitate easier statistical development.

17.4.1.2 Shape Space and Tangent Space

The Procrustes alignment procedure brings the shape set into shape space. Shape space is the set of all possible shapes in question, for example, scapula shapes and humerus shapes. Formally, the shape space $\Sigma_k^n S$ is the orbit shape of the non-coincident n-point set configurations in the $\mathbb{R}^k$ under the action of the Euclidean similarity transformations. The dimensionality of this space is:

$$M = kn - k - 1 - \frac{k(k-1)}{2} \tag{17.8}$$

If a relationship between the distance in shape space and Euclidean distance in the original plane can be established, the set of shapes actually forms a Riemannian manifold [64] containing the object class in question (e.g., scapula or humerus). This is also denoted as the Kendall shape space [60]. The Procrustes process produces a curved high dimensional space. Tangent space projection removes the nonlinearities induced by the Procrustes analysis. This facilitates the use of standard multivariate methods to model shape variation [59,60]. After projection into tangent space, the Procrustes residuals can be called Procrustes tangent vertices and these form the data for PCA.

17.4.1.3 Principal Component Analysis

Procrustes tangent vertices may be used to explore shape variability, just the same as Procrustes residuals, with the added benefit of being more amenable to linear multivariate methods [63]. Using the notation developed in Mutsvangwa, 2009 from [59], mathematically, $(k \times n)$ Procrustes tangent vertex variables can describe the variability of the object. Various statistical procedures can be carried out to investigate shape variation. One of them is PCA. PCA is a procedure that can decrease or summarize the $(k \times n)$ variables into uncorrelated linear combinations of these variables. The *rth* principal component PC(r) is given by:

$$PC(r) = \sum_{j}^{n}\left(\gamma_{rjx}S_{jx}^{PR} + \gamma_{rjy}S_{jy}^{PR} + \gamma_{rjz}S_{jz}^{PR}\right) \tag{17.9}$$

where the weights γ_{rjx}, γ_{rjy}, and γ_{rjz} describe the directions of variation in each Procrustes coordinate about the mean shape. A set of the same shape class of biological objects will always have some degree of inter-point correlation [60], especially since they belong to the same biological entity [58]. Thus, there exists a shape representation between points that accounts for the correlation between points. This can be used to reduce dimensionality in PCA. There are as many principal components as there are corresponding vertices. The principal components have the following properties [58]:

- All components are orthogonal to each other and so are statistically uncorrelated.
- Part of the variability of the sample is represented by each component in decreasing order, starting from the largest variability being represented by the first component and the second largest variability by the second and so on.
- Every component is a linear combination of the original variables.

To visualize the pattern of shape variability represented by each principal component, the mean shape can be warped by moving points according to the weights on the principal component. The decreasing order of significance of the components in accounting for variability means that only some need to be retained to account for a significant part of shape variability. The mean shape configuration of a sample has, by definition, all principal components equal to 0. For PC(r), shapes with the following vertices can be plotted: $\left(\overline{\alpha_{jx}} + c\sqrt{\lambda_r}, \gamma_{rjx}, \overline{\alpha_{jy}} + c\sqrt{\lambda_r}, \gamma_{rjy}, \overline{\alpha_{jz}} + c\sqrt{\lambda_r}, \gamma_{rjz}\right)$. The variance λ_r is captured by PC(r) so that configurations between −3 and +3 standard deviations either side of the mean shape may be obtained. A succinct, low-dimensional generative model can, thus, be derived from which valid shape new instances S_{new} can be generated:

$$S_{new} = \bar{\alpha} + \sum_{r=1}^{c} \lambda_r \gamma_r \tag{17.10}$$

17.4.2 Probabilistic Principal Component Analysis

This PCA model assumes that all the probability mass is concentrated on the subspace spanned by the principal components. This means that the probability of an object that does not strictly lie in this subspace is zero, which may not be the case in practice, since an object might be a valid and likely instance of the class and still not lie exactly on that subspace [31]. This may, on one hand, happen, since this subspace is only approximated from a finite number of examples, but also because the data itself is usually noisy. To alleviate this problem, Probabilistic PCA (PPCA) is a suggested substitute [31]. PPCA provides a well-defined probabilistic interpretation, by assuming a small amount of noise on the examples. The new generative model then becomes:

$$S_{new} = \bar{\alpha} + \sum_{r=1}^{c} \lambda_r \gamma_r + \varepsilon \tag{17.11}$$

where $\varepsilon \sim \aleph(0,\sigma^2)$. By virtue of the added noise term, this model now defines a valid probability distribution on the whole of high-dimensional subspace of Procrustes tangent vertices. The probability of an instance now depends on the distance from this subspace, that is, the objects that are far away are less likely than those instances that are closer to the space. It can be shown that when σ^2 goes to 0 then a standard PCA model is obtained [31].

17.4.2.1 Methods Employed

17.4.2.1.1 Data

The data was in the form of the mean-virtual estimates of the original instances. These are all in correspondence, as they were generated from a deformation of the mean-virtual objects, as described in the previous section. The reference shapes were the mean-virtual shapes themselves, which define the discretization and mesh topology.

17.4.2.1.2 Rescaling the Mean-Virtual Shape

The mean-virtual shape is still at a very small size (relatively) compared to the estimates because of the normalization, with centroid size at the beginning of IMCP. This may have an effect on the model created, depending on the number of instances that are used to create the model. This is because size is already a dominant mode in the PCA results, and using a mean-virtual shape, which is really small, can make this effect even more pronounced if there are not many instances in the model. In order to mitigate against this, it is best to bring the mean-virtual shape to more or less the same scale of size as the instance estimates. This was easily done in Amira using the alignSurface module. One of the estimates was taken as a reference, and the mean-virtual shape as the moving object in a rigid alignment, using correspondence as a control, and uniform scale for the final size of the mean-virtual shape.

17.4.2.2 With- and Without-Size SSM Corrections

17.4.2.2.1 With Size Correction

The estimates were pre-aligned to the rescaled mean-virtual shape. This was also done in Amira using the alignSurface module. The "reference_surface" was the rescaled mean-virtual shape. This procedure also completed the pre-alignment of all the estimates to the mean-virtual shape. After all the estimates were aligned, then the estimates and the rescaled mean-virtual shape constituted the input for the with-size model creation in STATISMO.

17.4.2.2.2 Without-Size Correction

In order to remove size from the models, all the estimates were rescaled to the rescaled mean-virtual shape in Amira using the alignSurface module. Rigid registration of the estimates on the mean-virtual shape using correspondence and constrained to uniform scale achieved the required size. The "reference_surface" was the rescaled mean-virtual shape. This also achieved the pre-alignment of all the estimates to the mean-virtual shape. After all the estimates are aligned, then all of them and the rescaled mean-virtual shape constituted the input for the with-size model creation in STATISMO.

17.4.2.2.3 Scapula

17.4.2.2.3.1 The 27 Sample Scapula Model In the interests of brevity, only selected results are presented in this section. These were chosen to highlight specific concepts of the results of the modeling process. In the "with-size" model of the scapula, the contribution by the first principal component alone claims close to 60% of total variation (Figure 17.23). The first seven principal components account for 90% of the variation. For the without-size model for the scapula, the first component contributes 27% of the total variation. The first 10 principal components account for 90% of the total variation. Figure 17.24 shows the without-size patterns of variation of some of the significant principal components of the without-size scapula model. For the scapula, the first three principal components are shown.

17.4.2.2.3.2 The 81 Sample Scapula Model As mentioned in Section 17.2.1.1, additional scapulae data was provided by Imascap. These data were used (54 healthy scapulae) to develop a larger instance-shape model. The first 10 principal components account for 90% of the variation (Figure 17.25). This represents a very compact model, although there are no reports in the literature to compare these results. The patterns of variation of the big model followed the trend of the 27-sample model, but were more pronounced.

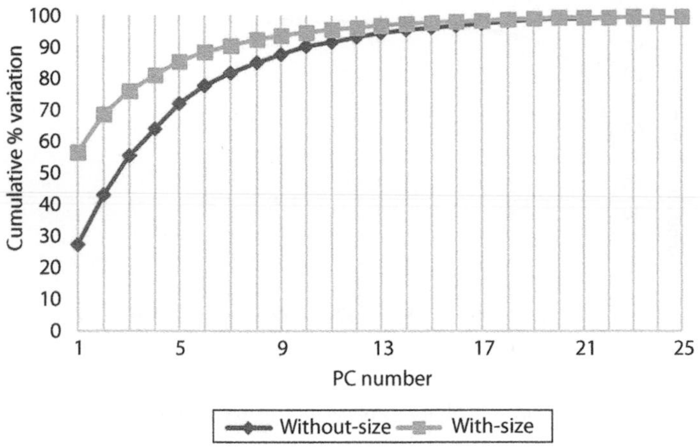

FIGURE 17.23 Cumulative variation contributions of each principal component for the with-size and without-size scapula SSM.

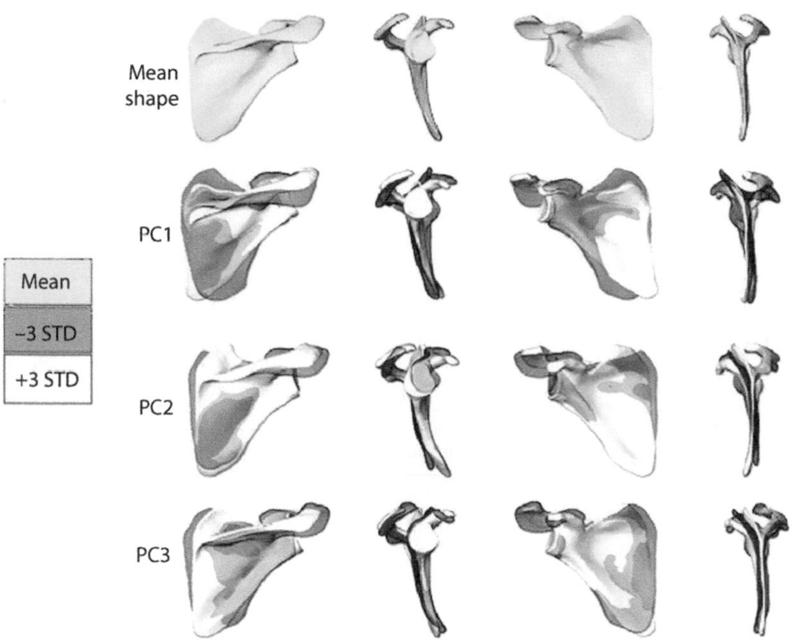

FIGURE 17.24 The pattern of shape variation for the first three principal components of the without-size models for the scapula.

17.4.2.2.4 Humerus

17.4.2.2.4.1 The Humerus Model For the humerus with-size model, the first principal component alone claims 70% of total variation (Figure 17.26). The first two principal components account for more than 90% of the variation. Unlike the scapula, the first component of the without-size humerus model has a relatively large contribution to the total variation, at close to 70% (Figure 17.27). The first four principal components represent close to 90% of the total variation. As expected, the with-size models, the first principal component that corresponds with the largest variation contributions represent scale-related changes in the size of both bone structure.

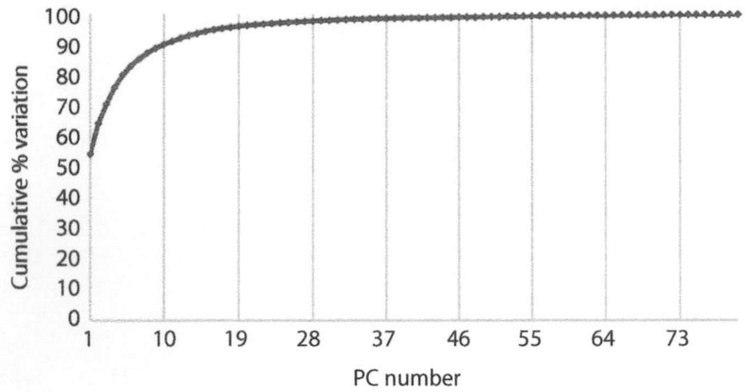

FIGURE 17.25 Cumulative variation contributions of each principal component for the without-size 81 sample scapula SSM.

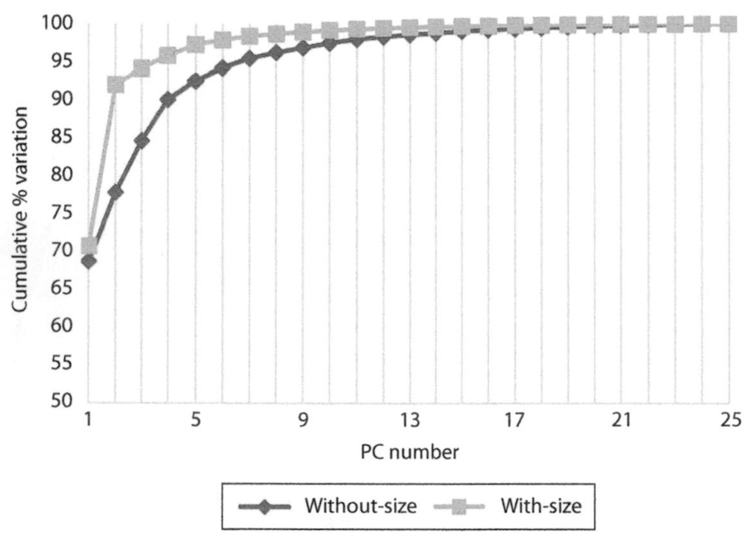

FIGURE 17.26 Cumulative variation contributions of each principal component for the with-size and without-size humerus SSM.

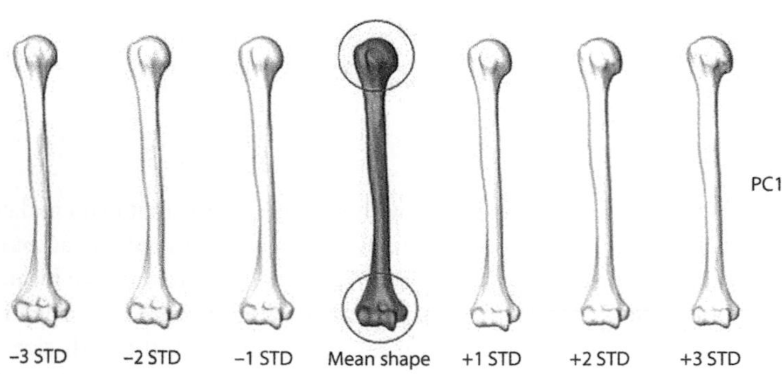

FIGURE 17.27 Without-size patterns of variation of the first principal component of the humerus model.

17.5 Theoretical Evaluation - SSM Robustness

To facilitate the comparison of the proposed methodology and other state-of-the-art methods, publicly available 42-segmented hippocampi data from [65] was obtained, as in [28]. The well-established measures of compactness, specificity and generality were used to measure the quality of shape models [26,28,30,56,66]. Briefly, compactness C measures a model's ability to use as few parameters as possible to cover the same variance and is calculated as:

$$C(R)=\sum_{r=1}^{R}\lambda^{r} \tag{17.12}$$

where:

λ^r is the rth largest eigenvalue
$C(R)$ is the cumulative variance of the Rth principal component

The generality $G(R)$ measures a model's ability to represent unseen instances of the class of object. This property is fundamental, as the goal of building a model is to fit the model to new examples. If a model is over-fitted to the training set, it will be unable to generalize to unseen examples. Generality is given by:

$$G(R)=\frac{1}{N}\sum_{i=1}^{N}\left|s_i'(R)-s_i\right|^2 \tag{17.13}$$

where:

N is the number of instances
$s_i'(R)$ is the best model reconstruction of the instance s_i from the model built excluding s_i with R principal components

Finally, specificity measures the model's ability to generate instances of the species of objects similar to those in the training set and is given by:

$$S(R)=\frac{1}{N}\sum_{j=1}^{N}\left|srandom_j(R)-s_j'\right|^2 \tag{17.14}$$

where:

$srandom_j(R)$ are shape examples randomly generated using R principal modes
s_j' is the nearest member of the training set to $srandom_j(R)$

For comparative purposes, in the work presented here, the range of the probability of the random sample $\mathbf{P}(srandom_j)$ had to be greater than probability of a sample drawn from within three times the standard deviation of the first principal component in a full instance model. This limited valid samples to those in the same range used in [28].

Further, the above measures were evaluated using the same metrics of RMS and Hausdorff distances between instantiated model objects and target objects. The results of the evaluation were then superimposed on those extracted from [28]. Through this process, a direct comparison of model quality between the Group-wise GMM and EM-ICP (both selected for best results in the mode quality evaluation) with the method presented here was possible on the same data. The results of all three methods are presented in Figure 17.28a–d. The model developed with our current method has a better generality for both distance metrics than those from the Group-wise GMM and EM-ICP methods. In addition, the current method

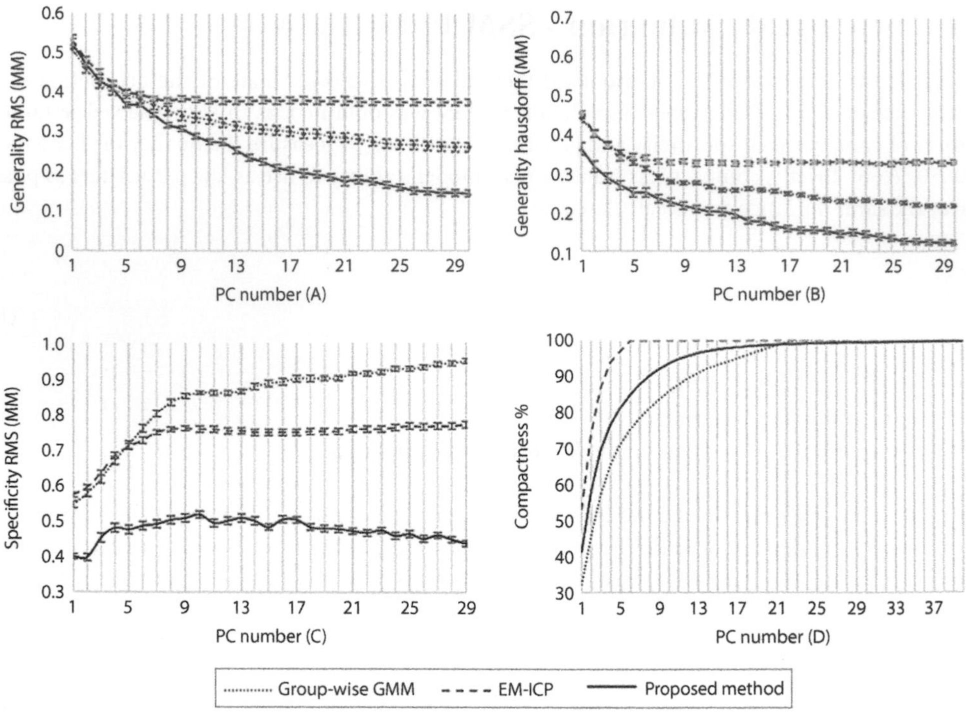

FIGURE 17.28 (a) Generality (RMS), (b) generality (Hausdorff), (c) specificity and finally, (d) compactness of the models from three methods for the hippocampi data. The standard errors were calculated as described in Davies et al. 2015.

produces a hippocampus model with a better specificity than the other two. In terms of compactness, the current method produced a model which lies in between the two other methods. The methodology presented here, thus, outperformed the other two methods in generality and specificity and performed slightly better than Group-wise GMM and slightly worse than EM-ICP in the test for compactness.

17.6 Clinical Evaluation - Validity Using Augmented SSMs

17.6.1 Need for Clinical Validity and Current Literature

Clinicians typically use external palpation strategies, along with range of motion tests, to determine the extent of the injury/deformity [67,68]. Surgeons on the other hand rely mostly on virtual palpations using 3D imaging for pre-surgical planning [69–71]. In events such as recurrent posterior shoulder dislocation due to excessive glenoid retroversion, glenoid osteotomy is the choice of intervention. However, surgical processes for glenohumeral deformity corrections are becoming more and more computerized [72] and, thus, need patient-specific shoulder joint models created from imaging modalities in order to analyze the morphology, as well as to pre-determine the course of action during the actual surgical procedure. In all these processes, developing a quick, accurate, reliable, and patient-specific shoulder model becomes very important [73,74]. With the advent of statistical methods in medical image analysis, it is now possible to build and use statistical shape and appearance models for biological shapes [3,10]. However, the computational ability to use these models in a clinical setting for patient-specific outcomes is not yet explored.

Previously, SSM has been used in a few biomechanical studies involving knee and hip joints [75,76], but has not been used in shoulder joints. Previous research on shoulder SSMs was focused on building methodological pipelines to build SSMs and did not evaluate these models in terms of their clinical robustness and reliability [19,28,29,77,78]. Computational robustness of an SSM can be checked through the measures of specificity, generality and compactness [19,79,80]. However, this does not automatically prove the clinical robustness and reliability of the model as these measures do not specifically target clinically relevant areas. Clinical validity on the other hand can be determined by evaluating the SSM validation in the region of clinical relevance (e.g., anatomical landmarks) [67]. Thus, in order to prove their clinical use, the tests of model robustness and clinical robustness are complementary and must be mutually exhaustive.

This study focused on illustrating the clinical validity of healthy adult scapula SSMs augmented with anatomical landmarks. As a reminder, for the scope of this chapter "clinical validity" is defined as the accuracy with which the SSM can predict the goodness of correspondence in the manually selected, clinically relevant anatomical landmarks, as defined in this study. First, a scapula SSM was built using previously published SSM building pipeline [19]. Computational robustness of the scapula SSM built using this pipeline had already been proven through the measures of generality, specificity and compactness [19]. A set of anatomical landmarks was selected from the clinically relevant regions for each of the training base scapulae by multiple observers. The intra- and inter-observer reliability of selecting the anatomical landmarks was then determined. The SSM was then augmented using a set of mean anatomical landmark positions transferred from anatomical landmark sets. Clinical validity of the SSM was determined by measuring the quality of correspondence of the augmented scapula SSM with either its training sample or an outside sample using anatomical landmark locations.

17.6.2 Anatomical Landmark Selection

As described earlier, the scapular bone data consisted of two sets, one with 27 bones and another with 54 healthy bones. The test of anatomical validity required manual selection of clinically significant anatomical landmarks on scapula SSM and checking its correspondence quality with scapulae models within or outside the training set. However, scapula SSM, being the probabilistic representation of all the instances, was too smooth to exhibit the correct location of anatomical landmarks, and, thus, selecting landmarks directly on the SSM was not realistic. Instead, landmarks were selected on the MV estimates of the training scapulae models, and later, the mean of these landmarks was transferred to MV shape to build augmented SSM(s). This was achieved by selecting a set of 16 anatomical landmarks on the MV estimates of the scapula models. Specifically, (1) six landmarks were located on glenoid rim surrounding glenoid cavity (Figure 17.29a), (2) four landmarks were located on the medial and superior edges of subscapular fossa (scapula blade) (Figure 17.29b), (3) four landmarks were located on the acromion (Figure 17.29c), and (4) one landmark each was located on the coracoid process and the notch made by scapular spine with supraspinous fossa (Figure 17.29d). Reliability of this manual selection process was determined using intra- and inter-observer reliability tests. A landmark selection guide (Appendix 17B) was prepared in order to train future users and to guide the observers of this study.

Landmarks were selected on all 27 training scapulae that were used to build the SSM (internal instances). Based on the methodology, landmarks were also selected on 20 external scapulae (randomly selected from the set of 54 healthy scapulae) that were not included in building the SSM (external instances). For the internal instances, MV estimates from each scapula were used for the landmark selection as they were already in correspondence with the SSM. For external instances, each instance was first brought in correspondence with the SSM by running the CPD algorithm, and then the estimated model of that instance was used for anatomical landmark selection. To ensure that each MV estimate truly represented its original instance, a Hausdorff distance [55] between original instance and its MV estimate was calculated. For the internal instances, if the Hausdorff distance was above 1.0 mm, the location of maximum distance was determined to make sure it did not fall within the vicinity of 16 anatomical landmarks. For 20 external instances, any MV estimate that either has a maximum Hausdorff distance above 2.0 mm at any location or has Hausdorff distance above 1.0 mm in the vicinity of anatomical landmarks was discarded from this study.

For 27 the internal instances, the maximum Hausdorff distance between an original instance and its MV estimate was found to be 1.99 mm, and a total of five MV estimates had Hausdorff distance greater than

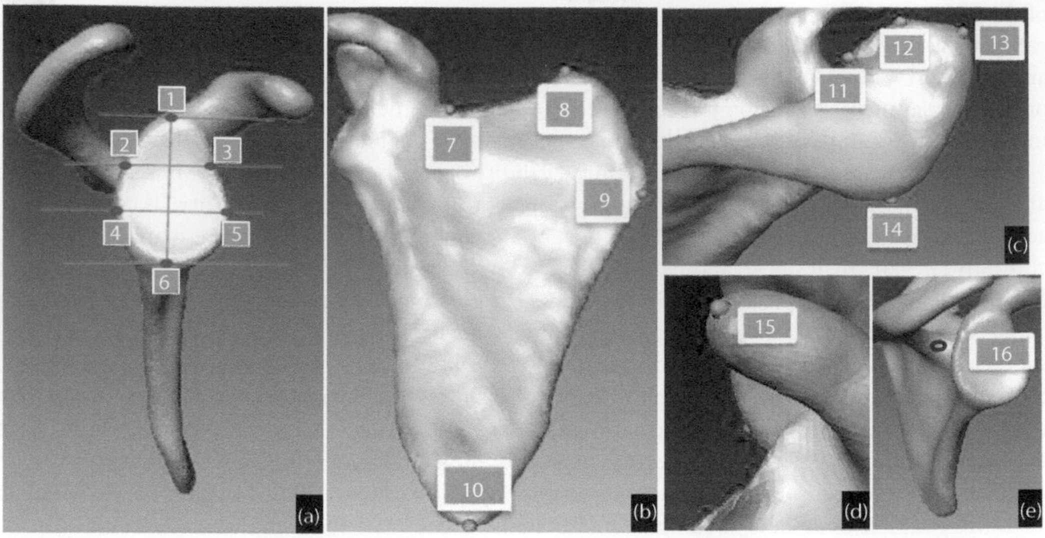

FIGURE 17.29 Anatomical landmark locations on the scapular bone. Sixteen landmarks were manually selected by multiple observers in clinically relevant regions of scapula. These include: (a) six landmarks on the glenoid rim, (b) four landmarks on the medial, superior and inferior edges of subscapular fossa, (c) four landmarks on the acromion, (d) one landmark on the coracoid process, and (e) one on the notch made by scapular spine with supraspinous fossa. A detailed landmark-selection guide is made available as a supplementary material (Appendix 17B).

1.0 mm. However, none of the maximum Hausdorff distance locations were within the vicinity of landmarks, and, thus, all 27 internal MV estimates were included in the internal validity measures. Similarly, for 20 external MV estimates, the maximum Hausdorff distance was found to be 1.86 mm and total three MV estimates had Hausdorff distance greater than 1.0 mm. After carefully checking the locations of maximum Hausdorff distances, all 20 external MV estimates were included in the external validity measures.

17.6.3 Intra- and Inter-Observer Reliability

For intra-observer reliability, anatomical landmarks were acquired on all 27 internal instances by three independent observers. Each observer followed the landmark selection guide (Appendix 17A) and repeated the landmark selection process five times. In order to minimize the recall bias, a time interval of 60–72 h was allowed to expire between consecutive landmark selection trials, and the order of instances was also randomized within each trial. Thus, each of the five datasets of landmark selection included x-, y-, and z-coordinates of 16 landmarks on 27 scapulae. Intra-observer reliability was defined by Intraclass Correlation Coefficients (ICCs), using a two-way analysis of variance (ANOVA) and considering the choice of observer as fixed effects [81]. For inter-observer reliability measurements, five independent observers (3 engineers, 2 clinicians) acquired anatomical landmarks on 27 internal instances. Inter-observer reliability was also defined by ICCs, using a two-way mixed effects (choice of observers) ANOVA [81]. Furthermore, standard error of measurement ($SEM = SD \times \sqrt{(1 - ICC)}$, where SD is standard deviation of whole set of measures) was also quantified for each set of landmark measurements. All the ICCs were obtained using Statistica Advanced module (StatSoft, Inc., Paris, France). All the observers successfully completed the reliability tests with the help of the anatomical landmark selection guide (Appendix 17B). Furthermore, all the intra-observers repeated the landmark selection process by respecting the recommended time gap and randomly selecting scapulae within each trial. A high degree of intra- and inter-observer reliability was found for all x-, y- and z-coordinates (Tables 17.7 and 17.8).

The ICC for inter-observer reliability ranged from 0.81 to 0.99 for medial-lateral direction, 0.89 to 0.99 for anterior-posterior direction, and 0.88 to 0.99 for superior-inferior direction

TABLE 17.7 Inter-Observer Reliability Interclass Correlation Coefficients (ICCs) in X, Y, and Z Coordinates of Anatomical Landmarks for All the Five Observers Combined

Landmark	X: Medial-Lateral	Y: Anterior-Posterior	Z: Superior-Inferior
1	0.98	0.95	0.97
2	0.87	0.97	0.98
3	0.92	0.97	0.99
4	0.89	0.98	0.95
5	0.91	0.99	0.96
6	0.98	0.96	0.99
7	0.99	0.98	0.92
8	0.99	0.99	0.99
9	0.84	0.99	0.99
10	0.99	0.99	0.98
11	0.81	0.89	0.92
12	0.92	0.92	0.88
13	0.97	0.97	0.98
14	0.99	0.97	0.91
15	0.98	0.97	0.99
16	0.91	0.93	0.94

TABLE 17.8 Intra-Observer Reliability Interclass Correlation Coefficients (ICCs) Values in X, Y, and Z Coordinates of Anatomical Landmarks for All the Three Observers

	X: Medial-Lateral			Y: Anterior-Posterior			Z: Superior-Inferior		
Landmark	O – 1	O – 2	O – 3	O – 1	O – 2	O – 3	O – 1	O – 2	O – 3
1	0.91	0.95	0.94	0.87	0.91	0.91	0.89	0.94	0.86
2	0.81	0.75	0.71	0.93	0.92	0.92	0.94	0.91	0.91
3	0.84	0.77	0.85	0.95	0.88	0.92	0.88	0.93	0.93
4	0.92	0.70	0.89	0.86	0.97	0.97	0.98	0.88	0.92
5	0.77	0.84	0.87	0.91	0.98	0.97	0.85	0.92	0.90
6	0.80	0.95	0.94	0.87	0.91	0.94	0.83	0.99	0.98
7	0.84	0.97	0.97	0.84	0.95	0.90	0.80	0.80	0.84
8	0.99	0.99	0.99	0.99	0.99	0.98	0.97	0.98	0.98
9	0.90	0.85	0.94	0.97	0.96	0.94	0.94	0.96	0.98
10	0.92	0.97	0.96	0.97	0.98	0.97	0.95	0.95	0.95
11	0.67	0.73	0.76	0.72	0.83	0.80	0.88	0.87	0.88
12	0.92	0.89	0.93	0.81	0.91	0.91	0.78	0.84	0.89
13	0.94	0.94	0.94	0.85	0.92	0.85	0.92	0.91	0.95
14	0.97	0.98	0.93	0.93	0.97	0.87	0.90	0.95	0.80
15	0.97	0.97	0.94	0.95	0.94	0.93	0.94	0.98	0.99
16	0.83	0.88	0.82	0.84	0.74	0.84	0.80	0.87	0.87

(Table 17.7). The ICC for intra-observer reliability for all the observers ranged from 0.67 to 0.99 for medial-lateral direction, 0.72 to 0.99 for anterior-posterior direction, and 0.78 to 0.99 for superior-inferior direction (Table 17.8). The lowest ICC measure was found to be 0.81 (Table 17.7) for selecting landmark 11 in medial-lateral direction (x-coordinates) for inter-observer reliability measures and 0.67 (Table 17.8) for selecting landmark 11 in medial-lateral direction for intra-observer reliability measures. The SEM was extremely low and ranged from 0.03 to 0.30 mm for all intra- and inter-observer measurements (Tables 17.9 and 17.10). Specifically, inter-observer SEM ranged from 0.03–0.29 mm for medial-lateral direction, 0.03–0.16 mm for anterior-posterior direction, and 0.03–0.19 mm for superior-inferior direction (Table 17.9). The intra-observer ICC for all the observers ranged from 0.05–0.30 mm for medial-lateral direction, 0.05–0.24 mm for anterior-posterior direction, and 0.0–0.30 mm for superior-inferior direction (Table 17.10).

TABLE 17.9 Inter-Observer Standard Error of Measurement (SEM) in Millimeter

Landmark	X: Medial-Lateral	Y: Anterior-Posterior	Z: Superior-Inferior
1	0.03	0.07	0.05
2	0.16	0.05	0.04
3	0.10	0.05	0.03
4	0.15	0.04	0.09
5	0.11	0.03	0.07
6	0.05	0.08	0.03
7	0.03	0.04	0.09
8	0.07	0.08	0.08
9	0.29	0.05	0.03
10	0.04	0.05	0.11
11	0.20	0.10	0.16
12	0.11	0.16	0.18
13	0.07	0.10	0.07
14	0.09	0.05	0.19
15	0.08	0.08	0.05
16	0.08	0.07	0.07

TABLE 17.10 Intra-Observer Standard Error of Measurement (SEM) in Millimeter for Three Observers

	X: Medial-Lateral			Y: Anterior-Posterior			Z: Superior-Inferior		
Landmark	O − 1	O − 2	O − 3	O − 1	O − 2	O − 3	O − 1	O − 2	O − 3
1	0.08	0.05	0.06	0.10	0.08	0.08	0.09	0.07	0.10
2	0.14	0.16	0.16	0.07	0.08	0.08	0.07	0.08	0.08
3	0.14	0.15	0.12	0.07	0.09	0.07	0.12	0.08	0.08
4	0.10	0.20	0.12	0.13	0.05	0.05	0.05	0.12	0.09
5	0.21	0.13	0.13	0.10	0.05	0.06	0.13	0.09	0.09
6	0.16	0.08	0.09	0.10	0.11	0.11	0.12	0.03	0.05
7	0.12	0.06	0.06	0.10	0.06	0.08	0.12	0.11	0.10
8	0.07	0.05	0.07	0.07	0.09	0.11	0.15	0.10	0.12
9	0.19	0.30	0.16	0.08	0.09	0.11	0.08	0.06	0.05
10	0.12	0.07	0.09	0.08	0.07	0.09	0.17	0.16	0.16
11	0.24	0.25	0.18	0.18	0.23	0.13	0.19	0.20	0.17
12	0.12	0.14	0.09	0.23	0.15	0.18	0.22	0.18	0.16
13	0.09	0.10	0.09	0.24	0.16	0.24	0.13	0.15	0.11
14	0.16	0.12	0.24	0.07	0.05	0.10	0.19	0.13	0.30
15	0.09	0.09	0.14	0.09	0.11	0.10	0.12	0.07	0.04
16	0.11	0.09	0.10	0.11	0.13	0.10	0.11	0.08	0.10

17.6.4 Building Augmented Scapula SSM

Anatomical landmark data from all the observers, MV shape, 27 internal MV estimates, and 20 external scapula instances were used for evaluating anatomical correspondence validity. Using this data, two types of augmented SSMs were developed viz. locally augmented SSM and globally augmented SSM (Figure 17.30). The local and global distinctions were made to illustrate various real-life scenarios and analyze the clinical validity for each of these. The local scenario represented the involvement of a single observer in the landmark data selection for the augmented SSM building, whereas a global scenario represented the involvement of multiple observers in the landmark data selection for building the augmented SSM. While anatomical landmarks were selected on MV estimates, augmented SSM was built based on the mean landmark data transferred to the unbiased reference MV shape, as described below.

To build the locally augmented SSM, a single observer was randomly chosen and the anatomical landmark data selected on 27 internal instances by this observer was taken into consideration. This data was then processed to obtain a set of 16 "repositioned" anatomical landmarks for each of the 27 MV estimates. The manually selected anatomical landmarks were points floating in space that were in closest proximity with the mesh surface of the MV estimate. These landmarks were first repositioned to the nearest vertex of the mesh of the MV estimate. Subsequently, repositioned landmark location data was averaged over 27 MV estimates to create a dataset of 16 averaged landmarks. To form a locally augmented SSM, this averaged dataset was transferred and repositioned to the nearest vertex on the surface of the MV shape. For building the globally augmented SSM, anatomical landmark data from all the five observers was taken into consideration. For each observer, manual anatomical landmarks were repositioned to closest vertices on 27 MV estimates. The repositioned landmark location data from

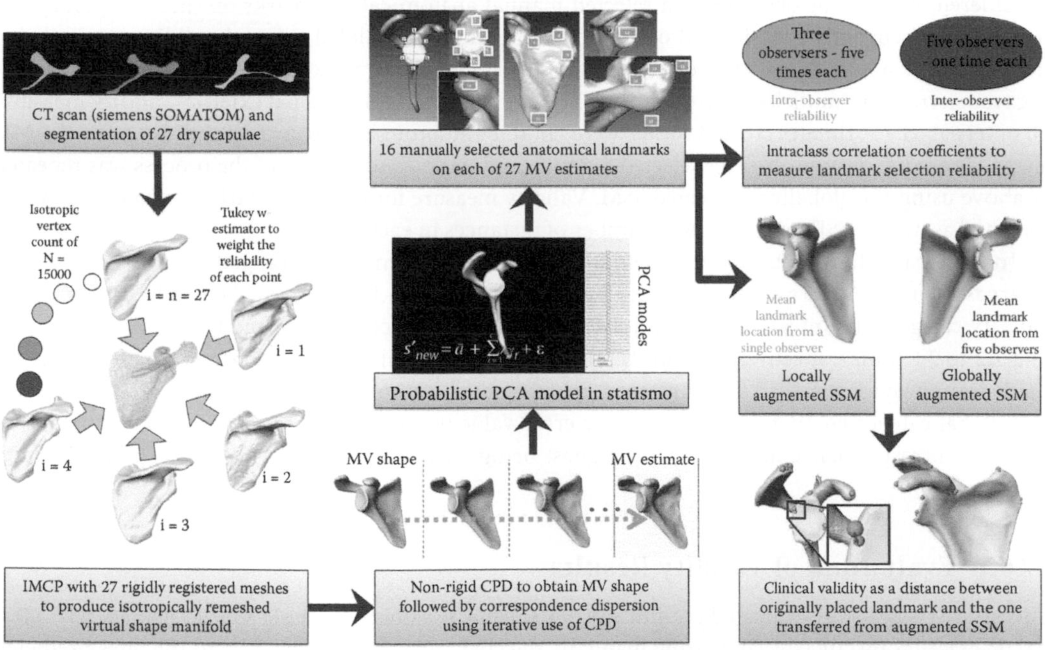

FIGURE 17.30 A schematic diagram showing the methodology used to determine clinical validity of an augmented statistical shape model of the adult scapula bone. First, a statistical shape model of an adult scapula bone was built using 27 dry scapulae, following a previously published methodology [19]. Sixteen anatomical landmarks in clinically relevant regions of scapula were selected by multiple observers, tested for their reliability of selection and used to augment the statistical shape model. Clinical validity was quantified as a distance between a manually selected anatomical landmark on the original scapula instance and a landmark transferred from an augmented (locally or globally) statistical shape model.

all the observers was then averaged (average of 27 × 5 measures for each landmark) to form a dataset of 16 averaged landmarks. A globally augmented SSM was formed by transferring and repositioning this average landmark dataset on the surface of the MV shape.

17.6.5 Local and Global Validity of Augmented SSM

Anatomical correspondence validity was a measure of how accurately one can estimate the anatomical locations using an SSM augmented with locally or globally placed landmarks. The location of the anatomical landmark on the given augmented SSM could be transferred to each of the MV estimates using the one-to-one correspondence between its vertices. Anatomical validity was determined as a distance between a manually placed landmark on MV estimates of the original scapula instances (internal or external) and its counterpart transferred from an augmented SSM using the correspondence coherence procedure as explained above. For each augmented SSM, the validity measure was determined on using both internal and external instances (Figure 17.30).

The local internal validity measure for each anatomical landmark was determined as the Euclidian distance between the repositioned manual landmark location and transferred landmark location for 27 internal instances. For determining the local external validity measure, the same observer manually selected the landmarks on the MV estimates of 20 external instances. Using the locally augmented SSM created previously, the local external validity measure for each anatomical landmark was determined as a Euclidian distance between the repositioned landmark location and the transferred landmark location on the external MV estimates.

For global internal validity measure, 27 MV estimates from corresponding internal instances were considered. An outside observer then selected manual anatomical landmarks on these scapulae. This data was subsequently repositioned on its MV estimate shape. Global internal validity measure was determined as a Euclidian distance between the repositioned landmark location data and corresponding transferred landmark location from globally augmented SSM. For global external validity measure, 20 external MV estimates created earlier were chosen and brought into correspondence with the MV shape. The same outside observer selected manual landmarks, and the rest of the process was repeated as above using the globally augmented SSM. Validity measure for each landmark in each category was reported as mean ± SD, averaged over number of instances in each category.

In the local validity category, the mean internal correspondence error ranged from 0.1 to 0.55 mm, whereas the mean external correspondence error ranged from 1.1 to 2.5 mm (Figure 17.31). For global validity category, the mean internal and external correspondence error ranged from 0.22 to 0.68 mm and 1.2 to 3.1 mm, respectively (Figure 17.31). The mean internal correspondence error for each landmark was always lower than its counterpart in external correspondence error measures, in both local and global categories. In general, the average error value was magnified by as much as eight times in external correspondence measures, with highest being 3.1 mm for landmark 8 in the global validity category.

17.6.6 Evaluation of Validity Results

This study successfully evaluated the strategy of using an augmented scapula SSM to deform any internal or external instance by comparing manually selected anatomical landmarks to the ones predicted by an augmented SSM. In doing so, this study advances our understanding of the role of augmented SSMs in scapular clinical morphometry, by providing validity measures pertaining to clinically relevant anatomical landmarks on the scapular bone surface. In clinical settings, skeletal morphology is either manually or virtually palpated for anatomical landmarks [82]. Although manual palpation helps in identification of painful areas or electrode positioning, virtual palpation is useful in quantifying morphological parameters or pre-surgical assessment and preparation. Inaccuracies in anatomical landmark selection may lead to surgical complications or wrong post-surgical clinical assessments [82].

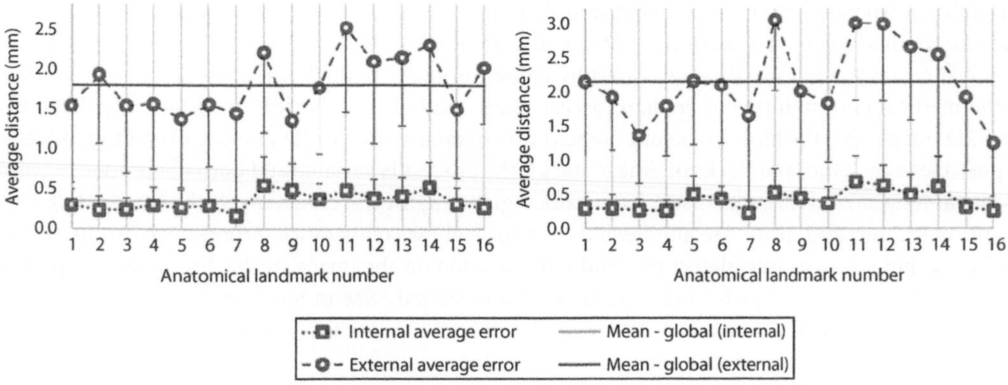

FIGURE 17.31 Clinical validity measures reported as mean error for the locally augmented statistical shape model (SSM) on left and globally augmented statistical shape model (SSM) on right. For each anatomical landmark, a 3D Euclidean distance between a manually selected anatomical landmark on original scapula instance and a landmark transferred from the locally augmented SSM was determined. Mean landmark locations on 27 training scapulae selected by a randomly chosen observer were used to form the locally augmented SSM. Mean landmark locations on 27 training scapulae selected by all five observers were used to form the globally augmented SSM. For internal mean error, for each landmark, mean distance from 27 training scapulae (used for building the SSM) was quantified. For external mean error, for each landmark, mean distance from 20 external scapulae (not used to build the SSM) was quantified. The solid line joining the landmarks in each graph represents internal mean error for 27 training scapulae. The dotted line joining the landmarks in each graph represents external mean error for 20 external scapulae. Error bars in each graph represent standard deviation of mean error.

Thus, accurate and reliable landmark selection is of utmost importance. As previously explained, Styner and colleagues [80] stated that landmark-based validity measure was not subjective, and, in fact, it was the only direct measure. The objectivity comes from selecting the anatomical landmarks manually and measuring the goodness of correspondence between the original manual landmark selection and the one predicted by deforming the augmented SSM to the original shape. The remaining three measures (generality, specificity and compactness) are linked to the modes of the SSM and, thus, are computed in the statistical space. But from the perspective of medical or clinical tool, we need to check and report if the SSM is able to transfer each anatomically relevant landmark in the patient-specific anatomy in the 3D space. We prove here that the augmented SSM is able to do that. Depending on the mesh deformation method used (constraint or not), we can check each clinically relevant area or landmark transferred with the augmented SSM. Moreover, since manual selection of landmarks was recognized by the clinical community, we selected this method for measuring clinical validity.

Findings from the intra- and inter-observer reliability indicated that anatomical landmarks can be specified with excellent repeatability and reproducibility on image-based 3D bone models; however, this may become a time consuming task. A 3D patient-specific scapular bone model derived from an augmented SSM could improve the accuracy and success ratio of shoulder surgeries to a great extent, as it would present the surgeon with an opportunity to visually understand the subject-specific morphology. Such a model, when clinically validated, may also be able to simulate the surgical procedure, when used in a surgical planning tool such as BLUEPRINT (Tornier surgical solutions, www.tornierblueprint.com). This study also lays a foundation for an accurate and reliable pipeline development of augmented SSMs for automatic segmentation of bone structures from medical images.

The manual selection of clinically relevant anatomical landmarks on scapular bone was not a known method; hence, it was necessary for the present study to undergo a rigorous reliability analysis through multiple users, including clinicians, researchers and engineers. The guideline created for the selection of manual landmarks complemented the integration of individual assessor's judgment by providing similar orientation of the scapula for all assessors for landmark selection. This helped in reducing the variability introduced by factors other than individual's assessment of the landmark location. It is true that the methods described here incorporate 3D, point-set based, non-rigid registration methods to establish correspondences. This was carefully chosen in order to apply the methods to a wider range of the medical community, rather than to keep them limited in the research community. Even if we have 2D image sets, the idea is to avoid the selection of landmarks on 2D images. Even though it is not impossible to pick the landmarks on image stacks, the process is variable and does not produce accurate results among mixed observers (surgeons, radiologists, engineers, clinicians, technicians, etc.). Using the proposed method, if we use augmented SSM for a global registration process, we can fit not only the image, but can also auto-detect the landmark location on the image stacks. From another perspective, we believe that if we prove the robustness of augmented SSM in holding the correspondences in landmarks across the training and outside samples, then it can be easily transformed to intensity or 2D images, where it is always difficult to locate bony landmarks.

A study performed directly on dry scapular bones quantified the reliability of using protractors for anatomical measurements around the glenoid and found excellent (> 97.5%) inter- and intra-assessor reliability [83]. In the current study, the reliability of glenoid rim landmark selection was > 70% for intra- and > 87% for inter-observer reliability. The lower reliability values could be attributed to the methods used to select the landmarks, as well as to calculate reliability measures. Reliability measures in the current study were based on two-way (mixed or random effects design) ANOVA, whereas Amadi and colleagues [83] used average error and average deviation measures to report reliability. Thus, the statistics in the current study used observers and scapula instances as factors and compensated for either the random or mixed effects due to these factors, which in turn reduced the reliability score.

Within the reliability measures, inter-observer reliability for all the coordinates of landmarks 11 and 12 was consistently lower than other landmarks (Table 6.1). These landmarks represented the acromion region of the scapula, and the lower reliability was primarily due to the fact that there was a relatively larger variability in the shape of the acromion, which was not adequately specified in the landmark selection guideline. We could not find any other study to compare these results with. Regardless, excellent reliability and low SEM measures indicated that we could reliably and precisely select anatomical landmarks and supported our strategy to use them in a globally augmented MV shape for clinical correspondence validity measures.

It has been shown that SSMs can be effectively used in situations during the surgery where it can be deformed to patient-specific bone shapes using sparse/partial bone digitization data and anatomical landmarks [84,85]. However, the accuracy of such deformable models in clinically relevant regions that are not covered by digitization needs to be carefully evaluated (before its implementation as an *in vivo* surgical planning tool). For example, in Zhu and Li [9,85] the overall differences between SSM-predicted distal femur surface and their native femur surface models were 0.16 ± 1.16 mm. However the maximum and minimum errors ranged from 3.62 to 3.31 mm, and it was not reported where on the distal femoral surface these errors occurred. These maximum errors could very well be occurring at the clinically relevant positions, and, thus, the average error metric did not provide clinical robustness of the method. The augmented scapula model proposed in the current study will be used further to register (rigid and non-rigid) to the patient-specific data (e.g., imaging), using the anatomical landmarks to set the initialization of rigid registration and to derive landmark-based constraints during non-rigid registration (deformation). Compared to using just an SSM for registration, an augmented SSM-based registration will provide the clinical robustness for the accuracy of fitting within the clinically relevant areas. For partial data, we are not able to check the accuracy in the missing region. But, with augmented SSM, it is possible to penalize or ignore the landmarks not in correspondence in the missing part and avoid the error propagation and, thereby, maintain good fit, which is more clinically relevant. Furthermore, biomechanical modeling studies typically use generic models to represent bone morphology, and using augmented SSMs built from patient-specific data would make these studies more clinically acceptable. To date, no data is available on the validity of clinical use of scapula SSMs, and the results of this study provided a much-needed evaluation for such techniques. To the knowledge of the authors, only one study [86] was found that directly validated the proximal femur (hip) bone SSM with original bone shape. In this study, nine morphometric parameters of proximal femur were measured and compared between the surface models of external femur instances and proximal femur SSM built using a training set. This study used distances, ratios and angles to define morphometry against the bony landmarks used in the current study. Since these morphometrics are determined using anatomical landmarks, it can be speculated that a direct validity using anatomical landmarks would automatically prove valid for morphometric measures. However, studies focused on both landmark and morphometry based SSM validation are warranted.

Predicting landmarks on the internal instances reported sub-millimeter correspondence errors, which illustrate the validity of using both locally and globally augmented SSMs in clinical settings. However, correspondence errors in predicting the landmarks on external instances in both local and global categories were in the magnitude of few millimeters questioning the validity of using the augmented scapula SSM for external instances. These higher errors could be attributed to the variation in size of the scapula, as well as the variation of scapular shape not being included in the training sample (internal instances) of the SSM. There was not much variation in the mesh size of 27 training scapulae with the average distance between the vertices being 0.23 mm ($\pm$ 0.005). A higher resolution mesh could yield even better validity for the internal instances, but it did not seem to significantly affect the correspondence error of external instances. A further analysis of correspondence errors revealed that errors were location specific. For glenoid rim landmarks 2–5, errors in medial lateral direction were highest; for acromion landmarks 11–13, errors in anterior-posterior direction were highest; and for scapular blade landmarks 7,8, and 10, superior-inferior errors were highest. It would be intuitive to assume that errors in different coordinates were correlated. But this was not the case, partly due to the landmark selection process, which constrained a 3D manifold in a 2D space for a given landmark selection. The axis of highest error corresponded with the plane of view of the observer for respective landmark selections in scapular blade and acromion regions; however, this was not true for glenoid rim landmark errors. It can be speculated that the process of repositioning the glenoid rim landmarks may have selected more medially or laterally placed vertices leading to larger errors in this direction.

17.7 Potential Applications and Future Research

This work introduces an SSM-building pipeline for creating statistical shape models, which has been shown to produce promising results for three structures of varying complexity. The main advantages in the pipeline include automation in that the method does not rely on manual landmarks or a regionalization step. Furthermore, there is no bias in the choice of reference due to the robust correspondences group-wise registration and the use of the probabilistic principal component analysis framework to increase the domain of the shape variability. A comparison of the methodology has also been performed with publicly available data, and the methodology has proven to be equal to or better than two state-of-the-art methods. This work also proposed a method to clinically validate the SSMs using scapula bone as an example and, in doing so, developed an augmented SSM for the same. This novel approach in using an augmented SSM enabled a better control over clinically relevant areas that were typically overlooked and under-evaluated in a standard SSM quality checking criterion viz. generality, specificity and compactness. In the local category, augmented SSM (formed using a single observer from available set of observers) represented a typical localized clinical scenario where a radiologist (or a medical student) normally handles such operations, and both the augmented SSM and internal/external instances will be processed by the same radiologist. In the global category, augmented SSM (formed using the mean from many observers) represented a larger collaboration group (mostly in international collaboration) that used a random set of observers from each establishment, and internal/external instances would be processed by any observers not involved in generating the augmented SSM. The internal and external correspondence errors exemplified whether the use of an augmented SSM provides acceptable errors in clinically relevant areas. In general, the results of correspondence error suggest that current SSM built from 27 scapulae may not have the required sample sufficiency. Thus, further evaluations in statistical stability of the SSM and completeness in terms of sample sufficiency is warranted before using it for external instances and in the clinical settings. This study was limited by its number of scapulae used for the construction of the SSM. The 27 instances included in the SSM did not represent the global scapular variability, which was evident in the external correspondence errors. Although we had additional data for 20 scapulae, it was important to keep them as external instances for the design of this study. Future research is directed toward building scapular SSM from higher training samples and also studying the sample sufficiency of the SSM, which was not in the scope of this study. Future work is also directed toward including pathological cases as training sets and understanding the clinical validity of predicting external pathological instances.

One can think of many possible applications for SSM. One important application is finite element (FE) modeling of bones and joints. Since every patient has a certain anatomy within the population described by SSM, one could pre-calculate the results of FE models for different values of mode parameters and use a look-up table approach for obtaining the results of FE analysis, thereby eliminating the need for the presence of a high-end computational facility in every hospital and reducing computation time. It should be, however, noted that the results of FE models generated based on SSM need to be corroborated against the simulation results of the FE models generated using standard segmentation techniques. The material properties of bones can be mapped based on the empirical relationships between the appearance and bone density, on the one hand, and the relationships between bone density and elastic properties of bones, on the other hand. Whether or not the statistical appearance model (SAM) provides accurate appearance and, thus, density distribution is dependent, among other factors, on the training population and whether it can well represent the individual for whom the patient-specific FE model is being built. Moreover, it is not clear whether statistical models of shape and appearance can be used for more specialized segmentation jobs, such as segmentation of cortical bone from trabecular bone particularly when only 2D information is available.

The other possible area of application is grafting of large bony defects. Modern additive manufacturing (3D printing) techniques have enabled the production of highly porous patient-specific bone substitutes that are considered very promising biomaterials for bone substitution [87–89]. However, one needs an accurate representation of bone shape so that the missing part of bone can be manufactured using additive manufacturing techniques. In such applications, active shape models could be used for estimating the original shape of bone, for example, by fitting the SSM to the intact parts of the bone. Even in the diagnosis and treatment of pathologies that involve bone deformities, many possible applications of SSMs and SAMs can be envisioned. Examples of such pathologies are cam deformities, genu valgum and varus deformities. In such application, SSM and SAM can be used for two main purposes. First, one could use SSM or SAM to estimate how the unreformed anatomy of the patient might have looked like. Second, new SSM and/or SAM can be trained using the data from pathological population to study, among others, the main modes of deformity, the shape changes caused by the deformity and the modifications caused by deformities in musculoskeletal loading.

The potential applications of SSM and SAM may increase as imaging modalities improve. For example, it may be possible to enrich SSM and SAM using imaging data acquired at the micro-scale, such as data acquired using micro-CT regarding the shape, orientation and distribution of individual trabeculae. That kind of information could, for instance, be used for representing anisotropy in SSM and SAM. In summary, statistical models of shape and appearance have been already used in different areas of bone-related research. However, more efforts in the generation of statistical models of shape and appearance and testing their accuracy are needed to allow for better exploitation of their potential in orthopedics research, diagnosis and treatment. That requires intense collaboration between engineers, radiologists and orthopedic surgeons.

References

1. Batchelor, G.P., Edwards, P.J., King, A.P., 3D medical imaging, in *3D Imaging, Analysis and Applications*, N. Pears, Liu, Y., Bunting, P., Eds. 2012, Springer-Verlag: London. 445–495.
2. Sarkalkan, N., et al., Statistical shape and appearance models for fast and automated estimation of proximal femur fracture load using 2D finite element models. *Journal of Biomechanics*, 2014. **47**(12): 3107–3114.
3. Sarkalkan, N., H. Weinans, and A.A. Zadpoor, Statistical shape and appearance models of bones. *Bone*, 2014. **60**: 129–140.
4. Rajamani, K.T., et al., Statistical deformable bone models for robust 3D surface extrapolation from sparse data. *Medical Image Analysis*, 2007. **11**(2): 99–109.
5. Stindel, E., et al., Bone morphing: 3D reconstruction without pre- or intraoperative imaging - concept and applications, in *Navigation and Robotics in Total Joint and Spine Surgery*. 2004, Springer: Berlin Heidelberg. 39–45.
6. Gomes, G., et al., Patient-specific modelling in orthopedics: From image to surgery, *Biomedical Imaging and Computational Modelling in Biomechanics*. 2013, Springer. 109–129.
7. Zheng, G., et al., A 2D/3D correspondence building method for reconstruction of a patient-specific 3D bone surface model using point distribution models and calibrated X-ray images. *Medical Image Analysis*, 2009. **13**(6): 883–899.
8. Lamecker, H., T.H. Wenckebach, and H.-C. Hege. Atlas-based 3D-shape reconstruction from X-ray images. 2006.
9. Zhu, Z. and G. Li, Construction of 3D human distal femoral surface models using a 3D statistical deformable model. *Journal of Biomechanics*, 2011. **44**(13): 2362–2368.
10. Heimann, T. and H.P. Meinzer, Statistical shape models for 3D medical image segmentation: A review. *Medical Image Analysis*, 2009. **13**(4): 543–563.
11. Pratt, P., et al., An effective visualisation and registration system for image-guided robotic partial nephrectomy. *Journal of Robotic Surgery*, 2012. **6**(1): 23–31.
12. Stindel, E., et al., Bone morphing: 3D morphological data for total knee arthroplasty. *Computer Aided Surgery*, 2002. **7**(3): 156–168.
13. Thompson, S., et al., Use of a CT statistical deformation model for multi-modal pelvic bone segmentation. *in Proceeding of SPIE, the International Society for Optical Engineering*. 2008. Society of Photo-Optical Instrumentation Engineers.
14. Barratt, D.C., et al., Instantiation and registration of statistical shape models of the femur and pelvis using 3D ultrasound imaging. *Medical Image Analysis*, 2008. **12**(3): 358–374.
15. Seim, H., et al., Automatic segmentation of the pelvic bones from CT data based on a statistical shape model. *VCBMe*, 2008: p. 93–100.
16. Becker, M., et al., Automatic construction of statistical shape models for vertebrae, in *Medical Image Computing and Computer-Assisted Intervention–MICCAI 2011*. 2011, Springer. p. 500–507.
17. Elias de Oliveira, M., et al. Statistical shape modeling of pathological scoliotic vertebrae: A comparative analysis. In *2010 Annual International Conference of the IEEE Engineering in Medicine and Biology Society (EMBC)*, 2010.
18. Mirzaalian, H., M. Wels, T. Heimann, B.M. Kelm, M. Suehling, Fast and robust 3D vertebra segmentation using statistical shape models. In *35th Annual International Conference of the IEEE Engineering in Medicine* and *Biology Society (EMBC)*. 2013.
19. Mutsvangwa, T., V. Burdin, C. Schwartz, C. Roux, An automated statistical shape model developmental pipeline: Application to the human scapula and humerus. *IEEE Transactions on Biomedical Engineering*, 2015. **62**(4): 1098–1107.
20. Boileau, P., R.J. Sinnerton, C. Chuinard, G. Walch, Arthroplasty of the shoulder. *Journal of Bone and Joint Surgery-British Volume*, 2006. **88**(5): 562.

21. Schwartz, C., et al., Using merged kinematic and anatomical data to evaluate humeral motion estimation: A pilot study. *Computer Methods in Biomechanics and Biomedical Engineering*, 2013(ahead-of-print): 1–5.
22. Schwartz, C., et al., Detection of incoherent joint state due to inaccurate bone motion estimation. *Computer Methods in Biomechanics and Biomedical Engineering*, 2013. **16**(2): 165–174.
23. Lebailly, F., L.V.P.C. Lima, A. Clairemidi, B. Aubert, S. Guerard, Y. Chaibi, ... W. Skalli, Semi-automated stereoradiographic upper limb 3D reconstructions using a combined parametric and statistical model: A preliminary study. *Surgical and Radiologic Anatomy*, 2012. **34**(8): 757–765.
24. van Kaick, O., H. Zhang, G. Hamarneh, D. Cohen-Or, A survey on shape correspondence. in *Proceeding of Eurographics State-of-the-art Report*. 2010. Citeseer.
25. Heimann, T. and H.-P. Meinzer, Statistical shape models for 3D medical image segmentation: A review. *Medical Image Analysis*, 2009. **13**(4): 543–563.
26. Styner, M.A., K.T. Rajamani, L-P. Nolte, G. Zsemlye, G. Székely, C.J. Taylor, R.H. Davies, Evaluation of 3D correspondence methods for model building, in *Information Processing in Medical Imaging*. 2003, Springer. p. 63–75.
27. Zhang, K.Y., A. Wiktorowicz-Conry, J.R. Hutchinson, M. Doube, M. Klosowski, S. Shefelbine, A.M.J. Bull, 3D morphometric and posture study of felid scapulae using statistical shape modelling. *PloS One*, 2012. **7**(4): e34619.
28. Rasoulian, A., R. Rohling, and P. Abolmaesumi, Group-wise registration of point sets for statistical shape models. *IEEE Transactions on Medical Imaging*, 2012. **31**(11): 2025–2034.
29. Hufnagel, H., J. Ehrhardt, X. Pennec, N. Ayache, H. Handels, Computation of a probabilistic statistical shape model in a maximum-a-posteriori framework. *Methods of Information in Medicine*, 2009. **48**(4): 314–319.
30. Su, Z., Statistical shape modelling: Automatic shape model building, in *Department of Medical Physics and Bioengineering*, 2011, London: University College.
31. Luthi, M., T. Albrecht, T. Gass, O. Goksel, M. Kistler, H. Bousleiman, ... T. Vetter, Statismo-A framework for PCA based statistical models. *The Insight Journal*, 2012. **1**: 1–18.
32. Yang, Y., Shape modelling of bones: application to the primate shoulder, in *Department of Bioengineering and Department of Computing*. 2008, London: Imperial College London.
33. Mayya, M., S. Poltaretskyi, C. Hamitouche, J. Chaoui, Scapula statistical shape model construction based on watershed segmentation and elastic reegistration. In *2013 IEEE 10th International Symposium on Biomedical Imaging (ISBI)*. 2013.
34. Styner, M., I. Oguz, S. Xu, C. Brechbühler, D. Pantazis, J.J. Levitt, M.E. Shenton, G. Gerig, Framework for the statistical shape analysis of brain structures using SPHARM-PDM. *Insight Journal*, 2006. (1071): 242–250.
35. Stalling, D., M. Westerhof, and H. Hege, eds. Amira: A highly interactive system for visual data analysis. *The Visualization Handbook*, ed. C.D. Hansen and C.R. Johnson. 2005, Elsevier. 749–767.
36. Ibanez, L., et al., The ITK software guide: The insight segmentation and registration toolkit. Kitware Inc, 2003. 5.
37. Schwartz, C., Contribution to the development of a common representation space for morphofunctional analysis of the upper limb: Application to the glenohumeral joint. (French), in *Department Image and Information Processing*. 2009, Ecole Nationale Supérieure des Télécommunications de Bretagne-ENSTB.
38. Jacq, J.J., T. Cresson, V. Burdin, C. Roux, Performing accurate joint kinematics from 3-D in vivo image sequences through consensus-driven simultaneous registration. *IEEE Transactions on Biomedical Engineering*, 2008. **55**(5): 1620–1633.
39. Jacq, J.J., C. Schwartz, V. Burdin, R. Gérard, C. Lefèvre, C. Roux, O. Rémy-Néris, Building and tracking root shapes. *IEEE Transactions on Biomedical Engineering*, 2010. **57**(3): p. 696–707.
40. Cootes, T.F., S. Marsland, C.J. Twining, K. Smith, C.J. Taylor, Groupwise diffeomorphic non-rigid registration for automatic model building, *Computer Vision-ECCV, 2004* 2004, Springer. 316–327.

41. Balci, S.K., P. Golland, M.E. Shenton, W.M. Wells, Free-form B-spline deformation model for groupwise registration. *In Medical Image Computing and Computer-Assisted Intervention: MICCAI. International Conference on Medical Image Computing and Computer-Assisted Intervention*. 2006. NIH Public Access.
42. Granger, S. and X. Pennec, Multi-scale EM-ICP: A fast and robust approach for surface registration, in *Computer Vision — ECCV 2002*, A. Heyden, et al., Editors. 2002, Springer Berlin Heidelberg. p. 418–432.
43. Besl, P.J. and N.D. McKay. Method for registration of 3-D shapes. in *Robotics-DL Tentative*. 1992. International Society for Optics and Photonics.
44. Rangarajan, A., H. Chui, and F.L. Bookstein, The softassign procrustes matching algorithm, in *Information Processing in Medical Imaging*. 1997, Springer Berlin Heidelberg. p. 29–42.
45. Nishino, K. and K. Ikeuchi, Robust simultaneous registration of multiple range images comprising a large number of points. *Electronics and Communications in Japan (Part II: Electronics)*, 2004. **87**(8): 61–74.
46. Jacq, J.-J., T. Cresson, V. Burdin, C. Roux, Performing accurate joint kinematics from 3-D emphasis emphasistype, *IEEE Transactions on Biomedical Engineering*, 2008. **55**(5): 1620–1633.
47. Jacq, J.-J., C. Schwartz, V. Burdin, R. Gérard, C. Lefèvre, C. Roux, O. Rémy-Néris, Building and tracking root shapes. *IEEE Transactions on Biomedical Engineering* , 2010. **57**(3): 696–707.
48. Tukey, J.W., *Exploratory Data Analysis*. Reading, MA, 1977. 231.
49. Stalling, D., H. Hege, and M. Zöckler, Amira-an advanced 3D visualization and modeling system. URL: http://amira.zib.de, 2007.
50. Myronenko, A., Non-rigid image registration: Regularization, algorithms and applications. in *Department of Science and Engineering, School of Medicine* 2010, Oregon: Oregon Health and Science University, 176.
51. Koch, M., S. Bauer, J. Hornegger, N. Strobel, Towards deformable shape modeling of the left atrium using non-rigid coherent point drift registration, in *Bildverarbeitung für die Medizin 2013*. 2013, Springer. p. 332–337.
52. Schwartz, C., Contribution to the development of a common representation space for morphofunctional analysis of the upper limb: Application to the glenohumeral joint., in *Ecole Nationale Supérieure des Télécommunications de Bretagne-ENSTB*. 2009, Brest, France: Matisse, Rennes.
53. Myronenko, A. and X. Song, Point set registration: Coherent point drift. *IEEE Transactions on Pattern Analysis and Machine Intelligence*, 2010. **32**(12): 2262–2275.
54. Combès, B. and S. Prima, A New eEficient EM-ICP Algorithm for Non-Linear Registration of 3D Point Sets. 2012, VISAGES: Vision Action et Gestion d'Informations en Santé - VISAGES.
55. Aspert, N., D. Santa-Cruz, and T. Ebrahimi. MESH: Measuring errors between surfaces using the Hausdorff distance. *In 2002 IEEE International Conference on Multimedia and Expo, 2002. ICME '02. Proceedings*. 2002.
56. Hufnagel, H., X. Pennec, J. Ehrhardt, N. Ayache, H. Handels, Generation of a statistical shape model with probabilistic point correspondences and the expectation maximization-iterative closest point algorithm. *International Journal of Computer Assisted Radiology and Surgery*, 2008. **2**(5): 265–273.
57. Slice, D.E., *Modern Morphometrics in Physical Anthropology*. **6**. 2005: Springer.
58. Halazonetis, D.J., Morphometrics for cephalometric diagnosis. *American Journal of Orthodontics and Dentofacial Orthopedics*, 2004. **125**(5): 571–581.
59. Mutsvangwa, T., Characterization of the facial phenotype associated with fetal alcohol syndrome using stereo-photogrammetry and geometric morphometrics, in *Human Biology*. 2009, Cape Town, South Africa: University of Cape Town.
60. Stegmann, M.B. and D.D. Gomez, A brief introduction to statistical shape analysis. *Informatics and Mathematical Modelling, Technical University of Denmark, DTU* 2002: p. 15.

61. Robinson, D., P.G. Blacwell, E.C. Stillman, A.H. Brook, Planar procrustes analysis of tooth shape. *Archives of Oral Biology*, 2001. **46**(3): 191–199.
62. McIntyre, G.T. and P.A. Mossey, Size and shape measurement in contemporary cephalometrics. *The European Journal of Orthodontics*, 2003. **25**(3): 231–242.
63. Hennessy, R. and J. Moss, Facial growth: Separating shape from size. *The European Journal of Orthodontics*, 2001. **23**(3): 275–285.
64. Burstall, F.E., *Basic Riemannian Geometry in Spectral Theory and Geometry*. London Mathematical Society Lecture Notes Series, 273, ed. E.B. Davies and Y. Safarov. 1999: Cambridge University Press.
65. Styner, M., I. Oguz, S. Xu, C. Brechbühler, D. Pantazis, J.J. Levitt, M.E. Shenton, G. Gerig, Framework for the statistical shape analysis of brain structures using SPHARM-PDM. *The Insight Journal*, 2006. (1071): 242.
66. Davies, R.H., et al., Building 3-d statistical shape models by direct optimization. *IEEE Transactions on Medical Imaging* , 2010. **29**(4): 961–981.
67. Gomes, G., Van Cauter, S., De. Beule, M., Vigneron, L., Pattyn, C., Audenaert, E., Patient-specific modelling in orthopedics: From image to surgery., in *Lecture Notes in Computational Vision and Biomechanics*. 2013, Springer. 109–129.
68. Paley, D., Normal lower limb alignment and joint orientation, in *Principles of Deformity Correction*. 2002, Springer: Berlin Heidelberg. p. 1–18.
69. Nizard, R., Computer assisted surgery for total knee arthroplasty. *Acta Orthopaedica Belgica*, 2002. **68**(3): 215–230.
70. Siston, R.A., N.J. Giori, S.B. Goodman, S.L Delp, Surgical navigation for total knee arthroplasty: A perspective. *Journal of Biomechanics*, 2007. **40**(4): 728–735.
71. Yoon, Y.S., A.J. Hodgson, J. Tonetti, B.A. Masri, C.P. Duncan, Resolving inconsistencies in defining the target orientation for the acetabular cup angles in total hip arthroplasty. *Clinical Biomechanics*, 2008. **23**(3): 253–259.
72. Verborgt, O., M. Vanhees, S. Heylen, P. Hardy, G. Declercq, R. Bicknell, Computer navigation and patient-specific instrumentation in shoulder arthroplasty. *Sports Medicine and Arthroscopy*, 2014. **22**(4): e42–49.
73. Zheng, G., M.A.G. Ballester, M. Styner, L-P. Nolte, Reconstruction of patient-specific 3D bone surface from 2D calibrated fluoroscopic images and point distribution model, in *Medical Image Computing and Computer-Assisted Intervention – MICCAI 2006*, R. Larsen, M. Nielsen, and J. Sporring, Editors. 2006, Springer, Berlin Heidelberg. 25–32.
74. Zheng, G. and S. Schumann, 3D reconstruction of a patient-specific surface model of the proximal femur from calibrated x-ray radiographs: A validation study. *Medical Physics*, 2009. **36**(4): 1155–1166.
75. Baka, N., M. de Bruijne, T. van Walsum, B.L. Kaptein, J.E. Giphart, M. Schaap, W.J. Niessen, B.P.F. Lelieveldt, Statistical shape model-based femur kinematics from biplane fluoroscopy. *IEEE Transactions on Medical Imaging*, 2012. **31**(8): 1573–1583.
76. Rao, C., C.K. Fitzpatrick, P.J. Rullkoetter, L.P. Maletsky, R.H. Kim, P.J. Laz, A statistical finite element model of the knee accounting for shape and alignment variability. *Medical Engineering and Physics*, 2013. **35**(10): 1450–1456.
77. Yang, Y.M., D. Rueckert, and A.M. Bull, Predicting the shapes of bones at a joint: application to the shoulder. *Computer Methods in Biomechanics and Biomedical Engineering*, 2008. **11**(1): 19–30.
78. Mayya, M., S. Poltaretskyi, C. Hamitouche, J. Chaoui, Mesh correspondence improvement using regional affine registration: Application to statistical shape model of the scapula. IRBM, 2015. **36**(4): 220-232.
79. Davies, R.H., *Learning Shapes: Optimal Models for Analyzing Natural Variability*, 2002, University of Manchester.

80. Styner, M.A., K.T. Rajamani, L-P. Nolte, G. Zsemlye, G. Székely, C.J. Taylor, R.H. Davies, Evaluation of 3D correspondence methods for model building. *Information Processing in Medical Imaging*, 2003. **18**: 63–75.
81. Shrout, P.E. and J.L. Fleiss, Intraclass correlations: Uses in assessing rater reliability. *Psychological Bulletin*, 1979. **86**(2): 420–428.
82. Van Sint Jan, S. and U. Della Croce, Identifying the location of human skeletal landmarks: Why standardized definitions are necessary--a proposal. *Clinical Biomechanics (Bristol, Avon)*, 2005. **20**(6): 659–660.
83. Amadi, H.O., Q.A. Fogg, U.C. Ugbolue, R.J.H. Emery, A.M.J. Bull, Reliability of a set of protractors for direct anatomical measurements around the glenoid and humeral head rims. *Journal of Anatomy*, 2012. **220**(5): 525–528.
84. Fleute, M., S. Lavallee, and R. Julliard, Incorporating a statistically based shape model into a system for computer-assisted anterior cruciate ligament surgery. *Medical Image Analysis*, 1999. **3**(3): p. 209–222.
85. Zhu, Z. and G. Li, An automatic 2D-3D image matching method for reproducing spatial knee joint positions using single or dual fluoroscopic images. *Computer Methods in Biomechanics and Biom edical Engineering*, 2012. **15**(11): p. 1245–1256.
86. Schumann, S., M. Tannast, L-P. Nolte, G. Zheng, Validation of statistical shape model based reconstruction of the proximal femur–A morphology study. *Medical Engineering and Physics*, 2010. **32**(6): 638–644.
87. Amin Yavari, S., T. Wauthle, J. van der Stok, A.C. Riemslag, M. Janssen, M. Mulier, … A.A. Zadpoor, Fatigue behavior of porous biomaterials manufactured using selective laser melting. *Materials Science and Engineering: C*, 2013. **33**(8): 4849–4858.
88. Van der Stok, J., O.P. van der Jagt, S.A. Yavari, M.F. De Haas, J.H. Waarsing, E.M. Van Hieshout, … H. Weinans, Selective laser melting-produced porous titanium scaffolds regenerate bone in critical size cortical bone defects. *Journal of Orthopaedic Research*, 2013. **31**(5): 792–799.
89. van der Stok, J., H. Wang, S.A. Yavari, M. Siebelt, M. Sandker, J.H. Waarsing, … H. Weinans, Enhanced bone regeneration of cortical segmental bone defects using porous titanium scaffolds incorporated with colloidal gelatin gels for time- and dose-controlled delivery of dual growth factors. *Tissue Engineering Part A*, 2013. **19**(23–24): 2605–2614.

Appendix I STATISMO software

The software package used was the Statismo model building toolkit [1]. The toolkit is developed in C++ with wrappers for VTK and ITK. From the documentation, Statismo confers the following advantages:

- High level: Most applications using statistical models exploit the fact that these models define a probability distribution over the objects of interest. The interface of Statismo is designed to reflect this fact and provides high-level methods to work with the represented distribution, rather than exposing the eigenvalues and eigenvectors of the model.
- Independence of data representation: Statistical models are used to model the variation in various types of objects, using a variety of different toolkits to represent the data. Independently of the concrete representation of the data, the ideas and interpretation of PCA-based statistical models is always the same. Statismo is designed to work for any kind of PCA models, independently of the data representation or the toolkit used.
- Reproducibility and easy data exchange: It is often fruitful to exchange data and models with colleagues or other researchers. An important design goal is to facilitate this exchange by providing a platform-independent data format, which contains all the information needed to use the model from an application. Furthermore, to help reproduce an experiment, the stored model should contain all the parameters that were used to build the model.
- Flexibility and extensibility: Statismo aims to simplify the integration of new algorithms for model building. This will not only increase the usefulness of Statismo itself, but will also allow such algorithms to benefit from the functionalities implemented in Statismo.
- PPCA: In addition, Statismo is based on the probabilistic principal component analysis paradigm described earlier.

The data was entered in Statismo in ".vtk" polydata format. The conversion from the ".surf" was performed by a custom-built code. The models produced were stored in a format based on HDF5, which includes all the information necessary to use the model, as well as metadata about the model creation, which helps to make model-building reproducible. The variance of noise added to the data was 1%.

Reference

1. Luthi, M., et al., Statismo-A framework for PCA based statistical models. *The Insight Journal*, 2012. 1: 1–18.

Appendix II How to Pick Landmarks on the Scapula

This document is a guideline for selecting anatomical landmarks on the scapula. All the users are requested to strictly adhere to this guideline in order to remove any bias due to the observer/user. A description of anatomical terminologies used for scapula bone is provided at the end of this document.

A Opening the Scapula Surface File in Amira

Although this guideline is made specific for 27 healthy adult scapulae that were CT-scanned and used in building the SSM model, it can be extended to any instances outside of these 27 scapulae.

For 27 scapulae:

- Open Amira.
- Open data and choose the folder.
    - We will use the SSM of 27 instances because we would like to test the validity on external instances as well.
    - We will use the CPD output file for each of the 27 instances because we need to verify the correspondences.
- Open the file "solutionInitiale 'i'.ply.ply_transformated_4.smooth.ply.surf," where i is the scapula number.
- Important Note: You must follow the order of landmarks as specified here. This order (from 1 to 16) should be maintained to make sure of the correct batch processing of landmark files.

B Preparing Scapula Surface File for Landmark Selection

To easily select the landmarks, we use the information of the mean curvature. After loading the .surf file, right click on scapula model (green bar in the pool panel showing its name) compute GetCurvature (Figure AII.1a).

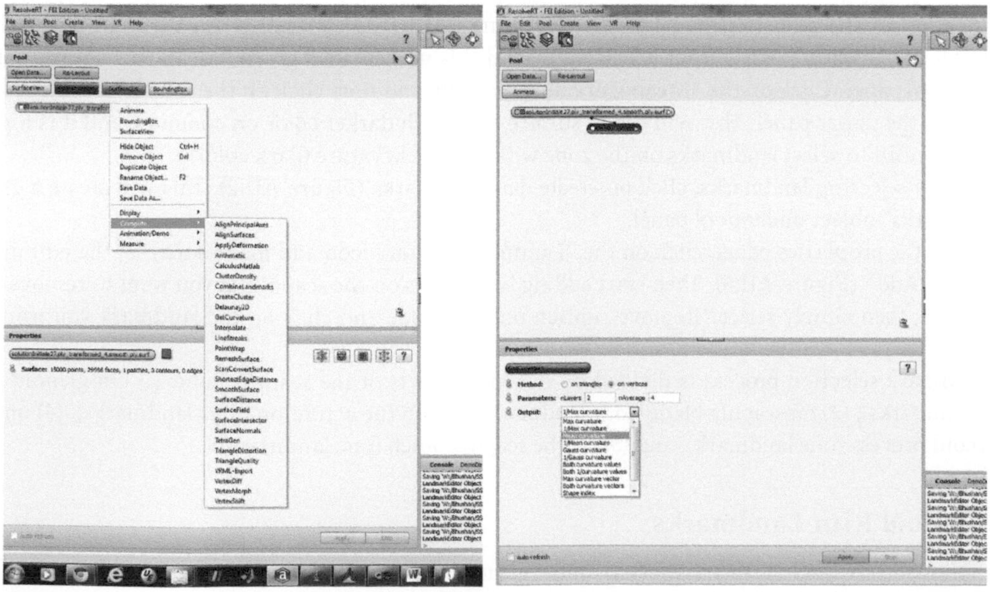

FIGURE AII.1 (a) Get curvature figure (b) Mean curvature.

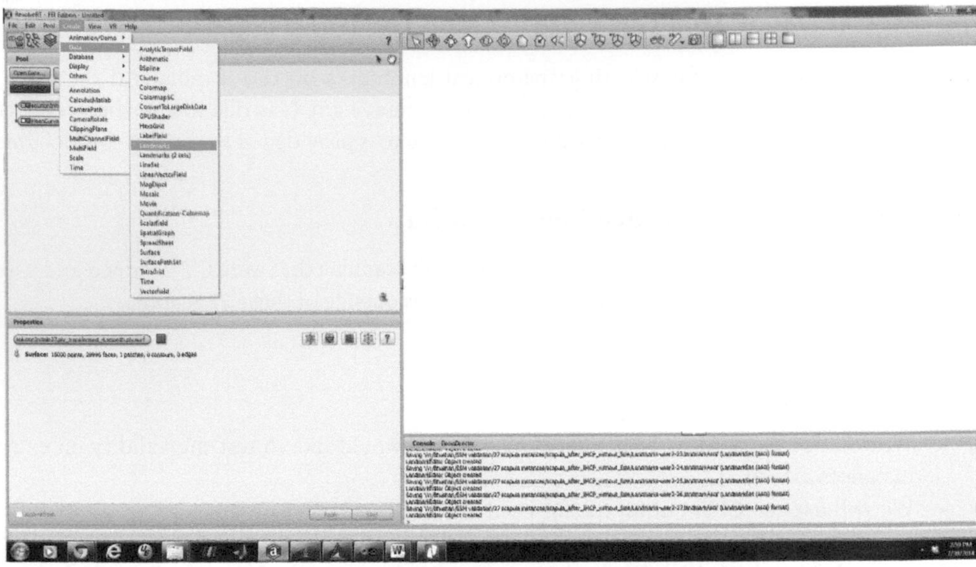

FIGURE AII.2 Create landmark object.

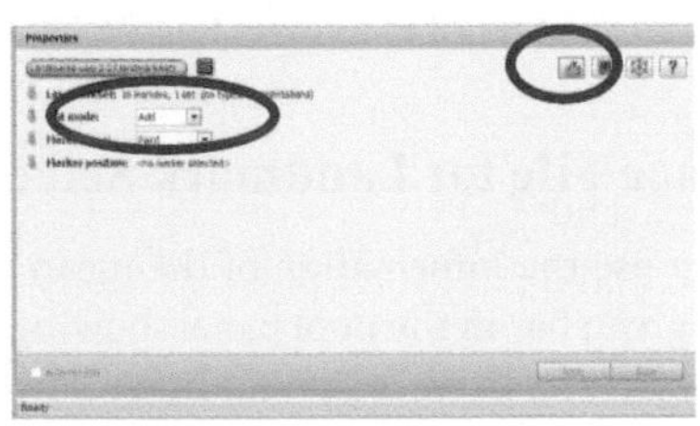

FIGURE AII.3 Landmark editor.

Then, under the properties panel for GetCurvature (Figure AII.1b), select "on Vertices" for Method and "Mean Curvature" for output. Click Apply and this will create a green "Mean Curvature" object under pool panel. Click on the "Mean Curvature" button, and then click on the yellow "Surface View" button in the upper panel. This will create surface view with darker color on contours, and it is highly recommended to select landmarks on the zone with a high curvature (dark color).

To start selecting landmarks, click on create data landmarks (Figure AII.2). This will create a green "Landmarks" object under pool panel.

Under the properties panel, click on the "Landmark Editor" icon and make sure that the edit mode is set on "Add" (Figure AII.3). Then start adding landmarks on the scapula. If you want to remove any landmark, then simply select "Remove" option on edit mode and click on the landmark you want to remove.

Landmark selection process is divided on five major parts of the scapula bone: (1) On glenoid rim (six landmarks), (2) on scapula blade (four landmarks), (3) on the acromion (four landmarks), (4) on the coracoid process (one landmark), and (5) on the scapula notch (one landmark).

C Glenoid Rim Landmarks

The following six landmarks are selected on the glenoid surface (Figure AII.4):

To select these six landmarks, orient the scapula in a lateral position in which the entire glenoid surface is revealed to the observer. Position the scapular blade in as vertical orientation as possible, zoom in

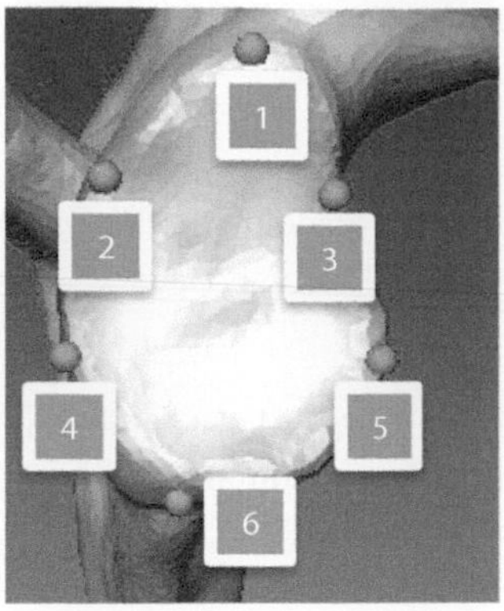

FIGURE AII.4 Glenoid landmarks.

on the glenoid rim. Press button, and then click and hold the left mouse button to rotate the scapula. Press button, and then click and hold the left mouse button to zoom in or out. You can also use the middle mouse wheel to zoom in or out.

C.1 Landmark 1

Superior middle point of the supraglenoid tubercle: the highest point on the glenoid rim. This is the most superior point on the glenoid rim. Select this point based on the superior-most point of the glenoid rim, defined by a dark colored mean curvature contour (Figure AII.5).

C.2 Landmark 2

For this point, you need to first visually locate the anterior point of the neck of the glenoid rim (Figure AII.6)—this is landmark 3 (Figure AII.5), but you **do not select it**.

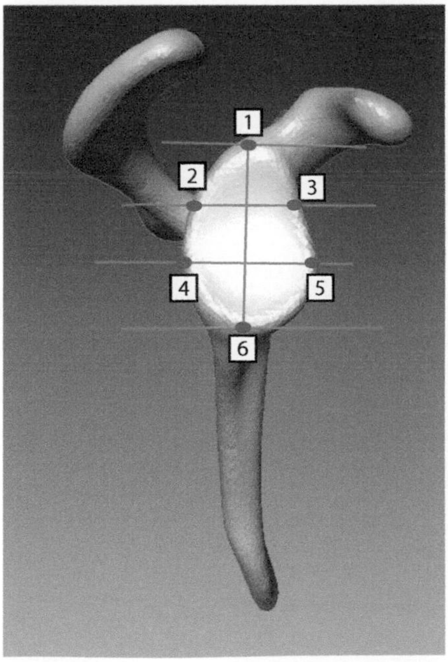

FIGURE AII.5 Scapula orientation for landmark selection.

Instead, you will draw an imaginary horizontal line from this landmark 3 to the posterior rim of the glenoid (Figure AII.5) and then select landmark 2.

C.3 Landmark 3

This is the anterior point of the neck of the glenoid rim (Figure AII.6). Select this point as landmark number 3.

C.4 Landmark 4

Again, for this point, you first need to visually locate the most anterior point of the glenoid rim (Figure AII.7) and then draw an imaginary horizontal line from this point to the posterior edge of the glenoid rim, and then select it as landmark number 4 (Figure AII.5).

C.5 Landmark 5

This is the most anterior point of the glenoid rim (Figure AII.7). Select this point as landmark 5.

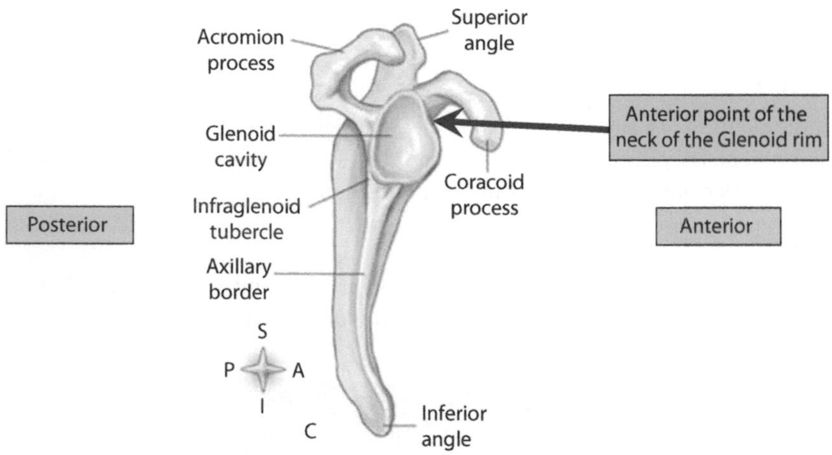

FIGURE AII.6 Anterior point of the neck of the glenoid rim.

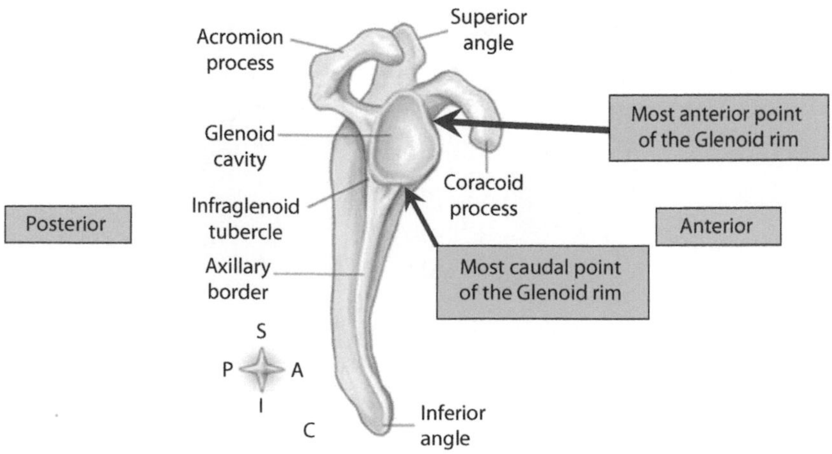

FIGURE AII.7 Anterior point of the neck of the glenoid rim

C.6 Landmark 6

This is the most caudal point of the glenoid rim (Figure AII.7). Select this point as landmark 6.

D Scapula Blade Landmarks

The following four landmarks are selected on the scapula blade (Figure AII.8):

To select these landmarks, first rotate and orient the scapula to view its anterior plane. This will expose the scapular blade in anterior view (Figure AII.8).

D.1 Landmark 7

This landmark is located in the suprascapular notch, which is located at the intersection of coracoid process and the superior border of the blade (Figure AII.8).

D.2 Landmark 8

Locate the superior angle of the scapular blade and select the point (visually the highest point) that makes this angle (Figure AII.8).

D.3 Landmark 9

This point is located in the region where the spine of the scapula intersects the medial border of the blade. If you see the posterior view of the blade, it will show the spine of the scapula (Figure AII.9). As this spine intersects the medial border, it fans out. You will select the superior point in this region. In order to make it simpler, you can see that from this point onwards, the lateral border curves in leading to the superior angle. You can remain in the anterior view and select this point easily. In the anterior view, this is the point that makes a distinctive angle on the medial border, and you select the point that makes this angle.

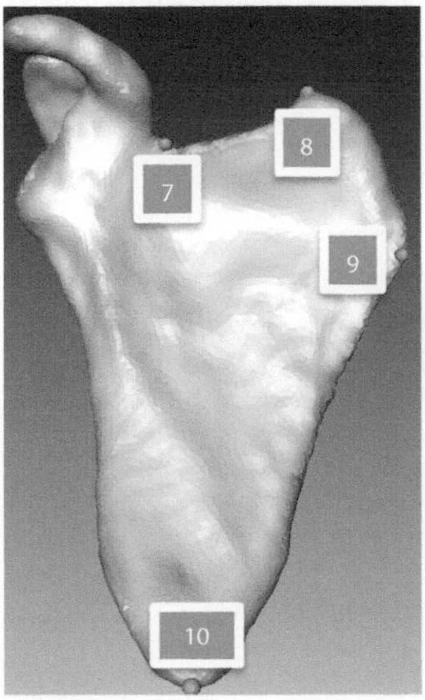

FIGURE AII.8 Scapula blade landmarks.

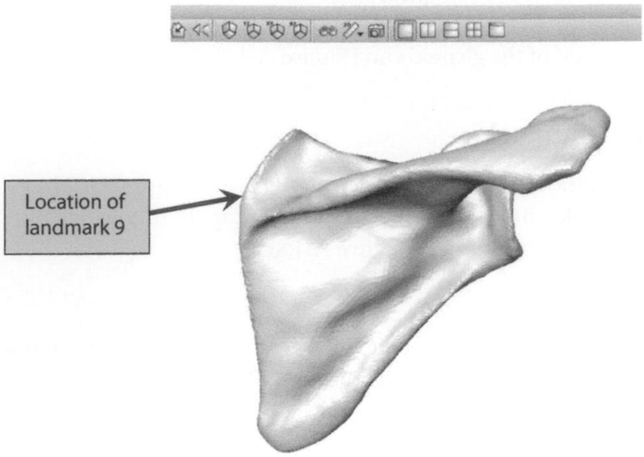

FIGURE AII.9 Posterior view of the scapula blade showing scapula spine and location of landmark 9.

D.3 Landmark 10

This is the point on the scapular blade that makes inferior angle. Visually, this is the lowest point on the blade.

E Acromion Landmarks

The following four landmarks are selected on the acromion (Figure AII.10):

In order to select these landmarks, first rotate the scapula in the posterior view (Figure AII.9). Then tilt the scapular blade with its inferior side going backward until you expose the plane of the acromion. In most of the cases, in this view, the spine makes a 45°angle with our viewing panel (Figure AII.11). Now, in this view, you will have to visually divide the acromion in different sections, starting from the superior edge of the spine. These sections define different angles of acromion, as explained in Figure AII.11. Specifically, we can imagine drawing six lines (Figure AII.11) from a to f and selecting landmarks on the intersection of these lines.

E.1 Landmark 11

This is the intersection of lines **a** and **b.**

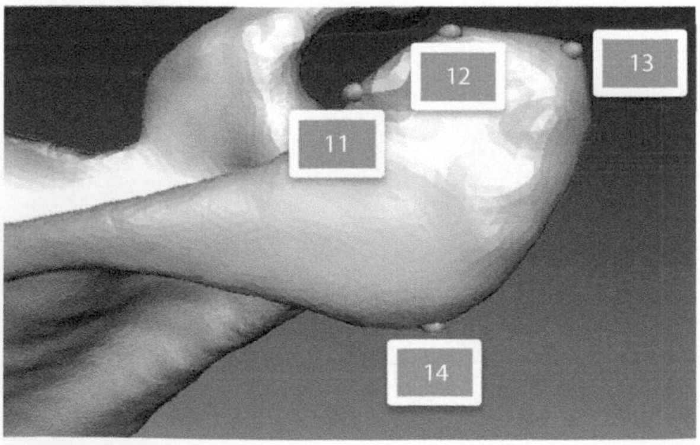

FIGURE AII.10 Acromion landmarks.

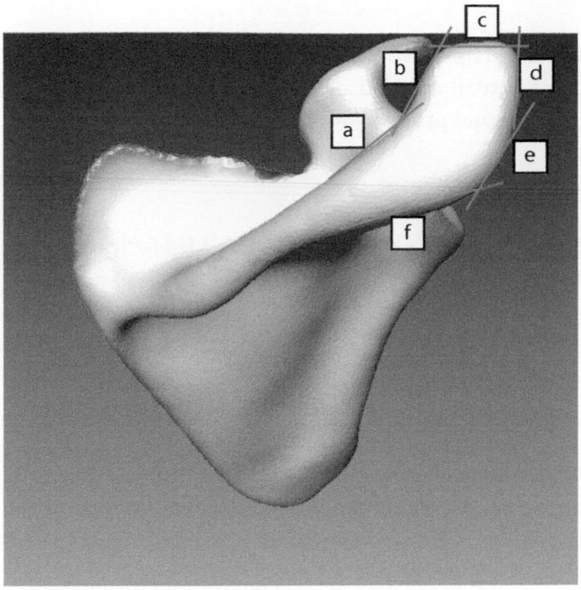

FIGURE AII.11 Sectioning the acromion using six lines.

E.2 Landmark 12

This is the intersection of lines **b** and **c**.

E.3 Landmark 13

This is the intersection of lines **c** and **d**.

E.4 Landmark 14

This is the intersection of lines **e** and **f**.

F. Coracoid Process Landmark

A single landmark is selected on the coracoid process (Figure AII.12).

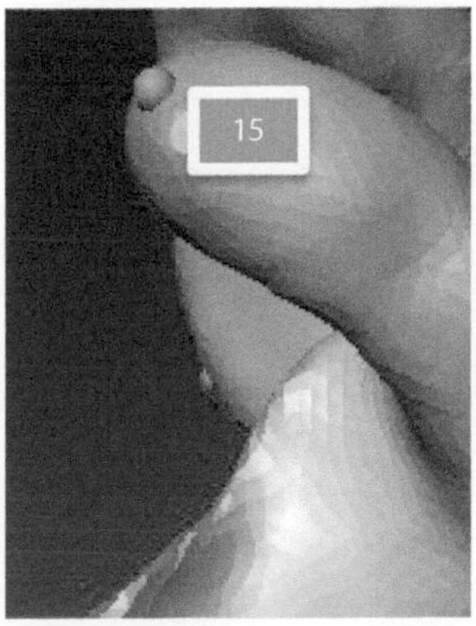

FIGURE AII.12 Coracoid process landmark.

F.1 Landmark 15

To select this landmark, first rotate and orient the scapula in such a way that its supraspinous fossa of the scapular blade, superior border and coracoid process are in one line (Figure AII.13). Then, add the landmark on the tip of the coracoid process with a high a level of curvature.

G Scapula Notch Landmark

A single landmark is selected on the scapula notch (Figure AII.14). Scapula notch is defined as a curved region between the posterior aspect of glenoid cavity and scapula spine. Notch is defined as the lowest location on this curvature, as seen in Figure AII.14. To select this landmark, orient the scapula in the lateral-posterior view so that the notch is exposed, then simply select the landmark.

G Saving Landmark File

Save the landmarks in this format: landmarks-user "j"-"i".landmarkAscii,

Where j represents the number of the user/observer and i the number of the scapula instance (from .surf file). Repeat these guidelines for each scapula.

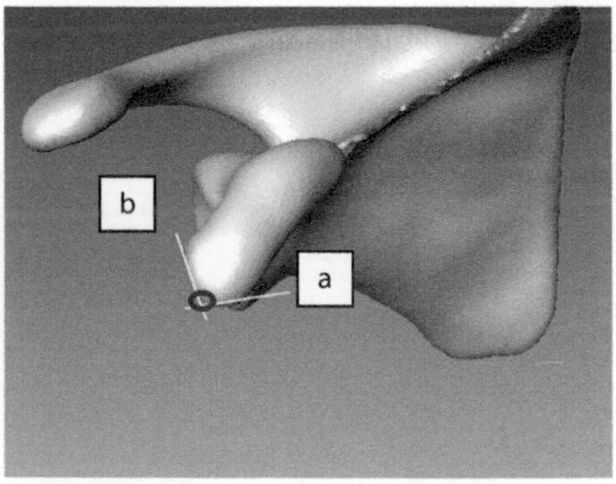

FIGURE AII.13 Landmark selection on coracoid process.

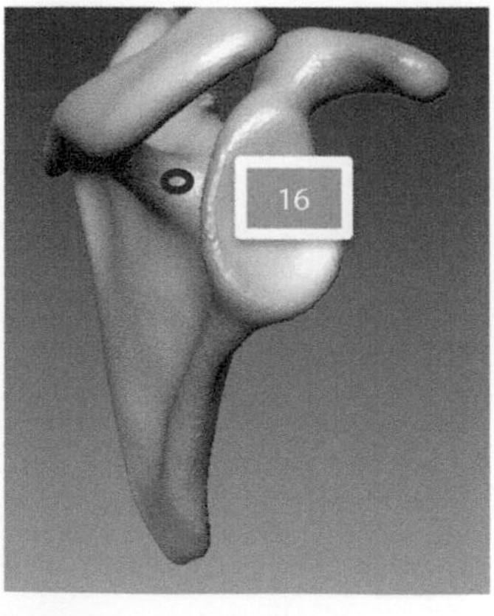

FIGURE AII.14 Scapular notch landmark.

Scapula Bone Anatomical Description

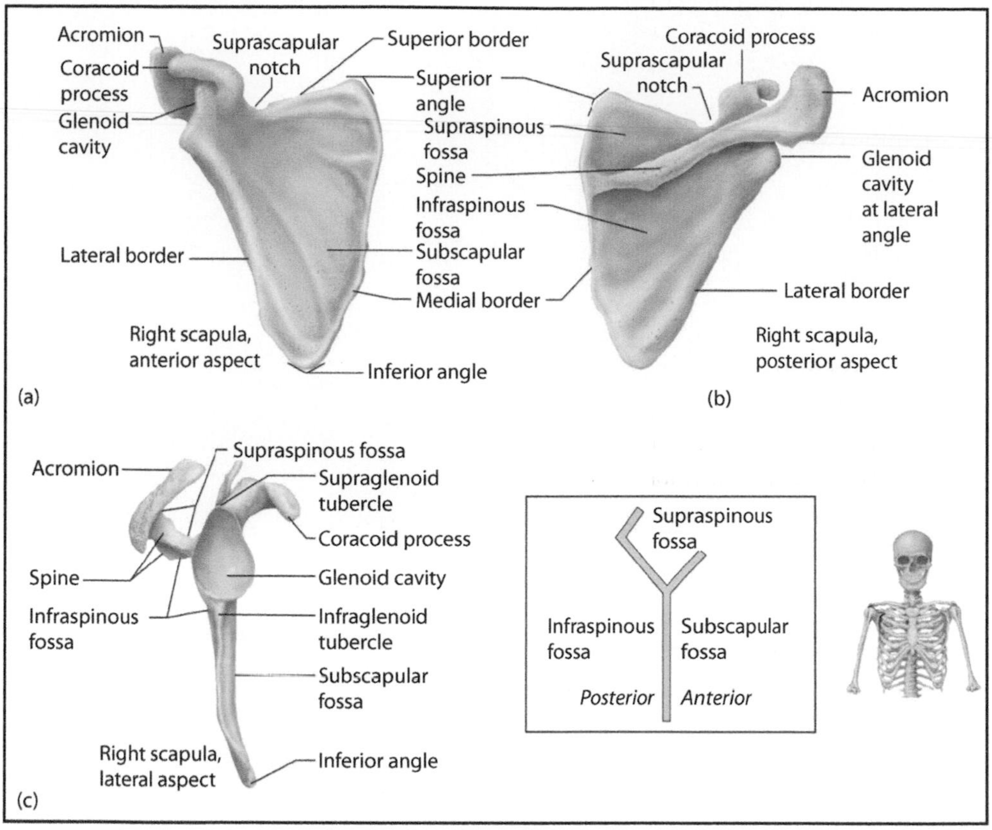

18

Disease-Inspired Feature Design for Computer-Aided Diagnosis of Breast Cancer Digital Pathology Images

Jesse Knight

April Khademi

18.1 Introduction

Breast cancer is the second leading cause of cancer death in Canada. One in nine women will be affected by the disease, and at least one in 30 will eventually die from it [1]. Breast cancer typically develops in the milk-producing tissues, known as the terminal ductal lobular units. The four most common breast cancers are Ductal Carcinoma *in situ* (DCIS), Lobular Carcinoma *in situ* (LCIS), Invasive Lobular carcinoma (ILC), and Infiltrating Ductal carcinoma (IDC). IDC is the most invasive and accounts for almost 75% of cases [2]. As invasive cancers are characterized by the propensity to metastasize—to spread to other parts of the body—they are more dangerous for the patient and usually result in worse outcomes.

Diagnosis of IDC occurs in two stages. First, the patient is sent for a mammogram, and the reviewing radiologist uses the Breast Imaging Reporting and Data System (BI-RADS) to note the existence and suspiciousness of any lesions. Usually, if a score of 4 or higher is reported for any of the lesions, the patient is sent for a biopsy to sample the tissue for further analysis. The sampled tissues are then examined under magnification by an anatomic pathologist. To quantify sample characteristics, grading and scoring systems are used by the pathologist to determine whether the tissue is diseased (diagnosis), what subclassifications of the disease are applicable [3], the level of tumor aggressiveness, as well as possible treatment options. The majority of these data are not discernable from imaging methods like mammograms alone. Therefore, pathology is the most informative modality for breast cancer diagnosis and provides critical information for patient treatment decisions.

However, the quality of analysis in some pathology centers has been called into question recently [4], and there is a large body of evidence indicating mid-to-high discordance rates between pathologists, even when using a consistent scoring system [5–11]. The net result of such inconsistency is a reduction in the quality of care, since different treatments can be prescribed depending on the slide reviewer.

To combat these challenges, many components of image interpretation can be automated through image analysis techniques, in order to provide objective, quantitative, reliable, and efficient results for pathology images. These methods can improve the concordance rates among pathologists, as well as enable large-scale research efforts that investigate treatment efficacies and disease survival.

In this chapter, an automated algorithm for the classification of breast cancer histopathology images is presented. The algorithm focuses on a small number of disease-inspired, intelligently designed features in order to differentiate between benign and malignant digital pathology images. The structure of the chapter is as follows. First, the challenges to quality in all aspects of the breast cancer pathology workflow are discussed. Next, the algorithm, which addresses analytic quality specifically, is presented. Detailed sections on preprocessing methods, feature extraction, image classification tools, and validation frameworks follow, concluding with the results of the presented method. The final section summarizes the topics of current research and future directions.

18.2 Background

Pathology is the study of bodily tissues and cells under magnification in order to make a judgment on the state of disease. Histological pathology analysis (histopathology) is done using high-power microscopes and thinly sliced tissue samples, extracted from the region(s). Samples are colored using a complementary set of stains, highlighting components of the tissue which are pertinent to the examination. The pathologist then identifies regions of the section which are diseased, if any, and employs semi-quantitative scoring schemes to quantify the severity of disease. In particular, two main types of slides are used commonly in breast cancer diagnosis, treatment planning, and patient management: immunohistochemistry (IHC) and hematoxylin and eosin (H&E).

18.2.1 Immunohistochemical Stains

Immunohistochemical slides are assays that investigate biomarker presence or absence in a cellular structure. The identification of biomarkers, or proteins, is important, as their presence is an indication of cell-state, which could be related to cancer aggressiveness, treatment efficacy, and prognosis, [5–8]. For example, the Ki-67 biomarker is a protein that indicates cell proliferation, which is important in determining whether cells are dividing at a non-normal rate (i.e., in aggressive tumors) [7]. Another biomarker, known as p53, is thought to be a tumor suppressor and indicates whether the natural cancer-fighting mechanisms of the body are functioning [8]. Other more popular and clinically used biomarkers are the Her2/neu, estrogen receptor (ER), and progesterone receptor (PR) proteins. Her2/neu is a membrane biomarker, and its presence (in high concentration) is an indication of a patient's likelihood to respond to Herceptin (a targeted chemotherapy drug) [5]. Similarly, the presence of high concentrations of ER/PR indicates a high likelihood that a patient would respond to hormone therapy [6].

To visualize these biomarkers, a DAB (3,3'-Diaminobenzidine) stain is used, which turns brown when the antibody of the stain binds to the antigen of the cell/tissue. Since tissue is essentially invisible under magnification, to visualize cells and tissue that do not have the biomarker(s), a counterstain is used. Once stained, pathologists use scoring systems to quantify the presence or absence of a particular biomarker.

In stains such as Ki-67 and p53, the number of cells that are positively stained brown (which indicates the presence of a biomarker) are simply counted. Therefore, the response is binary for every cell: either the cell has this protein, or it does not. Sometimes these binary scores are taken as a ratio, which requires counting the number of positively stained cells as well as the number of negatively stained cells. Other scoring methods, such as those used for Her2/neu, ER and PR, are not binary, as the intensity (darkness) of the stain is indicative of the concentration of the biomarker. For example, the Allred score, typically used for ER and PR, aims to quantify the proportion of positively stained cells, as well as the intensity level, or concentration of the protein present in the cell [12]. These scoring systems have a sliding scale usually quantized to a small number of bins.

18.2.2 Hematoxylin and Eosin Stains

More commonly used staining protocols, in all types of cancers and tissue analysis, include those that employ the H&E stains. Instead of considering specific proteins or biomarkers, H&E stains make visible the structural characteristics of cells and tissue [13]. Hematoxylin is a bluish (or purple) stain that has an affinity to acidic structures like nucleic acids (nuclei). Eosin usually has an affinity to the background stroma and is generally colored pink.

To diagnose and analyze breast cancer morphology, H&E slides are used to grade the cancer. The grade assigned for a breast carcinomas is an important surrogate marker of aggressiveness and metastatic potential that is used by oncologists when making treatment decisions [3]. The most commonly used grading system for invasive ductal carcinoma is the Nottingham Grading System and its variates (NGS) [10,14]. The NGS uses three morphological features: tubule formation (the percentage of tubules or ducts that occupy the tumor area); nuclear grade (shape, size, presence/lack of nucleoli, cellularity, and other nuclear features); and mitotic counts (the number of cells undergoing cell division per high-power field (HPF)). The NGS grading system is summarized in Table 18.1, where the total score for a given sample is the sum of scores in all three categories [14].

18.2.3 Challenges in Pathology

Pathology is the definitive diagnosis and is critical to patient care and management. As a result, the accuracy, quality, and reliability of pathological analyses are of utmost importance. Unfortunately, there has been some scrutiny around the quality of care and treatment of patients in some pathology centers [5–11]. Such deficiencies have stimulated the development of quality assurance protocols and governing bodies to ensure excellence in the pathology discipline [15]. Programs are being developed around the world, including in Canada [15], Europe [16], and the US [17].

TABLE 18.1 Nottingham Grading System Semi-Quantitative Scoring Criteria for Breast Cancer Histology

	Score		
Feature	1	2	3
Tubule formation	>75%	10%–75%	<10%
Nuclear pleomorphism	Small, regular uniform cells	Moderate increase in size and variability	Marked variation
Mitotic counts (per mm^2)	0–37	38–72	>73

Quality assurance programs and improvement plans in surgical pathology seek to "assure" and "improve" surgical pathology "products." The three quality areas within the discipline of pathology have been identified; these include a variety of factors along the pathology workflow. These three quality areas are summarized as follows [16,18]:

- **Pre-Analytic Factors:** All processes before the slide reaches the pathologist.
  - Quality issues include whether the right test has been ordered for the patient, whether specimen fixation occurs in a timely manner, and the quality of the staining.
  - Broadly grouped into: Test ordered, delivery, and accessioning.
- **Analytic Factors:** All processes involving the interpretation of the slide by the pathologist and the resulting diagnosis and/or treatment recommendation.
  - Quality issues include the diagnostic accuracy and precision, scoring/grading reproducibility, and bias.
- **Post-Analytic Factors:** All processes that occur after the pathologist has completed their examination.
  - Quality issues include challenges that arise during transcription, report delivery, follow up, monitoring, and maintenance.

One area that has recently been receiving much attention is the accuracy and reliability of the pathology analysis (analytic quality), and technologies are being designed, developed, and sought among the clinical and research communities to combat these challenges. Pathology slide interpretation is an inherently subjective process since it is based on visual examination. While semi-quantitative grading and scoring systems have been proposed and integrated into clinical practice with the aim of reducing subjectivity, reviewer disagreement persists [5–11].

In this chapter, the focus is on the analytic factors of the pathology workflow, the challenges associated with interpretation of histopathological images, and how automatic image analysis methods can reduce the variability of pathology slide interpretation.

18.2.4 Analytic Quality: Diagnostic Reproducibility, Precision, and Bias

There are challenges associated with the analysis of both breast cancer IHC and H&E-stained slides. As previously stated, the scoring methods for IHC assays rely on counts of cells which stain positively for each marker. For example, the Ki-67 proliferation index (PI) counts the proportion of malignant cells which are marked by the stain. Unfortunately, the concordance rates for Ki-67 PI have still been shown to be mid-low; in one study of 5 pathologists, grade 2 breast tumors were found to only have poor concordance ($k=0.17–0.49$), while grade 1/grade 3 tumors fared better, at $k=0.56–0.72$ [7]. In another study, a comparison of expert and non-expert pathologists scoring p53 expression in endometrial carcinomas found poor concordance ($k=0.13–0.25$) [8].

In these cases, lack of lab specialists and scoring experience were blamed for the undesirable results. However, there are also challenges with the histopathological methodology. For instance, what determines the threshold of stain darkness required to count a cell as positive? No specific definition is given. Additionally, since analysis of the entire tissue sample is laborious, at times, only a subset of the high powered fields (HPF, one visible area under magnification) are considered for scoring. Yet the number and location of HPFs in the sample is subjectively selected by each reviewer. Furthermore, some reviewers use a rough estimation of stained cell proportions, or "eyeballing," in lieu of physical counting; this could have significant impacts on the reported results. Efforts have been made to regulate Ki-67 scoring, however the recommendations continue to evolve [19].

Perhaps the most commonly discussed breast cancer pathology discordance rate is for the scoring of Her2/neu overexpression. The H-score, as it is called, measures the percentage of positively stained cells using a Her2/neu IHC assay, on four different intensity levels [20]. A Brazilian study analyzing agreement between 149 local and reference laboratories regarding Her2/neu overexpression in breast carcinomas found that in only 171 of 500 cases (34.2%) the same result was given [5]. Other studies have found similar, though less severe, weaknesses in estrogen and progesterone receptor (ER/PR) intensity and proportion scoring [6–11].

Evidently, visual estimation of antigen concentration level (i.e., in Her2/neu, ER/PR) entails subjectivity in making the distinctions between adjacent intensity levels, which is a greater challenge compared to the binary labeling issue in Ki-67 scoring. Moreover, the task is confounded by stain concentration variation between centers, meaning the same level of darkness or color could indicate different antigen concentration levels depending on the laboratory where it was prepared. In the absence of standardization, the observed discordance in pathological scoring might be expected.

While scoring systems consider a biomarker concentration and or presence in IHC slides, grading systems consider an aggregate of features based on morphological and structural characteristics of the tissue in H&E slides. As discussed, the most popular breast cancer grading tool is the NGS [9], and it considers three equally weighted features for classification: (1) tubule formation, (2) nuclear grade, and (3) mitotic count.

There is extensive literature on the agreement (or lack thereof) of pathologists using NGS-based systems. One study of six pathologists, considering each feature independently, found poor-to-moderate agreement for all three ($k=0.64$, 0.52, 0.40, for the tubule formation, nuclear pleomorphism, and mitotic count features, respectively). A more comprehensive and recent review of several studies on this topic concluded that there is large variability in grade reproducibility for the grading results overall [10].

Again, this is not surprising, as features like the proportion of tubule formation cannot be accurately and reproducibly quantified without computer assistance for counting pixels. Moreover, the nuclear pleomorphism features lack numerical definitions altogether, so the translation to a score is entirely dependent on the pathologist's interpretation of the criteria. Similarly, though mitotic counts are numerical, the identification of dividing cells is challenging, and the low number of these events reduces the statistical power of the feature for limited HPFs [21].

To conclude, there are a large number of challenges to providing a consistent and quantitative definition of image features. This results in variability in the accuracy and reliability of pathological slide interpretation and indicates the potential for breast cancer patient mismanagement. These factors, often compounded by limited experience and access to specialists in the lab, can be summarized as follows (as first reported by Khademi [22]):

- Establishing antigen-positivity based on visual perception of color
- Determining antigen-concentration based on visual perception of chromogen intensity
- Visual, qualitative definitions of thresholds to differentiate between intensity scores
- Choosing the most representative HPF for scoring and grading according to image understanding and experience
- Selecting the adequate number of HPF based on evolving quality assurance protocols

- Using qualitative descriptors to explain visual cues, such as cell and tissue morphology
- Manual counting or "eyeballing" to estimate the number of cells and objects
- Visual estimation of measurements, such as area, volume, size, and so on.

18.2.5 Digital Pathology

Evidently, there are many challenges to the analytic phase of the pathology workflow. However, recall that there are also difficulties in the pre- and post-analytic stages of pathology. The most comprehensive solutions to these challenges have come through technological developments and engineering. In fact, a movement toward digitization of pathology promises to improve all three quality phases and the entire pathology workflow through end-to-end solutions. Specifically, whole-slide image scanners, digital archives, and image viewing software are being integrated into hospital networks. The result is a completely digital workflow, from test ordering, to reporting, and follow up, which streamlines many of the processes that contributed to quality challenges. Some of the main digital pathology applications that address quality issues are summarized below [23]:

- *Database management*: Histological images and the associated clinical data of thousands of patients can be stored, transferred, and accessed more efficiently and less expensively, in contrast to the analog methods. Moreover, annotations and links to patient records can be consolidated in a common location.
- *Telemedicine and remote consultations*: With online access to digital images, obstacles to obtaining a second opinion are removed, and remote specialist and expert reviews are made possible, allowing underserved regions access to pathology expertise and opinions.
- *Potential for automated image analysis*: Now that pathology images are stored in a digital format, image analysis algorithms and software systems can be designed to automatically process the images. Such tools to aid pathologists in quantification and decision-making tasks are already under development and hold much promise to improve the efficiency and reliability of analysis and diagnosis.

This third aspect is, in fact, an exciting frontier in medical imaging research and its potential remains largely untapped. The goal of such research is a set of software modules that automatically perform image analysis and give prognostic quantities.

Automated image analysis systems for digital pathology images can generally be classified into preprocessing, segmentation, feature extraction, and decision-making (classification) tools. The goals of preprocessing modules are to denoise images, normalize the appearance of stains [24], or sometimes separate the images according to their stain concentrations [25,26]. These steps facilitate more simple or robust techniques downstream, by standardizing the images before further analysis. Standardization is an especially important objective, since histology sample preparation protocols are still not standardized, resulting in wide variability in the appearance of tissues stained by the same chromogens.

Perhaps the most ubiquitous, yet unsolved, challenge in automated pathology analysis is that of nucleus segmentation. Accurate segmentation of nuclei not only allows extraction of features for pleomorphic characterization like shape, texture, size, and color, but also precludes many metrics used to quantify cellular structure in samples [27,28]. Popular approaches have employed various combinations of contour-based object detection [29], thresholding, and morphological operations [27]. More ambitious approaches aim also to classify the segmented nuclei, since there are familiar flaws in manual performance of this task [27,30].

Other detection tasks, more commonly called rare event detection methods, have also received significant attention, such as the task of mitosis detection, since this feature is both clinically very useful and also extremely challenging to identify [31]. The most successful algorithm to date is a deep convolutional neural network (NN) trained on the MITOS dataset [21]. Deep NNs are powerful tools for object detection, but require large datasets for training supervision.

Finally, many methods, building on the above, have been proposed as tools for computer-aided diagnosis (CAD) or prognosis (CAP), estimating disease diagnosis or tumor grade based on image features [32]. Direct components of manual grading systems can be used, like mitotic counts, though more creative translations of qualitative features are often required. For instance, graph-based structural features are frequently derived from Voronoi diagrams, Delaunay triangulation maps, and minimum spanning trees, while texture features are computed using Gabor analysis, Harlack features, and gray-level co-occurrence matrices [22,24,32]. Feature sets are then computed for each image in a database and subsequently fed into a trainable model in order to resolve discriminatory features or combinations. The model is then able to predict diagnosis or prognosis for unseen images.

Overall, the goal of image analysis systems for digital pathology is to improve reproducibility, reduce or remove operator bias, and increase throughput. This will not only improve patient care, but facilitate large-scale pathology research in a way which is not feasible for manual methods.

To combat the associated analytical quality challenges for breast cancer diagnosis, the remainder of this chapter focuses on digital image analysis solutions for breast cancer digital pathology images. In particular, an automated classification system is proposed that is used to differentiate between malignant breast cancer and benign histopathology images, using a few judiciously designed features.

18.3 Breast Cancer Histopathology Diagnostic Tool

To combat the challenges associated with analytic quality in breast cancer grading systems, the remainder of this chapter is focused on the design and development of an algorithm for identifying cancer in H&E-stained digital breast pathology images. While automated methods for IHC-stained image analysis are also of clinical interest, they are a subject for future work. In particular, the focus is on intelligent, disease-inspired features in order to differentiate images of malignant breast cancer from benign histology.

Existing methods for CAD and CAP of breast cancer histopathology images have been proposed in the literature [28,33–35]. These usually aim to stratify tumor grade automatically, and likely many of the same features which distinguish low and high grade cancer can be used to denote images of benign versus malignant pathology. Approaches can be broadly classified into those which extract only a few, relatively complex features (like tubule formation), and those which employ a large number of more simple features (like texture measures). Using a supervised predictive or regressive model, features are then used to label images with a particular grade or diagnosis. For example, in the work by Doyle et al. [33], almost 3,500 features are considered for prognostic tasks using support vector machine (SVM) classification, including textural features, wavelet-filtered images, gray-level co-occurrence features, triangulation statistics, and graph based features. In contrast, only three features are constructed in the work by Dalle et al. [35] for giving prognostic grade, drawn from the NGS criteria.

18.3.1 Motivation of the Approach

The algorithm presented here is also inspired by the NGS (see criteria in Table 18.1), but uses three entirely novel features, which robustly emulate those used by pathologists for grading breast cancer tumors. Specifically, two features attempt to quantify tubule formation (and cellular organization in general), and one feature computes a measure of nucleic pleomorphism. These metrics are helpful in differentiating benign and malignant cases because in aggressive tumors, the organized arrangement of cells is disrupted by deregulation of normal cell-to-cell signaling pathways. The result is that cells are distributed randomly throughout the tissue, rather than in rings and layers (as in normal or benign samples). Similarly, cancerous cells tend to have enlarged, textured nuclei, which distinguishes them from normal cells. The features are derived mainly from the hematoxylin image, as computed using color deconvolution and a binary nuclei segmentation mask. No attempt is made to identify or quantify mitotic events; while this feature is extremely useful, mitotic detection is a very challenging task worthy of its own investigations [21].

Overall, the emphasis is on a small number (only three) of intelligently designed features, drawn from the established hallmarks of cancerous pathology. This model is more difficult to design, but will maintain good generalization performance on new datasets since it inherently describes the manifestations of the disease itself. The proposed method contrasts high-dimensionality approaches, which consider hundreds of features [33]. Given the small size of typical validation databases in this field, such high-dimensional approaches have a large (often unmentioned) capacity to over-fit the training database, leading to poor performance on new images. A more detailed treatment of this topic can be found in recent machine-learning texts [36].

Intelligent feature design typically requires several modular steps of image processing. For this task of classifying breast cancer histopathology slides, these include preprocessing, stain deconvolution, nuclei segmentation, extraction of disease-specific features, and supervised training a classification tool. This pipeline is shown in Figure 18.1. The remainder of the chapter will present each step in detail and conclude with the results of the model.

18.3.2 Color Space Transformations

Many of the processing steps will involve manipulating the representation of color in the digitized image. Therefore, it is helpful to first describe some common color spaces and their relationships.

The universality of red-green-blue (RGB) digital displays, means that almost all color images are stored in the RGB color space. This color space is represented by orthogonal dimensions *R*, *G*, and *B*, each spanning the discrete range of values [0, *L*], where *L*+1 is defined by the bit-precision of the data type. An arbitrary color vector in the RGB space, *f*, then has components in each dimension *r*, *g*, and *b*:

$$f = \begin{bmatrix} r \\ g \\ b \end{bmatrix}, \tag{1.1}$$

where:

$f = [0,0,0]^T$ is black

$f = [L,L,L]^T$ is white

A given color image with size $N \times M$, is then simply a set of color vectors in *xy* space:

$$f(x,y) = \begin{bmatrix} r(x,y) \\ g(x,y) \\ b(x,y) \end{bmatrix}, \tag{1.2}$$

Indexed by discrete spatial dimensions (pixels) $x \in [1,N]$ and $Y \in [1,M]$.

Color image data can be transformed between color spaces using linear transformation matrices. One useful color space considered in this work is the HSV (hue, saturation, value) space. Value, or brightness,

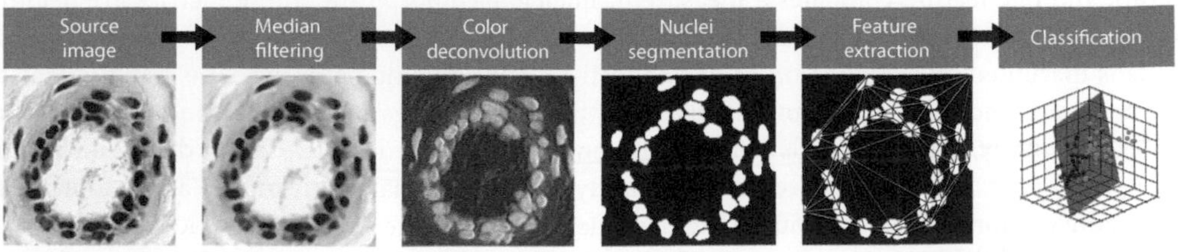

FIGURE 18.1 Block diagram overview of algorithm showing preprocessing and feature extraction steps.

is the proximity of the color to black or white, along the line from $[0,0,0]^T$ to $[L,L,L]^T$ in the RGB space. Hue is the dominant color component, or the polar angle around the value line, while saturation is the intensity of the color, or the magnitude of the polar arm stemming from the value line. Another useful space that is used in this work is the L*a*b* space, comprising of a similar brightness dimension L^*, and two color dimensions a^* and b^*, spanning red to green, and yellow to blue, respectively. The L*a*b* space was designed to cover a wider range of perceptible colors, is a perceptually uniform (linear) color space, and is device-independent, unlike the RGB or HSV color spaces. The equations to achieve the discussed color space transformations can be found in the text by Plataniotis and Venetsanopuolos [37].

18.4 Preprocessing

The image analysis pipeline begins with two important preprocessing steps. The goals of preprocessing are to reduce the impacts of image degradations like noise and stain variability so that downstream processing can give more robust results. The two steps are denoising and color deconvolution.

18.4.1 Denoising

During the slide digitization process, acquisition noise can be introduced, which degrades the quality of the images. Correction of these artifacts facilitates more reliable processing of pixels by rejecting outlier pixel values and strengthening consistent intensities within objects. Since this noise is usually high-frequency, denoising filters are typically low-pass. However, because object edges are also important for delineating structures and linear low-pass filters blur edges, nonlinear filters are preferred. A simple, yet effective nonlinear filter is the median filter; for all pixels in the input image, the median filter computes the median value of the pixels in a (3×3) neighborhood surrounding the pixel as the output pixel value. Each RGB color channel is processed independently:

$$C(x,y) = \text{Median}\{c(x-i, y-j) \mid i = [-1:+1], j = [-1:+1]\}, c = \{r,g,b\} \tag{1.3}$$

18.4.2 Stain Deconvolution

Despite advances in whole slide imaging and other engineering solutions in the pathology workflow, histopathology slide preparation has yet to be standardized. The result is that different stain vendors and application protocols yield digital images with varying stain hues and intensities [28]. Figure 18.2 demonstrates this range of characteristics in four digitized slides. Evidently, these variations present challenges for automated analysis algorithms, since a particular hue of purple may dictate nuclei in one image, but in another image, this color may not even be present. This is even more apparent in software applications that use static vectors for eosin or hematoxylin stain (ImageJ), which undoubtedly fail because the hues in the image depend on how the slide was prepared. Other variables that create color variability in digital pathology images include the stain concentration itself, illumination of the image, and the histological content of the slide.

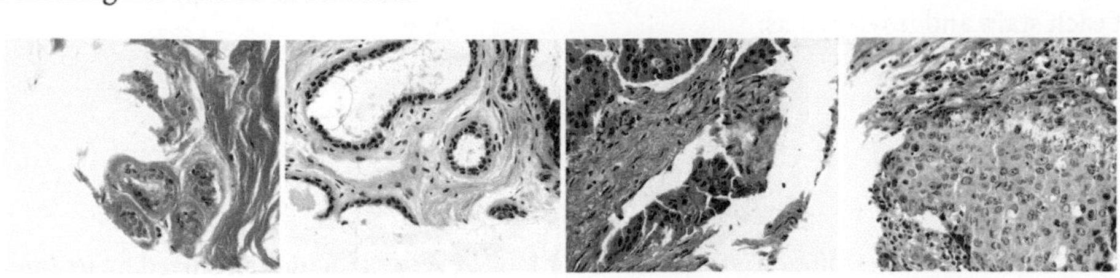

FIGURE 18.2 H&E-stained digital slides showing variability of stain hue and concentration across images due to different stain vendors and application protocols.

Stain deconvolution is a powerful tool for separating specific chromatic components of an image, permitting isolated analysis of the structures stained by each chromogen. Furthermore, if the stain hues are estimated on a per-image basis, the resulting stain concentration images are *independent* of the stain preparation protocol; that is, color normalization procedures are rendered unnecessary.

In H&E-stained images, stain deconvolution aims to un-mix the hematoxylin and eosin stains in the image into separate gray-scale images, effectively isolating the nuclei and background stromal structures, respectively. In fact, stain deconvolution is no more than a color space transformation; the data in the original RGB dimensions are transformed to the new space spanned by hematoxylin (*H*), eosin (E), and one "no stain" (*NS*) dimensions [25]. A bright pixel in one of the three resulting stain images represents a high concentration of that stain at this location in the original image. Importantly, pixels are assigned a mixture of stain components, instead of hard class memberships, providing a more complete representation of stain interactions, unlike other methods [24].

The model for stain deconvolution is based on Beer's Law of absorbance, which states that the transmitted intensity of monochromatic light passing through an absorbing medium has the following relationship:

$$I = I_o \times e^{-\sum s_i \alpha_i} \tag{1.4}$$

where:

I_0 is the incident light
I is the transmitted light
s_i is the stain concentration of stain i
α_i is the absorption coefficient of the corresponding stain

To linearize this relationship, the optical density space is often used by taking the logarithm of the ratio of the incident and transmitted light:

$$D = -\ln\left(\frac{I}{I_o}\right) = \sum s_i \alpha_i \tag{1.5}$$

For RGB images (three wavelengths of light) with three stains, this relationship may be written in matrix notation as in:

$$D = SV \tag{1.6}$$

where:

$D = [D_R, D_G, D_B]^T$ are the three optical density values of a single RGB (intensity) pixel
$S = [s_1, s_2, s_3]^T$ are the stain concentrations for each of the three stains at this pixel

V is the matrix that contains three stain vectors, which are vectors $(\vec{v_1}, \vec{v_2}, \vec{v_3})$ that characterize the color of each stain and is written as:

$$V = \begin{bmatrix} v_{1R} & v_{1G} & v_{1B} \\ v_{2R} & v_{2G} & v_{2B} \\ v_{3R} & v_{3G} & v_{3B} \end{bmatrix} \tag{1.7}$$

If the stain vectors can be estimated, stain concentration at every pixel may be solved by using [25]:

$$S = inv(V) \times D \tag{1.8}$$

In the case of H&E deconvolution, s_1 corresponds to the amount of hematoxylin stain at every pixel, denoted *h(x,y)*, and the stain color is characterized by v_1. Similarly, s_2 describes the amount of eosin stain at every pixel, denoted *e(x,y)*, and the stain color is characterized by v_2. The last "stain" actually refers to no-stain, and the color can be defined by $[0,0,0]^T$. The result of solving for S is three new stain image channels:

$$\begin{bmatrix} h(x,y) \\ e(x,y) \\ ns(x,y) \end{bmatrix} \tag{1.9}$$

The challenging aspect of color deconvolution is the estimation of per-image stain vectors before stain un-mixing [26]. In the proposed algorithm, this is done by first estimating the hue of each stain and then combining these with typical values for the value and saturation components (in HSV space).

The hues of the stain vectors are estimated using a saturation-weighted hue histogram; this minimizes the impact of noisy hue values in nearly white pixels, which represents pixels with low stain concentration [38]. This histogram, shown in Figure 18.3, typically demonstrates a bimodal shape. However this bimodal nature is not a reliable feature, and bimodal peaks do not represent true stain hues, since stain mixing skews the distributions. Therefore, to estimate the stain hues, a single Gaussian distribution is fitted to the saturation-weighted histogram using expectation–maximization, and the hues are selected as the hue values at the mean ±1.5 standard deviation (black diamonds, Figure 18.3, left). See also Figure 18.3, right, which shows the distribution of pixels in the RGB space, as well as the selected stain vectors.

Hues are the most discriminatory characteristic of stains, but do not fully define the stain vectors. To complete the HSV-space representation of each stain color, typical values of saturation (0.9) and intensity (0.7) are used. The third, "no stain" vector is then defined as pure white in the HSV space, $[0,0,1]^T$, and the color vectors are transformed back to the RGB space. The three stain vectors, $(\vec{v_1}, \vec{v_2}, \vec{v_3})$, are then used to compute the stain deconvolution matrix *inv*(*V*); applying this to the optical density (negative log image) yields hematoxylin, eosin, and background images, *h*(*x,y*), e(*x,y*), and ns(*x,y*).

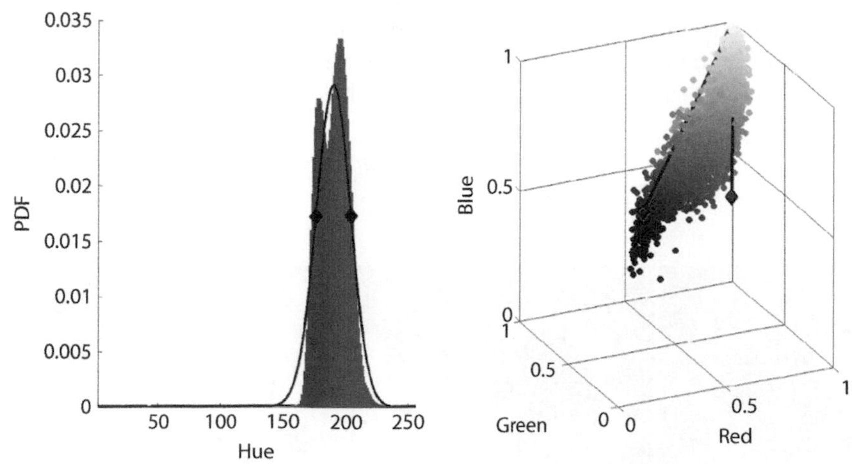

FIGURE 18.3 Saturation weighted histogram (left) and corresponding distribution of colors in the image (right). Histogram black curve shows best-fit Gaussian distribution and selected hues marked with black diamonds; resulting color vectors are shown in the point cloud with black vectors with diamond caps.

18.5 Nuclei Segmentation

As noted previously, nuclei are the most important objects for identifying malignancy. Analysis of their shape and texture requires that these objects be first segmented. Since hematoxylin is a nucleophilic stain, the extraction of the hematoxylin image greatly simplifies the challenge of identifying and delineating nuclei. However, variability in the appearance of nuclei, especially between normal and cancerous cells, makes segmentation difficult regardless [28]. In fact, a plethora of nuclei segmentation algorithms have been proposed [28]. After an initial nuclei estimate, many of these employ morphological operators to clean up segmentations after a first pass, based on reasonable assumptions about nucleus size and shape [28]. In this work, we employ a thresholding-based approach using a novel feature image and morphological postprocessing.

18.5.1 Novel Feature Image and Thresholding for Robust Nuclei Segmentation

To account for intensity variations in the hematoxylin image, image normalization is first performed using the image mean μ_h and standard deviation σ_h:

$$\check{h}(x,y)=\frac{h(x,y)-\mu_h}{\sigma_h} \tag{1.10}$$

Next, a combined feature image $\eta(x,y)$ is defined using the weighted sum of two filtered versions of the normalized hematoxylin image:

$$\eta(x,y)=\alpha_1 h_1(x,y)+\alpha_2 h_2(x,y) \tag{1.11}$$

where:

$h_1(x,y)$ is a nonlinearly smoothed version of $\hat{h}(x,y)$

$h_2(x,y)$ negatively highlights edges in $\hat{h}(x,y)$

The smoothed image $h_1(x,y)$ identifies where nuclei are expected to exist, while the edge image $h_2(x,y)$ emphasizes boundaries around nuclei. A pixel will, therefore, be predicted to be nucleus when h_1 has a high value (nuclei interior) and h_2 is also high (non-edgy region), or at least not low. Figure 18.4 gives an example of a hematoxylin image and the corresponding feature images.

The smoothed image is computed using gray-scale morphological opening [39]:

$$h_1(x,y)=\left(\hat{h}(x,y)\circ SE(x,y)\right) \tag{1.12}$$

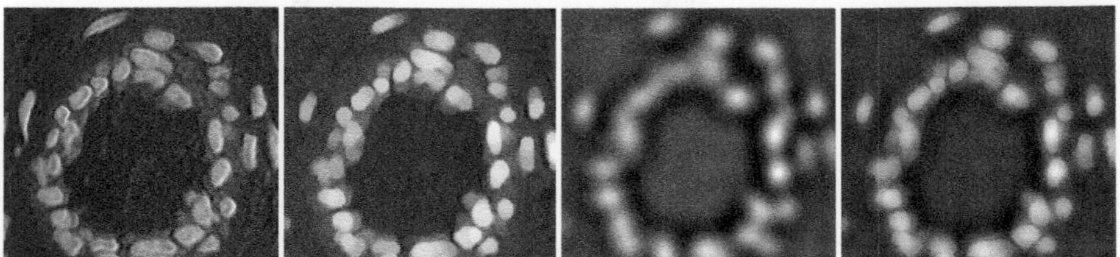

FIGURE 18.4 Original hematoxylin image $h(x,y)$, low-pass filtered image $h_1(x,y)$, edge-eroding image $h_2(x,y)$, and the resulting feature image $\eta(x,y)$ showing crisp outline of nuclei objects.

With a circular structuring element, $SE(x,y)$, having a radius of 4 pixels (area 13 μm^2). This size was selected empirically during development after showing best results. For bright structures in the hematoxylin image larger than $SE(x,y)$, this operation smooths the object interior (such as textured malignant nuclei), while maintaining crisp boundaries with the background. Objects smaller than $SE(x,y)$ are removed.

The edge image $h_2(x,y)$ is computed using the positive Laplacian of a Gaussian operator. Usually, it is negative; however, this positively emphasizes edges, whereas it is desired that edges are negative here.

$$LoG(x,y) = +\frac{1}{\pi\sigma^4}\left[1 - \frac{x^2+y^2}{2\sigma^2}\right]e^{-\left(\frac{x^2+y^2}{2\sigma^2}\right)} \tag{1.13}$$

$$h_2(x,y) = \left(\check{h}(x,y) \star LoG(x,y)\right) \tag{1.14}$$

The LoG operator acts as a band-pass filter, capable of simultaneously emphasizing edges and blurring the result to suppress noise. Choosing =8 pixels (4 μm), nucleus-sized objects have their edges darkened, but are also highlighted as bright focal points corresponding to the positive peak of the positive $LoG(x,y)$ $LoG(x,y)$ operator.

Finally, the combined feature image $\eta(x,y)$ is defined as the weighted sum of the images $\hat{h}_1(x,y)$ and $\hat{h}_2(x,y)$ (also normalized), as shown above, using equal weights $\alpha_1=1$, and $\alpha_2=1$. These weight values were validated empirically and have demonstrated good results. A threshold of $\eta > 2$ (corresponding to one standard deviation in both images, or an equivalent combination) is then used to segment nuclei in the image, yielding a binary mask, $N_0(x,y)$ indicating whether each pixel is part of an estimated nucleus or not.

18.5.2 Object-Size Assertion

To analyze nuclei independently, the binary objects in $N_0(x,y)$ should correspond to single nuclei. Unfortunately, adjacent nuclei are not adequately separated using this feature image $\eta(x,y)$ and thresholding alone. Therefore, segmented objects in $N_0(x,y)$ are refined using minimum and maximum size specifications of 100 pixels (25 μm^2) and 400 pixels (100 μm^2), respectively [40]. This removes objects which are too small (e.g., blotches of hematoxylin), and objects which are too large (representing multiple nuclei) are split using the watershed transformation [39]. The image used to define watershed boundaries is $h_1(x,y)$, with inverse contrast (yielding dark nuclei). While the distance transform image is often used for watershed segmentation [28], the distance transform relies exclusively on the contours of the initial segmentation. As such, it ignores additional feature image information within the objects being considered for splitting.

The size assertion routine is also recursive, such that the watershed segmentation results are themselves tested for size conformity. Excessive recursion is guarded against with a maximum self-call allowance of 20, after which the offending object is removed from the nuclei mask (an extremely rare occurrence resulting from large blotches of hematoxylin).

The result of this algorithm is a binary nucleus segmentation mask, $N(x,y)$, containing exclusively objects sized 100–400 pixels, derived from the hematoxylin image.

18.6 Disease-Inspired Feature Design

Using the stain separated images from Section 18.4 and nuclei masks from Section 18.5, features inspired by the NGS can be extracted. The features are then used to classify images in the database as benign or cancerous.

18.6.1 Structural Feature I: Interstitial Distance (F_{S1})

Structural objects like tubule cross sections are extremely useful for characterizing the malignancy of breast cancer pathology slides; however, reliably quantifying the presence of these features is challenging, and algorithms targeting this feature are not extensive [28]. A novel approach, presented here, considers the interstitial spaces between nuclei—defined as any area between three or more nuclei. Specifically, a measure that quantifies the relative amount of spaces between nuclei is designed, making the feature robust to scale, shift, and rotation.

The fundamental principle surrounding the design of this feature is that cancerous nuclei lack spatial structure and are, therefore, distributed more chaotically throughout the slide. Conversely, tubules in low-grade or benign samples cluster nuclei tightly together, forming structures and leaving large gaps (such as the interiors of tubules). To quantify the size of these interstitial spaces, the distance transform of the nuclei segmentation mask is used:

$$\mathcal{D}(x,y\,|\,N)=\min\left(\sqrt{(x-x_n)^2+(y-y_n)^2}\;|\;x_N,y_N\in N(x,y)=1\right) \tag{1.15}$$

In nuclei masks from non-malignant tissues, there will be many large interstitial spaces, while in cancerous images, there will be many small interstitial spaces, and fewer large spaces, since there are essentially no structures. In the distance transform image, these manifest as a high proportion of high-amplitude peaks in images of non-malignant pathology (Figure 18.5, top), versus only small-amplitude peaks in chaotic images (Figure 18.5, bottom).

To quantify these proportions globally, the histogram of the distance transform image, $p(D)$, can therefore be used. Specifically, images of benign histology would be expected to maintain a more evenly distributed histogram, while malignant cases would have a high proportion of smaller distance values, peaking sooner. To model these differences continuously, the cumulative distribution function (CDF) can be considered, where a more gradual increase, versus a sharp increase, would be observed for the healthier (structured) versus malignant (chaotic) nuclei. Finally, the area under this CDF curve then quantifies the difference between these spatial arrangements, since it will be higher for the quickly-peaking disordered nuclei and lower for a more gradual CDF. This area is therefore the feature of interest and is computed using trapezoidal approximation.

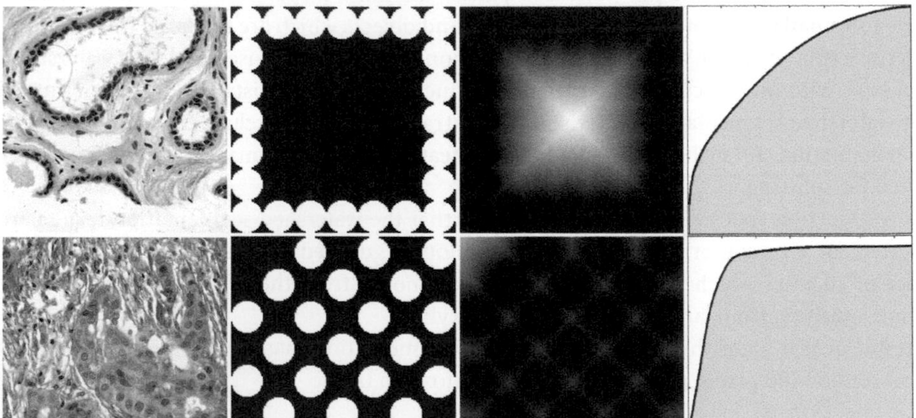

FIGURE 18.5 Toy example of the interstitial space feature for structured nuclei (top) and even or chaotic nuclei (bottom); left-to-right: example histopathology image, toy representative nuclei mask, distance transform image, and the CDF of distance image histogram. Large area under the curve indicates a more even distribution of nuclei in the image.

18.6.2 Structural Feature II: Triangulation Edge Lengths (F_{S2})

Another indicator of nuclei organization is the distribution of distances between them. For instance, if many nuclei in the image form epithelial structures, the distances between them will be smaller. These types of features can be quantified using the following method.

First, the centroid of each nucleus object is identified, and Delaunay triangulation computes the map of edge connections between the centroids. In Figure 18.6, it is shown how a healthy, structured arrangement yields many short edge lengths, while chaotic nuclei exhibit more medium length edges. Next, the lengths of all edges are calculated, and the probability distribution of the edge lengths for each image is approximated using the histogram (with kernel density estimation), giving $p(l)$. As before, the CDF of edge lengths is quantified from this probability estimate, and the area under the curve is considered as the feature. Here, however, a high proportion of small edge lengths (indicating epithelial organization, as in healthy tissues) will yield a sharper increase in CDF and more area under the curve. Conversely, disparate nuclei in malignant cases will result in longer edge lengths, a shallower rise in CDF, and less area under the curve (Figure 18.6, right).

18.6.3 Nucleus Homogeneity Feature (F_N)

The second component of the NGS aims to quantify nuclear pleomorphism—the degree of nuclear abnormality. A major feature of nuclei pleomorphism is lack of internal homogeneity; healthy nuclei tend to have darker, concentrated centers while cancerous nuclei usually appear grainy and transparent toward the center. Therefore, the distance from the nucleus edge should be positively correlated with the

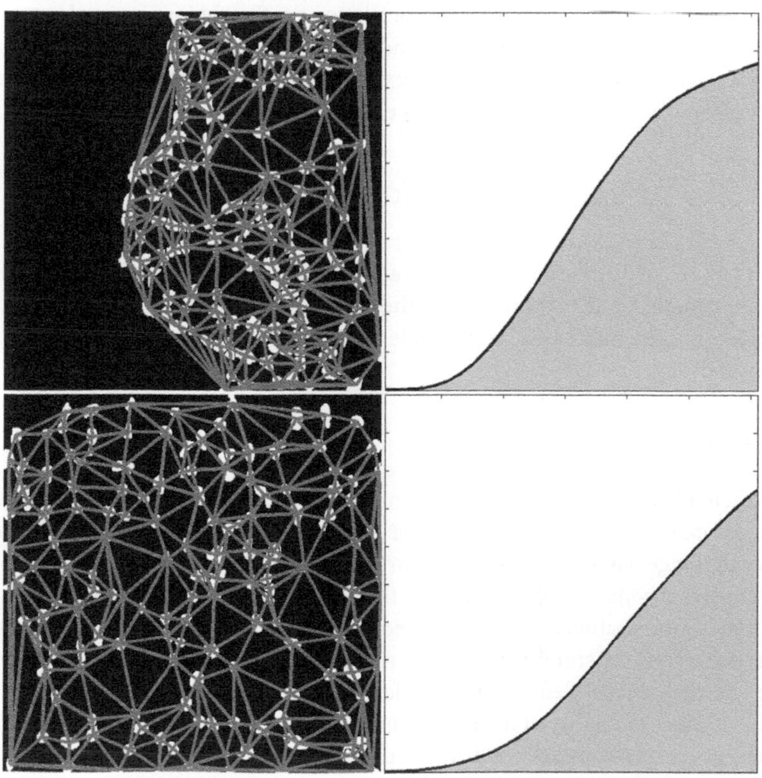

FIGURE 18.6 Delaunay triangulation and edge length CDF for an image of benign (top), and malignant (bottom) histology. Larger area under the curve indicates more tightly packed (organized) nuclei.

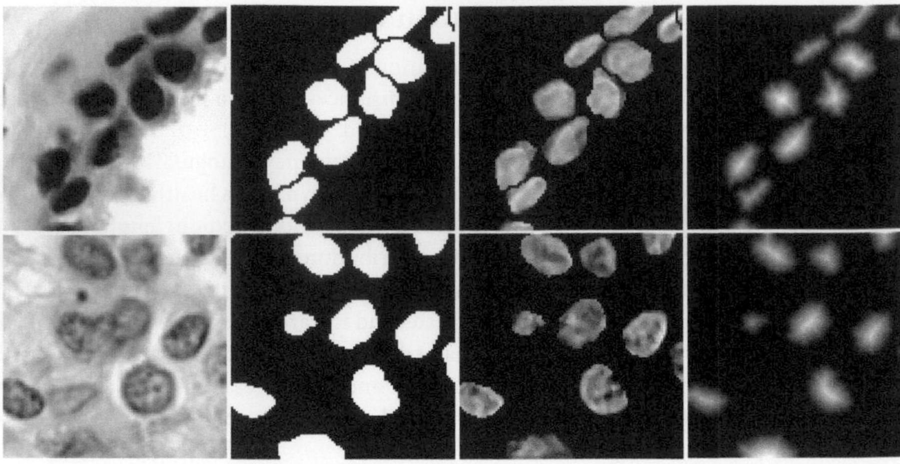

FIGURE 18.7 Benign (top) and malignant (bottom) nuclei showing difference in nuclear stain density and correlation with distance transform image; left-to-right: original image, segmentation, masked hematoxylin image, and distance transform image.

stain density in healthy nuclei and negatively correlated in cancerous ones. Figure 18.7 shows characteristic healthy and cancerous nuclei and the corresponding hematoxylin image and distance transform of the segmentation. These two images should depict similar intensity values in healthy nuclei and opposite values in malignant samples.

For a single nucleus object $O(x,y)$ in $N(x,y)$, a correlation coefficient is computed between the internal distance transform of a nucleus mask $D(O(x,y))$ and the hematoxylin image over the same area $h(x,y) \times O(x,y)$. In order to define the global feature, however, all nucleus objects should be considered, so the average of correlation coefficients for all objects is used:

$$F_{NH} = \frac{1}{\#n}\sum_{n}^{\#n} corr\left(h\left(O_n\left(x,y\right)\right),\ \mathcal{D}\left(O_n\left(x,y\right)\right)\right) \tag{1.16}$$

This approach is more robust than other texture analysis methods in that it leverages the spatial constraints of the problem to quantify the feature, rather than a specific texture type. For instance, methods like Gabor analyses require multiscale, multi-angle wavelet definitions, which are expensive to compute and complex to analyze.

18.7 Classifier Tools

Finally, the extracted features are used to train a model for predicting whether an image is cancerous. Intuitively, images can be thought of as points in a space, where each dimension of the space is one feature. Since each image has a unique set of feature values, it exists as a unique point in the space. The task is then to construct some decision surface in the space which maximally separates the benign and malignant examples. Constraints are usually placed on the boundary, so that new images (test data) are classified in a reliable way; generally, this involves reducing the level of detail in the surface.

In fact, the choice of model defines the complexity of the decision surface, so should therefore be informed by the number of quality of features identified. For fewer features with high correlation with the output labels, linear or smooth boundaries should be used, as these will easily maintain good generalization performance [36]. For approaches with many features (high dimensionality), nonlinear classifiers like decision trees or random forests can be used; however, with small databases, it is more risky to use classifiers with higher capacity for detail, as they are likely to over-fit the training data.

Given the judiciously designed features outlined in this method, two simplistic classifiers are considered, a Naïve Bayesian classifier (NB) and an SVM. The only difference between these models is that the SVM boundary is linear with all dimensions, while the NB surface represents the interface of two Gaussian probability distributions in the feature space. It should also be noted that each feature (space dimension) is normalized to have zero mean and unit standard deviation before training the model.

18.8 Validation Framework

Despite extensive work on digital pathology analysis recently, few investigations employ robust validation frameworks. This goal includes a large number of images, originating from multiple centers using different stain protocols, and appropriate cross-validation (CV) techniques. It should be hard to promote a method which is validated on only single-center data for anything more than in-house work, given the level of image variability described. For example, in a well-known paper [33], one model achieves an impressive accuracy of 95.8%; yet this model had over 2,300 features for only 48 images, all from the same center. Moreover, there is no mention of the number of cross-validation randomizations. The validation framework employed here attempts to combat such downfalls by considering both image variability and a large number of cross-validation trials.

18.8.1 Image Database

The H&E-stained breast cancer histopathology image database used contains 58 images, with 26 examples of malignant pathology, and 32 examples of benign, as identified by an expert pathologist [41]. The images were acquired at 20 × magnification, covering an area of 0.17 mm^2 each (768 × 896 pixels, 0.5 μm/pixel). The database also has a number of challenging features, including large variations in stain hue and concentration, small overall field size, and a large number of tissue tears resulting from the fixation process. These reduce the consistency, quantity, and quality of histopathological information presented in the images, respectively. Strong performance on these images, therefore, would indicate good generalization of the algorithm on higher quality images.

18.8.2 Cross-Validation

To ensure that the model does not perform well simply by exploiting a chance partition through the training data, training and testing are repeated using randomized, three-fold cross-validation. Specifically, 1,000 iterations are computed using a random one-third of the image data for training the classifier tool and the remaining two-thirds for testing; in each iteration, all three partitions are rotated through for the training, yielding 116,000 total test classifications.

18.8.3 Validation Metrics

Performance for this binary classification problem is evaluated using three standard statistical measures: accuracy (*Acc*), sensitivity (*Sen*), and specificity (*Spe*). Images of malignant pathology (*M*, as identified by the expert), which are classified as malignant by the model, are considered true positives (*TP*), while true negatives (*TN*) are defined as images of benign histology, which are correctly labeled benign. False negatives (*FN*) and false positives (*FP*) follow accordingly, and the performance measures are defined as shown:

$$Acc = \frac{TP + TN}{TP + FP + TN + FN} \tag{1.17}$$

$$Sen = \frac{TP}{TP + FN} \tag{1.18}$$

$$Spe = \frac{TN}{TN + FP} \tag{1.19}$$

18.9 Experimental Results and Discussion

In this section, intermediate results from each processing step are presented, including preprocessed images and nuclei segmentation results. This is followed by an analysis of the features and a summary of the classification performance.

18.9.1 Stain Deconvolution

Following median filtering, the second step in the algorithm aims to un-mix the stains in each RGB image in the database. The stain deconvolution is computed using per-image optical density vector estimation, yielding separate stain images. In Figure 18.8, the gray-scale hematoxylin and eosin images are shown for the source image at left. The stain distributions appear to correlate well with the perceived distributions of stains in the original images. In the work by Gavrilovic et al. [26], stain ground truth images were created by coloring two adjacent sections with single-stains only; this method of validation could be considered in future works.

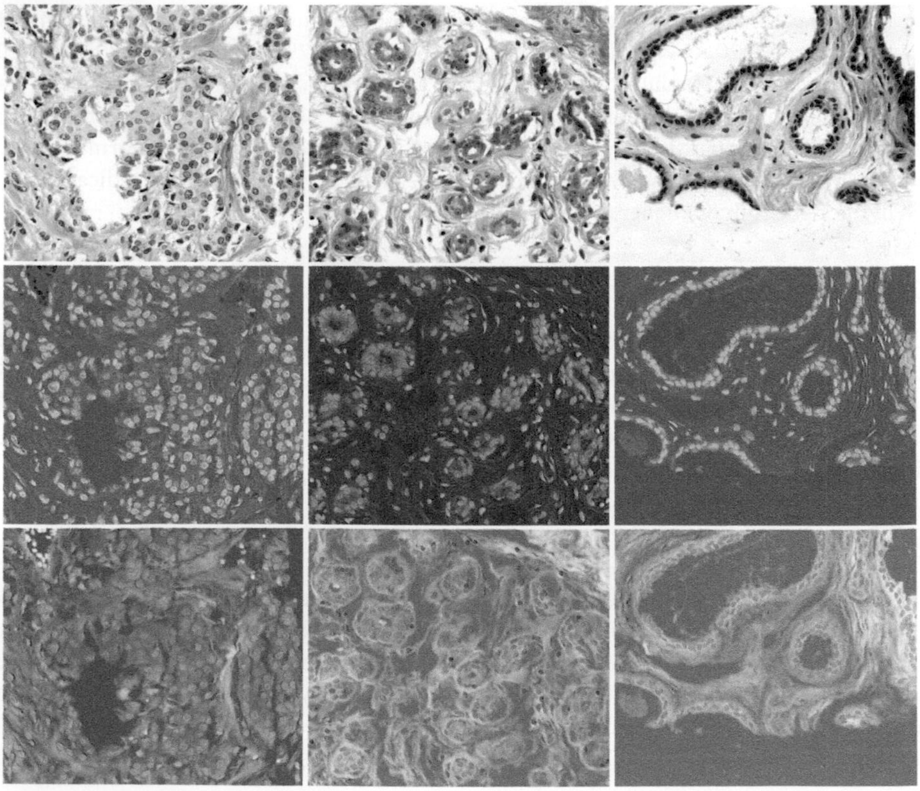

FIGURE 18.8 Stain deconvolution source images (top) and resulting hematoxylin (middle) and eosin images (bottom), showing reliable separation despite varying stain appearance due to per-image stain estimation.

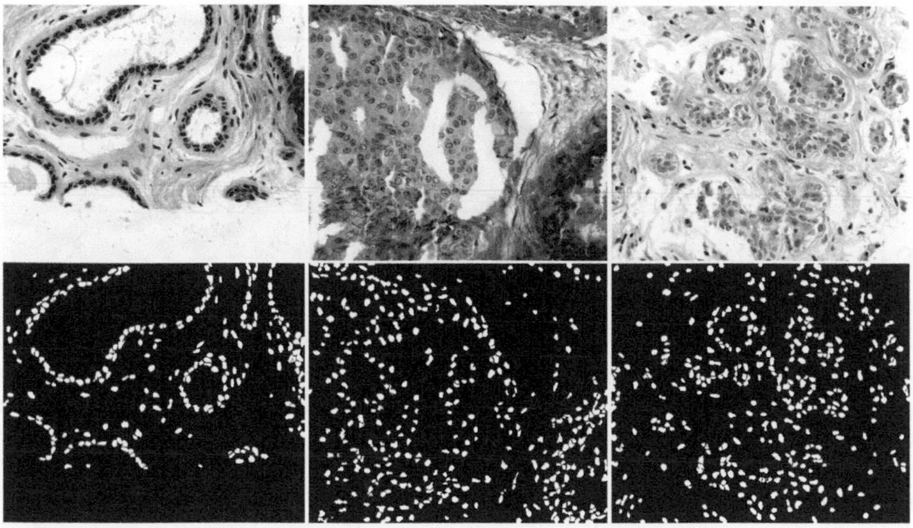

FIGURE 18.9 Nuclei segmentation: original images (top) and final masks (bottom).

18.9.2 Nuclei Segmentation

The next step in the algorithm is the segmentation of nuclei objects. This step is of utmost importance, since all ensuing features rely on the segmentation mask $N(x,y)$.

Several example results are shown in Figure 18.9, where it can be seen that the novel feature image $\eta(x,y)$ highlights nuclei well and that thresholding could then provide a very good initial segmentation. This paradigm of extensive preprocessing, followed by relatively simple thresholding can often give more reliable results, especially compared to techniques which try to incorporate all assumptions and operations in a few sweeping steps.

Following initial thresholding, the recursive separation of contiguous multi-nuclei objects using the watershed transform successfully asserts the assumed size constraints. However, this technique is relatively sensitive to the image used to define the transform. During the development stage, the gray-level morphological opening image, $\hat{h}_1$, was found to give better results than edge- or contour-based watershed transformations.

Validation of the nuclei segmentation results is not presented here; however, visual inspection confirms that the majority of nuclei are identified. Robust frameworks for nuclei segmentation can be found in the works by Veta et al. [13] and Wienert et al. [29]. The principal error in the method presented is the underestimation of nuclei area, especially in pale, textured nuclei (i.e., malignant), since the assumption is that nuclei are represented by hematoxylin image peaks. This limitation is also compounded by object edge erosion by the *LoG* operator in $\eta(x,y)$. However, a specific, rather than sensitive, identification of nuclei objects is preferred for the ensuing definition of features, since false positive nuclei objects may have large influence on the structural features.

18.9.3 Feature Quality

The ideal feature set has minimal cross correlation and maximum correlation with the output, thereby maximizing the separation of data used for classification. For the features employed here, all three are correlated with the binary label of malignancy, M as shown in Table 18.2. It can also be seen that cross correlations with other features have similar magnitudes as correlations with M (mean absolute values 0.58 and 0.62, respectively). The last column of Table 18.2 also gives the classification accuracy for each feature in isolation (in which case, the model is irrelevant), using the same cross-validation framework as described above.

TABLE 18.2 Correlations of Features with Each Other and with Malignancy (M)

Feature	Correlation (r^2)				1-Feature Accuracy
	F_{S1}	F_{S2}	F_N	M	
Interstitial distance F_{S1}	1	−0.64	−0.73	+0.60	0.80
Triangulation edge length F_{S2}	−0.64	1	+0.39	−0.66	0.81
Nucleus homogeneity F_N	−0.73	+0.39	1	−0.59	0.79

18.9.4 Classification Performance

This set of three Nottingham-inspired features achieves accuracies of 84.4% and 85.7% using SVM and NB classifiers, respectively. These results correspond to 116,000 image classifications using challenging test-heavy, three-fold cross-validation, as described in Section 18.8.2. Since the features employed here have direct correlation with malignancy, quasi-linear classifiers like NB and SVMs would be expected to perform well. Conversely, nonlinear classifiers like random forests would not be expected to improve accuracy significantly by identifying nonlinearly separable subspaces. The benefits of NB and SVM surfaces are their more reliable generalization performances.

Additional performance data, including sensitivity, specificity, and the total number of true/false positive/negative classifications, are shown in Table 18.3. The decision surfaces for each model (using full-batch training) are also illustrated in Figure 18.10. It can be seen that the feature representation of most images (points in the feature space) separates the two classes very well; however, a few images end up on the "wrong" side of the space.

In fact, the majority of classification errors generally arise from the same eight images, including four images comprising 12,280/15,921 (77%) of false positives, and another four responsible for 11,637/18,914 (62%) of false negatives (aggregate values from both classifiers). These images represent the few inseparable points in the feature space using quasi-linear classifiers, and this result implies they are simply challenging images. From Figure 18.11, it is easy to see why: many images have either a high proportion of background and/or tissue tear artifacts. The malignant examples, which are falsely labeled benign (bottom), have nuclei which are mostly clustered together, potentially instances of lower-grade cancer. Conversely, the false positive images (top) have very small tubules and nuclei which are distributed throughout the slide, which could be confused for chaotic nuclei, as considered by the structural features.

In fact, these errors highlight the major weakness of the algorithm currently—a reliance on the level of organization of nuclei in the image. From Table 18.2, it is clear than while the two structural features are most helpful, they are also partially redundant, having the largest cross-correlation (0.73). Therefore, one of these might be removed in future works.

It is expected that the presently described model would benefit most from the inclusion of a mitosis detection feature, since this NGS criterion is currently ignored. Due to the field size of the images employed in this investigation and the difficulty of this task in general, it was not feasible to include mitosis quantification here.

TABLE 18.3 Classification Performance of the Two Tools Using Cross-Validation

Classifier	Predicted	Truth		Accuracy (%)	Sensitivity (%)	Specificity (%)
		M	B			
SVM	M	53906	7953	84.4	84.2	84.7
	B	10094	44047			
NB	M	55030	7676	85.7	86.0	85.2
	B	8970	44,324			

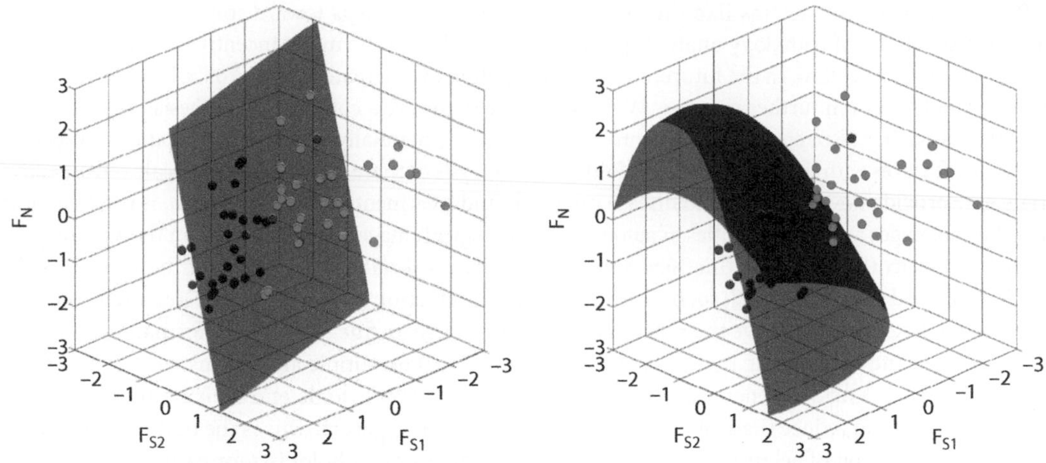

FIGURE 18.10 Feature space and decision surface for SVM and NB classifiers using and showing all training data (58 single-HPF images); separability of benign cases (blue) and malignant (red) is good even in the low dimensional feature space.

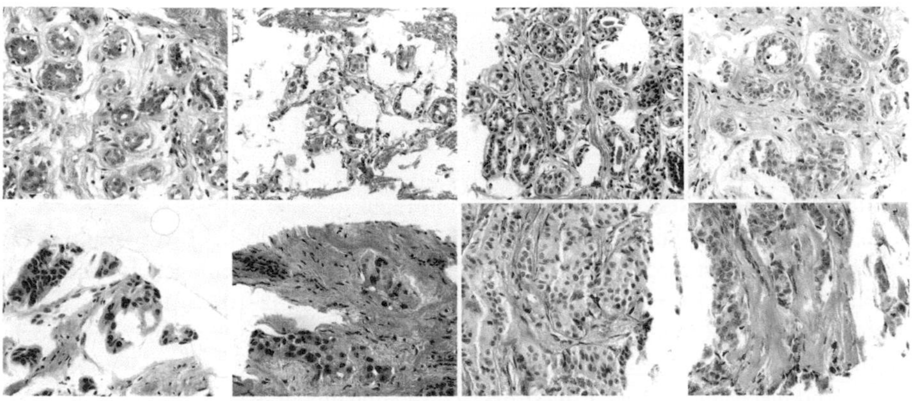

FIGURE 18.11 Commonly misclassified images: false positives (top) with small tubule structures and evenly distributed nuclei and false negatives (bottom) with relatively low numbers of nuclei overall; also shown is the high degree of sectioning artifacts.

Conclusions and Future Directions

In this chapter, a method for automated diagnosis of H&E stained breast cancer pathology images is presented. The endeavor is motivated by a review of the many challenges and pitfalls of manual histopathological analysis, including subjective interpretations of grading systems, biased estimation of tissue sample features, and arbitrary selection of HPFs for analysis. Extensive literature suggests that these factors are currently undermining the quality of patient care.

The algorithm is inspired by criteria already used by pathologists for making disease prognosis, namely the Nottingham Grading System. Extensive preprocessing and careful construction of only three features facilitates image classification with relatively simple Naïve Bayes and SVM decision tools. This approach is contrasted against many recently developed algorithms, which use large numbers of features and nonlinear classifier models, limiting the certainty of their generalization performance; yet the current method achieves competitive accuracy (~85%) on an extremely challenging image database using robust validation methodology.

The development of systems like this one represent important steps toward improving the accuracy and reproducibility of pathology analysis, particularly considering the many incentives for digitizing the entire pathology workflow. In the future, complete integration of automated image analysis tools with other patient data holds even further promise. While scoring systems are a convenient synthesis of information for physicians, they also represent an information bottleneck, potentially obscuring correlations in specific pathology features with patient outcomes. Multivariate analysis of these features in large-scale retrospective trials may provide novel insights into disease prognosis and treatment efficacies. As such, it would appear that the challenges and opportunities of automated pathological image analysis are still in their infancy.

In looking to the future of this new and exciting field, special attention should be drawn to the MICCAI Grand Challenges [42], such as the 2008 mitosis detection competition [31]. Research competition frameworks like these have the potential to simultaneously define rigorous performance standards through large and challenging image databases and to compare multiple algorithms on the same data. Researchers in this field are encouraged to participate in such events to defend their claims of algorithmic superiority and reduce the ambiguity in defining the state of the art. This is the best way to accelerate the dissemination of helpful automated tools into the hands of pathologists for patient care.

References

1. Canadian Breast Cancer Society, Canadian Breast Cancer Society, 2016. [Online]. Available: http://www.cbcf.org/ [Accessed: 09-Feb-2016].
2. C O'Connell, and V L Dickey, *Hematology and Oncology*. Malden, Massachusetts: Blackwell Publishing Ltd, 2005.
3. S. S. Cross, Grading and scoring in histopathology, *Histopathology*, 33, 2, 99–106, 1998.
4. B. McLellan, R. McLeod, and J. Srigley, Report of the Investigators of Surgical and Pathology Issues at Three Essex County Hospitals: Hôtel-Dieu Grace Hospital, Leamington District Memorial Hospital and Windsor Regional Hospital, Toronto, 2010.
5. S. C. L. Wludarski, L. F. Lopes, T. R. Berto, E. Silva, F. M. Carvalho, L. M. Weiss, and C. E. Bacchi, HER2 testing in breast carcinoma: Very low concordance rate between reference and local laboratories in Brazil, *Appl. Immunohistochem. Mol. Morphol.*, 19, 2, 112–118, 2011.
6. F. Bischoff, T. Pham, K. Wong, E. Villarin, X. Xu, K. Kalinsky, and J. Mayer, Immunocytochemistry staining for estrogen and progesterone receptor in circulating tumor cells: Concordance between primary and metastatic tumors, *Cancer Res.*, 72, 24 Supplement, P2–01–10–P2–01–10, 2014.
7. Z. Varga, J. Diebold, C. Dommann-Scherrer, et al., How reliable is Ki-67 immunohistochemistry in grade 2 breast carcinomas? A QA study of the Swiss Working Group of Breast- and Gynecopathologists, *PLoS One*, 7, 5, e37379, 2012.
8. K. Garg, M. M. Leitao, C. A. Wynveen, G. L. Sica, J. Shia, W. Shi, and R. A. Soslow, p53 overexpression in morphologically ambiguous endometrial carcinomas correlates with adverse clinical outcomes, *Mod. Pathol.*, 23, 1, 80–92, 2010.
9. H. F. Frierson, R. A. Wolber, K. W. Berean, D. W. Franquemont, M. J. Gaffey, J. C. Boyd, and D. C. Wilbur, Interobserver reproducibility of the Nottingham modification of the Bloom and Richardson histologic grading scheme for infiltrating ductal carcinoma, *Am. J. Clin. Pathol.*, 103, 2, 195–8, 1995.
10. E. A. Rakha, J. S. Reis-Filho, F. Baehner, et al. Breast cancer prognostic classification in the molecular era: The role of histological grade, *Breast Cancer Res.*, 12, 4, 207, 2010.
11. E. Kornaga, A. Klimowicz, M. Konno, N. Guggisberg, T. Ogilvie, R. Cartun, D. Morris, M. Webster, and A. Magliocco, Comparison of three commercial ER/PR assays on a single clinical outcome series, *Cancer Res.*, 72, 24 Supplement, P1–07–10–P1–07–10, 2014.
12. M. E. H. Hammond, D. F. Hayes, M. Dowsett et al., American Society of Clinical Oncology/College of American Pathologists Guideline Recommendations for Immunohistochemical Testing of Estrogen and Progesterone Receptors in Breast Cancer (Unabridged Version), 2010.

13. M. Veta, P. J. van Diest, R. Kornegoor, A. Huisman, M. A. Viergever, and J. P. W. Pluim, Automatic nuclei segmentation in H&E stained breast cancer histopathology images, *PLoS One*, 8, 7, e70221, 2013.
14. C. W. Elston and I. O. Ellis, Pathological prognostic factors in breast cancer. I. The value of histological grade in breast cancer: Experience from a large study with long-term follow-up, *Histopathology*, 19, 5, 403–410, 1991.
15. Institute for Quality Management in Healthcare, About, 2013. [Online]. Available: https://iqmh.org/About. [Accessed: 09-Feb-2016].
16. Royal College of Pathologists, What is Quality in Pathology? Report of a Meeting to Discuss the Development of Laboratory Accreditation in the UK, London, 2009.
17. G. Konstantakos and D. A. Novis, Reducing errors in the practices of pathology and laboratory medicine, *Am. J. Clin. Pathol.*, 126, 1 Supplement, S30–S35, 2006.
18. R. E. Nakhleh, What is quality in surgical pathology? *J. Clin. Pathol.*, 59, 7, 669–672, 2006.
19. M. Dowsett, T. O. Nielsen, R. A'Hern et al., Assessment of Ki67 in breast cancer: Recommendations from the International Ki67 in Breast Cancer working group, *J. Natl. Cancer Inst.*, 103, 22, 1656–1664, 2011.
20. S. Detre, G. Saclani Jotti, and M. Dowsett, A 'quickscore' method for immunohistochemical semi-quantitation: Validation for oestrogen receptor in breast carcinomas, *J. Clin. Pathol.*, 48, 9, 876–878, 1995.
21. D. C. Ciresan, A. Giusti, L. M. Gambardella, and J. Schmidhuber, Mitosis detection in breast cancer histology images with deep neural networks, *Med. Image Comput. Comput. Assist. Interv.*, 16, Pt 2, 411–418, 2013.
22. A. Khademi. Image Analysis Solutions for Automatic Scoring and Grading of Digital Pathology Images, *Canadian Journal of Pathology*, 5(2), pp. 51–55, June 2013.
23. L. Pantanowitz, Digital images and the future of digital pathology, *J. Pathol. Inform.*, 1, 1, 15, 2010.
24. S. Kothari, J. H. Phan, T. H. Stokes, and M. D. Wang, Pathology imaging informatics for quantitative analysis of whole-slide images, *J. Am. Med. Inform. Assoc.*, 20, 6, 1099–1108, 2013.
25. A. C. Ruifrok and D. A. Johnston, Quantification of histochemical staining by color deconvolution, *Anal. Quant. Cytol. Histol.*, 23, 4, 291–299, 2001.
26. M. Gavrilovic, J. C. Azar, J. Lindblad, C. Wahlby, E. Bengtsson, C. Busch, and I. B. Carlbom, Blind color decomposition of histological images, *IEEE Trans. Med. Imaging*, 32, 6, 983–994, 2013.
27. H. Irshad, A. Veillard, L. Roux, and D. Racoceanu, Methods for nuclei detection, segmentation, and classification in digital histopathology: A review-current status and future potential, *IEEE Rev. Biomed. Eng.*, 7, 97–114, 2014.
28. M. Veta, Breast cancer histopathology image analysis: A review, *IEEE Trans. Biomed. Eng.*, 61, 5, 1400–1411, 2014.
29. S. Wienert, D. Heim K., Saeger A. Stenzinger, M. Beil P., Hufnagl M. Dietel, C. Denkert and F. Klauschen, Detection and segmentation of cell nuclei in virtual microscopy images: A minimum-model approach, *Sci. Rep.*, 2, 503, 2012.
30. T. J. Fuchs and J. M. Buhmann, Computational pathology: Challenges and promises for tissue analysis, *Comput. Med. Imaging Graph.*, 35, 7–8, 515–530, 2011.
31. M. Veta, P. J. van Diest, S. M. Willems, H. Wang, A. Madabhushi, A. Cruz-Roa, F. Gonzalez, A. B. L. Larsen, J. S. Vestergaard, A. B. Dahl, D. C. Ciresan, J. Schmidhuber, A. Giusti, L. M. Gambardella, F. B. Tek, T. Walter, C.-W. Wang, S. Kondo, B. J. Matuszewski, F. Precioso, V. Snell, J. Kittler, T. E. de Campos, A. M. Khan, N. M. Rajpoot, E. Arkoumani, M. M. Lacle, M. A. Viergever, and J. P. W. Pluim Assessment of algorithms for mitosis detection in breast cancer histopathology images, *Med. Image Anal.*, 20, 1, 237–248, 2015.
32. M. N. Gurcan, L. E. Boucheron, A. Can, A. Madabhushi, N. M. Rajpoot, and B. Yener, Histopathological image analysis: A review, *IEEE Rev. Biomed. Eng.*, 2, 147–171, 2009.

33. S. Doyle, S. Agner, A. Madabhushi, M. Feldman, and J. Tomaszewski, Automated grading of breast cancer histopathology using spectral clusteringwith textural and architectural image features, in *2008 5th IEEE International Symposium on Biomedical Imaging: From Nano to Macro*, 496–499, Paris, France 2008.
34. B. Weyn, G. van de Wouwer, A. van Daele, P. Scheunders, D. van Dyck, E. van Marck, and W. Jacob, Automated breast tumor diagnosis and grading based on wavelet chromatin texture description, *Cytometry*, 33, 1, 32–40, 1998.
35. J.-R. Dalle, W. K. Leow, D. Racoceanu, A. E. Tutac, and T. C. Putti, Automatic breast cancer grading of histopathological images, *Conf. Proc. Annu. Int. Conf. IEEE Eng. Med. Biol. Soc. IEEE Eng. Med. Biol. Soc. Annu. Conf.*, 2008, 3052–3055, 2008.
36. I. Goodfellow, Y. Bengio, and A. Courville, Machine Learning Basics, in *Deep Learning*, Cambridge, Massachusetts: MIT Press, 98–165, 2016.
37. K. Plataniotis and A. Venetsanopuolos, *Color Image Processing and Applications*. Berlin: Springer-Verlag, 2000.
38. X. Li and K. N. Plataniotis, Blind stain decomposition for histo-pathology images using circular nature of chroma components, in *2015 IEEE International Conference on Acoustics, Speech and Signal Processing (ICASSP)*, 2015, 877–881.
39. R. C. Gonzalez and R. E. Woods, *Digital Image Processing* (3rd edn), Upper Saddle River, New Jersey: Prentice Hall, 2006.
40. P. H. Tan, B. B. Goh, G. Chiang, and B. H. Bay, Correlation of nuclear morphometry with pathologic parameters in ductal carcinoma in situ of the breast, *Mod. Pathol.*, 14, 10, 937–941, 2001.
41. University of California Santa Barbara, Breast Cancer Dataset, *Bio-Segmentation Benchmark dataset*, 2016. [Online]. Available: http://bioimage.ucsb.edu/research/bio-segmentation.
42. Consortium for Open Medical Image Computing, Grand Challenges in Biomedical Image Analysis, 2016. [Online]. Available: http://grand-challenge.org/.

19

Medical Microwave Imaging and Analysis

Rohit Chandra

Ilangko Balasingham

Huiyuan Zhou

Ram M. Narayanan

19.1 Introduction

Medical imaging is usually a noninvasive tool to visualize the interior of the body of both healthy and diseased subjects. The imaging of the body is done for clinical analysis and detection of any abnormality. The most common examples of medical imaging are imaging done in different phases of cancer management, imaging for detecting bone fracture, and identifying tumors. Depending upon the application, clinical data for imaging can be obtained through a variety of techniques. These techniques rely on x-rays, magnetic resonance imaging (MRI), computed tomography (CT), ultrasound (US), and positron emission tomography (PET). Different techniques provide images with different resolution. Some of the techniques are costly due to the high maintenance and implementation cost of the equipment. Mostly, the equipment are bulky and are not portable. Moreover, some may use ionizing radiation that may have health risks. For example, MRI can provide a good resolution image that can be used for applications such as brain imaging for tumor detection, making the analysis easier. However, MRI equipment is costly and has high maintenance costs. Similarly, CT has a good spatial resolution, but uses x-rays that may ionize the tissue, posing a possible health risk. Moreover, CT cannot be used reliably for imaging soft tissues [1]. PET can be more informative when it comes to soft tissue imaging or providing functional information, but suffers from poor spatial resolution [1]. X-rays are most commonly used in mammography for breast cancer detection. Apart from being ionizing in nature, for x-rays, high false alarm has also been observed in mammography [2].

Microwave Imaging is an imaging technique using nonionizing electromagnetic (EM) signals in the frequency range of hundreds of megahertz to a few gigahertz. It is emerging as an alternative imaging technique to the aforementioned medical imaging techniques due to several advantages. As it uses nonionizing, low power EM signals, it is a low health-risk method. Moreover, the microwave imaging equipment consists of a microwave source, a receiver, an antenna-array for transmitting the signals, and a radiofrequency switch to switch between different antenna elements in the antenna-array. This equipment usually costs a fraction of the cost of the equipment for other diagnostic methods, making the microwave imaging a cost-effective technique. Moreover, the equipment is portable and can fit inside an ambulance for fast diagnosis of life-threatening conditions, like stroke while a patient is still on the way to the hospital [3]. However, there are some disadvantages and challenges for medical microwave imaging. The biggest disadvantage is that the images obtained from microwave imaging are low spatial resolution images.

Microwave imaging for medical applications started as a feasibility study by imaging organs as a perfused canine kidney by Larsen and Jacobi [4] in the 1980s. Over the years, medical microwave imaging has advanced significantly with the development of robust imaging algorithms and simple, but fast data acquisition hardware supported by the advancements made in microelectronics, material science, and embedded systems. This has helped in moving the medical microwave imaging from imaging of perfused organs to more application-based imaging for different pathological conditions. To this end, proof-of-concept of medical microwave imaging for various applications has been reported in the literature. Imaging for various pathological conditions that have been investigated are brain imaging for strokes [3,5], cerebral edema [6], breast cancer [7], bone imaging [8,9], heart imaging [10–15], and joint tissues [16].

The physical basis of medical microwave imaging is the tissue-dependent dielectric contrast that is used to reconstruct signals and images using radar-based or tomographic imaging algorithms. Different tissues have different dielectric properties that are characterized by relative permittivity and conductivity. Moreover, tissue of the same kind, if tumorous, will have different dielectric properties. Due to this difference, the interaction of the EM signals will be different for different tissues. This can be exploited by different reconstruction algorithms to construct a map or an image, either in 2D or 3D, that shows different tissue dielectric properties or the location of a tumor inside the body. Further, the reconstruction algorithms can be categorized into quantitative and qualitative imaging algorithms. A map or the image of the distribution of the various tissues with the values of dielectric properties in the body can be generated by quantitative imaging algorithms. Quantitative algorithms are usually based on the inverse EM scattering problem. On the other hand, qualitative imaging algorithms use radar-like techniques for image generation to differentiate the tumorous tissues from the normal tissues. Tumorous or the malignant tissues usually scatter EM signals stronger than other normal tissues and can be detected by qualitative imaging algorithms.

This chapter discusses the principle behind medical microwave imaging techniques by highlighting various medical applications. A brief discussion will be presented for various imaging algorithms. An application of microwave imaging for detection of brain tumor will be discussed. The chapter is based on our previous research papers, [17] and [18].

19.2 Dielectric Properties of Tissues

As explained in the introduction, the fundamental notion of microwave imaging is the tissue-dependent dielectric contrast that is used to reconstruct images using radar-based or tomographic imaging techniques. Hence, before the details of the microwave imaging setup and algorithms are introduced, it is beneficial to understand the tissue dielectric properties.

Different types of tissues, including both normal and malignant tissues of the same kind, have distinct electrical properties. The dielectric properties of the tissue consist of the relative permittivity and the conductivity. The difference occurs primarily due to the difference in the water content in the tissue [19]. For example, fat is a low water content tissue having low relative permittivity and conductivity. On the other hand, muscle is a high water content tissue having high relative permittivity and conductivity at microwave frequencies. Further, a tumorous or a malignant tissue has a higher rate of metabolism, resulting in more blood flowing through it. This changes the electrical properties of the tissue from its normal healthy kind. Apart from the difference in the water content in healthy and malignant tissues, a variety of other factors have been found that explain the difference in electrical properties between healthy tissues and malignant tissues. These include necrosis and inflammation causing breakdown of the cell membrane [20], charging of the cell membrane [21], change in the dielectric relaxation time [22], and difference in the sodium content [20]. This difference in the electrical properties of the tissues results in the variation in the scattered field by the different tissues. This forms the basis of the microwave imaging. A summary of measurements for various healthy tissues is provided by Gabriel et al. in [23]. The measured electrical properties of various healthy and malignant tissues from 50 to 900 MHz are presented in [24] and shown in percentage change in Figure 19.1. However, this very basis of electrical properties of the tissues that forms the fundamental notion of the microwave imaging gives rise to several challenges. For example, for some tissues, the difference between the normal and the malignant tissue may be low. This may result in a very small difference in the scattered field from the normal and the malignant tissues. For example, there exists a large contrast between cancerous breast and healthy fatty tissues. This difference is small when the cancerous tissue is compared with the healthy glandular or the fibro-connective breast tissue [25]. Hence, the detection of the cancerous tissue in a close vicinity of the glandular tissue may be a challenging task with microwave imaging. Moreover, the presence of multiple tissues with different properties results in a complex scattering environment. To solve these challenges, either use of a high dynamic range system is required to capture the small difference in the scattered field or the use of contrast agents is proposed to enhance the electrical properties of the malignant tissues [26–29]. The idea is to administer a contrast agent to the body by methods like

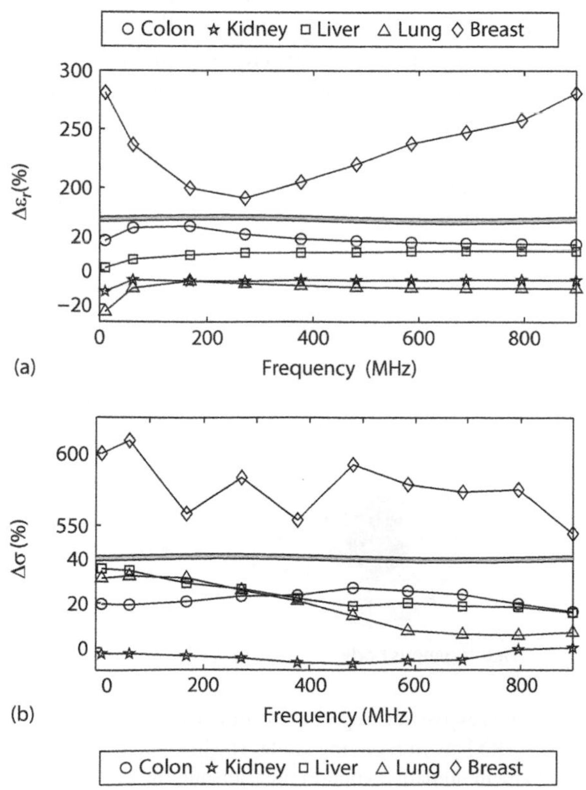

FIGURE 19.1 Percentage change in the dielectric properties of malignant tissues, with respect to the healthy tissue from 50 to 900 MHz: (a) percentage change in the relative permittivity (b) percentage change in the conductivity [17,24] © 2017, IEEE. Reprinted with permission.

intravenous injection. Some volume of the contrast agent will then reach and bind the cancerous tissues enhancing their electrical properties. Some research in this field has been done by using microbubbles, single-walled carbon nanotubes [27], or magnetic nanoparticles [29] as contrast agents for breast cancer detection. These feasibility studies show the potential of using contrast agents in microwave imaging. Another challenge is the frequency dispersion of dielectric properties of the tissues. This means that the dielectric properties of the tissues vary with the frequency of the microwave used. This is a challenge when wide bandwidth, as in the radar-based microwave imaging methods, is used, as it results in the distortion of the wideband pulses. Usually, the Debye model or the Cole-Cole model [30] is used to model the frequency dispersive nature of the tissues. Moreover, the attenuation of the microwave signal increases with the frequency due to increase in the conductivity resulting in a lower penetration depth. Hence, choice of frequency is critical for imaging.

19.3 Microwave Imaging Setup

A simple microwave imaging setup is shown in Figure 19.2. It consists of several antennas located around the body to be imaged. The antenna can form a 3D array. One antenna transmits at a time: it is connected to the transmitter and the scattered signal from the body is collected by the rest of the antennas that are connected to the receiver. Usually, there is an RF switch, that quickly switches between different antennas, such that all or some antennas act as a transmitter while the rest of the antennas are connected to the receiver. Between the antennas and the body, there is a homogeneous matching medium. The matching medium helps in reducing the reflections by coupling the signal to the body. This is because of the fact that, in the absence of the matching medium, a high reflection may occur between the tissue and the medium, which is the air in which the antenna system is kept. This would result in a weak signal that could penetrate the body. Hence, matching medium helps in reducing this reflection, and a relatively higher strength of the signal can penetrate the body. A simple matching medium that has been used is water. After the scattered signal is collected by the antenna and processed by the receiver to get meaningful signal data, the data are transferred to a computer. The computer executes the imaging algorithm to generate the image of the body. By the analysis of the generated image, the position of the tumor can be detected.

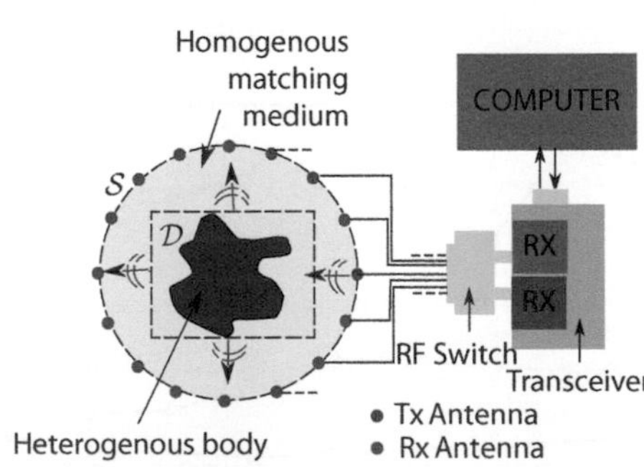

FIGURE 19.2 Basic setup for microwave imaging in a two-dimensional case when one antenna (TX) transmits the signal and all other antennas (RX) are in receiving mode. The imaging domain $\mathcal{D}$ contains the heterogeneous body to be imaged. The measurement domain $\mathcal{S}$ contain the transmit and the receive antennas. The whole system is in a homogeneous medium that acts as a matching medium. The RF switch switches between different antennas, such that all or some antennas acts a transmitter while the rest of the antennas are connected to the receiver. The transceiver is connected to a computer to which the collected signal data is transferred. The computer also executes the imaging algorithm.

19.4 Microwave Imaging Algorithms

Microwave Imaging algorithms can be divided into two categories, namely, quantitative and qualitative. Quantitative imaging algorithms generate the image with the distribution of the electrical properties of the body, either in 2D or 3D. On the other hand, qualitative imaging algorithms generate the 2D or the 3D image of the intensity of the scattered signal that shows the location of the strong scatterer as the tumor. In this section, both types of the algorithms are described.

19.4.1 Quantitative Imaging Algorithms

Quantitative imaging is also called tomography. The microwave tomography problem is formulated in terms of electric fields. To formulate the problem, two domains are defined: bounded domain *(D)* and the measurement domain *(S)*. The bounded domain is a domain in which the body to be imaged is enclosed. The antennas are located on the measurement domain. These are shown in Figure 19.2. Three electric fields are defined for the purpose: the incident field $\mathbf{E}^{\text{inc}}$ on $\mathcal{D}$, the total field $\mathbf{E}^{\text{total}}$ in $\mathcal{D}$, and the scattered field $\mathbf{E}^{\text{scatt}}$ on $\mathcal{S}$. A time harmonic dependence $\exp(j\omega t)$ with $j^2 = -1$, and $\omega = 2\pi f$ where f is frequency, is assumed. The scattered field on $\mathcal{S}$ is defined by the following *data equation:*

$$\mathbf{E}^{\text{scatt}}(\mathbf{p}) = k_b^2 \int_{\mathcal{D}} \mathcal{G}(\mathbf{p}, \mathbf{r}')\chi(\mathbf{r})\mathbf{E}^{\text{total}}(\mathbf{r})dv(\mathbf{r}'), \tag{19.1}$$

where:

- $\mathbf{p}\in\mathcal{S}$, $\mathbf{r}'$, and $\mathbf{r}$ are position vectors
- k_b is the wavenumber of the matching medium
- $\mathcal{G}$ is the Green's function that has a different expression for transverse magnetic (TM), transverse electric (TE), or 3D full vectorial case illumination [32]
- $\chi(\mathbf{r})$ is the contrast function containing the permittivity $(\mathbf{r})$ of the body in a matching medium with the permittivity ε_b and is defined as:

$$\chi(r) = \frac{\varepsilon(\mathbf{r}) - \varepsilon_b}{\varepsilon_b}. \tag{19.2}$$

The total electric field $\mathbf{E}^{\text{total}}$ satisfies the *domain equation* in $\mathcal{D}$:

$$\mathbf{E}^{\text{total}}(\mathbf{r}) = \mathbf{E}^{\text{inc}}(\mathbf{r}) + k_b^2 \int_{\mathcal{D}} \mathcal{G}(\mathbf{r}, \mathbf{r}')\chi(\mathbf{r}')\mathbf{E}^{\text{total}}(\mathbf{r}')dv(\mathbf{r}'). \tag{19.3}$$

The objective is to determine the contrast by solving these two equations. This is a nonlinear and ill-posed inverse EM scattering problem. Hence, the best way to solve them is through iterations where the domain $\mathcal{D}$ is discretized into N cells or pixels. In each of these pixels, the electrical properties are assumed to be constant. Many quantitative reconstruction algorithms have been developed [31]. These can be further categorized into two groups depending upon the cost-function used. These are either Newton-type iterative algorithms [32,33–39] or Modified Gradient (MGM) [40,41] and Contrast Source Inversion (CSI) algorithms [42]. The cost-function for the Newton-type algorithms is of least square type as:

$$\mathcal{C}_{Newton} = \frac{\sum_t \left\| \mathbf{E}_t^{\text{scatt}} - \mathbf{E}_{\text{meas},t}^{\text{scatt}} \right\|_{\mathcal{S}}^2}{\sum_t \left\| \mathbf{E}_{\text{meas},t}^{\text{scatt}} \right\|_{\mathcal{S}}^2}, \tag{19.4}$$

where:

The summation is done over the transmit antennas

$E^{scatt}_{meas,t}$ is the computed scattered field at all the receiver antennas, from the data and the domain equations for a particular transmit antenna t

$E^{scatt}_{meas,t}$ is the measured scattered field for that transmit antenna

The cost-function is minimized iteratively by updating the contrast as $\chi_{n+1} = \chi_n + \nu_n \Delta\chi_n$, ν_n being the appropriate step length and $\Delta\chi_n$ being the correction at the nth iteration. As the cost-function is nonlinear in χ and is ill-posed, it is regularized before performing the minimization. A common regularization method is Tikhonov regularization [36]. Newton-Kontorovich [32], Gauss-Newton inversion [39], Distorted Born Iterative method [33], Levernberg-Marquardt [36], and Log-Magnitude and Phase Reconstruction (LMPR) [38] are some of the Newton-type algorithms that have been used for microwave imaging of different human body parts.

The disadvantage of the Newton-type algorithms is that the data equation or the forward equation has to be solved at each iteration. This results in a high computational demand that may limit the applicability in a full 3D vectorial case [42]. Hence, MGM and CSI methods have been proposed that do not require solving the forward equation. This can be done by formulating the cost-function in terms of unknown contrast and unknown contrast sources. The contrast source is defined as $\mathbf{w}(\mathbf{r}) = \chi(\mathbf{r})\, \mathbf{E}^{\text{total}}(\mathbf{r})$ [42]. Consequently, the cost function is written as [42]:

$$\mathcal{C}_{\text{CSI}} = \frac{\sum_t \left\| \mathbf{E}^{\text{scatt}}_{\text{meas},t} - \mathcal{G}_S(\mathbf{w}_t) \right\|_S^2}{\sum_t \left\| \mathbf{E}^{\text{scatt}}_{\text{meas},t} \right\|_S^2} + \frac{\sum_t \left\| \chi \mathbf{E}^{\text{inc}}_t - \mathbf{w}_t + \chi \mathcal{G}_{\mathcal{D}}(\mathbf{w}_t) \right\|_{\mathcal{D}}^2}{\sum_t \left\| \chi \mathbf{E}^{\text{inc}}_t \right\|_{\mathcal{D}}^2}, \tag{19.5}$$

where $\mathcal{G}_S$ and $\mathcal{G}_{\mathcal{D}}$ are Green function's operator in $\mathcal{S}$ and $\mathcal{D}$, respectively. The minimization of the cost-functional iteratively constructs sequences of the contrast sources and the contrast, assuming one to be constant and updating the other. Another variant of CSI is formulated by multiplying the cost-function by a regularization term and the algorithm is called the multiplicative regularized contrast source inversion (MR-CSI). The MGM is similar to CSI, where the fields and the contrast are updated at each step instead of contrast sources and the contrast.

19.4.2 Qualitative Imaging Algorithms

In general, quantitative algorithms are computationally and memory-wise demanding. Moreover, in some applications such as in breast cancer detection, the main objective is not to know the electrical properties of the tissues, but to determine the existence and location of the tumor. In such cases, qualitative imaging algorithms can be used. Qualitative imaging algorithms are similar to radar-based algorithms where the objective is to detect strong scattering objects. In case of medical microwave imaging, the malignant or the tumorous tissue is a strong scatterer due to higher dielectric properties than the surrounding tissues. Various radar-based imaging algorithms to focus the tumor, such as confocal microwave imaging [2], beamforming [43], and tissue sensing adaptive radar [44] are used. Ultra-wideband (UWB) signal is used for qualitative imaging to have a good time resolution.

In qualitative imaging, each antenna transmits a short pulse at a time (UWB in frequency domain), and the backscatter response is received by the same antenna. The backscatter response consists of the tumor response, scatter from the skin, and backscatter from other tissues. Signal processing is used to reduce the effect of the skin and the backscatter from other tissue, to enhance the signal backscattered by the tumor. For example, in confocal microwave imaging, the processed backscattered waveform at each of the antennas is integrated over time to obtain B_t integrated waveforms, where t is the number of the transmit antennas. The reconstructed image is then created by time-shifting and summing data points from these integrated waveforms for each synthetic focal point **r** in the imaged body. The intensity $I(\mathbf{r})$ of a pixel in the image at **r** is the square of the coherently summed values [2]:

$$I(\mathbf{r}) = \left[\sum_t c_t B_t\left(\tau_t(\mathbf{r})\right) \right]^2, \tag{19.6}$$

where:

$\tau_t(r) = 2\lvert r - r_t \rvert / v\Delta T$	is the time delay from the tth antenna at $\mathbf{r}_t$ to the synthetic focal point at **r** in the body
υ	is the propagation velocity of the signal inside the body, assuming a homogeneous medium
c_t	are weights to compensate for the radial spreading of the cylindrical waves as they propagates outward from the transmit antenna

19.5 An Example: Brain Tumor Detection

This section is based on the paper [18] and presents an example showing the use of microwave imaging for detection of brain tumor of size 5 mm at two different locations in the brain. An anatomically realistic numerical phantom with a model of a brain tumor is used, where scattered electromagnetic signals are generated using the phantom by finite-difference-time-domain (FDTD) simulations. The microwave imaging technique based on the Levernberg-Marquardt iterative scheme is used to solve the inverse scattering problem for the head of the phantom in the 403.5 MHz medical radio (MedRadio) band. Two-dimensional quantitative images having the electrical properties of the brain are reconstructed. Differential images are obtained by taking the difference between the reconstructed images with and without the tumor model, and with and without the expected effect of the contrast agents.

An anatomically realistic numerical phantom called Billie [45], provided by ITIS foundation, is used.* Simulations are done in a full wave EM solver SEMCAD-X [46]. As the imaging algorithm is implemented for a 2D case, a 1 mm thick slice of the phantom head was obtained by terminating the phantom with perfect electric conductor boundary conditions along the height. The simulation domain in other directions is terminated by a perfectly matched layer with the absorbing boundary condition. The relative permittivity ε_r and the conductivity σ at 403.5 MHz of the sliced phantom at the level of the antennas are shown in Figure 19.3a,b, respectively. Figure 19.3a also shows the positions of the

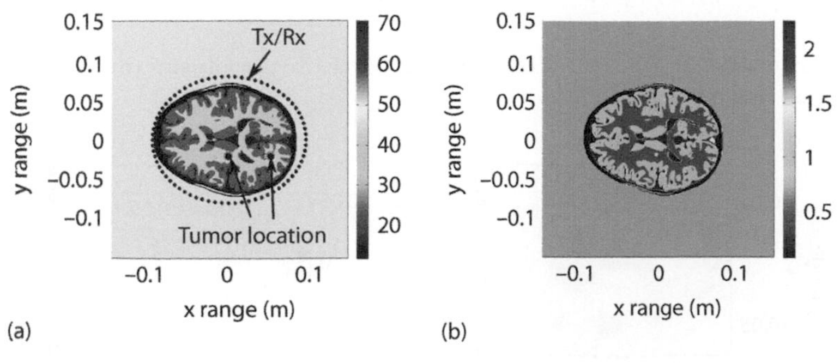

FIGURE 19.3 Dielectric properties of the head of the phantom Billie, (a) relative permittivity, (b) conductivity. The positions of the transmit and the receive antennas are also shown. The two locations of the tumor considered are also marked. Taken from [18], © EurAAP, used with permission.

* www.itis.ethz.ch

72 antennas, which are located around the head on an elliptical path at an angular separation of 5°. One antenna transmits at a time, and the other 71 antennas act as receivers. The external matching medium is assigned the electrical properties of the skin. The two locations where the tumor of diameter 5 mm was placed are shown in Figure 19.3a. It should be noted that, though this figure shows the two tumor locations in the same figure, for brevity, two separate simulations are done by placing the tumor model, one at each location. These two locations are cases of a deep brain and a superficial brain tumor location. The dielectric properties of the tumor are taken as $\varepsilon_r = 65$ and $\sigma = 1.01$ S/m from [47], which presents the electrical properties of the carcinoma brain tissue of a rat. The phantom is illuminated by a sinusoidal signal transmitted by a current source, oriented along the height of the phantom with an amplitude of 1 A. A 2D field sensor is considered the receiver antenna. To get the measured scattered field for the minimization of the cost-functional C (Equation 19.4), simulations are done with and without (but with the external medium) the phantom. The electric fields recorded by the field sensors are then subtracted in the two cases to obtain the scattered field.

Figure 19.4 shows the relative permittivity (Figure 19.4a) and the conductivity (Figure 19.4b) of the brain obtained from the imaging algorithm. The microwave imaging problem is divided into $N = 4841$ discrete cells or pixels where the electrical properties are assumed to be constant within a cell. The dimension of a cell is approximately $\lambda_{ext}/50 \times \lambda_{ext}/50$, where λ_{ext} is the wavelength in the exterior medium that is chosen to have electrical properties of the skin. Reconstructed images with a simulated tumor of 5 mm diameter are shown in Figure 19.5. By visually investigating the reconstructed image, it is difficult to detect the tumor. In [16], a method of differential imaging is presented to detect the meniscal tear in the knee joint, where the images were reconstructed using radar-based imaging method. Similarly, here we create a differential image by using the image without tumor as a reference image. The differential image is shown in Figure 19.6. As can be seen in the differential image, a tumor can be detected both in the relative permittivity image as well as in the conductivity image.

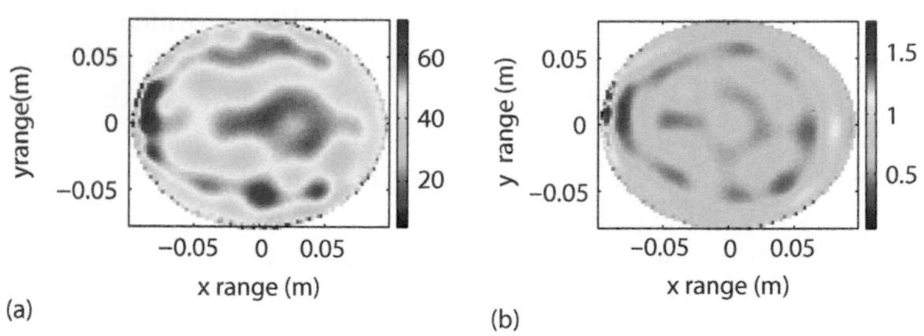

FIGURE 19.4 Microwave imaged brain of the phantom, (a) relative permittivity, (b) conductivity. Taken from [18], © EurAAP, used with permission.

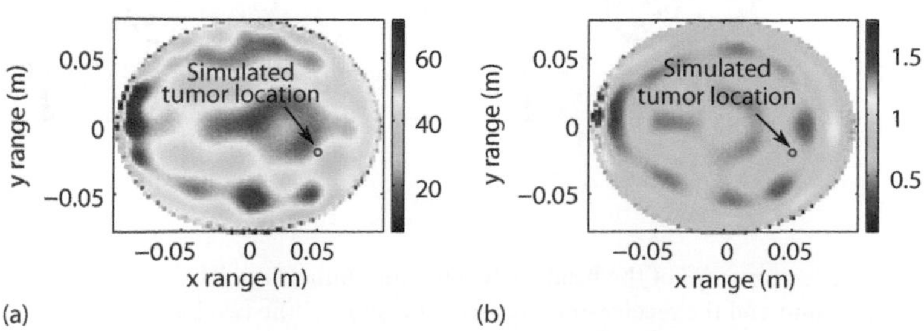

FIGURE 19.5 Microwave imaged brain of the phantom with 5 mm tumor, (a) relative permittivity, (b) conductivity. The location of the simulated tumor is also shown. Taken from [18], © EurAAP, used with permission.

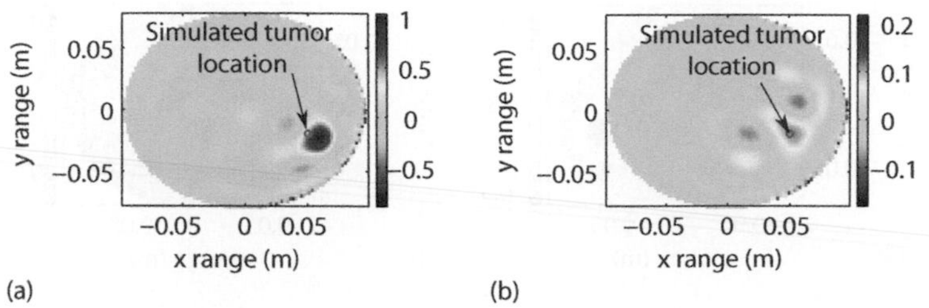

FIGURE 19.6 Differential image of the brain of the phantom with 5 mm tumor, (a) differential relative permittivity, (b) differential conductivity. The simulated location of the tumor is also shown. Taken from [18], © EurAAP, used with permission.

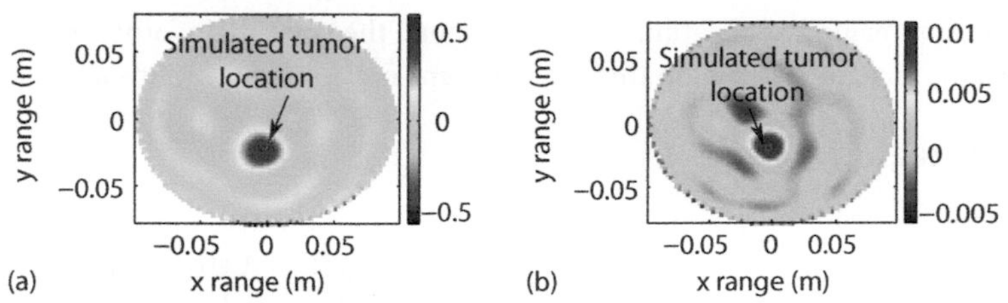

FIGURE 19.7 Differential image of the brain of the phantom with 5 mm tumor in deep brain (a) differential relative permittivity (b) differential conductivity. The real location of the tumor is also shown. Taken from [18], © EurAAP, used with permission.

The error distance between the actual position of the tumor and the detected position is calculated as $d_{\text{error}} = \|\hat{\mathbf{r}} - \mathbf{r}\|$, where $\mathbf{r} = (x_{tumor}, y_{tumor})$ is the real position of the center of the tumor and $\hat{\mathbf{r}}$ is the estimated position that is the cell center of the image inside the brain having highest differential value of ε_r or σ. d_{error} is found to be 11.35 mm from the relative permittivity differential image and 1.13 mm from the conductivity image. Next, the 5 mm tumor model is placed deep into the brain of the phantom. The differential image in such a case is shown in Figure 19.7. d_{error} is 5.7 and 1.44 mm calculated from the differential relative permittivity image and the differential conductivity image, respectively.

One of the limitations with the above approach for practical cases would be to get a reference image, the one without the tumor. This could be done by obtaining an image by microwave imaging of a head of a healthy person with same gender, age, and approximately same head size. However, even a small anatomical difference between the healthy person and the person with the tumor might make the detection of the tumor as small as 5 mm a challenging task. Thus, a more practical way to obtain the reference image can be through microwave imaging of the head of the person with a tumor before injecting the contrast agent. As discussed in the Section 19.2, the contrast agent can enhance the dielectric properties of the tumor. The differential image can then be obtained by the subtraction of the image before contrast agent from the image obtained after the effect of the contrast agent. In [26], it was shown that the contrast agent, like single walled carbon nanotubes (SWCT), can enhance the relative permittivity by 22% and the conductivity by 66% of the breast tumor. Assuming a similar effect on the brain tumor at 403.5 MHz, resulting in the new electrical properties as $\varepsilon_r' = 80$ and $\sigma' = 1.68$ S/m, differential imaging was done. Figure 19.8 shows the differential relative permittivity images of the two locations of the simulated tumor for a high SNR case. As can be seen from the figure, the use of contrast agent can facilitate detection of the tumor. However, to actually detect the tumor, one would need additional image processing algorithms.

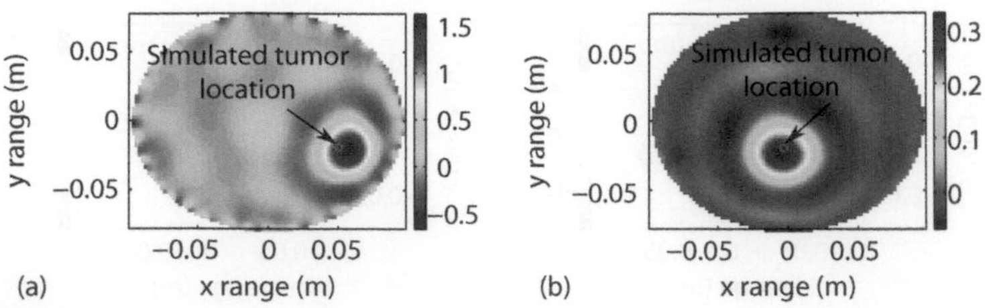

FIGURE 19.8 Differential relative permittivity image of the brain of the phantom with 5 mm simulated tumor at the two locations created by taking difference with and without the expected effect of the contrast agent. Taken from [18], © EurAAP, used with permission

19.6 Other Reported Applications of Microwave Medical Imaging

Apart from the brain imaging application to detect tumors, there are many other medical applications for which microwave imaging is used. In this section, some of these applications are discussed.

19.6.1 Brain Imaging

Microwave imaging of the brain can be done to detect and locate the area of the damaged brain tissues due to injuries or conditions such as ischemic or hemorrhagic stroke. A stroke may result in accumulation of a blood clot within the brain. As there is a difference between the electrical properties of the blood and normal brain tissues, microwave imaging can be used to detect the blood clot. Research reported in [5,48,49–52] focus on such stroke detection in the brain using microwave imaging. Quantitative methods used so far include a Newton-type iterative scheme for 2D tomography [5], the Born iterative method [49], and a multiplicative regularized Gauss-Newton inversion [53]. Qualitative approach as confocal imaging has also shown to detect stroke [48,50].

19.6.2 Breast Imaging

One of the most widely investigated medical application of microwave imaging is detection of breast tumor. The breast tumor has relatively high contrast when compared with the prominent fat tissue in a breast making the tumor as a significant scatterer. Hence, radar-based techniques can be effectively applied to locate the tumor, as reported in [2,43,44,54–60]. The tomographic methods can also be used where the tumor can be distinguished easily from the rest of the fat tissues in the reconstructed image as discussed in [39,61–66]. A review of extensive research work done for breast cancer imaging is provided in [7] and [67].

19.6.3 Bone Imaging

Microwave imaging for bone has been done to detect leukemia in the bone marrow [8]. Another application that is reported is the determination of the bone density for detection of osteoporosis [9]. Leukemia causes the cellular population in the bone to increase, which, in turn, increases the relative permittivity and decreases the conductivity up to a factor of 2. Both of these applications require the electrical properties of the bone to be known, and, thus, quantitative approaches are used. The Levenberg-Marquardt method has been used for leukemia detection and Gauss-Newton iterative method for the detection of osteoporosis.

19.6.4 Soft Tissues and Joint Imaging

During a bone fracture, soft tissues may also get injured. The commonly used method to detect a bone fracture is using x-rays. However, x-rays cannot detect the injured soft tissues [1]. Microwave imaging can be used to detect the soft tissue injury as well. In [68], a feasibility study using a Newton and the MR-CSI method for functional imaging of the extremity soft tissues of a pig's hind leg is presented.

Salvador et al. [16] presented lesions of clinically relevant sizes and shapes in the menisci, ligaments, and tendons in the knee joint. Images were reconstructed with and without the model of the meniscal tear in the knee. A tear as thin as 1.3 mm × 8 mm was detected in the difference image (but not in the reconstructed image), and two tears separated by a minimum distance of 9 mm were detected.

19.6.5 Heart Imaging

The objective of heart imaging using microwaves can be to detect any pathological conditions, such as myocardial infarction [12], as the dielectric properties of myocardial tissue have a strong dependence on coronary blood flow. It can also be used to obtain the temporal images of a beating heart for heartbeat extraction [15], due to millisecond temporal resolution of microwave imaging. Semenov et al. [10] reconstructed the quantitative images of an excised canine perfused heart in a static, as well as in a beating, case. Further, Semenov et al. [11] were able to successfully reconstruct the internal structure of an excised canine heart, including the left and the right ventricles, using the 3D MGM and in [12] for the detection of the myocardial infarction. Using the same 3D gradient method in [13], Semenov et al. were able to reconstruct the heart in an intact swine. An ultra-wideband, radar-based approach to image the human heart in real time is presented by Brovoll et al. in [14] and [15].

19.6.6 Miscellaneous Biomedical Body Imaging

Apart from the presented medical applications of microwave imaging, microwave imaging has also been used to image various body parts. For example, human arm imaging is done using the CSI method by Gilmore et al. [69]. Human thorax and arm imaging using the Newton-Kantorovich algorithm to obtain their quantitative permittivity image is presented by Mallorqui et al. [70] at 434 MHz. The Levenberg-Marquardt method [36] for imaging the human torso at the level of the small intestine is used in the 403.5 MHz MedRadio band to obtain permittivity images by Chandra et al. [71]. The reconstructed image was used as a basis for the localization of wireless capsule endoscopes in a 2D case [71] and in a simple 3D body [73]. A whole-body imaging of a dog using a 3D gradient based iterative algorithm is shown by Semenov et al. [72]. A comparison study of different quantitative image reconstruction methods is done on the experimental data of a phantom and a pig's hind leg in the frequency range 0.9–2.05 GHz by Semenov et al. [31]. Different methods employed are Newton-based 2D method and 3D gradient method and MR-CSI for 2D and 3D cases. This research shows the feasibility of imaging large biological bodies using microwaves and opens up a new area where microwave imaging can be employed for medical applications.

19.7 Open Challenges and Future Directions

During the last two decades, feasibility studies for various applications using microwave imaging as presented in the previous sections have been done by various research groups. During this period, medical application using microwave has advanced from a single antenna sensing application to multi-antenna imaging applications. More sophisticated 3D antenna array is also being used. Early stage commercial start-up companies, like EMTensor [74] for brain stroke and Micrima [75] for breast cancer, have been setup. The ultimate goal for microwave imaging is to be used in a real clinical setting as a cost-effective alternative to currently used imaging modalities, like MRI and CT-scan. However, most of the feasibility study of microwave imaging for medical imaging is done through computer simulations using computer models of biomedical bodies. Though in some research, actual biomedical bodies or phantoms have been used, there are several open challenges that remain. These open challenges need to be addressed and more research has to be done before microwave imaging can be effectively used in a real clinical environment. Some of these open challenges are discussed in this section.

Coupling the microwave signal to the body: there are large differences in the electrical properties between the air and tissues of the biomedical body. Due to this difference, in the absence of any coupling or matching medium, microwave signals transmitted from the antenna undergo reflection at the air-tissue boundary, leaving a fraction of the transmitted microwave signal to be coupled to the body. The coupled signal is further attenuated by the lossy tissues, leaving a very weak scattered signal to be used for imaging. Hence, matching medium is used to reduce the reflections at the air-tissue boundary. Water (not pure) is usually used as a matching medium. However, water itself has relatively high losses. Therefore, a low-loss matching medium is required. Some research in this regard has been done by Hamsakutty et al. [76], where sodium meta silicate gel for 2.45 GHz imaging frequency is proposed. Another solution for low strength of the scattered signal is to use a high dynamic range system.

Contrast agents: In MRI, contrast agents are usually used to enhance the magnetic properties of the tumor that makes it clearly visible in the image. Similar to this, development of a biocompatible contrast agent for microwave imaging of the body is an open challenge. Contrast agents can be used in cases where the difference in the electrical properties of the malignant tissue and the healthy tissue is small enough to be detected by the microwave imaging system. The idea is to administer a contrast agent to the body by methods like intravenous injection, as previously discussed. Some volume of the contrast agent will then reach and bind with the cancerous tissues enhancing their electrical properties. Use of carbon nanotubes and microbubbles as contrast agents for breast cancer has been proposed by Shea et al. [26]. Contrast agents can be specifically used for the qualitative imaging algorithms.

Advancement in the imaging algorithms: At the core of microwave imaging are the imaging algorithms. Computational and memory efficiency are the properties needed for a robust imaging algorithm. Quantitative imaging algorithms used are EM inverse problem and, hence, suffer from open challenges of the inverse problem like nonlinearity and ill-posedness. Usually methods employed to tackle these challenges of an inverse problem result in a computationally and memory demanding imaging algorithms. Thus, there is a scope to develop robust, computational, and memory-effective imaging algorithms.

Antennas and Measurement System: Usually, microwave imaging system employs a large number of the antennas in an array. This is done in order to reduce the non-uniqueness and the ill-posedness of nonlinear inverse problems to some extent. However, using a large number of antennas increases the processing complexity of the system and adds to the cost. Moreover, a large number of the antennas means closely spaced antennas that may couple with each other and introduce error in the measured data. Development of an effective antenna system with an optimum number of antennas is a future research direction.

A usual measurement system for microwave imaging uses the vector network analyzer (VNA). The VNA-based system records the path-gain between the antennas, rather than the electric-field that is required by the imaging algorithms. To convert the path-gain to the electric field calibration is done. Developing a calibration method that is less prone to errors is also needed.

Frequency band and resolution: One of the open challenges of microwave imaging is the usage of an optimum frequency band. The reason for this is that the attenuation of the microwave signal increases at higher frequency. On the other hand, using low attenuation, low frequency bands result in a lower image resolution. Moreover, a frequency band that is optimum for one body part may not be suitable for another part due to variation in the tissue properties and size. Several frequency bands, such as the 403.5 MHz MedRadio band, 900 MHz, UWB (3.1–10.6 GHz), and so on, have been used, depending upon the application. Thus, a thorough investigation is required before using a particular frequency band for particular application to obtain an acceptable resolution of the image.

Conclusions

This chapter presented an alternative imaging technique called microwave imaging for medical applications that uses nonionizing electromagnetic (EM) waves for imaging unlike ionizing x-rays. Moreover, microwave imaging equipment is low-cost and can be easily maintained, unlike costly and high-maintenance equipment of other imaging modalities like MRI. Various research reports have shown the feasibility of using microwave imaging for applications such as brain stroke and tumor detection, breast cancer detection, heart imaging, and bone imaging. At least two commercial companies, Micrima and EMTensor, have started. It was discussed that the physical basis for microwave imaging is the variation in the dielectric properties of various tissues, either healthy or malignant. Further, it was discussed that various microwave imaging algorithms can be broadly categorized into qualitative and quantitative imaging algorithms. A quantitative imaging technique solves the inverse EM problem to obtain images that show the distribution of the dielectric properties of different tissues. On the other hand, qualitative imaging algorithms are radar-based algorithms that detect a strong scatterer, like a tumor. An example of microwave imaging was presented for the detection of brain tumors of size 5 mm at two different locations in the brain. It was shown that such a small tumor can be detected using microwave imaging with a millimeter accuracy.

With benefits like nonionizing nature, low health risk, low cost, and portability, microwave imaging can emerge as a potential alternative for currently used imaging modalities, like costly MRI or the ionizing, x-ray-based CT-scan. However, before microwave imaging can be successfully implemented in a clinical setting, various challenges that it faces have to be solved. These challenges and future directions for microwave imaging were reported in the chapter.

References

1. S. Semenov, Microwave tomography: Review of the progress towards clinical applications, *Philos. Trans. A Math. Phys. Eng. Sci*, 367, 1900, 3021–3042, 2009.
2. E. C. Fear, X. Li, S. C. Hagness, and M. A. Stuchly, Confocal microwave imaging for breast cancer detection: Localization of tumors in three dimensions, *IEEE Trans. Biomed. Eng.*, 49, 8, 812–822, 2002.
3. M. Persson, A. Fhager, H. D. Trefna, and Y. Yu, Microwave-based stroke diagnosis making global prehospital thrombolytic treatment possible, *IEEE Trans. Biomed. Eng.*, 61, 11, 2806–2817, 2014.
4. L. E. Larsen and J. H. Jacobi, *Medical Application of Microwave Imaging*, New York, IEEE Press, 1986.
5. S. Y. Semenov, and D. R. Corfield, Microwave tomography for brain imaging: Feasibility assessment for stroke detection, *Int. J. Antennas Propag.*, Article ID 254830, 2008.
6. J. C. Lin and M. J. Clarke, Microwave imaging of cerebral edema, *Proc. IEEE*, 70, 5, 523–524, 1982.
7. N. K. Nikolova, Microwave imaging for breast cancer, *IEEE Microw. Mag.*, 12, 7, 78–94, 2011.
8. D. Colton and P. Monk, The detection and monitoring of leukemia using electromagnetic waves: Numerical analysis, *Inverse Prob.*, 11, 2, 329–342, 1995.
9. P. M. Meaney, D. Goodwin, A. H. Golnabi, and T. Zhou, Clinical microwave tomographic imaging of the calcaneus: A first-in-human case study of two subjects, *IEEE Trans. Biomed. Eng.*, 59, 12, 3304–3313, 2012.
10. S. Y. Semenov, R. H. Svenson, A. E. Boulyshev, A. E. Souvorov, V. Y. Borisov, Y. Sizov, A. N. Starostin, K. R. Dezern, G. P. Tatsis, and V. Y. Baranov, Microwave tomography: Two-dimensional system for biological imaging, *IEEE Trans. Biomed. Eng.*, 43, 9, 869–877, 1996.
11. S. Y. Semenov, Three-dimensional microwave tomography: Experimental imaging of phantoms and biological objects, *IEEE Trans. Microw. Theory Tech.*, 48, 6, 1071–1074, 2000.
12. S. Y. Semenov, A. E. Bulyshev, V. G. Posukh, Y. E. Sizov, T. C. Williams, and A. E. Souvorov, Microwave tomography for detection/imaging of myocardial infarction. I. Excised canine hearts, *Ann. Biomed. Eng.*, 31, 3, 262–270, 2003.
13. S. Y. Semenov, V. G. Posukh, A. E. Bulyshev, T. C. Williams, Y. E. Sizov, P. N. Repin, A. Souvorov, and A. Nazarov, Microwave tomographic imaging of the heart in intact swine, *J. Electromag. Waves Appl.*, 20, 7, 873–890, 2006.
14. S. Brovoll, T. Berger, Y. Paichard, O. Aardal, T. S. Lande, and S.-E. Hamran, Time-lapse imaging of human heartbeats using UWB radar, *IEEE Biomed. Circuits Sys. Conf. (BioCAS)*, 142–145, 2013.
15. S. Brovoll, T. Berger, Y. Paichard, O. Aardal, T. S. Lande, and S.-E. Hamran, Time-lapse imaging of human heart motion with switched array UWB radar, *IEEE Trans. Biomed. Circuits Sys.*, 8, 5, 704–715, 2014.

16. S. M. Salvador, E. C. Fear, M. Okoniewski, and J. R. Matyas, Exploring joint tissues with microwave imaging, *IEEE Trans. Microw. Theory Tech.*, 58, 8, 2307–2313, 2010.
17. R. Chandra, H. Zhou, I. Balasingham, and R. M. Narayanan, On the opportunities and challenges in microwave medical sensing and imaging, *IEEE Trans. Biomed. Eng.*, 62, 7, 1667–1682, 2015.
18. R. Chandra and I. Balasingham, Detection of brain tumor and localization of a deep brain RF-source using microwave imaging, 9th European Conference *on* Antennas *and* Propagation *(EuCAP)*, Lisbon, April 2015.
19. J. L. Schepps, and K. R. Foster, The UHF and microwave dielectric properties of normal and tumor tissues: Variation in dielectric properties with tissue water content, *Phys. Med. Biol.*,. 25, 6, 1149–1159, 1980.
20. L. Sha, E. R. Ward, and B. Stroy, A review of dielectric properties of normal and malignant breast tissue, *Proc. IEEE Southeast Con.*, 457–462, 2002.
21. R. Pethig, Dielectric properties of biological materials: Biophysical and medical applications, *IEEE Trans. Elec. Insul,* EI-19, 5, 453–474, 1984.
22. M. Lazebnik, M. Okoniewski, J. H. Booske, and S. C. Hagness, Highly accurate debye models for normal and malignant breast tissue dielectric properties at microwave frequencies, *IEEE Microw. Wireless Comp. Lett.*, 17, 12, 822–824, 2007.
23. C. Gabriel, S. Gabriel, and E Corthout, The dielectric properties of biological tissues. I. literature survey, *Phys. Med. Bio.l*, 41, 11, 2231–2249, 1996.
24. W. T. Joines, Y. Zhang, C. Li, and R. L. Jirtle, The measured electrical properties of normal and malignant human tissues from 50 to 900 MHz, *Med. Phys.*, 21, 4, 547–550, 1994.
25. M. Lazebnik, D. Popovic, L. McCartney, et al., A large-scale study of the ultrawideband microwave dielectric properties of normal, benign and malignant breast tissues obtained from cancer surgeries, *Phys. Med. Biol.*, 52, 20, 6093–6115, 2007.
26. J. D. Shea, P. Kosmas, B. D. Van Veen, and S. C. Hagness, Contrast-enhanced microwave imaging of breast tumors: A computational study using 3D realistic numerical phantoms, *Inverse Prob.*, 26, 7, 2010.
27. C. Yifan, I. J. Craddock, and P. Kosmas, Feasibility study of lesion classification via contrast-agent-aided UWB breast imaging, *IEEE Trans. Biomed. Eng.*, 57, 5, 1003–1007, 2010.
28. A. Mashal, B. Sitharaman, X. Li, P. K. Avti, A. V. Sahakian, J. H. Booske, and S. C. Hagness, Toward carbon-nanotube-based theranostic agents for microwave detection and treatment of breast cancer: Enhanced dielectric and heating response of tissue-mimicking materials, *IEEE Trans. Biomed. Eng.*, 57, 8, 1831–1834, 2010.
29. G. Bellizzi, O. M. Bucci and I. Catapano, Microwave cancer imaging exploiting magnetic nanoparticles as contrast agent, *IEEE Trans. Biomed. Eng.*, 58, 9, 2528–2536, 2011.
30. S. Gabriel, R. W. Lau, and C. Gabriel, The dielectric properties of biological tissues. III. Parametric models for the dielectric spectrum of tissues, *Phys. Med. Biol.*, 41, 11, 2271–2293, 1996.
31. S. Y. Semenov, A. E. Bulyshev, A. Abubakar, V. G. Posukh, Y. E. Sizov, A. E. Sou-vorov, P. M. van den Berg, and T. C. Williams, Microwave-tomographic imaging of the high dielectric-contrast objects using different image-reconstruction approaches, *IEEE Trans. Microw. Theory Tech.*, 53, 7, 2284–2294, 2005.
32. N. Joachimowicz, C. Pichot, and J. P. Hugonin, Inverse scattering: An iterative numerical method for electromagnetic imaging, *IEEE Trans. Antennas Propag.,* 39, 12, 1742–1753, 1991.
33. W. C. Chew, and Y. M. Wang, Reconstruction of two-dimensional permittivity distribution using the distorted Born iterative method, *IEEE Trans. Med. Imaging,* 9, 2, 218–225, 1990.
34. A. Franchois and A. G. Tijhuis, A quasi-Newton reconstruction algorithm for a complex microwave imaging scanner environment, *Radio Sci.,* 38, 2, VIC 12-1-VIC 12-13, 2003.
35. A. E. Souvorov, A. E. Bulyshev, S. Y. Semenov, R. H. Svenson, A. G. Nazarov, Y. E. Sizov, and G. P. Tatsis, Microwave tomography: A two-dimensional Newton iterative scheme, *IEEE Trans. Microw. Theory Tech.*, 46, 11, 1654–1659, 1998.

36. A. Franchois, and C. Pichot, Microwave imaging-complex permittivity reconstruction with a Levenberg-Marquardt method, *IEEE Trans. Antennas Propag.*, 45, 2, 203–215, 1997.
37. J. De Zaeytijd, A. Franchois, C. Eyraud, and J. M. Geffrin, Full-wave three-dimensional microwave imaging with a regularized Gauss-Newton method-theory and experiment, *IEEE Trans. Antennas Propag.*, 55, 11, 3279–3292, 2007.
38. P. M. Meaney, K. D. Paulsen, B. W. Pogue, and M. I. Miga, Microwave image reconstruction utilizing log-magnitude and unwrapped phase to improve high-contrast object recovery, *IEEE Trans. Med. Imaging,* 20, 2, 104–116, 2001.
39. T. Rubæk, P. M. Meaney, P. Meincke, and K. D. Paulsen, Nonlinear microwave imaging for breast-cancer screening using Gauss-Newton's method and the CGLS inversion algorithm, *IEEE Trans. Antennas Propag.*, 55, 8, 2320–2331, 2007.
40. R. E. Kleinman, and P. M. van den Berg, A modified gradient method for two-dimensional problems in tomography, *J. Comput. Appl. Math.*, 42, 17–35, 1992.
41. A. E. Bulyshev, Three-dimensional microwave tomography: Theory and computer experiments in scalar approximation, *Inverse Probl.*, 16, 863–875, 2000.
42. A. Abubakar, P. M. van den Berg, and J. J. Mallorqui, Imaging of biomedical data using a multiplicative regularized contrast source inversion method, *IEEE Trans. Microw. Theory Tech.*, 50, 7, 1761–1771, 2002.
43. J. E. Bond, X. Li, S. C. Hagness, and B. D. Van Veen, Microwave imaging via space-time beamforming for early detection of breast cancer, *IEEE Trans. Antennas Propag.*, 51, 8, 1690–1705, 2003.
44. J. M. Sill and E. C. Fear, "Tissue sensing adaptive radar for breast cancer detection–Experimental investigation of simple tumor models," *IEEE Trans. Microw. Theory Tech.*, 53, 11, 3312–3319, 2005.
45. A. Christ, W. Kainz, E. G. Hahn, et al., The virtual family development of anatomical CAD models of two adults and two children for dosimetric simulations, *Phy. Med. Biol.*, 55, 2, N23–N38, 2010.
46. Schmid & Partner Engineering AG, Zeughausstrasse 43, 8004 Zürich, Switzerland. Website: https://www.speag.com/.
47. D. S. Yoo, The dielectric properties of cancerous tissues in a nude mouse xenograft model, *Bioelectromagnetics*, 25, 7, 492–497, 2004.
48. S. Mustafa, B. Mohammed, and A. Abbosh, Novel preprocessing techniques for accurate microwave imaging of human brain, *IEEE Antennas Wireless Propag. Lett.*, 12, 460–463, 2013.
49. D. Ireland, K. Bialkowski, and A. Abbosh, Microwave imaging for brain stroke detection using Born iterative method, *IET Microw. Antennas Propag.*, 7, 11, 909–915, 2013.
50. B. J. Mohammed, A. M. Abbosh, S. Mustafa, and D. Ireland., Microwave system for head imaging, *IEEE Trans. Instrum. Meas.*, 63, 1, 117–123, 2014.
51. M. Jalilvand, X. Li, T. Zwick, W. Wiesbeck, and E. Pancera, Hemorrhagic stroke detection via UWB medical imaging, *Proc. of 5th Eur. Conf. Antennas Propag.*, 2911–2914, 2011.
52. R. Scapaticci, L. Di Donato, I. Catapano, and L. Crocco., A feasibility study on microwave imaging for brain stroke monitoring, *Prog. Electromagn. Res. B*, 40, 305–324, 2012.
53. P. Mojabi, and J. LoVetri, Microwave biomedical imaging using the multiplicative regularized Gauss-Newton inversion, *IEEE Antennas Propag. Lett.*, 8, 645–648, 2009.
54. S. C. Hagness, A. Taflove, and J. E. Bridges, Two-dimensional FDTD analysis of a pulsed microwave confocal system for breast cancer detection: Fixed-focus and antenna-array sensors, *IEEE Trans. Biomed. Eng.*, 45, 12, 1470–1479, 1998.
55. S. C. Hagness, A. Taflove, and J. E. Bridges, Three-dimensional FDTD analysis of a pulsed microwave confocal system for breast cancer detection: Design of an antenna-array element, *IEEE Trans. Antennas Propag.*, 47, 5, 783–791, 1999.
56. Xu Li, and S. C. Hagness, A confocal microwave imaging algorithm for breast cancer detection, *IEEE Microw. Wireless Compon. Lett.*, 11, 3, 130–132, 2001.

57. H. B. Lim, N. Thi Tuyet Nhung, E. P. Li, and N. Duc Thang, Confocal microwave imaging for breast cancer detection: Delay-multiply-and-sum image reconstruction algorithm, *IEEE Trans. Biomed. Eng.*, 55, 6, 1697–1704, 2008.
58. D. J. Kurrant, E. C. Fear, and D. T. Westwick, Tumor response estimation in radar-based microwave breast cancer detection, *IEEE Trans. Biomed. Eng.*, 55, 12, 2801–2811, 2008.
59. X Li, S. K. Davis, S. C. Hagness, D. W. van der Weide, and B. D. Van Veen, Microwave imaging via space-time beamforming: Experimental investigation of tumor detection in multilayer breast phantoms, *IEEE Trans. Microw. Theory Tech.*, 52, 8, 1856–1865, 2004.
60. Y. Xie, B. Guo, L. Xu, J. Li, and P. Stoica, Multistatic adaptive microwave imaging for early breast cancer detection, *IEEE Trans. Biomed. Eng.*, 53, 8, 1647–1657, 2006.
61. P. M. Meaney, M. W. Fanning, D. Li, S. P. Poplack, and K. D. Paulsen, A clinical prototype for active microwave imaging of the breast, *IEEE Trans. Microw. Theory Tech.*, 48, 11, 1841–1853, 2000.
62. A. E. Souvorov, A. E. Bulyshev, S. Y. Semenov, R. H. Svenson, and G. P. Tatsis, Two-dimensional computer analysis of a microwave flat antenna array for breast cancer tomography, *IEEE Trans. Microw. Theory Tech.*, 48, 8, 1413–1415, 2000.
63. A. E. Bulyshev, S. Y. Semenov, A. E. Souvorov, R. H. Svenson, A. G. Nazarov, Y. E. Sizov, and G. P. Tatsis, Computational modeling of three-dimensional microwave tomography of breast cancer, *IEEE Trans. Biomed. Eng.*, 48, 9, 1053–1056, 2001.
64. D. Li, P. M. Meaney, and K. D. Paulsen, Conformal microwave imaging for breast cancer detection, *IEEE Trans. Microw. Theory Tech.*, 51, 4, 1179–1186, 2003.
65. M. J. Burfeindt, N. Behdad, B. D. Van Veen, and S. C. Hagness, Quantitative microwave imaging of realistic numerical breast phantoms using an enclosed array of multiband, miniaturized patch antennas, *IEEE Antennas Wireless Propag. Lett.*, 11, 1626–1629, 2012.
66. T. U. Gürbüz, B. Aslanyürek, A. Yapar, H. Sahintürk. and I. Akduman, A nonlinear microwave breast cancer imaging approach through realistic body-breast modeling, *IEEE Trans. Antennas Propag.*, 62, 5, 2596–2605, 2014.
67. E. C. Fear, S. C. Hagness, P. M. Meaney, M. Okoniewski, and M. A. Stuchly, Enhancing breast tumor detection with near-field imaging, *IEEE Microw. Mag.*, 3, 1, 48–56, 2002.
68. S. Semenov, Microwave tomography for functional imaging of extremity soft tissues: Feasibility assessment, *Phys. Med. Biol*, 52, 5705–5719, Article ID 673027, 2007.
69. C. Gilmore, A. Zakaria, S. Pistorius, and J. LoVetri, Microwave imaging of human forearms: Pilot study and image enhancement, *Int. J. Biomed. Imaging*, 17, 2013.
70. J. J. Mallorqui, N. Joachimowicz, A. Broquetas, and J. C. Bolomey, Quantitative images of large biological bodies in microwave tomography by using numerical and real data, *Electron. Lett.*, 32, 23, 2138–2140, 1996.
71. R. Chandra, A. J. Johansson, M. Gustafsson, and F. Tufvesson, A microwave imaging based technique to localize an in-body RF-source for biomedical applications, *IEEE Trans. Biomed. Eng.*, 62, 5, 1231–1241, 2015.
72. S. Y. Semenov, R. H. Svenson, A. E. Bulyshev, A. E. Souvorov, A. G. Nazarov, Y. E. Sizov, V. G. Posukh, A. Pavlovsky, P. N. Repin, A. N. Starostin, B. A. Voinov, M. Taran, G. P. Tatsis, V. Y. Baranov, Three-dimensional microwave tomography: Initial experimental imaging of animals, *IEEE Trans. Biomed. Eng.*, 49, 1, 55–63, 2002.
73. R. Chandra, and I. Balasingham, A microwave imaging-based 3D localization algorithm for an in-body RF source as in wireless capsule endoscopes, *37th Annual International Conference of the IEEE Engineering in Medicine and Biology Society (EMBC)*, 4093–4096, Milan, Italy 2015.
74. EMTensor, TechGate Tower, Donau-City-Strasse 1, 1220 Vienna, Austria. Website: http://www.emtensor.com/
75. Micrima Limited, One Glass Wharf, Temple Quay, Bristol, BS2 0EL. Website: http://micrima.com/
76. V. Hamsakutty, A. Lonappan, V. Thomas, G. Bindu, J. Jacob, J. Yohannan, and K. T. Mathew, Coupling medium for microwave medical imaging applications, *Electron. Lett*, 39, 21, 1498–1499, 2003.

20

Making Content-Based Medical Image Retrieval Systems Worth for Computer-Aided Diagnosis: From Theory to Application

Agma Juci Machado Traina

Marcos Vinícius Naves Bedo

Lucio Fernandes Dutra Santos

Luiz Olmes Carvalho

Glauco Vítor Pedrosa

Alceu Ferraz Costa

Caetano Traina Jr.

20.1 Introduction

The amount of medical data generated in hospitals and medical centers has been growing at a very fast pace, due to the increased usage of different modalities of digital medical exams, as well as the widespread acceptance of the Picture Archiving and Communication Systems (PACS). Medical images are usually stored with text-based descriptions of their content, which are limited due to a lack of systematization in the annotation process. Consequently, as the volume of stored images grows, searching for relevant medical images is becoming less efficient and more tiresome. These shortcomings have been addressed by the *Content-Based Image Retrieval* (CBIR) systems and, specifically, by the *Content-Based Medical Image Retrieval* (CBMIR) for the medical domain (Datta et al. 2008; Müller and Deserno 2011). Given a query image as an example with and without text annotations, a CBMIR finds the most similar images.

Potential applications of CBMIR systems range from clinical decision support to medical education and research. Retrieving similar images can add more information and confidence to the analysis and evaluation process, which supports the specialists in the final diagnostic. However, when searching for similar medical images, it is important that the CBMIR system analyzes the same anatomical structures and visual patterns in the image exam that are relevant to the medical specialists. Simple CBMIR systems essentially employ a predefined search strategy, using a fixed combination of feature extraction methods and distance functions (Güld et al. 2007; El-Naqa et al. 2004). This limits the image comparisons to a single search space, limiting the system capability to capture the large variety of distinct visual patterns that may exist between different images in the same domain (Marques and Rangayyan 2013; Kumar et al. 2013).

In this chapter, we discuss the potential and the limitations of using CBMIR systems in clinical environments and how they can be further improved to become a powerful asset for Computer-Aided Diagnosis (CAD). We present the development of a CBMIR system for a real medical environment: the Higiia system. Higiia is a CBMIR system specialized for mammograms, which is able to encode *perceptual parameters* (i.e., pairs of feature extraction methods and distance functions) according to an initial evaluation defined by the medical specialist. This is a powerful asset to enhance the effectiveness of CBMIR systems, once it directs the similarity search regarding the user perception and avoids the retrieval of unwanted cases. Our experiments were conducted together with radiologists and resident physicians of the Center of Images and Medical Physics with the Clinical Hospital of the School of Medicine at the University of São Paulo in Ribeirão Preto, Brazil. The results show that Higiia improves the expert confidence as the relevance feedback cycles are performed, highlighting its importance to help medical diagnoses and education.

Although much work has been done trying to develop effective and efficient CBMIR systems, the use of these systems over massive datasets may have drawbacks that impact their usability in real clinical applications. These problems are related to the fact that many images returned are closer to each other than to the query element (near-duplicates). To narrow down this drawback, we also present a technique developed to bridge the *usability gap* in CBMIR systems exploring both similarity and diversity. Usability gaps are related to how easy a software tool is to use from the radiologist's perspective, while diversity refers to elements not only similar to the query element, but also diverse among themselves. The proposed approach called BridGE (*Better result with influence diversification to Group Elements*) aims at adding new relevant information to the user specialist, reducing the need of further query refinement or relevance feedback cycles. The results are displayed to the user specialist as a traditional CBMIR result, whereas the radiologists are able to expand the clusters and navigate through them. The results support our claim that a CBMIR system empowered with diversity is able to bridge the usability gap, grouping near-duplicates and being at least two orders of magnitude faster than its main competitors.

The remainder of this chapter is structured as follows: Section 20.2 summarizes the main concepts related to the retrieval of similar medical images. In Section 20.3, we present the Higiia system, which is a CBMIR system implemented and evaluated in a clinical environment. We detail the Higiia architecture, as well as the experiments performed to evaluate the main contribution of this system in a real application within Clinical Hospital with our university. Section 20.4 presents the BridGE technique, aiming at narrowing down the problem of near-duplicate images, which brings potential benefits to construct CBMIR systems more effective in real medical scenarios with large datasets. Finally, Section 20.5 presents the conclusions.

20.2 Background and Concepts

A CBMIR system is centered on the notion of *similarity* between images: given a dataset of images, the user wants to retrieve the most similar images based on a query image (Datta et al. 2008; Müller and Deserno 2011). The concept of *similarity* in CBMIR systems can be explored by *image context* and/or *image content* (Ghosh et al. 2011; Akgül et al. 2011). The *context-based approach* relies on textual attributes to each image, such as keywords, tags, or ontologies (Hu et al. 2003). These textual attributes, referred to as *annotations*, are employed to retrieve medical images. Examples of this approach are the systems MEhS and RadSem (Díaz-Galiano et al. 2008; Möller et al 2009). The disadvantages of this textual-based approach are the large human effort required to tag a large image dataset and the subjectivity of the annotation of each image, which may derail the entire retrieval strategy (Liu et al. 2007; Muramatsu et al. 2012).

To overcome textual-based drawbacks, the *content-based approach* relies on extracting low-level characteristics from the images using feature extraction methods (El-Naqa et al. 2004; Alto et al. 2005), which is the main interest of our work. In this section, we present some definitions related to CBMIR systems based on a fully content-based approach for an effective implementation of these systems within a medical domain.

20.2.1 Architecture of CBMIR Systems

The CBMIR architecture is divided into an offline and an online module (Güld et al. 2007; Marques and Rangayyan 2013; Müller and Deserno 2011). Figure 20.1 illustrates the main modules and the interactions. In the offline stage, the diagnosed images are stored in a Database Management System (DBMS), creating the *knowledge database*, which stores both the predefined set of images for classification purposes and their diagnoses, which are pieces of information usually found in a Computer-Aided Diagnosis Atlas. Inserting a new image triggers the feature extraction methods that represent each image in a particular feature extractor domain. Alongside the DBMS is the Feature Indexing Module, which relies on a distance function to organize the stored images and improves the query processing. Therefore, for each pair of available distance function and domain extractor, the CBMIR keeps an index structure over the images stored into the DBMS.

Online processing is triggered whenever the user submits an image as a query element. At this point, an expert user can define the query parameters, whereas default values are expected to be available for non-expert users. The query parameters generally consist of: (1) the image feature extractor, (2) the distance function employed, (3) the similarity search operation, and (4) a model for a relevance feedback cycle.

The images are retrieved from the dataset according to the distances to the query image and the *retrieval operation* requested by the user. The fundamental retrieval operations are the *range* and the *k-nearest neighbor* queries. Nevertheless, sophisticated retrieval criteria may be employed, such as

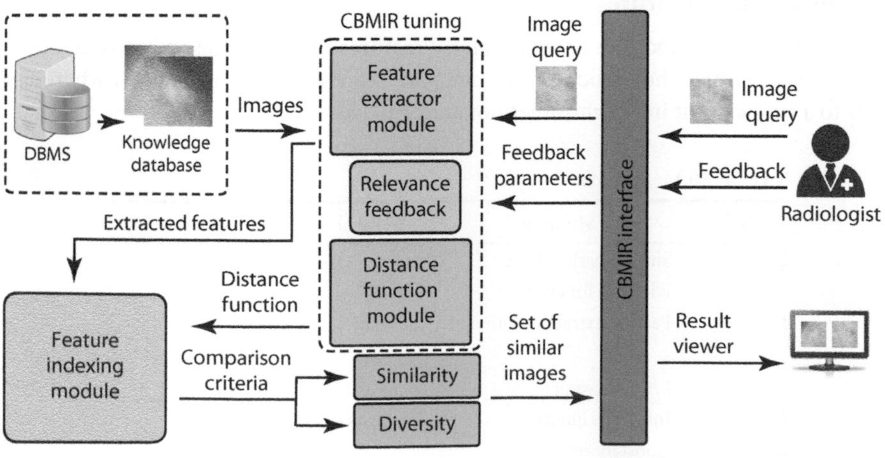

FIGURE 20.1 The architecture of a CBMIR system.

diversity queries, as we will discuss in Section 20.4. The set of images complying with the search criterion are returned as the answer to the similarity search.

After a query processing, the CBMIR displays the answer to the radiologist, which may now decide to further explore the search space, by providing feedback to re-execute the query. In this case, the CBMIR refines the query following a relevance feedback cycle. There are a number of cycles available. For instance, using the corner stone Rocchio relevance feedback method, the user selects some images as relevant or non-relevant (Rocchio 1971). The selected images are evaluated to define a new query element by composing the original center image with the relevant and non-relevant images selected. The new query element is used to execute a new similarity search, closing the first relevance feedback cycle.

By storing the previously diagnosed images into a DBMS, this generic architecture provides a portable way to integrate CBMIR, CAD, and PACS into a single clinical suite application (Lehmann et al. 2003). However, the content-based retrieval operations are still performed at the application level, as similarity queries are not naturally supported by the current commercial DBMS. A significant improvement here is to take advantage of similarity-extended DBMS systems (Kaster et al. 2011; Bedo et al. 2014) to perform similarity searching using an SQL extension. In this way, such core-extended systems are employed to handle the range and the *k*-nearest neighbor similarity criteria. Exploiting this concept, in Section 20.3 we will introduce the Higiia system, a clinical suite framework that benefits from similarity queries performed on the similarity-extended DBMS. Before that, we will review important concepts regarding to CBMIR in the next subsections.

20.2.2 Feature Extraction Module

A crucial module of CBMIR systems is the feature extractor, which encodes the medical images into an n-dimensional array (the feature vector) denoted by $f=(v_1, v_2, \ldots, v_n)$, where each value v_i describes an intrinsic feature of the image's visual content. Formally, a feature extractor method is given by Definition 20.1.

Definition 20.1 (Feature Extractor Method). *Given a data domain* $\mathbb{S}$ *and a feature extractor domain* $\mathbb{F} \in \mathbb{R}^n$, *a feature extractor method* $\mathfrak{F}: \mathbb{S} \mapsto \mathbb{F}$ *is a computational function that represents the original data as a summarization in* $\mathbb{F}$.

There are several feature extractor methods proposed in the literature. They can be divided into *global features* and *local features*. The former (Section 20.2.2.1) considers the image structure as a whole, whereas the latter (Section 20.2.2.2) computes the features based on local regions-of-interest in each image. Table 20.1 summarizes the main symbols used in this chapter.

20.2.2.1 Global Image Features

Many feature extractor methods are based on global features that describe an entire image. Such features are attractive because they produce very compact representations of images, where each image corresponds to a single point in a high-dimensional feature space.

TABLE 20.1 Symbols

Symbol	Meaning	Symbol	Meaning
ζ	Similarity threshold	L_p	Minkowski distance
D	Distance function	PP	Perceptual parameter
$\mathbb{F}$	Feature extractor domain	Rq	Range query
		$\mathbb{S}$	Data domain
F	Feature vector	S	Set of elements
I	Influence intensity	s_i, s_j, s_u	Elements in S
K	Constant number	s_q	Query center
$kNNq$	k-nearest neighbor query	v_i	Feature value

Color is a widely used visual feature for CBMIR due to its invariance with respect to image scaling, translation, and rotation. Most of the CBMIR systems employ color or gray intensity as the basic image feature descriptor. Usually, each pixel is represented as a point in a color-scale system (either gray or a true-color scale). Histogram-based methods evaluate the relative frequency of each color in the image. The *equi-width histogram* (Ioannidis 2003) sorts the pixel values into equally partitioned color groups, where the sum of frequencies represents the cumulative probability function. Another histogram-based approach is the metric histogram (Traina et al. 2003), which summarizes an *equi-width histogram* joining similar frequencies into consecutive intensities. The Color Layout Descriptor (CLD) (Bay et. al. 2008) is designed to capture the spatial distribution of color in an image, by using a grid-based representative color selection and discrete cosine transform with quantization. There are many examples of this kind of extractor in the literature.

Texture features are intended to capture the granularity and repetitive patterns within an image. From a statistical point of view, textures can be seen as complicated visual patterns from which sets of statistics can be obtained for characterization purposes. A statistical method of examining texture that considers the spatial relationship of pixels is the gray-level co-occurrence matrix (GLCM) (Srinivasan and Shobha 2008). The GLCM is a matrix that counts the number of occurrences of pairs of gray levels at a given displacement and angle. Statistics such as contrast, energy, and entropy are computed from the GLCM to obtain texture features (Haralick et al. 1973). The color and edge directivity descriptor (CEED) (Chatzichristofis and Boutalis 2008) incorporates color and texture information in a histogram. The feature vector dimensionality of CEDD is limited to 54 bytes per image, rendering this descriptor suitable for use in large image databases. One of the most important attributes of the CEDD is the low computational power needed for its extraction. Wavelet transforms such as Haar and Daubechies wavelets (Mallat 2008) are also an important tool to texture analysis. Other useful techniques for image feature extractor methods are based on the Zernike moments, which are able to gather the distribution of shapes in an image (Sonka et al. 2014).

Shape features have been increasingly adopted as a way for quickly finding the same anatomical region of the same disease in different exam images, such as mammography and melanomas (Sonka et al. 2014).

20.2.2.2 Local Image Features

Another approach for feature extraction uses local features, which are computed from local image regions based on points of interest. One advantage of using local features is that they may be used to recognize objects despite significant clutter and occlusion. They also do not require segmenting the object from the background, unlike many texture and shape features.

The Scale Invariant Feature Transform (SIFT) (Lowe 1999) and Speed Up Robust Features (SURF) (Bay et. al. 2008) are the two popular approaches to encode local image descriptors, being invariant to image scale and rotation, viewpoint change, noise, and illumination changes in the image content. SIFT is the earliest and classic descriptor, which uses difference of Gaussians over various scales of an image to detect interest points. SURF uses Haar wavelets and is a very fast feature detector and descriptor. SIFT presents an overall accuracy higher than SURF, whereas SURF is faster than SIFT for computing local features.

One of the key issues in dealing with local features is that there may be differing numbers of feature points in each image, which increases the cost of comparing images. A popular strategy to overcome this problem is the Bag-of-Visual-Words (BoVW) representation (Boreau et. al. 2010; Jgou et al. 2010). This model encodes each local feature vector as a *visual word*. A visual word is generated by clustering local feature vectors detected in a set of training images. Each cluster is considered as a visual word, and a set of visual words is considered a visual dictionary. This representation has a final feature vector of fixed sized, making the task of computing the similarity between images based on local features easier. A drawback of the BoVW is that different images may have identical histograms of visual words, although some works have been proposed to address this problem by encoding *spatial information* into the BoVW representation (Lazebnik et al. 2006; Penatti et.al 2014; Tao et.al 2014; Savarese et al. 2006).

20.2.3 Distance Function Module

The similarity between two feature vectors is evaluated by a function (*metric*) that quantifies the similarity as the distance between these vectors, expressed as a real number, according to Definition 20.2 (Deza and Deza 2013; Zezula et al. 2006).

Definition 20.2 (Metric). *Given a data domain* $\mathbb{S}$, *a distance function* $d:\mathbb{S}\times\mathbb{S}\mapsto\mathbb{R}_+$ *is called a metric on* $\mathbb{S}$ *iff, for all* $a,b,c\in\mathbb{S}$, *there holds:*

- $d(a,b)>0$ (non-negativity)
- $d(a,b)=0 \Longleftrightarrow a=b$ (identity of indiscernibles)
- $d(a,b)=d(b,a)$ (symmetry)
- $d(a,b)\leq d(a,c)+d(b,c)$ (triangle inequality)

Following Definition 20.2, the smaller the distance value, the more similar the elements are, whereas a distance zero indicates that the compared elements are the totally identical.

A well-known family of distance functions is the Minkowski family, which includes the Manhattan (L_1) and the Euclidean (L_2) distances (Deza and Deza 2013). The distance function members of the Minkowski family compare a pair of feature vectors dimension by dimension, assuming that each dimension i is independent from the others, according to Equation 20.1, where $X=\{x_1,x_2,\ldots,x_n\}$ and $Y=\{y_1,y_2,\ldots,y_n\}$ denote feature vectors obtained by the same feature extractor method, the operator $\{-\}$ is defined for each dimension, and p is the distance characterization.

$$L_p(X,Y)=\sum_{i=1}^{n}\left|x_i-y_i\right|^{p1/p} \tag{20.1}$$

A variation of L_1 is the Canberra Distance, which is expressed in Equation 20.2. When $x_i=y_i=0$, the result is defined as zero.

$$C(X,Y)=\sum_{i=1}^{n}\frac{\left|x_i-y_i\right|}{\left|x_i\right|+\left|y_i\right|} \tag{20.2}$$

Another widely employed member of the Minkowski family is the Chebyshev metric: L_∞. There are distance functions able to compare feature vectors that are not defined in $\mathbb{R}^n$ (Zezula et al. 2006). For instance, the metric histogram distance is able to obtain the similarity between two metric histograms by calculating the area difference under their piecewise linear approximation curve (Traina et al. 2003).

There are other important functions that can measure similarity, although not every one of them respects the requirements of Definition 20.2 (Deza and Deza 2013). For example, the *Kullback-Leibler* distance does not satisfy the triangular inequality nor the symmetry property, while the *Jeffrey's Divergence* does not satisfy the triangular inequality property (Kullback and Leibler 1951). In such cases, those functions can still be employed to measure the distance between two elements, but they lack on enabling using the metric access methods.

20.2.4 Perceptual Parameters

The pair $\mathfrak{F},d$, which associates a feature extractor method $\mathfrak{F}$ to a distance function d, composes the *perceptual parameters* (*PP*). In a CBMIR system, $\{PP_1, PP_2, \ldots, PP_n\}$ is the set of all pairs composed of available feature extractor methods and distance functions. The choice of perceptual parameters impacts the search in different ways (Felipe et al. 2009; Chadha et al. 2012). For instance, a feature extractor method based on texture may be more suitable to a given analysis than a color extractor, while the distance function is used to define the relationship among the elements in the search space. Therefore, the CBMIR should represent the image query according to the same feature extractor method that represents the stored images and must also employ the user-defined distance function to measure the similarity between the query image and the database elements.

20.2.5 Similarity Queries

Similarity search is the information retrieval process where a query is given by an object (e.g., a medical image), referred to as the *query center* or *query element*, and the answer is a set of elements similar to the query object (Hetland 2009; Beecks et al. 2013). There are two main types of similarity queries: the *similarity range* and the *k-nearest neighbor* queries. A *range query* retrieves all elements of a dataset that are dissimilar to the query center by at most a maximum threshold, where the dissimilarity is computed by a distance function or a metric. A range query is formally expressed according to the Definition 20.3.

Definition 20.3 (Range query: *Rq*). *Let* $\mathbb{S}$ *be a data domain, S be a set of elements in the domain* $\mathbb{S}$, *d be a distance function and* $\zeta \in \mathbb{R}_+$ *be a similarity threshold. A range query over the query center* $s_q \in \mathbb{S}$ *is given by* Equation 20.3.

$$Rq(S,d,\xi,s_q) = \{s_i \in S \mid d(s_q,s_i) \leq \xi\} \tag{20.3}$$

An example of a range query is: "*retrieve all the images that are similar to the image <brain_tissue_1.jpg> up to 10 units of distance.*" Although a range query is the most basic kind of similarity query, it may not be intuitive to the user to properly define the threshold ζ, since small changes may produce a great impact on the result set cardinality. The suitable values of ζ are often in a narrow range, and defining it properly may require a deeper understanding of the data domain.

Aimed at solving this issue, another kind of similarity query—the *k-nearest neighbor*—enables the user to directly specify the number of elements in the answer set. A *k*-nearest neighbor query retrieves an amount *k* of elements that are the most similar to the query center, evaluating the similarity by a distance function. For example, "*retrieve the 5 images most similar to the image <brain_tissue_1.jpg>.*" A *k*-nearest neighbor query is formally expressed by Definition 20.4.

Definition 20.4 (*k*-nearest neighbor query: *kNNq*). *Let* $\mathbb{S}$ *be a data domain, S be a set of elements in domain* $\mathbb{S}$, *d be a distance function and* $k \in \mathbb{Z}_+^*$ *be a positive integer. A k-nearest neighbor query over the query center* $s_q \in \mathbb{S}$ *is given by* Equation 20.4.

$$kNNq(S,d,k,s_q) = A = \{s_i \in S \mid \forall s_j \in S - A : d(s_q,s_i) \leq d(s_q,s_j)\}, |A| = k \tag{20.4}$$

A *k*-nearest neighbor query is often more useful in a clinical environment, since retrieving the most similar cases increases the physician reliability in presenting a diagnostic (Carvalho et al. 2014).

20.3 Higiia: A CBMIR System for Mammography

In this section, we present a CBMIR system that employs the concepts aforementioned. Higiia is a similarity-enabled system that supports perceptual retrieval, that is it enables setting up perceptual parameters on-the-fly, adjusting the similarity evaluation to user-chosen feature extractor methods and distance functions. Although default settings are available, users can manually tune the search when they know the semantics of the perceptual parameters regarding both system and the image domain. Higiia natively provides support for several file formats, including the standard DICOM. It was also

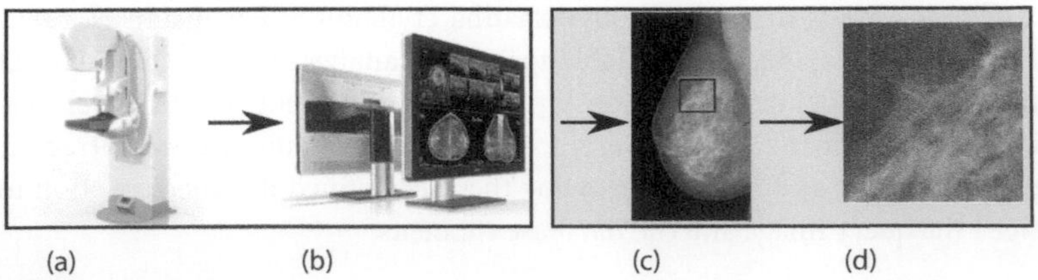

FIGURE 20.2 Image flow from the digital mammogram systems to the CBMIR Higiia. (a) The digital mammogram is taken from a patient. (b) The digital mammogram is managed by the PACS system. (c) The radiologist selects regions with potential abnormalities (ROIs). (d) The ROI analysis becomes available for the Higiia system. Components (a) and (b) are standard to any PACS, while components (c) and (d) are specific to Higiia.

fully coded over a DBMS-oriented architecture. Therefore, it can handle subsets of an enterprise medical image database in an efficient and scalable manner, allowing building distinct medical contexts through filters over DICOM or traditional attributes.

Let us describe the architecture of the Higiia system assuming a CBMIR tool specialized for mammograms (Bedo et al. 2012). Higiia implements all of the modules previously shown in Figure 20.1. The system was defined according to the requirements given by radiologists, which includes a clean interface and image processing tool as the windowing operation for medical images. Figure 20.2 presents the relationship between other digital mammogram systems and Higiia.

Initially, a digital mammogram from the patient (Figure 20.2a) is processed according to the implementation of the local PACS system (Figure 20.2b) and submitted to Higiia. Then, the radiologist defines the regions with suspected abnormalities by selecting specific *Regions of Interest* (ROIs) from the digital mammogram (Figure 20.2c). Each ROI (Figure 20.2d) is inserted into the Higiia pool list, which becomes available to the experts alongside with the similarity searching options.

Regarding the offline stage, the Feature Extractor Module represents the image query using visual features of color, texture, and shape, which are indexed and stored into the extended-DBMS, according to every distance function available. In the online stage, the query parameters (including the perceptual parameters) are applied over the indexes, allowing the retrieval of the image identifiers according to the comparison criteria. The Higiia tool, using the extended-DBMS query capabilities, internally performs this step. For instance, consider that the image query was already stored in the database. Then, given the query parameters, Higiia builds an extended-SQL expression to represent the similarity query, as, for example, in the following selection command:

```
SELECT image FROM knowledge_database_table
WHERE image NEAR(SELECT image FROM query_table WHERE idImage = 1)
BY perceptual_parameter STOP AFTER k;
```

where `image` is the medical image attribute of the `knowledge _ database _ table` that stores related and diagnosed images, and `perceptual _ parameter` defines the search space. The operator `NEAR` refers to the *k*-nearest neighbor operation and *k* is the expected number of images to be returned. Every search is executed employing indexes, which were created during the offline stage. A related case occurs when the query image is not stored into the DBMS. Higiia saves the query image in a temporary buffer, enabling the SQL query to be rewritten as follows:

```
SELECT image FROM knowledge_database_table
WHERE image NEAR 'query_image.dcm'
BY perceptual_parameter STOP AFTER k;
```

After building the SQL statement, Higiia invokes the similarity-extended DBMS for its execution. Higiia also provides a relevance feedback module as a query refinement option. This module collects

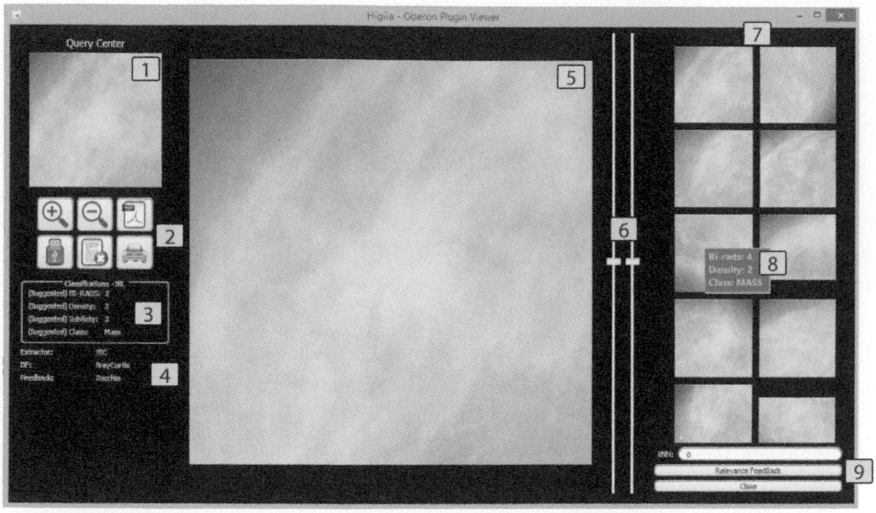

FIGURE 20.3 The Higiia viewer interface. The numbers indicate system functionalities for similarity searching.

information from the users' interactions and employs the Rocchio strategy to generate a new query element, which is employed to perform a relevance feedback cycle. The expert may request as many relevance feedback cycles as necessary. Finally, Higiia requests the user diagnosis including BI-RADS, density, and report, which are stored in the database. After the expert finishes the diagnosis section, the ROI is removed from the Higiia study pool.

Higiia has a graphical user interface designed to provide detailed information about the ROIs retrieved. For each similar image recovered, the expert can visualize its diagnosis. The Higiia viewer interface is depicted in Figure 20.3. It presents the query image at the top-left of the screen (Figure 20.3(1)), while in the remainder of the left column the user can check the automatic system classification (Figure 20.3(3)). The perceptual parameters employed are listed in Figure 20.3(4). The buttons depicted in Figure 20.3(2) provide extra functionalities, such as exporting images in PDF file format or printing them. The screen center presents the current image under analysis, which can be either the query center or a retrieved image (Figure 20.3(5)). Initially, the image under analysis is set as the query image, but the expert can proceed choosing any of the retrieved images by just double-clicking it, as the user explores the result set.

The result of the current similarity query is presented in the right column (Figure 20.3(7)) as well as the stored diagnosis related to each retrieved ROI, as shown in Figure 20.3(8). After analyzing the diagnosis retrieved, the experts can remove non-relevant images by right-clicking on them, or just selecting the relevant ones through a left-click on the mouse over the image. Also, using the bar controls depicted in Figure 20.3(6), it is possible to perform adjustments in the analyzed image, such as framing, zooming, and translation. Subsequently, a new query employing relevance feedback can also be issued, which allows providing a new number of images, as depicted in Figure 20.3(9). The relevance feedback module performs a new similarity query and the right column of the result (Figure 20.3(8)) is updated, enabling the user to continue the analysis by exploring new result sets.

Each ROI can be reported with a new diagnosis, which is stored into the DBMS. Such information can be uploaded in the institution's hospital information system.

20.3.1 The Higiia Knowledge Database

Higiia takes advantage of previous studies (e.g., Pereira Jr. et al. 2007) to build a default knowledge database. Although newly analyzed ROIs can be easily added to the system by inserting images into the knowledge database table, the system's distribution default knowledge database is pre-loaded with 2919 ROIs, extracted

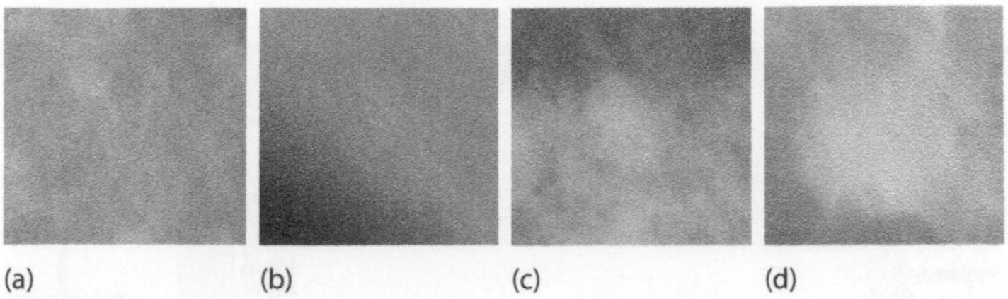

FIGURE 20.4 Examples of the knowledge database: (a) Benign calcification, (b) malignant calcification, (c) benign mass, and (d) malignant mass.

from the Digital Database for Screening Mammography* (DDSM). The ROIs in this knowledge base are divided in two classes: mass and calcification lesions. The images were labeled according to the first edition of the Breast Imaging Reporting and Data System (BI-RADS) standard as proposed by the American College of Radiologists† (ACR). Figure 20.4 outlines some of the Higiia knowledge database images.

20.3.2 Higiia in Action: Potential and Limitations

To illustrate the potential and limitations of the CBMIR technology, this section presents an experiment with the current release of the Higiia system. The experiment was performed employing 10 specialists, including radiologists and resident physicians from the Center of Images and Medical Physics of the Clinical Hospital of the School of Medicine of the University of São Paulo at Ribeirão Preto. The experts were asked to classify a number of images using Higiia.

Each expert reviewed in average six images, and each image was queried by, at least, two different professionals. At the end of the experiments, 65 classifications about 30 ROIs employed as query images were collected. To enable a fair comparison of the CBMIR process, the queried ROIs were also diagnosed by other physicians (non-involved with Higiia) in an evaluation process with other resources, such as biopsy and medical records. Those diagnoses were stored in a separated table as the ground truth.

For this experiment, the Higiia knowledge database images had their gray-level histograms extracted, and the Euclidean metric (L_2) was employed to perform the similarity search operations. This perceptual parameter (*Histogram*, *Euclidean*) was tested using tenfold cross-validation and obtained an accuracy of 62% regarding ROIs containing mass and 40% for ROIs with calcification. In other words, the perceptual parameter (*Histogram*, *Euclidean*) seems more suitable to search ROIs containing mass than calcification. The hypothesis here is that the perceptual parameter depends on the diagnostic hypothesis to deliver good-quality image retrieval.

To perform the analysis, a set of four steps between the expert and the system were defined as follows. In the first step, a query image was randomly picked from the Higiia pool and presented to an expert. The experts were expected to provide an initial diagnosis and a degree of certainty about their own classification. Next, Higiia performed a similarity query using the query image and presented the retrieved images to the experts. Then, the experts checked the diagnoses of past cases and indicated the non-relevant images, which were removed from the system interface. In the third step, the system asked the users to provide a second diagnosis alongside the certainty degree. Notice that the experts were allowed to maintain the first diagnosis as well as the certainty degree. If the experts were not sure enough about the diagnosis, the system enabled a fourth step, triggering a relevance-feedback query. The relevance-feedback query performs a new similarity query, guiding the user interactively back to the third step until the user is confident about the diagnosis.

* <http://marathon.csee.usf.edu/Mammography/Database.html> Access: November 14, 2015.

† <http://www.acr.org> Access: November 14, 2015.

TABLE 20.2 Diagnosis Accuracy with and without the CBMIR Higiia

	No-System	Similarity Query	Relevance Feedback
	Classification of Mass		
Expert hit ratios (%)	89	93	93
System precision (%)	—	61	62
	Classification of Calcification		
Expert hit ratios (%)	47	31	27
System precision (%)	—	39	41

The interactions between Higiia and the 10 users were stored into a log file and recorded three main measures: (1) the system precision, (2) the expert hit ratio, and (3) the expert average confidence level. All measures were calculated according to a given instant of interaction with the system. The tool precision was the fraction of retrieved images marked as "relevant," while the experts hit ratio was calculated as the average expert accuracy. Table 20.2 presents the comparison for the first two measures for the sequential expert classifications.

Notice that, for a system precision level of 61%, the experts' hit ratios were up to 93% regarding ROIs containing mass. Even for system precision levels higher than 50%, the diagnosis hit ratio showed noticeable impact. Naturally, that was the expected behavior for the system, which improved the overall diagnosis process in the two-phase query process.

On the other hand, for ROIs containing calcifications, the results showed that the system precision scales up for relevance-feedback cycles. However, the experts' hit ratio drops with the use of similarity searching. In order to understand why the system presented such a different behavior for the two classes of ROIs, a posterior analysis has been performed over the expert confidence level through the interactions. Figure 20.5 summarizes the average confidence of the experts regarding the three main interactions with Higiia, considering four feature extractors (Histogram, Haralick, Haar wavelets, Daubechies wavelets) and five distance functions (City-Block—L_1, Euclidean—L_2, Chebyshev, Jeffrey-Divergence, Canberra).

For the mass ROIs images, the experts increased the confidence level by means of similarity querying and at least one relevance feedback cycle. Moreover, 90% of the classification changes were related to previous diagnoses with low levels of confidence. Still, the classification ratios marked as "completely confident" decreased through the user and system interaction. This fact indicates that Higiia returned interesting findings that potentially brought further insights to the analyzed image.

Regarding the calcification of ROIs, the majority of experts reported "some level of not confident," which, in practical terms, would lead to a second analysis by other means than just the CBMIR. Notwithstanding,

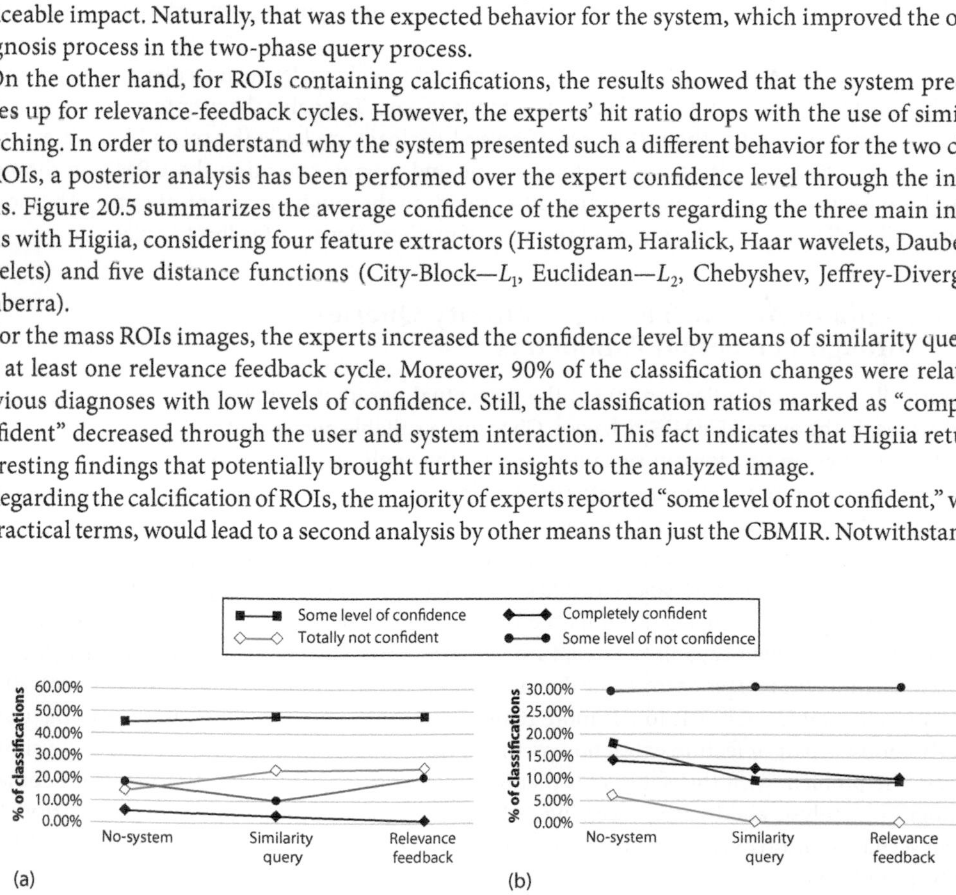

FIGURE 20.5 Experts confidence level using the CBMIR for ROIs with (a) mass and (b) calcification.

the system precision increased through relevance feedback cycles (Table 20.2), but the hit ratio dropped substantially, as well as the expert confidence level on the new diagnosis (Silva et al. 2013).

Those results corroborate the initial hypothesis that the perceptual parameter depends on the diagnostic hypothesis to deliver good-quality image retrieval. In this case, it can be explained as the gray-scale histogram feature extractor method is neither capable of shape nor size detection, representing the query image only by the color-frequency in a reduced gray-scale spectrum (Traina et al. 2003). Therefore, the histogram extractor did not produce a faithful expression for calcification ROIs. For mass ROIs, on the other hand, the representation seems good enough to not compromise the similarity retrieval.

Complementarily, the second part of the perceptual parameter—the distance function—also plays an important role to define the behavior of the search space. The choice of the distance function may compromise the entire similarity searching process (Felipe et al. 2009; Muramatsu et al. 2012). The need for properly setting the perceptual parameter is the first key observation to make a CBMIR technology suitable to efficiently act as a Computer-Aided Diagnosis tool in the future. Based on that, Section 20.3.3 details this situation and discusses proper strategies to handle the perceptual parameter choice for Higiia.

The obtained results also enabled an important discussion with respect to the relevance feedback cycles. Figure 20.5 shows that the diagnosis confidence level has monotonically increased or decreased, except for one case. It means that Higiia contributes to the expert confidence as the relevance feedback cycles are performed. The drawback of this approach is that many cycles are necessary to allow the user to navigate through the search space, which requires a lot of time from the expert and consume more computational resources.

Therefore, another search mechanism can be employed to delay relevance feedback cycles as much as possible, bringing similar but useful past cases to the expert. To enable the user to explore the search space using similarity and without the need of several feedback cycles is the second key observation, which improves the similarity search and saves time and money regarding the CBMIR technology. Taking this observation into account, Section 20.4 provides a thorough discussion about another search criterion, called *diversity*, which may avoid the need of further relevance feedback cycles.

20.3.3 Higiia in Action: Tuning Similarity Queries through Perceptual Parameters

As empirically evaluated in Section 20.3.2, the perceptual parameter employed in a similarity query strongly impacts the overall CBMIR behavior. Choosing a suitable perceptual parameter is not straightforward, as it relies on the domain semantics and on the applications. For instance, do the extracted features represent the relevant visual patterns? Or does the distance function correctly highlight the differences and define a proper search space?

Basic CBMIR methods essentially employ a predefined search strategy, using a fixed perceptual parameter (Güld et al. 2007; El-Naqa et al. 2004). Although such a perceptual parameter is conceivably well-suited for a specific scenario, it limits the image comparisons to a single search space. In other words, it limits the system capability to capture the large variety of distinct visual patterns that may exist between different images of the same domain (Marques and Rangayyan 2013; Kumar et al. 2013).

A solution to enable a CBMIR to use more than one perceptual parameter is to rely on the user to manually choose a pair of feature extraction method and distance function before executing a similarity query. The problem with this solution is that it requires the user to be an expert on both domains: mammograms analysis and image processing techniques, which is not the case for the majority of the practical scenarios. Moreover, the training cost and time for those professionals to just set the CBMIR would be prohibitive.

Additionally, new perceptual parameters could be added to a CBMIR system, what would potentially increase the complexity of tuning the system. To illustrate this, consider a scenario where a CBMIR implements five distance functions and three feature extraction methods. The amount of available

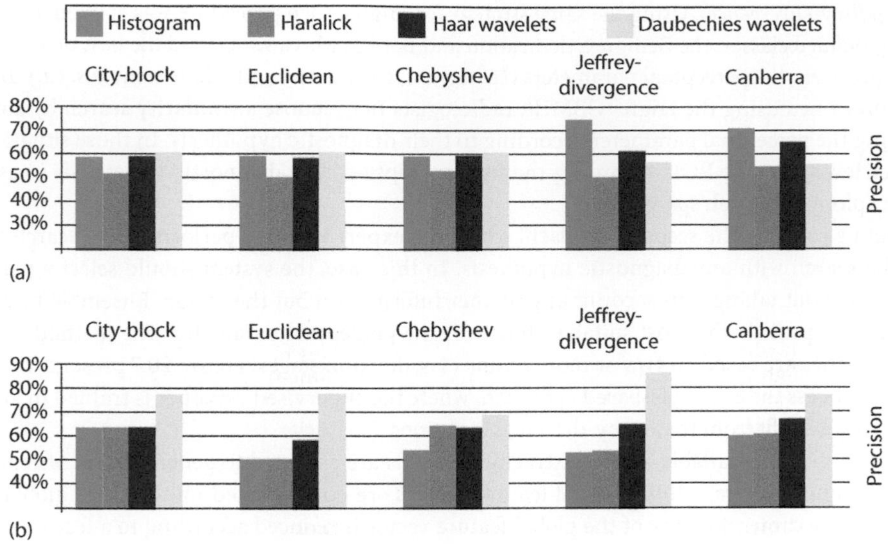

FIGURE 20.6 Average precision exploring four feature extractors and five distance functions, totaling 20 perceptual parameters over the Higiia knowledge database and its evaluation on: (a) Malignant calcification and (b) malignant mass.

perceptual parameters already adds up to 15. Assume now that the CBMIR developing team designs a new feature extractor and adds it to the system. According to this example, as shown in Figure 20.6, now the user must choose among 20 possible perceptual parameters. This occurs because the number of perceptual parameters increases linearly with respect to the number of new extractors or distance functions. For modern CBMIR applications, the number of possible perceptual parameters is much larger, turning this expert-driven choice quite hard some.

To tackle this problem, two complementary approaches can be used. Both rely on the hypothesis that the overall classification precision defines the most suitable perceptual parameter for each diagnostic hypothesis. Therefore, a similarity query should depend on a previous definition of the diagnostic classes. The first approach employs frequency-driven strategies whereas the second approach is related to ensemble models to determine the most suitable perceptual parameter (Keysers et al. 2003; Rahman et al. 2007; Yildizer et al. 2012; Bedo et al. 2015).

The first approach is to employ *frequency-driven* strategies, which are offline-only methods that associate a diagnosis class to a perceptual parameter. Such association relies on classifying a query element by using a classifier such as IB1 (Aha et al. 1991). The idea is that a query image can be labeled according to the majority class of its set of most similar images that have been selected according to the perceptual parameter. By applying this reasoning to the entire CBMIR knowledge database and relying on some model validation (Picard and Cook 1984), such as the well-known ten-fold cross-validation or leave-one-out, it is possible to generalize this statistical analysis and determine the most precise perceptual parameter regarding each diagnosis class.

To illustrate the behavior of a frequency-driven method, Higiia was evaluated with four feature extractor methods, namely *Histogram*, *Haralick*, *Haar* and *Daubechies wavelets* and 5 distance functions, which are *City-Block*, *Euclidean*, *Chebyshev*, *Jeffrey-Divergence*, and *Canberra*. Figure 20.6 presents the precision of each possible perceptual parameter regarding two diagnosis classes: Malignant Calcification and Malignant Mass. The ten-fold cross-validation approach was employed to obtain the average precision for each possible perceptual parameter.

For Malignant Calcification ROIs, the combination (*Histogram*, *Jeffrey-Divergence*) obtained the highest precision, whereas for the Malignant Mass class the most suitable perceptual parameter was (*Daubechies wavelets, Jeffrey-Divergence*). Using the same strategy for Higiia, such procedure was applied to other two previously defined classes: the Benign Calcification and Benign Mass, which had the best precision when using, respectively, the perceptual parameters (*Histogram, Chebyshev*) and (*Haar Wavelets, City-Block*).

Therefore, when using the Higiia CBMIR, radiologists may choose a similarity search by automatically setting the perceptual parameter according to their diagnostic hypothesis. In those situations, the system retrieves relevant ROIs regarding the user perception. The diagnostic hypothesis is used as a guide to explore the search space.

Now, let us consider the second approach, where the expert wants to perform a free search without biasing the search with any diagnostic hypothesis. In this case, the system should select a perceptual parameter without taking into account any further information but the image. Ensemble models can be used here to provide the most suitable choice for the perceptual parameter. The methods based on those approaches are based on two or more online classification stages. Figure 20.7 presents the overall pipeline to process the ensemble-based approach, where the supervised classifier is trained in an offline stage and may benefit from frequency-driven associations.

In the first step, all available feature extractor methods are executed to generate a set of representations for the query image. All individual feature vectors are concatenated into a single global feature vector. Next, the dimensionality of the global feature vector is reduced according to a feature selection method as, for instance, the Principal Component Analysis (PCA) (Jolliffe 1986) or the Fast Correlation-Based Filter (FCBF) (Yu and Liu 2003). The next step is to resort to a supervised classifier to automatically select the most suitable parameter regarding the resulting feature vector.

At this point, a traditional similarity query is triggered using the perceptual parameter selected by the system as having the best precision, using indexes whenever possible. Thus, the images retrieved are those most similar according to the system's best perception of similarity. The pipeline of Figure 20.7 was implemented on Higiia to experiment whether the perceptual parameters choice following the ensemble model in fact improves the query precision. The experiment considers the same feature extractor methods and distance functions of the frequency-driven experiment reported in Section 20.3.2 employing the FCBF as the feature selector method. The resulting feature vector has 32 dimensions.

Higiia employs the ensemble-based version of the Online Sequential Extreme Learning Machine (OS-ELM) (Liang et al. 2006) supervised classifier. It was chosen because it scales well for increasingly large datasets. Moreover, when compared to other state-of-the-art classifiers such as support vector machines (Cortes and Vapnik 1995) or artificial neural networks, OS-ELM presents equivalent accuracy

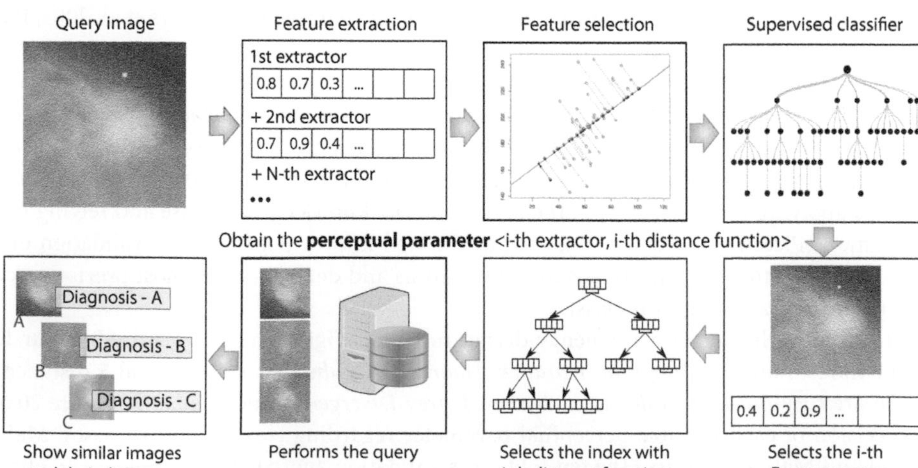

FIGURE 20.7 Overall pipeline of the ensemble-based strategy used to choose a suitable perceptual parameter and perform a similarity query.

TABLE 20.3 Precision Ratio Achieved by Higiia when Using the Ensemble Model to Perform Similarity Queries Compared to Other Approaches

	Theoretical Best Choice	Automatic Higiia Choice	Basic CBMIR Approach	Theoretical Worst Choice
Benignant calcification (%)	67.5	**64.5**	61	50
Malignant calcification (%)	73	**70**	58	50
Benignant mass (%)	69	**67.5**	61	50.5
Malignant mass (%)	86	**78.5**	62	51

Note: The Higiia results are shown in bold.

ratio for most of the evaluated domains whereas requiring much less computational power in the training stage (Huang et al. 2012; Liang et al. 2006). Once the perceptual parameter is chosen, Higiia starts the regular similarity query processing. Query results are shown to the experts with automatic ROI classification regarding the similarity perception provided by the OS-ELM classifier.

To evaluate the resulting precision for the relevant images retrieved, another experiment using the four distinct ROI classes (Section 20.3.2) was performed, as well as a ten-fold cross-validation. Thus, the performance of Higiia was compared to the theoretical upper and lower bounds in terms of system precision. The theoretical upper bound is reached only when the Higiia ensemble model selects the best perceptual parameter for every evaluated query element in the cross-validation process. Likewise, the theoretical lower bound is reached only when Higiia picks the worst perceptual parameters for every query image.

The precision of Higiia was compared to the basic CBMIR approach, which is, basically, the setting used for the experiments shown in Section 20.3.2, that is the precision was taken using Higiia with the same perceptual parameter for every query element. Table 20.3 summarizes the precision ratio regarding the four possible scenarios. As it can be seen, for Malignant Mass and Malignant Calcification classes, Higiia improved the hit ratio over the basic CBMIR approach in up to 14%. Likewise, for the Benign Calcification class, the accuracy gain was 3.5%. Regarding the images containing Benign Mass, Higiia was 6.5% more accurate than the standard approach.

The automatic selection of the perceptual parameter increased the overall similarity queries precision an average of more than 10%, when compared to the basic CBMIR. Although the implemented feature extractor methods are also valid to represent general purpose images, the ensemble-model approach boosted the Higiia capabilities in such a way that it increased the precision more than twice in relation to the traditional CBMIR approach. The experiments confirm that properly addressing the problem of choosing perceptual parameters enables CBMIR tools to improve the quality of the retrieved images. Moreover, Higiia outputs enable the expert to interactively explore the result set and, at the same time, provide a "*second opinion*" in easily understandable way, as it relies on the retrieved image set. Following, we investigate how CBMIR tools may reduce the number of interactions while enabling the user to navigate faster through the search space.

20.4 Is Being Similar Really Enough for CBMIR?

Intuitively, content-based queries should retrieve the images most relevant (most similar) to the query element. As discussed in Section 20.3.3, selecting adequate perceptual parameters associated to relevance feedback techniques improves the definition of similarity in a CBMIR system, making the results nearer to the users' expectation of what is similarity among the images.

However, nowadays, the volume of images produced and gathered in health centers grows exponentially, and querying massive database employing *k*-nearest neighbor and/or range criteria may often retrieve images too much similar among themselves. Often, the retrieved images are more similar to each other than to the query element, called *near-duplicates* (Banda et al. 2014). Having *near-duplicates* in the result requires query reformulations or several relevance feedback cycles and often induces users to think that the database does not store their desired answer, which leads the users to give up the use of a CBMIR system.

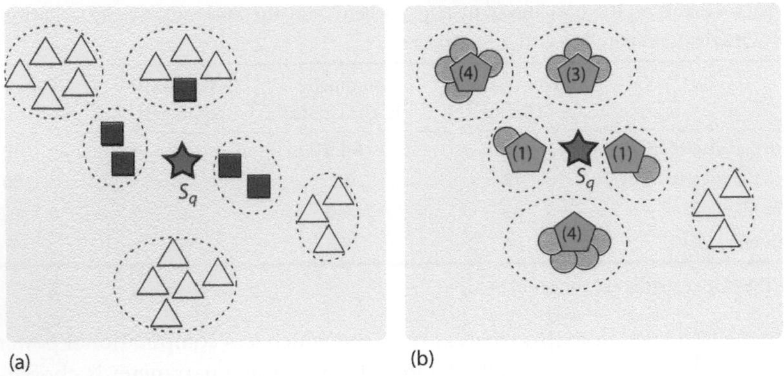

FIGURE 20.8 Element selection for similarity queries in an Euclidean bi-dimensional space. The squares are the elements selected while the triangles are the elements not returned to answer the query. (a) The solution space for the plain *k*-nearest neighbor operation centered at element s_q, including near-duplicates, which are inside the dotted balls. (b) A diversified result: clustered near-duplicates (circles) and its representative elements (pentagons), with the number of grouped elements indicated.

Figure 20.8a illustrates the result obtained by the traditional k-nearest neighbor algorithm in a search space where dotted balls show the relation between the near-duplicates group. As it can be seen, the plain k-nearest neighbor operation returns two near-duplicates, which may be unsatisfactory to the expert since they probably bring the same information. Near-duplicates require more effort from the specialist to navigate and analyze the images (Santos et al. 2013).

A solution to reduce the near-duplicates problem is adding a diversity degree to the similarity query process, taking advantage of the relationship among the retrieved images to obtain a holistic vision of the database. The intuition of diversity is to generate a result set that includes elements not only similar to the query element, but also diverse among themselves. This property has been tackled in a number of ways in several research areas (Santos et al. 2013; Drosou and Pitoura 2010), including medical images systems (Dias et al. 2013). The most common way to include diversity is solving a bi-criteria optimization problem, where similarity and diversity compete with each other ruled by a trade-off parameter defined by an expert (Carbonell and Goldstein 1998; Gil-Costa et al. 2011; Dias et al. 2013; Vieira et al. 2011). Those methods receive the result of a basic similarity query configured to retrieve more elements than requested by the user and, thereafter, apply the bi-criteria objective function to re-rank the solution inducing diversity among the elements sent to the user. This basic approach results in an NP-hard problem (Vieira et al. 2011), and despite efforts to reduce the computational costs, the solution to this problem remains costly. Moreover, setting the *trade-off* parameter among similarity and diversity for each query is difficult and at least unintuitive (Angel and Koudas 2011). Therefore, it was not adopted in CBMIR systems before.

On the other hand, diversity can be considered in CBMIR systems by using the similarity searching process to automatically group near-duplicates, which provides more information to the specialist right in the first result and improves the query performance. For instance, the answer illustrated in Figure 20.8b shows that diversified answers explore more elements than the traditional *k*-nearest neighbor (Figure 20.8a). In addition, on the front-end application of a CBMIR, the user can hierarchically visualize each near-duplicate group in the result set and proceed interactively, performing either the shifting of the query element in an intuitive way or even proceeding to a relevance feedback interaction. The next section details diversity and shows how it can enhance a CBMIR.

20.4.1 Applying Diversity to CBMIR to Group Near-Duplicate Elements

The Result Diversification based on Influence (RDI) (Santos et al. 2013) is a diversity approach based on element separation distances. It assumes that if two elements s_i and s_j are closer than a minimum distance

ζ, they ought to bring the same amount of information and only one of them should be returned in the final answer. Such minimum distance is estimated using the concept of "influence intensity" (Definition 20.5) using only the position among the elements relative to the query element s_q.

Definition 20.5 (Influence). *Given $s_i, s_j \in \mathbb{S}$ and the distance $d(s_i, s_j)$ between them, the influence among the elements is defined as the inverse of the distance, such as in Equation 20.5.*

$$I(s_i, s_j) = \frac{1}{d(s_i, s_j)} \tag{20.5}$$

The concept of influence is applied to select elements s_j based on $I(s_i, s_q)$ in such a way that every element s_j having $I(s_j, s_i) \geq I(s_j, s_q)$ is considered to exert more "influence" over s_i than s_q. Figure 20.9a depicts an Euclidean search space where image s_j is more influenced by a selected image s_i than by the query element s_q, as $I(s_j, s_i) \geq I(s_j, s_q)$. In this example, it illustrates the fact that image s_j provides more information (is closer) to s_i, (an image already selected as part of the final result) than to the query element: a near-duplicate situation. Aimed at avoiding near-duplicates, it is safe to assume that as s_q distinctly recognizes s_i from the other elements distant at most $d(s_i, s_q)$, then if metric d holds the symmetry property, s_i also recognizes s_q at the same distance. In other words, s_q "influences" the importance of s_i for the query with the same intensity that s_i does regarding s_q. Thus, the distance $d(s_i, s_q)$ can be used to obtain the set of all elements that can surely be considered influenced by every s_i closer than the minimum distance $\epsilon = d(s_i, s_j)$, as represented in Figure 20.9b.

Here, the BridGE technique—Better Result with Influence Diversification to Group Elements (Santos et al. 2014)—is introduced. This method uses the influence concept to cluster medical images. The strategy is to select the most similar k images that are not influenced by each other, and, for each one of them, it groups the corresponding influenced images. The subset of k images retrieved provides a new hierarchical way to present the results, showing the neighborhood around elements with respect to the query element. It is important to highlight that near-duplicate images are retained to enable the user to ask more information about them. Even though the images are similar among themselves, users may be interested due to other data possible associated to them, distinct from the image itself, such as its associated medical record or other metadata.

BridGE builds the result set incrementally, selecting non-influenced elements (called representatives) to group near-duplicates. Therefore, the representative image summarizes the entire set of near-duplicates regarding the query element. This structure brings more information when compared to the plain similarity approach, as the user can navigate through the neighborhood of each representative element. This operation

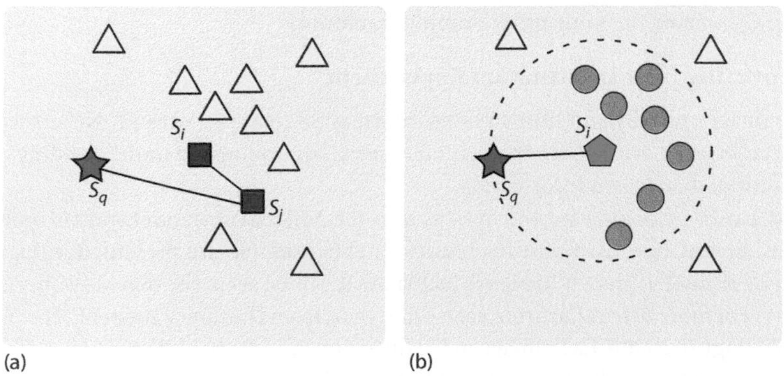

FIGURE 20.9 Influence intensity in an Euclidean bi-dimensional space. Squares are the selected elements and triangles are not-influenced elements. (a) Element S_j is more influenced by S_i than by S_q. (b) The diversified set generated by the influence of S_i for a query centered at S_q, near-duplicates (circle) and S_i as a representative element (pentagon).

is called a "locally expanded query." Figure 20.8a,b illustrate the difference between the traditional result produced by a similarity query and by the BridGE approach when searching for the five most similar cases.

The underlying intuition of BridGE is that the additional information associated to each element can both impact the decision support process and reduce the time needed by the user to analyze the images. Thus, it also addresses two problems commonly happening in tools that analyze several database elements to diversify a query: (1) maintain the search space around the elements similar to the query element (query focus), and (2) retrieve the answer in acceptable time (computational cost). The query focus is related to the semantic of the returned elements, as too many dissimilar elements from the query element may induce users to think that the query was misinterpreted. The computational cost is related to the system acceptability, because the user will not wait minutes to receive the answer from a query that may yet need to be further reformulated or refined.

20.4.2 Evaluating Similarity with Diversity

This section evaluates the BridGE algorithm, comparing it to the traditional (non-diverse) *k*-nearest neighbor, to a diversity algorithm based on the *k*-medoids clustering method (CLT) (Vieira et al. 2011) and to the diversification algorithm based on the optimization approach (OPT) (Dias et al. 2013). The CLT algorithm employs a concept similar to that used in BridGE, as it also groups similar elements and returns a set of representative elements. OPT represents the concept of diversity on similarity queries already applied to medical images. The evaluation followed two strategies. The first determines which method has higher probability to empower a similarity query to recover new information. In other words, it measures the number of images retrieved by BridGE and by the traditional *kNNq*. The second point evaluates the computational performance, measuring the cost to provide diversity to a similarity query.

The algorithms were evaluated using the real image dataset MRIBalan (Dias et al. 2013). It is composed of 704 images of magnetic resonance imaging (MRI) obtained from the Clinical Hospital of the School of Medicine of the University of São Paulo at Ribeirão Preto. The images feature vectors were obtained by the method proposed in (Balan et al. 2005) and were compared using the Euclidean distance function (L_2), evaluated over the 30 features extracted. The parameter "balance among similarity and diversity" required by CLT and OPT was set to 0.5, as it was reported by their authors to achieve, on average, the best diversification (Dias et al. 2013; Vieira et al. 2011). Regarding BridGE, a parameter-free technique, it does not need to set any query parameter besides the number k of elements to retrieve.

The experiments were performed on a computer with an Intel Core i7 processor and 8 GB of main memory, on the GNU Linux distribution Ubuntu 11.10. To obtain fair comparisons, every algorithm was implemented in C++ using the same programming framework.

20.4.2.1 Recovering New Information Experiment

Near-duplicate images usually add little to none information to a query result. Nevertheless, the near-duplicates should be made available to the user on request, improving the understanding of the answers as it includes additional, relevant information.

Suppose that a user is looking for the five cases in the MRIBalan dataset most similar to the MRI image from the current case. Two possible results for this example are presented in Figure 20.10. The result obtained by BridGE is shown in Figure 20.10a. As it can be seen, the diversity query retrieved five elements, all of them more dissimilar from each other than from the query element. The plain similarity query is shown in Figure 20.10b. Each image in the diversified result shows the number of near-duplicate images it represents, which indicates if grouped images do exist. If near-duplicates are interesting to the user, she can expand the query result without re-submitting a query or requesting the metadata associated with the grouped images. Notice that the query coverage was improved to naturally include diversity without missing the context of retrieving the most similar images without requesting new parameters from the user.

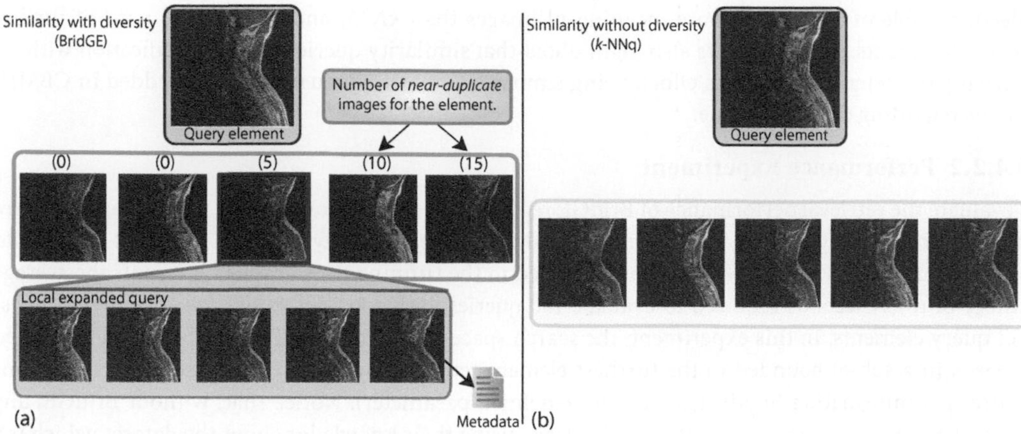

FIGURE 20.10 Result sets for a query over the MRIBalan dataset centered at the query element shown at the top left and considering $k = 5$. (a) BridGE result, highlighting the near-duplicate images grouped to a result element (b) Result of a traditional *kNNq*. Notice that the images returned by the *kNNq* are part in BridGE result and part in its third near-duplicate group.

In order to evaluate the information retrieval power of BridGE, we measured the number of images retrieved by BridGE and by the traditional *kNNq*. Figure 20.11a shows the measurements obtained to answer similarity queries over the MRIBalan dataset. The elements retrieved by BridGE are sorted and the similar elements grouped, as presented in Figure 20.10. The images that have more probability to add new information over the previous ones are shown on the top. Thus, BridGE retrieves the number of images required by the specialist, as it can be seen in Figure 20.11b. The *kNNq* query linearly fetches the images with a slight slope, often returning less than 50% of the images not near-duplicated from each other. That is, the traditional *kNNq* brings less relevant information to the user.

To evaluate the user acceptability when using diversity, BridGE was embedded into the extended-DBMS used by Higiia (Section 20.3) and then presented to specialists as a CBMIR operation. They performed similarity queries using both the traditional *kNNq* and BridGE. They stated that BridGE is,

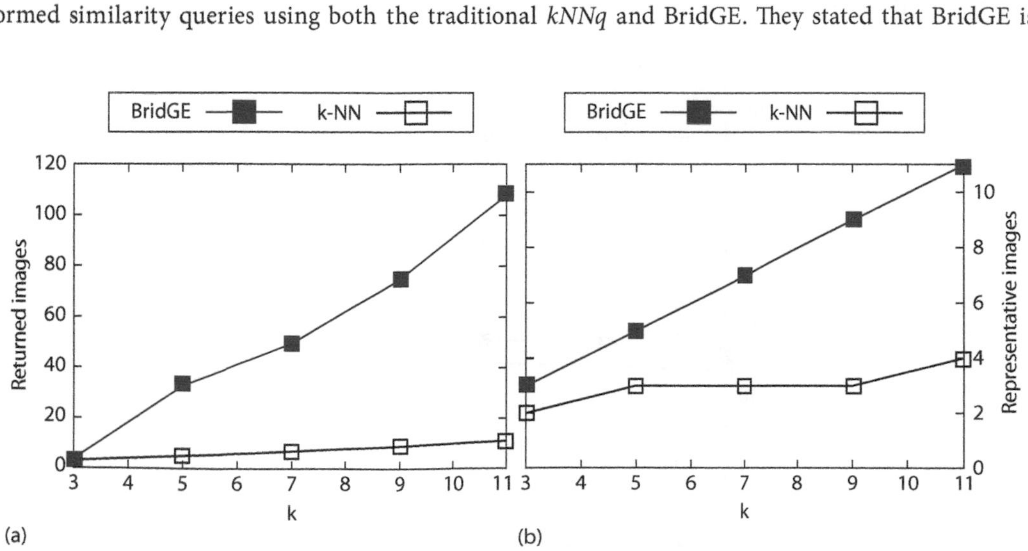

FIGURE 20.11 Relationship among the number of images retrieved and of representative images retrieved by BridGE and *kNNq*. (a) Total number of images retrieved; (b) number of representative images.

indeed, capable of presenting more meaningful images than *kNNq* and reduces the need of further query refinements. The specialists also highlighted that similarity queries with diversification without external parameters demand less effort, being simpler and more adequate to be embedded in CBMIR systems regarding clinical routine.

20.4.2.2 Performance Experiment

To evaluate the retrieval performance of BridGE regarding its nearest competitors, batched queries were performed on the MRIBalan dataset. For each evaluation, we randomly chose 100 different elements to be employed as query centers. Each point measured in the running time graphs represents the average number of microseconds required to evaluate 100 queries with constant values for *k*, centered at distinct query elements. In this experiment, the search space of CLT and OPT algorithms were restricted to search in a subset bounded to the furthest element retrieved by BridGE, because those algorithms require this information in advance (as a user-defined parameter). Notice that, without BridGE, any diversity parameter must be estimated by the users, using their knowledge about the dataset, which is a factor that hurts CLT and OPT usability.

Figure 20.10 shows, in log scale, the average time required by each technique to answer queries when *k* varies from 3 to 11 over the MRIBalan dataset. Since including diversity adds to the computational costs, the traditional non-diverse *kNNq* was the fastest technique and a theoretical lower bound. However, BridGE reduces the number of interactions cycles required to achieve users' satisfaction, by reducing the total time of each section. Moreover, the experiments showed that BridGE is faster than any other experimental diversity approach to execute similarity queries with diversity. Particularly, BridGE was always around two orders of magnitude faster than OPT, and around four orders of magnitude faster than CLT. Such results support our claim that BridGE can be seamlessly integrated to a CBMIR system at the cost of just doubling the time required for a similarity query, while the competitors are at least 100 times slower. Moreover, BridGE does not require new information from the specialist to retrieve relevant images reducing the relevance feedback cycles, turning the use of diversity on medical images transparent and intuitive.

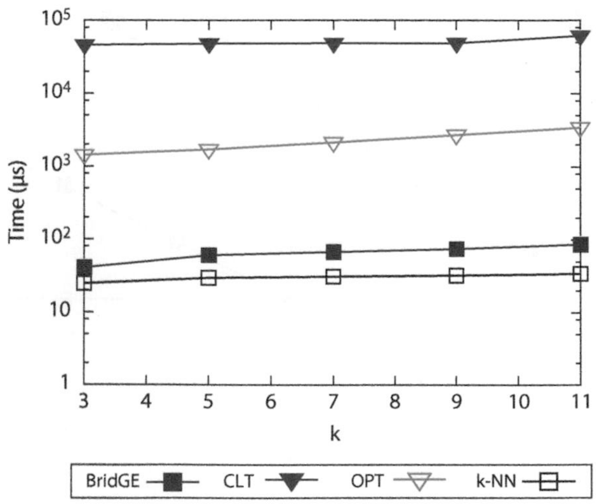

FIGURE 20.12 Average running time of queries over the MRIBalan dataset, for *k* varying from 3 to 11.

Conclusions

In this chapter, we addressed the task of retrieving similar medical images based on their visual content using content-based medical image retrieval (CBMIR) systems. We discussed the concepts and challenges associated with the implementation of a CBMIR system.

The first challenge that we addressed was the selection of perceptual parameters. In a real scenario, users may have to choose from a large number of feature extraction methods and distance functions to use a CBMIR system. To tackle this problem, we present a CBMIR system, called Higiia (Section 20.3), which was evaluated in a real medical environment. Higiia uses machine learning to automatically suggest perceptual parameters to the user.

We also discussed how near duplicate images increases the effort required from the users to find useful information. To solve this problem, we presented BridGE (Section 20.4), an effective technique that bridges usability by better expressing similarity queries that consider the relationship among the elements in the result set.

References

Aha, D. W., D. Kibler, and M. K. Albert. 1991. Instance-based learning algorithms. *Machine Learning* 6 (1): 37–66.

Akgül, C., D. L. Rubin, S. Napel, C. F. Beaulieu, H. Greenspan, and B. Acar. 2011. Content-based image retrieval in radiology: Current status and future directions. *Journal of Digital Imaging* 24 (2): 208–222.

Alto, H, R M. Rangayyan, and J. E. L Desautels. 2005. Content-based retrieval and analysis of mammographic masses. *Journal of Electronic Imaging* 14 (2): 1–17.

Angel, A., and N. Koudas. 2011. Efficient diversity-aware search. *In International Conference on Management*, 781–792. San Francisco, CA.

Balan, A. G. R., A. J. M. Traina, C. Traina Jr., and P. M. Azevedo-Marques. 2005. Fractal analysis of image textures for indexing and retrieval by content. In *International Conference on Computer-Based Medical Systems*, 581–586. Dublin, Ireland.

Banda, J. M., M. A. Schuh, T. Wylie, P. McInerney, and R. A. Angryk. 2014. When too similar is bad: A practical example of the solar dynamics observatory content-based image-retrieval system. In *New Trends in Databases and Information Systems*, edited by B. Catania, T. Cerquitelli, S. Chiusano, G. Guerrini, M. Kämpf, A. Kemper, B. Novikov, T. Palpanas, J. Pokorný, and A. Vakali, 241: 87–95. Cham, Switzerland: Springer.

Bay, H., A. Ess, T. Tuytelaars, and L. V. Gool. 2008. Speeded-up robust features (SURF). *Computer Vision and Image Understanding* 110 (3): 346–359.

Bedo, M. V. N., D. P. Santos, M. P. Silva, P. M. A. Marques, A. P. L. F. Carvalho, and C. Traina Jr. 2015. Endowing a content-based medical image retrieval system with perceptual similarity using ensemble strategy. *Journal of Digital Imaging* 29 (1): 22–37.

Bedo, M. V. N., M. P. Silva, D. Kaster, P. H. Bugatti, C. Traina Jr., and A. J. M. Traina. 2012. Higiia: A perceptual medical CBIR system applied to mammography classification. In *Brazilian Symposium on Databases*, 13–18. São Paulo, Brazil: Brazilian Computer Society.

Bedo, M. V. N., A. J. M. Traina, and C. Traina Jr. 2014. Seamless integration of distance functions and feature vectors for similarity queries processing. *Journal of Information and Data Management* 5 (3): 308–320.

Beecks, C., M. S. Uysal, P. Driessen, and T. Seidl. 2013. Content-based exploration of multimedia databases. In *International Workshop on Content-Based Multimedia Indexing*, 59–64. Veszprém, Hungary.

Boureau, Y.-L., F. Bach, Y. LeCun, and J. Ponce. 2010. Learning mid-level features for recognition. In Trevor Darrell, David Hogg, David Jacobs (Eds.), *Computer Vision and Pattern Recognition*, 2559–2566. San Francisco, CA: IEEE Computer Society.

Carbonell, J., and J. Goldstein. 1998. The use of MMR, diversity-based reranking for reordering documents and producing summaries. In *SIGIR Conference on Research and Development*, 335–336. Melbourne, Australia.

Carvalho, L. O., E. Seraphim, T. F. P. Seraphim, A. J. M. Traina, and C. Traina Jr. 2014. MedInject: A general-purpose information retrieval framework applied in a medical context. In *International Conference on Computer Based Medical Systems*, 308–313. New York, NY.

Chadha, A., S. Mallik, and R. Johar. 2012. Comparative study and optimization of feature-extraction techniques for content based image retrieval. *International Journal of Computer Applications* 52 (20): 35–42.

Chatzichristofis, S. A. and Y. S. Boutalis. 2008. CEDD: Color and edge directivity descriptor: A compact descriptor for image indexing and retrieval. In *International Conference on Computer Vision Systems*, 312–322. Santorini, Greece.

Cortes, C. and V. Vapnik. 1995. Support-vector networks. *Machine Learning* 20 (3): 273–297.

Datta, R., D. Joshi, J. Li, and J. Z. Wang. 2008. Image retrieval: Ideas, influences, and trends of the new age. *ACM Computing Surveys* 40 (2): 1–60.

Deza, M. M and E. Deza. 2013. *Encyclopedia of Distances*. New York, NY: Springer.

Dias, R. L., R. Bueno, and M. X. Ribeiro. 2013. Reducing the complexity of k-nearest diverse neighbor queries in medical image datasets through fractal analysis. In *International Conference on Computer-Based Medical Systems*, 101–106. Porto, Portugal.

Díaz-Galiano, M. C., M. Á. García-Cumbreras, M. T. Martín-Valdivia, A. Montejo-Ráez, and L. A. Ureña-López. 2008. Integrating MeSH ontology to improve medical information retrieval. In *Advances in Multilingual and Multimodal Information Retrieval*, edited by C. Peters, V. Jijkoun, T. Mandl, H. Müller, D. W. Oard, A. Peñas, V. Petras, and D. Santos, 601–606. Berlin, Germany: Springer.

Drosou, M., and E. Pitoura. 2010. Search result diversification. *SIGMOD Record* 39 (1): 41–47.

El-Naqa, I., Y. Yang, N. P. Galatsanos, R. M. Nishikawa, and M. N. Wernick. 2004. A similarity learning approach to content-based image retrieval: Application to digital mammography. *IEEE Transactions on Medical Imaging* 23 (10): 1233–1244.

Felipe, J. C., C. Traina Jr. and A. J. M. Traina. 2009. A new family of distance functions for perceptual similarity retrieval of medical images. *Journal of Digital Imaging* 22 (2): 183–201.

Ghosh, P., S. Antani, L. R. Long, and G. R. Thoma. 2011. Review of medical image retrieval systems and future directions. In *International Symposium on Computer-Based Medical Systems*, 1–6. Bristol, UK.

Gil-Costa, V., R. L. T. Santos, C. Macdonald, and I. Ounis. 2011. Sparse spatial selection for novelty-based search result diversification. *In International Conference on String Processing and Information Retrieval*, 344–355. Pisa, Italy.

Güld, M. O., C. Thies, B. Fischer, and T. M. Lehmann. 2007. A generic concept for the implementation of medical image retrieval systems. *International Journal of Medical Informatics* 76 (2–3): 252–259.

Haralick, R., K. Shanmugam, and I. Dinstein. 1973. Texture features for image classification. *IEEE Transactions on Systems, Man, and Cybernetics* 3 (6): 610–621.

Hetland, M. L. 2009. The basic principles of metric indexing. *In Swarm Intelligence for Multi-Objective Problems in Data Mining*, edited by C. A. Coello Coello, S. Dehuri, and S. Ghosh, 199–232. Chennai, India: Springer.

Hu, B., S. Dasmahapatra, P. Lewis, and N. Shadbolt. 2003. Ontology-based medical image annotation with description logics. In *International Conference on Tools with Artificial Intelligence*, 77–82. Sacramento, CA.

Huang, G.-B., H. Zhou, X. Ding, and R. Zhang. 2012. Extreme learning machine for regression and multiclass classification. *IEEE Transactions on Systems, Man, and Cybernetics, Part B: Cybernetics* 42 (2): 513–529.

Ioannidis, Y. 2003. The history of histograms (abridged). In *International Conference on Very Large Data Bases*, 19–30. Berlin, Germany.

Jgou, H., M. Douze, and C. Schmid. 2010. Improving bag-of-features for large scale image search. *International Journal of Computer Vision* 87 (3): 316–336.
Jolliffe, I. T. 1986. *Principal Component Analysis*. Newbury Park, CA: Springer.
Kaster, D. S., P. H. Bugatti, M. P Silva, A. J. M. Traina, P. M A. Marques, A. C. Santos, and C. Traina Jr. 2011. MedFMI-SiR: A powerful DBMS solution for large-scale medical image retrieval. In *Information Technology in Bio and Medical Informatics*, edited by C. Böhm, S. Khuri, L. Lhotská, and N. Pisanti, 16–30. Berlin: Springer.
Keysers, D., J. Dahmen, H. Ney, B. B. Wein, and T. M. Lehmann. 2003. Statistical framework for model-based image retrieval in medical applications. *Journal of Electronic Imaging* 12 (1): 59–68.
Kullback, S., and R. A. Leibler. 1951. On information and sufficiency. *The Annals of Mathematical Statistics* 22 (1): 79–86.
Kumar, A., J. Kim, W. Cai, M. Fulham, and D. Feng. 2013. Content-based medical image retrieval: A survey of applications to multidimensional and multimodality data. *Journal of Digital Imaging* 26 (6): 1025–1039.
Lazebnik, S., C. Schmid, and J. Ponce. 2006. Beyond bags of features: Spatial pyramid matching for recognizing natural scene categories. In *IEEE Conference on Computer Vision and Pattern Recognition*, 2169–2178. New York, NY.
Lehmann, T. M., M. O. Guld, C. Thies, B. Fischer, D. Keysers, M. Kohnen, H. Schubert, and B. B. Wein. 2003. Content-based image retrieval in medical applications for picture archiving and communication systems. *Proceedings of the Spie* 5033: 109–117.
Liang, N.-Y., G.-B. Huang, P. Saratchandran, and N. Sundararajan. 2006. A fast and accurate online sequential learning algorithm for feedforward networks. *IEEE Transactions on Neural Networks* 17 (6): 1411–1423.
Liu, Y, D Zhang, G Lu, and W.-Y. Ma. 2007. A survey of content-based image retrieval with high-level semantics. *Pattern Recognition* 40 (1): 262–282.
Lowe, D. G. 1999. Object recognition from local scale-invariant features. In *International Conference on Computer Vision*, 1150–1157. Toronto, Canada.
Mallat, S. 2008. *A Wavelet Tour of Signal Processing*. Burlington, VT: Academic Press.
Marques, P. M. A., and R. M. Rangayyan. 2013. Content-based retrieval of medical images: Landmarking, indexing, and relevance feedback. *Synthesis Lectures on Biomedical Engineering* 8 (1): 1–143.
Möller, M., S. Regel, and M. Sintek. 2009. RadSem: Semantic annotation and retrieval for medical images. In *The Semantic Web: Research and Applications*, edited by L. Aroyo, P. Traverso, F. Ciravegna, P. Cimiano, T. Heath, E. Hyvönen, R. Mizoguchi, E. Oren, M. Sabou, and E. Simperl, 21–35. Berlin, Germany: Springer.
Müller, H., and T. M. Deserno. 2011. Content-based medical image retrieval. In *Biomedical Image Processing*, edited by T. M. Deserno, 471–494. London, UK: Springer.
Muramatsu, C, K. Nishimura, M. Oiwa, M. Shiraiwa, T. Endo, K Doi, and H Fujita. 2012. Correspondence among subjective and objective similarities and pathologic types of breast masses on digital mammography. In *Breast Imaging*, edited by A. D. A. Maidment, P. R. Bakic, and S. Gavenonis, 450–457. Chennai, India: Springer Berlin Heidelberg.
Penatti, O. A., F. B. Silva, E. Balle, V. Gouet-Brunet, and R. S. Torres. 2014. Visual word spatial arrangement for image retrieval and classification. *Pattern Recognition* 47:705–720.
Pereira Jr., R. R. P. M. A. Marques, M. O. Honda, S. K. Kinoshita, R. Engelmann, C. Muramatsu, and K. Doi. 2007. Usefulness of texture analysis for computerized classification of breast lesions on mammograms. *Journal of Digital Imaging* 20 (3): 248–255.
Picard, R. R., and R. D. Cook. 1984. Cross-validation of regression models. *Journal of the American Statistical Association* 79 (387): 575–583.
Rahman, M. M., P. Bhattacharya, and B. C. Desai. 2007. A framework for medical image retrieval using machine learning and statistical similarity matching techniques with relevance feedback. *Transactions on Information Technology in Biomedicine* 11 (1): 58–69.

Rocchio, J. J. 1971. Relevance feedback in information retrieval. In *The Smart Retrieval System -Experiments in Automatic Document Processing*, edited by G. Salton, 313–323. Upper Saddle River, Englewood Cliffs, NJ: Prentice-Hall.

Santos, L. F. D., M. V. N. Bedo, M. P. Silva, A. J. M. Traina, and C. Traina Jr. 2014. Being similar is not enough: How to bridge usability gap through diversity in medical images. *In International Conference on Computer Based Medical Systems*, 287–293. New York, NY.

Santos, L. F. D., W. D. Oliveira, M. R. P. Ferreira, A. J. M. Traina, and C. Traina Jr. 2013. Parameter-free and domain-independent similarity search with diversity. In *International Conference on Scientific and Statistical Database Management*, 5:1–5:12. Baltimore, Maryland.

Savarese, S., J. Winn, and A. Criminisi. 2006. Discriminative object class models of appearance and shape by correlatons. *In IEEE Computer Vision and Pattern Recognition*, 2033–2040. New York, NY.

Silva, M. P., J. P. Souza, P. H. Bugatti, M. V. N. Bedo, D. S. Kaster, R. T. V. Braga, A. D. Bellucci, P. M. A. Marques, C. Traina Jr., and A. J. M. Traina. 2013. Does a CBIR system really impact decisions of physicians in a clinical environment? In *International Symposium on Computer-Based Medical Systems*, 41–46. Porto, Portugal.

Sonka, M., V. Hlavac, and R. Boyle. 2014. *Image Processing, Analysis, and Machine Vision*. Stamford, CT: Cengage Learning.

Srinivasan, G. N., and G. Shobha. 2008. Statistical texture analysis. *Proceedings of World Academy of Science, Engineering and Technology* 36:1264–1269.

Tao, W., Y. Zhou, L. Liu, K. Sun, and Z. Zhang. 2014. Spatial adjacent bag of features with multiple superpixels for object segmentation and classification. *Information Sciences* 281:373–385.

Traina, A. J. M., C. Traina Jr., J. M. Bueno, F. J. T. Chino, and P. M. A. Marques. 2003. Efficient content-based image retrieval through metric histograms. *World Wide Web* 6 (2): 157–185.

Vieira, M. R., H. L. Razente, M. C. N. Barioni, M. Hadjieleftheriou, D. Srivastava, C. Traina Jr., and V. J. Tsotras. 2011. On query result diversification. In Serge Abiteboul, Klemens Böhm, Christoph Koch, Kian-Lee Tan (Eds.), *International Conference on Data Engineering*, 1163–1174. Hannover, Germany: IEEE Computer Societ.

Yildizer, E., A. M. Balci, M. Hassan, and R. Alhajj. 2012. Efficient content-based image retrieval using multiple support vector machines ensemble. *Expert Systems with Applications* 39 (3): 2385–2396.

Yu, L., and H. Liu. 2003. Feature selection for high-dimensional data: A fast correlation-based filter solution. *In International Conference on Machine Learning*, 856–863. Washington, DC.

Zezula, P., G. Amato, V. Dohnal, and M. Batko. 2006. *Similarity Search: The Metric Space Approach*. New York, NY: Springer.

21

Health Informatics for Research Applications of CAD

Thomas M. Deserno

21.1 Summary

Controlled clinical trials (CCT) are majorly performed in the western world to proof safety, efficiency, and efficacy of novel drugs or medical devices. Still, the comprehensive documentation, as well as data capturing, is majorly paper-based. Electronic systems are now available, but they stand alone, not providing any standardized interface for data exchange and system interconnections. However, image- or signal-based surrogates (imaging biomarkers) are acquired increasingly in those trials approaching big data volumes, which cannot be handled manually.

We present an already implemented approach on system interconnection that is based on existing standards, such as digital imaging and communications in medicine (DICOM), clinical data interchange standards consortium (CDISC) operational data model (ODM), and provides simple object access protocol (SOAP), as well as representational state transfer (REST) web services. The core idea is taken from HL7-based hospital information systems, where the particular information system components are interconnected via a centralized communication server. Using this approach, electronic data capture (EDC) in CCTs is enhanced seamlessly with binary image or signal data. Furthermore,

MATLAB®- or Java-based data processing is integrated. We demonstrate full automation in two case studies: (1) photography of skin lesions to monitor wound healing, and (2) Holter-based electrocardiography (ECG) monitoring over 24 h and seven days, which have been recorded for full 12 leads at a sample rate of 1000 Hz and 10 bit.

In the first example, smartphone cameras are used to capture the scene, where a special calibration card has been placed next to the lesion of interest. The card is enhanced with a barcode that identifies the subject. After image acquisition, the image is transferred via the Internet (WLAN or G4) to the analysis server, where it is processed automatically for card detection, normalization with respect to geometry and colors, and barcode reading. The subject identifier is returned to the smart device and allows immediate interconnection with the EDC system, pointing the research nurse directly to the corresponding electronic case report form (eCRF).

In the second example, up to 8.5 GB of uncompressed data per subject must be handled. Here, the memory card is detached from the Holter device, and the files are transferred to the EDC system, manually. Then, automatic processing is started instantaneously, where the measurements are taken after separating all heart cycles (about 600,000 up to 900,000 cycles per seven day recording) and representing each individual cycle by Gaussian-based models (99 parameters per cycle for a 12 lead recording). The model parameters are estimated using the Levenberg-Marquardt algorithm. They are used to track medically relevant ECG descriptors over the time automatically. In both cases, the binary signal and image data can be enveloped according to the DICOM standard and persistently stored in a picture archiving and communication system (PACS), which is also interconnected via the communication server. This system architecture is able to cope with the increasing number of trials, but decreasing number of subjects per trial, which forms medical research for personalized medicine.

21.2 Introduction

Translational medicine aims at transferring results from novel research immediately out of the laboratory to improve patient treatment and health care, that is from bench to bedside (B2B). This holds for drugs as well as medical devices. Figure 21.1 visualizes the steps required to transform research to patient benefit, and clinical trials are a central component here.

Controlled clinical trials are performed in order to gather reliable data on safety, efficiency, and efficacy of drugs, therapeutic procedures, medical devices, or surgical assistance systems. For instance at Uniklinik RWTH Aachen, a German hospital of highest level of supply with approx. 1,200 beds, 45,000 stationary inpatients and 230,000 outpatient cases per year, there are currently 340 trials running, and more than 11,500 subjects have been enrolled into clinical trials within the last five years. Despite these large numbers, subject handling, documentation, and data collection is still majorly paper-based [1].

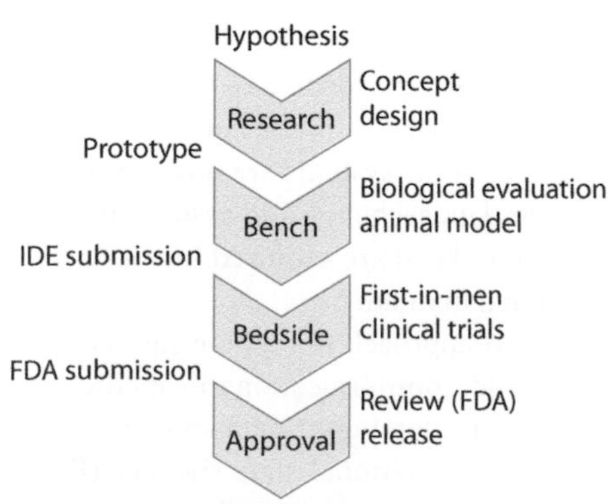

FIGURE 21.1 Major steps of translational research.

Some electronic systems are available to partially support tasks, but automation is lacking since these systems are stand-alone, without any interconnections. In particular, for investigator initiated trails (IIT), the small budget and lack of funding hinders expensive information technology (IT) solutions. However, in a university, IITs are by far the majority of the conducted clinical trials.

Another aspect is the type of data that is collected in such trials. Today, quantitative imaging biomarkers, also referred to as image-based surrogates, play an important role [2]. Biosignals, such as electrocardiography (ECG) or electroencephalography (EEG), and medical images, such as radiographs, photographs, three-dimensional (3D) data (e.g., computed tomography (CT), magnetic resonance imaging (MRI), or positron emission tomography (PET)) are acquired frequently for research in clinical trials. Therefore, there is a need for complete automation of work- and dataflow in clinical trials, including the medical data, biosignals, and images.

21.3 State of the Art

In this section, we briefly review existing IT systems supporting research and clinical trials and healthcare information systems (HIS) architectures for research and patient care.

21.3.1 Clinical Trial Management System (CTMS)

In order to manage the high number of trials, subjects, and persons involved, a CTMS is required [3]. Such systems aim at supporting one or more of the tasks, which are regularly performed when conducting a trial (Figure 21.2) by automation (red line in Figure 21.2) and by fostering inter-disciplinary collaboration. There are several instances involved in trials and, hence, several components to be controlled, for example, using dashboarding techniques or graphical data aggregations [4]:

- *Actors:* A lot of persons collaborate in conducting trials. They are best described by their role within a trial, where one human entity may have several roles, and some roles are rather institutional than personal. These roles are: sponsor, ethics, authorities, sites, investigators, research nurses, monitors, data manager, statistician, and so on.

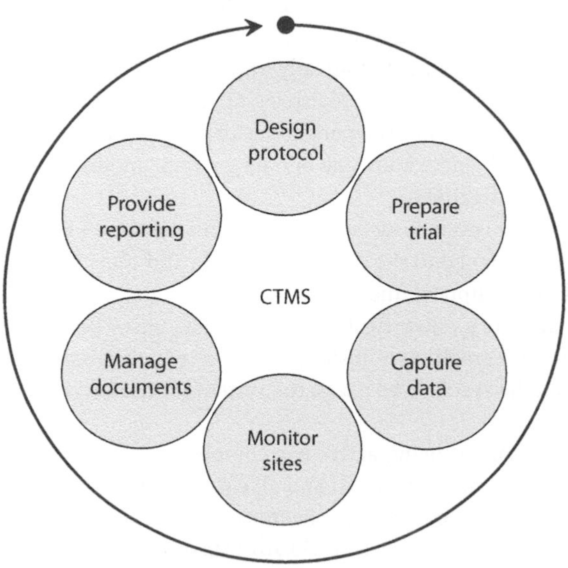

FIGURE 21.2 Tasks conducting clinical trials.

- *Documents:* Several documents are required to conduct a trial, and document management is a major component of any CTMS. Some documents, such as standard operation procedures (SOPs), may apply to several trials, while others are trial specific: investigators brochure (IB), study protocol, informed consent form, and so on.
- *Status:* The life cycle of a study is usually divided into: (1) preparation, which is before any patients are enrolled; (2) recruiting, which starts when the first patient is sought to be enrolled; (3) active, not recruiting, as long as enrolled patients have not passed the entire study program and follow-up visits are scheduled, and (4) completed.
- *Finances:* Control of efforts, costs, and billing is also important for successful trials. CTMS offers logs for tracking efforts of the study personnel, calculating costs for trials, and triggering the liquidation with the trials sponsor.
- *Reporting:* An adverse event (AE) is any untoward medical occurrence in a patient or clinical investigation subject, disregarding whether that event is related directly to the trial or not (e.g., car accident). Such AE is considered as serious (serious adverse event, SAE) if it results in death, hospitalization, persistent or significant disability, congenital anomaly, or birth defect. AE and SAE need special treatment for reporting to the authorities. After the trial, statistical analysis is performed and the final results are reported.

Since the market for such systems is small, the systems are rare and expensive. Most academia use paper-based trial management, accompanied by some proprietary components that do not cover all of the required functionality.

21.3.2 Electronic Data Capture System (EDCS)

Prospective, randomized, controlled clinical trials are organized by visits and events. For each visit, a well-defined set of data is recorded for each subject that has been enrolled in the trial. The schedule is following a strict protocol. In addition, events such as AEs or SAEs may trigger further data collection and management. Such events may be used also for photographic wound documentation, if the images are not linked to a certain visit but may be acquired by any time during the trial protocol. The main components of any EDCS are [5]:

- *eCRF:* Electronic case report forms are used to collect the study-specific data. Improving paper-based CRFs, an eCRF allows range checks, rules, and other data evaluations directly when attempting to manually enter or automatically import data.
- *Audit trail:* In CCTs, as well as medical registries, all data entry and modification has to be logged by recording the identifier (ID) of the person performing the entry, the old and new values that have been entered, and the date and time of data entry or modification. Using EDCS, the audit trail is generated automatically.
- *Query:* If the monitor, for instance, detects some inconsistencies or has some questions, he invokes a query. This query is assigned to the data entry person and must be resolved.
- *Communication:* EDCSs support query management by integrated messaging between the users according to their roles (e.g., investigator, monitor, data entry person). Some EDCS are providing more comprehensive emailing to foster the communication between involved persons. Mails may be sent between the system and users, the systems and user groups, or in between the users or user groups.
- *Export:* After the last patient being active has completed the last visit or has been dropped out (i.e., the last patient went out of the trial) and all entries into the eCRF have been completed and approved by the monitor, the data is exported for statistical assessment. Interim evaluations usually are not part of the study protocol to avoid any bias in the investigators. Statistical analysis and reporting, however, are not part of EDCS.

There is a set of systems available, commercially as well as non-commercially. Some commercial systems are BioClinica,* Medrio-edu,† OpenClinica,‡ REDCap,§ SecuTrial,¶ and Viedoc.** Partly, they are available freely for the academia and IITs, and some of them are open source, for example, OpenClinica.

21.3.3 Medical Registry

Medical registries are also EDCS. Here, however, data is collected partly in retrospect, not following a certain protocol. Mostly, registers are built to enable research on rare diseases: rare disease registry (RDR). The National Institutes of Health (NIH) Office of Rare Diseases Research (ORDR) urgently recommends the development of a minimal common registry model that should be open source and broadly available. In particular, ORDR has developed specific recommendations for [6]:

- *Standardized vocabulary,* terminology, codes, and diagnoses, which aim at finding commonalities across all rare diseases and at developing a minimal common registry model;
- *Technology and informatics,* which aim at developing an open-source software/hosted registry solution;
- *Biorepositories and biospecimens,* which aim at establishing disease biospecimen repositories using patient registries as sources for donors;
- *Clinical research,* patient care, and disease management, which aim at developing centralized registries;
- *Patient participation,* outreach activities, and patient advocacy, which aim at writing a Registry-building for Dummies handbook; and
- *Bioethical and legal issues,* which aim at bringing in ethical as well as regulatory expertise.

Still, however, these goals have rarely been reached [7]. Regarding academia, neither a standard on database design, nor on functionality, nor on user or data interfaces has been established yet [8,9].

21.3.4 Healthcare Information System (HIS)

HISs provide a common source of information about a patient's health history. In the hospital, these systems enhance the ability of healthcare professionals to coordinate care by providing a patient's health information and visit history at the place and time that it is needed.

Therefore, the main function of any HIS is data warehousing, a rather static model of information management. An HIS is often composed of a large variety of sub-systems in medical specialties from a multi-vendor market. Specialized implementations are named according to their specialty, for example, the laboratory information system (LIS), radiology information system (RIS), or PACS.

For internal and external communication among healthcare providers, HIS provides interfaces, where all communication is usually following the HL7 protocol. Regarding interconnection of the internal sub-systems, a communication server is often applied, since it centralizes and standardizes the communication and allows sub-modules operations to connect on different versions of HL7 standard or to provide only proprietary interfaces (Figure 21.3).

* www.bioclinica.com /
† www.medrio.com /
‡ www.openclinica.com /
§ www.project-redcap.org/
¶ www.secutrial.com /
** www.viedoc.com /

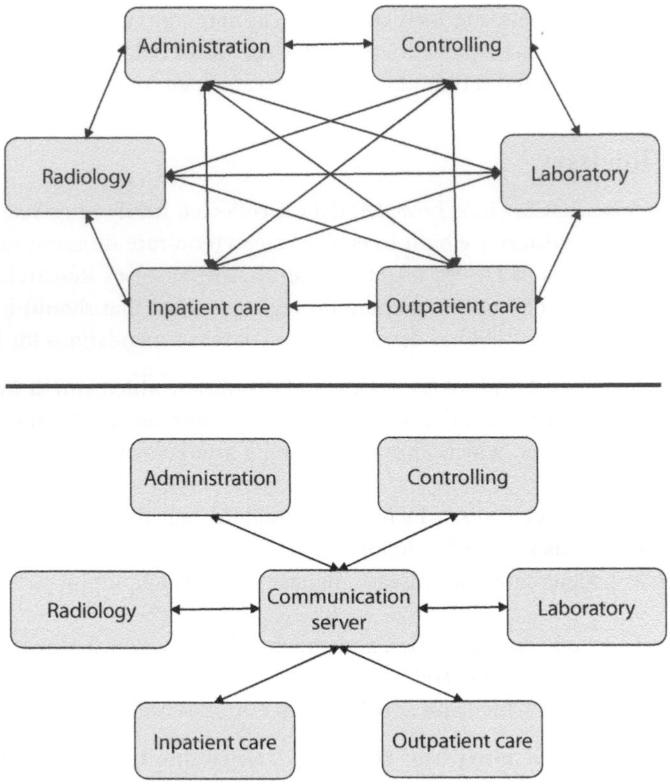

FIGURE 21.3 Architectures for healthcare information systems.

21.3.5 System Interconnection

There are only a few protocols aiming at interconnecting systems in clinical research:

- The clinical data interchange standards consortium (CDISC) operational data model (ODM) is designed to facilitate the regulatory-compliant acquisition, archive, and interchange of metadata and data for clinical research studies. ODM is a vendor-neutral, platform-independent format for interchange and archival of clinical study data.
- The fast healthcare interoperability resources (FHIR) standard is a next-generation framework created by HL7. FHIR combines the best features of HL7's different versions, while leveraging the latest web standards (REST). FHIR solutions are built from a set of modular components called "Resources." These resources can easily be assembled into working systems that solve real world clinical and administrative problems at a fraction of the price of existing alternatives. FHIR is suitable for use in a wide variety of contexts, such as mobile phone apps, cloud communications, EHR-based data sharing, server communication in large institutional healthcare providers, and much more.

Despite the CDISC protocols or the FHIR initiative, so far, CTMS, EDCS, and RDR are stand-alone systems and are not communicating to each other. Actually, trials, sites, and users are double defined in CTMS and EDCS, and single sign on (SSO) mechanisms are not available.

21.4 Our Previous Work

In this section, we summarize our previous work at Uniklinik RWTH Aachen to establish IT support for the Clinical Trial Center Aachen (CTC-A). The CTC-A has been founded to support clinical research and to improve planning, conduction, and reporting of clinical trials.

21.4.1 Study Management Tool

A CTMS has been developed for the Clinical Trial Center Aachen (CTC-A) [10]. It is a web-based application based on the Google Web Toolkit (GWT) and Java programming language, Sencha ExtGWT graphical user interfaces (GUI), a relational database (MySQL) that is fed via Hibernate and Gilead libraries. JUnit tests are performed. Jaspers reports are used to produce graphics-enhanced PDF-documents and dashboards. The main relations are:

- *Study:* Title and all relevant descriptions and identifiers such as the ClinicalTrial.Gov reference number.
- *Person:* All persons in their study-specific role (principal investigator, monitor, statistician, data manager, research nurse, etc.).
- *Department:* Every person is associated to only one department. This allows easy access levels (i.e.: none, own, department, all, master). The level department allows access to all trial data, where persons from the same department are involved, disregarding their particular roles.
- *Patient:* Subject-identifying data is separated, and the identification number is generated from the person's name, date of birth, and gender.
- *Efforts:* There are several categories (study preparation, IT implementation, informed consent, etc.), which are used by all persons to track their study-specific efforts. Since 2010, more than 170,000 h have been recorded by CTC-A stuff.
- *Documents:* SOPs and all other study-specific documents are archived in the system. A versioning is performed and several rights are attached to the documents.

Based on that system, CTC-A has been granted the ISO 9001 certificate.

21.4.2 OpenClinica

OpenClinica (OC) is an open source EDCS. It allows data acquisition in clinical trials by user- defined eCRFs maintaining industry standards. OC is approved by regulatory authorities such as the Food and Drug Administration (FDA). The web application is based on a PostgreSQL database and developed using Java server pages (JSP). This architecture allows access through modern web browsers and supports distributed data acquisition in multi-site trials. OpenClinica is structured in several packages, such as OC Core and OC WS for core and web service functionality, respectively. In previous works, we have integrated extended functionality:

- *OC Big:* Binary large objects (BLOBs) are robustly integrated into OC.* It is developed using JavaScript (jQuery) and the PHP hypertext preprocessor (PHP), and it is based on other open source projects, namely jQuery file upload and zip.js. Disregarding the bandwidth, it ensures data transfer without time outs and links the data into the subjects eCRF accordingly [11]. OC Big is open source, too.†
- *OC Tab:* This smartphone-based application (App) allows data entry into OCs eCRFs. The eCRF structure is gathered via CDISC ODM and web services from the OC instance. All visualizations

* https: / / community.openclinica.com / extension / openclinica-big-data-oc-big
† http: //idmteam.github.io/oc-big/

are optimized for data entry on tablets, considering validation rules (e.g., range checks). OC Tab supports off- and online mode, whereby data is directly submitted as soon as an Internet connection is available [12].

- *OC ToGo:* Using this mobile App, images captured with a smartphone-integrated camera are directly transferred into OCs eCRFs. The identifiers for study, event, and subject are queried from OC via web services and selected by the user to set the context of the image [13]. Barcode-enhanced color cards can also be placed in the camera's field of view and automatically extracted by an optical character recognition (OCR) and barcode reader. This has reduced manual interaction at a patient's bed side.

21.4.3 RDR Framework

The RDR Framework is a software architecture and library for the development of web-based medical registries [7]. The RDR Framework is based on a similar technology stack as the Study Management Tool (primarily GWT) and defines data objects used in medical registries (e.g., subject, visit). The web application includes core and optional modules, which are activated by configuration [7]:

- *RDR Core:* Provides core functionality such as instantiation of eCRFs (EDCs), user levels and rights (Access Control), standardized data objects (Terminology), and logging of data modifications (Audit Trail).
- *PID:* Securely connects a database of subjects personal and identifying data with the RDR core, where only de-identified medical data is archived. This module generates a unique patient identifier (PID) using name, date of birth, and gender.
- *Bioreg:* Interfaces data sharing with centralized biomaterial banks.
- *BLOB:* Stable and user-friendly integration of binary biosignals (e.g., ECG recordings), or images (e.g., radiographs), which can be handled as digital imaging and communications in medicine (DICOM) objects and in other data formats.
- *BLOB Analysis:* Integrates algorithms for signal and image processing. The algorithms themselves are treated as black box, and only the input/output data and parameters are defined in an extensible markup language (XML)-based format.
- *SOP:* Provides a document archive for SOPs and other important documents, as well as subject-specific data captured in the portable document format (PDF). All documents can be modified, but all versions stay registered.

21.5 System Interconnection

In this section, the system architecture supporting control and automation of clinical trials, the protocols in use, and the resulting potentials for integrated signal and image analysis are described.

21.5.1 System Architecture

The system architecture is adopted from the well-established structure of information systems that are already in use for medical care. Hospital information systems (HIS) are usually composed of several modules, which are interconnected by a central communication server (Figure 21.4).

Accordingly, we suggest an application server that is designed in the center of the research information system. It connects our CTMS with services such as de-identification (integration of the PID module provided by the German TMF), randomization (modules based on the statistical software R), a full-grown PACS to archive persistently image and signal data of study subjects, OpenClinica as open source EDCS, and our RDR Framework.

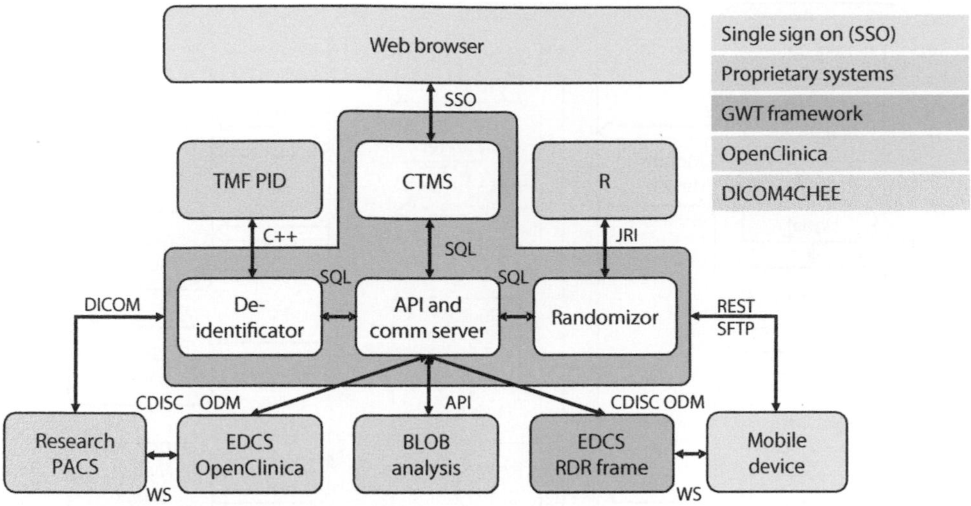

FIGURE 21.4 System architecture.

21.5.2 Protocols

Whenever possible, external components and services are integrated using the simple object access protocol (SOAP) and representational state transfer (REST) web service (WS) protocols containing CDISC ODM data. If required, an interface is implemented to translate into any (proprietary) format. The De-identificator and the Randomizer (Figure 21.4), for instance, translate remote procedure calls and data into the desired formats, C++ and the Java R Interface (JRI), respectively. Here, such modules can be integrated closely using the GWT-based software framework. The research PACS is connected according to the DICOM standard [14].

21.5.3 Image and Signal Data Management

Non-DICOM data objects, such as Holter ECG recordings or digital photographs, which may be recorded using mobile devices, can be enveloped as a DICOM object and stored within the PACS, too. Alternatively, they may be stored with the other patient data in the EDCS.

Data analysis (BLOB analysis) is integrated via an application programming interface (API), see Figure 21.4. So far, we have integrated MATLAB®, Java (ImageJ) and C++, as well as R for signal and image processing, as well as randomization, respectively. The automated analysis of BLOB data is depicted in Figure 21.5. The communication server is connected to a database where the modified BLOBS are stored. This cloud-based repository is mounted by all of the servers to avoid data exchange based on the secure file transfer protocol (SFTP), as it has been used in previous versions [15].

21.6 Cad in Research Applications

The architecture and communication protocols have been implemented and they are being used in several clinical trials, supervised by the CTC-A. In this section, full of signal and image analysis is demonstrated by describing two case studies.

21.6.1 Case Study I: Wound Imaging

Calciphylaxis is a rare disease, and a highest-level hospital such as the Uniklinik RWTH Aachen may see one or two patients a year. Therefore, subject data needs integration from multiple sites. Along several

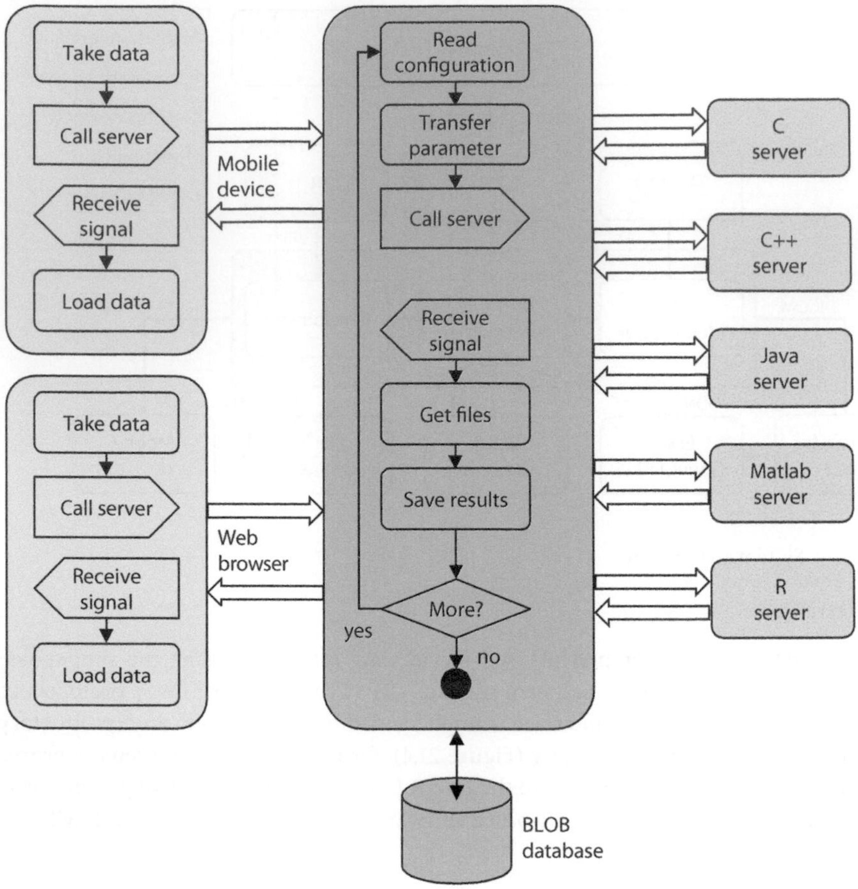

FIGURE 21.5 Communication for signal and image analysis.

clinical parameters, necrotic tissue and skin lesions are documented photographically. In this example, a mobile camera equipped with Internet access and processor (a smart device) is used, and the images are generated using a special app.

Next to the lesion of interest, a color calibration card is positioned into the field of view. The card is specially designed to support its automatic localization in the image, color, and geometry calibration of the image, and subject identification. For the latter, a unique code is printed on each card twice, as a numerical number and as barcode. Optical character recognition (OCR) and barcode readers are integrated to robustly extract the subject identifier. This supports full automation of data entry.

Figure 21.6 visualizes the automation process. In the upper left, digital wound imaging is shown as it has been acquired using low-cost consumer hardware [16]. A lattice detection algorithm is used to locate the reference card, disregarding where it is placed in the image and disregarding its size and projective distortions [17]. The upper right panel in Figure 21.6 depicts the lattice points. The yellow lines indicate the reference points on the card, which are used to feed a least-squares algorithm for inverse perspective projection. Thereafter, the card is transformed into its standard rectangular position, and the identifiers (both barcode and characters) are extracted. The redundant code ensures robustness and that the images may be registered correctly into the eCRF of a different subject. In addition, a checksum is integrated into the bar code, providing further reliability.

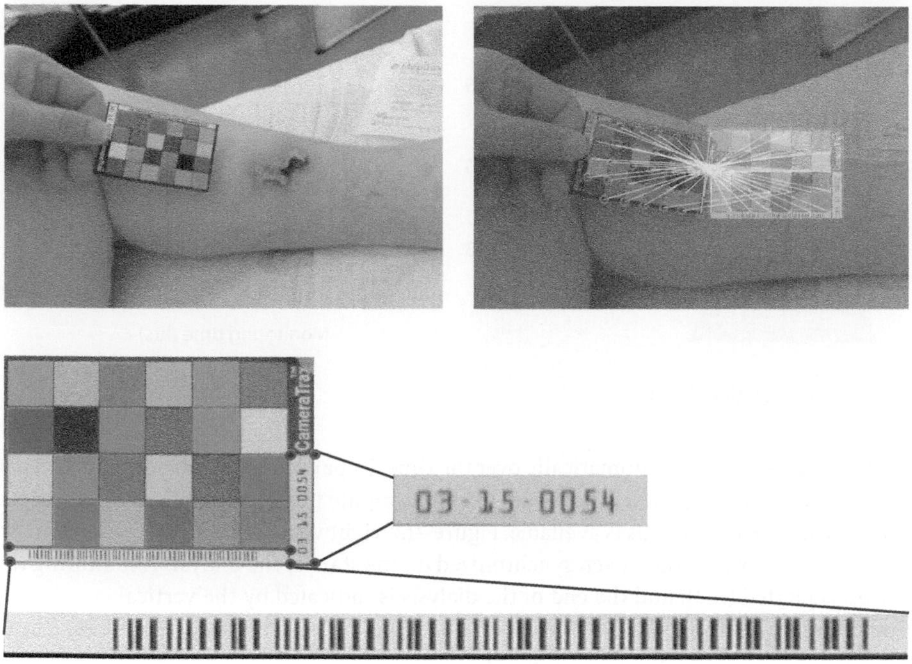

FIGURE 21.6 Wound imaging.

21.6.2 Case Study II: ECG Recording

In this controlled clinical trial, patients with diabetes mellitus and severe heart diseases are enrolled. The study aims at identifying ECG-based surrogates for the indication of sudden cardiac death (SCD), which remains a major public health problem worldwide, causing more than 300,000 deaths per year in the United States [18]. SCD occurs frequently during the dialysis sequence. Therefore, the subjects are equipped with a 12 lead ECG Holter monitor, and data is recorded over a period of seven days and 24 h a day.

The recording period covers three dialysis cycles and delivers about 600,000–900,000 heart cycles per person. If this data would bc printed, 15.1 km of paper would be required for each study subject. Clearly, this is far beyond being assessable manually. Therefore, automated CAD analysis is required. However, automation is difficult here, too, due to the large data volumes: the sample rate of 1000 Hz and 10 bit quantization yields up to 8.5 GB uncompressed signal data per subject.

For automation of signal analysis, the data is decomposed into distinct heart cycles. A vector cardiogram-based R-spike detection is applied [19]. The ECG model fitting is based on Gaussians, yielding a novel paradigm of ECG analysis, where the entire recording is processed rather than some isolated events are detected and counted [20]. In particular, computational ECG aims at predicting and preventing serious events, such as SCD.

Computational ECG is visualized in Figure 21.7. Wellness industry and sportive consumers who like monitoring their body performance and sharing it on social networks will open economical markets that allow low-cost hardware, for instance, embedded in smart clothes (Figure 21.7, left). More meaningful use of this technology will start with high-risk patients, for instance, in ambient assisted living (AAL) or as security methods in public transportation on roads, railways, or airways, before applied to all humans [20].

The ECG model fitting is based on Gaussians with 99 parameters per cycle for a 12 lead recording. The model parameters are estimated using the Levenberg-Marquardt algorithm [19]. They are used to

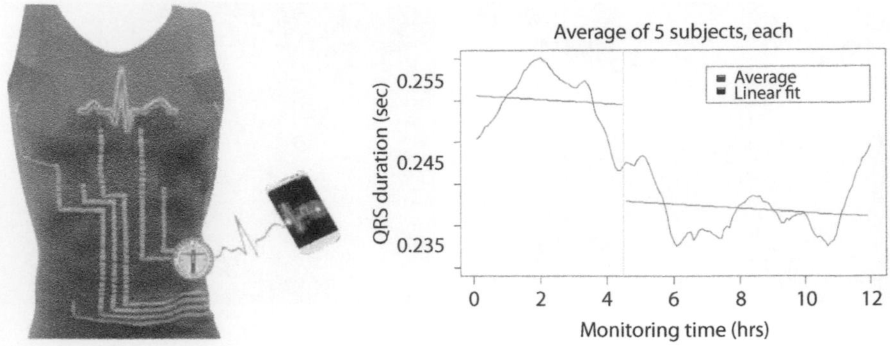

FIGURE 21.7 Computational ECG.

track relevant ECG descriptors automatically over the time. In particular, mean and standard deviation of the Gaussians describing Q- and S-waves are used to compute the QRS-duration in each of the cycle.

So far, data from five individuals is available. Figure 21.7, right visualizes the QRS- duration averaged over 15 measurements, which have been synchronized on the start of the dialysis (monitoring time = 0). A 12 h period is plotted (red), and the end of the dialysis is indicated by the vertical gray line. In both periods (during and after the dialysis), a linear regression model has been fitted (blue lines). The offset in both parts differ significantly (Students t-test, $p \ll 0.001$) [20].

21.7 Discussion

So far, CAD in clinical trials is not automated. The data is recorded manually, disregarding whether a paper-based or electronic CRF is used. In this chapter, we have presented a system architecture, interfaces, and protocols for automation of such processes. This automation affects the workflow, the dataflow, and the data analysis.

21.7.1 Automation of Workflow

Figure 21.8 depicts the workflow in clinical trials recording BLOB data before (top) and after (bottom) automation. Initially, recordings have been copied to a compact disc and sent by mail to the center for data analysis. After weeks, the surrogates that have been measured from the data usually are faxed back to the site (in order to save time) and the research nurse enters the values into the eCRF, manually (Figure 21.8, upper part). Using the communication server, the workflow is automated, and measures based on BLOB recordings are entered instantaneously into the eCRF. This reduces errors and improves the data quality. Therefore, the trial's sample size, conducting time, and financial efforts are reduced.

21.7.2 Automation of Data Flow

Applying CAD allows also for identifying the subject (barcode reading and OCR) and hence, automatic assignment of BLOB data into the correct eCRF fields. This essentially simplifies the data flow in clinical trials. Furthermore, it supports automation of data processing and analysis.

21.7.3 Automation of Data Processing

Our system architecture is open for any API. So far, we have implemented MATLAB® and Java interfaces for signal and image analysis, respectively. Examples have been given. The advantages are time savings and instantaneous approval that recorded data is processable. Depending on the type of data,

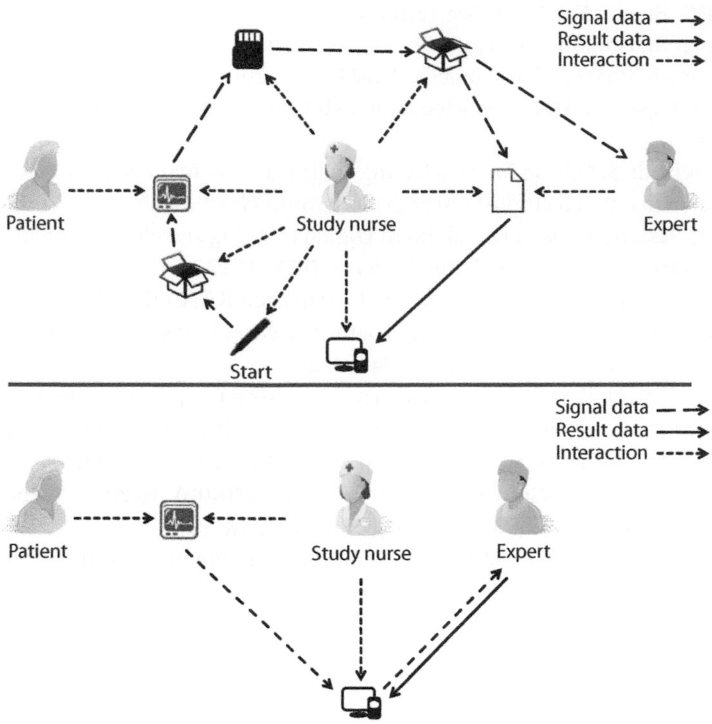

FIGURE 21.8 Automation of workflow and data flow.

for example, wound photography, the research nurse is still at the patient's bedside when she is notified that there were some problems in automatic image analysis. Hence, she might easily repeat the imaging and restart the CAD process.

Conclusion

We have presented a system architecture and implementation to automate the CAD in clinical trials, in particular for signal- and image-based biomarkers. The concept has been applied to several use cases, where impact on both processing time and processing quality has been shown. System integration reduces the variation of measurements and, hence, the number of required subjects, the costs of trials, and the duration of the trials. Therefore, novel drugs and devices become available earlier to the healthcare systems, which is the goal of the B2B-paradigm.

Acknowledgments

The ECG trial has been funded by the European Foundation for the Study of Diabetes (EFSD), grant number 74550-94555. The clinical trial has been registered under NCT02001480 and was approved by the Ethics Committee of the RWTH Aachen University.

References

1. Patwardhan S, Gogtay N, Thatte U, Pramesh CS. Quality and completeness of data documentation in an investigator-initiated trial versus an industry-sponsored trial. *Indian J Med Ethics*. 2014 Jan-Mar;11(1):19–24.

2. Abramson RG, Burton KR, Yu JP, Scalzetti EM, Yankeelov TE, Rosenkrantz AB, Mendiratta-Lala M, Bartholmai BJ, Ganeshan D, Lenchik L, Subramaniam RM. Methods and challenges in quantitative imaging biomarker development. *Acad Radiol.* 2015 Jan;22(1):25–32.
3. Reboussin D, Espeland MA. The science of web-based clinical trial management. *Clin Trials.* 2005;2(1):1–2.
4. Leroux H, McBride S, Gibson S. On selecting a clinical trial management system for large scale, multi-centre, multi-modal clinical research study. *Stud Health Technol Inform.* 2011;168:89–95.
5. Deserno TM, Deserno V, Haak D, Kabino K. Digital Imaging and electronic data capture in multi-center clinical trials. *Stud Health Technol Inform.* 2015;216:930.
6. Rubinstein YR, Groft SC, Bartek R, Brown K, Christensen RA, Collier E, Farber A et al. Creating a global rare disease patient registry linked to a rare diseases biorepository database: Rare Disease-HUB (RD-HUB). *Contemp Clin Trials* 31:394404, 2010.
7. Deserno TM, Haak D, Brandenburg V, Deserno V, Classen C, Specht P. Integrated image data and medical record management for rare disease registries: A general framework and its instantiation to the German calciphylaxis registry. *J Digit Imaging* 2014; 27(3): 702–713.
8. Stausberg J, Altmann A, Antony G, Drepper J, Sax U, Schiitt A. Register for networked medical research in Germany. *Appl Clin Inform* 1(4):408418, 2010.
9. Drepper J, Semler SC (ed). IT infrastructure in patient-centered research. State of the art and required actions (German). Technical Report. IT-Reviewing Board of TMF. Akademische Verlagsgesellschaft AKA GmbH, Berlin. 2013 (ISBN 978-389838-690-6).
10. Deserno TM, Deserno V, Legewie V, Schafhausen J, Eisert A, Schmidt-Kotsas A, Kirstein S, Willems J, Spitzer K, Schulz JB. IT support for translational management of clinical trials using the Google web toolkit. GMS Publishing House 2011; 56: 023. [in German].
11. Haak D, Samsel C, Gehlen J, Jonas S, Deserno TM. Simplifying electronic data capture in clinical trials: Workflow embedded image and biosignal file integration and analysis via web services. *J Digit Imaging* 2014; 27(5): 571–80.
12. Haak D, Dovermann J, Kramer C, Merkelbach K, Deserno TM. Data recording in clinical trials by ODM supporting tablets and smartphones. *Proceedings 59th GMDS Annual Conference in Gttingen*, 2014, DocAbstr. 210.
13. Haak D, Gehlen J, Jonas S, Deserno TM. OC ToGo. Bed site image integration into OpenClinica with mobile devices. *Proceedings SPIE* 2014; 9039: 091-6.
14. Haak D, Page CE, Reinartz S, Krger T, Deserno TM. DICOM for clinical research: PACS-integrated electronic data capture in multi-center trials. *J Digit Imaging* 2015; 28(5): 558–6.
15. Deserno TM, Haak D, Samsel C, Gehlen J, Kabino K. Integration image management and analysis into OpenClinica using web services. *Proceedings SPIE* 2013; 8674: 0F1-10.
16. Jose A, Haak D, Jonas SM, Brandenburg V, Deserno TM. Human wound photogrammetry with low-cost hardware based on automatic calibration of geometry and color. *Proceedings of SPIE* 2015; 9414: 3J1-8.
17. Park M, Brocklehurst K, Collins RT, Liu Y. Deformed lattice detection in real- world images using mean-shift belief propagation. *IEEE Trans Pattern Anal Mach Intell* 2009; 31(10): 1804–16.
18. Mitrani RD, Myerburg RJ. Ten advances defining sudden cardiac death. *Trends Cardiovasc Med.* 2015 [Epub ahead of print].
19. Sartor M, Jonas S, Wartzek T, Leonhardt S, Wanner C, Marx N, Deserno TM. Non-linear time normalization in long term ECG: Generation of unique pseudo images from multi-lead ECG. In: Deserno TM, Handels H, Meinzer HP, Tolxdorff T (ed). *Image Processing for Medical Applications.* Springer, Berlin 2014; 300–5. [in German].
20. Deserno TM, Marx N. Computational electrocardiography. Revisiting Holter ECG monitoring. *Methods Inf Med* 2016; 55(4): 305–11.

Concluding Remarks

Paulo Mazzoncini de Azevedo-Marques

Arianna Mencattini

Marcello Salmeri

Rangaraj Mandayam Rangayyan

In the preparation of this book, our aim was to present a wide range of topics and applications to demonstrate the impressive impact of CAD and related fields of study on health care of the entire human body. Our wish is to motivate and attract more researchers and innovators to the related fields of research and development to continue to address the current challenges and lead to further advances. We believe the chapters in this book set the stage for these purposes.

Reiche et al. have demonstrated, in Chapter 1, the robustness of their segmentation method and investigated quantitative shape features to discriminate between two classes of brain lesions in FLAIR MR images using a novel and exploratory noise analysis approach.

In Chapter 2, Hatanaka and Fujita have shown that analysis of retinal fundus images is effective in the diagnosis of not only diseases of the eye, such as glaucoma, diabetic retinopathy, and age-related macular degeneration, but also systemic hypertension.

Oloumi et al. have demonstrated, in Chapter 3, that a measure of vascular tortuosity, derived by applying image processing methods to retinal fundus images, can facilitate CAD of retinopathy in premature infants with high accuracy.

Roychowdhury et al., in Chapter 4, have shown how image processing methods may be applied to measure sub-retinal layer thickness distributions in OCT images. The thickness maps can aid analysis of disorganization in sub-retinal layers and assessment of diabetic macular edema.

Chapter 5 by Muramatsu et al. presented CAD methods for analysis of dental panoramic radiographs for screening and efficient diagnosis of oral as well as systemic diseases, such as osteoporosis.

In Chapter 6, Pérez-Carrasco et al. discussed the problem of burn diagnosis in clinical practice. They highlighted the importance of using CAD methods to derive an accurate initial assessment, which is crucial to the evolution and response to treatment of the burn wound.

Gutierrez et al. demonstrated, in Chapter 7, the role of cardiac image segmentation in a wide range of applications, including quantification of volume, localization of pathology, CAD, and image-guided interventions.

In Chapter 8, Pezeshk et al. have described methods to augment databases for training and testing CAD systems by inserting lesions or tumors from one image into another. The variations in the replicated lesions provided by various transformations facilitate realistic simulation of additional cases and lead to improved training of CAD systems.

Koenigkam Santos and Weinheimer presented, in Chapter 9, clinical and technical aspects related to CAD and QCT in the evaluation of diffuse lung diseases, focusing on emphysema, airway diseases such as asthma, and cystic fibrosis.

Chapter 10 by Mencattini et al. demonstrated that CAD techniques can be designed to detect bilateral asymmetry in mammograms. Such an approach could facilitate detection of breast cancer even in the absence of well-recognized signs such as masses and calcifications.

In Chapter 11, Chan et al. introduced the DBT imaging technique and its applications in breast cancer detection. The authors compared, in detail, CAD techniques for DBT and standard digital mammography and described their potential capabilities and limitations in the detection of signs of breast cancer, such as calcifications and masses.

Nogueira-Barbosa and Azevedo-Marques presented, in Chapter 12, a review of CAD methods for spinal abnormalities, with focus on vertebral body fracture and intervertebral disc degeneration. They indicated that MRI is currently the preferred option for evaluation of spinal abnormalities.

Chapter 13 by Yuan and Meng has shown how capsule endoscopy images may be obtained with advanced instrumentation and analyzed using image processing, feature extraction, and classification techniques to detect ulcers.

Virmani and Kumar described, in Chapter 14, a computerized tissue classification system to detect focal liver disease based on texture analysis.

In Chapter 15, Cipriani Frade et al. presented computational aspects of CAD and CBIR via color image analysis applied to images of dermatological lesions on the lower limbs.

Chapter 16 by Boyd and Lagacé showed how imaging techniques, image analysis procedures, and finite element modeling methods could be used to derive parameters that could assist in the diagnosis of osteoporosis. Various in vivo imaging techniques were used to derive a number of parameters, such as bone mineral density, trabecular volume fraction, and cortical thickness, that could be useful in the assessment of fracture risk.

Borotikar et al. presented, in Chapter 17, techniques for augmented statistical shape modeling, mesh registration, and derivation of patient-specific anatomical references with applications in shoulder arthroplasty and rehabilitation. The methods were integrated in a pipeline to facilitate pre-surgical planning and computer-aided surgery.

In Chapter 18, Knight and Khademi have shown how image processing and pattern recognition techniques can be designed to detect breast cancer in histopathology images. Such approaches extend CAD techniques from medical images of patients to histological images of tissue samples and can improve the accuracy of the most decisive step in clinical diagnosis.

Chandra et al. have shown, in Chapter 19, that diagnostically useful images may be obtained by using nonionizing microwaves. If the associated technical difficulties are resolved, microwave imaging could provide a low-cost alternative to CT and MR imaging.

In Chapter 20, Traina et al. discussed the potential applications and limitations of CBIR systems in the clinical environment. They presented a CBIR system specialized for mammography that is able to encode perceptual parameters using feature extraction methods and distance functions. The methods were shown to reduce the need for query refinement or relevance feedback cycles.

Chapter 21 by Deserno and Reichertz has shown how signal and image processing algorithms for biomedical and CAD applications can be integrated with models and schema drawn from the domain of informatics for improved management of patient care.

An important point to note from our discussion in the preface and from the chapters of the book is that quantitative analysis becomes possible by the application of computational procedures to medical images. The logic of medical or clinical diagnosis via image analysis can then be objectively encoded and consistently applied in routine or repetitive tasks. However, we emphasize that the end-goal of medical image analysis should be computer-*aided* diagnosis and not automated diagnosis. A physician or medical specialist typically uses a large amount of information in addition to images and measurements, including the general physical and mental state of the patient, family history, and socioeconomic factors affecting the patient, many of which are not amenable to quantification and logical rule-based processes. Medical images are, at best, indirect indicators of the state of the patient; the results of image analysis need to be integrated with other clinical signs, symptoms, and nonmedical information by a physician. The general background knowledge and intuition of the medical specialist play important roles in arriving at the final diagnosis. Keeping in mind the realms of practice of various licensed and regulated professions, liability, and legal factors, the final diagnostic decision and communication with the patient are best left to the physician or medical specialist. It is expected that quantitative and objective analysis facilitated by the application of medical image analysis, medical image informatics, CBIR, and CAD will lead to improved diagnostic decision by the physician.

CAD has become a part of the clinical workflow in the detection of breast cancer with mammograms, but is still in the infancy of its full potential for application to many other types of diseases or lesions imaged with various modalities. CBIR is an alternative and complementary approach for image retrieval based on key words and metadata. Initial results are promising for the use of CBIR as a diagnostic support tool. In the future, it is likely that CAD and CBIR will be incorporated into PACS and will be employed as useful tools for diagnostic examinations in routine clinical work.

It is evident from the material presented in this book that digital image processing techniques can assist in quantitative analysis of medical images, that pattern recognition and classification techniques can facilitate CAD, and that CAD systems can assist in achieving efficient diagnosis, in designing optimal treatment protocols, in analyzing the effects of or response to treatment, and in clinical management of various abnormal conditions. Medical imaging, medical image analysis, medical image informatics, CBIR, and CAD are proven, as well as essential techniques for health care.

Index

Printed and bound by CPI Group (UK) Ltd, Croydon, CR0 4YY
01/11/2024
01782603-0012